Principles of Engineering Physics 1

This is a textbook for an introductory course in engineering physics. It provides a coherent treatment of the basic principles and theories of engineering physics and offers a balance between theoretical concepts and their applications. Beginning with a comprehensive discussion on oscillations and waves with applications in the field of mechanical and electrical engineering, it goes on to explain basic concepts such as Huygen's principle, Fresnel's biprism, Fraunhofer diffraction and polarization.

All chapters are interspersed with rich pedagogical features such as solved problems, unsolved exercises and multiple choice questions with answers. It will help undergraduate students of engineering acquire skills for solving difficult problems in quantum mechanics, electromagnetism, nanoscience, energy systems and other engineering disciplines.

Md. N. Khan is Associate Professor at the Department of Physics, Indira Gandhi Institute of Technology (IGIT), Odisha. He has more than 22 years of teaching experience and has taught courses on engineering physics, physics of semiconductor devices and materials science. His areas of interest include X-ray scattering and materials science.

S. Panigrahi is Senior Professor at the Department of Physics and Astronomy, National Institute of Technology (NIT), Rourkela. He has more than two decades of teaching and research experience in the field of solid state physics, materials science and ferroelectrics.

W0081111

Principles of
Engineering Physics 1

Md. N. Khan
S. Panigrahi

CAMBRIDGE
UNIVERSITY PRESS

CAMBRIDGE
UNIVERSITY PRESS

University Printing House, Cambridge CB2 8BS, United Kingdom

One Liberty Plaza, 20th Floor, New York, NY 10006, USA

477 Williamstown Road, Port Melbourne, vic 3207, Australia

4843/24, 2nd Floor, Ansari Road, Daryaganj, Delhi – 110002, India

79 Anson Road, #06–04/06, Singapore 079906

Cambridge University Press is part of the University of Cambridge.

It furthers the University's mission by disseminating knowledge in the pursuit of education, learning and research at the highest international levels of excellence.

www.cambridge.org
Information on this title: www.cambridge.org/9781316635643

First published 2016

Printed in India by International Print- O -Pac Limited, Noida, U.P.

A catalogue record for this publication is available from the British Library

ISBN 978-1-316-63564-3 Paperback

Additional resources for this publication at www.cambridge.org/9781316635643

To all our beloved people who have sacrificed their lives for the betterment of the world through science, technology and social service.

Acknowledgment

Contents

Preface

Science in general may be described as organized common sense. In the real world of science, nothing prevails except rationality and logics. Science does not believe in miracles. Clear understanding of the basic principles of science is essential for technological and social development. Once upon a time, the base of engineering was mainly empirical; however, now it is completely scientific. Physics is a fundamental aspect of science on which all engineering sciences have been built upon. Nowadays, more stress is given to the understanding of the basic principles rather than on remembering specific procedures. The fundamental concepts of physics have paved the way for the development of technologies. All modern technological advances from laser micro surgery to television, from computers to dishwashers to mobile phones, from remote controlled toys to space vehicles, trace back directly to the principles of physics. Accordingly, the syllabus of engineering courses includes physics as an essential ingredient.

This book, entitled *Principles of Engineering Physics 1,* is designed as a textbook keeping in view the engineering physics course curricula prescribed by most technical universities of India. The present book begins with oscillations and waves and ends with holography, containing altogether fourteen chapters. This book is written in a logical and coherent manner for easy understanding. The concepts of physics are mathematized without losing the beauty of the physical ideas involved. Emphasis has been given to an understanding of the basic concepts and their applications to a number of engineering problems. Each topic has been discussed in detail, both conceptually and mathematically, so that students do not face any kind of difficulties. All the derivations and solutions of numerical examples are given in detail. Each chapter contains a large number of solved numerical examples, unsolved numerical problems with answers, practical applications, theoretical questions, and multiple choice questions with answers. Certain topics and derivations that are not included directly in the syllabi have also been included in the book for the sake of continuity and completeness. The scope of the book thus has been expanded beyond the basic needs of undergraduate engineering students. We hope this book will be of immense help not only to the students but also to the teachers.

The authors sincerely request the readers for their constructive criticisms via emails *mdnkhan1964@yahoo.com* and *spanigrahi@nitrkl.ac.in* for future modification of the book.

Acknowledgment

It is a pleasure to express our deep appreciation to the engineering students (both continuing and passed out) of IGIT Sarang and NIT Rourkela who have borne with us in our class teachings. Many suggestions from our colleagues, students and reviewers have gone a long way in the development of this book. Our sincere thanks are due to them. We gratefully acknowledge the ideas received from a number of standard books on physics as given in the bibliography. We sincerely thank the editorial team at Cambridge University Press, India, for their keen interest in publishing this book in a nice format. We particularly wish to thank Gauravjeet Singh Reen for many helpful suggestions and improvements.

1 Oscillations and Waves

1.1 Introduction

Objects subjected to restoring forces when displaced from their normal positions and released, perform to and from or vibrating motions. They move back and forth along a path, repeating over and over again, a series of motions. Such motion of constant frequency is called periodic motion or harmonic motion and objects performing such type of motion are called harmonic oscillators. In this book, it is tacitly assumed that there is a linear relationship between force and displacement; frequency remains constant throughout the motion. In real systems however, the linear behavior, implicit in simple harmonic motion, is rarely obeyed. If the frequency of the oscillatory system is not constant, then it is called anharmonic motion – its study is beyond the scope of this book due to its mathematical complexities.

In oscillatory systems, it is not necessarily the bodies themselves who execute oscillations; bodies may be at rest. If the physical properties of a system undergo changes in an oscillatory manner, the system will also be called an oscillatory system. The electromagnetic energy transfer between the capacitor and inductor in a tank circuit used in electronic gadgets, variation of pressure in air due to propagation of sound waves, vibration of the diaphragm of a speaker in sound systems, flow of alternating current, variation of electric and magnetic vectors during propagation of electromagnetic waves, etc., are examples of oscillatory systems.

1.1.1 Parameters of an oscillatory system

i. *Mean position*   The position of the oscillating body when there is no oscillation is called the mean position or equilibrium position. This is the rest position of the oscillating body.

ii.   *Amplitude* (r) It is the absolute value of the maximum displacement of the oscillating particle from its mean position or equilibrium position.

iii.   *Time period* (T) It is the time required for one complete oscillation.

iv.   *Frequency* (v) It is the number of complete oscillations made by the oscillating body in one second. The relation between frequency 'v' and time period 'T' from definition is $T = \dfrac{1}{v}$.

1.2   Simple Harmonic Oscillation (SHO)

Let a body of mass 'm' be placed on a frictionless plane with a massless spring attached to it (Fig. 1.1). The other end of the spring is fixed to a rigid support. The spring–body system is in the relaxed state, i.e., the spring is neither compressed nor extended. Notice the position of the body – it is called the mean position or equilibrium position. Now, the body is pulled through a displacement 'x'. The spring exerts a restoring force on the body tending to pull it backwards. This restoring force 'F' is proportional to the displacement (i.e., elongation of the spring) 'x' and is opposite in direction to the displacement.

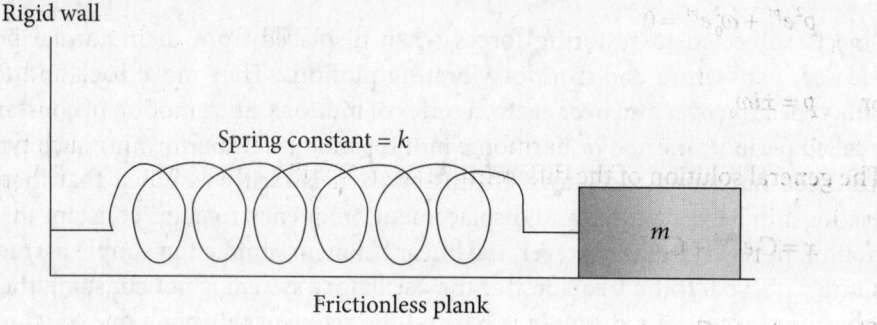

Figure 1.1 | Simple harmonic oscillator. Spring–body system is placed on a frictionless plane

Mathematically, $F \propto -x$ (negative sign appears since the restoring force and displacement 'x' are in opposite directions)

$$F = -kx \tag{1.1}$$

where k is the proportionality constant and is known as spring constant. This equation is called Hooke's law of elasticity. Applying Newton's laws of motion, Hooke's law can be written as

$$ma = -kx$$

or $$m\frac{d^2x}{dt^2} = -kx \tag{1.2}$$

Equation (1.2) is the differential equation of motion of a simple harmonic oscillator in the absence of other forces. It can also be written as

$$\frac{d^2x}{dt^2} + \omega_0^2 x = 0 \tag{1.3}$$

where $\omega_0^2 = \dfrac{k}{m}$ (1.4)

Here m is the mass of the body attached at the end of the free end of the spring and ω_0 is called the natural angular frequency of oscillation. The general solution of the differential Eq. (1.3) is determined in the following way. Other methods for the purpose are also available.

Assume a solution of the form

$$x = e^{pt} \tag{1.5}$$

Putting (1.5) into (1.3), we get

$$p^2 e^{pt} + \omega_0^2 e^{pt} = 0$$

or $p = \pm i\omega_0$

The general solution of the differential Eq. (1.3) will be given by

$$x = C_1 e^{i\omega_0 t} + C_2 e^{-i\omega_0 t} \tag{1.6}$$

The constants C_1 and C_2 must be complex in order for Eq. (1.6) to be the general solution. Since $e^{i\omega_0 t}$ and $e^{-i\omega_0 t}$ are complex conjugates of each other, the constants C_1 and C_2 must be complex conjugate of each other so that x is real. For this, we set

$$C_1 = C = Ae^{i\theta}$$

and $C_2 = C^* = Ae^{-i\theta}$

Upon this substitution into Eq. (1.6), we get

$$x = Ce^{i\omega_0 t} + C^* e^{-i\omega_0 t} = Ae^{i\theta} e^{i\omega_0 t} + Ae^{-i\theta} e^{-i\omega_0 t}$$

or $x = 2A\cos(\omega_0 t + \theta) = r\cos(\omega_0 t + \theta),\ 2A = r$ (1.7)

$(x = r\sin(\omega_0 t + \theta)$ may also be a solution).

Here

   r = amplitude of the oscillation.

   θ = the initial phase of the oscillation.

   $\omega_0 t + \theta$ = phase of the motion.

The constants r and θ can be obtained from the initial conditions of the simple harmonic motion.

   The velocity 'v' of the body executing SHO is determined by differentiating Eq. (1.7) with respect to time.

$$v = -r\omega_0 \sin(\omega_0 t + \theta) \tag{1.8}$$

The acceleration 'a' of the body executing SHO is determined by differentiating Eq. (1.8) with respect to time.

$$a = -\omega_0^2 r \cos(\omega_0 t + \theta). \tag{1.9}$$

Putting the value of $r\cos(\omega_0 t + \theta) = x$ from (1.7) into Eq. (1.9), we get

$$a = -\omega_0^2 x \tag{1.10}$$

Equation (1.10) shows that in the case of a simple harmonic motion, acceleration is directed opposite to displacement.

   If T is the time period, then in T seconds, the number of complete oscillations is 1. So, in 1 second, the number of oscillations will be $1/T$ which by definition is the frequency v. Hence, we have

$$v = \frac{1}{T} \tag{1.11}$$

Frequency and time period are inversely proportional to each other.

   From the definition of angular frequency, we have

$$\omega_0 = \frac{2\pi}{T} = 2\pi v \tag{1.12}$$

1.2.1 Energy of a simple harmonic oscillator

By the definition of potential energy, the potential energy of a simple harmonic oscillator at any instant or at any position is given by

$$E_p = \int_0^x F dx = \int_0^x kx \, dx = \frac{1}{2} kx^2 \tag{1.13}$$

Here 'x' has been measured from the mean position $x = 0$. Putting the value of x from Eq. (1.7) into Eq. (1.13), we have

$$E_P = \frac{1}{2}kr^2 \cos^2(\omega_0 t + \theta) \qquad (1.14)$$

Putting the value of k from Eq. (1.4) into Eq. (1.14), we get

$$E_P = \frac{1}{2}m\omega_0^2 r^2 \cos^2(\omega_0 t + \theta) \qquad (1.15)$$

Since the maximum value of $\cos^2(\omega_0 t + \theta)$ is 1, the maximum value of potential energy-$E_{P\max}$ from Eq. (1.15) is found out to be

$$E_{P\max} = \frac{1}{2}m\omega_0^2 r^2 \qquad (1.16)$$

By the definition of kinetic energy, the kinetic energy of a simple harmonic oscillator at any instant or at any position is given by

$$E_K = \frac{1}{2}mv^2$$

Putting the value of v from Eq. (1.8) into the previous equation, we get

$$E_K = \frac{1}{2}mr^2\omega_0^2 \sin^2(\omega_0 t + \theta) \qquad (1.17)$$

Since the maximum value of $\sin^2(\omega_0 t + \theta)$ is 1, the maximum value of kinetic energy $E_{K\max}$ from Eq. (1.17) is found out to be

$$E_{K\max} = \frac{1}{2}m\omega_0^2 r^2 \qquad (1.18)$$

The total energy of a simple harmonic oscillator at any instant or at any position is given by

$$E = E_P + E_K$$

$$= \frac{1}{2}m\omega_0^2 r^2 \cos^2(\omega_0 t + \theta) + \frac{1}{2}mr^2\omega_0^2 \sin^2(\omega_0 t + \theta)$$

$$= \frac{1}{2}m\omega_0^2 r^2 = E_{K\max} = E_{P\max} \qquad (1.19)$$

Equation (1.19) shows that the total energy of a simple harmonic oscillator at any instant or at any position is constant.

1.2.2 Characteristics of SHO

From the discussions in the previous section, we can infer the following characteristics of SHO.

When the body is released, it moves to and fro about the mean position with constant amplitude 'r'. As the body moves towards the mean position, its speed increases but the force and hence, its acceleration decreases; both become zero at the mean position. Due to the inertia of motion, the body overshoots the mean position, but at the same time, a retarding force comes into action to oppose the motion. This retarding force increases until the body reaches the largest distance from the mean position. Here, it stops and begins its return journey. This process goes on continuously for an indefinite period if there is no dissipative force. Throughout the motion, the force as well as the acceleration is directly proportional to displacement and directed towards the mean position. The distances travelled by the vibrating body on the two sides of the mean position are equal. This type of motion is called simple harmonic oscillation. The SHO is graphically represented in Fig. 1.2.

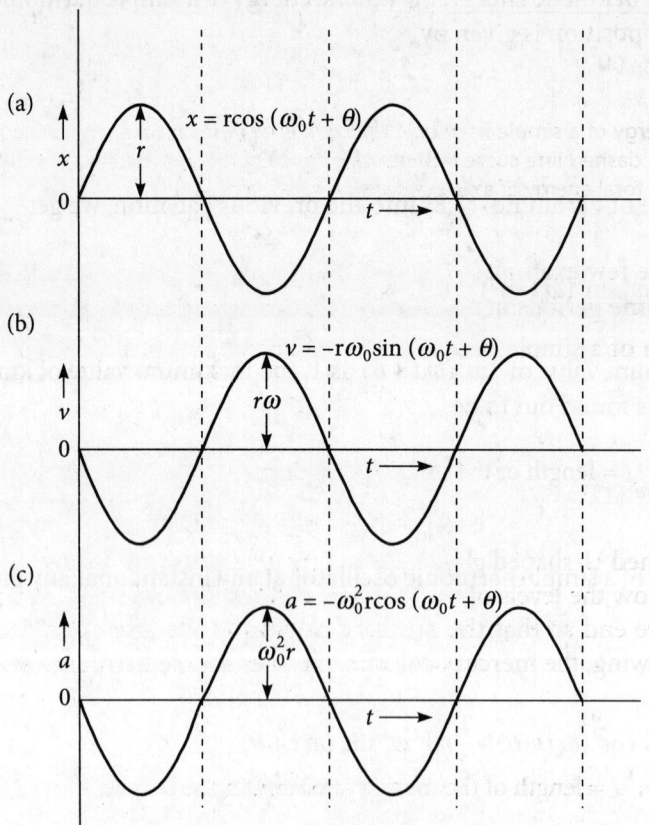

Figure 1.2 Graphical representation of SHO. (a) displacement-time plot, (b) velocity–time plot, (c) acceleration–time plot

For a particular simple harmonic oscillator, mass, amplitude and angular frequency are constants with respect to time. Therefore, we can conclude that the total energy of a simple harmonic oscillator at any instant or position is constant. Moreover, the total energy of a simple harmonic oscillator at any instant or position is equal to the maximum values of potential or kinetic energy. The simultaneous variation of potential energy and kinetic energy of a simple harmonic oscillator according to Eqs (1.15) and (1.17) is depicted in Fig. 1.3(a) and in Fig. 1.3(b).

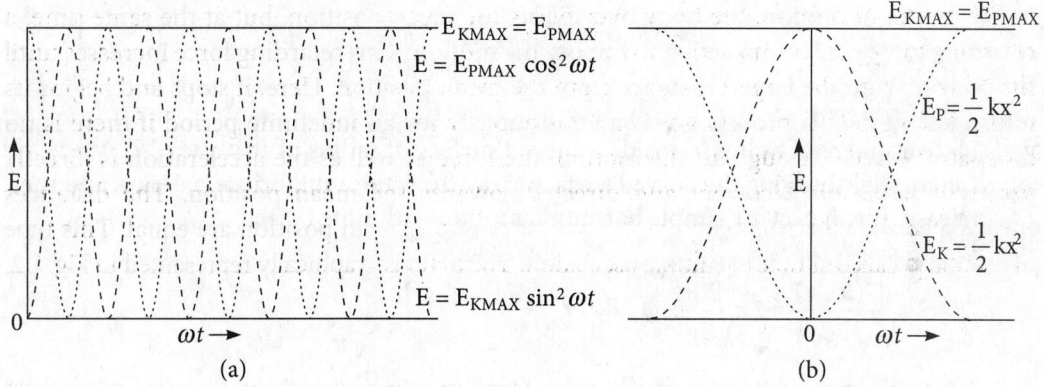

Figure 1.3 | Energy of a simple harmonic oscillator. The dotted curve represents potential energy, the dashed line curve represents kinetic energy and the continuous curve represents the total energy of a simple harmonic oscillator

We will now cite few examples of simple harmonic motion under ideal conditions. The derivations for time periods in all the examples are left as exercises to the students.

i.   The motion of a simple pendulum in vacuum is simple harmonic. Its time period is given by

$$T = 2\pi \sqrt{\frac{\ell}{g}}, \quad \ell = \text{length of the simple pendulum.}$$

ii.   Take a cleaned U-shaped glass tube and fix it vertically on a stand. Partially fill it with mercury. Now the levels of the mercury column in both sides are equal. Blow slowly through one end so that the mercury column in the other end rises slightly. When we stop blowing, the mercury column executes simple harmonic motion having time period

$$T = 2\pi \sqrt{\frac{\ell}{2g}}, \quad \ell = \text{length of the mercury column in the U-tube.}$$

iii. A body dropped into a hole dug through the centre of earth executes simple harmonic motion with time period

$$T = 2\pi \sqrt{\frac{R}{g}}, \quad R = \text{radius of Earth.}$$

iv. Place a small block inside a smooth curved surface having radius of curvature 'R'. Move the sphere slightly along the curved surface away from the equilibrium position and then release. It will perform simple harmonic motion with time period

$$T = 2\pi \sqrt{\frac{R}{g}}.$$

v. Place a sphere of radius 'r' inside a curved surface of radius of curvature 'R'. Move the sphere slightly along the curved surface away from the equilibrium position and then release. It will execute simple harmonic motion with time period

$$T = 2\pi \sqrt{\frac{7(R-r)}{5g}} \approx 2\pi \sqrt{\frac{7R}{5g}} \quad \text{if} \quad R >> r$$

vi. If a small iron cylinder, partially submerged in mercury vertically, is pressed slightly and then released, it will execute simple harmonic motion with time period

$$T = 2\pi \sqrt{\frac{\ell \rho}{\rho' g}}.$$ Here ℓ = length of the cylinder, ρ = density of the cylinder, ρ' = density of mercury.

Example 1.1

A spring 15 cm in length is fixed to the ceiling. When a body of mass 1 kg is hung at the free end, its length becomes 17 cm. Calculate the spring constant of the spring.

Solution

The increase in length 'x' of the spring when a 1kg body is hung = 17 cm – 15 cm = 2 cm = 0.02 m

The downward force acting on the spring = weight of the body hung. At the equilibrium position, the restoring force = downward force acting on the spring.

$$kx = mg$$

or $$k = \frac{mg}{x} = \frac{1 \times 9.8 \text{ N}}{0.02 \text{ m}} = 490 \text{ N/m}$$

The spring constant of the spring is calculated to be 490 N/m.

Example 1.2

A 5 kg body extends a spring 15 cm from its relaxed position. The body is removed and a 1 kg body is hung from the same spring. The 1 kg body is pulled and released. Calculate the time period of oscillation of the body.

Solution

The increase in length 'x' of the spring when a 5 kg body is hung = 15 cm = 0.15 m

The downward force acting on the spring = weight of the body hung. At the equilibrium position, the restoring force = downward force acting on the spring.

$$kx = mg$$

or $k = \dfrac{mg}{x} = \dfrac{5 \times 9.8 \text{ N}}{0.15 \text{ m}} = 326.67 \text{ N/m}$ is the spring constant of the spring.

The natural angular frequency of the oscillation $\omega_0 = \sqrt{\dfrac{k}{m}}$.

The time period of oscillation $T = 2\pi\sqrt{\dfrac{m}{k}} = 2\pi\sqrt{\dfrac{1}{326.67}}$ s $= 0.35$ s

Example 1.3

A body oscillates with simple harmonic motion obeying the equation $y = 12 \cos\left(0.7\pi t + \dfrac{\pi}{5}\right)$ m.

Calculate the velocity and acceleration of the body at time $t = 3$ s. Also calculate the natural frequency and time period of the harmonic motion.

Solution

The velocity 'v' of the body at any time t is

$$v = \frac{dy}{dt} = \frac{d}{dt}\left[12\cos\left(0.7\pi t + \frac{\pi}{5}\right)\right]\text{m/s} = -8.4\pi \sin\left(0.7\pi t + \frac{\pi}{5}\right)\text{m/s}$$

So the velocity of the body at $t = 3$ s will be

$$v = -8.4\pi\sin\left(0.7\pi \times 3 + \frac{\pi}{5}\right)\text{m/s} = -21.35 \text{ m/s}$$

The velocity is directed opposite to the displacement.

The acceleration 'a' of the body at any time t is

$$a = \frac{dv}{dt} = \frac{d}{dt}\left[-8.4\pi\sin\left(0.7t\pi + \frac{\pi}{5}\right)\right]\text{m/s}^2 = -5.88\pi^2\cos\left(0.7\pi t + \frac{\pi}{5}\right)\text{m/s}^2$$

So the acceleration of the body at $t = 3$ s will be

$$a = -5.88\pi^2 \cos\left(0.7\pi \times 3 + \frac{\pi}{5}\right) \text{m/s}^2 = -34.11 \text{ m/s}^2.$$

The acceleration is directed opposite to the displacement.

The initial phase constant $= \dfrac{\pi}{5}$.

The angular frequency $\omega = 0.7\pi \text{s}^{-1}$

Hence, the frequency of the oscillation $v = \dfrac{0.7\text{s}^{-1}}{2} = 0.35\text{s}^{-1}$

The time period of the oscillation $T = \dfrac{1}{v} = 2.86\text{s}$

1.3 Damped Harmonic Oscillation (DHO)

In case of simple harmonic oscillation, the amplitude of oscillation does not decrease. However in reality, in the case of simple pendulums, amplitude decreases with the passing of time. This is due to the viscosity of the medium. The force is the dissipative force F_d. Harmonic oscillation under the influence of a spring–like restoring force in a viscous medium is called damped harmonic oscillation.

For a low velocity dissipative force, F_d is directly proportional to the velocity of the oscillating body and is always directed opposite to the velocity of the body. Mathematically, we have

$$F_d = -b\frac{dx}{dt} \tag{1.20}$$

The minus sign appears because the direction of the dissipative force F_d is opposite to that of velocity. The restoring force on the oscillating body from Eq. (1.1) is $-kx$. Therefore, the total force on the vibrating body in a dissipative medium is

$$F_{\text{total}} = -b\frac{dx}{dt} - kx \tag{1.21}$$

or $\quad ma = -b\dfrac{dx}{dt} - kx$

or $\quad \dfrac{d^2x}{dt^2} + \dfrac{b}{m}\dfrac{dx}{dt} + \dfrac{k}{m}x = 0 \tag{1.22}$

Let $\dfrac{b}{m} = 2\gamma$ and $\dfrac{k}{m} = \omega_0^2$

γ and $\dfrac{\omega_0}{2\pi}$ are the damping coefficient of the dissipative medium and the natural frequency of the damped harmonic oscillator. Putting these substitutions into Eq. (1.22), we get

$$\frac{d^2x}{dt^2} + 2\gamma \frac{dx}{dt} + \omega_0^2 x = 0 \qquad (1.23)$$

Equation (1.23) is the differential equation of motion of a damped harmonic oscillation in a dissipative medium. The solutions of this differential equation depend upon the relative values of γ and ω_0. The general solution of the differential Eq. (1.23) is determined in the following way. Other methods for the purpose are also available.

Assuming a solution of the form $x = e^{pt}$, from Eq. (1.23), we have

$$p^2 + 2\gamma p + \omega_0^2 = 0$$

or $\qquad p = -\gamma \pm \sqrt{\gamma^2 - \omega_0^2}$ $\qquad\qquad\qquad\qquad\qquad$ (1.24)

Depending upon the values of γ, three different situations can arise for a damped harmonic oscillator. If the medium and frequency of the oscillator is such that the $\gamma < \omega_0$ condition is satisfied, the oscillator is said to be underdamped. If the medium and frequency of the oscillator is such that the $\gamma > \omega_0$ condition is satisfied, the oscillator is said to be overdamped. If the medium and frequency of the oscillator is such that the $\gamma = \omega_0$ condition is satisfied, the oscillator is said to be critically damped.

Case 1: $\gamma < \omega_0$ (Under damped)

From Eq. (1.24), we have

$$p = -\gamma \pm i\omega_1 \quad \text{where} \quad \omega_1 = \sqrt{\omega_0^2 - \gamma^2}$$

The general solution of the differential Eq. (1.23) will be given by

$$x = C_1 e^{-\gamma t + i\omega_1 t} + C_2 e^{-\gamma t - i\omega_1 t} \qquad (1.25)$$

The constants C_1 and C_2 must be complex in order that Eq. (1.25) is the general solution. Since $e^{-\gamma t + i\omega_1 t}$ and $e^{-\gamma t - i\omega_1 t}$ are complex conjugates of each other, the constants C_1 and C_2 must be complex conjugates of each other, so that x is real. For this, we set

$$C_1 = C = A e^{i\theta}$$

and $C_2 = C^* = Ae^{-i\theta}$

Upon this substitution in Eq. (1.25), we get

$$x = Ce^{-\gamma t + i\omega_1 t} + C^* e^{-\gamma t - i\omega_1 t} = Ae^{i\theta} e^{-\gamma t + i\omega_1 t} + Ae^{-i\theta} e^{-\gamma t - i\omega_1 t}$$

or $x = Ae^{-\gamma t} \left(e^{i(\omega_1 t + \theta)} + e^{-i(\omega_1 t + \theta)} \right)$

or $x = 2Ae^{-\gamma t} \cos\left(\omega_1 t + \theta\right)$

Putting $2A = r$ into the aforementioned equation, we get

$$x = re^{-\gamma t} \cos\left(\omega_1 t + \theta\right) \tag{1.26}$$

Here the values of the constants 'r' and 'θ' depend upon the initial conditions. The frequency of oscillation in this case is $\dfrac{\omega_1}{2\pi}$ and the amplitude is $re^{-\gamma t}$ which decreases exponentially with time. The phenomenon is plotted in the Fig. 1.4. This frequency does not change as the oscillations decay.

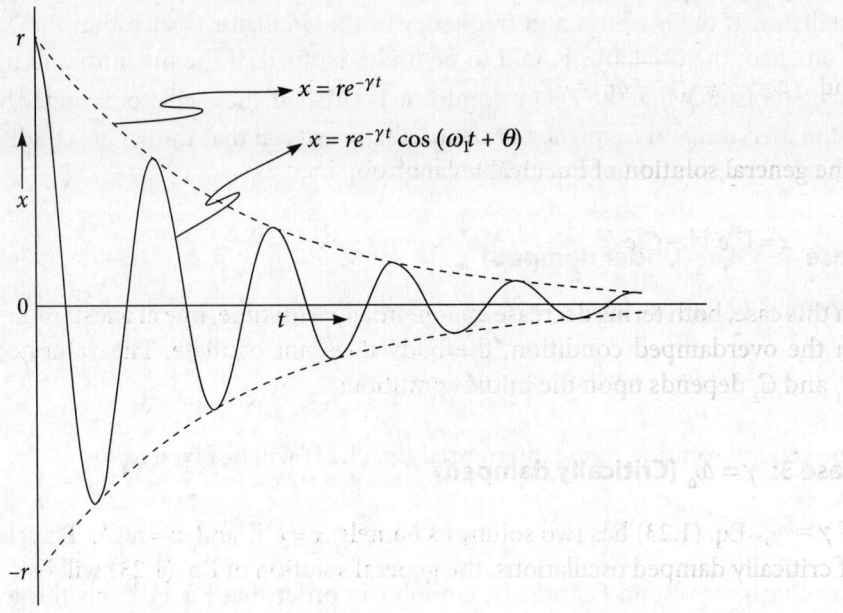

Figure 1.4 | Motion of a damped harmonic oscillator. The solid curve is the plot of $x = re^{-\gamma t} \cos \omega t$ and the dashed curve is the plot of $x = \pm re^{-\gamma t}$

The total energy of the damped harmonic oscillator according to Eq. (1.19) may be given as

$$E = E_{P\,max} = \left(\frac{1}{2}kx^2\right)_{max} = \left(\frac{1}{2}kr^2 e^{-2\gamma t}\cos^2(\omega_1 t + \theta)\right)_{max}$$

$$= \frac{1}{2}m\omega_0^2 r^2 e^{-2\gamma t}$$

or $E = E_0 e^{-2\gamma t}$ (1.27)

Here $E_0 = \frac{1}{2}m\omega_0^2 r^2$ is the total initial energy of the damped harmonic oscillator at $t = 0$. Thus, the total energy decreases more rapidly than the amplitude with time.

Case 2: $\gamma > \omega_0$ (Over damped)

If $\gamma > \omega_0$, $\sqrt{\gamma^2 - \omega_0^2}$ is real and the two solutions of p are

$$p = -\gamma - \sqrt{\gamma^2 - \omega_0^2} = -\gamma_1$$

and $p = -\gamma + \sqrt{\gamma^2 - \omega_0^2} = -\gamma_2$

The general solution of Eq. (1.23) will be obtained as

$$x = C_1 e^{-\gamma_1 t} + C_2 e^{-\gamma_2 t}$$ (1.28)

In this case, both terms decrease exponentially with time, one at a faster rate than the other. In the overdamped condition, the body does not oscillate. The values of the constants C_1 and C_2 depends upon the initial conditions.

Case 3: $\gamma = \omega_0$ (Critically damped)

If $\gamma = \omega_0$, Eq. (1.23) has two solutions namely $x = e^{-\gamma t}$ and $x = te^{-\gamma t}$. Therefore, in this case of critically damped oscillations, the general solution of Eq. (1.23) will be

$$x = (C_1 + C_2 t)e^{-\gamma t}$$ (1.29)

and declines exponentially at a faster rate than that of Case 2. In this case, the body comes to rest in finite time without oscillation.

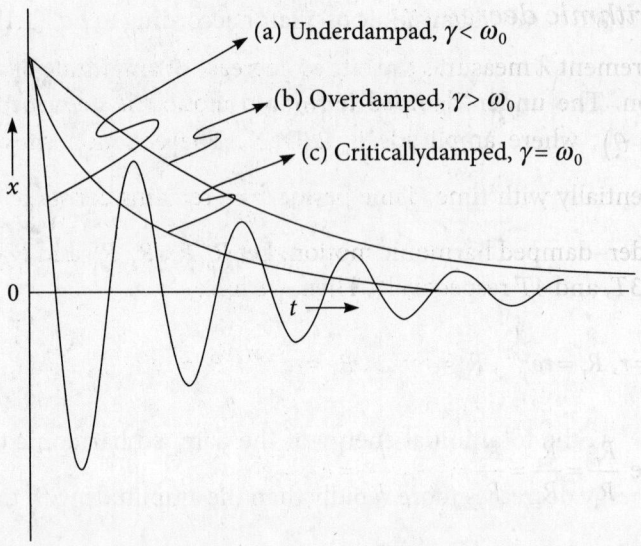

Figure 1.5 | Return of a harmonic oscillator to equilibrium position. (a) Under damped (b) Over damped (c) Critically damped

All the three cases are plotted in Fig. 1.5.

The pointer reading meters, hydraulic and pneumatic spring returns for door are so designed that they come to equilibrium/rest position quickly without over shoots or oscillation. These devices are so designed that the condition of critical damping $\gamma = \omega_0$ is satisfied. If the condition of over damping $\gamma > \omega_0$ is satisfied in these devices, it will take a long time to come to equilibrium position. If the condition of under–damping $\gamma < \omega_0$ is satisfied in these devices, it will over shoot the equilibrium position creating noise and will thus affect the life span of the devices.

1.3.1 Damping of an oscillator

Damping is an effect that reduces the amplitude of oscillations in an oscillatory system, particularly the harmonic oscillator. In a simple harmonic oscillator, total energy $\left(E_0 = \dfrac{1}{2} m\omega_0^2 r^2 \right)$ which is proportional to square of amplitude remains constant throughout time and positions. Due to the presence of damping force $-2m\gamma \dfrac{dx}{dt}$ in the case of under–damped harmonic motion, the amplitude R and energy E of the oscillator becomes $R = re^{-\gamma t}$ and $E = E_0 e^{-2\gamma t}$ respectively. The larger the value of the damping coefficient γ, the more rapidly does the amplitude and energy decrease. Measuring natural decay in terms of the fraction e^{-1} of the original value is a very common practice in physics. Thus, we can make use of the exponential factor to express the rates at which the amplitude and energy are decreased. The time for a natural decay process to reach zero is of course theoretically infinite. We shall discuss the damping of an oscillator in three different ways.

1.3.1(a) Logarithmic decrement

Logarithmic decrement λ measures the rate of decrease of amplitude of an under–damped harmonic motion. The under–damped harmonic motion is described by the equation $x = re^{-\gamma t}\cos(\omega_1 t + \theta)$, where amplitude is $R = re^{-\gamma t}$. The equation shows that amplitude decreases exponentially with time. Time period $\dfrac{2\pi}{\omega_1}$ remains constant. Let T be the time period of the under–damped harmonic motion. Let R_0, R_1, R_2, R_3, and R_4 be the amplitudes at time 0, T, $2T$, $3T$, and $4T$ respectively. Then, we have

$$R_O = re^{-\gamma 0} = r,\ R_1 = re^{-\gamma T},\ R_2 = re^{-\gamma 2T},\ R_3 = re^{-\gamma 3T},\ R_4 = re^{-\gamma 4T},\text{ etc.}$$

Now we can have $\dfrac{R_0}{R_1} = \dfrac{R_1}{R_2} = \dfrac{R_2}{R_3} = \dfrac{R_3}{R_4} = \dfrac{R_4}{R_5} = e^{\gamma T} = e^{\lambda}$

where $\lambda = \gamma T$ is called logarithmic decrement. Taking the natural logarithm of the aforementioned equation, we get

$$\ln\frac{R_0}{R_1} = \ln\frac{R_1}{R_2} = \ln\frac{R_2}{R_3} = \ln\frac{R_3}{R_4} = \ln\frac{R_4}{R_5} = \ln e^{\gamma T} = \ln e^{\lambda} = \lambda$$

In general, we can have

$$\ln\frac{R_{n-1}}{R_n} = \lambda = \gamma T \tag{1.30}$$

Now we can define the logarithmic decrement in the following way. Logarithmic decrement is defined as the natural logarithm of the ratio of two consecutive and successive amplitudes which are separated by one time period. In Eq. (1.30), amplitudes R_n and R_{n-1} are separated by one time period.

Example 1.4

A point performs a damped harmonic oscillation with frequency ω_1 and damping coefficient γ. Find the initial amplitude 'r' and the initial phase θ if at the moment $t = 0$ the displacement of the point and its velocity are $y(0) = 0$ and $v(0) = b$ respectively.

Solution

The equation of an oscillatory damped harmonic motion is given by

$$y(t) = re^{-\gamma t}\cos(\omega_1 t + \theta) \tag{A}$$

The velocity of the oscillating point is given by

$$v(t) = \frac{dy}{dt} = r\omega_1 e^{-\gamma t} \sin(\omega_1 t + \theta) - r\gamma e^{-\gamma t} \cos(\omega_1 t + \theta)$$ (B)

Putting $t = 0$ into (A), we get

$$y(0) = re^{-\gamma 0} \cos(\omega_1 \times 0 + \theta) = r \cos\theta$$

or $0 = r\cos\theta$

or $\theta = \dfrac{\pi}{2}$ or $-\dfrac{\pi}{2}$

Case 1: Suppose $\theta = \dfrac{\pi}{2}$

Putting $t = 0$ and $\theta = \dfrac{\pi}{2}$ into (B), we get

$$v(0) = r\omega_1 \ e^{-\gamma 0} \sin\left(\omega_1 \times 0 + \frac{\pi}{2}\right) - r\gamma e^{-\gamma 0} \cos\left(\omega_1 \times 0 + \frac{\pi}{2}\right)$$

$$b = r\omega_1 \times 1$$

$$r = \frac{b}{\omega_1} \text{ when } \theta = \frac{\pi}{2}$$

Case 2: Suppose $\theta = -\dfrac{\pi}{2}$

Putting $t = 0$ and $\theta = -\dfrac{\pi}{2}$ into (B), we get

$$v(0) = r\omega_1 \ e^{-\gamma 0} \sin\left(\omega_1 \times 0 - \frac{\pi}{2}\right) - r\gamma \ e^{-\gamma 0} \cos\left(\omega_1 \times 0 - \frac{\pi}{2}\right)$$

$$b = r\omega_1 \times (-1)$$

$$r = \frac{-b}{\omega_1} \quad \text{when} \quad \theta = -\frac{\pi}{2}$$

Example 1.5

The amplitude of an under–damped oscillator falls to 1/10 of its initial value after 10 oscillations. If the time period is 2 s, calculate the (a) damping coefficient, (b) logarithmic decrement, (c) time during which energy falls to 1/10 of its original value.

Solution

The amplitude of the under–damped harmonic oscillation after n number of oscillations is given by

$$R_n = re^{-\gamma n T}$$

where T is the time period.

or $\quad \dfrac{r}{R_n} = e^{\gamma n T}$

(a) According to the problem, $n = 10$, $T = 2$ s, $\dfrac{r}{R_{10}} = 10$

So we have

$$10 = e^{\gamma \times 10 \times 2}$$

or $\quad 20\gamma = \ln 10$

or $\quad \gamma = \dfrac{2.3}{20} = 0.115$

(b) Logarithmic decrement λ is given by

$$\lambda = \gamma T = 0.115 \times 2 = 0.23$$

(c) The energy of an under–damped harmonic oscillator is given by

$$E = E_0 e^{-2\gamma t}$$

or $\quad \dfrac{E}{E_O} = e^{-2\gamma t}$

According to the given data $\dfrac{E}{E_0} = \dfrac{1}{10}$

or $\dfrac{1}{10} = e^{-2\gamma t}$

or $e^{2\gamma t} = 10$

or $2\gamma t = \ln 10$

or $t = \dfrac{\ln 10}{2 \times 0.115} = 10\text{s}.$

Example 1.6

A body performs damped harmonic oscillation according to the equation $x = re^{-\gamma t} \sin \omega t$. Calculate the (a) amplitude of oscillation and (b) velocity of the body at time $t = 0$; (c) the moments of time at which body reaches the extreme positions.

Solution

(a)   The equation of motion of the damped harmonic oscillator is given as $x = re^{-\gamma t} \sin \omega t$. The amplitude of oscillation at any time is

$$R_t = re^{-\gamma t}.$$

So the amplitude of oscillation at time $t = 0$ will be

$$R_O = re^{-\gamma \times 0} = r$$

(b)   The instantaneous velocity v of the damped harmonic oscillator at any time is given by

$$v = \frac{d}{dt}re^{-\gamma t} \sin \omega t = -r\gamma e^{-\gamma t} \sin \omega t + r\omega e^{-\gamma t} \cos \omega t \qquad \text{(A)}$$

So the velocity v of the damped harmonic oscillator at time $t = 0$ will be

$$v(0) = r\omega$$

Since the body is performing damped harmonic motion, the velocity of the body at extreme positions is zero. If the body reaches the extreme positions at time t_n, then from (A), we have

$$r\gamma e^{-\gamma t_n} \sin \omega t_n = r\omega e^{-\gamma t_n} \cos \omega t_n$$

or $t_n = \dfrac{1}{\omega}\left(\tan^{-1} \dfrac{\omega}{\gamma} + n\pi \right)$

Example 1.7

A body performs damped harmonic oscillation according to the equation $x = re^{-\gamma t}\cos(\omega t + \theta)$. Calculate the (a) initial amplitude r of oscillation and (b) initial phase θ if at time $t = 0$, velocity of the body is u and $x = 0$.

Solution

According to the problem, at $t = 0$, $x = 0$. So putting $t = 0$ and $x = 0$ in the equation $x = re^{-\gamma t}\cos(\omega t + \theta)$, we get

$$0 = re^{-\gamma 0}\cos\left(\omega \times 0 + \theta\right)$$

or $\cos\theta = 0$

or $\theta = \pm\dfrac{\pi}{2}$

The instantaneous velocity v of the damped harmonic oscillator at any time is given by

$$v = \frac{d}{dt}re^{-\gamma t}\cos(\omega t + \theta)$$

$$= -r\gamma e^{-\gamma t}\cos\left(\omega t + \theta\right) - r\omega e^{-\gamma t}\sin\left(\omega t + \theta\right) \tag{A}$$

According to the problem, at $t = 0$, $v = u$. So putting $t = 0$ and $v = u$ in (A), we get

$$u = -r\gamma\cos\theta - r\omega\sin\theta \tag{B}$$

If $\theta = \dfrac{\pi}{2}$ is put in (B), u becomes negative and if $\theta = -\dfrac{\pi}{2}$ is put in (B), u becomes positive. Thus, we have

$$\theta = \frac{\pi}{2} \text{ if } u < 0$$

and

$$\theta = -\frac{\pi}{2} \text{ if } u > 0$$

Putting $\theta = \pm\dfrac{\pi}{2}$ in (B), we get

$$|u| = r\omega$$

or   Initial amplitude of oscillation r will be $r = \dfrac{|u|}{\omega}$

1.3.1(b) Modulus of decay

The amplitude of an under–damped harmonic oscillator is given as

$$R_t = R_0 e^{-\gamma t}$$

The time interval during which amplitude decreases to $\dfrac{1}{e}$ of the original amplitude is called modulus of decay or relaxation time τ. At $t = \dfrac{1}{\gamma}$ seconds, the amplitude becomes $\dfrac{R_0}{e}$ and hence, $\dfrac{1}{\gamma}$ is the modulus of decay or relaxation time i.e.,

$$\tau = \frac{1}{\gamma} \tag{1.31}$$

1.3.1(c) Quality factor of an under damped harmonic oscillator

This measures the rate at which the energy decays. The energy of an under damped harmonic oscillator at any instant of time is given as

$$E = E_0 e^{-2\gamma t} \tag{1.32}$$

The quality factor Q is defined as the number of radians through which an under–damped system goes so that its energy decreases to $\dfrac{E_0}{e}$. During time interval $\dfrac{1}{2\gamma}$, energy decreases to $\dfrac{E_0}{e}$.

During time interval T seconds (time period), the oscillator traces 2π radians. Then during the time interval $\dfrac{1}{2\gamma}$ seconds, the oscillator traces $\dfrac{2\pi}{T} \times \dfrac{1}{2\gamma} = \dfrac{\omega_1}{2\gamma}$ radians. By definition, quality factor Q is given by

$$Q = \frac{\omega_1}{2\gamma} = \frac{1}{2}\omega_1 \tau \tag{1.33}$$

In the aforementioned expression, ω_1 is the angular frequency of the under–damped oscillator.

From Eq. (1.32), we get

$$-dE = 2\gamma E_0 e^{-2\gamma t} dt$$

So the energy lost per cycle

$$-dE = 2\gamma E_0 e^{-2\gamma t} \times T = E_0 e^{-2\gamma t} \times \frac{2\pi}{Q}$$

or $$\frac{-dE}{E_0 e^{-2\gamma t}} = \frac{2\pi}{Q}$$

or $$Q = 2\pi \frac{E_0 e^{-2\gamma t}}{-dE}$$

Thus, quality factor can be defined as

$$Q = 2\pi \frac{\text{Energy stored in the system}}{\text{Energy lost per cycle}}$$

(1.34)

1.4 Forced Vibrations

The real life phenomena of forced vibrations can be observed in the vibration of a bridge under the influence of marching soldiers, vibration of an electric motor due to the periodic impulses from an irregularity in the shaft, vibration of a tuning fork when subjected to the periodic force of a sound wave, and so on. The harmonic oscillations under the influence of an externally applied sinusoidally varying force are called forced vibrations.

Let the externally applied sinusoidally varying force be

$$F_0 \cos(\omega t + \theta_0)$$

(1.35)

where
F_0 = Amplitude of the externally applied sinusoidally varying force.

ω = Angular frequency of the externally applied sinusoidally varying force.

θ_0 = Initial phase of the externally applied sinusoidally varying force.

From Eq. (1.20), the dissipative force on the oscillator is $F_d = -b\dfrac{dx}{dt}$ and from Eq. (1.1), the restoring force on the oscillating body is $-kx$. Therefore, the total force on the vibrating body in this case is

$$F_{\text{Total}} = -b\frac{dx}{dt} - kx + F_0 \cos(\omega t + \theta_0)$$

By applying Newton's law of motion to the aforementioned equation, the equation of motion of a forced harmonic oscillator is obtained as

$$\frac{d^2x}{dt^2} + 2\gamma\frac{dx}{dt} + \omega_0^2 x = \frac{F_0}{m}\cos(\omega t + \theta_0)$$

(1.36)

where $\frac{b}{m} = 2\gamma$ and $\frac{k}{m} = \omega_0^2$. γ and $\frac{\omega_0}{2\pi}$ are the damping coefficients of the dissipative medium and the natural frequency of an undamped harmonic oscillator respectively. Equation (1.36) is the differential equation of motion of a forced vibration in a dissipative medium. The general solution of the differential Eq. (1.36) is determined in the following way using complex functions. Other methods for the purpose are also available. Equation (1.36) is modified to

$$\frac{d^2x}{dt^2} + 2\gamma\frac{dx}{dt} + \omega_0^2 x = \frac{1}{m}F_0 e^{i\theta_0} e^{i\omega t}$$

(1.37)

where $\frac{F_0}{m}\cos(\omega t + \theta_0)$ is the real part of $\frac{1}{m}F_0 e^{i\theta_0} e^{i\omega t}$, i.e.,

$$\text{Re}\left(\frac{1}{m}F_0 e^{i\theta_0} e^{i\omega t}\right) = \frac{F_0}{m}\cos(\omega t + \theta_0)$$

Assume a complex solution of the differential Eq. (1.37) of the form

$$x = x_0 e^{i\omega t}$$

(1.38)

Hence, the complex velocity of the forced oscillator is

$$\frac{dx}{dt} = i\omega x_0 e^{i\omega t}$$

(1.39)

and the complex acceleration of the forced oscillator is

$$\frac{d^2x}{dt^2} = -\omega^2 x_0 e^{i\omega t}$$

(1.40)

Putting Eq. (1.38) into Eq. (1.40) and Eq. (1.37), we get

$$-\omega^2 x_0 e^{i\omega t} + i2\gamma\omega x_0 e^{i\omega t} + \omega_0^2 x_0 e^{i\omega t} = \frac{1}{m}F_0 e^{i\theta_0} e^{i\omega t}$$

or $\quad x_0 = \dfrac{\dfrac{F_0}{m}e^{i\theta_0}}{\omega_0^2 - \omega^2 + 2i\gamma\omega} = \dfrac{\dfrac{F_0}{m}e^{i\theta_0}}{\left[\left(\omega_0^2 - \omega^2\right) + 4\gamma^2\omega^2\right]^{\frac{1}{2}} e^{i\tan^{-1}\frac{2\gamma\omega}{\omega_0^2 - \omega^2}}}$ (1.41)

Let

$$\tan^{-1}\frac{2\gamma\omega}{\omega_0^2 - \omega^2} = \frac{\pi}{2} - \beta$$ (1.42)

Upon this substitution in Eq. (1.41), we get

$$x_0 = \dfrac{\dfrac{F_0}{m}e^{i\theta_0}}{\left[\left(\omega_0^2 - \omega^2\right) + 4\gamma^2\omega^2\right]^{\frac{1}{2}} e^{i\left(\frac{\pi}{2} - \beta\right)}} = \dfrac{\dfrac{F_0}{m}e^{i\theta_0} e^{-i\frac{\pi}{2}} e^{i\beta}}{\left[\left(\omega_0^2 - \omega^2\right) + 4\gamma^2\omega^2\right]^{\frac{1}{2}}}$$

$$= \dfrac{-i\dfrac{F_0}{m}e^{i(\theta_0 + \beta)}}{\left[\left(\omega_0^2 - \omega^2\right) + 4\gamma^2\omega^2\right]^{\frac{1}{2}}}$$

or $\quad x_0 = \dfrac{\dfrac{F_0}{m}e^{i(\theta_0 + \beta)}}{i\left[\left(\omega_0^2 - \omega^2\right) + 4\gamma^2\omega^2\right]^{\frac{1}{2}}}$ (1.43)

Putting this value of x_0 in Eq. (1.38), we obtain the complex solution of the differential Eq. (1.38) as

$$x = \dfrac{\dfrac{F_0}{m}e^{i(\omega t + \theta_0 + \beta)}}{i\left[\left(\omega_0^2 - \omega^2\right)^2 + 4\gamma^2\omega^2\right]^{\frac{1}{2}}}$$ (1.44)

This is the expression for a complex position. The real position will be the real part of the aforementioned equation. Hence, the real position of the forced harmonic oscillator is given by

$$x = \frac{F_0}{m\left[\left(\omega_0^2 - \omega^2\right)^2 + 4\gamma^2\omega^2\right]^{\frac{1}{2}}} \sin\left(\omega t + \theta_0 + \beta\right) \tag{1.45}$$

The general solution of the differential equation of motion (1.37) of a forced harmonic oscillation is

$$x = re^{-\gamma t}\cos\left(\omega_1 t + \theta\right) + \frac{F_0}{m\left[\left(\omega_0^2 - \omega^2\right)^2 + 4\gamma^2\omega^2\right]^{\frac{1}{2}}} \sin\left(\omega t + \theta_0 + \beta\right) \tag{1.46}$$

This above solution contains two constants namely r and θ which can be determined from the initial boundary conditions. The first term dies out exponentially in time and is called a transient term. After some time, the driven body starts oscillating with the frequency of the driving force or the applied force. This state is said to be the steady state of the forced vibrations. The second term is called the steady state term and oscillates with constant amplitude $\dfrac{F_0}{m\left[\left(\omega_0^2 - \omega^2\right)^2 + 4\gamma^2\omega^2\right]^{\frac{1}{2}}}$. In steady state, Eq. (1.46) becomes

$$x = \frac{F_0}{m\left[\left(\omega_0^2 - \omega^2\right)^2 + 4\gamma^2\omega^2\right]^{\frac{1}{2}}} \sin\left(\omega t + \theta_0 + \beta\right) \tag{1.47}$$

The phase of oscillation of the forced oscillator is $\left(\omega t + \theta_0 + \beta\right)$ and that of the applied force is $\left(\omega t + \theta_0\right)$. Hence, the phase difference between oscillation and the applied force is β and is defined mathematically in Eq. (1.42) as

$$\tan^{-1}\frac{2\gamma\omega}{\omega_0^2 - \omega^2} = \frac{\pi}{2} - \beta$$

or $\quad \beta = \tan^{-1}\dfrac{\omega_0^2 - \omega^2}{2\gamma\omega} \tag{1.48}$

This equation shows that the phase difference between oscillation and the applied force depends upon the frequency of the applied force and the damping coefficient of the medium.

1.4.1 Velocity of the forced harmonic oscillator

The complex velocity of the forced harmonic oscillator is obtained by differentiating Eq. (1.44) with respect to time as

$$v = i\omega \frac{\dfrac{F_0}{m} e^{i(\omega t + \theta_0 + \beta)}}{i\left[\left(\omega_0^2 - \omega^2\right)^2 + 4\gamma^2\omega^2\right]^{\frac{1}{2}}} = \omega \frac{\dfrac{F_0}{m} e^{i(\omega t + \theta_0 + \beta)}}{\left[\left(\omega_0^2 - \omega^2\right)^2 + 4\gamma^2\omega^2\right]^{\frac{1}{2}}}$$

The real velocity will be the real part of the aforementioned equation. Hence, the real velocity of the forced harmonic oscillator will be

$$v = \frac{\omega F_0}{m\left[\left(\omega_0^2 - \omega^2\right)^2 + 4\gamma^2\omega^2\right]^{\frac{1}{2}}} \cos\left(\omega t + \theta_0 + \beta\right) \tag{1.49}$$

1.4.2 Total energy of the forced harmonic oscillator

The instantaneous kinetic energy E_K of the forced harmonic oscillator is given as

$$E_K = \frac{1}{2}mv^2 .$$

Putting the value of v from Eq. (1.49) into the aforementioned equation, we get

$$E_K = \frac{1}{2} \frac{\omega^2 F_0^2}{m\left[\left(\omega_0^2 - \omega^2\right)^2 + 4\gamma^2\omega^2\right]} \cos^2\left(\omega t + \theta_0 + \beta\right) \tag{1.50}$$

The instantaneous potential energy E_P of the forced harmonic oscillator is given as

$$E_P = \frac{1}{2}kx^2$$

Putting the value of x from Eq. (1.47) into the aforementioned equation, we get

$$E_P = \frac{1}{2}m\omega_0^2 \times \frac{F_0^2}{m^2\left[\left(\omega_0^2 - \omega^2\right)^2 + 4\gamma^2\omega^2\right]} \sin^2\left(\omega t + \theta_0 + \beta\right)$$

or $E_P = \dfrac{1}{2} \dfrac{\omega_0^2 F_0^2}{m\left[\left(\omega_0^2 - \omega^2\right)^2 + 4\gamma^2\omega^2\right]} \sin^2\left(\omega t + \theta_0 + \beta\right)$ (1.51)

The total energy E of the forced harmonic oscillator will be given by

$E = E_K + E_P$

Putting the values of E_K and E_P from Eqs (1.50) and (1.51) into the aforementioned equation, we have

$$E = \dfrac{1}{2} \dfrac{\omega^2 F_0^2}{m\left[\left(\omega_0^2 - \omega^2\right)^2 + 4\gamma^2\omega^2\right]} \cos^2\left(\omega t + \theta_0 + \beta\right)$$

$$+ \dfrac{1}{2} \dfrac{\omega_0^2 F_0^2}{m\left[\left(\omega_0^2 - \omega^2\right)^2 + 4\gamma^2\omega^2\right]} \sin^2\left(\omega t + \theta_0 + \beta\right)$$

The total average energy $< E >$ of the forced harmonic oscillator over a complete cycle will be given by

$$< E > = \dfrac{1}{2} \dfrac{\omega^2 F_0^2}{m\left[\left(\omega_0^2 - \omega^2\right)^2 + 4\gamma^2\omega^2\right]} < \cos^2\left(\omega t + \theta_0 + \beta\right) >$$

$$+ \dfrac{1}{2} \dfrac{\omega_0^2 F_0^2}{m\left[\left(\omega_0^2 - \omega^2\right)^2 + 4\gamma^2\omega^2\right]} < \sin^2\left(\omega t + \theta_0 + \beta\right) >$$

or $< E > = \dfrac{1}{2} \dfrac{\omega_0^2 F_0^2}{m\left[\left(\omega_0^2 - \omega^2\right)^2 + 4\gamma^2\omega^2\right]} \times \dfrac{1}{2}$

$$+ \dfrac{1}{2} \dfrac{\omega_0^2 F_0^2}{m\left[\left(\omega_0^2 - \omega^2\right)^2 + 4\gamma^2\omega^2\right]} \times \dfrac{1}{2}$$

or $<E>= \dfrac{1}{4} \dfrac{F_0^2}{m\left[\left(\omega_0^2 - \omega^2\right)^2 + 4\gamma^2\omega^2\right]}\left(\omega^2 + \omega_0^2\right)$ (1.52)

1.4.3 Power of the forced harmonic oscillator

In steady state, the power dissipated by the oscillator due to a dissipative medium is compensated by the power absorption from the driving source. The power of a body 'P' moving under the action of force F with velocity v is defined as Fv. Therefore, the power dissipated by the forced harmonic oscillator will be given by

$P = Fv$

Putting the value of v from Eq. (1.49) and $F = F_0 \cos(\omega t + \theta_0)$, which is the externally applied sinusoidally varying force, we get the power of the forced harmonic oscillator as

$$P = \dfrac{F_0^2}{m}\dfrac{\omega}{\left[\left(\omega_0^2 - \omega^2\right)^2 + 4\gamma^2\omega^2\right]^{\frac{1}{2}}}\cos\left(\omega t + \theta_0\right)\cos\left(\omega t + \theta_0 + \beta\right)$$

$$= \dfrac{F_0^2}{m}\dfrac{\omega}{\left[\left(\omega_0^2 - \omega^2\right)^2 + 4\gamma^2\omega^2\right]^{\frac{1}{2}}}\cos\left(\omega t + \theta_0\right)$$

$$\times\left[\cos\left(\omega t + \theta_0\right)\cos\beta - \sin\left(\omega t + \theta_0\right)\sin\beta\right]$$

$$= \dfrac{F_0^2}{m}\dfrac{\omega}{\left[\left(\omega_0^2 - \omega^2\right)^2 + 4\gamma^2\omega^2\right]^{\frac{1}{2}}}\left[\cos\beta\cos^2\left(\omega t + \theta_0\right) - \dfrac{1}{2}\sin2\left(\omega t + \theta_0\right)\sin\beta\right]$$

or $P = \dfrac{F_0^2}{m}\dfrac{\omega\cos\beta\cos^2\left(\omega t + \theta_0\right)}{\left[\left(\omega_0^2 - \omega^2\right)^2 + 4\gamma^2\omega^2\right]^{\frac{1}{2}}} - \dfrac{F_0}{2m}\dfrac{\omega\sin\beta\sin2\left(\omega t + \theta_0\right)}{\left[\left(\omega_0^2 - \omega^2\right)^2 + 4\gamma^2\omega^2\right]^{\frac{1}{2}}}$ (1.53)

Now the average value of $\cos^2(\omega t + \theta_0)$ and $\sin 2(\omega t + \theta_0)$ over a complete cycle of the applied force is $\dfrac{1}{2}$ and zero respectively. Therefore, the average value of power $<P>$ over a complete cycle will be

$$<P> = \frac{F_0^2}{2m} \frac{\omega \cos \beta}{\left[\left(\omega_0^2 - \omega^2 \right)^2 + 4\gamma^2 \omega^2 \right]^{\frac{1}{2}}} \tag{1.54}$$

Now either from Eq. (1.48) or from Fig. 1.6, we can have

$$\cos \beta = \frac{2\gamma \omega}{\left[\left(\omega_0^2 - \omega^2 \right)^2 + 4\gamma^2 \omega^2 \right]^{\frac{1}{2}}}$$

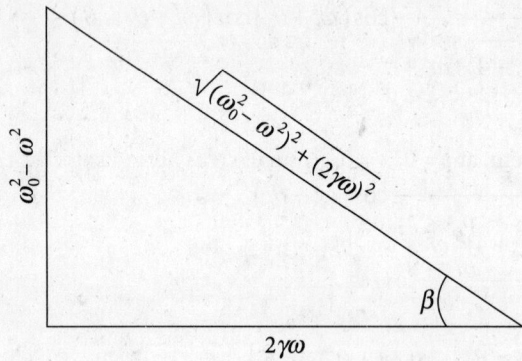

Figure 1.6 | Evaluation of trigonometric angles. This right triangle is drawn in accordance with Eq. (1.48)

Putting this value of $\cos \beta$ into Eq. (1.54), we get

$$<P> = \frac{F_0^2}{m} \frac{\gamma \omega^2}{\left[\left(\omega_0^2 - \omega^2 \right)^2 + 4\gamma^2 \omega^2 \right]^{\frac{1}{2}}} \tag{1.55}$$

This expression gives the amount of average power dissipated by the forced oscillator in the dissipative medium over a complete cycle of oscillation. In steady state, this must be equal to the power absorbed by the forced oscillator from the driving source.

Example 1.8

A ball of mass m performs undamped harmonic oscillations about the point $x = 0$ with natural frequency ω_0. At the moment $t = 0$, when the ball is in equilibrium position, a force $F_x = F_0 \cos \omega t$ along the x-axis was applied to it. Find the position of the ball at any time.

Solution

In this problem $\gamma = 0$ because the motion is undamped and $\theta_0 = 0$. So with the help of Eq. (1.46), we can determine the position of the ball by

$$x(t) = r \cos \omega t + \frac{\frac{F_0}{m}}{\omega_0^2 - \omega^2} \sin(\omega_0 t + \beta) \tag{A}$$

Putting $\beta = \dfrac{\pi}{2}$ into the Eq. (A), we get

$$x(t) = r \cos \omega t + \frac{\frac{F_0}{m}}{\omega_0^2 - \omega^2} \sin\left(\omega_0 t + \frac{\pi}{2}\right) = r \cos \omega t + \frac{\frac{F_0}{m}}{\omega_0^2 - \omega^2} \cos \omega_0 t \tag{B}$$

According to the problem, at $t = 0$, $x = 0$. Applying this boundary condition to (B), we get

$$x(0) = r \cos(\omega \times 0) + \frac{\frac{F_0}{m}}{\omega_0^2 - \omega^2} \cos(\omega_0 \times 0)$$

or $\quad r = -\dfrac{\frac{F_0}{m}}{\omega_0^2 - \omega^2}$

Putting the value of r into (B), we get

$$x(t) = -\frac{\frac{F_0}{m}}{\omega_0^2 - \omega^2} \cos \omega t + \frac{\frac{F_0}{m}}{\omega_0^2 - \omega^2} \cos \omega_0 t$$

or $\quad x(t) = \dfrac{\frac{F_0}{m}}{\omega_0^2 - \omega^2}\left(\cos \omega_0 t - \cos \omega t\right)$

1.5 Displacement Resonance

Forced vibrations occur if a system is continuously driven by an external agency. A simple example is a child's swing that is pushed on each downswing. Of special interest are systems undergoing harmonic oscillation and driven by sinusoidal force. This leads to the important phenomenon of resonance. Resonance occurs when the driving frequency approaches the natural frequency of free vibrations. The result is a rapid take-up of energy by the vibrating system, with a continuous growth of the vibration amplitude. A simple disturbance can set a harmonic oscillator into motion. Repeated disturbances can increase the amplitude of the oscillations if they are applied in phase with the natural frequency. Even a very small disturbance, repeated periodically at just the right frequency, can cause a very large amplitude motion to build up. In general, whenever a system capable of oscillation is acted upon by a periodic series of impulses having frequencies equal to or nearly equal to the natural frequencies of oscillation of the system, the system is set into oscillation with a relatively large amplitude. This phenomenon is known as resonance. Therefore we can define resonance as the phenomenon in which a body is set to violent vibrations by a strong periodic force whose frequency coincides exactly or nearly with the natural frequency of the body.

The amplitude of forced vibration in the steady state from Eq. (1.45) is

$$\frac{F_0}{m}\frac{1}{\sqrt{\left(\omega_0^2 - \omega^2\right)^2 + \left(2\gamma\omega\right)^2}}$$

and, as is obvious, it depends on the frequency of the driving force ω. This amplitude is maximum when the denominator is minimum, i.e., when

$$\frac{d}{d\omega}\left[\left(\omega_0^2 - \omega^2\right)^2 + \left(2\gamma\omega\right)^2\right]^{\frac{1}{2}} = 0$$

or $\omega_0^2 - \omega^2 = 2\gamma^2$

or $\omega = \omega_0 \sqrt{1 - 2\left(\dfrac{\gamma}{\omega_0}\right)^2}$ (1.56)

Thus, the amplitude is maximum when ω is given by Eq. (1.56). The frequency at which resonance occurs (i.e., amplitude of oscillation become maximum) is called resonant frequency. Thus, Eq. (1.56) gives the expression for resonant frequency. When damping is extremely small (i.e., $\gamma \to 0$), the resonance occurs at a frequency very close to the natural frequency of the system (i.e., $\omega \to \omega_0$ when $\gamma \to 0$). The variation of amplitude with ω is depicted in Fig. 1.7.

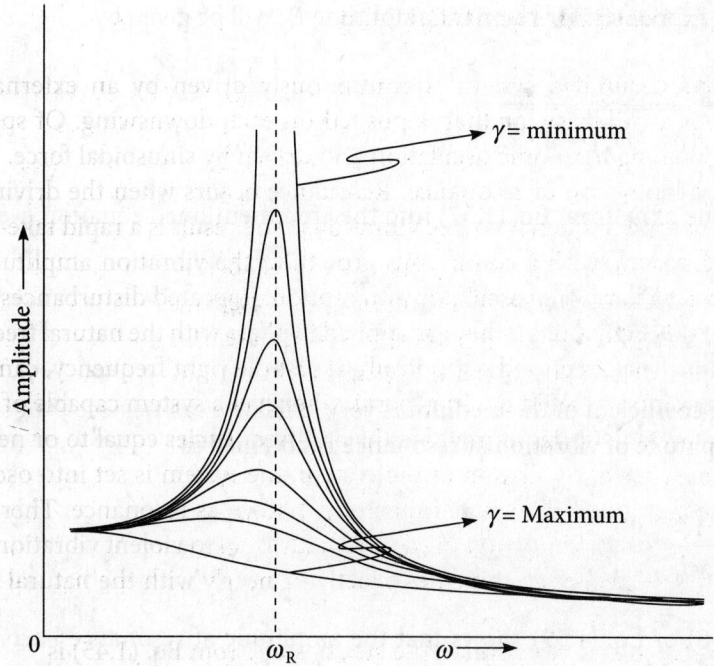

Figure 1.7 | The variation of amplitude with frequency of the applied force for various values of γ. Observe that with increase of damping, the resonance occurs at smaller values of ω

It is interesting to observe that when damping decreases, maximum amplitude becomes very sharp and the amplitude falls off rapidly as we go away from the resonant frequency.

1.5.1 Resonant amplitude

The amplitude of the forced oscillation at resonance is called resonant amplitude R_R and the corresponding frequency is called resonant frequency ω_R. The expression for resonant frequency ω_R is obtained from Eq. (1.56) as

$$\omega_R = \omega_0 \sqrt{1 - 2\left(\frac{\gamma}{\omega_0}\right)^2} \tag{1.57}$$

The amplitude of forced vibration in the steady state from Eq. (1.45) is

$$\frac{F_0}{m} \frac{1}{\sqrt{\left(\omega_0^2 - \omega^2\right)^2 + \left(2\gamma\omega\right)^2}}$$

and hence, the expression for resonant amplitude R_R will be given by

$$R_R = \frac{F_0}{m} \frac{1}{\sqrt{\left(\omega_0^2 - \omega_R^2\right)^2 + 4\gamma^2 \omega_R^2}}$$

Putting the value of ω_R from Eq. (1.57) into the aforementioned equation, we get

$$R_R = \frac{F_0}{m} \frac{1}{2\gamma\sqrt{\omega_0^2 - \gamma^2}} \qquad (1.58)$$

If the damping coefficient of the medium is very small, we can neglect γ^2 in comparison to ω_0^2 and the amplitude of vibration at resonance is obtained as

$$R_R = \frac{F_0}{m} \frac{1}{2\gamma\omega_0} \qquad (1.59)$$

Equations (1.58) or Eq. (1.59) shows that the amplitude at resonance becomes infinitely large when $\gamma \to 0$. Moreover, the same aforementioned equations shows that if the damping coefficient is very large (i.e., $\gamma \to \infty$), resonance cannot occur (i.e., $R_R \to 0$). The resonant frequency is inversely proportional to the damping coefficient for small damping.

1.5.2 Sharpness of resonance

The amplitude of vibration of an oscillatory system is maximum when the angular frequency of the sinusoidally varying applied force is equal to $\sqrt{\omega_0^2 - 2\gamma^2}$. The amplitude of vibration decreases when the angular frequency of the sinusoidally varying applied force is less than or more than $\sqrt{\omega_0^2 - 2\gamma^2}$. This has been depicted in Fig. 1.7. If the amplitude decreases rapidly for a slight departure of the frequency of the applied force from $\sqrt{\omega_0^2 - 2\gamma^2}$ the resonance is more sharp. If the amplitude decreases slowly for a large departure of the frequency of the applied force from $\sqrt{\omega_0^2 - 2\gamma^2}$, the resonance is less sharp. How rapidly the amplitude decreases on either side of $\sqrt{\omega_0^2 - 2\gamma^2}$ is represented by the sharpness of resonance.

The sharpness of resonance depends upon the damping coefficient. For small damping $(\gamma \to 0)$ at resonance, we can have

$$\omega_R \approx \omega_0$$

or $\quad \omega_R = \omega_0 \pm \gamma \qquad (\because \quad \gamma \text{ is very small })$

or $\quad \omega_1 = \omega_0 + \gamma$

and

$$\omega_2 = \omega_0 - \gamma$$

From the last two equations here, we have

$$\omega_1 - \omega_2 = 2\gamma = \frac{\omega_R}{Q} \qquad (1.60)$$

Hence, $\omega_1 - \omega_2 = 2\gamma$ is the width of the resonance curve. The width of the resonance curve $\omega_1 - \omega_2$ is approximately proportional to the sharpness of resonance. Therefore, the sharpness of resonance is approximately proportional to the damping coefficient. More less the damping, more sharp is the resonance curve and vice versa. The more is the quality factor Q, the more sharp is the resonance curve.

1.5.3 Quality factor of a forced harmonic oscillator

According to relation (1.34), the quality factor Q of a forced harmonic oscillator can be defined as

$$Q = 2\pi \frac{\text{average energy stored per cycle}}{\text{average energy dissipated per cycle}} \qquad (1.61)$$

or $$Q = 2\pi \frac{<E>}{T<P>},$$

T = time period of the applied force

or $$Q = \omega \frac{<E>}{<P>} \qquad (1.62)$$

Putting the values of $<E>$ and $<P>$ from Eqs (1.52) and (1.55) respectively into Eq. (1.62) at resonance, we get

$$Q = \omega_R \frac{\dfrac{F_0^2\left(\omega_R^2 + \omega_0^2\right)}{4m\left[\left(\omega_0^2 - \omega_R^2\right)^2 + 4\gamma^2\omega_R^2\right]}}{\dfrac{\gamma\omega_R^2 F_0^2}{m\left[\left(\omega_R^2 - \omega_0^2\right)^2 + 4\gamma^2\omega^2\right]}}$$

or
$$Q = \frac{\omega_R^2 + \omega_0^2}{4\gamma\omega_R} \qquad (1.63)$$

For small damping, $\omega_R \approx \omega_0$. Therefore, for small damping, the quality factor from Eq. (1.63) becomes

$$Q = \frac{2\omega_R^2}{4\gamma\omega_R} = \frac{\omega_R}{2\gamma} \qquad (1.64)$$

Equations (1.63) and (1.64) show that larger the value of quality factor (i.e., lesser is γ), sharper is the resonance.

At displacement resonance from Eq. (1.58), we have

$$R_R = \frac{F_0}{m} \frac{1}{2\gamma\sqrt{\omega_0^2 - \gamma^2}}$$

At low frequency from Eq. (1.47), we have

$$R_0 = \frac{F_0}{m\omega_0^2}$$

or
$$\frac{R_R}{R_0} = \frac{F_0}{m} \frac{1}{2\gamma\sqrt{\omega_0^2 - \gamma^2}} \times \frac{m\omega_0^2}{F_0} = \frac{\omega_0}{2\gamma\left(1 - \dfrac{\gamma^2}{\omega_0^2}\right)^{1/2}}$$

or
$$\frac{R_R}{R_0} = \frac{Q}{\left(1 - \dfrac{1}{4Q^2}\right)^{1/2}} \approx Q\left(1 + \frac{1}{8Q^2}\right) \approx Q$$

The aforementioned equation predicts that the value of Q can be used as an amplification factor in different branches of science and technology.

1.5.4 Examples of resonance

i.   When a tuning fork is struck and its stem is placed on the sonometer, the air inside the sonometer is set into resonant vibration.

ii. Soldiers marching over a suspension bridge are always advised to break their steps. If incidentally the frequency of the stepping of the soldiers coincide with the natural frequency of the bridge, violent oscillations set in, making the bridge to collapse. This was what happened to Tacoma Narrows Bridge (Washington, U.S.A.). In 1940, the newly constructed bridge collapsed by a mild storm just four months after it was built.

iii. Dancing of a small object placed on sound box.

iv. We are able to listen to a particular radio station because the tuned circuit in the radio set resonate at the frequency of the incoming electromagnetic wave which nearly coincide with its own natural frequency.

v. A stationary tuning fork vibrates when another vibrating tuning fork having a natural frequency equal to that of the stationary one is brought near it.

Example 1.9

Forced harmonic oscillations have the same displacement amplitudes at the natural frequencies $\omega_1 = 400/s$ and $\omega_2 = 800/s$. Calculate the resonant frequency at which the displacement is maximum.

Solution

The amplitude of oscillation in case of forced vibration is given by

$$\frac{\dfrac{F_0}{m}}{\sqrt{\left(\omega^2 - \omega_0^2\right)^2 + \left(2\gamma\omega_0\right)^2}}$$

According to the problem,

$$\frac{\dfrac{F_0}{m}}{\sqrt{\left(\omega^2 - \omega_1^2\right)^2 + \left(2\gamma\omega_1\right)^2}} = \frac{\dfrac{F_0}{m}}{\sqrt{\left(\omega^2 - \omega_2^2\right)^2 + \left(2\gamma\omega_2\right)^2}}$$

or $2\omega^2 - \omega_1^2 - \omega_2^2 = 4\gamma^2$ (A)

From Eq. (1.56), the resonant frequency is given by

$$\omega_{res} = \sqrt{\omega^2 - 2\gamma^2}$$

Putting the value of $2\gamma^2$ from (A) into this equation, we have

$$\omega_{res} = \sqrt{\omega^2 - \frac{2\omega^2 - \omega_1^2 - \omega_2^2}{2}}$$

or $\omega_{res} = \sqrt{\dfrac{\omega_1^2 + \omega_2^2}{2}}$

According to the problem, $\omega_1 = 400/s$ and $\omega_2 = 800/s$. Putting these values into the aforementioned equation, we get the resonant frequency at which the displacement is maximum as

$$\omega_{res} = \sqrt{\frac{400^2 + 800^2}{2}} = 632.46/s \quad \text{or} \quad v_{res} = \frac{\omega_{res}}{2\pi} = 100 \text{ Hz.}$$

1.6 Coupled Oscillators

In the section on simple harmonic oscillators, the motion of a single particle held in place by springs was considered. In this section, the motion of a group of particles bound by spring-like forces to one another is discussed.

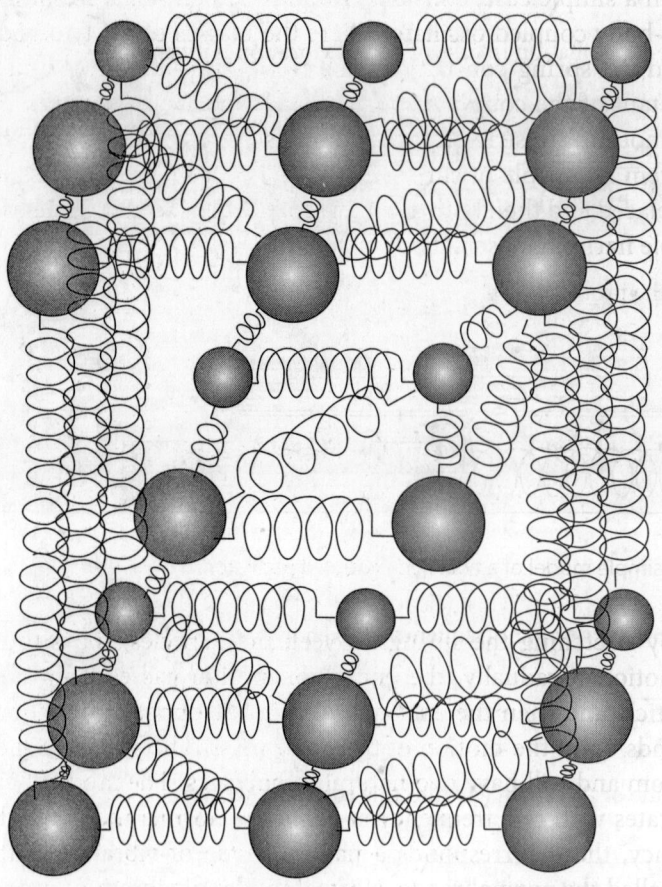

Figure 1.8 | Cubic arrangement of atoms in a crystal as inferred from X-ray data. The spring like forces act between neighbouring atoms

Simple harmonic oscillators interacting with each other by any type of influence are called coupled oscillators. The understanding and manipulation of these phenomena has far-reaching implications in many fields of physics and engineering. For example, a system of particles held together by springs turns out to be a useful model of the behavior of atoms mutually bound in a crystalline solid. The atoms of a crystal are held in place by mutual forces of interaction that oppose any disturbance from equilibrium positions, similar to the forces in a spring (Fig. 1.8). For small displacements of the atoms, they behave mathematically just like spring forces – that is, they obey Hooke's law. Each atom is free to move in three dimensions rather than one and therefore, each atom added to a crystal adds three normal modes. In a typical crystal at ordinary temperature, all these modes are always excited by random thermal energy.

The lower-frequency, longer-wavelength modes may also be excited mechanically. These are called sound waves. The behavior of oscillatory electrical circuits inductively coupled to each other can be explained using the concepts of coupled oscillators.

To begin with a simple case, consider two particles in a line, as shown in Fig. 1.9. The system is a two-body coupled oscillator. Here the masses of the two bodies are same and equal to 'm' and the spring constants of the two springs are same and equal to 'k'. k_c is the spring constant of the coupling spring connecting the two bodies. For mathematical simplicity, the motion is restricted to one dimension only, i.e., along the x-axis. Even this elementary system is capable of surprising behavior. If one particle is held in place while the other is displaced and then both are released, the displaced particle immediately begins to execute simple harmonic oscillation.

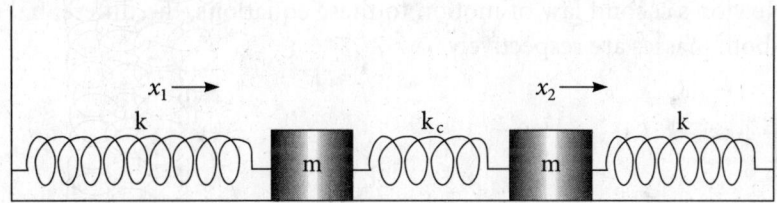

Figure 1.9 | A simple model of a two-body coupled oscillator.

This motion, by stretching the spring between the particles, starts to excite the second particle into motion. Gradually, the energy of motion passes from the first particle to the second particle and then the energy passes in the opposite direction. The motion of one mass depends upon the motion of the other mass. The system will oscillate with two degree of freedom and will have two natural frequencies. The mode of vibration in which the system vibrates with one frequency is called the normal mode of vibration. For each natural frequency, there corresponds a natural mode of vibration with a displacement configuration called the normal mode. Normal modes of vibrations are the free vibrations that depend only on the mass and stiffness of the system and how they are distributed.

To analyze the possible motions of the system, one may write the equations of motion for the two bodies in differential form. Again for mathematical simplicity, the system is made ideal, neglecting all dissipative forces like friction, air resistance, etc. The masses of the springs are also neglected. Assume that all the springs are initially at their natural lengths. We shall consider oscillations along the horizontal line (*x*-axis). This corresponds to longitudinal oscillations.

The force on each particle depends on two factors – first is on its displacement from its equilibrium position and second is on its distance from the other particle, because the spring between them stretches or compresses according to that distance. For this reason, the motions are coupled and the solution of each equation (i.e., the motion of each particle) depends on the solution of the other (i.e., the motion of the other).

Move the left mass a distance x_1 to the right and the right mass a distance x_2 to the right side. The coupling spring is therefore compressed by x_1 and stretched by x_2. The forces acting on the left mass are:

i.   The force kx_1 towards the left side due to the stretching of the left spring by x_1.

ii.   The force $k_C x_1$ to the left due to the compression of the coupling spring by x_1.

iii.   The force $k_C x_2$ to the right due to the stretching of the coupling spring by x_2.

The resultant force F_1 acting on the left mass is

$$F_1 = -kx_1 - k_C x_1 + k_C x_2$$

Similarly, the resultant force F_2 acting on the right mass is

$$F_2 = -kx_2 - k_C x_2 + k_C x_1$$

Applying Newton's second law of motion to these equations, the differential equation of motion for both masses are respectively

$$m\frac{d^2 x_1}{dt^2} + kx_1 - k_C(x_2 - x_1) = 0 \tag{1.65}$$

$$m\frac{d^2 x_2}{dt^2} + kx_2 - k_C(x_1 - x_2) = 0 \tag{1.66}$$

Equations (1.64) and (1.66) can be respectively rewritten as

$$\frac{d^2 x_1}{dt^2} + \frac{k + k_C}{m}x_1 - \frac{k_C}{m}x_2 = 0 \tag{1.67}$$

$$\frac{d^2 x_2}{dt^2} + \frac{k + k_C}{m}x_2 - \frac{k_C}{m}x_1 = 0 \tag{1.68}$$

Addition and subtraction of Eqs (1.67) and (1.68) gives

$$\frac{d^2(x_1 + x_2)}{dt^2} + \frac{k}{m}(x_1 + x_2) = 0 \tag{1.69}$$

$$\frac{d^2\left(x_1 - x_2\right)}{dt^2} + \frac{k + 2k_c}{m}\left(x_1 - x_2\right) = 0 \tag{1.70}$$

Equations (1.69) and (1.70) indicate that $x_1 + x_2$ and $x_1 - x_2$ are the normal coordinates of the motion. Also $\sqrt{\dfrac{k}{m}}$ and $\sqrt{\dfrac{k + 2k_c}{m}}$ are the normal frequencies of the system. It turns out that these normal frequencies are the frequencies at which the masses oscillate in their normal modes of vibration. This will be further explained later. With the following substitution into the aforementioned equations

$$q_1 = x_1 + x_2 \tag{1.71}$$

$$q_2 = x_1 - x_2 \tag{1.72}$$

$$\omega_1 = \sqrt{\frac{k}{m}} \tag{1.73}$$

$$\omega_2 = \sqrt{\frac{k + 2k_c}{m}} \tag{1.74}$$

we have

$$\frac{d^2 q_1}{dt^2} + \omega_1^2 q_1 = 0 \tag{1.75}$$

$$\frac{d^2 q_2}{dt^2} + \omega_2^2 q_2 = 0 \tag{1.76}$$

Now we have our equations in a much simpler form. In fact, these equations are exactly the same as that of the simple harmonic motion. The solutions are

$$q_1(t) = C_1' \cos \omega_1 t + C_2' \sin \omega_1 t \tag{1.77}$$

$$q_2(t) = C_3' \cos \omega_2 t + C_4' \sin \omega_2 t \tag{1.78}$$

From Eqs (1.71) and (1.72), we have

$$x_1 = \frac{q_1 + q_2}{2} \quad \text{and} \quad x_2 = \frac{q_1 - q_2}{2} \tag{1.79}$$

Thus, general solutions in terms of x_2 and x_1 are given as

$$x_1(t) = C_1 \cos \omega_1 t + C_2 \sin \omega_1 t + C_3 \cos \omega_2 t + C_4 \sin \omega_2 t \tag{1.80}$$

$$x_2(t) = C_1 \cos \omega_1 t + C_2 \sin \omega_1 t - C_3 \cos \omega_2 t - C_4 \sin \omega_2 t \tag{1.81}$$

Equations (1.80) and (1.81) show that motion of both masses is a superposition of two harmonic vibrations having angular frequencies ω_1 and ω_2. The constants C_1, C_2, C_3 and C_4 are evaluated from initial boundary conditions.

The coordinates q_1 and q_2 in terms of which the equations of motion take the form of a set of linear ordinary differential equations with constant coefficients involving just one of the coordinates, are called normal coordinates. A vibration (solution) involving only one dependent variable (normal coordinate) is called a normal mode of vibration and has its own normal frequency. In other words, we can define normal mode vibration as a vibration in which both masses harmonically oscillate at the same frequency. In such a normal mode, all the components of the system oscillate with the same normal frequency. Note that the normal mode corresponding to q_1 has a frequency ω_1, which is equal to the frequency of the individual oscillators. The normal mode corresponding to q_2 has a frequency ω_2, which is greater than the frequency of the individual oscillators. Since $\omega_2 > \omega_1$, the oscillation with frequency ω_1 is referred to as the normal mode of vibration of the coupled oscillator with lowest mode and the oscillation with frequency ω_2 is referred to as the normal mode of vibration of the coupled oscillator with highest mode.

Case 1: Normal mode of oscillation at higher frequency (anti-symmetric mode)

According to Eqs (1.80) and (1.81), the normal mode of vibration of higher frequency ω_2 occurs when $C_1 = 0$ and $C_2 = 0$. Under this condition, Eqs (1.80) and (1.81) become

$$x_1(t) = C_3 \cos \omega_2 t + C_4 \sin \omega_2 t \tag{1.82}$$

$$x_2(t) = -C_3 \cos \omega_2 t - C_4 \sin \omega_2 t \tag{1.83}$$

In Eqs (1.82) and (1.83), x_1 and x_2 are opposite in sign and equal in magnitude. Therefore, we conclude that in the normal mode of vibration of higher frequency ω_2, the two masses are vibrating in opposite directions or are out of phase. The normal mode of vibration of higher frequency is also known as the anti-symmetric mode.

Case 2: Normal mode of oscillation at lower frequency (symmetric mode)

Similarly, the normal mode of vibration at lower frequency ω_1 is obtained when $C_3 = 0$ and $C_4 = 0$. Under this condition, Eqs (1.80) and (1.81) become

$$x_1(t) = C_1 \cos \omega_1 t + C_2 \sin \omega_1 t \tag{1.84}$$

$$x_2(t) = C_1 \cos \omega_1 t + C_2 \sin \omega_1 t \tag{1.85}$$

In Eqs (1.84) and (1.85), x_1 and x_2 have the same sign and are equal in magnitude. Therefore, we conclude that in the normal mode of vibration of lower frequency ω_1, the two masses are vibrating in the same direction or are in phase. The normal mode of vibration of lower frequency is also known as the symmetric mode.

Linear combination of normal modes

Suppose the left-side mass is displaced through a distance 'r' from its rest position keeping the right-side mass at rest. Both masses are released at time $t = 0$. Mathematically, this case can be described as

$$x_1(0) = r \quad \text{and} \quad \left.\frac{dx_1(t)}{dt}\right|_{t=0} = 0$$

$$x_2(0) = 0 \quad \text{and} \quad \left.\frac{dx_2(t)}{dt}\right|_{t=0} = 0$$

Putting these conditions into Eqs (1.80) and (1.81), we obtain

$$C_1 = \frac{r}{2}, \ C_3 = \frac{r}{2}, \ C_2 = 0, \ \text{and} \ C_4 = 0$$

Putting $C_1 = \dfrac{r}{2}, \ C_3 = \dfrac{r}{2}, \ C_2 = 0, \ \text{and} \ C_4 = 0$ into Eqs (1.80) and (1.81), we obtain

$$x_1(t) = \frac{r}{2}\left(\cos \omega_1 t + \cos \omega_2 t\right) = r \cos\left(\frac{\omega_2 + \omega_1}{2}\right)t \cos\left(\frac{\omega_2 - \omega_1}{2}\right)t \tag{1.86}$$

$$x_2(t) = \frac{r}{2}\left(\cos \omega_1 t - \cos \omega_2 t\right) = r \sin\left(\frac{\omega_2 + \omega_1}{2}\right)t \sin\left(\frac{\omega_2 - \omega_1}{2}\right)t \tag{1.87}$$

Unlike the case of pure normal modes, neither mass performs simple harmonic motion. The amplitude of the left-side mass is $r\cos\left(\dfrac{\omega_2 + \omega_1}{2}\right)t$, whereas amplitude of the right-side mass is $r\sin\left(\dfrac{\omega_2 + \omega_1}{2}\right)t$. Since there is a phase difference of $\pi/2$, the crest of x_1 corresponds to the troughs of x_2. Each mass oscillates between the maximum amplitude 'r' and an amplitude of zero This has been illustrated in Fig. 1.10

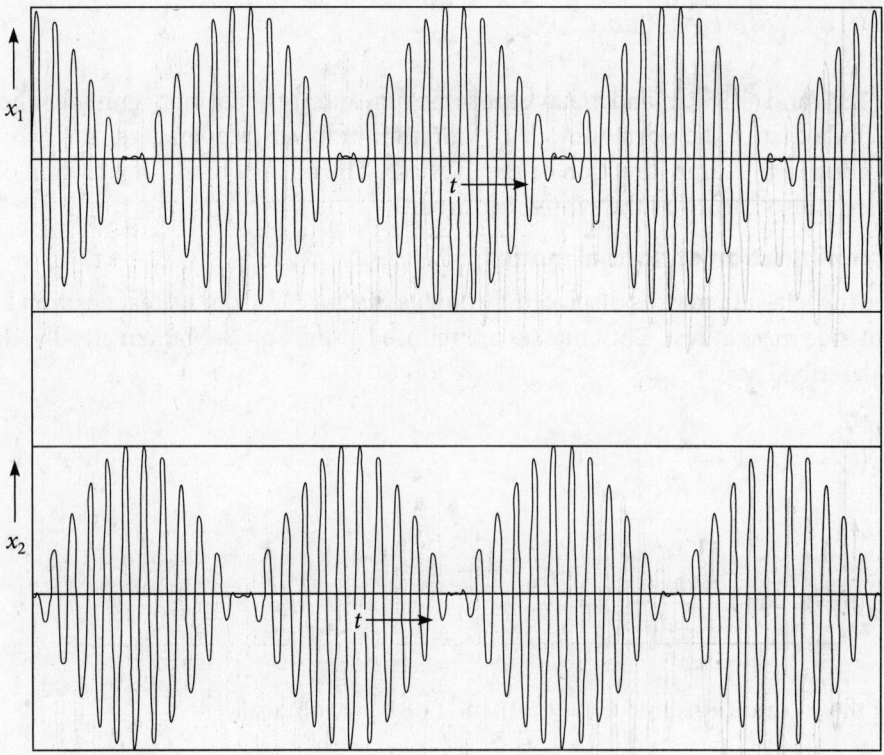

Figure 1.10 Linear combination of normal mode when $x_1(0) = r$, $x_2(0) = 0$ and the left-side mass is released from rest

If we give the system initial conditions of $x_1(0) = r_1$, $x_2(0) = r_2$ (i.e., two bodies have been displaced through distances of r_1 and r_2),

$$\frac{dx_1(0)}{dt} = 0 \text{ and } \frac{dx_2(0)}{dt} = 0$$

(i.e., the two masses are released from rest), we can see a similar behavior. This has been illustrated in Fig. 1.11.

However, neither mass's oscillations ever reach an amplitude of zero. This is because neither mass started at its equilibrium position. Here each mass oscillates between the maximum amplitude r_1 and the minimum amplitude r_2. The energy of the system is continuously oscillating between the two masses, while the total energy of the system remains constant. The mathematics of the problem is left as an exercise to the students.

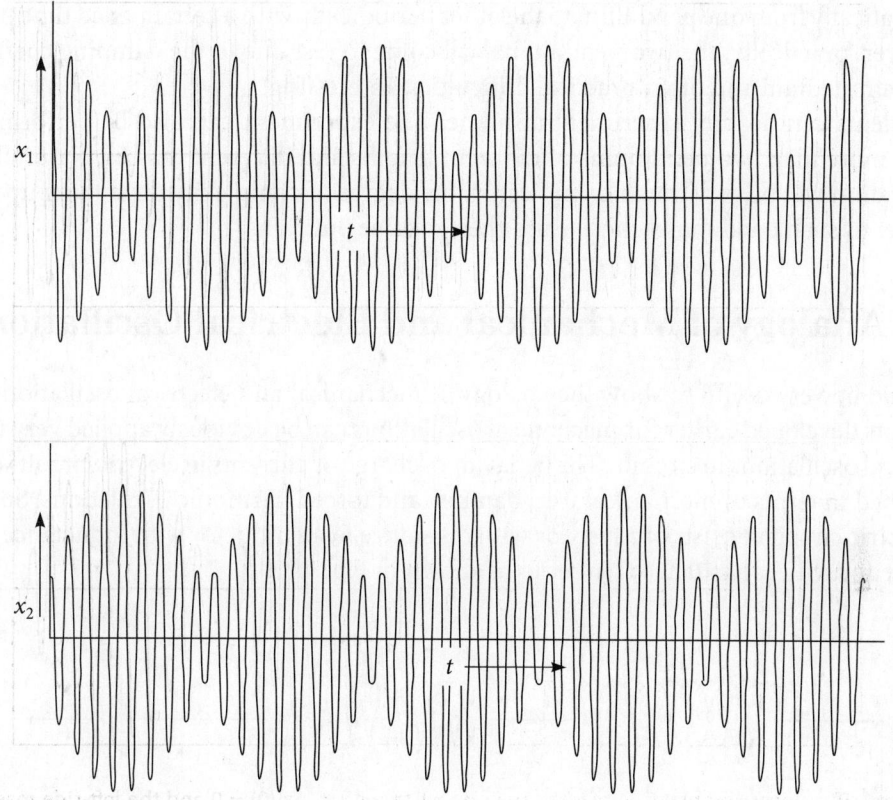

Figure 1.11 | Linear combination of normal mode when $x_1(0) = r_1$, $x_2(0) = r_2$ and the two masses are released from rest

1.6.1 Experiment on a two-body coupled oscillator

An easy and interesting experiment can be performed on the coupled oscillator. A thick cotton string is fixed tightly between two rigid support separated by a distance of less than one meter. Two identical (i.e., same length and same mass) simple pendulums of length approximately half meter are taken – they can be named A and B. These two simple pendulums are knotted firmly to the tightly fixed string with a separation of around half meter. After knotting the two pendulums, the cotton string will sag slightly. Let the two pendulums become stationary. Without disturbing the other pendulum (B), oscillate one pendulum (A) in a plane perpendicular to the tightly fixed string.

Now observe the two pendulums. After some time, the stationary pendulum B will start oscillating with a slowly increasing amplitude and the oscillating pendulum A oscillates with decreasing amplitude. Again, after some time, pendulum A comes to rest/equilibrium position and pendulum B oscillates with the largest amplitude. Then the phenomenon will start to reverse. This process will continue for infinite time. Energy transfer takes place automatically from one pendulum to the other periodically with a certain fixed time period. However, practically, the two pendulums will come to rest due to the damping coefficient of the air medium and other practical difficulties. Interesting!

Students can do this experiment at home. The experiment can also be performed by taking more than two pendulums of different lengths and different masses. Naturally, the observation will be very much complicated, the degree of complexity depending upon the number of such pendulums chosen and their relative positions.

1.7 Analogy of Mechanical and Electrical Oscillations

It would be very useful to show the analogy of mechanical and electrical oscillations. The concepts developed earlier for mechanical oscillations can be obviously applied very well to electrical oscillations in circuits. The behavior of charge or currents in electric circuits can be described in terms of mechanical free, damped and forced harmonic oscillations. Suppose an electric circuit consists of a resistor with resistance R, an inductor with inductance L and a capacitor with capacitance C in series as shown in Fig. 1.12.

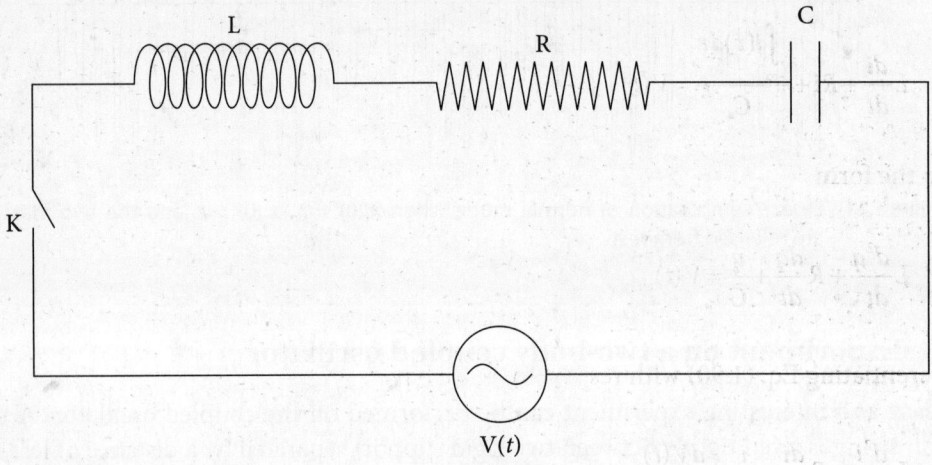

Figure 1.12 | LCR series circuit with applied voltage $V(t)$

If V_R, V_L and V_C are the instantaneous voltages across resistor, inductor, and capacitor respectively, then by applying Kirchhoff's voltage law to the circuit, we have

$$V_R + V_L + V_C = V(t) \tag{1.88}$$

Here

t = time

$V(t)$ = time varying applied voltage

$$V_R = iR = R\frac{dq}{dt} = \text{voltage across resistor}$$

$$V_L = L\frac{di}{dt} = L\frac{d^2q}{dt^2} = \text{voltage across inductor}$$

$$V_C = \frac{q}{C} = \frac{\int\limits_{-\infty}^{t} i(\tau)d\tau}{C} = \text{voltage across capacitor}$$

$$q = \text{charge}$$

$$i = \text{current}$$

Thus, Eq. (1.88) can be re-written either in the form

$$L\frac{di}{dt} + Ri + \frac{\int\limits_{-\infty}^{t} i(\tau)d\tau}{C} = V(t) \tag{1.89}$$

or in the form

$$L\frac{d^2q}{dt^2} + R\frac{dq}{dt} + \frac{q}{C} = V(t) \tag{1.90}$$

Differentiating Eq. (1.90) with respect to 't', we have

$$L\frac{d^2i}{dt^2} + R\frac{di}{dt} + \frac{i}{C} = \frac{dV(t)}{dt} \tag{1.91}$$

For a sinusoidally varying voltage source, like $V(t) = V_0 \sin \omega t$, Eqs (1.90) and (1.91) becomes respectively

$$L\frac{d^2q}{dt^2} + R\frac{dq}{dt} + \frac{q}{C} = V_0 \sin \omega t \tag{1.92}$$

$$L\frac{d^2i}{dt^2} + R\frac{di}{dt} + \frac{i}{C} = V_0 \omega \cos \omega t \tag{1.93}$$

Under different boundary conditions, these differential equations can be solved for the current and charges at any instant of time.

A glance at Eqs (1.90)–(1.93) along with that of mechanical problems shows the equivalence of electrical and mechanical oscillations as shown in the following table.

Equivalence of Electrical and Mechanical Oscillations

Mechanical Oscillation	Electrical Oscillation
Mass	Inductance
Damping coefficient	Resistance
Spring constant	Reciprocal of Capacitance
Applied Force	Applied voltage
Displacement	Charge
Velocity	Current
Potential Energy in spring = $\dfrac{1}{2}kx^2$	Energy in Capacitor = $\dfrac{1}{2}\dfrac{q^2}{C}$
Kinetic Energy = $\dfrac{1}{2}mv^2$	Energy in inductor = $\dfrac{1}{2}Li^2$

For equivalent expressions in electrical oscillations we have to replace the variables and other parameters according to the aforementioned table. In this way, we can get voltage equations from force equations as shown in the following text.

For example, let us discuss briefly the damped harmonic electrical oscillation produced when a charged capacitor is connected to a resistor and inductor in series as shown in Fig. 1.13.

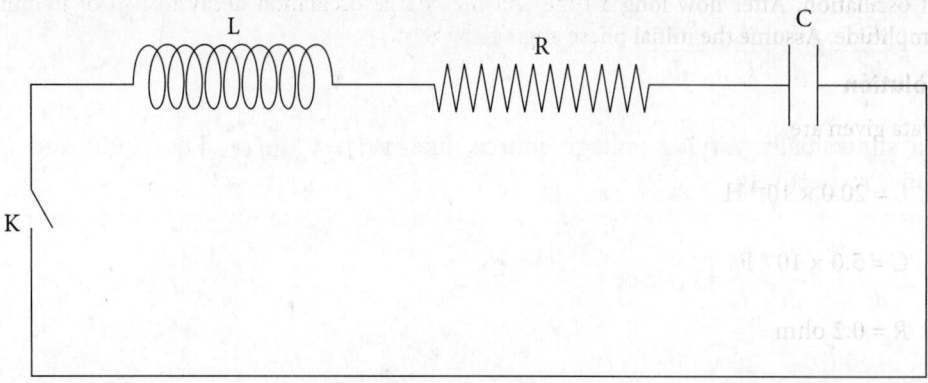

Figure 1.13 | LCR series circuit without applied voltage

In this case, since there is no externally applied voltage, from Eq. (1.90), we get

$$\frac{d^2q}{dt^2} + \frac{R}{L}\frac{dq}{dt} + \frac{1}{LC}q = 0$$

Comparing this equation with Eq. (1.23) for $\frac{R^2}{4L^2} < \frac{1}{LC}$, we can have

$$q = q_0 e^{-\gamma t} \cos\left(\omega_1 t + \theta\right) \qquad (1.94)$$

Here $\omega_1 = \sqrt{\omega_0^2 - \gamma^2}$, $\omega_0 = \frac{1}{\sqrt{LC}}$, $\gamma = \frac{R}{2L}$, $q_0 e^{-\gamma t}$, and θ are the angular frequency, natural angular frequency, attenuation factor, instantaneous amplitude, and initial phase angle of oscillation respectively. The parameters q_0 and θ can be determined from initial boundary conditions.

Similarly, by comparing Eq. (1.93) with Eq. (1.36), we have

$$i = \frac{V_0 \omega}{\sqrt{\left(\omega^2 L - \dfrac{1}{C}\right)^2 + R^2 \omega^2}} \sin(\omega t + \beta)$$

with $\beta = \tan^{-1}\left(\dfrac{1 - \omega^2 LC}{RC\omega}\right)$

Example 1.10

An LCR circuit as shown in Fig. 1.13 contains an inductor of inductance 20.0 mH, a capacitor of capacitance 5.0 μF and a resistor of resistance 0.2 ohm. Calculate the angular frequency of oscillation. After how long a time will the charge oscillation decay to half of its initial amplitude. Assume the initial phase angle to be zero.

Solution

Data given are

$L = 20.0 \times 10^{-3}$ H

$C = 5.0 \times 10^{-6}$ F

$R = 0.2$ ohm

$\dfrac{q}{q_0} = \dfrac{1}{2}$

The angular frequency of the electrical oscillation is

$$\omega_1 = \sqrt{\frac{1}{LC} - \frac{R^2}{4L^2}} = \sqrt{1.0 \times 10^7 - 2.5} \text{ rad/s} = 3162.3 \text{ rad/s}$$

$$\gamma = \frac{R}{2L} = 1.58 \text{ rad/s}$$

Putting the given conditions in Eq. (1.94), we get

$$\frac{1}{2} = e^{-1.58t}$$

or $t = \dfrac{\ln 2}{1.58} \text{ s} = 0.44 \text{ s}$

Thus, the charge oscillation decays to half of its initial amplitude in 0.44 s.

1.8 Wave as a Periodic Variation Quantity in Space and Time

When a pebble is thrown into the still water of a pond, disturbance is created at the point where the pebble enters the water. The disturbance created thus is not confined to that point alone; it spreads out and reaches the boundary of the pond. The water does not move or flow towards the edge of the pond; only the disturbance moves. The water particles move up and down over a short distance in a vertical direction about their mean position as a result of which disturbance travels in a horizontal direction. At any time, the amount of disturbance is not the same at all points on the water surface. Also, at any point, the amount of disturbance varies with time. In non-technical language, we call this disturbance, water waves. This behavior is characteristic of all wave motions.

Now we can define the wave in the following way. A wave is defined as a disturbance that moves through a medium in such a manner that at any position, the displacement of the particles of the medium is a function of time while at any instant the displacement of the particles of the medium is a function of the position of the point. The medium as a whole does not move in the direction of the motion of the wave. The wave is also called a progressive wave or a travelling wave (see stationary waves).

1.8.1 Wave equation

Sinusoidal waves can be represented by an equation of the form

$$\psi(x,t) = r\sin(\omega t \pm kx) \quad \text{or} \quad \psi(x,t) = r\sin\left[k(vt \pm x)\right] \tag{1.95}$$

The '+' sign is used if the wave is travelling towards the left side ($-x$ direction) and the '$-$' sign is used if the wave is travelling towards the right-side ($+x$ direction).

Here

r = amplitude of the wave

ω = angular frequency of the wave

$k = \dfrac{2\pi}{\lambda}$ = wave number of the wave = magnitude of propagation vector.

v = phase speed of the wave $= \dfrac{\omega}{k}$

x and t = position and time functions.

The speed of the wave (wave speed) and the speeds of the particles of the medium (particle speed) are not the same. The particle speed is found out by differentiating the aforementioned equation. The particle speed is not the same for all the particles. The relation between maximum particle speed V_{max} and wave speed 'v' is given by

$$V_{max} = \frac{2\pi r}{\lambda} v \qquad (1.96)$$

1.8.2 Wave equation in differential form

The general wave equation in one dimension is given by

$$\frac{\partial^2 \Psi(x,t)}{\partial x^2} = \frac{1}{v^2}\frac{\partial^2 \Psi(x,t)}{\partial t^2} + \frac{1}{\xi}\frac{\partial \Psi(x,t)}{\partial t} \qquad (1.97)$$

and in three dimension, it is given by

$$\nabla^2 \Psi = \frac{1}{v^2}\frac{\partial^2 \Psi}{\partial t^2} + \frac{1}{\xi}\frac{\partial \Psi}{\partial t}$$

or $\qquad \nabla^2 \Psi - \dfrac{1}{v^2}\dfrac{\partial^2 \Psi}{\partial t^2} - \dfrac{1}{\xi}\dfrac{\partial \Psi}{\partial t} = 0 \qquad (1.98)$

where $\nabla^2 = \dfrac{\partial^2}{\partial x^2} + \dfrac{\partial^2}{\partial y^2} + \dfrac{\partial^2}{\partial z^2}$ and is called a Laplacian operator. In these equations, ξ called diffusivity, takes care of the energy loss of the wave in the medium. Here, Ψ is called wave function and v is the speed of the wave. The wave function plays the role of disturbance in wave propagation. It contains all the information regarding the wave propagation.

The solution of this wave Eq. (1.97) gives the direction of propagation, the wave speed and many other properties of waves. In case of a stretched string, Ψ is the displacement of the string from the x-axis; in case of a sound wave, Ψ is the pressure difference; in case of light wave, Ψ is either an electric field or a magnetic field. Solutions of this differential wave equation are of many kinds, reflecting a variety of waves that can occur.

1.9 Longitudinal and Transverse Waves

Depending upon the direction of oscillation of the particles of the medium, progressive waves can be divided into two categories of waves; one is longitudinal waves and the other is transverse waves.

1.9.1 Longitudinal waves

A longitudinal wave is defined as a wave in which particles of the medium oscillate in the direction of propagation of the wave. A beautiful example of a longitudinal wave is the sound wave. Sound waves propagate in the air medium. In the propagation of sound waves, air particles oscillate (but do not move) about their mean position in the direction in which sound travels. In this case, air particles get compressed at one region and at the adjacent region, air particles get rarefied. This process goes on alternately. The two compression and rarefaction regions are pictorially represented in Fig. 1.14.

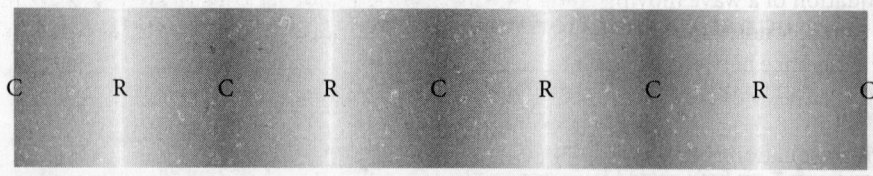

Figure 1.14 | Propagation of longitudinal waves. *C* represents a compression region and *R* represents a rarefaction region

1.9.2 Transverse waves

The transverse wave is defined as the wave in which particles of the medium oscillate in a direction perpendicular (i.e., transverse) to the direction of propagation of the wave. One of the visible examples of a transverse wave is the water wave. Water wave propagates in the water medium. In the propagation of a water wave, water particles oscillate about their mean position in a direction perpendicular to the direction in which the water wave travels.

In this case, water particles get raised above the normal level of the water surface at one region and at the adjacent region, water particles get depressed below the normal level of the water surface. The raised region is called 'crest' and the depressed region is called 'trough'. This process goes on alternately and is represented in Fig. 1.15.

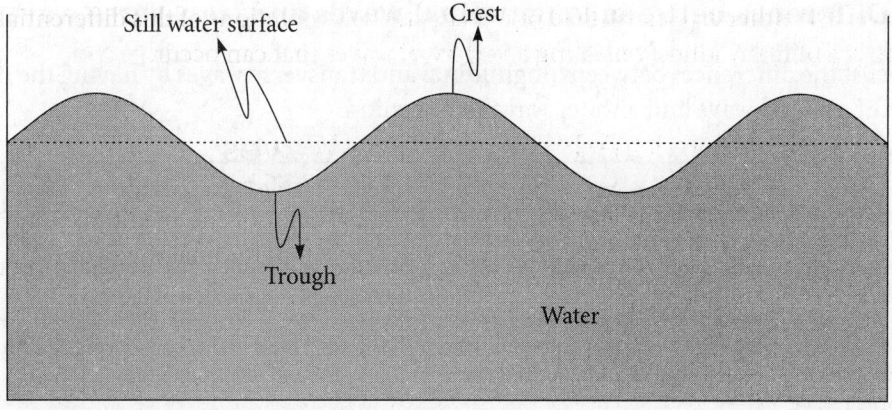

Figure 1.15 | Water waves or ripples

All electromagnetic waves belong to the transverse wave category. In this case, the electromagnetic field oscillates and no material medium is necessary.

Example 1.11

The equation of a wave moving along a string is $\Psi = 4 \sin \pi(0.04\,x - 8\,t)$. Here Ψ, x are in cm and t is in seconds. Find the amplitude, frequency, and speed of the wave. Also calculate the phase difference between two points situated 25 cm apart.

Solution

The given equation is $\Psi = 4 \sin \pi(0.04x - 8t) = 4 \sin 2\pi(0.02x - 4t)$

Comparing this equation with the standard equation $\Psi = r \sin 2\pi\left(\dfrac{x}{\lambda} - \dfrac{t}{T}\right)$, we get $r = 4$ cm, $\dfrac{1}{\lambda} = 0.02/\text{cm}$, and $\dfrac{1}{T} = 4/\text{s}$. Thus, we have amplitude $r = 4$ cms, wavelength $\lambda = 50$ cms, time period $T = 0.25$ s and frequency $\dfrac{1}{T} = 4$ Hz. The velocity of the wave is given by

$$\text{Speed} = \text{frequency } x. \text{ Wavelength} = 4 \times 50 \text{ cm/s} = 200 \text{ cm/s}$$

The phase difference between the two points is given by

$$\text{Phase difference} = \frac{2\pi}{\lambda} \times \text{path difference}$$

$$= \frac{2\pi}{50} \times 25 = \pi \text{ radian}$$

1.9.3 Difference between longitudinal waves and transverse waves

We can list the differences between longitudinal and transverse waves by having the mental picture of a sound wave and a water wave in our mind.

	Longitudinal wave	Transverse wave
1	Here, the particles oscillate about their mean position in the direction of propagation of the wave	Here, the particles oscillate about their mean position in the direction perpendicular to the direction of propagation of the wave
2	The wave travels in the form of compression and rarefaction. One compression and one rarefaction form a wave.	The wave travels in the form of crest and troughs. One crests and one trough form a wave.
3	The production of this type of wave is possible in all type of media	The production of this type of wave is possible in any type of media which have elasticity.
4	The longitudinal wave cannot be polarized	The transverse wave can be polarized
5	Examples: sound wave, seismic primary waves, shock waves in air (shock waves are created by violent explosion), etc.	Examples: All type of electromagnetic waves, waters waves, seismic shear waves, seismic Love waves, etc.

1.9.4 Characteristics of progressive waves

The following are the characteristics of progressive waves.

i.   A progressive wave is the disturbance produced in the medium by the repeated periodic motion of the particles of the medium, motion being handed over from particle to particle.

ii. It is the progressive wave which advances. The particles of the medium do not travel but oscillate about their mean position.

iii. There is a continuous phase difference between the particles of the medium when progressive wave advances.

iv. The velocity of the progressive wave is uniform through out the medium.

v.   When a progressive wave propagates, each particle of the medium oscillates about their mean position with the same amplitude and time period. The velocity of the particle is different at different points. It is maximum at the mean position and zero at the extreme position.

vi. The particle velocity is different from the wave velocity.

vii. There is transmission of energy across every plane in the direction of propagation waves

1.10 Stationary Waves

A stationary wave is defined as a wave produced by the superposition of two identical progressive waves (i.e., progressive waves having the same wavelength, the same time period, the same frequency and the same speed) propagating through a medium in a line but in opposite directions. The new wave produced by this method is called a stationary wave because there is no transmission of energy across any plane. That means stationary waves do not carry energy. There are certain points in the medium at which particles of the medium are permanently at rest. These points are called nodes. There are certain points in the medium at which particles of the medium are oscillating with maximum amplitude. These points are called anti nodes. The stationary wave can not carry energy because the particles of the medium at the nodes are permanently at rest without having any energy. Stationary waves are also called standing waves.

1.10.1 Formation of stationary waves

A stationary wave is formed when the incident wave and the corresponding wave reflected by a rigid wall are superimposed on each other. In Fig. 1.16, the displacement graphs of both incident and reflected waves are plotted.

The incident wave is supposed to be travelling from left to right. The reflected wave (Fig. 1.17) is travelling from right to left. These incident and reflected waves superimpose on each other and thus give rise to stationary waves. The continuous curve in 3 of Fig. 1.16 depicts the resultant stationary waves. The mode of stationary waves depends upon the phase difference of the two waves. The nature of stationary waves is shown graphically in Fig. 1.17 at different intervals. Here, the incident wave (I) is represented by dotted curves and the reflected wave (II) is represented by dashed line curves.

Case 1:

At $t = 0$, the second crest of the incident wave coincides with the trough of the reflected wave. Since the two waves have equal amplitude, the resultant stationary wave is a straight line represented by a solid line. At all the points $P_1, P_2, P_3, P_4, P_5, P_6, P_7, P_8, P_9$, the particles of the medium have zero amplitude. See Fig. 1.17(a)

Case 2:

At $t = T/4$, the second incident and reflected wave have individually advanced towards each other by a distance of $\lambda/4$ and the relative displacement between them is $\lambda/2$. $\left(\dfrac{\lambda}{4} + \dfrac{\lambda}{4} = \dfrac{\lambda}{2}\right)$.

As a result of this, the crest of the incident wave coincides with the crest of the reflected wave giving a resultant stationary wave of maximum amplitude. Points P_1, P_3, P_5, P_7 and P_9 acquire maximum displacements while the points $P_2, P_4, P_6,$ and P_8 have zero amplitude. See Fig. 1.17(b).

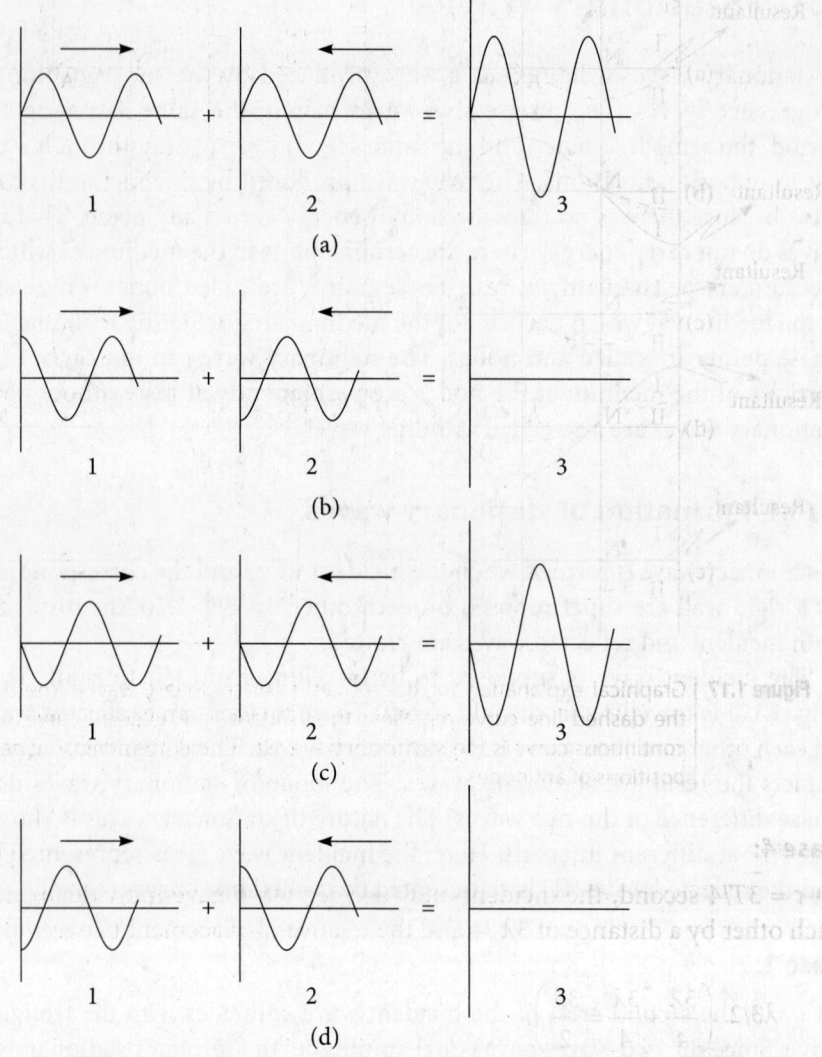

Figure 1.16 | Stationary waves as superposition of right and left going waves. 1 and 2 are components, 3 the resultant

Case 3:

At $t = T/2$ second, the incident and the reflected wave have individually advanced towards each other by a distance of $\lambda/2$ and the relative displacement between them is $\lambda \left(\dfrac{\lambda}{2} + \dfrac{\lambda}{2} = \lambda \right)$.

As a result, the crest of the incident wave coincides with the trough of the reflected wave giving a resultant stationary wave of zero amplitude. In this case, again, the resultant stationary wave is a straight line. See Fig. 1.17(c).

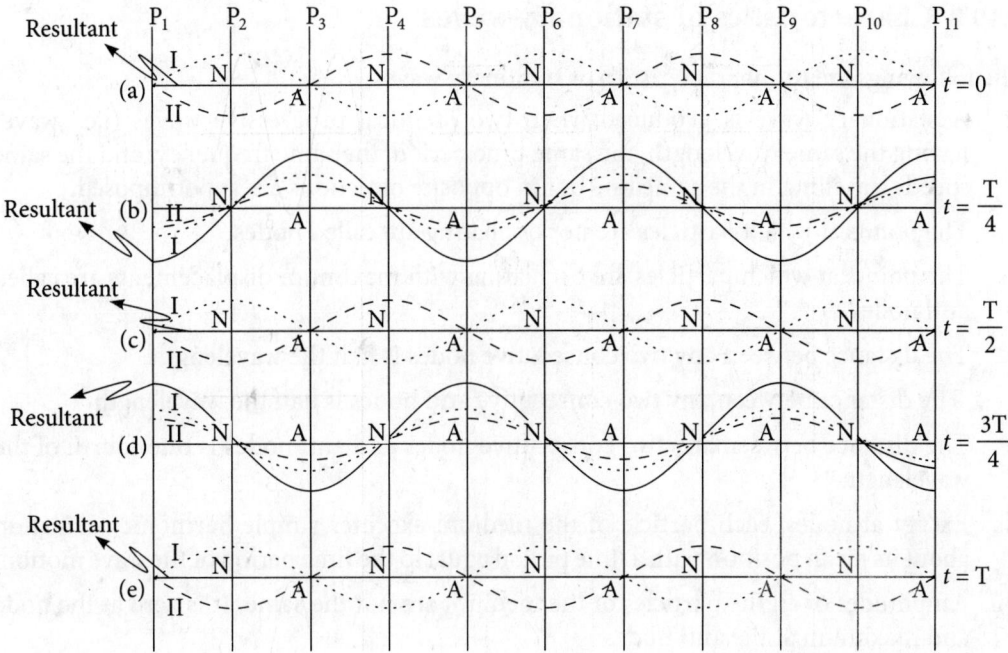

Figure 1.17 Graphical explanation for the formation of stationary waves. The dotted curve and the dashed line curve represent the incident and reflected curves respectively. The continuous curve is the stationary wave. Ns are the positions of nodes and As are the positions of antinodes

Case 4:

At $t = 3T/4$ second, the incident and reflected wave have individually advanced towards each other by a distance of $3\lambda/4$ and the relative displacement between them is

$$\lambda 3/2. \left(\frac{3\lambda}{4} + \frac{3\lambda}{4} = \frac{3\lambda}{2} \right).$$

As a result of this, the trough of the incident wave coincides with the trough of the reflected wave giving a resultant stationary wave of maximum amplitude but opposite to that of Case 2. Points P_1, P_3, P_5, P_7 and P_9 acquire maximum displacements while the points P_2, P_4, P_6, and P_8 have zero amplitude. See Fig. 1.17(d).

Case 5:

At $t = T$ second, the crest of the incident wave coincides with the trough of the reflected wave. Since the two waves have equal amplitude, the resultant stationary wave is a straight line represented by a solid line. At all the points, P_1, P_2, P_3, P_4, P_5, P_6, P_7, P_8, P_9, the particles of the medium have zero amplitude. See Fig. 1.17(e).

1.10.2 Characteristics of stationary waves

The followings are the characteristics of stationary waves.

i.   A stationary wave is produced when two identical progressive waves (i.e., waves having the same wavelength, the same time period, the same frequency, and the same speed) travelling in the same line but in opposite directions are superimposed.

ii.   The points at which particles are not oscillating are called nodes.

iii. The points at which particles are oscillating with maximum displacements are called anti nodes

iv.   The distance between any two consecutive nodes is half the wavelength.

v.   The distance between any two consecutive anti nodes is half the wavelength

vi.   The distance between any two consecutive nodes and anti nodes is one-fourth of the wavelength.

vii. Except at nodes, each particle of the medium executes simple harmonic oscillation about its mean position with a time period equal to the time period of the wave motion.

viii. Amplitudes of all the particles of the medium are not the same. It is zero at the node and maximum at the anti node.

1.10.3 Differences between progressive and stationary waves

	Progressive waves	Stationary waves
1	A progressive wave is produced due to the oscillation of the particles of the medium.	A stationary wave is produced when two identical progressive waves travelling in the same line but opposite directions are superimposed.
2	The waves move with a velocity depending upon the properties of the medium.	The waves remain stationary and do not move.
3	Each particle of the medium executes periodic motion about their mean position with the same amplitude.	Except the node, all the particles of the medium execute SHO with varying amplitude.
4	There is a continuous change of phase from particle to particle.	All the particles between two consecutive nodes are at the same phase, but differ in phase by π from those in the preceding as well as succeeding similar segments.
5	At any instant all the particles do not come together in the mean position, they pass their mean position in succession but with the same velocity.	All the particles pass their mean position at a time, but with different velocities.

6	Each particle of the medium undergoes similar change of pressure and density	There is no change of pressure and densities at the antinodes while there is maximum change of pressure and densities at the nodes.
7	There is transmission of energy across every plane in the direction of propagation of waves.	There is no flow of energy across any plane.
8	A complete wavelength contains a compression and rarefaction in the case of longitudinal waves and crest and trough in the case of transverse waves.	The wavelength is the distance between two alternate nodes and anti nodes.
9	Compression and rarefaction move from point to point throughout the medium.	The compression and rarefaction do not move from point to point; they simply appear at and disappear at certain equidistance fixed points.
10	No particle of the medium is permanently at rest.	Particles at the nodes are permanently at rest.
11	The equation of a progressive wave is given by $$\psi(x,t) = r \sin (kx \mp \omega t)$$	The equation of a stationary wave is given by $$\psi(x,t) = 2r \sin kx \cos \omega t$$

1.11 Reflection of a Wave at the Boundary of Two Media

When a wave is allowed to fall on the interface separating two media, the wave is returned to the same medium completely or partially. This phenomenon is called reflection. Reflection is defined as the phenomenon by virtue of which wave (actually energy contained in the wave) is sent back to the same medium by an interface separating the two media.

1.11.1 Reflection of transverse waves

Case 1: Passing from rarer medium to denser medium

Let a transverse wave travel along a string fixed to a point 'P' on the rigid wall. Suppose a crest is formed in the string just before the wall. The string will exert an upward force on the point P as a result of which a downward force of same magnitude will act (Newton's third law) on the string at the point 'P'. The result will be the formation of a trough just before the point 'P'. See Fig. 1.18. Therefore, there is a phase change π when a transverse wave is reflected back by the surface of a denser medium.

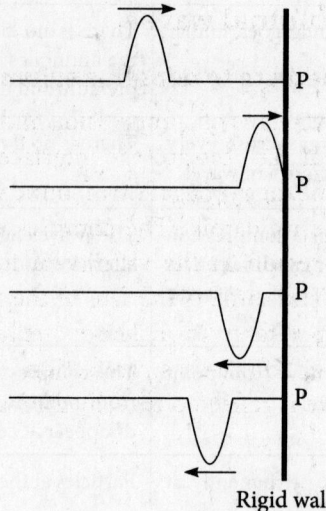

Rigid wall

Figure 1.18 | Reflection of transverse waves from a rigid (hard) wall

Case 2: Passing from denser medium to rarer medium

Let a transverse wave travel along a string fixed to a point 'P' on a mass-less ring capable of sliding freely in the vertical direction without any friction on the rigid support. Suppose a crest is formed in the string just before the support. The string will exert an upward force on the point P as a result of which the ring will slide in the upward direction. The result will be the formation of a crest just before the point 'P'. See Fig. 1.19. Therefore, there is no phase change when a transverse wave is reflected back by the surface of a rarer medium.

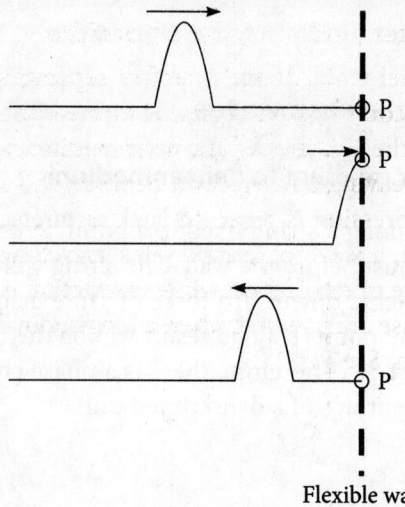

Flexible wall

Figure 1.19 | Reflection of transverse waves from a flexible (soft) wall

1.11.2 Reflection of longitudinal waves

Case 1: Passing from rarer medium to denser medium

Longitudinal waves propagate by means of compression and rarefaction. Let a longitudinal wave travel towards the rigid wall, i.e., toward the interface separating the rarer medium from the denser medium. (imagine the propagation of sound waves). Suppose a compression region is formed just before the rigid wall. The compressed particles will rebound in a group forming a compression region in the vicinity of the wall. Thus, compression is reflected back as compression. The same is the fate of the rarefaction. Therefore, a wave of compression travelling from a rarer to a denser medium is reflected as a wave of compression while rarefaction is reflected as rarefaction. That means there is no phase change when a longitudinal wave is reflected back by the surface of a denser medium. See Fig. 1.20.

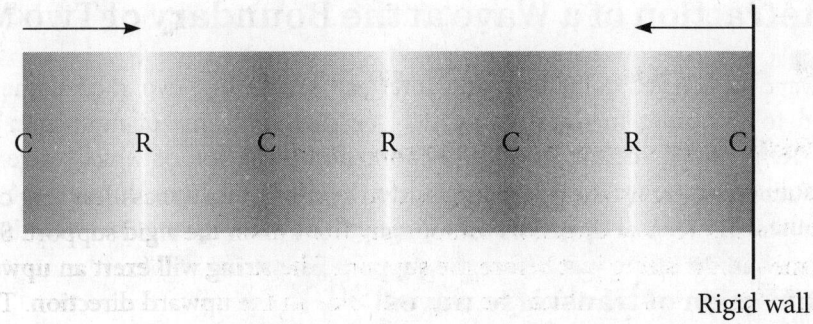

Rigid wall

Figure 1.20 │ Reflection of longitudinal waves from a rigid (hard) wall

Case 2: Passing from denser medium to rarer medium

Let a longitudinal wave travel towards the interface separating a denser medium from a rarer medium. Suppose a compression region is formed just before the interface. The compression region presses the particles of the rarer medium which yield to it and move forward with higher speed, leaving a rarefaction behind. This rarefaction travels back as a reflected wave. Thus, compression is reflected back as rarefaction. The same is the fate of the rarefaction. Therefore, a wave of compression travelling from a denser to a rarer medium is reflected as a wave of rarefaction while rarefaction is reflected as compression. That means that there is phase change of π when a longitudinal wave is reflected back by the surface of a rarer medium. See Fig. 1.21.

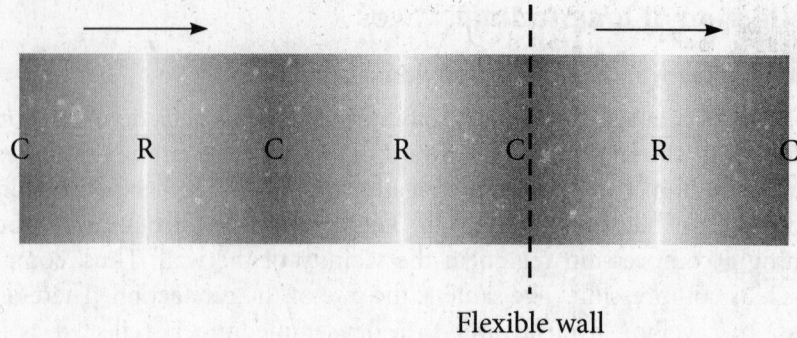

Flexible wall

Figure 1.21 | Reflection of transverse waves from a flexible (soft) wall

1.12 Refraction of a Wave at the Boundary of Two Media

When a wave is allowed to fall on the interface separating two media, the wave is transmitted to the other medium completely or partially. This phenomenon is called refraction. Refraction is defined as the phenomenon by virtue of which wave (actually energy contained in the wave) is transmitted to the other medium through the interface separating the two media. The speed and wavelength of waves change in refraction.

1.12.1 Refraction of transverse waves

Case 1: Passing from rarer medium to denser medium

Let a transverse wave travel along a thin string attached to a thick string at a point 'P' on the interface. Suppose a crest is formed in the thin string just before the interface. The thin string will exert an upward force on the point P as a result of which a downward force of the same magnitude will act (Newton's third law) on the light string. The result will be the formation of a trough just before the interface and a crest just after the interface. See Fig. 1.22. Hence, a crest is refracted or transmitted as a crest. Therefore, there is no change of phase when a transverse wave is refracted going from a rarer medium to a denser medium.

Case 2: Passing from denser medium to rarer medium

Let a transverse wave travel along a thick string attached to a thin string at a point 'P' on the interface. Suppose a crest is formed in the thick string just before the interface. The thick string will exert an upward force on the point P as a result of which light string will yield to the upward force resulting in the formation of a crest just after and before the interface. See Fig. 1.23. Hence, a crest is refracted or transmitted as a crest. Therefore, there is no change of phase when a transverse wave is refracted going from a denser medium to a rarer medium.

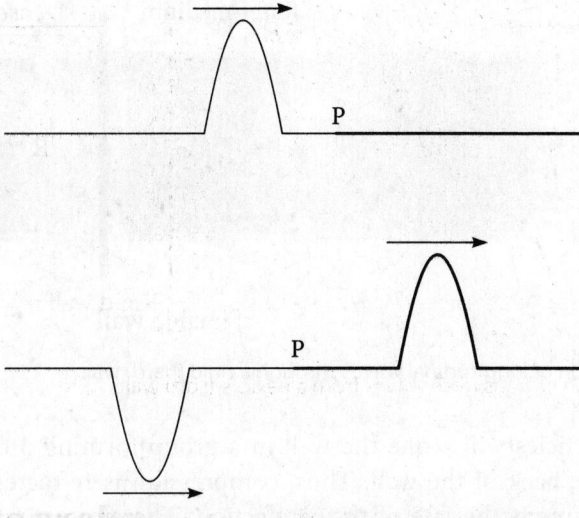

Figure 1.22 │ Refraction of transverse waves through a flexible (soft) wall

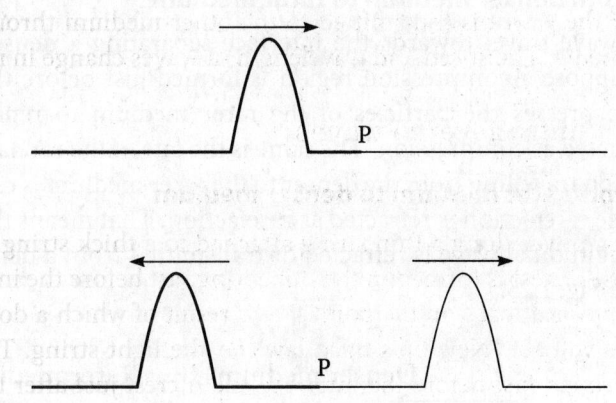

Figure 1.23 │ Refraction of transverse waves through a rigid (hard) wall

1.12.2 Refraction of longitudinal waves

Case 1: Passing from rarer medium to denser medium

As mentioned earlier, longitudinal waves propagate in the form of compression and rarefaction. Let a longitudinal wave travel towards the rigid wall, i.e., toward the interface separating a rarer medium from a denser medium. Suppose a compression region is formed just before the rigid wall. See Fig. 1.24.

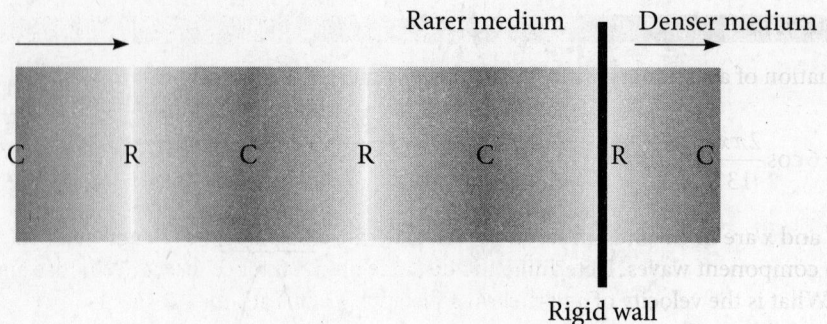

Figure 1.24 | Refraction of longitudinal waves through a rigid (hard) wall

The compressed particles will strike the wall in a group forming a compression region in the vicinity of the back of the wall. Thus, compression is refracted or transmitted as compression. The same is the fate of the rarefaction. Therefore, a wave of compression travelling from the rarer to the denser medium is refracted as a wave of compression while rarefaction is refracted as rarefaction. That means there is no phase change when a longitudinal wave is transmitted or refracted from a rarer medium to a denser medium.

Case 2: Passing from denser medium to rarer medium

Let a longitudinal wave travel towards the interface separating a denser medium from a rarer medium. Suppose a compression region is formed just before the interface. The compression region presses the particles of the rarer medium to replace them. Thus, compression is refracted as compression. The same is the fate of the rarefaction. Therefore, a wave of compression travelling from the denser to the rarer medium is refracted as a wave of compression while rarefaction is refracted as rarefaction. That means there is no change of phase when a longitudinal wave is refracted or transmitted from a denser medium and a rarer medium. See Fig. 1.25.

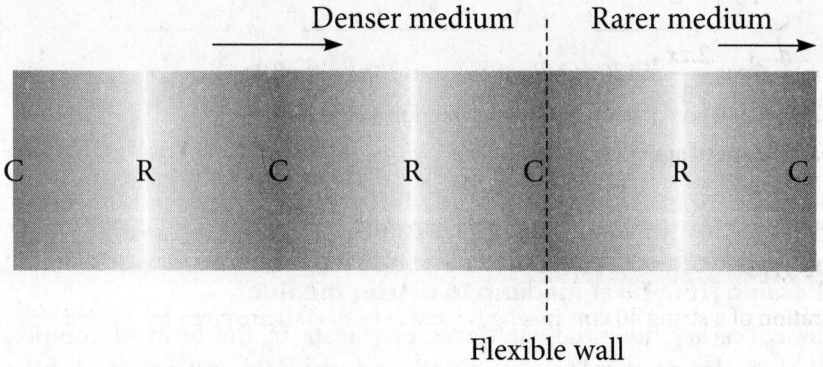

Figure 1.25 | Refraction of longitudinal waves through a flexible (soft) wall

Example 1.12

The equation of a stationary wave in a medium is given as

$$\Psi = 6\cos\frac{2\pi x}{13}\sin 20\pi t.$$

Here Ψ and x are in cm and t in seconds. Calculate the amplitude, wavelength, and velocity of the two component waves. Determine the distance between two consecutive nodes and anti-nodes. What is the velocity of a particle at a distance 13 cms at time 0.2 s?

Solution

The given equation

$$\Psi = 6\cos\frac{2\pi x}{13}\sin 20\pi t$$

can be written as

$$\Psi = 2\times 3\cos\frac{2\pi x}{13}\sin 2\pi\times 10t$$

Comparing this equation with the standard stationary wave equation

$$\Psi = 2r\cos\frac{2\pi x}{\lambda}\sin 2\pi\frac{t}{T},$$

we get

amplitude = 3 cms, wavelength $\lambda = 13$ cm time period $T = 0.1$ s, frequency = 10/s, velocity = wavelength × frequency = 13×10 cm/s = 130 cm/s.

The distance between two consecutive nodes and anti nodes = 13 cms/2 = 6.5 cm.
The particle speed is given by

$$\frac{d\Psi}{dt} = \frac{d}{dt}6\cos\frac{2\pi x}{3}\sin 20\pi t = 6\cos\frac{2\pi x}{13}\cos 20\pi t\times 20\pi$$

So the particle speed at $x = 13$ cm and $t = 0.2$ s will be $= 6\times\cos 2\pi\times\cos 4\pi\times 20\pi = 120\pi$ cm/s

Example 1.13

The vibration of a string 40 cms in length fixed at both ends are given by the equation

$$\Psi = 4\sin\frac{4\pi x}{25}\cos(4\pi t).$$

Here Ψ, x are in cm and t in seconds. What is the maximum displacement of a particle at 2.5 cm? Write down the equation of the component waves whose superposition gives this stationary wave.

Solution

The given equation

$$\Psi = 4\sin\frac{4\pi x}{25}\cos(4\pi t)$$

can be written as

$$\Psi = 2\times 2\sin\frac{2\times 2\pi x}{25}\cos(2\pi\times 2t)$$

Comparing this equation with a standard equation of the form

$$\Psi = 2r\sin\frac{2\pi x}{\lambda}\cos\left(2\pi\frac{t}{T}\right),$$

we have amplitude = 2 cm, wavelength = 25/2 = 12.5 cm, time period = 1/2 = 0.5 s, frequency = 2/s, speed = 12.5 × 2 = 25 cm/s

The maximum displacement of the particle of the medium is

$$\Psi_{max} = 4\sin\frac{4\pi x}{25}.$$

The maximum displacement of a particle at $x = 2.5$ cms is

$$4\times\sin\left(\frac{4\pi}{25}\times 2.5\right) = 4\times 0.95 = 3.80 \text{ cm}.$$

The equations of the component waves are

$$\Psi_1 = r\sin\frac{2\pi}{\lambda}(x - vt) = 2\sin\frac{2\pi}{12.5}(x - 25t)$$

and $\Psi_1 = r\sin\dfrac{2\pi}{\lambda}(x + vt) = 2\sin\dfrac{2\pi}{12.5}(x + 25t).$

1.13 Wave Packet

The concept of a wave packet is purely quantum mechanical. A pure sine wave extends from $-\infty$ to $+\infty$. That means it is completely unlocalized. A classical particle is approximately localized. We do not know the exact position vector of an electron in an atom. Such a situation can be described by the concept of wave packets.

A wave packet is defined as a group of waves of slightly different wavelengths, with phases and amplitudes chosen such that they are superposed to interfere constructively over only a small region of space; outside of this space they produce an amplitude that reduces to zero rapidly as a result of destructive interference. Thus, a wave packet is a localized wave, amplitude of which is zero except over a small region. When a wave packet is localized in a small region, it can be regarded as a point. Therefore, the motion of a point particle can be described by the motion of a wave packet. It also has a periodic structure that is characteristic of a wave. We can represent a wave packed in one dimension by an expression such as

$$\Psi(x) = \int\limits_0^\infty r(\lambda) \, \cos\frac{2\pi x}{\lambda} \, d\lambda \tag{1.99}$$

where $\cos\dfrac{2\pi x}{\lambda}$ is a sinusoidal wave with wavelength λ. The integral represents a superposition in which we add a very large number of such waves with slightly different values of λ, each with an amplitude $r(\lambda)$ that depends upon λ. The wave packet is also called a wave group. See Fig. 1.26.

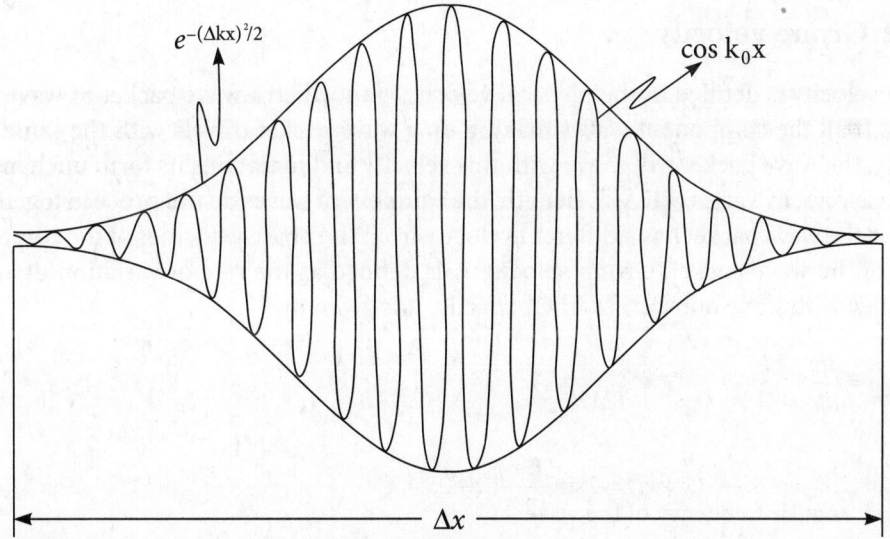

$e^{-(\Delta k x)^2/2}$

$\cos k_0 x$

Δx

Figure 1.26 | The resultant of the addition of many sine waves of different amplitudes and wavelengths

1.14 Phase Velocity and Group Velocity

1.14.1 Phase velocity

The velocity of an individual wave is defined as the phase velocity. It is also the velocities with which the component waves of a wave packet move inside the wave packet. The phase velocity is not the velocity of propagation of any physical quantity; it simply represents the velocity of a point of constant phase on the wavefront. Let x be the position of a point on the wave front of the wave. The phase velocity v_p is defined as

$$v_p = \frac{dx}{dt} \qquad (1.100)$$

From Section 1.10, we have

$$\omega t \pm kx = \text{constant or } \frac{dx}{dt} = \frac{\omega}{k}.$$

or $\quad v_p = \dfrac{\omega}{k} \qquad (1.101)$

We can also get Eq. (1.101) by using the relations $k = \dfrac{2\pi}{\lambda}$, $\omega = 2\pi v$, and $v = \lambda v$. Here v is the linear frequency of the wave.

1.14.2 Group velocity

Group velocity is defined as the physical velocity with which a wave packet or wave group travels. If all the component waves making up a wave packet travels with the same phase velocity, the wave packet will move with this velocity and maintains its form unchanged. If the phase velocity varies with wavelength, the component waves do not proceed together. In this case the wave packet has a different velocity from the phase velocities of the component waves of the wave packet. Group velocity v_g is defined as the rate of variation of angular frequency with wave number. Mathematically, it is given by

$$v_g = \frac{d\omega}{dk} \qquad (1.102)$$

Here

ω = angular frequency of the wave

$k = \dfrac{2\pi}{\lambda}$ = magnitude of propagation vector

The phenomenon of variation of phase velocity with wavelengths of the component waves of a wave packet is called dispersion. Depending upon how the phase velocity varies with wave number k in a particular situation, the group velocity may be less or more than the phase velocities of the constituent waves of the wave packet. If the constituent waves of a wave packet travel with the same velocities as that of the wave packet, then the phase velocity and group velocity are the same. This happens in case of light waves moving in vacuum. The wave packet moves with group velocity where as within the wave packet, constituent waves may move with phase velocities. The phase velocity cannot be defined for a wave packet. It is meaningful only for component waves making up the wave packet. Electromagnetic signals are transmitted with group velocity and not with phase velocity. Phase speed and group speed would be better terms in comparison to the terms phase velocity and group velocity respectively.

Example 1.14

Two harmonic waves given by $\psi_1 = 12\sin(4\pi t - 5x)$ and $\psi_2 = 16\sin(8\pi t - 7x)$ are superposed to produce a single wave. Find the group speed of the resultant wave.

Solution

Comparing the given two equations with a standard wave equation, the angular frequencies ω's of the two waves are obtained as

$$\omega_1 = 4\pi \tag{A}$$

$$\omega_2 = 8\pi \tag{B}$$

Again comparing the given two equations with a standard wave equation, the propagation vectors of the two waves are obtained as

$$k_1 = 5 \tag{C}$$

$$k_2 = 7 \tag{D}$$

Hence, the group speed of the resultant wave is given by

$$v_g = \frac{d\omega}{dk}$$

$$\text{or} \quad v_g = \frac{\Delta\omega}{\Delta k} = \frac{\omega_2 - \omega_1}{k_2 - k_1} \tag{E}$$

Putting the values of ω_1, ω_2, k_1 and k_2 from equations (A), (B), (C), and (D) into equation (E), we have

$$v_g = \frac{(8\pi - 4\pi)/\text{s}}{(7-5)/\text{cm}}$$

$$\text{or} \quad v_g = 6.28 \text{ cm/s}$$

1.14.3 Relation between phase velocity and group velocity

The phase velocity is the average velocity of component waves making up the wave packet and is defined in Eq. (1.101) as

$$v_p = \frac{\omega}{k}$$

or $\quad \omega = v_p k$

The group velocity is the velocity of the wave packet and is defined in Eq. (1.102) as

$$v_g = \frac{d\omega}{dk} = \frac{d(v_p k)}{dk}$$

or $\quad v_g = v_p + k\dfrac{dv_p}{dk}$ $\hfill$ (1.103)

Since $k = \dfrac{2\pi}{\lambda}$, Eq. (1.103) becomes

$$v_g = v_p + k\frac{dv_p}{d\lambda} \times \frac{d\lambda}{dk} = v_p + k\frac{dv_p}{d\lambda} \times \frac{d}{dk}\left(\frac{2\pi}{k}\right)$$

or $\quad v_g = v_p - \lambda\dfrac{dv_p}{d\lambda}$ $\hfill$ (1.104)

This shows that the group velocity may be equal to, less than or more than the phase velocity of a wave depending upon the variation of phase velocity with respect to the wavelength.

Case 1: $\quad \dfrac{dv_p}{d\lambda} = 0 \Rightarrow v_g = v_p$

If the phase velocity of the component waves is independent of the wavelength, i.e., $\dfrac{dv_p}{d\lambda} = 0$, then group velocity and phase velocity are equal as in the case of electromagnetic waves propagating in vacuum. In this case, wave packet moves with the velocity of the component waves making up the wave packet. $\dfrac{dv_p}{d\lambda} = 0$ in a non-dispersive media.

Case 2: $\dfrac{dv_p}{d\lambda} > 0 \Rightarrow v_g < v_p$

If the phase velocity of the component waves increases with increase of the wavelength, i.e., $\dfrac{dv_p}{d\lambda} > 0$, then group velocity is less than the phase velocity as in the case of an electromagnetic wave propagating in a transparent medium like water or glass. In this case, the wave packet moves with a velocity less than that of the component waves making up the wave packet. When white light passes through a glass prism, constituent wavelengths are separated out and travel through the prism with different phase velocities. This phenomenon is called normal dispersion and the medium is called a normal dispersive medium.

Case 3: $\dfrac{dv_p}{d\lambda} < 0 \Rightarrow v_g > v_p$

If the phase velocity of the component waves increases with decrease of the wavelength, i.e., $\dfrac{dv_p}{d\lambda} < 0$, then group velocity is more than the phase velocity as in the case of an electromagnetic wave propagating in an ionized media. In this case, the wave packet moves with a speed more than that of the component waves making up the wave packet. This phenomenon is called an anomalous dispersion and the medium is called anomalous dispersive medium.

Example 1.15

The de Broglie wavelength of a moving electron is 5×10^{-12} m. Find its kinetic energy, phase speed and group speed.

Solution

de Broglie's famous relation is $p = \dfrac{h}{\lambda}$

or $pc = \dfrac{hc}{\lambda} = \dfrac{6.63 \times 10^{-34} \times 3 \times 10^8}{5.0 \times 10^{-12}} \text{J} = 3.98 \times 10^{14} \text{J}$

or $pc = 637$ keV

The rest energy of the electron $E_0 = m_e c^2 = 511$ keV

The kinetic energy KE of the electron is

$$\text{K.E.} = E - E_0 = \sqrt{E_0^2 + (pc)^2} - E_0 = \sqrt{(551)^2 + (637)^2} \text{ keV} - 511 \text{ keV} = 305.63 \text{ keV}$$

We know that

$$E = \frac{E_0}{\sqrt{1 - \dfrac{v^2}{c^2}}}$$

Hence, the speed of the electron can be found out from

$$v = c\sqrt{1 - \frac{E_0^2}{E^2}} = c\sqrt{1 - \frac{511}{816.63}} = 0.780\ c = \text{The group speed of the electron} = v_g = 0.780\ \text{m/s}$$

The phase speed of the electron is given by

$$v_p = \frac{c^2}{v} = \frac{c^2}{0.780c} = 1.28\ c$$

The phase speed is more than the light speed in vacuum!

1.15 Uncertainty Principle

1.15.1 Uncertainty principle for classical waves

The sinusoidal wave $\Psi = r_1 \sin k_1 x$ extends from $-\infty$ to $+\infty$, and hence it is located everywhere. It is called unlocalized (i.e., there is uncertainty in position $\Delta x = \infty$). Since there are a large number of waves, the wavelength can be accurately determined (i.e., uncertainty in wave number $\Delta k = 0$). However, it cannot describe a particle. The sinusoidal wave $\psi = r_1 \sin k_1 x + r_2 \sin k_2 x$ (one more wave is added) also extends from $-\infty$ to $+\infty$; however, we know a bit more about the location of the wave. In case of sound wave, it is the well known phenomenon of beats. See Fig. 1.27.

The particles of the medium oscillate with much less amplitude at certain points. Unlike the first case, the determination of the location of the wave improves, i.e., uncertainty in position Δx decreases at the cost of the accurate determination of wavelength, i.e., uncertainty in wave number Δk increased. If we superpose a very large number of waves, the resultant wave will be like shown in Fig. 1.24. Here, uncertainty in position Δx decreases and uncertainty in wave number Δk increases, i.e., they are inversely related. So approximately, we have

$$\Delta x \cdot \Delta k \approx 1 \tag{1.105}$$

This is the wave number and position uncertainty relationship for classical waves which states that for any type of wave, the position can be determined accurately at the cost of our knowledge of its wavelength.

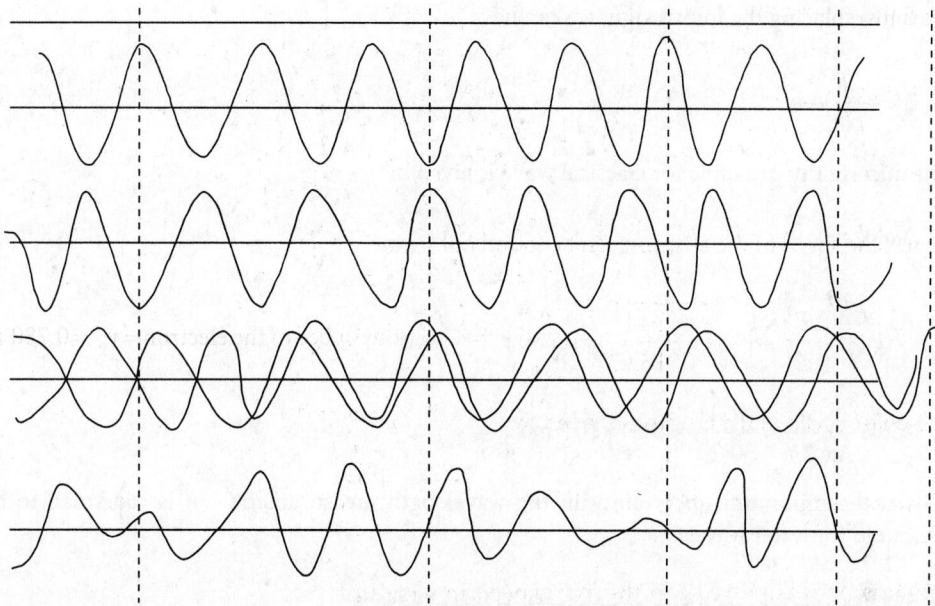

Figure 1.27 | Two waves of different frequency combined to cause beats

One of the other uncertainty relationships for classical waves between angular frequency and time is

$$\Delta\omega \cdot \Delta t \approx 1. \tag{1.106}$$

It states that for any type of wave, the angular frequency can be determined accurately at the cost of our knowledge of its time interval.

Example 1.16

One day standing on the Puri sea beach (Odisha), you observe that within a distance of 30 m there are 30 number of water wave crests. Calculate the minimum uncertainty in the wavelength measurement.

Solution

The wavelength λ of the water wave is found out as

$$\lambda = \frac{30\text{m}}{30} = 1 \text{ m}$$

$$k = \frac{2\pi}{\lambda}$$

$$\text{or} \quad dk = -\frac{2\pi}{\lambda^2} d\lambda$$

Thus by replacing the minus sign, we can have

$$\Delta k = \frac{2\pi}{\lambda^2} \Delta\lambda$$

The uncertainty principle for classical wave is given by

$$\Delta k \Delta x \approx 1$$

or $\quad \Delta x \left(\frac{2\pi}{\lambda^2} \Delta\lambda \right) \approx 1$

or $\quad \Delta\lambda \approx \frac{1}{\Delta x} \frac{\lambda^2}{2\pi} \approx \frac{1}{30} \frac{1^2}{2\pi} \approx 5.31 \times 10^{-3}\,\text{m}.$

This is the minimum uncertainty in the wavelength measurement – it is too small to be measured for water waves.

1.15.2 Heisenberg's uncertainty principle

This principle is also called the indeterminacy principle. German physicist Werner Heisenberg, in the year 1927, made the historical statement that even in theory, the position and the velocity of an object cannot both be measured with cent percent accuracy simultaneously. The concepts of exact position and exact velocity together, have no meaning in nature. Ordinary experience provides no clue of this principle. It is easy to measure both the position and the velocity of an automobile, because the uncertainties implied by this principle for ordinary objects are too small to be observed. The complete rule dictates that the product of the uncertainties in position and velocity is equal to or greater than a small physical quantity

$$1.054 \times 10^{-34}\,\text{Js}, \quad \left(= \frac{h}{2\pi} = \hbar \right),$$

where h is Planck's constant. It is only for atoms and subatomic particles that the product of the uncertainties becomes significant due to their very small masses. Any attempt to measure precisely the position of an electron, will knock it about in an unpredictable way, so that a simultaneous measurement of its velocity has no validity. This result has nothing to do with inadequacies in the measuring instruments, the technique, or the observer. It arises out of the intimate connection in nature between particles and waves in the realm of subatomic dimensions. Every particle has a wave associated with it and it exhibits wavelike behavior. The particle is most likely to be found in those places where the undulations of the wave are most intense. The more intense the undulations of the associated wave become, the more ill defined becomes the wavelength and so is the momentum of the particle $\left(p = \frac{h}{\lambda} \right)$. Hence, a completely localized wave has an indeterminate wavelength.

The corresponding particle has a definite position but no definite velocity. A particle wave having a well-defined wavelength is spread out. The particle, while having precise velocity may be present almost anywhere.

A quite accurate measurement of one observable results in large uncertainty in the measurement of the other conjugate observable. The uncertainty principle is alternatively expressed as the product of the uncertainties in the momentum and the position of a particle is equal to or more than $\hbar$. The principle applies to other related conjugate pairs of observables, such as energy–time, angular momentum–angle. The product of the uncertainty in an energy measurement and the uncertainty in the time interval during which the measurement is made also equals to or is more than $\hbar$. Three famous uncertainty principles, mathematically, are as follows

i. $\Delta x\,\Delta p \geq \dfrac{\hbar}{2}.$ (1.107)

(Heisenberg's uncertainty principle for position and momentum; see Fig. 1.28.)

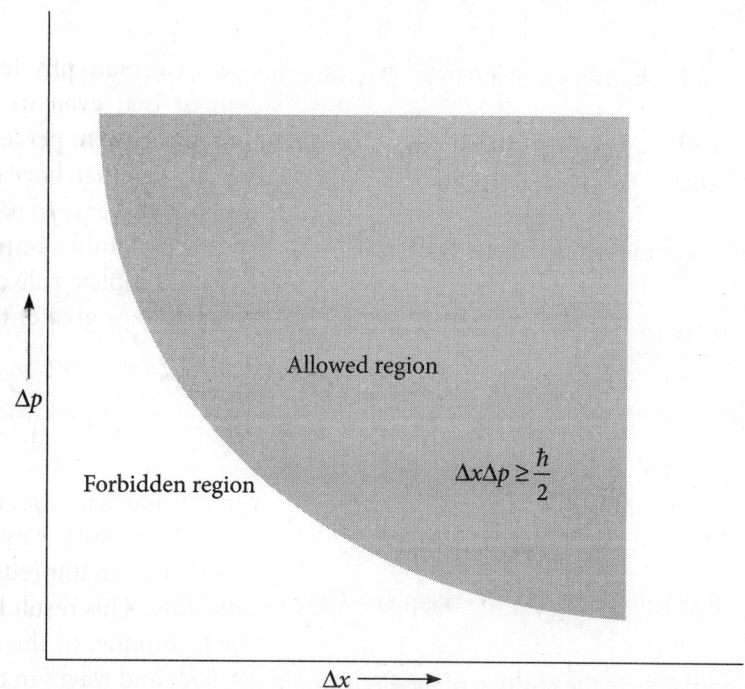

Figure 1.28 | Heisenberg's uncertainty principle for position and momentum. The allowed region has values such that the product $\Delta x\Delta p$ is greater than or equal to $\dfrac{h}{4\pi}$. In the forbidden region, $\Delta x\Delta p$ is less than $\dfrac{h}{4\pi}$.

ii. $\Delta E\,\Delta t \ge \dfrac{\hbar}{2}.$ (1.108)

(Heisenberg's uncertainty principle for energy and time)

iii. $\Delta J\Delta\theta \ge \dfrac{\hbar}{2}$ (1. 109)

(Heisenberg's uncertainty principle for angular momentum and angle)

Heisenberg's uncertainty principle and its applications are described in great detail in Section 7.7. Philosophically, we may say that, 'We cannot know the future for sure because we do not know the present for sure.'

Example 1.17

An electron has a speed 2×10^4 cm/s accurate to 0.01%. With what accuracy can we locate position of this electron?

Solution

The accuracy of the electron speed is 0.01%. Then, uncertainty in speed will be

$$= \Delta v = v \times \frac{0.01}{100} = 2 \times 10^4 \text{ cm/s} \times \frac{0.01}{100} = 0.02\,\text{m/s}$$

The uncertainty in momentum will be $\Delta p = m\Delta v.$

$$\therefore \quad \Delta p = 9.11 \times 10^{-31}\,\text{kg} \times 0.02\,\text{m/s} = 1.8 \times 10^{-32}\,\text{kgm/s}$$

We know that

$$\Delta p \Delta x \approx \frac{h}{4\pi} = 5.28 \times 10^{-35}\,\text{J.s}$$

$$\therefore \quad \Delta x = 5.28 \times 10^{-35}\,\text{J.s}/(1.8 \times 10^{-32}\,\text{kgm.s}) = 2.93 \times 10^{-3}\,\text{m}.$$

This is the uncertainty in the position of the electron.
Thus, the accuracy with which we can locate the electron is 2.93×10^{-3} m.

Example 1.18

A cricket ball of mass 150 gm is moving at a speed of 160 km/hour. The speed can be measured with an accuracy of 1%. With what precision can we measure its position simultaneously?

Solution

The speed of the cricket ball is 160 km/hour = 44.44 m/s. The accuracy of the electron speed is 1%. Then uncertainty, in the speed of the ball is

$$\Delta v = v \times \frac{1}{100} = 44.44 \times 0.01 \text{ m/s} = 0.44 \text{ m/s}$$

The uncertainty in momentum will be $\Delta p = m \Delta v$.

$$\therefore \quad \Delta p = 0.15 \text{ kg} \times 0.44 \text{ m/s} = 0.07 \text{ kgm/s}$$

We know that

$$\Delta p \Delta x \approx \frac{h}{4\pi} = 5.28 \times 10^{-35} \text{ J.s}$$

$$\therefore \quad \Delta x = 5.28 \times 10^{-35} \text{ Js}/(0.07 \text{ kgm.s}) = 7.5 \times 10^{-34} \text{ m}.$$

This is the uncertainty in the position of the cricket ball.
Thus, the precision with which we can locate the cricket is 7.5×10^{-34} m.
The uncertainty in the position of the cricket ball is 20 orders of magnitude smaller than the size of an atomic nucleus.

Example 1.19

If the average period that elapses between the excitation of an atom and the time it radiates energy is 10^{-10} s, then calculate the uncertainty in the energy of the emitted photon and the limit of accuracy with which the frequency of the emitted radiations can be determined.

Solution

The uncertainty relation between energy and time is given by

$$\Delta E \Delta t \approx \frac{h}{4\pi} = 5.28 \times 10^{-35} \text{ Js}.$$

The uncertainty in time is $\Delta t = 10^{-10}$ s. The uncertainty in the energy is given by

$$\Delta E = 5.28 \times 10^{-35} \text{ Js}/(\Delta t)$$

$$\text{or} \quad \Delta E = 5.28 \times 10^{-35} \text{ Js}/\left(10^{-10} \text{ s}\right) = 5.28 \times 10^{-25} \text{ J}$$

The energy of the emitted photon in terms of frequency is given by

$$E = h\nu$$

or $\quad \Delta E = h\Delta\nu$

or $\quad \Delta\nu = \dfrac{\Delta E}{h}$ is the uncertainty in frequency of the emitted photon.

or $\quad \Delta\nu = \dfrac{5.28 \times 10^{-25}\,\text{J}}{6.63 \times 10^{-34}\,\text{Js}}$

or $\quad \Delta\nu = 7.96 \times 10^{-8}\,\text{s}^{-1}$

The limit of accuracy with which the frequency of the emitted radiations can be determined is $7.96 \times 10^{-8}\,\text{s}^{-1}$.

1.16 Superposition of Waves

In our day-to-day life, there are lots of phenomena in which a large number of waves pass through a single point at the same time. Particles cannot pass through each other; they collide and their course of action changes. However, waves can pass through each other freely without being modified. Your friend's voice quality does not change in the presence of other sounds although they have passed through the same space simultaneously. The phenomenon involving a combination of two or more waves travelling in a medium simultaneously is called superposition of waves. The superposition phenomenon is governed by very simple law, as long as the wave function is linear, that is, it contains the function $\Psi(x, t)$ only to the first power. The wave function is linear for the waves producing small disturbances.

1.16.1 Basis for the principle of superposition

The general differential wave equation

$$\nabla^2\Psi = \frac{1}{v^2}\frac{\partial^2\Psi}{\partial t^2}$$

is a homogeneous linear differential equation. It is the property of the linear differential equation that the algebraic sum of all its solutions is also one of its solutions. Since all the harmonic wave functions (i.e., mathematical representation of disturbances) are the solutions of this differential wave equation, their summation, that is, the resultant wave function (resultant disturbance) will satisfy the differential wave equation. This means that if $\Psi_1, \Psi_2, \Psi_3, \Psi_4, ..., \Psi_N$ are the N solutions, then $\Psi = \Psi_1 + \Psi_2 + \Psi_3 + \Psi_4 + ... + \Psi_N$ will be a solution of the differential wave equation. Thus, Ψ is the resultant wave function satisfying the general differential wave equation.

1.16.2 Principle of superposition

Based on the discussions in the previous section, we can now enunciate the principle of superposition. The principle of superposition states that, 'when two or more waves of small disturbance pass through a point simultaneously, the resultant disturbance at the point is equal to the sum of all the disturbances each wave would produce in the absence of other waves.' In other words, when a non-dispersive medium is disturbed simultaneously by any number of waves, the instantaneous resultant displacement of the medium at every point at any instant is the sum of all the displacements taken independently.

Mathematically, if $\Psi_1, \Psi_2, \Psi_3, \Psi_4, ..., \Psi_N$ are individual disturbances of a particle of the medium at a point, produced by N waves separately, then the resultant displacement/disturbance Ψ, when they pass simultaneously through the point will be given by

$$\Psi = \Psi_1 + \Psi_2 + \Psi_3 + \Psi_4 + ... + \Psi_N \tag{1.110}$$

1.16.3 Two beams superposition in one direction

Case 1: Two waves having the same frequency travelling in the same direction (Production of interference pattern)

Let two harmonic waves Ψ_1 and Ψ_2 having the same frequency travelling in the same $+x$ direction in a medium be represented mathematically by

$$\Psi_1 = r_1 \sin(\omega t - kx) \text{ and } \Psi_2 = r_2 \sin(\omega t - kx + \delta).$$

By the principle of superposition, the resultant displacement will be given by

$$\Psi = \Psi_1 + \Psi_2 = r_1 \sin(\omega t - kx) + r_2 \sin(\omega t - kx + \delta). \tag{1.111}$$

$$= r_1 \sin(\omega t - kx) + r_2 \left[\sin(\omega t - kx)\cos\delta + \cos(\omega t - kx)\sin\delta \right]$$

$$= r_1 \sin(\omega t - kx) + r_2 \sin(\omega t - kx)\ \cos\delta + r_2 \cos(\omega t - kx)\sin\delta$$

or $$\Psi = \sin(\omega t - kx)(r_1 + r_2 \cos\delta) + r_2 \sin\delta \cos(\omega t - kx) \tag{1.112}$$

Let $r_1 + r_2 \cos\delta = R\cos\theta$ (1.113)

and $\quad r_2 \sin \delta = R \sin \theta$ $\hfill (1.114)$

Putting these substitutions into Eq. (1.112), we have

$$\Psi = R \cos \theta \sin(\omega t - kx) + R \sin \theta \cos(\omega t - kx)$$

or $\quad \Psi = R \left[\cos \theta \; \sin(\omega t - kx) + \sin \theta \cos(\omega t - kx) \right]$

or $\quad \Psi = R \sin \theta \sin(\omega t - kx + \theta)$ $\hfill (1.115)$

The values of R and θ are found out by the following way.

Using the Eqs (1.113) and (1.114), we have

$$\left(R \sin \theta\right)^2 + \left(R \cos \theta\right)^2 = \left(r_1 + r_2 \cos \delta\right)^2 + \left(r_2 \sin \delta\right)^2$$

or $\quad R^2 \left(\sin^2 \theta + \cos^2 \theta\right) = r_1^2 + r_2^2 \left(\cos^2 \delta + \sin^2 \delta\right) + 2 r_1 r_2 \cos \delta$

or $\quad R = \sqrt{r_1^2 + r_2^2 + 2 r_1 r_2 \cos \delta}$ $\hfill (1.116)$

Thus, the amplitude of the resultant wave is maximum ($= r_1 + r_2$) when $\cos \delta = 1$ and minimum ($= r_1 - r_2$) when $\cos \delta = -1$. The same conditions are valid for the distribution of an intensity function (since intensity is proportional to square of the amplitude] with respect to the phase difference of the two superposing waves.

Again from Eqs (1.113) and (1.114), we have

$$\tan \theta = \frac{r_2 \sin \delta}{r_1 + r_2 \cos \delta}$$ $\hfill (1.117)$

Equation (1.115) is of a similar form as that of the two superposing waves and it also satisfies the general differential wave equation. Thus, the resultant is a wave with a new amplitude and a phase angle whose values can be calculated from Eqs (1.116) and (1.117) respectively. From Eqs (1.116) and (1.117), it is clear that the phase angle and the amplitude R of the resultant wave changes with the initial phase difference δ of the superposing waves because $\tan \theta$ and R is a function of δ.

Case 2: Two waves having the same frequency travelling in opposite direction (Production of stationary waves)

Let two harmonic waves Ψ_1 and Ψ_2 having the same frequency and amplitude travelling, in opposite directions in a line in a medium be superimposed on each other. This is the superimposition of the incident wave and the reflected wave moving along a string, which gives rise to stationary waves. The incident wave and the corresponding reflected wave may be represented mathematically by

$$\Psi_I = r \sin(\omega t - kx) \text{ and } \Psi_R = r \sin(\omega t + kx) \text{ respectively.}$$

By the principle of superposition, the resultant displacement will be given by

$$\Psi = \Psi_I + \Psi_R = r \sin(\omega t - kx) + r \sin(\omega t + kx)$$

or $\quad \Psi = r\left[\sin(\omega t - kx) + \sin(\omega t + kx)\right] = 2r \cos kx \sin \omega t \qquad (1.118)$

Putting $R = 2r \cos kx$ in this equation, we have

$$\Psi = R \sin \omega t \qquad (1.119)$$

Equation (1.119) is also a wave equation representing stationary wave or standing wave with amplitude $R = 2r \cos kx$. Thus, amplitude is a function of the position 'x' and varies from 0 to $2r$. The amplitude is $2r$ when

$$|2r \cos kx| = 1$$

or $\quad kx = n\pi$, n is a whole number, i.e., $n = 0, 1, 2, 3, \ldots$

or $\quad x = \dfrac{n\pi}{k} = \dfrac{n\pi}{2\pi\big/\lambda} = n\pi \times \dfrac{\lambda}{2\pi} = \dfrac{n\lambda}{2}$

The amplitude is 0 when

$$|2r \cos kx| = 0$$

or $\quad kx = (2n+1)\dfrac{\pi}{2}$

or $\quad x = \dfrac{(2n+1)\pi}{2k} = (2n+1)\dfrac{\lambda}{4}$

Thus amplitude is maximum (i.e., displacements of the particles of the medium are at maximum distance from the mean position) at $x = 0, \dfrac{\lambda}{2}, \dfrac{2\lambda}{2}, \dfrac{3\lambda}{2}, \dfrac{4\lambda}{2} \ldots$, etc and amplitude is zero (i.e., displacements of the particles of the medium are zero) at $x = \dfrac{1\lambda}{4}, \dfrac{3\lambda}{4}, \dfrac{5\lambda}{4}, \dfrac{7\lambda}{4} \cdots$,

etc. The points at which amplitude is maximum are called antinodes and the points at which amplitude is minimum are called nodes. Thus nodes occur at $x = \dfrac{1\lambda}{4}, \dfrac{3\lambda}{4}, \dfrac{5\lambda}{4}, \dfrac{7\lambda}{4} \cdots$, etc and anti-nodes occur at $x = 0, \dfrac{\lambda}{2}, \dfrac{2\lambda}{2}, \dfrac{3\lambda}{2}, \dfrac{4\lambda}{2} \cdots$, etc

The stationary wave is graphically represented in the Fig. 1.27. We also see from Eq. (1.119) that at a time when sin $\omega t = 1$, all the particles for which cos kx is positive reach their positive maximum displacement. At this particular instant, all the particles for which cos kx is negative, reach their negative maximum displacement. At a time when sin $\omega t = 0$, all the particles of the medium crosses their mean position. See Fig. 1.29.

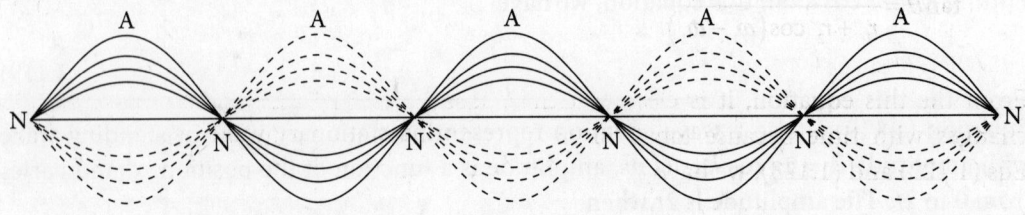

Figure 1.29 | External appearance of a string when stationary wave is produced on it. As and Ns are positions of antinodes and nodes respectively

Case 3: Two waves having NEARLY the same frequency (Beat phenomenon)

Consider two waves of the same frequency having nearly the same frequency (i.e., $\omega_1 - \omega_2$ is very small.). For the sake of simplicity, it is assumed that two waves are in phase at any point in the medium at time $t = 0$. The two waves are represented by

$$\Psi_1 = r_1 \sin \omega_1 t \quad \text{and} \quad \Psi_2 = r_2 \sin \omega_2 t \tag{1.120}$$

The resultant wave according to the superposition principle will be given by

$$\Psi = \Psi_1 + \Psi_2 = r_1 \sin \omega_1 t + r_2 \sin \omega_2 t$$

or $\quad \Psi = r_1 \sin \omega_1 t + r_2 \left[\sin\left\{ \omega_1 t - \left(\omega_1 t - \omega_2 t \right) \right\} \right]$

or $\quad \Psi = r_1 \sin \omega_1 t + r_2 \left[\sin \omega_1 t \cos\left(\omega_1 t - \omega_2 t \right) - \cos \omega_1 t \sin\left(\omega_1 t - \omega_2 t \right) \right]$

or $\quad \Psi = \sin \omega_1 t \left[r_1 + r_2 \cos\left(\omega_1 t - \omega_2 t \right) \right] - r_2 \cos \omega_1 t \left[\sin\left(\omega_1 t - \omega_2 t \right) \right] \tag{1.121}$

Now let $R \cos \theta = r_1 + r_2 \cos\left(\omega_1 t - \omega_2 t \right)$ \tag{1.122} and $R\sin\theta = r_2\sin\left(\omega_1 t - \omega_2 t\right)$ $\qquad\qquad$ (1.123)

Substituting Eqs (1.122) and (1.123) into Eq. (1.121), we get the resultant wave as

$\qquad \Psi = R\cos\theta\sin\omega_1 t - R\sin\theta\cos\omega_1 t$

or $\quad \Psi = R\sin\left(\omega_1 t - \theta\right)$ $\qquad\qquad$ (1.124)

Taking the ratio of Eqs (1.123) and (1.122), we have

$$\tan\theta = \frac{r_2\,\sin\left(\omega_1 - \omega_2\right)t}{r_1 + r_2\,\cos\left(\omega_1 - \omega_2\right)t}$$ $\qquad\qquad$ (1.125)

From the this equation, it is clear that the phase angle θ of the resultant wave (1.124) changes with time because $\tan\theta$ is a function of time. Again squaring and adding Eqs (1.122) and (1.123), we have

$$R^2 = \left[r_1 + r_2\,\cos\left(\omega_1 - \omega_2\right)t\right]^2 + \left[r_2\,\sin\left(\omega_1 - \omega_2\right)t\right]^2$$

or $\quad R = \sqrt{r_1^2 + r_2^2 + 2r_1 r_2\,\cos\left(\omega_1 - \omega_2\right)t}$

$\qquad = \sqrt{r_1^2 + r_2^2 + 2r_1 r_2\,\cos 2\pi\left(v_1 - v_2\right)t}$ $\qquad\qquad$ (1.126)

From Eq. (1.126) it is clear that the amplitude of the resultant wave changes with time because R is a function of time.

The amplitude of the resultant wave is maximum when $\cos 2\pi\left(v_1 - v_2\right)t = 1$ in Eq. (1.126).

or $\quad 2\pi\left(v_1 - v_2\right)t = 2n\pi$, where $n = 0, 1, 2, 3, \ldots$, etc.

or $\quad t = \dfrac{n}{v_1 - v_2}$ $\qquad\qquad$ (1.127)

The amplitude of the resultant wave is maximum at time

$\qquad 0s, \dfrac{1}{v_1 - v_2}s, \dfrac{2}{v_1 - v_2}s, \dfrac{3}{v_1 - v_2}s, \dfrac{4}{v_1 - v_2}s, \ldots$ etc.

The maximum value of the amplitude R is $R_{max} = \sqrt{r_1^2 + r_2^2 + 2r_1r_2} = r_1 + r_2$

The amplitude of the resultant wave is minimum when $\cos 2\pi (v_1 - v_2)t = -1$.

or $\quad 2\pi (v_1 - v_2)t = (2n+1)\pi$, where $n = 0, 1, 2, 3, \ldots$ etc.

or $\quad t = \dfrac{2n+1}{2(v_1 - v_2)}$ $\hspace{4cm}$ (1.128)

The amplitude of the resultant wave is minimum at time

$$\frac{1}{2(v_1 - v_2)}s, \frac{3}{2(v_1 - v_2)}s, \frac{5}{2(v_1 - v_2)}s, \frac{7}{2(v_1 - v_2)}s, \ldots \text{etc.}$$

The minimum value of the amplitude R is

$$R_{min} = \sqrt{r_1^2 + r_2^2 + 2r_1r_2 \times (-1)} = r_1 - r_2.$$

Thus, the maxima and minima occur alternately after equal intervals of time. The time interval between any two consecutive minima and maxima (difference between nth maxima time and nth minima time) is $\dfrac{1}{2(v_1 - v_2)}s$.

The time interval between any two consecutive minima or two consecutive maxima is $\dfrac{1}{v_1 - v_2}s$. Therefore, the number of maxima (i.e., beats in sound wave or optical beats in light wave) produced per second is $v_1 - v_2$. The difference between the frequencies of the two superposing waves is the number of maxima produced per second.

1.16.4 Multiple beam superpositions

Let us consider the superposition of n number of waves having the same frequency. The displacements of the particles of the medium due to the n waves are here taken to be along the same line. Mathematically, we can represent the n waves by the following equations.

$$\Psi_1 = r_1 \sin(\omega t - \delta_1)$$

$$\Psi_2 = r_2 \sin(\omega t - \delta_2)$$

$$\Psi_3 = r_3 \sin(\omega t - \delta_3)$$

$$\ldots\ldots\ldots\ldots\ldots$$

....

....

$$\Psi_n = r_n \sin(\omega t - \delta_n)$$

Here we have assumed all the waves to have the same frequency ω.

The resultant displacement of the particles of the medium under the influence of the first two waves are given by

$$\Psi = \Psi_1 + \Psi_2 = r_1 \sin(\omega t - \delta_1) + r_2 \sin(\omega t - \delta_2)$$

or $\Psi = (r_1 \cos\delta_1 + r_2 \cos\delta_2) \sin\omega t - (r_1 \sin\delta_1 + r_2 \sin\delta_2) \cos\omega t.$ (1.129)

Since rs and δs are constants, we can make the following substitutions.

$$r_1 \cos\delta_1 + r_2 \cos\delta_2 = R\cos\theta \tag{1.130}$$

and

$$r_1 \sin\delta_1 + r_2 \sin\delta_2 = R\sin\theta. \tag{1.131}$$

With these substitution, Eq. (1.125) becomes

$$\Psi = R\cos\theta \sin\omega t - R\sin\theta \cos\omega t.$$

or $\Psi = R\sin(\omega t - \theta)$ (1.132)

Equation (1.132) is the mathematical representation of the resultant wave with amplitude R and phase angle θ. The values of R and θ can be calculated by squaring and adding both sides of Eqs (1.130) and (1.131).

$$R^2 = (r_1 \cos\delta_1 + r_2 \cos\delta_2)^2 + (r_1 \sin\delta_1 + r_2 \sin\delta_2)^2$$

or $R = \sqrt{r_1^2 + r_2^2 + 2r_1 r_2 \cos(\delta_1 - \delta_2)}$ (1.133)

Taking the ratio of Eqs (1.131) and (1.130), we have

or $\tan\theta = \dfrac{r_1 \sin\delta_1 + r_2 \sin\delta_2}{r_1 \cos\delta_1 + r_2 \cos\delta_2}$ (1.134)

Thus, when two waves are superposed on each other, the resultant wave is of the form $\Psi = R\sin(\omega t - \theta)$. The resultant wave and the two superposing waves have the same form. In a similar way, if the third wave is superposed on $\Psi = R\sin(\omega t - \theta)$, the resultant will also be of the form $\Psi = R\sin(\omega t - \theta)$ (of course, here the numerical values of R and θ will be different but the equations will be similar). Thus, we conclude that when a large number of harmonic waves are superposed on each other, the resultant wave will be of the form $\Psi = R\sin(\omega t - \theta)$, where R is the amplitude and θ is the phase angle. The equations for them are given by

$$R^2 = \left(r_1 \cos\delta_1 + r_2 \cos\delta_2 + r_3 \cos\delta_3 + \ldots\ldots + r_N \cos\delta_N\right)^2 +$$

$$\left(r_1 \sin\delta_1 + r_2 \sin\delta_2 + r_3 \sin\delta_3 + \ldots\ldots + r_N \sin\delta_N\right)^2 \tag{1.135}$$

$$\tan\theta = \frac{r_1 \sin\delta_1 + r_2 \sin\delta_2 + r_3 \sin\delta_3 \ldots r_N \sin\delta_N}{r_1 \cos\delta_1 + r_2 \cos\delta_2 + r_3 \cos\delta_3 \ldots r_N \cos\delta_N} \tag{1.136}$$

Example 1.20

Two simple harmonic oscillations acting in a line simultaneously on a particle are given by

$$\Psi_1 = 3\sin\left(\omega t + \frac{\pi}{5}\right) \text{ and } \Psi_2 = 5\sin\left(\omega t + \frac{\pi}{3}\right).$$

Calculate the resultant equation, phase, and amplitude of the resultant motion.

Solution

The resultant motion will be controlled by the equation

$$\Psi = \Psi_1 + \Psi_2 = 3\sin\left(\omega t + \frac{\pi}{5}\right) + 5\sin\left(\omega t + \frac{\pi}{3}\right)$$

$$= \sin\omega t \left[3\cos\frac{\pi}{5} + 5\cos\frac{\pi}{3}\right] + \left[3\sin\frac{\pi}{5} + 5\sin\frac{\pi}{3}\right]\cos\omega t$$

$$= \sin\omega t\left[R\cos\theta\right] + \left[R\sin\theta\right]\cos\omega t$$

or $\quad \Psi = R\sin(\omega t + \theta)$

where $R^2 = \left(3\cos\frac{\pi}{5} + 5\cos\frac{\pi}{3}\right)^2 + \left(3\sin\frac{\pi}{5} + 5\sin\frac{\pi}{3}\right)^2$

$$R^2 = (2.43 + 2.50)^2 + (1.76 + 4.33)^2$$

or $R = 7.84$ m is the amplitude of the resultant motion

and $\tan\theta = \dfrac{R\sin\theta}{R\cos\theta} = \dfrac{3\sin\dfrac{\pi}{5} + 5\sin\dfrac{\pi}{3}}{3\cos\dfrac{\pi}{5} + 3\cos\dfrac{\pi}{5}} = 1.24$

or $\theta = 51°$ is the phase angle of the resultant motion.

1.16.5 Coherent and incoherent superposition

Two waves are said to be coherent waves if the phase difference between the two waves at a point in space remains constant throughout the observation time. The waves are said to be incoherent waves if the phase difference between the two waves at a point in space varies randomly throughout the observation time. Coherent superposition is defined as the superposition of coherent waves. In the superposition of coherent waves, the resultant intensity I is directly proportional to the square of the vector sum of amplitudes of individual waves, i.e. $I = k(r_1 + r_2 + r_3 \ldots)^2$. Incoherent superposition is defined as the superposition of incoherent waves. In the incoherent superposition of waves, the resultant intensity is directly proportional to the sum of the squares of the amplitudes of individual waves, i.e. $I = k(r_1^2 + r_2^2 + r_3^2 \ldots)$. That means in the case of incoherent superposition, the resultant intensity is equal to the sum of the intensities of individual waves which is not the case with coherent superposition.

Mathematical treatment of coherent superposition

In Section 1.16, we have discussed in detail superposition principle taking different cases. In coherent superposition, the resultant amplitude does not change with respect to time or space at a point. The resultant amplitudes in Eq. (1.116) or Eq. (1.133) remain constant with time. The superposition of waves always having constant phase difference (i.e., phases of all the waves do not change with lapse of time) is called coherent superposition. In case of coherent superposition, if δ is the phase difference between the two waves then,

$$\frac{d\delta}{dt} = 0.$$

From Eq. (1.116), we have

$$R^2 = r_1^2 + r_2^2 + 2r_1r_2\cos\delta \tag{1.137}$$

where

R = the resultant amplitude of two superposing waves.

r_1 = the amplitude of first wave.

r_2 = the amplitude of second wave.

δ = the phase difference between the two superposing waves.

In coherent superposition δ, the phase difference between the superposing waves is independent of time. Therefore, the time average of $\cos \delta$, at a point will be $\cos \delta$, i.e.,

$$\langle \cos \delta \rangle = \cos \delta \qquad (1.138)$$

Hence, for coherent superposition Eq. (1.138) does not change. From Eq. (1.137), the intensity of the resultant wave will be

$$I = I_1 + I_2 + 2\sqrt{I_1}\sqrt{I_2}\cos \delta \qquad (1.139)$$

The interference pattern produced due to coherent superposition of two coherent light sources having the same amplitudes (i.e., $I_1 = I_2 = I_0$) is visible and is completely described by the equation

$$I = 4I_0 \cos^2 \left(\frac{\varphi}{2} \right). \qquad (1.140)$$

Mathematical treatment of incoherent superposition

In incoherent superposition, the resultant amplitude changes with respect to time at a point. The resultant amplitudes in Eq. (1.116) or Eq. (1.133) does not remain constant with time. The superposition of waves always having time-varying phase difference (i.e., phases of all the waves do change with lapse of times.) is called incoherent superposition. In case of incoherent superposition, if δ is the phase difference between two waves then,

$$\frac{d\delta}{dt} \neq 0.$$

From Eq. (1.116), we have

$$R^2 = r_1^2 + r_2^2 + 2r_1 r_2 \cos \delta \qquad (1.141)$$

In incoherent superposition δ, the phase difference between the two superposing waves varies with time. The value of $\cos \delta$ will vary from -1 to $+1$ with respect to time. Therefore, the time average of $\cos \delta$ at a point will be 0, i.e.,

$$\langle \cos \delta \rangle = 0$$

Hence, for incoherent superposition, Eq. (1.141) becomes

$$R^2 = r_1^2 + r_2^2 + 2r_1 r_2 \times 0$$

or $\quad R^2 = r_1^2 + r_2^2$

From this equation, the intensity of the resultant wave in case of incoherent superposition will be

$$I = I_1 + I_2 \tag{1.142}$$

The interference pattern produced due to incoherent superposition of two coherent light sources having the same amplitudes (i.e., $I_1 = I_2 = I_0$) may be described by

$$I = 4I_0 \left\langle \cos^2 \left(\frac{\varphi}{2} \right) \right\rangle. \tag{1.143}$$

Here I is the intensity at any point, I_0 is the maximum intensity on the screen and $\left\langle \cos^2 \left(\frac{\varphi}{2} \right) \right\rangle$ is the time average of $\cos^2 \left(\frac{\varphi}{2} \right)$. The maximum and minimum values of $\cos^2 \left(\frac{\varphi}{2} \right)$ with the passing of time are 1 and 0. Therefore, the value of $\left\langle \cos^2 \left(\frac{\varphi}{2} \right) \right\rangle$ is 0.5.

$$I = 4I_0 \times 0.5 = 2I_0 = I_0 + I_0$$

This equation shows that if two superposing waves are incoherent, then intensity will not vary from point to point on the screen and the interference pattern will not be visible. Thus, the resultant intensity is equal to the sum of two individual intensities of the superposing waves as in Eq. (1.142).

Questions

1.1 What is an oscillatory system? Give few examples.

1.2 Define the parameters of an oscillatory system.

1.3 What is simple harmonic oscillation? Describe an example.

1.4 Derive the differential equation of motion of a simple harmonic oscillation.

1.5 Derive the expression of frequency of a simple harmonic oscillation.

1.6 Derive the equation to describe the position of a simple harmonic oscillator at any instant of time.

1.7 What are the characteristics of simple harmonic oscillation?

1.8   Draw the graphs of displacement-time, speed–time and acceleration–time of a simple harmonic oscillator on a single sheet.

1.9   Derive the expression for the potential energy of a simple harmonic oscillator.

1.10   Derive the expression for the kinetic energy of a simple harmonic oscillator.

1.11   Prove that the total energy of a simple harmonic oscillator is constant through out the motion.

1.12   Prove that the average potential energy and average kinetic energy of a simple harmonic oscillator are the same during one complete oscillation.

1.13   Prove that the maximum potential energy and maximum kinetic energy of a simple harmonic oscillator are the same.

1.14   Plot in a single plot the kinetic, potential and total energy of a simple harmonic oscillator.

1.15   Prove that the change of phase when a simple harmonic oscillator moves from equilibrium position to extremum position is $\pi/2$.

1.16   What is the nature of the displacement–acceleration graph of a simple harmonic oscillator?

1.17   What is the nature of the speed–acceleration graph of a simple harmonic oscillator?

1.18   Show that the displacement and velocity graph of a simple harmonic oscillator is elliptical.

1.19   A loaded bus is more comfortable than an empty one. Explain.

1.20   What do you mean by a damped harmonic oscillation ? Give few examples.

1.21   Describe in detail the different parameters required to measure the damping of an under damped harmonic oscillation.

1.22   Explain the role of a dissipative force and a restoring force during the motion of an oscillator.

1.23   What is the coefficient of damping? What is its units?

1.24   Derive the differential equation of motion of a damped harmonic oscillation.

1.25   Derive the equation to describe the position of a damped harmonic oscillator at any instant of time.

1.26   What do you mean by underdamped harmonic oscillation?

1.27   What are the characteristics of underdamped harmonic oscillation?

1.28   Calculate the rate of energy loss by an underdamped harmonic oscillator.

1.29   What do you mean by overdamped harmonic oscillation?

1.30   What are the characteristics of overdamped harmonic oscillation?

1.31   What do you mean by critically damped harmonic oscillation?

1.32   What are the characteristics of critically damped harmonic oscillation?

1.33   Prove that the time period of an underdamped harmonic oscillator remains constant during the motion.

1.34   Differentiate between underdamped harmonic oscillation, overdamped harmonic oscillation and critically damped harmonic oscillation.

1.35   Show graphically underdamped, overdamped, and critically damped harmonic oscillation.

1.36   Give examples of underdamped, overdamped, and critically damped harmonic oscillation.

1.37   Derive an expression for the amplitude of underdamped harmonic oscillation.

1.38   Differentiate between simple harmonic oscillation, and damped harmonic oscillation.

1.39   Derive the expression for the potential energy of an underdamped harmonic oscillator.

1.40   Derive the expression for the kinetic energy of an underdamped harmonic oscillator.

1.41   Prove that the total energy of an underdamped harmonic oscillator is not constant through out the motion.

1.42   Prove that the total energy of an overdamped harmonic oscillator is not constant throughout the motion.

1.43   Prove that the total energy of a critically damped harmonic oscillator is not constant throughout the motion.

1.44   Derive an expression for the amplitude of critically damped harmonic oscillation. Plot this amplitude versus time.

1.45   Prove that the average potential energy and average kinetic energy of an underdamped harmonic oscillator are the same during one complete oscillation.

1.46   Prove that the maximum potential energy and maximum kinetic energy of an underdamped harmonic oscillator are the same.

1.47   Show graphically with equations the variation of total energy of an underdamped harmonic oscillator with time when $\gamma << \omega_0$.

1.48   What is logarithmic decrement?

1.49   What do you mean by forced harmonic oscillation?

1.50   Establish the differential equation of motion of a forced harmonic oscillator subjected to a sinusoidal force.

1.51   What is the steady state oscillation of forced vibration? What is the expression for amplitude of steady state oscillation of forced vibration?

1.52   Derive the expression for the kinetic energy of a forced harmonic oscillator.

1.53   Derive the expression for the potential energy of a forced harmonic oscillator.

1.54   Derive the expression for the total energy of a forced harmonic oscillator.

1.55   Derive the condition for amplitude resonance of a system in a dissipative medium. Reduce the condition in vacuum.

1.56   Prove that the larger the value of quality factor, the sharper is the resonance.

1.57   Derive the condition for velocity resonance.

1.58   Derive the condition for average power resonance.

1.59 Prove that velocity resonance and power resonance of a forced harmonic oscillator occurs under the same condition

1.60 Compare simple harmonic, damped harmonic and forced harmonic oscillations.

1.61 What do you mean by coupled oscillators? Give few examples.

1.62 Enumerate the importance of the concept of coupled oscillators.

1.63 Explain the motion of a coupled oscillator when two bodies of equal masses are displaced by equal distances in the same direction and then released from rest.

1.64 Prove that the motion of either mass of a coupled oscillator is a superposition of two harmonic vibrations.

1.65 What are the characteristics of wave motion?

1.66 Write down the differential equation of wave motion stating clearly the meaning of each symbol.

1.67 Show that wave function $\psi(x,t) = r \sin(\omega t - kx)$ satisfy the wave equation $\dfrac{\partial^2 \psi}{\partial x^2} = \dfrac{1}{v^2} \dfrac{\partial^2 \psi}{\partial t^2}$.

1.68 What are the characteristics of transverse waves?

1.69 What are the characteristics of longitudinal waves?

1.70 Differentiate between longitudinal waves and transverse waves.

1.71 Define progressive wave. What are its characteristics?

1.72 Define stationary wave. What are its characteristics?

1.73 Differentiate between progressive wave and transverse wave.

1.74 Differentiate between progressive wave and stationary wave.

1.75 Explain how a stationary wave is produced.

1.76 Explain the phenomenon of reflection of transverse waves in passing from a rarer medium to a denser medium.

1.77 Explain the phenomenon of reflection of transverse waves in passing from a denser medium to a rarer medium.

1.78 Explain the phenomenon of reflection of longitudinal waves in passing from a rarer medium to a denser medium.

1.79 Explain the phenomenon of reflection of longitudinal waves in passing from a denser medium to a rarer medium.

1.80 Explain the phenomenon of refraction of transverse waves in passing from a rarer medium to a denser medium.

1.81 Explain the phenomenon of refraction of transverse waves in passing from a denser medium to a rarer medium.

1.82 Explain the phenomenon of refraction of longitudinal waves in passing from a rarer medium to a denser medium.

1.83 Explain the phenomenon of refraction of longitudinal waves in passing from a denser medium to a rarer medium.

1.84 What is a wave packet? What are its characteristics?

1.85 What do you mean by phase speed of a wave?

1.86   What do you mean by group speed of a wave?

1.87   Define phase speed and group speed of a wave mathematically.

1.88   Differentiate between phase speed and group speed of a wave.

1.89   Derive the relation between phase speed and group speed of a wave.

1.90   What are the conditions under which phase speed is equal to, less than, and more than group speed of a wave?

1.91   Derive the equation $v_g = \dfrac{dE}{dp}$; v_g, E and p are group speed, energy and momentum of a particle respectively.

1.92   Uncertainty principle is a measure of the accuracy of measurement or inherent properties of nature? Explain

1.93   Write down the uncertainty principle for a classical wave and a quantum mechanical wave explaining each term.

1.94   Represent uncertainty principle graphically.

1.95   Can the product in the uncertainty principle be made equal to zero? Explain.

1.96   State the principle of superposition of waves.

1.97   Two waves of the same frequency and travelling in the same direction are superposed. Find the resultant wave.

1.98   Two waves of the same frequency and travelling in opposite directions are superposed. Find the resultant wave.

1.99   Explain how beat is produced.

1.100 Prove that in incoherent superposition of waves, the resultant intensity is equal to the summation of the intensities of individual waves.

1.101 Using Eqs (1.123) and (1.124), derive Eqs (1.125), (1.132), and (1.133).

1.102 What do you mean by coherent superposition? What are its characteristics?

1.103 What do you mean by incoherent superposition? What are its characteristics?

1.104 Differentiate between coherent and incoherent superposition.

1.105 The propagation of light in free space is with phase speed or group speed? What is the difference between phase speed and group for electromagnetic wave propagating in vacuum?

Problems

1.1   A spring of 10 cm length is fixed to a ceiling. When a body of mass 1 kg is hung at the free end, its length becomes 32 cm. Calculate the time period of its natural vibration.
[Ans   0.03 s]

1.2   A body oscillates with simple harmonic motion obeying the equation

$$y = 10\cos\left(0.5\pi t + \frac{\pi}{3}\right)\text{m.}$$

Calculate the velocity and acceleration of the body at time $t = 4$ s and the frequency and time period of the harmonic motion.   [Ans   13.6 m/s, 12.34 m/s^2, 0.25 s^{-1}, 4 s.]

1.3   A harmonic wave is represented by $\psi = A \cos m(x - ct)$. Determine the wavelength and time period of the wave.   [Ans $\dfrac{2\pi}{m}, \dfrac{2\pi}{mc}$]

1.4   A point performs a damped harmonic oscillation with frequency ω and damping coefficient γ. Find the initial amplitude r and the initial phase θ if at the moment $t = 0$, the displacement of the point and its velocity are $y(0) = y_0$ and $v(0) = 0$.

$$[\text{Ans} \quad r = |y_0|\sqrt{1 + \frac{\gamma^2}{\omega^2}}, \quad \tan\theta = -\frac{\gamma}{\omega} \text{ when } -\frac{\pi}{2} < \theta < 0 \text{ if } y_0 > 0$$

$$\text{and } \frac{\pi}{2} < \theta < \pi \text{ if } y_0 < 0]$$

1.5   The amplitude of an underdamped oscillator of frequency 100 per second falls to $\dfrac{1}{10}$ of its initial value after 1000 cycles. Calculate the (a) damping coefficient, (b) logarithmic decrement and (c) time in which energy falls to $\dfrac{1}{10}$ of its original value.

       [Ans   23/s, 2.3×10^{-3}, 5 s]

1.6   A body performs damped harmonic oscillation according to the equation $x = 2e^{-0.2t} \sin 3\pi t$ m. Calculate the (a) amplitude of oscillation and (b) velocity of the body at time $t = 2$s; (c) the moments of time at which the body reaches the extreme positions.

       [Ans   1.34 m, 12.64 m/s, $\approx \dfrac{2n+1}{6}$ with $n = 0, 1, 2, 3, \ldots$]

1.7   A body performs damped harmonic oscillation according to the equation $x = re^{-0.4t} \cos(2\pi t + \theta)$ m. Calculate the (a) initial amplitude r of oscillation and (b) initial phase θ if at time $t = 0$, velocity of the body is 20 cm/s and $x = 0$.

       [Ans   3.2 cm, $-\dfrac{\pi}{2}$]

1.8   A ball of mass 250 gm performs undamped harmonic oscillation about the point $x = 0$ with natural frequency 3 Hz. At the moment $t = 0$, when the ball is in equilibrium position, a force $F = 100 \cos 3\pi t$ Newton along the x-axis was applied to it. Find the position and the speed of the ball at time 1 s.   [Ans   3.0 cm, 0]

1.9   A body performs damped harmonic oscillation according to the equation $x = re^{-\gamma t} \cos \omega t$. Calculate the (a) amplitude of oscillation and (b) velocity of the body at time $t = 0$; (c) the moments of time at which body reaches the extreme positions.

       [Ans r, $-\gamma r$, $\dfrac{1}{\omega}\left(n\pi - \tan^{-1}\dfrac{\gamma}{\omega}\right)$]

1.10   Forced harmonic oscillations have same amplitudes at frequencies 200/s and 300/s. Find the frequency at which amplitude resonance takes place.   [Ans   2265/s]

1.11   Two waves each of equal amplitude and equal frequency pass through a point in the medium in the same direction with a phase difference of 120°. Calculate the amplitude of the resultant wave at this point.   [Ans   Amplitude of either wave]

1.12 The ratio of amplitude of two waves is 1 : 4. If these two waves superpose on each other, find the ratio of minimum and maximum amplitudes. [Ans 3 : 5]

1.13 Two waves of the same frequency and same amplitude are reaching a point simultaneously. What should be the phase difference of the two waves so that the amplitude of the resultant wave will be r where r is the amplitude of a wave? [Ans 120°]

1.14 The equation of a stationary wave produced in a stretched string is $\Psi = 6\cos\left(\dfrac{\pi x}{3}\right)$ sin(6πt), where x and t are measured in centimeters and seconds respectively. Calculate the frequency, amplitude and velocity of its constituent waves. [Ans 3 Hz, 3 cm, 18 cm/s]

1.15 Three harmonic waves are represented by $\Psi_1 = 2\sin(\omega t - 30)$, $\Psi_2 = 5\sin(\omega t + 60)$ and $\Psi_3 = 4\sin(\omega t + 30)$. Find the resultant wave equation if they are superposed. [Ans $\psi = 9.36\sin(\omega t + 34.72)$]

1.16 Two sources vibrating according to the equation $\Psi_1 = 4\sin 2\pi t$, and $\Psi_1 = 3\sin 2\pi t$ send out waves in all directions with speed 2.40 m/s. Find the equation of motion of a particle 5 m from the first source and 3 m from the second source when angular speed of both waves is 2π rad/s [Ans $\psi = 3.08\sin(2\pi t - 25.3)$]

1.17 Two coherent beams of intensities I_1 and I_2 interfere. What will be the maximum and minimum intensity? [Ans $\left(\sqrt{I_1}+\sqrt{I_2}\right)^2$, $\left(\sqrt{I_1}-\sqrt{I_2}\right)^2$]

1.18 An electron moves in the x-direction with a speed of 4.8×10^5 m/s. We can measure its speed with a precision of 1%. With what precision can we simultaneously measure its position? With what precision can we measure its velocity along the y-direction? [Ans $\Delta p_X = 4.4\times10^{-24}$ kgm/s, $\Delta x = 2.4\times10^{-11}$ m, $\Delta p_Y = 0$ kgm/s, $\Delta y = \inf$. That is we know nothing about the y coordinate]

1.19 A point moves on the XY plane obeying the equation $x = a\sin \omega t$ and $y = b\cos \omega t$, where a, b, and ω are positive constants. Calculate the trajectory of motion y(x) of the particle and the direction of its motion along the trajectory. [Ans $\dfrac{x^2}{a^2}+\dfrac{y^2}{b^2}=1$, clockwise]

1.20 Two cubes of mass m_1 and m_2 are interconnected by a weightless spring of force constant k and placed on a frictionless plane. Two cubes were compressed towards each other and then released simultaneously. Find the natural frequency of the oscillation. See Fig. 1.9. [Ans $\dfrac{1}{2\pi}\sqrt{\dfrac{k(m_1+m_2)}{m_1 m_2}}$]

1.21 Two harmonic waves given by $\psi_1 = 3\sin(5\pi t - 4x)$ and $\psi_2 = 4\sin(7\pi t - 8x)$ are superposed to produce a single wave. Find the group speed of the resultant wave. [Ans 1.57 cm/s]

Multiple Choice Questions

1. What is the magnitude of restoring force on an oscillator when the oscillator is at rest at its mean position.

    (i)   zero   (ii) infinity

    (iii) any value   (iv) cannot be said

2. Which of the following is correct? In simple harmonic motion, the total energy is maximum only at

    (i)   equilibrium position

    (ii) extreme position

    (iii) any position in between the extremum and minimum position

    (iv) it is constant; independent of time or position

3. Which of the following is correct? In a simple harmonic motion, kinetic energy is maximum at

    (i)   equilibrium position

    (ii) extreme position

    (iii) any position in between the extremum and minimum position

    (iv) none of the above statements

4. Which of the following is correct? In simple harmonic motion potential energy is maximum at

    (i)   equilibrium position

    (ii) extreme position

    (iii) any position in between the extremum and minimum position

    (iv) none of the above statements

5. How many times does the potential energy of a simple harmonic oscillator attain minimum value during one complete oscillation?

    (i)   1   (ii) 2

    (iii) 3   (iv) 4

6. How many times does the kinetic energy of a simple harmonic oscillator attain minimum value during one complete oscillation?

    (i)   1   (ii) 2

    (iii) 3   (iv) 4

7. How many times does the potential energy of a simple harmonic oscillator attain maximum value during one complete oscillation?

    (i)   1   (ii) 2

    (iii) 3   (iv) 4

8. How many times does the kinetic energy of a simple harmonic oscillator attain maximum value during one complete oscillation?

    (i)   1   (ii) 2

    (iii) 3   (iv) 4

9. What is the magnitude of dissipative force on an oscillator when the oscillator is at rest.

    (i)   zero   (ii) infinity

    (iii) any value   (iv) can not be said

10. Which of the following is correct? In damped harmonic motion

    (i)   amplitude as well as time period changes

    (ii) only time period changes

    (iii) only amplitude changes

    (iv) none of the above statements

11. In damped harmonic motion, frequency does not change.

    (i)   True   (ii) False

12. Which of the following is correct?

    (i) $\nabla^2 \psi = \dfrac{1}{v^2}\dfrac{\partial^2 \psi}{\partial x^2}$ (ii) $\nabla^2 \psi = \dfrac{1}{v^2}\dfrac{\partial^2 \psi}{\partial t^2}$

    (iii) $\nabla^2 \psi = v^2 \dfrac{\partial^2 \psi}{\partial x^2}$ (iv) $\nabla^2 \psi = v^2 \dfrac{\partial^2 \psi}{\partial t^2}$

13. Phase change of π occurs due to

    (i)   reflection of transverse waves in passing from a rarer medium to a denser medium

    (ii) reflection of a transverse wave in passing from a denser medium to a rarer medium

    (iii) reflection of longitudinal waves in passing from a rarer medium to a denser medium

    (iv) reflection of longitudinal waves in passing from a denser medium to a rarer medium

14. The phase speed and group speed for an electromagnetic wave propagating in vacuum are equal.

    (i)   True   (ii) False

15. The angle between the propagation vector and the normal to the wave front is

    (i)   0   (ii) π

    (iii) $\dfrac{\pi}{2}$ (iv) $\dfrac{\pi}{4}$

16. What type of superposition produces standing waves?

    (i)   Two waves having the same frequency travelling in the same direction

    (ii) Two waves having the same frequency travelling in opposite directions

(iii) Two waves having the same amplitude travelling in the same direction

(iv) Two waves having different amplitudes travelling in the same direction

17. What type of superposition produces interference patterns?

(i) Two waves having the same frequency travelling in the same direction

(ii) Two waves having the same frequency travelling in the opposite direction

(iii) Two waves having the same amplitude travelling in the same direction

(iv) Two waves having different amplitudes travelling in the same direction

18. What type of superposition produces beats

(i) Two waves having the same frequency travelling in the same direction

(ii) Two waves having the same frequency travelling in the opposite direction

(iii) Two waves having the same amplitude

(iv) Two waves having different frequency

19. The differential equation of motion of a freely oscillating body is given by $\dfrac{d^2x}{dt^2} + \omega x = 0$. The natural frequency of the body will be

(i) $\sqrt{\omega}$

(ii) $\dfrac{2\pi}{\omega}$

(iii) $\dfrac{2\pi}{\sqrt{\omega}}$

(iv) $\dfrac{\sqrt{\omega}}{2\pi}$

20. Which of the following properties does not change when a wave passes from one medium to another?

(i) speed

(ii) wavelength

(iii) frequency

(iv) amplitude

21. Which of the following properties of a wave is independent of the other?

(i) speed

(ii) wavelength

(iii) frequency

(iv) amplitude

22. In a stationary wave in air, the variation of pressure at a node is

(i) maximum

(ii) minimum

(iii) initially maximum then minimum

(iv) cannot be said

23. In a stationary wave

(i) strain is maximum at nodes

(ii) strain is minimum at nodes

(iii) strain is maximum at antinodes

(iv) strain is minimum at antinodes

24. Which of the following is incorrect regarding phase speed and group speed

(i) $v_p = \dfrac{\omega}{k}$

(ii) $v_p = \dfrac{d\omega}{dk}$

(iii) $v_g = \dfrac{d\omega}{dk}$

(iv) $v_g = \dfrac{dE}{dP}$, $P = $ Momentum, $E = $ Energy

25. Two simple harmonic oscillators of masses m_1 and m_2 oscillate with frequency v_1 and v_2 respectively under the same restoring force. Which of the following is correct

(i) $\dfrac{v_1}{v_2} = \dfrac{m_1}{m_2}$

(ii) $\dfrac{v_1}{v_2} = \dfrac{m_2}{m_1}$

(iii) $\dfrac{v_1}{v_2} = \sqrt{\dfrac{m_2}{m_1}}$

(iv) $\dfrac{v_1}{v_2} = \sqrt{\dfrac{m_1}{m_2}}$

Answers

1 (i)	2 (ii)	3 (i)	4 (ii)	5 (ii)	6 (ii)	7 (ii)	8 (ii)
9 (i)	10 (iii)	11 (i)	12 (ii)	13 (i)	14 (i)	15 (i)	16 (ii)
17 (i)	18 (ii)	19 (iv)	20 (iii)	21 (i)	22 (i)	23 (i)	24 (ii)
25 (iii)							

2 Interference

2.1 Introduction

In Section 1.16 of the previous chapter we learned that two beams of light waves can cross each other without either one producing any modification on the other after passing beyond the region of crossing. However, from the concepts explained in Section 1.16.2, we expect some modifications in the amplitudes or intensity (since intensity $\propto$ amplitude²) of the two waves inside the region of crossing. The intensity of the resultant wave becomes a function of the position of the point. At certain points intensity is maximum and at other points it is minimum. In other words, we say that the two waves interfere with each other inside the region of crossing. This modification of intensity obtained by the superposition of two or more beams of light waves is called interference of light. The phenomenon of interference of light complements the validity of the concept that light is a wave. As a result of the short wavelength and disordered phase relationships of the interfering light waves, the interference pattern is not visible to the naked eye without special arrangements. It was in the year 1801 that Thomas Young for the first time demonstrated the interference of sunlight experimentally. Before discussing the interference phenomenon, let us discuss Huygens' principle, a helpful tool and an early concept in favour of the wave theory of light when the scientific world was mesmerized by Newton's corpuscular theory of light.

2.2 Huygens' Principle

Huygens, a Dutch mathematician, in 1678, propounded a theory regarding the propagation of light wave in any medium. According to this theory, light is a sort of disturbance in the medium in which it propagates in all direction from a point source. To explain the propagation of light in vacuum, he postulated an all-pervading medium called 'ether' (Later on, in the year 1881, Michelson and Morley, American scientists, performed a high precision optical experiment and completely disapproved the presence of this ether medium).

Huygens' principle governs the propagation of light waves in any medium. The principle in original form is stated in his book '*Traité de la Lumière*' (Treatise on light), which says, 'In considering the propagation of waves, we must remember that each particle of the medium through which wave spread does not only communicate its motion to that neighbour which lie in the straight line drawn from the luminous point, but shares also with all particles which touch it and resist its motion. Each particle is thus to be considered as the centre of a wave'.

2.2.1 Explanation

Huygens assumed light to be a longitudinal wave (Actually, light is a transverse wave as concluded from Maxwell's electromagnetic theory as well as from concepts of light polarization) which propagates in pulses with high speed in the hypothetical ether medium in the form of compression and rarefaction (shown roughly in Fig. 2.1).

Figure 2.1 | Propagation of light wave as visualized by Huygens. According to Huygens light is a longitudinal wave which propagates in pulses in the hypothetical ether medium in the form of compression and rarefaction

Suppose in a homogeneous isotropic ether medium, a point source O emits light in the form of pulses which propagates in all direction with equal speed v so that after time t the light pulses will reach all the points on the spherical surface $A_1A_2A_3A_4 \ldots A_N \ldots$ simultaneously as shown in Fig. 2.2(a). The ether particles present at all the points on the spherical surface $A_1A_2A_3A_4 \ldots A_N \ldots$ will vibrate simultaneously in phase. Therefore, this spherical surface $A_1A_2A_3A_4 \ldots A_N \ldots$ will be the wavefront of the light wave at time t and its radius will be vt. Thus, at time t, the shape of the wavefront is a sphere of radius vt. After lapses of dt time from this moment, the light pulses will have reached another spherical surface

$A'_1A'_2A'_3A'_4...A'_N...$ of radius $v(t + dt)$ and this spherical surface will be the position of the wavefront after time $t + dt$. Thus, we have after a time t $A_1A_2A_3A_4...A_N...$ as the position of a wavefront and after time $t + dt$ $A'_1A'_2A'_3A'_4...A'_N...$ is another position of the same wavefront.

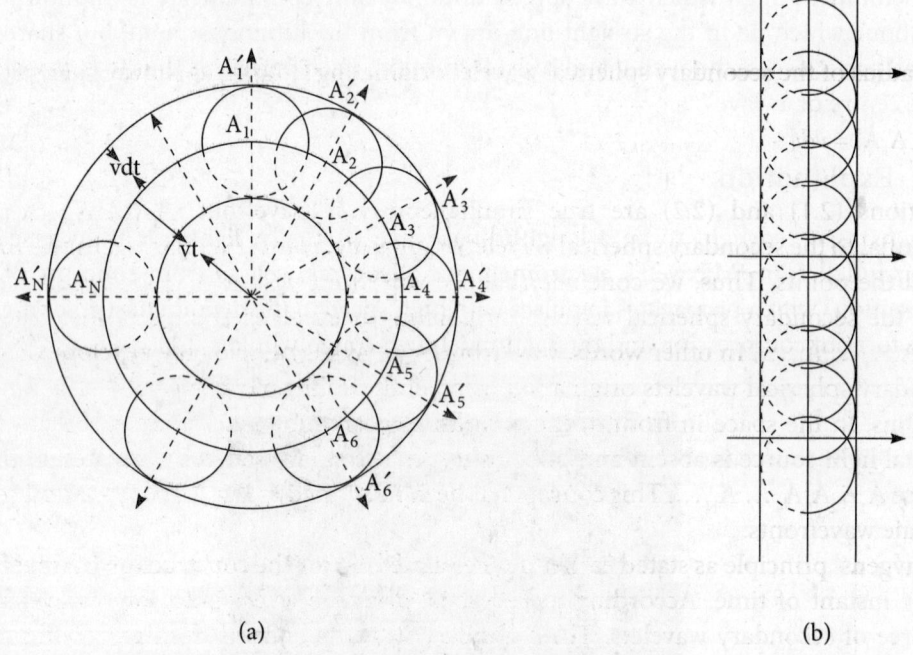

(a)

(b)

Figure 2.2 | (a) Explanation of Huygens' principle of propagation and construction of spherical wavefronts. (b) Explanation of Huygens' principle of propagation and construction of plane wavefronts

The light pulses originating from O propagate through the ether medium in the form of disturbances all the ether particles on the wavefront $A_1A_2A_3A_4...A_N...$ are disturbed by the light pulses originating from O. The disturbance at every point on the wavefront $A_1A_2A_3A_4...A_N...$ is exactly the same as that which originated at O t time earlier. Thus, all the points on this wavefront become the origin of ether disturbance which is exactly the same to the ether disturbance originated at O at t time earlier. Each of this ether disturbance originated from the wavefront is called secondary spherical wavelets or simply secondary wavelets. The ether disturbance originating from all the points of the wavefront $A_1A_2A_3A_4...A_N...$ will be propagated in all directions with speed v and will reach spherical surfaces each of radii vdt in time dt. The radius of the secondary spherical wavelets is thus vdt.

Now we shall show that the wavefront $A'_1A'_2A'_3A'_4...A'_N...$ is tangential to these secondary spherical wavelets. The radius of the wavefront $A_1A_2A_3A_4...A_N...$ is

$$OA = vt$$

The radius of the wavefront $A_1'A_2'A_3'A_4'...A_N'...$ as shown in Fig. 2.2 is

$$OA_1' = v(t+dt) = vt + vdt$$

or $\quad OA + A_1A_1' = vt + vdt$ $\hfill$ (2.1)

The radius of the secondary spherical wavelet originating from A_1 as shown in Fig. 2.2 is

$$A_1A_1' = vdt \hfill (2.2)$$

Equations (2.1) and (2.2) are true simultaneously if wavefront $A_1'A_2'A_3'A_4'...A_N'...$ is tangential to the secondary spherical wavelet originating from A. Similarly, it can be proved for all the points. Thus, we conclude that the wavefront $A_1'A_2'A_3'A_4'...A_N'...$ is tangential to all the secondary spherical wavelets originating from all the points of the wavefront $A_1A_2A_3A_4...A_N....$ In other words, wavefront $A_1'A_2'A_3'A_4'...A_N'...$ is the envelope of all the secondary spherical wavelets originating from all the points of wavefront $A_1A_2A_3A_4...A_N$ .... Thus, in the space in front of the wavefront, everything takes place exactly as if the original light source is absent and only a sheet of secondary sources were present on the surface $A_1A_2A_3A_4...A_N....$ This concept has been illustrated in Fig. 2.2 for spherical as well as plane wavefronts.

Huygens' principle as stated earlier defines a method for the construction of wavefronts at any instant of time. According to Huygens' principle, every point on a wavefront is a source of secondary wavelets. The secondary wavelets emitted from each point of the wavefront propagate in all directions with speed equal to the speed of the wave. The new wavefront at a later stage is obtained by drawing a surface tangential to all the secondary wavelets.

2.2.2 Construction of a new wavefront

Let us consider the construction and position of a new wavefront if a wavefront at a slightly earlier instant of time is known. Suppose in a homogeneous isotropic medium, a point source O as shown in Fig. 2.2 emits light which propagates in all directions with equal speeds v. $A_1A_2A_3A_4...A_N...$ is the position of the spherical wavefront after time t. According to Huygens' principle, all points on this wavefront emit secondary wavelets which again propagate in all directions with the same speed v as that of the original wave. Let us select few points such as $A_1, A_2, A_3, A_4...A_N...$ on this spherical wavefront. Our aim is to determine the new wavefront after $t + dt$ time. During the time interval dt the radii of all the secondary spherical wavelets emitted from the points $A_1, A_2, A_3, A_4...A_N...$ will be vdt as shown in Fig. 2.2(a). According to Huygens' principle, the new wavefront will be the surface which is tangential to all these spherical secondary wavelets. Hence, we can construct a surface which is tangential to all the secondary spherical wavelets. This surface is the shape and position of the new wavefront as it satisfies the conditions of a wavefront.

2.2.3 Absence of backward waves

According to Fresnel, the intensity at any point on the spherical secondary wavelets is given by

$$I = k(1 + \cos\theta)^2 \tag{2.3}$$

where

    k = proportionality constant.

    θ = angle between the normal and the central line passing through the source S.

In Eq. (2.3), $\frac{1}{2}(1 + \cos\theta)$ is called Fresnel's obliquity factor or inclination factor which follows automatically from the rigorous theory of diffraction.

When we move the point on the secondary wavelet away from the central line, the angle θ increases from zero and becomes equal to π for the opposite point as shown in Fig. 2.3. For opposite points on all the secondary spherical wavelets, intensity becomes

$$I = k(1 + \cos\theta)^2 = k(1 + \cos\pi)^2 = 0$$

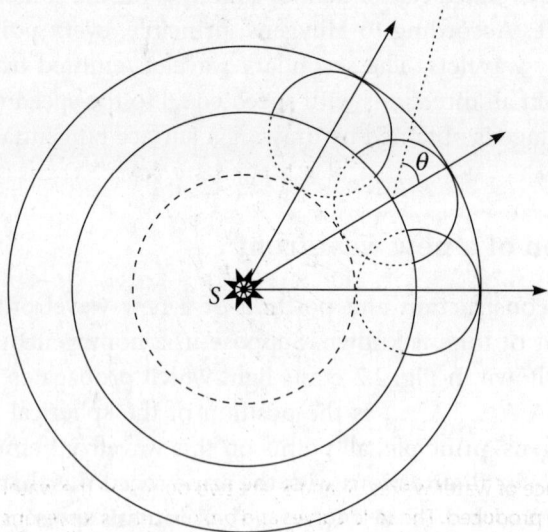

Figure 2.3 | Explanation of absence of backward waves in Huygens' wave theory of light

Thus, intensity at back point zero implies that there is no secondary wavelet and hence no wavefront is present in the backward direction. The aforementioned analysis shows the absence of backward wavefronts in Huygens' wave theory of light.

2.2.4 Applications

Huygens' principle can be applied successfully to explain the phenomena of rectilinear propagation of light waves, specular reflection of plane waves by a plane smooth surface, diffuse reflection of plane waves by a rough surface, image formation by mirrors, refraction of plane waves by plane and curved surfaces, refraction of spherical waves by plane and curved surfaces, total internal reflection. Almost all the phenomena of geometrical optics can be explained by Huygens' principle. Huygens' principle in conjunction with Fresnel's theory finds applications in explaining the phenomena of interference and diffraction.

2.3 Interference of Water Waves

Let us consider the interference pattern produced on the water surface due to interference of water waves. In the Fig. 2.4, S_1 and S_2 are two points on the water surface where circular water waves of equal amplitude and constant phases are produced. These waves travel in the water medium in the form of crests and troughs. As the circular water waves from the two sources S_1 and S_2 advance in the forward direction, they interfere with each other.

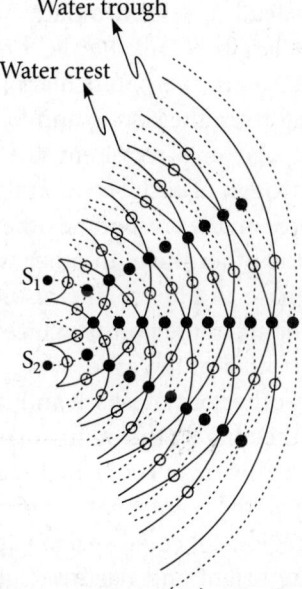

Figure 2.4	Interference of water waves. S_1 and S_2 are two points on the water surface where water waves are produced. The solid curves and broken curves represent crests and troughs of the water waves respectively. At the points shown by the empty circle the crest of one wave superimposes on the trough of other wave and hence have zero displacements. At the points shown by the black circle two crests or two troughs of the two waves superimpose on each other and hence, have maximum displacements

At any instant, water particles will be under the action of two waves originating from S_1 and S_2. When the waves have equal amplitudes, the net displacement of the water particles will be zero if the crest of one wave superimposes on the trough of the other wave. The net displacement of the water particles will be maximum (two times the amplitude of one wave if they have the same amplitude) if the crest of one wave superimposes on the crest of the other wave and the trough of one wave superimposes on the trough of the other wave. There is no loss of energy during interference; only re-distribution occurs. The law of conservation of energy holds good in the interference phenomenon.

2.4 Young's Double Slit Experiment

In Young's original setup, shown in Fig. 2.5, sunlight was first allowed to pass through a pin hole S and then at a considerable distance away through two pin holes S_1 and S_2 situated on an opaque screen. The pin hole S was situated on the perpendicular bisector of $S_1 S_2$. Finally, sunlight was incident on a far away screen placed parallel to the pin holes S_1 and S_2. Two sets of spherical waves emerging from the two pin holes interfered with each other and the interference pattern was obtained on the screen. Young found that the illumination on the screen was not uniform but a symmetric pattern of varying intensity.

In today's laboratory set up, pin holes are replaced by rectangular fine slits and monochromatic light is used instead of polychromatic sunlight. According to Huygens' principle, cylindrical wavelets emerging from the slit S reach S_1 and S_2 at the same time since $SS_1 = SS_2$; the slit S is situated on the perpendicular bisector of $S_1 S_2$. Therefore, at first, the cylindrical wavelets emerging from the slits S_1 and S_2 have the same phases and same amplitudes. The continuous circular arcs represent the wave crest whereas the broken circular arcs represent the wave trough of each wave. At the points shown by a black circle, the crest of one wave superimposes on the crest of the other wave or the trough of one wave superimposes on the trough of the other wave. In other words, at these points, two waves superimpose on each other in phase and hence, the resultant amplitude is twice that of a component wave. On the other hand, at the points marked by an empty circle, the crest of one wave superimposes on the trough of the other wave. In other words, at these points, two waves superimpose on each other out of phase and hence, the resultant amplitude is zero. The straight solid lines connecting all the points marked by black circles touches the screen at P_0 and P_1. The bright narrow regions of equal spacing appear at these points. The straight broken lines connecting all the points marked by empty circles touches the screen at D_1 and D_2. The dark narrow regions of equal spacing appear at points D_1 and D_2. Thus, on screen, a number of alternate bright and narrow regions of equal spacing known as interference fringes are observed parallel to the slits S_1 and S_2. To be most accurate, the loci of black circles or empty circles are confocal hyperbolae. Due to the small field of view of the eye-piece, these confocal hyperbolae appear to be straight.

The fact that the pattern on the screen is only due to interference of the two waves emerging from the two slits S_1 and S_2 can be proved by closing either of the two slits. When one slit is closed, well-defined interference fringes are replaced by coarser fringes of unequal spacings. This is due to the diffraction phenomenon of light by the unclosed slit. The coarse

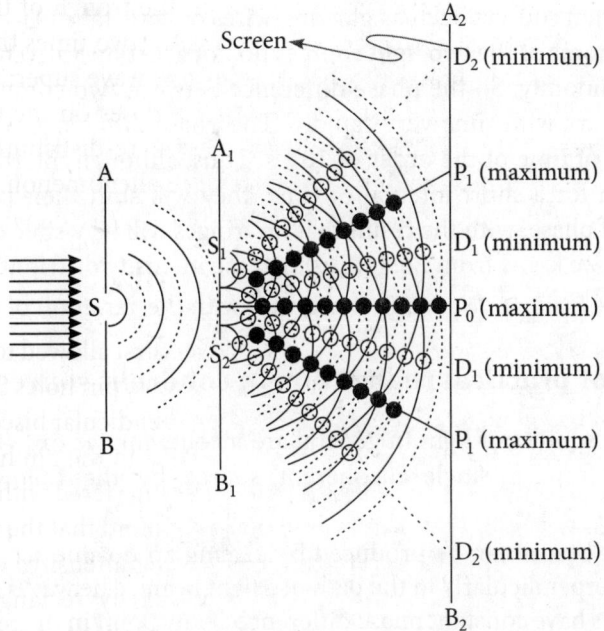

Figure 2.5 Young's double slit experiment. The screen AB containing slit S, the screen $A_1 B_1$ containing slits S_1 and S_2 and the screen $A_2 B_2$ are perpendicular to the plane of the page. At the points shown by the empty circles, the crest of one wave superimposes on the trough of the other wave and hence, have zero intensity at the points D_1 and D_2. At the points shown by the black circles, two crests or two troughs of two waves superimpose on each other and hence, have maximum intensity at the points P_1 and P_0

bright fringe changes to fine dark fringe and the coarse dark fringe changes to fine bright fringe when the closed slit is opened. The corpuscular theory of light totally failed to explain, whereas, the wave theory of light completely explained this phenomenon.

Young's double slit experiment was a turning point in the acceptance of the wave theory of light.

2.5 Coherent Sources

Two sources of waves are said to be coherent only if the phase difference between the two interfering waves do not change with the passage of time, i.e., $\frac{\partial \delta}{\partial t} = 0$. This is possible if either the phase of the two sources remain constant with respect to time or changes by equal amounts so that the difference of phases remains constant. If this condition is satisfied, a stable-well-defined fringe pattern will be visible on the screen. For completely independent light sources such as glowing wires or stars this condition of coherence cannot be satisfied. In any light source, light is emitted due to the electronic transitions that occur among the energy levels of atoms. The chances that all the electronic transitions in the atoms between two independent light sources such as glowing wires or stars take place simultaneously, is very remote. The atoms of the two light sources do not act cooperatively in emitting lights. They emit light randomly. So the phase difference between two completely independent light sources will vary with time very rapidly. The phase difference will change within a very short interval of time of the order of 10^{-8} s. Thus, although interference fringes may exist on the screen for a short interval of time, they will shift their positions each time there is a change of phase, with the result that no fringes will be visible on the screen at all. Mathematically, as we know from Eq. (1.139) that the intensity distribution pattern follows the equation $I = I_1 + I_2 + 2\sqrt{I_1}\sqrt{I_2}\cos\delta$, it will vary with the variation of phase difference δ.

2.5.1 Methods of practical realization of coherent sources

Although ordinary sources of light in general are incoherent, we can obtain two perfectly coherent sources from a single incoherent source by the following experimental manipulation.

i. Two coherent beams can be produced by placing an opaque screen containing two parallel slits perpendicularly in the path of a light beam. The waves emerging from the slits will always have constant phase difference at any point in the region in which they overlap to produce an interference pattern. This is what happens in Young's double slit experiment.

ii. A narrow beam of light coming from a source can be divided into two coherent beams by the phenomenon of complete reflection through suitable experimental arrangements. The interference effects produced in Lloyd's mirror is due to the superposition of a direct beam and a completely reflected beam of light. The interference effects produced in Fresnel's double mirror is due to the superposition of two reflected beams of light.

iii. A narrow beam of light coming from a source can be divided into two coherent beams by the phenomenon of refraction through suitable experimental arrangements. The interference effects produced in Billet's split lens is due to the superposition of two refracted beams of light.

iv. A narrow beam of light coming from a source can be divided into two coherent beams by the phenomenon of partial reflection/refraction through suitable experimental arrangements. The interference effects produced in thin liquid films is due to the superposition of two partially reflected rays – one from the upper surface and the other from the bottom surface of the liquid films.

2.6 Classification of the Interference Phenomenon

The interference phenomenon can be broadly divided into the following two categories depending upon the methods by which two coherent interfering waves are produced.

Division of wavefront

The setup which divides the incident wavefront into two parts by using the phenomenon of reflection, refraction or diffraction in such a manner that they superpose on each other to produce interference effects, come under this category of interference phenomenon. The interference effects produced in Young's double slits, Fresnel's biprism, Lloyd's single mirror, Fresnel's double mirror, and so on are examples of interference phenomena produced by the division of wavefronts methods.

Division of amplitude

The setup which divides the amplitude of the incident wave into two parts by using the phenomenon of partial reflection or refraction in such a manner that they superpose on each other to produce interference effects, come under this category of interference phenomenon. The interference effects produced in liquid films, Newton's ring and interferometers are examples of interference phenomena produced by division of amplitude methods. In this case, an extended source of light is used.

2.7 Theory of Interference

In the Fig. 2.6, a monochromatic source of light is placed behind the slit S. The monochromatic wave emerging from the slit S is incident simultaneously on the two slits S_1 and S_2 which are equidistance from the slit S. The separation $2d$ between the two slits S_1 and S_2 is very small of the order of 10^{-2} cm. Now we can take the light waves emerging from the two slits S_1 and S_2 to have the same amplitudes, same wavelengths and same frequencies. Let the phase difference between the waves emerging from the two slits S_1 and S_2 have constant phase difference so that the interference pattern observed on the screen is independent of time.

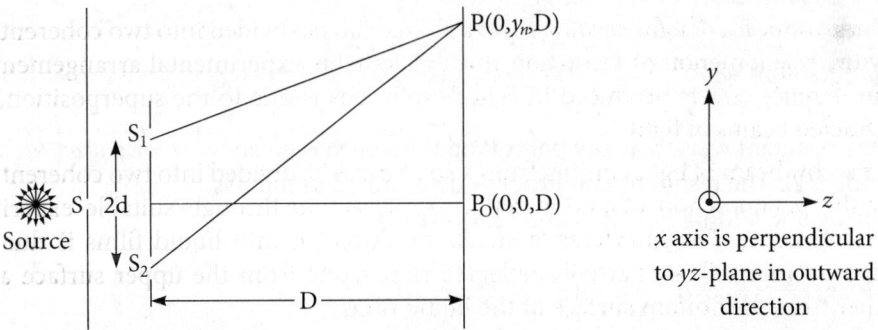

Figure 2.6 | Explanation of interference phenomenon. S_1 and S_2 are two rectangular slits perpendicular to the plane of the page. $2d$ is the separation between slits S_1 and S_2. D is the perpendicular distance between the plane of slits and the screen. P is any point on the screen. According to our coordinate systems, $S_1(0, d, 0)$, $S_2(0, -d, 0)$ and $P(0, y_n, D)$. Fig. 2.7 may be consulted

Let the wave emitted from the slit S_1 be represented by

$$\psi_1 = r \sin \omega t \tag{2.4}$$

Let the wave emitted from the slit S_2 be represented by

$$\psi_2 = r\sin(\omega t + \delta) \qquad (2.5)$$

According to our assumption, two waves ψ_1 and ψ_2 have the same amplitudes r, same angular frequencies ω and constant phase difference δ. The resultant wave ψ obtained by the principle of superposition at any point P is given by

$$\psi = \psi_1 + \psi_2$$

$$= r\sin\omega t + r\sin(\omega t + \delta)$$

$$= r\sin\omega t + r\sin\omega t\cos\delta + r\cos\omega t\sin\delta$$

or $\quad \psi = r\sin\omega t(1 + \cos\delta) + r\cos\omega t\sin\delta \qquad (2.6)$

Let $\quad R\cos\alpha = r(1 + \cos\delta) \qquad (2.7)$

and $\quad R\sin\alpha = r\sin\delta \qquad (2.8)$

Putting this substitution into Eq. (2.6), we have

$$\psi = \sin\omega t \times R\cos\alpha + \cos\omega t \times R\sin\alpha$$

$$= R(\sin\omega t\cos\alpha + \cos\omega t\sin\alpha)$$

or $\quad \psi = R\sin(\omega t + \alpha)$

Thus, the resultant wave ψ at any point P on the screen is given by $\psi = R\sin(\omega t + \alpha)$, whose amplitude is R. The resultant amplitude is determined as follows.

Squaring and adding Eqs (2.7) and (2.8), we get

$$(R\sin\alpha)^2 + (R\cos\alpha)^2 = (r\sin\delta)^2 + [r(1+\cos\delta)]^2$$

or $\quad R^2 = r^2 + r^2 + 2r^2\cos\delta$

or $\quad R^2 = 2r^2(1 + \cos\delta) \qquad (2.9)$

The intensity I of a wave is defined as the amount of energy flowing perpendicularly through unit area in unit time. Mathematically, the expression for the intensity of a wave is

$$I = 2\pi^2\rho v r^2, \ \rho = \text{density of the medium}, \ v = \text{speed of the wave}$$

or $I = k(\text{amplitude})^2$

where $k = 2\pi^2 \rho v$ is constant for a particular medium and particular wave. Therefore, the intensity is directly proportional to the square of the amplitude of the resultant wave. The intensity of the resultant wave is thus found out to be

$$I = 2kr^2(1 + \cos\delta) \tag{2.10}$$

The relation between optical path difference Δ (optical path = geometrical path × refractive index) and phase difference δ is

$$\delta = \frac{2\pi}{\lambda}\Delta \tag{2.11}$$

or $$\Delta = \frac{\lambda}{2\pi}\delta \tag{2.12}$$

2.7.1 Constructive interference ($I = I_{\text{max}}$)

The interference is said to be constructive (formation of bright fringe) if the intensity of the resultant wave I is maximum. From Eq. (2.10), the maximum value of intensity $I = 2kr^2(1 + \cos\delta)$ is found out as follows.

$I = 2kr^2(1 + \cos\delta)$ is maximum when $\cos\delta$ is maximum. The maximum value of $\cos\alpha$ is 1. Hence, the maximum value of I is

$$I = I_{\text{max}} = 2kr^2(1+1) = 4kr^2 \quad \text{if} \quad \cos\delta = +1$$

or $\delta = 2n\pi, n = 0, 1, 2, 3, \ldots$ (maxima) $\tag{2.13}$

Constructive interference occurs if the phase difference between the interfering waves is an even multiple of π.

We know that for constructive interference, $\delta = 2n\pi$. Hence, the constructive interference condition in terms of the path difference will be given by

$$\Delta = \frac{\lambda}{2\pi} \times 2n\pi$$

or $\Delta = n\lambda = 2n\left(\dfrac{\lambda}{2}\right).$ (maxima) $\tag{2.14}$

Therefore, we conclude that bright fringes are formed at those points for which the optical path difference between the waves emerging from the two slits S_1 and S_2 are even multiples of half the wavelengths.

The maximum intensity of the resultant wave is $4kr^2$, whereas the maximum intensity of the component wave is kr^2 since the amplitude of each component wave is r. Hence, the maximum intensity of the resultant wave is four times the maximum intensity of the component wave in case of constructive interference.

2.7.2 Destructive interference ($I = I_{min}$)

The interference is said to be destructive (formation of dark fringe) if the intensity of the resultant wave I is minimum. From Eq. (2.10), the minimum value of intensity $I = 2kr^2(1 + \cos \delta)$ is found out as follows.

$I = 2kr^2(1 + \cos \delta)$ is minimum when $\cos \delta$ is minimum. The minimum value of $\cos \alpha$ is -1. Hence, the minimum value of I is

$$I = I_{min} = 2kr^2(1 - 1) = 0 \quad \text{if} \quad \cos \delta = -1$$

or $\quad \delta = (2n + 1)\pi, n = 0, 1, 2, 3, \dots$ (minima) $\qquad\qquad$ (2.15)

Destructive interference occurs if the phase difference between the interfering waves is an odd multiple of π.

We know that for destructive interference $\delta = (2n + 1)\pi$. Hence, the destructive interference condition in terms of the optical path difference will be given by

$$\Delta = \frac{\lambda}{2\pi} \times (2n+1)\pi$$

or $\quad \Delta = (2n+1)\left(\dfrac{\lambda}{2}\right).$ (minima) $\qquad\qquad$ (2.16)

Therefore, we conclude that dark fringes are formed at those points for which the optical path difference between the waves emerging from the two slits S_1 and S_2 are odd multiples of half the wavelength.

The minimum intensity of the resultant wave is 0, whereas the minimum intensity of the component wave is kr^2 since the amplitude of each component wave is r. Hence, the minimum intensity of the resultant wave is kr^2 less than the minimum intensity of the component wave in case of destructive interference.

2.7.3 Fringe spacing β

Let P be any point on the screen. The formation of a bright fringe or a dark fringe at P depends upon the optical path difference Δ between the waves emerging from the two slits S_1 and S_2. According to Eq. (2.14), a bright fringe is formed at P if

$$\Delta = 2n\left(\frac{\lambda}{2}\right) \quad \text{(Bright fringe)}$$

and according to Eq. (2.16), a dark fringe is formed at P if

$$\Delta = (2n+1)\left(\frac{\lambda}{2}\right) \quad \text{(Dark fringe)}$$

Let the distance between the plane of the slits and the screen be D; O is a point on the screen so that SO as shown in Fig. 2.7 is the perpendicular bisector of S_1S_2. We can take O as the origin of our coordinate system as shown in Fig. 2.7.

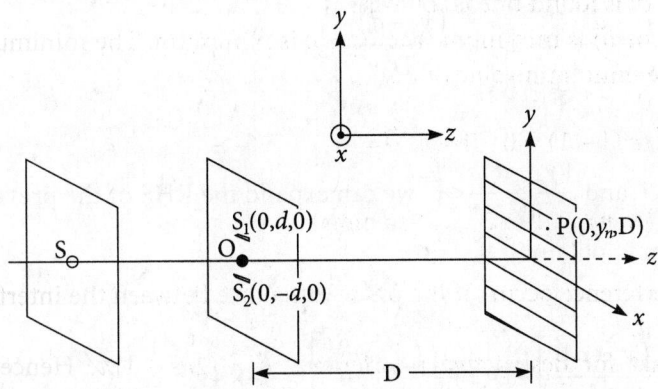

Figure 2.7 Our coordinate system, along with the experimental setup for Young's double slit experiment. The coordinates of the two rectangular slits S_1 and S_2 are S_1 (0, d, 0) and S_2(0, −d, 0) P is any point on the screen along a straight line parallel to the Y-axis and its coordinates will be P(0, y_n, D). As shown in the figure, D is along the Z-axis

P(0, y_n, D) is any point on the screen along a straight line parallel to the Y-axis. The separation D between the plane of the slits and the screen is along the Z-axis as shown in Fig. 2.7. According to our coordinate system, the coordinates of the two rectangular slits S_1 and S_2 are S_1(0, d, 0) and S_2(0, −d, 0). Suppose the nth order fringe appears at P(0, y_n, D). Hence, we have

$$S_1P = \sqrt{(0-0)^2 + (d-y_n)^2 + (0-D)^2} \tag{2.17}$$

$$S_2P = \sqrt{(0-0)^2 + (-d-y_n)^2 + (0-D)^2} \tag{2.18}$$

The optical path difference Δ between the two interfering waves emerging from S_1 and S_2 in vacuum as shown in Fig. 2.6 is

$$\Delta = S_2P - S_1P$$

As our experimental setup is immersed in air medium having refractive index 1, the optical path and the geometrical path are the same.

Putting Eqs (2.17) and (2.18) into the aforementioned equation, we have

$$\Delta = \sqrt{(0-0)^2 + (-d-y_n)^2 + (0-D)^2} - \sqrt{(0-0)^2 + (d-y_n)^2 + (0-D)^2}$$

$$= \sqrt{(y_n+d)^2 + D^2} - \sqrt{(y_n-d)^2 + D^2}$$

or $\quad \Delta = D\left[1 + \frac{(y_n+d)^2}{D^2}\right]^{\frac{1}{2}} - D\left[1 + \frac{(y_n-d)^2}{D^2}\right]^{\frac{1}{2}}$

Since $\dfrac{(y_n+d)^2}{D^2} < 1$ and $\dfrac{(y_n-d)^2}{D^2} < 1$, we can expand the RHS of the previous equation by binomial theorem to obtain

$$\Delta = D\left[1 + \frac{1}{2}\frac{(y_n+d)^2}{D^2}\right] - D\left[1 + \frac{1}{2}\frac{(y_n-d)^2}{D^2}\right] = \frac{1}{2D} \times 4y_n d$$

The optical path difference Δ between two interfering waves emerging from S_1 and S_2 is thus

$$\Delta = \frac{2y_n d}{D}$$

The position of the nth order fringe is obtained from this equation as

$$y_n = \frac{D\Delta}{2d} \tag{2.19}$$

For the nth order bright fringe, Δ from Eq. (2.14) is given as $\Delta = 2n\left(\dfrac{\lambda}{2}\right)$. Putting this value of Δ into Eq. (2.19), the position of the nth order bright fringe is obtained as

$$y_n = \frac{Dn\lambda}{2d} \tag{2.20}$$

Similarly, the position of the $(n + 1)$th order bright fringe is given by

$$y_{n+1} = \frac{D(n+1)\lambda}{2d}$$

In between the nth and $(n + 1)$th order bright fringe there must exists the nth order dark fringe. Here, the distance between two consecutive bright fringes or two consecutive dark fringes is defined as fringe spacing. Therefore, the separation $y_{n+1} - y_n$ gives the spacing of the nth order dark fringe β. Hence, we have

$$\beta = y_{n+1} - y_n = \frac{D\lambda}{2d} \tag{2.21}$$

as the spacing of the nth order dark fringe.

For the nth order dark fringe, Δ from Eq. (2.16) is given as $\Delta = (2n+1)\left(\dfrac{\lambda}{2}\right)$. Putting this value of Δ into Eq. (2.19), the position of the nth order dark fringe is obtained as

$$y_n = \frac{D(2n+1)\lambda}{4d} \tag{2.22}$$

Similarly, the position of the $(n + 1)$th order dark fringe is given as

$$y_{n+1} = \frac{D(2n+3)\lambda}{4d}$$

In between the nth and $(n+1)$th order dark fringe there must exists the nth order bright fringe. Therefore, the separation $y_{n+1} - y_n$ gives the spacing of the nth order bright fringe β. Hence, we have

$$\beta = y_{n+1} - y_n = \frac{D\lambda}{4d} \times 2 = \frac{D\lambda}{2d} \tag{2.23}$$

Equations (2.21) and (2.23) prove that the spacing of the nth order bright fringe and the dark fringe are the same. Therefore, we simply say fringe spacing β instead of bright fringe spacing or dark fringe spacing. Hence, the nth order fringe spacing is

$$\beta = \frac{D\lambda}{2d} \tag{2.24}$$

Equation (2.24) does not contain n. Therefore, we conclude that the spacing of any order fringe have the same value. From Eq. (2.24), we can have

i. $\beta \propto D$ when γ and d remains unchanged.

ii. $\beta \propto \lambda$ when D and d remains unchanged.

iii. $\beta \propto \dfrac{1}{d}$ when λ and D remains unchanged.

These conclusions are in perfect agreement with the experimental interference pattern on the screen.

2.7.4 Intensity distribution curve

The intensity distribution of interference fringes on screen produced due to coherent superposition of two monochromatic light waves of the same amplitude and same frequency follows the equation $I = 2kr^2(1 + \cos\delta)$ (Eq. (2.10)). Again according to equation

$$\beta = \frac{D\lambda}{2d}$$

(Equation (2.24)), spacing of the fringes on the screen is constant for given values of D, d and λ. The intensity I is taken along the Y-axis and phase difference δ is taken along the X-axis. As phase difference δ gradually increases from 0 to π, $\cos\delta$ decreases gradually from +1 to –1 via 0 and as a consequence, intensity I diminishes from $4kr^2$ to 0. The shape of the intensity distribution curve is therefore as shown in Fig. 2.8 for different positive and negative values of δ.

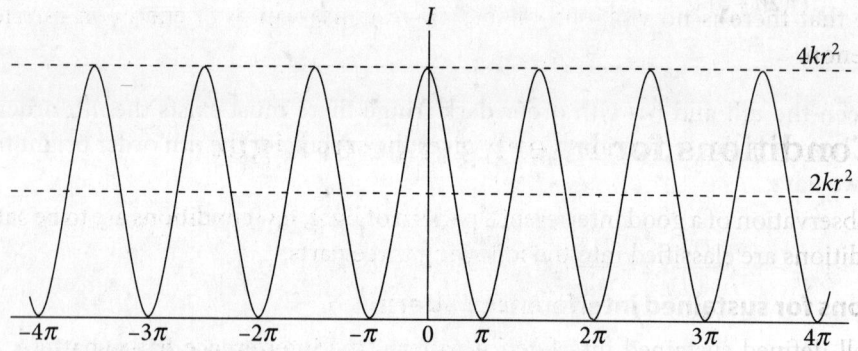

Figure 2.8 The intensity distribution $I = 2kr^2(1 + \cos\delta) = 4kr^2\cos^2\left(\dfrac{\delta}{2}\right)$ of interference fringes on a screen produced due to coherent superposition of two monochromatic light waves of the same amplitude and same frequency

2.8 Conservation of Energy in Interference

No destruction or creation of energy occurs in interference phenomenon; there is only re-distribution of energy. The transformation of energy from one form to another form does not occur in interference phenomenon. The law of conservation of energy should hold good in interference phenomenon.

The energy which disappears from the dark fringe appears in the bright fringe. The average value of energy or intensity $<I>$ over any range of phase difference is the same as if interference effects are absent. We can show it mathematically in the following way.

$$<I>= \frac{\int\limits_{0}^{2\pi} I d\delta}{\int\limits_{0}^{2\pi} d\delta} = \frac{\int\limits_{0}^{2\pi} 2kr^2(1+\cos\delta)d\delta}{\int\limits_{0}^{2\pi} d\delta}$$

$$= \frac{\int\limits_{0}^{2\pi} 2kr^2 d\delta + \int\limits_{0}^{2\pi} \cos\delta d\delta}{\int\limits_{0}^{2\pi} d\delta} = \frac{2kr^2 \times 2\pi}{2\pi}$$

or $<I>= 2kr^2 = \frac{1}{2}4kr^2$

Thus, the average value of intensity is half of the maximum intensity $4kr^2$. Therefore, we conclude that there is no violation of the law of conservation of energy in interference phenomenon.

2.9 Conditions for Interference of Light

For the observation of a good interference pattern of light, few conditions are to be satisfied. The conditions are classified into the following three parts.

Conditions for sustained interference pattern

For a well-defined sustained interference pattern, the interference fringe pattern should remain stable throughout the observation time. To accomplish this, the experimental arrangements must satisfy the following conditions.

i. The interfering two waves of light must originate from the same source of light so that they have constant phase difference at all times.

As we know that the intensity distribution pattern follows from the equation $I = I_1 + I_2 + 2\sqrt{I_1}\sqrt{I_2}\cos\delta$, it will vary with the variation of phase difference δ. If the phase difference δ varies with time, the interference pattern will change with time.

Therefore, the two interfering light waves should be coherent, i.e., $\frac{d\delta}{dt} = 0$.

ii. The interfering two waves of light must have the same wavelength, i.e., they must be monochromatic waves (A monochromatic light source emits light of a single color, i.e., single wavelength; mono = single, chroma = color). If a source of white light (polychromatic) is used, no fringes will be seen at all except for some at the central region. The central dark fringe will be accompanied by a few colored fringes on both sides.

iii. The interfering two waves of light must have the same period or same frequency. If frequencies are different, the maximum due to one wave may coincide with the minimum due to the other wave.

Conditions for observation of interference pattern

For a very clear observation of interference pattern, the experimental arrangements must satisfy the following conditions.

i. The separation between the two sources (2d) should be small. The expression for fringe spacing β as obtained in Eq. (2.24) is

$$\beta = \frac{D\lambda}{2d}.$$

If 2d is small, fringe spacing β will be large and fringes will be separately visible. If 2d is large, fringe spacing β will be small and fringes will not be visible clearly due to the limited resolving power of our eyes.

ii. The distance D between the two sources and screen should be large. The expression for fringe spacing β as obtained in Eq. (2.24) is

$$\beta = \frac{D\lambda}{2d}.$$

If D is small fringe, spacing β will be small and the fringes will not be visible clearly due to the limited resolving power of our eyes. If D is large, fringe spacing β will be large and fringes will be visible clearly.

iii. The background of the observation should be dark in order to observe the interference fringe pattern clearly.

Conditions for good contrast interference pattern

In a well-defined good contrast interference pattern, the dark fringe should be fully dark ($I_{min} = 0$) and the bright fringe should be fully bright (I_{max} = maximum) throughout the observation time. To accomplish this, the experimental arrangements must satisfy the following conditions.

i. The interfering two waves of light must have the same or nearly equal amplitudes. If the two interfering waves have different amplitudes, the dark fringe will not be fully dark as is evident from equation $I = k\left(r_1^2 + r_2^2 + 2r_1 r_2 \cos\delta\right)$. $I = 0$ only when $r_1 = r_2$ with $\cos\delta = -1$.

ii.   The sources should be narrow. If sources are not narrow, interference between the waves from doifferent parts of the same source will take place and the contrast will be poor.

iii. The two interfering light waves must be in the same state of polarization.

iv.   The two interfering light waves must be propagated almost in the same direction.

v.   The original source from which the two interfering light waves are originating must be monochromatic or nearly monochromatic. If instead of a monochromatic source a polychromatic source is used, then each separate colored light may produce its own interference pattern with its own spacing. The different interference pattern due to different wavelengths will overlap each other and the net effect will be a blurred interference pattern.

Example 2.1

The wavelength of light in vacuum is 5890 Å. Calculate the wavelength of the same light in a medium having absolute refractive index 1.4.

Solution

The absolute refractive index is defined by

$$\mu = \frac{c}{v}$$

or $\mu = \frac{v\lambda_o}{v\lambda_m}$ $\qquad [\because \quad v = v\lambda]$

or $\lambda_m = \frac{\lambda_o}{\mu}$

Putting the values of λ_o and μ into this equation, we get

$$\lambda_m = \frac{5890\text{Å}}{1.4} = 4207\text{Å}$$

Example 2.2

In Young's double slit experiment, a 2 cm space on the screen placed at 200 cm contains 20 fringes. Find the slit separation if the wavelength of light used is 5100 Å.

Solution

Twenty fringes occupy 2 cm space on the screen. Hence, the distance occupied by one fringe on the screen, i.e., fringe spacing β will be

$$\beta = \frac{2 \text{ cm}}{20} = 0.1 \text{ cm} = 0.001 \text{ m}$$

Data given are $D = 200$ cm $= 2$ m and $\lambda = 5100 \overset{\circ}{A} = 5100 \times 10^{-10}$ m

The distance between the two slits, i.e., slit separation, $2d$, is given as

$$2d = \frac{D\lambda}{\beta}$$

Putting the values of D, λ and β into this equation, we get

$$2d = \frac{2 \times 5100 \times 10^{-10}}{0.001} \text{ m} = 1.02 \times 10^{-3} \text{ m} = 0.102 \text{cm}$$

Example 2.3

In Young's double slit experiment, two coherent sources are 0.02 cm apart and fringes are observed on a screen 100 cm away. It is found that the 6th bright fringe is situated at a distance of 1.2 cm from the central fringe. Calculate the wavelength of the monochromatic light used.

Solution

The data given are $2d = 0.02$ cm, $D = 100$ cm, $n = 6$, and $y_n = 1.2$ cm

The position of the nth order bright fringe from the central fringe is

$$y_n = \frac{Dn\lambda}{2d} \quad \text{or} \quad \lambda = \frac{y_n 2d}{Dn}$$

Putting the values of $2d$, D, n, and y_n into this equation, we get

$$\lambda = \frac{1.2 \times 0.02}{100 \times 6} \text{ cm} = 4000 \text{ Å}$$

2.10 Shape of Interference Fringes

We can very easily form a mental picture regarding the shape of interference fringes by deriving the equation of the locus of the points having a given path difference Δ between the two interfering waves originating from the two slits S_1 and S_2. Interference fringes are nothing but the loci of the points having a given path difference Δ between the two interfering waves.

2.10.1 Shape of interference fringes on the XY-plane (Hyperbolic)

The coordinate system we have chosen is shown in Fig. 2.7. According to our coordinate system, the coordinates of any point P on the screen are (x, y, D), the coordinates of the slit S_1 are $(0, d, 0)$ and that of S_2 are $(0, -d, 0)$. Let Δ be the optical path difference between the two interfering waves originating from the two slits S_1 and S_2. Therefore, from the Fig. 2.6, we have

$$S_2P - S_1P = \Delta \tag{2.25}$$

However, $S_2P = \sqrt{(0-x)^2 + (-d-y)^2 + (0-D)^2}$

and $S_1P = \sqrt{(0-x)^2 + (d-y)^2 + (0-D)^2}$

So Eq. (2.25) becomes

$$\sqrt{x^2 + (y+d)^2 + D^2} = \sqrt{x^2 + (y-d)^2 + D^2} + \Delta$$

Squaring both sides of this equation, we have

$$(y+d)^2 = (y-d)^2 + \Delta^2 + 2\Delta\sqrt{x^2 + (y-d)^2 + D^2}$$

or $\quad 4yd - \Delta^2 = 2\Delta\sqrt{x^2 + (y-d)^2 + D^2}$

Again squaring both sides of this equation, we have

$$\left(4yd - \Delta^2\right)^2 = 4\Delta^2\left[x^2 + \left(y-d\right)^2 + D^2\right]$$

or $\quad 16y^2d^2 + \Delta^4 - 8yd\Delta^2 = 4x^2\Delta^2 + 4y^2\Delta^2 + 4d^2\Delta^2 - 8yd\Delta^2 + 4D^2\Delta^2$

or $\quad 16y^2d^2 + \Delta^4 = 4x^2\Delta^2 + 4y^2\Delta^2 + 4d^2\Delta^2 + 4D^2\Delta^2$

or $\quad 16y^2d^2 - 4y^2\Delta^2 - 4x^2\Delta^2 = -\Delta^4 + 4d^2\Delta^2 + 4D^2\Delta^2$

or $4\left(4d^2 - \Delta^2\right)y^2 - 4\Delta^2 x^2 = \Delta^2\left(4D^2 + 4d^2 - \Delta^2\right)$ (2.26)

Dividing both sides of this equation by $\Delta^2\left(4D^2 + 4d^2 - \Delta^2\right)$, we have

$$\frac{4\left(4d^2 - \Delta^2\right)y^2}{\Delta^2\left(4D^2 + 4d^2 - \Delta^2\right)} - \frac{4\Delta^2 x^2}{\Delta^2\left(4D^2 + 4d^2 - \Delta^2\right)} = 1$$

or $$\frac{y^2}{\left(\dfrac{\Delta^2 D^2}{4d^2 - \Delta^2} + \dfrac{\Delta^2}{4}\right)} - \frac{x^2}{\left(D^2 + d^2 - \dfrac{\Delta^2}{4}\right)} = 1$$ (2.27)

In optical experiments, to obtain interference fringes, we need the optical path difference $\Delta \approx 10^{-8}$ cm and $2d \approx 10^{-2}$ cm. Therefore, neglecting $\dfrac{\Delta^2}{4}$ from Eq. (2.27), we get

$$\frac{y^2}{\dfrac{\Delta^2 D^2}{4d^2 - \Delta^2}} - \frac{x^2}{D^2 + d^2} = 1$$ (2.28)

Equations (2.27) or (2.28) are of the form $\dfrac{y^2}{a^2} - \dfrac{x^2}{b^2} = 1$ with

$$a = \frac{\Delta D}{\sqrt{4d^2 - \Delta^2}}$$ (2.29)

and $b = \sqrt{D^2 + d^2}$ (2.30)

The hyperbola defined by the equation

$$\frac{y^2}{a^2} - \frac{x^2}{b^2} = 1$$

is shown in the Fig. 2.9. Equations (2.27) or (2.28) define the loci of a point on the $Z = D$ plane for which the path difference $S_2 P - S_1 P = \Delta$ is constant.

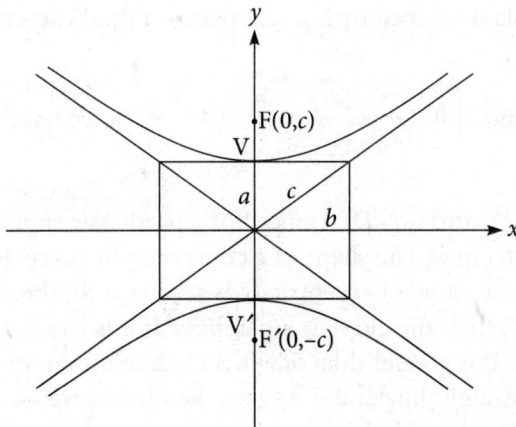

Figure 2.9 | The hyperbola defined by the equation $\dfrac{y^2}{a^2}-\dfrac{x^2}{b^2}=1$ with $a^2 + b^2 = c^2$. The vertices are $V(0, a)$ and $V'(0, -a)$; foci are $F(0, c)$ and $F(0, -c)$

The hyperbola has the vertices at $(0, \pm a)$ and foci at $(0, \pm c)$. For a hyperbola.

$$c^2 = a^2 + b^2$$

Putting the values of a and b from Eqs (2.29) and (2.30) into this equation, we get

$$c^2 = \frac{\Delta^2 D^2}{4d^2 - \Delta^2} + D^2 + d^2$$

or $\quad c^2 = \dfrac{4d^2 D^2 + 4d^4 - \Delta^2 d^2}{4d^2 - \Delta^2}$

Without hampering physics, we can neglect $4d^4$ and $\Delta^2 d^2$ in this equation and obtain

$$c = \frac{2dD}{\sqrt{4d^2 - \Delta^2}} \tag{2.31}$$

The vertices of the hyperbola described by Eq. (2.28) are on the Y-axis at

$$\left(0, \frac{D\Delta}{\sqrt{4d^2 - \Delta^2}}\right) \quad \text{and} \quad \left(0, -\frac{D\Delta}{\sqrt{4d^2 - \Delta^2}}\right)$$

or approximately at

$$\left(0, \frac{D\Delta}{2d}\right) \quad \text{and} \quad \left(0, -\frac{D\Delta}{2d}\right).$$

The foci of the hyperbola described by Eq. (2.28) are on the Y-axis at

$$\left(0, \frac{2dD}{\sqrt{4d^2 - \Delta^2}}\right) \quad \text{and} \quad \left(0, -\frac{2dD}{\sqrt{4d^2 - \Delta^2}}\right)$$

or approximately at $(0, D)$ and $(0, -D)$. Hence, for a particular experimental setup, the foci are the same for all the fringes. The shape of a conic section curve depends upon the value of the eccentricity e. If the value of eccentricity is zero ($e = 0$), then the curve is a circle; if it is equal to one ($e = 1$), then the curve is a parabola; if it is less than one ($e < 1$), then the curve is an ellipse; and if it is greater than one ($e > 1$), then the curve is a hyperbola. If $e \gg 1$, curves will be nearly a straight line and if $e \to \infty$, then the curve $\to$ straight line.

The eccentricity e of the hyperbola

$$\frac{y^2}{a^2} - \frac{x^2}{b^2} = 1$$

is defined by $e = \dfrac{c}{a}$. Putting the values of c and a from Eqs (2.31) and (2.29) into this equation, we have the eccentricity of the hyperbola defined by Eq. (2.28) as

$$e = \frac{\dfrac{2dD}{\sqrt{4d^2 - \Delta^2}}}{\dfrac{\Delta D}{\sqrt{4d^2 - \Delta^2}}}$$

or $\quad e = \dfrac{2d}{\Delta}$ \hfill (2.32)

Thus, eccentricity of the hyperbola described by Eqs (2.27) or (2.28) is obtained approximately as $e = \dfrac{2d}{\Delta}$. Putting the values of $2d$ and Δ into Eq. (2.28), the eccentricity of the hyperbola is found much greater than 1. Larger value of eccentricity means the hyperbola is more flat.

The eccentricity of a bright hyperbolic fringe is obtained by putting the value Δ from Eq. (2.13) (for bright fringe $\Delta = 2n\left(\dfrac{\lambda}{2}\right)$) into Eq. (2.32). Hence, the eccentricity of a bright hyperbolic fringe will be

$$e = \frac{2d}{n\lambda} \quad \text{(for bright fringe)} \hfill (2.33)$$

The eccentricity of a dark hyperbolic fringe is obtained by putting the value Δ from Eq. (2.15) (for dark fringe $\Delta = (2n+1)\left(\dfrac{\lambda}{2}\right)$) into Eq. (2.32). Hence, the eccentricity of a dark hyperbolic fringe will be

$$e = \frac{4d}{(2n+1)\lambda} \quad \text{(for dark fringe)} \tag{2.34}$$

Equations (2.33) and (2.34) show that for all types of fringes, eccentricity is much more than 1. Hence, hyperbolic fringes are nearly straight lines. Any plane parallel to the plane of the slits is a transverse section. Therefore, on any transverse plane, the interference fringes appear straight though they are actually confocal hyperbolas of large eccentricities. In Fig. 2.7, the XY-plane is the transverse section containing the straight interference fringe pattern.

2.11 Interference Fringes in 3-D Space

In Eq. (2.26), in place of D, let us put z, i.e., $D = z$. As D is along the Z-axis we have

$$4\left(4d^2 - \Delta^2\right)y^2 - 4\Delta^2 x^2 = \Delta^2\left(4z^2 + 4d^2 - \Delta^2\right)$$

or $\quad 4\left(4d^2 - \Delta^2\right)y^2 - 4\Delta^2 x^2 = 4\Delta^2 z^2 + \Delta^2\left(4d^2 - \Delta^2\right)$

or $\quad 4\left(4d^2 - \Delta^2\right)y^2 - 4\Delta^2 x^2 - 4\Delta^2 z^2 = \Delta^2\left(4d^2 - \Delta^2\right)$

Dividing both sides of this equation by $\Delta^2(4d^2 - \Delta^2)$, we have

$$\frac{4\left(4d^2 - \Delta^2\right)y^2}{\Delta^2\left(4d^2 - \Delta^2\right)} - \frac{4\Delta^2 x^2}{\Delta^2\left(4d^2 - \Delta^2\right)} - \frac{4\Delta^2 z^2}{\Delta^2\left(4d^2 - \Delta^2\right)} = 1$$

or $\quad \dfrac{y^2}{\dfrac{\Delta^2}{4}} - \dfrac{x^2}{d^2 - \dfrac{\Delta^2}{4}} - \dfrac{z^2}{d^2 - \dfrac{\Delta^2}{4}} = 1$

or $\quad \dfrac{x^2}{d^2 - \dfrac{\Delta^2}{4}} - \dfrac{y^2}{\dfrac{\Delta^2}{4}} + \dfrac{z^2}{d^2 - \dfrac{\Delta^2}{4}} = -1 \tag{2.35}$

Equation (2.35) is of the form $\dfrac{x^2}{a^2} - \dfrac{y^2}{b^2} + \dfrac{z^2}{a^2} = -1$ with

$$a^2 = d^2 - \frac{\Delta^2}{4}$$

and $\quad b^2 = \dfrac{\Delta^2}{4}$

Equation (2.34) describes a hyperboloid of two sheets.

The hyperboloid of two sheets defined by the equation $\dfrac{x^2}{a^2} - \dfrac{y^2}{b^2} + \dfrac{z^2}{a^2} = -1$ is shown in Fig. 2.10.

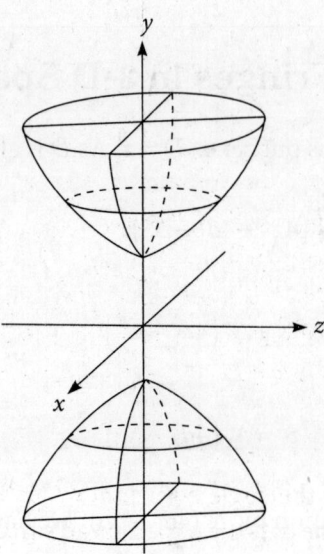

Figure 2.10 The hyperboloid of two sheets defined by the equation $\dfrac{x^2}{a^2} - \dfrac{y^2}{b^2} + \dfrac{z^2}{a^2} = -1$. The cross-sections by the $x = k$ plane is a hyperbola, by the $y = k$ plane is a circle and by the $z = k$ plane is a hyperbola

Equation (2.34) is the loci of points in space for which path difference $S_2P - S_1P = \Delta$ is constant. The hyperboloid surface described by Eq. (2.35) touches the Y-axis at the points

$$\left(0, \frac{\Delta}{2}, 0\right) \text{ and } \left(0, -\frac{\Delta}{2}, 0\right)$$

and does not touch the X-axis and the Z-axis.

2.11.1 Shape of interference fringes on the ZX-plane (Circular)

Now let us observe the interference pattern on the ZX-plane along the Y-axis (see Fig. 2.7). The $y = k$ plane (the ZX-plane at k distance from the origin) cuts the hyperboloid surface described by the equation

$$\frac{x^2}{d^2 - \dfrac{\Delta^2}{4}} - \frac{y^2}{\dfrac{\Delta^2}{4}} + \frac{z^2}{d^2 - \dfrac{\Delta^2}{4}} = -1$$

in a circle. Therefore, the shape of the interference fringes on the $y = k$ plane are circular. The radius of the circular fringe on the $y = k$ plane is determined by putting $y = k$ in Eq. (2.35). Hence, we have

$$\frac{x^2}{d^2 - \dfrac{\Delta^2}{4}} - \frac{k^2}{\dfrac{\Delta^2}{4}} + \frac{z^2}{d^2 - \dfrac{\Delta^2}{4}} = -1$$

or

$$\frac{x^2}{d^2 - \dfrac{\Delta^2}{4}} + \frac{z^2}{d^2 - \dfrac{\Delta^2}{4}} = \frac{4k^2}{\Delta^2} - 1$$

or

$$x^2 + z^2 = \left(\frac{4k^2}{\Delta^2} - 1 \right)\left(d^2 - \frac{\Delta^2}{4} \right)$$

This equation is in the form of the circle equation $x^2 + z^2 = r^2$ in the XZ-plane. Thus, we come to know that the $y = k$ plane cuts the hyperboloid surface in a circle. Hence, the loci of points having constant path difference on the $y = k$ plane (the XZ-plane at k distance from the origin) is a circular fringe of radius

$$\sqrt{\left(\frac{4k^2}{\Delta^2} - 1 \right)\left(d^2 - \frac{\Delta^2}{4} \right)}.$$

The approximate radius of the circular fringe on the $y = k$ plane can be found out in the following manner. The square of radius of the circular fringe r^2 is given by

$$r^2 = \left(\frac{4k^2}{\Delta^2} - 1 \right)\left(d^2 - \frac{\Delta^2}{4} \right)$$

$$= \frac{4k^2d^2}{\Delta^2} - d^2 - k^2 + \frac{\Delta^2}{4}$$

or $\quad r^2 = \frac{1}{\Delta^2}\left(4k^2d^2 - \Delta^2d^2 - \Delta^2k^2 + \frac{\Delta^4}{4}\right)$

As earlier, neglecting Δ^2d^2 and $\frac{\Delta^4}{4}$ from this equation, for the same reason, we obtain

$$r^2 = \frac{k^2}{\Delta^2}\left(4d^2 - \Delta^2\right)$$

By neglecting Δ^2 from the term $(4d^2 - \Delta^2)$ in this equation, we get the approximate value of the radius of the circular fringe as

$$r = \frac{2dk}{\Delta} \tag{2.36}$$

For bright fringe, $\Delta = n\lambda$. Hence, the radius of the nth order circular bright fringe on the $y = k$ plane is obtained as

$$r = \frac{2dk}{n\lambda} \quad \text{(Bright fringe)} \tag{2.37}$$

For dark fringe, $\Delta = (2n+1)\frac{\lambda}{2}$. Hence, the radius of the nth order circular dark fringe on the $y = k$ plane is obtained as

$$r = \frac{4dk}{(2n+1)\lambda} \quad \text{(Dark fringe)} \tag{2.38}$$

2.11.2 Shape of interference fringes on the XY-plane (Hyperbolic)

If we put $z = k$ in the hyperboloid Eq. (2.35), we get

$$\frac{x^2}{d^2 - \frac{\Delta^2}{4}} - \frac{y^2}{\frac{\Delta^2}{4}} + \frac{k^2}{d^2 - \frac{\Delta^2}{4}} = -1$$

or $\quad \dfrac{y^2}{\frac{\Delta^2}{4}} - \dfrac{x^2}{d^2 - \frac{\Delta^2}{4}} = 1 + \dfrac{k^2}{d^2 - \frac{\Delta^2}{4}} \tag{2.39}$

This is an equation of a hyperbola on the $z = k$ plane (the XY-plane at k distance from the origin) and has been discussed in Section 2.10.1. The interference fringes on the XY-plane are therefore hyperbolic in shape as discussed earlier.

2.11.3 Shape of interference fringes on the YZ-plane (Hyperbolic)

If we put $x = k$ in the hyperboloid Eq. (2.35), we get

$$\frac{k^2}{d^2 - \frac{\Delta^2}{4}} - \frac{y^2}{\frac{\Delta^2}{4}} + \frac{z^2}{d^2 - \frac{\Delta^2}{4}} = -1$$

or
$$\frac{y^2}{\frac{\Delta^2}{4}} - \frac{z^2}{d^2 - \frac{\Delta^2}{4}} = 1 + \frac{k^2}{d^2 - \frac{\Delta^2}{4}} \tag{2.40}$$

The Eq. (2.40) is similar to Eq. (2.39). Equation (2.40) represents a hyperbola on the $x = k$ plane (the YZ-plane at k distance from the origin). The interference fringes on the YZ-plane are therefore hyperbolic in shape.

2.12 Newton's Rings

Let a plano-convex lens of large focal length be placed on a plane glass plate so that the curved surface of the lens is in contact with the glass surface. It is assumed that the plano-convex lens touches the glass plate at a single point. Now the plano-convex lens and the glass plate system is in the air medium. Therefore, a thin air film of uniformly and slowly increasing thickness in the outward direction is formed around the point of contact. Concentric circular interference fringes are formed when a parallel beam of monochromatic light is incident normally on the plane surface of either the plano-convex lens or the glass plate. The thickness of these alternate bright and dark circular interference fringes decreases as their radii increases as shown in Fig. 2.12. Any one of these concentric circular interference fringes is actually the locus of the points of equal depths or thicknesses. Therefore, these concentric circular interference fringes are called fringes of constant thickness. These concentric circular interference fringes so formed are termed Newton's rings since this phenomenon was first analyzed by Newton, though it was first observed by Robert Hooke in the year 1635.

2.12.1 Experimental setup

The experimental setup is depicted in Fig. 2.11. S a source of monochromatic light is placed at the focus of a convex lens so that after refraction through the lens, the refracted beam becomes parallel. It is one of the properties of a convex lens. This parallel beam of light is allowed to fall on a glass plate G placed at 45° with the incident beam. The glass plate G so placed will reflect a part of the incident beam of light vertically downward towards the plane surface of the plano-convex lens placed on another glass plate as shown in Fig. 2.11.

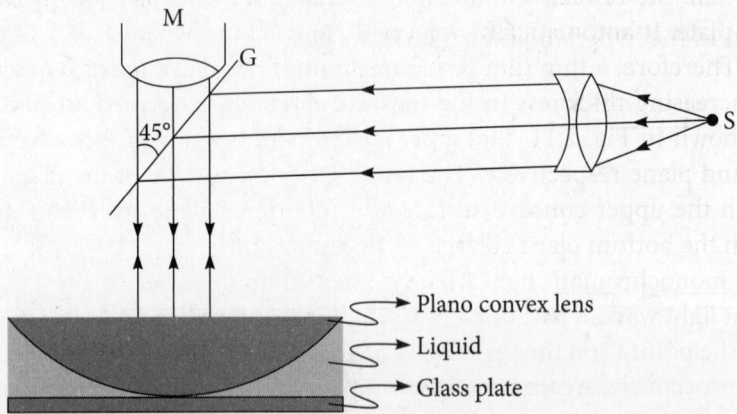

Plano convex lens
Liquid
Glass plate

Figure 2.11 | The experimental setup for observation of Newton's rings by reflected light. G is the glass plate, M is the travelling microscope and S is a monochromatic source of light

A microscope is placed over the glass plate G and is focused on the bottom surface of the thin film enclosed by the plano-convex lens and glass plate system. When viewed through the eye-piece of the properly focused microscope, alternate bright and dark concentric circular interference fringes are seen similar to that as depicted in Fig. 2.12.

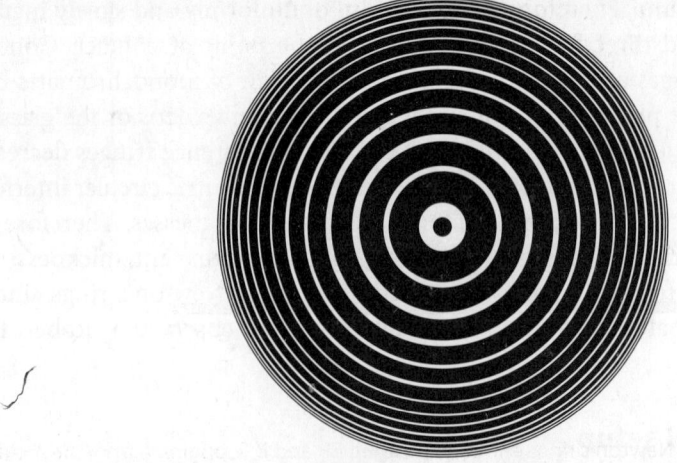

Figure 2.12 | Newton's rings by the reflected light as seen by a well-focused microscope when a thin air film is enclosed between the glass plate and the spherical surface of the plano-convex lens. The central fringe is dark

2.12.2 Theory

To generalize, we can assume that a transparent liquid of refractive index μ is inserted between the curved surface of the lens and the glass plate. Thus, the space between the curved surface of the lens and the glass plate contains a medium of refractive index μ which is less than the refractive indices of the material of the plano-convex lens and the bottom glass plate. It automatically reduces to air medium when $\mu \to 1$ in the following calculations. Therefore, a thin film of the medium of refractive index μ and of uniformly and slowly increasing thickness in the outward direction is formed around the point of contact. As shown in Fig. 2.11, the upper surface and bottom surface of this liquid film are concave and plane respectively. The lower spherical surface of the plano-convex lens coincides with the upper concave surface and the upper plane surface of the glass plate coincides with the bottom plane surface of the liquid film.

Let a ray of monochromatic light AB be incident at point B. Due to division of amplitude of the incident light wave, a part of it is reflected along BR and the other part is transmitted along BC. At the point C on the plane glass plate, reflection and transmission occurs along CB_1 and CT respectively. Again at the point B_1 on the bottom surface of the plano-convex lens, reflection and transmission occurs along B_1C_1 and B_1R_1 respectively. At the point C_1 on the plane glass plate, reflection and transmission occurs along C_1B_2 and C_1T respectively. This process of reflection and transmission goes on continuously as shown in Fig. 2.13. The two waves BR and B_1R_1 originate from a single light wave AB and hence should be capable of producing an observable interference pattern. Similarly, the two transmitted waves CT and C_1T originate from a single light wave AB and hence should be capable of producing an observable interference pattern.

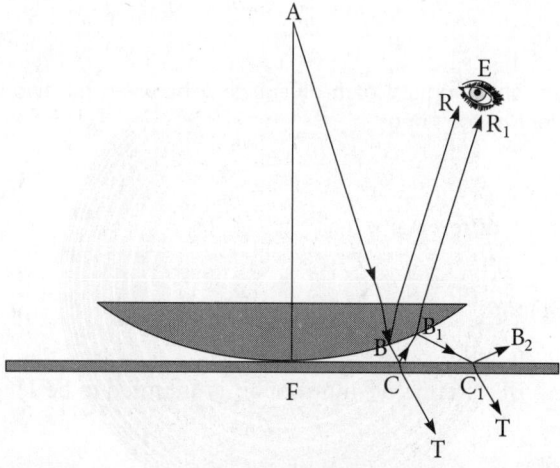

Figure 2.13 | Formation of Newton's rings by reflected light. BR and B_1R_1 originate from the incident light wave AB by partial reflections. They interfere to produce Newton's rings. E is the position of the eye

2.12.3 Calculations

Suppose e is the thickness of the liquid film at the point B_1. Since the focal length of the plano-convex lens is large (hence, the radius of curvature of the plano-convex lens is large), the upper concave surface of the liquid film can be assumed to be plane with a small inclination to the glass plate. Therefore, under this assumption, the thin liquid film formed between the plano-convex lens and the glass plate system is a wedge-shaped film. The optical path difference Δ between the waves BR and B_1R_1 can be calculated in the following manner with the help of Fig. 2.14.

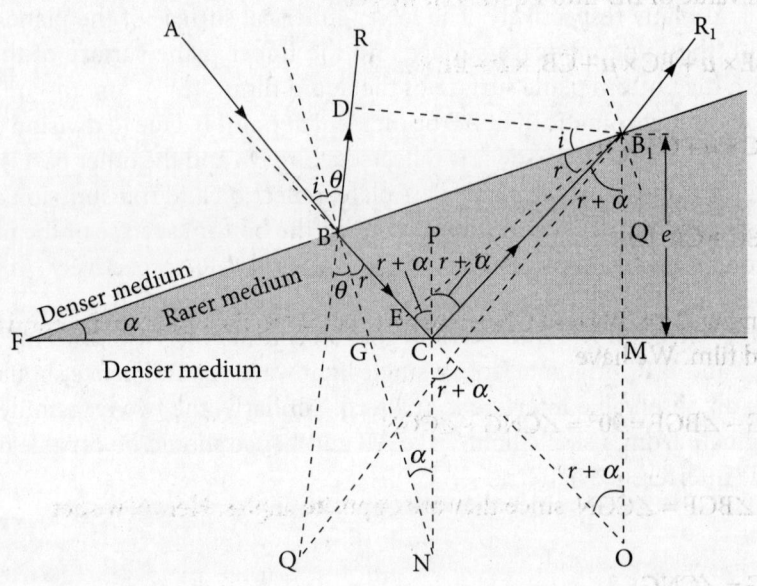

Figure 2.14 | Calculation of the optical path difference Δ between the two interfering waves BR and B_1R_1 for Newton's rings

$$\Delta = \text{geometrical path difference} \times \mu$$

$$= BCB_1 \times \mu - BD \times \mu_{air}$$

$$= (BC + CB_1)\mu - BD \text{ (Refractive index of air is assumed to be 1)}$$

$$= (BE + EC + CB_1)\mu - BD$$

or $\quad \Delta = BE \times \mu + EC \times \mu + CB_1 \times \mu - BD$ \hfill (2.41)

In the right triangle BB_1D $\sin i = \dfrac{BD}{BB_1}$ and in the right triangle BB_1E, $\sin r = \dfrac{BE}{BB_1}$. Hence, by applying Snell's law, we have

$$\mu = \frac{\sin i}{\sin r} = \frac{BD}{BE}$$

or $\quad BD = \mu BE$

Putting this value of BD into Eq. (2.41), we get

$$\Delta = BE \times \mu + EC \times \mu + CB_1 \times \mu - BE \times \mu$$

$$= EC \times \mu + CB_1 \times \mu$$

or $\quad \Delta = \left(EC + CB_1\right)\mu$ $\hfill$ (2.42)

As shown in Fig. 2.14, BN and CN are perpendicular to the upper surface and lower surface of the liquid film. We have

$$\angle BFG + \angle BGF = 90° = \angle CNG + \angle CGN$$

However, $\angle BGF = \angle CGN$ since they are opposite angles. Hence, we get

$$\angle BFG = \angle CNG$$

or $\quad \alpha = \angle CNG$

Again $\angle CNB + \angle CBN = \angle BCP$

or $\quad \alpha + r = \angle BCP$

Since the angle of incidence is equal to the angle of reflection, we have

$$\alpha + r = \angle B_1CP$$

Since $\angle BCP$ and $\angle NCO$ are opposite angles, we have

$$\alpha + r = \angle NCO$$

Since CN is parallel to MO, we have

$$\alpha + r = \angle COM$$

$$\alpha + r = \angle B_1 CP \Rightarrow \angle CB_1 M = \alpha + r$$

Thus, $B_1 CO$ is an isosceles triangle with

$$CB_1 = CO$$

Putting this value of CB_1 into Eq. (2.42), we get

$$\Delta = (EC + CO)\mu$$

or $\quad \Delta = EO \times \mu$ $\hfill$ (2.43)

In the right triangle $B_1 EO$,

$$EO = B_1 O \cos(\alpha + r) = (B_1 M + MO)\cos(\alpha + r) = 2B_1 M \cos(\alpha + r)$$

($B_1 CO$ is an isosceles triangle $\Rightarrow MO = B_1 M$)

or $\quad EO = 2e \cos(\alpha + r)$

where $B_1 M = e$ is the thickness of the liquid film at point B_1.
  Putting this value of EO into Eq. (2.43), we have

$$\Delta = 2e \cos(\alpha + r) \times \mu$$ $\hfill$ (2.44)

This equation shows that the optical path difference Δ between the two interfering waves BR and $B_1 R_1$ varies with the thickness of the film as well as with the angle of incidence. The phase difference δ' between the two interfering waves BR and $B_1 R_1$ due to optical path difference is

$$\delta' = \frac{2\pi}{\lambda} \times \Delta$$ $\hfill$ (2.45)

It has been assumed that the refractive index of the liquid is less than the refractive index of the glass. The wave BR as shown in Fig. 2.13 is obtained from the incident wave AB by the reflection of the incident wave from the upper surface of the liquid film. That means wave BR is obtained by the reflection of the incident wave from a surface backed by a rarer medium. Hence, no phase change occurs due to reflection and the phase difference due to reflection between the waves AB and BR is zero. Similarly, the wave $B_1 R_1$ as shown in the

Fig. 2.13 is obtained from the incident wave AB by the reflection of the incident wave from the glass surface. That means wave B_1R_1 is obtained by the reflection of the incident wave from a surface backed by a denser medium. Hence, the phase change of π occurs due to reflection and the phase difference due to reflection between the waves AB and B_1R_1 is π. In case of transverse waves, no phase change occurs during the refraction or transmission phenomenon. Therefore, the extra phase difference between the two interfering waves BR and B_1R_1 due to reflection is π. The total phase difference δ between the two interfering waves BR and B_1R_1 due to optical path difference and due to reflection is given by

δ = Phase difference due to optical path difference + phase difference due to reflection.

or $\quad \delta = \delta' + \pi$

or $\quad \delta = \dfrac{2\pi}{\lambda} \times \Delta + \pi$ $\hfill (2.46)$

According to Eq. (2.13), constructive interference occurs if the total phase difference between the interfering waves is an even multiple of π. Hence, the bright circular fringes appear at the points for which

$\delta = 2n\pi$

or $\quad \dfrac{2\pi}{\lambda} \times \Delta + \pi = 2n\pi$

or $\quad \Delta = (2n-1)\dfrac{\lambda}{2}$

$2n - 1$ and $2n + 1$ represent odd numbers. Hence, we can replace $2n - 1$ by $2n + 1$ in this equation without hampering anything and the equation becomes

$\Delta = (2n+1)\dfrac{\lambda}{2}$

$2\mu e \cos(\alpha + r) = (2n+1)\dfrac{\lambda}{2}$ (Bright rings) $\hfill (2.47)$

According to Eq. (2.14), destructive interference occurs if the total phase difference between the interfering waves is an odd multiple of π. Hence, the dark circular fringes appear at the points for which

$\delta = (2n+1)\pi$

or $\quad \dfrac{2\pi}{\lambda} \times \Delta + \pi = 2n\pi + \pi$

or $\quad \Delta = 2n\left(\dfrac{\lambda}{2}\right)$

$$2\mu e \cos(\alpha + r) = 2n\left(\dfrac{\lambda}{2}\right) \quad \text{(Dark rings)} \tag{2.48}$$

Equations (2.47) and (2.48) represent the condition for the formation of bright Newton's rings and dark Newton's rings respectively. In Eqs (2.47) and (2.48), α the angle between the lower surface of the plano-convex lens and the glass plate is exceedingly small. Hence, in comparison to r, we can neglect α. Therefore, the Eqs (2.47) and (2.48) become

$$2\mu e \cos r = (2n+1)\dfrac{\lambda}{2} \quad \text{(Bright rings)} \tag{2.49}$$

$$2\mu e \cos r = 2n\left(\dfrac{\lambda}{2}\right) \quad \text{(Dark rings)} \tag{2.50}$$

The experimental arrangements can be designed so that light wave is allowed to fall normally on the plane surface of the plano-convex lens which makes the angle of incidence zero. Hence, under these experimental conditions which makes $r = 0$, Eqs (2.49) and (2.50) boil down to

$$2\mu e = (2n+1)\dfrac{\lambda}{2} \quad \text{(Bright rings)} \tag{2.51}$$

$$2\mu e = 2n\left(\dfrac{\lambda}{2}\right) \quad \text{(Dark rings)} \tag{2.52}$$

If we join all the points on the glass plate below the plano-convex lens having the same thicknesses, the locus will be a perfect circle. The thickness of the liquid film at the points where the nth order bright fringe appears is obtained from Eq. (2.51) and is given by

$$e = (2n+1)\dfrac{\lambda}{4\mu} \tag{2.53}$$

Similarly, the thickness of the liquid film at the points where the nth order dark fringe appears is obtained from Eq. (2.52) and is given by

$$e = 2n\left(\frac{\lambda}{4\mu}\right) \tag{2.54}$$

2.12.4 Diameter of the nth order Newton's ring

The radius of the nth order fringe can be calculated looking at Fig. 2.15.

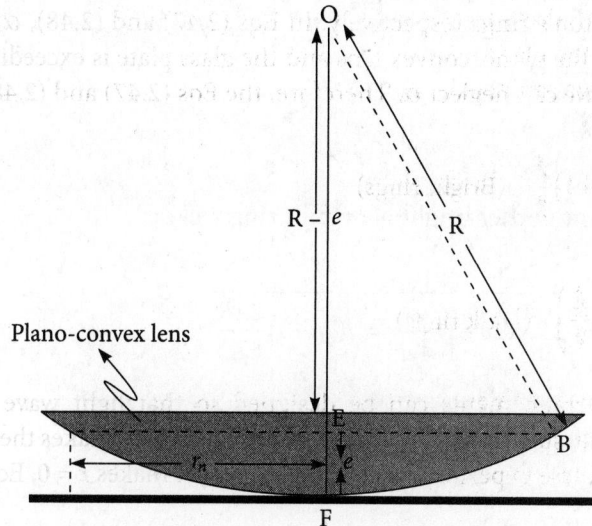

Figure 2.15 | Calculation of radii of nth order Newton's rings. R is the radius of curvature of the plano-convex lens and e is the thickness of the thin film at B

In Fig. 2.15, R is the radius of curvature of the curved surface of the plano-convex lens and r_n is the radius of nth order fringe. Since OEB_1 is a right triangle, we have

$$R^2 = r_n^2 + (R - e)^2$$

or $\quad r_n^2 = 2Re - e^2$

Since the radius of curvature of the curved surface of the plano-convex lens is very large, the thickness e will be a very small quantity, as a result of which e^2 is negligibly small in comparison to $2Re$. Hence, neglecting e^2 in the aforementioned equation, we get

$$r_n^2 = 2Re \tag{2.55}$$

or $r_n = \sqrt{2Re}$ (2.56)

2.12.5 Diameter of the nth order bright Newton's ring

Putting the value of e from Eq. (2.53) into Eq. (2.56), we shall get the radius of the nth order bright Newton's ring as

$$r_n = \sqrt{2R \times (2n+1)\frac{\lambda}{4\mu}}$$

$$r_n = \sqrt{(2n+1)\frac{\lambda R}{2\mu}}$$

The diameter of the nth order bright Newton's ring will be

$$D_n = 2r_n = \sqrt{4} \times \sqrt{2R \times (2n+1)\frac{\lambda}{4\mu}} = \sqrt{(2n+1)\frac{2\lambda R}{\mu}}$$ (2.57)

Putting $\dfrac{2\lambda R}{\mu} = K^2$ in this equation, we have

$$D_n \propto \sqrt{(2n+1)}$$

For any particular experimental setup, $\dfrac{2\lambda R}{\mu}$ is constant. Hence, this equation shows that the diameter of the nth order bright Newton's ring is directly proportional to the square root of odd numbers.

2.12.6 Diameter of the nth order dark Newton's ring

Putting the value of e from Eq. (2.54) into Eq. (2.56), we shall get the radius of the nth order dark Newton's ring as

$$r_n = \sqrt{\frac{n\lambda R}{\mu}}$$

The diameter of the nth order dark Newton's ring will be

$$D_n = 2r_n = \sqrt{4} \times \sqrt{\frac{n\lambda R}{\mu}} = \sqrt{2n\frac{2\lambda R}{\mu}} \qquad (2.58)$$

Putting $\dfrac{2\lambda R}{\mu} = K^2$ in this equation, we have

$$D_n \propto \sqrt{2n}$$

For any particular experimental setup, $\dfrac{2\lambda R}{\mu}$ is constant. Hence, this equation shows that the diameter of the nth order dark Newton's ring is directly proportional to the square root of even numbers.

2.12.7 Central fringe as seen by the reflected light

The plano-convex lens touches the glass plate only at the central point. The thickness of the liquid film at this point is zero. The optical path difference $\Delta = 2e\cos(\alpha + r) \times \mu$ for this point ($e = 0$), according to Eq. (2.45), reduces to zero, i.e., $\Delta = 0$. Therefore, the total phase difference from Eq. (2.52) will be

$$\delta = \frac{2\pi}{\lambda} \times 0 + \pi = \pi \quad \text{(for central fringe)}$$

According to Eq. (2.15), destructive interference occurs if the total phase difference between the interfering waves is an odd multiple of π. Hence, a dark fringe appears at the central point and the central point appears perfectly black as seen by the reflected monochromatic light. This is true when the space between the curved surface of the lens and the glass plate contains a medium of refractive index μ which is less than the refractive indices of the material of the plano-convex lens and the bottom glass plate.

2.13 Newton's Rings by Transmitted Light

In the previous section, we observed and discussed the formation of Newton's rings by reflected light due to interference of reflected waves BR and B_1R_1. Now, we shall discuss the formation of Newton's rings by transmitted light due to the interference of transmitted waves CT and C_1T as depicted in Fig. 2.16.

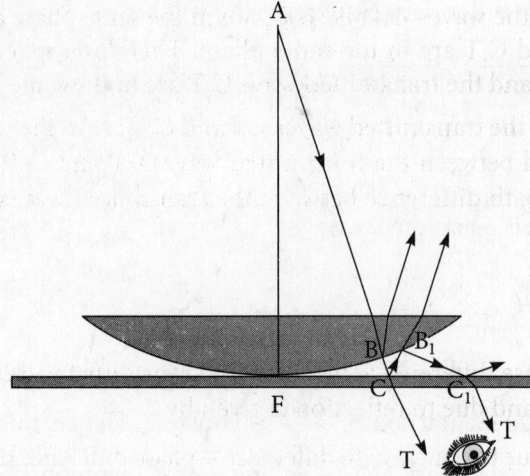

Figure 2.16 | Formation of Newton's rings by transmitted light. Waves CT and C_1T originate from the incident light wave AB by partial transmissions. They interfere to produce Newton's rings. E is the position of the eye

No phase change occurs during refraction or transmission transverse waves like light. Phase change of π occurs during reflection of transverse waves like light from a surface only backed by an optically denser medium. When light is reflected by a glass surface, the phase difference between incident wave and reflected wave is π as glass is optically denser than air. With these points in mind, wave CT is produced from the incident wave AB due to transmissions at the points B and C. Hence, the incident wave AB and the transmitted wave CT are in phase. However, C_1T is produced from the incident wave AB by two reflections at the two points C and B_1 and one transmission at the point C_1.

i.   The waves AB and BC are in the same phase as wave BC is produced from AB due to refraction at point B. The waves BC and CT are in the same phase as wave CT is produced from BC due to refraction at point C. Therefore, we can conclude that the incident wave AB and the refracted wave CT are in the same phase.

ii. The wave CB_1 is produced from the wave BC due to reflection from a surface backed by a denser medium (glass plate). Hence, there is a phase difference of π between the waves BC and CB_1 or between the waves AB and CB_1.

iii. The wave B_1C_1 is produced from the wave CB_1 due to reflection from a surface backed by a denser medium (plano-convex lens). Hence, there is a phase difference of π between the waves B_1C_1 and CB_1.

iv. Combining (ii) and (iii), the phase difference between the waves AB and B_1C_1 will be $\pi \pm \pi = 2\pi$ or $0°$. In other words, the wave AB and the wave B_1C_1 are in the same phase.

v.   The waves B_1C_1 and C_1T are in the same phase as the wave C_1T is produced from the wave B_1C_1 due to refraction at point C_1.

vi.   According to (iv), the waves AB and B_1C_1 are in the same phase and according to (v), the waves B_1C_1 and C_1T are in the same phase. Therefore, we can conclude that the incident wave AB and the transmitted wave C_1T are in the same phase.

Combining (i) and (vi), the transmitted waves CT and C_1T are in the same phase. No phase difference is introduced between the transmitted waves CT and C_1T due to reflections at C and B_1. The optical path difference between the transmitted waves CT and C_1T can be calculated as

$$\Delta = 2e\cos(\alpha + r) \times \mu \tag{2.59}$$

Therefore, the total phase difference δ between the transmitted waves CT and C_1T due to optical path difference and due to reflection is given by

δ = Phase difference due to optical path difference + phase difference due to reflection, i.e.,

$$\delta = \delta' + 0°$$

or $$\delta = \frac{2\pi}{\lambda} \times \Delta \tag{2.60}$$

According to Eq. (2.13), constructive interference occurs if the total phase difference between the interfering waves is an even multiple of π. Hence, bright circular fringes appear at the points for which

$$\delta = 2n\pi$$

or $$\frac{2\pi}{\lambda} \times \Delta = 2n\pi$$

or $$\Delta = 2n\left(\frac{\lambda}{2}\right)$$

or $$2\mu e\cos(\alpha + r) = 2n\left(\frac{\lambda}{2}\right) \quad \text{(Bright rings)} \tag{2.61}$$

According to Eq. (2.15), destructive interference occurs if the total phase difference between the interfering waves is an odd multiple of π. Hence, dark circular fringes appear at the points for which

$$\delta = (2n+1)\pi$$

or $\quad \dfrac{2\pi}{\lambda} \times \Delta = (2n+1)\pi$

or $\quad \Delta = (2n+1)\dfrac{\lambda}{2}$

or $\quad 2\mu e \cos(\alpha + r) = (2n+1)\dfrac{\lambda}{2}$ (Dark rings) $\hfill (2.62)$

Equations (2.61) and (2.62) represent the condition for the formation of bright Newton's rings and dark Newton's rings respectively. In Eqs (2.61) and (2.62), α the angle between the lower surface of the plano-convex lens and the glass plate is exceedingly small. Hence, in comparison to r, we can neglect α. Therefore, Eqs (2.61) and (2.62) become

$$2\mu e \cos r = 2n\left(\dfrac{\lambda}{2}\right) \quad \text{(Bright rings)} \hfill (2.63)$$

$$2\mu e \cos r = (2n+1)\dfrac{\lambda}{2} \quad \text{(Dark rings)} \hfill (2.64)$$

The experimental arrangements can be designed so that the light wave falls normally on the plane surface of the plano-convex lens which makes the angle of incidence. Hence, under these experimental conditions which makes $r = 0$, Eqs (2.63) and (2.64) boil down to

$$2\mu e = 2n\left(\dfrac{\lambda}{2}\right) \quad \text{(Bright rings)} \hfill (2.65)$$

$$2\mu e = (2n+1)\dfrac{\lambda}{2} \quad \text{(Dark rings)} \hfill (2.66)$$

If we join all the points on the glass plate below the plano-convex lens having the same thicknesses, the locus will be a perfect circle. The thickness of the liquid film at the points where the nth order bright fringe appears is obtained from Eq. (2.65) and is given by

$$e = 2n\left(\dfrac{\lambda}{4\mu}\right) \hfill (2.67)$$

Similarly, the thickness of the liquid film at the points where the nth order dark fringe appears is obtained from Eq. (2.66) and is given by

$$e = (2n+1)\dfrac{\lambda}{4\mu} \hfill (2.68)$$

2.13.1 Diameter of the *n*th order Newton's ring

The radius of the *n*th order bright fringe can be calculated looking at Fig. 2.15. In the figure, R is the radius of curvature of the curved surface of the plano-convex lens and r_n is the radius of the *n*th order bright fringe. Since OEB_1 is a right triangle, we have

$$R^2 = r_n^2 + (R-e)^2$$

or $r_n^2 = 2Re - e^2$ (2.69)

Since the radius of curvature of the curved surface of the plano-convex lens is very large, the thickness e will be a very small quantity as a result of which e^2 is negligibly small in comparison to $2Re$. Hence, we get

$$r_n^2 = 2Re$$ (2.70)

or $r_n = \sqrt{2Re}$ (2.71)

2.13.2 Diameter of the *n*th order bright Newton's ring

Putting the value of e from Eq. (2.67) into Eq. (2.70), we shall get the radius of the *n*th order bright Newton's ring as

$$r_n = \sqrt{n\frac{\lambda R}{\mu}}$$

The diameter of the *n*th order bright Newton's ring will be

$$D_n = \sqrt{2n\frac{2\lambda R}{\mu}}$$ (2.72)

Putting $\dfrac{2\lambda R}{\mu} = K^2$ in Eq. (2.72), we have

$$D_n \propto \sqrt{2n}$$

For any particular experimental setup, $\dfrac{2\lambda R}{\mu}$ is constant. Hence, this equation shows that the diameter of the *n*th order bright Newton's ring is directly proportional to the square root of even numbers.

2.13.3 Diameter of the *n*th order dark Newton's ring

Putting the value of e from Eq. (2.68) into Eq. (2.70), we shall get the radius of the *n*th order dark Newton's ring as

$$r_n = \sqrt{(2n+1)\frac{\lambda R}{2\mu}}$$

The diameter of the *n*th order dark Newton's ring will be

$$D_n = 2r_n = \sqrt{4} \times \sqrt{(2n+1)\frac{\lambda R}{2\mu}}$$

or $\quad D_n = \sqrt{(2n+1)\frac{2\lambda R}{\mu}}$ \hfill (2.73)

Putting $\dfrac{2\lambda R}{\mu} = K^2$ in this equation, we have

$$D_n \propto \sqrt{2n+1}$$

For any particular experimental setup, $\dfrac{2\lambda R}{\mu}$ is constant. Hence, this the above equation shows that the diameter of the *n*th order dark Newton's ring is directly proportional to the square root of odd numbers.

2.13.4 Central fringe as seen by the transmitted light

The plano-convex lens touches the glass plate only at the central point. The thickness e of the liquid film at this point is zero. The optical path difference $\Delta = 2e\cos(\alpha + r) \times \mu$ for this point ($e = 0$) according to Eq. (2.45) reduces to zero. Therefore, the total phase difference, from Eq. (2.46), will be

$$\delta = \frac{2\pi}{\lambda} \times 0 = 0 \quad \text{(for central fringe)}$$

According to Eq. (2.13), constructive interference occurs if the total phase difference between the interfering waves is zero. Hence, a bright fringe appears at the central point and the central point appears bright as seen by the transmitted monochromatic light.

2.13.5 Discussions

The law of conservation of energy holds good in Newton's ring experiment. The points at which light energy appears to be absent as seen by the reflected waves, are the same points at which light energy is present as seen by the transmitted waves. The regions at which bright Newton's rings are formed by the transmitted wave are, the same regions at which dark Newton's rings are formed by the reflected wave. The rings observed by the reflected waves are exactly complementary to those observed by the transmitted light.

i.   As discussed earlier, the central fringe appears dark when observed by the reflected light wave, whereas it appears bright when observed by the transmitted light wave.

ii.   From Eq. (2.57), the diameter of the nth order bright fringe as seen by the reflected light wave is

$$D_n = \sqrt{(2n+1)\frac{2\lambda R}{\mu}}.$$

From Eq. (2.73), the diameter of the nth order dark fringe as seen by the transmitted light wave is

$$D_n = \sqrt{(2n+1)\frac{2\lambda R}{\mu}}.$$

They are located at the same position.

iii.   From Eq. (2.58), the diameter of the nth order dark fringe as seen by the reflected light wave is

$$D_n = \sqrt{2n\frac{2\lambda R}{\mu}}.$$

From Eq. (2.72), the diameter of the nth order bright fringe as seen by the transmitted light wave is

$$D_n = \sqrt{2n\frac{2\lambda R}{\mu}}.$$

They are located at the same position.

Example 2.4

A plano-convex lens of radius of curvature 3.5 m is placed on an optically plane glass plate in an air medium and is illuminated by a parallel beam of monochromatic light. The diameter of the 6th bright ring as seen by the reflected light is 0.72 cm. Calculate the wavelength of the light.

Solution

Data given are $R = 3.5$ m $= 350$ cm, $n = 6$, $D_n = 0.72$ cm, $\mu = 1$, for the air medium.
The diameter of the nth bright ring as seen by the reflected light is given by

$$D_n = \sqrt{(2n+1)\frac{2\lambda R}{\mu}}$$

or $\lambda = \dfrac{D_n^2 \mu}{2R(2n+1)}$

Putting the values of the given data into this equation, we get

$$\lambda = \frac{0.72^2 \times 1}{2 \times 350 \times 13} Å = 5697 Å$$

Example 2.5

A plano-convex lens of radius of curvature 185 cm is placed on an optically plane glass plate. The space between the lens and the glass plate is filled with carbon tetrachloride having a refractive index of 1.461. It is illuminated by a parallel beam of monochromatic light. The diameter of the 12th dark ring as seen by the reflected light is 0.65 cm. Calculate the wavelength of the light.

Solution

Data given are $R = 150$ cm, $n = 12$, $D_n = 0.65$ cm, $\mu = 1.461$.
The diameter of the nth dark ring as seen by the reflected light is given by

$$D_n = \sqrt{\frac{4n\lambda R}{\mu}}$$

or $\lambda = \dfrac{D_n^2 \mu}{4nR}$

Putting the values of the given data into this equation, we get

$$\lambda = \frac{0.65^2 \times 1.461}{4 \times 12 \times 185} Å = 6951 Å$$

Example 2.6

A plano-convex lens of radius of curvature 3.5 m is placed on an optically plane glass plate in air medium and is illuminated by a parallel beam of monochromatic light. The diameter of the 6th bright ring as seen by the transmitted light is 0.7 cm. Calculate the wavelength of the light.

Solution

Data given are $R = 3.5$ m $= 350$ cm, $n = 6$, $D_n = 0.7$ cm, $\mu = 1$ for air medium.

The diameter of the nth bright ring as seen by the transmitted light is given by

$$D_n = \sqrt{\frac{4n\lambda R}{\mu}}$$

or $\lambda = \dfrac{D_n^2 \mu}{4nR}$

Putting the values of the given data into this equation, we get

$$\lambda = \frac{0.7^2 \times 1}{4 \times 6 \times 350} \text{Å} = 5833\text{Å}$$

Example 2.7

A plano-convex lens of radius of curvature 95 m is placed on an optically plane glass plate. The space between the lens and the glass plate is filled with water having a refractive index of 1.333. It is illuminated by a parallel beam of monochromatic light. The diameter of the 15th dark ring as seen by the transmitted light is 0.45 cm. Calculate the wavelength of the light.

Solution

Data given are $R = 95$ cm, $n = 15$, $D_n = 0.45$ cm, $\mu = 1.333$.

The diameter of the nth dark ring as seen by the transmitted light is given by

$$D_n = \sqrt{\frac{2(2n+1)\lambda R}{\mu}}$$

or $\lambda = \dfrac{D_n^2 \mu}{2(2n+1)R}$

Putting the values of the given data into this equation, we get

$$\lambda = \frac{0.45^2 \times 1.333}{2 \times 31 \times 95} \text{Å} = 4583\text{Å}$$

Example 2.8

A plano-convex lens of radius of curvature 3 m is placed on an optically plane glass plate in an air medium and is illuminated by a parallel beam of monochromatic light of wavelength 5893 Å. Calculate the change in diameter of the 5th bright ring when the air medium is replaced by water of refractive index 1.33 and is observed by the reflected light.

Solution

Data given are $R = 3$ m $= 300$ cm, $n = 5$, $\lambda = 5893$ Å.

The diameter of the nth bright ring as seen by the reflected light in a medium of refractive index μ is given by

$$D_n = \sqrt{\frac{(2n+1)2\lambda R}{\mu}}$$

Hence, the diameter of the nth bright ring as seen by the reflected light in the air medium is given by

$$D_{n-\mathrm{air}} = \sqrt{(2n+1)2\lambda R}$$

The change in diameter of the nth ring will be

$$D_{n-\mathrm{air}} - D_n = \sqrt{(2n+1)2\lambda R} - \frac{\sqrt{(2n+1)2\lambda R}}{\sqrt{\mu}}$$

or $\quad D_{n-\mathrm{air}} - D_n = \sqrt{(2n+1)2\lambda R} \times \left(1 - \frac{1}{\sqrt{\mu}}\right)$

Putting the values of n, λ, R and μ into this equation, we get

$$D_{n-air} - D_n = \sqrt{(11 \times 2 \times 5893 \times 10^{-8} \times 300} \times \left(1 - \frac{1}{\sqrt{1.33}}\right) \mathrm{cm} = 0.083 \mathrm{cm}$$

Example 2.9

A plano-convex lens of radius of curvature 280 cm is placed on an optically plane glass plate in an air medium and is illuminated by a parallel beam of monochromatic light of wavelength 5893 Å. Calculate the change in diameter of the 7th bright ring when the air medium is replaced by water of refractive index 1.33 and is observed by the transmitted light.

Solution

Data given are $R = 280$ cm, $n = 7$, $\lambda = 5893$ Å $= 5893 \times 10^{-8}$ cm.

The diameter of the nth bright ring as seen by the transmitted light in a medium of refractive index μ is given by

$$D_n = \sqrt{\frac{4n\lambda R}{\mu}}$$

Hence, the diameter of the nth bright ring as seen by the transmitted light in the air medium is given by

$$D_{n-air} = \sqrt{4n\lambda R}$$

The change in diameter of the nth ring will be

$$D_{n-air} - D_n = \sqrt{4n\lambda R} - \frac{\sqrt{4n\lambda R}}{\sqrt{\mu}}$$

or $\quad D_{n-air} - D_n = \sqrt{4n\lambda R} \times \left(1 - \frac{1}{\sqrt{\mu}}\right)$

Putting the values of n, λ, R and μ into this equation, we get

$$D_{n-air} - D_n = \sqrt{(4 \times 7 \times 5893 \times 10^{-8} \times 280}) \times \left(1 - \frac{1}{\sqrt{1.33}}\right) \text{cm} = 0.090 \text{ cm}$$

Example 2.10

A plano-convex lens of radius of curvature 3.5 m is placed on an optically plane glass plate in an air medium and is illuminated by a parallel beam of monochromatic light. The diameter of the 6th dark ring as seen by the reflected light is 0.72 cm. Calculate the wavelength of the light.

Solution

Data given $R = 3.5$ m $= 350$ cm, $n = 6$, $D_n = 0.72$ cm.

The diameter of the nth dark ring as seen by the reflected light is given by

$$D_n = \sqrt{\frac{4n\lambda R}{\mu}}$$

or $\lambda = \dfrac{D_n^2 \mu}{4nR}$

Putting the values of the given data into this equation, we get

$$\lambda = \frac{0.72^2 \times 1}{4 \times 350 \times 6}\text{cm} = 6171\text{Å}$$

Example 2.11

In a Newton's ring experiment in air, the diameter of the 10th bright ring was 0.272 cm and that of the 15th bright ring was 0.555 cm. If the radius of curvature of the plano-convex lens is 200 cm, calculate the wavelength of the monochromatic light used.

Solution

The data given are D_n = 0.272 cm, D_{n+p} = 0.555 cm, n = 10, $n + p$ = 15, p = 5, R = 200 cm. We know that

$$\lambda = \frac{D_{n+p}^2 - D_n^2}{4pR}$$

Putting the given values into this equation, we get

$$\lambda = \frac{0.555^2 - 0.272^2}{4 \times 5 \times 200}\text{cm} = 5851\text{Å}$$

Example 2.12

In a Newton's ring experiment in air, the diameter of the 10th bright ring was 0.272 cm and that of the 15th bright ring was 0.555 cm. If the wavelength of the monochromatic light used is 5893 Å, calculate the radius of curvature of plano-convex lens used.

Solution

The data given are D_n = 0.272 cm, D_{n+p} = 0.555 cm, n = 10, $n + p$ = 15, p = 5,

$\lambda = 5893\text{Å} = 5893 \times 10^{-8}\text{cm}.$

We know that

$$\lambda = \frac{D_{n+p}^2 - D_n^2}{4pR}$$

or $R = \dfrac{D_{n+p}^2 - D_n^2}{4p\lambda}$

Putting the given values into this equation, we get

$$R = \frac{0.555^2 - 0.272^2}{4 \times 5 \times 5893 \times 10^{-8}} \text{ cm} = 198.6 \text{ cm}$$

Example 2.13

In a Newton's ring experiment when a certain liquid is inserted in between the plano-convex and the glass plate, the radius of the 8th dark ring is measured to be 6 cm. If the wavelength of the monochromatic light used is 5893 Å and the radius of curvature of the plano-convex lens is 254 cm, calculate the refractive index of the liquid inserted.

Solution

The data given are $D_n = 0.6$ cm, $n = 8$, $\lambda = 5893$ Å $= 5893 \times 10^{-8}$ cm, $R = 254$ cm
The diameter of the nth dark ring as seen by the reflected light is given by

$$D_n = \sqrt{\frac{4n\lambda R}{\mu}}$$

or $\mu = \dfrac{4n\lambda R}{D_n^2}$

Putting the values of n, λ, R and D_n into this equation, we get

$$\mu = \frac{4 \times 8 \times 5893 \times 10^{-8} \times 254}{0.6^2} = 1.33$$

Example 2.14

In Newton's ring experiment, light containing two wavelengths λ_1 and λ_2 are incident normally on a plano-convex lens of radius of curvature R placed in an air medium. If the nth dark ring due to λ_1 coincides with the $(n + 1)$th dark ring due to λ_2, prove that the radius of the nth dark ring due to λ_1 is $\sqrt{\dfrac{\lambda_1 \lambda_2 R}{\lambda_1 - \lambda_2}}$.

Solution

The radius of the nth dark ring due to λ_1 is

$$\sqrt{Rn\lambda_1},\tag{A}$$

The radius of the $(n + 1)$th dark ring due to λ_2 is

$$\sqrt{R(n+1)\lambda_2},\tag{B}$$

According to the problem statement, the nth dark ring due to λ_1 coincides with the $(n + 1)$th dark ring due to λ_2, i.e., the two radii are the same. Hence, we have

$$\sqrt{Rn\lambda_1} = \sqrt{R(n+1)\lambda_2}$$

or $Rn\lambda_1 = R(n+1)\lambda_2$

or $n = \dfrac{\lambda_2}{\lambda_1 - \lambda_2}$ $$\tag{C}$$

Putting this value of n into Eq. (A), we get the radius of the nth dark ring due to λ_1 as

$$\sqrt{\frac{\lambda_1\lambda_2 R}{\lambda_1 - \lambda_2}}$$

Example 2.15

The light coming from a sodium vapor lamp generally used in Newton's ring experiment in the laboratory is not perfectly monochromatic it contains two wavelengths namely 5890 Å and 5896 Å. It is found that the nth dark ring due to 5896 Å coincides with the $(n + 1)$th dark ring due to 5890 Å. Calculate n. Also calculate the diameter of the nth dark ring due to 5896 Å if the radius of curvature of the plano-convex lens is 100 cm. The experiment is in general performed in an air medium.

Solution

The radius of the nth dark ring due to 5896 Å is

$$\sqrt{Rn5896\times10^{-8}} \text{ cm},\tag{A}$$

The radius of the $(n + 1)$th dark ring due to 5890 Å is

$$\sqrt{R(n+1)5890\times10^{-8}},$$

According to the question, the nth dark ring due to 5896 Å coincides with the $(n + 1)$th dark ring due to 5890 Å, i.e., the two radii are the same. Hence, we have

$$\sqrt{Rn5896\times10^{-8}}\ \text{cm} = \sqrt{R(n+1)5890\times10^{-8}}$$

or $6n = 5890$

or $n = 982$

Putting this value of n into Eq. (A), we get the radius of the nth dark ring due to 5896 Å as

$$\sqrt{100\times982\times5896\times10^{-8}}\ \text{cm} = 2.41\,\text{cm}$$

Hence, the diameter of the nth dark ring due to 5896 Å will be $= 2 \times 2.41$ cm $= 4.82$ cm. Obviously, this ring lies outside the field of view of the travelling microscope!

Example 2.16

In Newton's ring experiment in a laboratory, a source of light having two wavelengths 5760 Å and 4800 Å is used. It is found that the nth dark ring due to 5760 Å coincides with the $(n+2)$th dark ring due to 4800 Å. Calculate n. Also calculate the radii of the nth dark rings due to 5760 Å and 4800 Å if radius of curvature of the plano-convex lens is 90 cm.

Solution

The radius of nth dark ring due to 5760 Å is

$$\sqrt{Rn5760\times10^{-8}}\ \text{cm}, \quad \mu \text{ is assumed to be 1} \qquad\qquad\text{(A)}$$

The radius of the $(n + 2)$nd dark ring due to 4800 Å is

$$\sqrt{R(n+2)4800\times10^{-8}}\ \text{cm}, \quad \mu \text{ is assumed to be 1}$$

According to the question, the nth dark ring due to 5760 Å coincides with the $(n + 2)$nd dark ring due to 4800 Å i.e., the two radii are same. Hence, we have

$$\sqrt{Rn5760 \times 10^{-8}}\ \text{cm} = \sqrt{R(n+2)4800 \times 10^{-8}}$$

or $96n = 960$

or $n = 10$

Putting this value of n into Eq. (A), we get the radius of the nth dark ring due to 5760 Å as

$$\sqrt{90 \times 10 \times 5760 \times 10^{-8}}\ \text{cm} = 0.2277\,\text{cm}$$

Similarly, the radius of the nth dark ring due to 4800 Å is obtained as

$$\sqrt{90 \times 10 \times 4800 \times 10^{-8}}\ \text{cm} = 0.2078\,\text{cm}$$

Example 2.17

In the Newton's ring experiment by a light of wavelength 5890 Å, the diameter of the 10th dark ring is 0.532 cm as seen by the reflected light. Calculate the radius of curvature of the plano-convex lens and the thickness of the film at this point.

Solution

The data given are $n = 10$, $D_{10} = 0.532$ cm, $\lambda = 5890$ Å $= 5890 \times 10^{-8}$ cm, $\mu = 1$ since it is an air film.

The square of the diameter of the nth order dark fringe as seen by reflected light in the air medium is

$$D_n^2 = 4n\lambda R$$

or $R = \dfrac{D_n^2}{4n\lambda}$

Putting the data given into this equation, we get

$$R = \frac{0.532^2}{4 \times 10 \times 5890 \times 10^{-8}}\ \text{cm} = 120.13\,\text{cm}$$

In Newton's ring experiment, the thickness of the thin film e at the nth order dark fringe is as seen by the reflected beam obtained from the formula $2\mu e = n\lambda$ and is given by

$$e = \frac{n\lambda}{2\mu}$$

Putting the data given into this equation, we get

$$e = \frac{10 \times 5890 \times 10^{-8}}{2 \times 1} \text{ cm} = 2.945 \times 10^{-4} \text{ cm} = 2.945 \times 10^{-4} \text{ cm}$$

Example 2.18

In Newton's ring experiment, by a light of wavelength 7000 Å, the diameter of the 15th dark ring is 0.565 cm when a thin liquid film is formed between the lens and the glass plate. If the radius of curvature of the plano-convex lens is 102.9 cm, calculate the refractive index of the liquid.

Solution

The data given are $n = 15$, $D_{15} = 0.565$ cm, $\lambda = 7000$ Å $= 7000 \times 10^{-8}$ cm.

The square of the diameter of the nth order dark fringe as seen by reflected light in the air medium is

$$D_n^2 = \frac{4n\lambda R}{\mu}$$

or $\mu = \frac{4n\lambda R}{D_n^2}$

Putting the data given into this equation, we get

$$\mu = \frac{4 \times 15 \times 7000 \times 10^{-8} \times 102.9}{0.565^2} = 1.354$$

2.14 Determination of Wavelength of Light using Newton's Ring

The experimental arrangement to determine the wavelength of monochromatic light by using Newton's ring is shown in Fig. 2.11. S is the source of the monochromatic light of which wavelength is to be determined. Generally, a sodium light source is taken. The source of monochromatic light is placed at the focus of a convex lens so that after refraction through the lens, the refracted beam becomes parallel. This parallel beam of light is allowed to fall on a glass plate G placed at 45° with the incident beam. The glass plate G so placed will reflect a part of the incident beam of light vertically downward towards the plane surface of the plano convex lens placed on another glass plate as shown in Fig. 2.11. A thin air film ($\mu = 1$) is enclosed between the lower surface of the plano convex lens and the glass plate. A travelling microscope is placed over the glass plate G and when it is properly focused on the air film, alternate bright and dark concentric circular interference fringes are seen when viewed through the eye-piece of the microscope. These concentric circular interference fringes are called Newton's rings. The diameter of any Newton's rings can be measured with help of the travelling microscope. The cross-wire in the eye-piece of the travelling microscope is placed in contact with a ring; the microscope is then moved horizontally to the exact opposite end of the ring and the diameter of the ring is found out. In this manner, diameters of approximately ten number of rings are found out.

2.14.1 Theory for the experiment

From Eq. (2.57), square of the diameter of the nth number of bright rings is given by

$$D_n^2 = (2n+1)\frac{2\lambda R}{\mu}$$

The refractive index μ of thin air film is nearly 1. Hence, the above equation becomes

$$D_n^2 = (2n+1)2\lambda R \tag{2.74}$$

Similarly, the square of the diameter of the $(n + p)$th number bright ring will be given by

$$D_{n+p}^2 = \left[2(n+p)+1\right]2\lambda R \tag{2.75}$$

Subtracting Eq. (2.74) from Eq. (2.75), we get

$$D_{n+p}^2 - D_n^2 = \left[2(n+p)+1\right]2\lambda R - (2n+1)2\lambda R$$

or $\quad D_{n+p}^2 = 4\lambda Rp + D_n^2 \tag{2.76}$

This equation is in the form of

$y = mx + c$ with $y = D_{n+p}^2$, $x = p$, $c = y$-intercept $= D_n^2$, and $m = $ slope$= 4\lambda R$

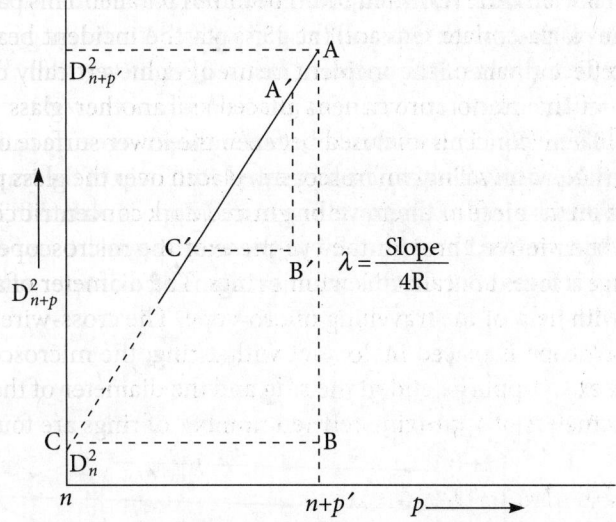

$$\lambda = \frac{\text{Slope}}{4R}$$

Figure 2.17 | The slope of this straight line is put in the working formula (2.77) to find out the wavelength of the monochromatic light

If we plot D_{n+p}^2 along the Y-axis and p along the X-axis with $p = 1, 2, 3, \ldots, 10$, we shall get a straight line as shown in Fig. 2.17. By determining the slope m of this straight line from the plot, we can get the wavelength λ by using the formula

$$\lambda = \frac{\text{slope}}{4R} \tag{2.77}$$

The radius of curvature of the curved surface R of the plano-convex lens can be found out accurately either by using a spherometer or by Boy's method. Knowing the value of R, by using this formula, we can easily calculate the wavelength of the monochromatic light.

2.15 Determination of Refractive Index of Liquids using Newton's Rings

The experimental arrangement for determining the wavelength of monochromatic light using Newton's ring is shown in Fig. 2.18. S is the source of the monochromatic light. Generally, a sodium light source is taken. The source of monochromatic light is placed at the focus of a convex lens so that after refraction through the lens the refracted beam becomes parallel. This parallel beam of light is allowed to fall on a glass plate G placed at 45° with the incident beam. The glass plate G so placed will reflect a part of the incident beam of light vertically downward towards the plane surface of the plano-convex lens placed on another glass plate as shown in Fig. 2.18. The given transparent liquid whose refractive index is to be found out is poured into a shallow container containing the plano-convex lens and the glass plate system. A thin liquid film of refractive index μ is filled between the lower surface of the plano-convex lens and the glass plate. A travelling microscope is placed over the glass plate G and when it is properly focused on the air film, alternate bright and dark concentric circular interference fringes are seen when viewed through the eye-piece of the microscope. These concentric circular interference fringes are called Newton's rings. The diameter of any Newton's ring can be measured with help of the travelling microscope. The cross-wire in the eye-piece of the travelling microscope is placed in contact with a ring; the microscope is then moved horizontally to the exact opposite end of the ring and the diameter of the ring is found out. In this manner, diameters of approximately ten number of rings are found out.

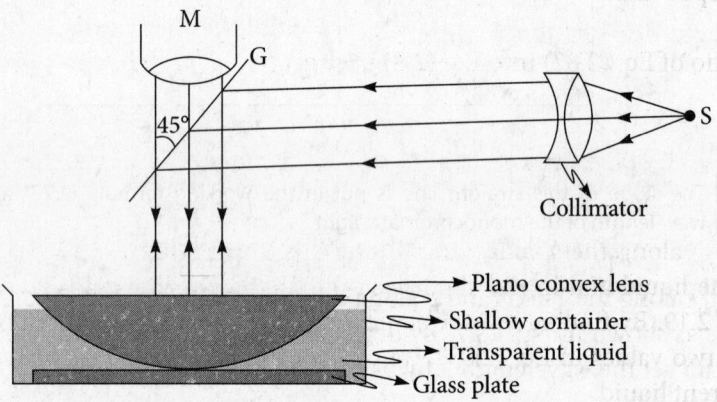

Figure 2.18 Experimental arrangement to measure the refractive index of a transparent liquid by using Newton's ring

2.15.1 Theory for the experiment

From Eq. (2.57), the square of the diameter of the nth number of bright ring is given by

$$D_n^2 = (2n+1)\frac{2\lambda R}{\mu} \tag{2.78}$$

Similarly, the square of the diameter of $(n + p)$th number of bright ring will be given by

$$D_{n+p}^2 = [2(n+p)+1]\frac{2\lambda R}{\mu} \tag{2.79}$$

Subtracting Eq. (2.78) from Eq. (2.79), we get

$$D_{n+p}^2 - D_n^2 = [2(n+p)+1]\frac{2\lambda R}{\mu} - (2n+1)\frac{2\lambda R}{\mu}$$

or $\quad D_{n+p}^2 = \dfrac{4\lambda Rp}{\mu} + D_n^2$ (2.80)

This equation is in the form of

$y = mx + c$ with $\; y = D_{n+p}^2, \; x = p, c = y\text{-intercept} = D_n^2, \;$ and $\; m_L = \text{slope} = \dfrac{4\lambda R}{\mu}$ (2.81)

In the absence of the thin liquid film (i.e., in the presence of the thin air film), the slope will be

$m_{\text{Air}} = \text{slope} = 4\lambda R$ (2.82)

Taking the ratio of Eq. (2.82) into Eq. (2.81), we get

$\mu = \dfrac{m_{\text{Air}}}{m_L}$ (2.83)

If we plot D_{n+p}^2 along the Y-axis and p along the X-axis with $p = 1, 2, 3, \ldots, 10$ in the presence of the liquid and in the absence of the liquid, we shall get two straight lines as shown in Fig. 2.19. By finding out the slopes of these two straight lines from the plot and putting these two values in Eq. (2.83), we can get the value of the refractive index of the given transparent liquid.

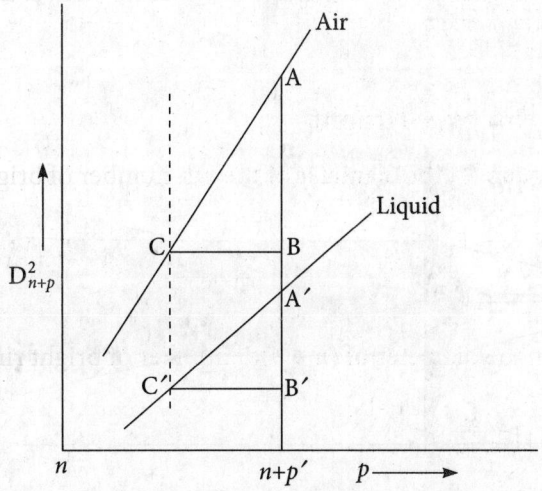

Figure 2.19 The slopes of these two straight lines are put in the working formula (Eq. 2.83) to find out the refractive index of the given liquid. $m_{\text{Air}} = \dfrac{AB}{CB}$ and $m_L = \dfrac{A'B'}{C'B'}$. Since as shown in the figure, $CB = C'B'$, we have $\mu = \dfrac{m_{\text{Air}}}{m_L} = \dfrac{AB}{A'B'}$

2.16 Fresnel's Biprism

Fresnel's biprism, as shown in Fig. 2.20(a), is essentially made of two prism, each of a very small refracting angle α placed base to base. Practically, a biprism is constructed from a single glass piece by grinding and polishing it so that its obtuse angle is slightly less than $180°$.

2.16.1 Determination of wavelength of light using a biprism

The experimental setup is shown schematically in Fig. 2.20(b). In this figure, slit S is perpendicular to the plane of the page. The obtuse angle of the biprism faces the monochromatic source of light S, i.e., the slit so that two virtual coherent sources S_1 and S_2 are created on the plane of the slit at equal distance from it due to refraction. This will be accomplished when the edge of the biprism at the obtuse angle is exactly parallel to the slit and plane surface of the biprism and the plane containing the slit are exactly parallel to each other.

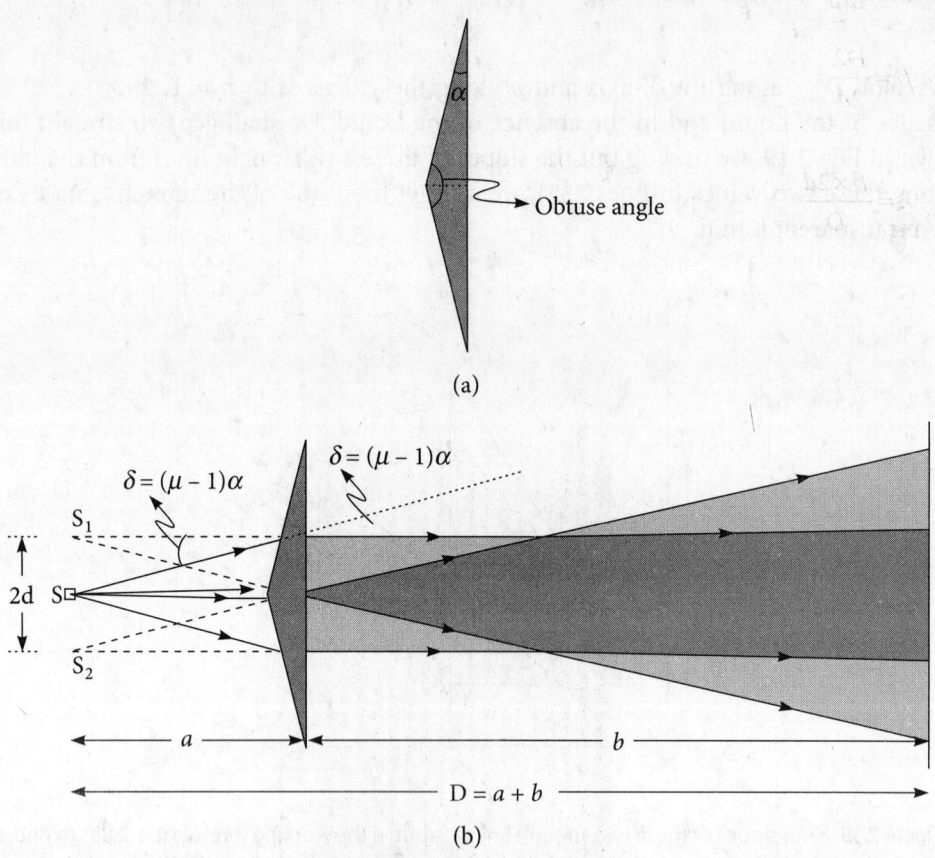

Figure 2.20 (a) Fresnel's biprism. (b) Ray diagram for determination of wavelength of light using a biprism

The wavefront which is incident on the plane surface of the biprism is divided into two parts as explained here. According to the laws of refraction, the portion of the wavefront incident on the upper part and the lower part of the biprism are bent towards the thicker part, i.e., towards the obtuse angle edge of the biprism after refraction. The two refracted wavefronts are thus derived from a single monochromatic source satisfying the fundamental condition of interference. As the two refracted wavefronts passes through common space at the right side, they produce interference patterns of equally spaced non-localized fringes. These fringes can be captured on a screen placed in the overlapping region of the two refracted wavefronts. Moreover, interference fringes can be seen through a powerful eye-piece it its focal plane. When they are extended towards the left side, they meet at two points on the plane of the slit producing two virtual sources S_1 and S_2 of equal brightness, i.e., equal amplitudes. Now the two refracted wavefronts which interfere to produce interference fringes appear to come from these two virtual coherent sources. The fringe pattern thus produced is illustrated in Fig. 2.21.

If $2d$ is the distance between S_1 and S_2 with $SS_1 = SS_2 = d$, Eqs (2.21) and (2.23) show that the spacings β of bright fringe and dark fringe are equal and are given by

$$\beta = \frac{D\lambda}{2d}$$

or $\quad \lambda = \dfrac{\beta \times 2d}{D}$ $\hfill (2.84)$

Figure 2.21 | Fringe pattern as obtained in Fresnel's biprism experiment

To determine of wavelengths of monochromatic light λ using this formula, we need to measure the bright fringe spacing β, the distance between the two virtual sources $2d$ and the distance between slit and eye-piece D.

2.16.1A Experimental setup

A heavy metallic optical bench is utilized to perform this experiment. A scale is attached to one of the two railings. It is provided with four uprights to support an adjustable slit, a biprism, a high accuracy micrometer, Ramsden's eye-piece and a convergent lens. These uprights are attached with vernier scales, their heights can be changed according to our requirement and are capable of moving freely over the rails. There are also arrangements in the uprights so that the slit and the biprism can be rotated in their own planes. These have been illustrated schematically in Fig. 2.22.

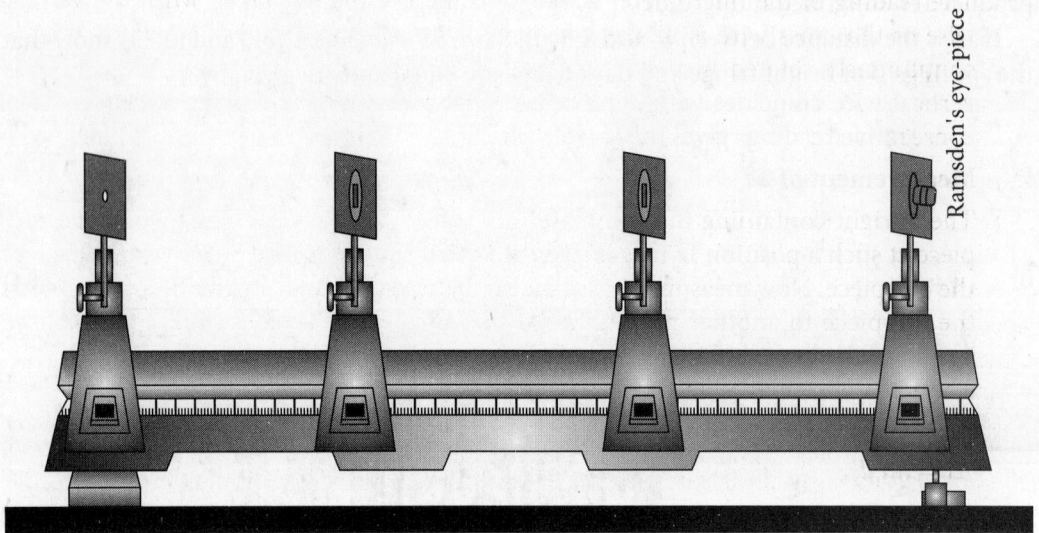

Ramsden's eye-piece

Figure 2.22 | Schematic diagram of an optical bench

2.16.1B Adjustments

i. The optical bench is made perfect horizontal with the help of a spirit level and levelling screws.

ii. The cross wires in the eye-piece are made perfectly vertical/horizontal.

iii. The slit and eye-piece are kept at the same height on two uprights. The slit is made exactly vertical by making its bright image coincide with the vertical wire of the eye-piece.

iv. The biprism is placed on another upright and is kept between the eye-piece and the slit at the same height. The edge at the obtuse angle of the biprism and slit length are made exactly parallel. This is accomplished when two equally bright images of the slit are visible through the eye-piece.

v. The line joining the centers of the slit, the edge at the obtuse angle of the biprism and cross wires and the optical bench should be perfectly parallel.

Sharp fringes are visible through the eye-piece when suitable separations are maintained among the four uprights.

2.16.1C Measurements

i. **Measurement of D**

The difference in the positions of uprights containing the slit and the eye-piece gives the value of D.

ii. **Measurement of β**

The reading of the micrometer attached to the eye-piece is taken when the vertical wire of the cross wire is made to coincide with a bright fringe. The micrometer is continuously rotated in the same direction and readings are taken each time the vertical wire coincides with consecutive bright fringes. The difference between two consecutive readings gives the fringe spacing β.

iii. **Measurement of 2d**

The upright containing the convex lens is placed between the biprism and the eye-piece at such a position L_1 that images of virtual sources S_1 and S_2 are visible through the eye-piece. Now measure the distance d_1 between S_1 and S_2. Move the lens towards the eye-piece to another position L_2 so that images of virtual sources S_1 and S_2 are again visible through the eye-piece. Again, measure the distance d_2 between S_1 and S_2. The actual distance $2d$ between the virtual sources S_1 and S_2 will be given by

$$2d = \sqrt{d_1 d_2}$$

There is another method of finding $2d$ in terms of the refracting angle α. As we know, the refractive index μ of the prism material in terms of the minimum angle of deviation δ is given by

$$\mu = \frac{\sin[(\alpha + \delta)/2]}{\sin(\alpha/2)}$$

For a small refracting angle α, this equation gives

$$\mu \approx 1 + \delta/\alpha$$

or $\delta \approx (\mu - 1)\alpha$

Here δ and α have been depicted in Fig. 2.20. δ and α are to be measured in radians only. Applying simple geometry in Fig. 2.20, we have

$$\frac{SS_1}{a} = \delta = (\mu - 1)\alpha$$

or $2d = 2a(\mu - 1)\alpha$ (2.85)

Putting the values of $D(= a + b)$, β, and $2d$ into $\lambda = \dfrac{\beta \times 2d}{D}$, the wavelength of monochromatic light can be obtained.

Example 2.19

A biprism is placed 5 cm from a slit illuminated by sodium light of wavelength 5890 Å Calculate the fringe spacing if the distance between the biprism and the screen is 75 cm and the two virtual sources are separated by 0.05 cm

Solution

Data given are $\lambda = 5890$ Å $= 5890 \times 10^{-8}$ cm, $a = 5$ cm, $b = 75$ cm, $2d = 0.05$ cm. The fringe spacing is calculated as

$$\beta = \frac{(a+b)\lambda}{2d} = \frac{(5 \text{ cm} + 75 \text{ cm}) \times 5890 \times 10^{-8} \text{ cm}}{0.05 \text{ cm}} = 9.424 \times 10^{-2} \text{ cm}$$

Example 2.20

A biprism of refractive index 1.5 and refracting angle 1° is placed 10 cm from a monochromatic light source of wavelength 5890 Å. Find the fringe spacing observed at a distance of 110 cm from the biprism.

Solution

Data given are $\lambda = 5890$ Å $= 5890 \times 10^{-8}$ cm, $a = 10$ cm, $b = 110$ cm,

$$\alpha = 1° = \frac{\pi}{180} \text{ rad} = \frac{22/7}{180} \text{ rad} \text{ and } \mu = 1.5.$$

The distance between the two virtual sources is calculated as

$$2d = 2a(\mu - 1)\alpha = 2 \times 10 \text{ cm} (1.5 - 1) \times \frac{22/7}{180} = 0.175 \text{ cm}$$

The fringe spacing is calculated as

$$\beta = \frac{(a+b)\lambda}{2d} = \frac{(10\text{cm} + 110\text{cm}) \times 5890 \times 10^{-8}\,\text{cm}}{0.175\text{cm}} = 4.04 \times 10^{-2}\,\text{cm}$$

2.17 Interferometers

Optical instruments that produce optical interference are called interferometers. Standardization of meter scale, determination of wavelengths of light, and thickness of thin films can be determined accurately by interferometers devised by utilizing the concepts of interference of light. The interferometer was first envisioned and constructed by Albert Abraham Michelson around 1880.

2.17.1 Michelson interferometer

In the Michelson interferometer, different types of interference patterns are produced due to the splitting and reunion of two coherent light beams by partial reflection and refraction. Michelson himself and Morley used this interferometer to prove the absence of hypothetical ethereal medium and paved the way for Einstein and his special theory of relativity. In this device, both equal thickness and equal inclination interference fringes can be produced.

Principle

Different types of interference patterns are produced in the instrument due to the superposition of two coherent light beams produced by the division of amplitude method. The two beams thus produced, after travelling different optical paths are superposed to give interference patterns. The basic principle is depicted in Fig. 2.23.

Construction

In a Michelson interferometer, division of the amplitude of incident light is done by means of arrangements of mirrors and glass plates in such a manner that each part is made to travel different paths and brought back together where they interfere according to their path difference. The essential parts of a Michelson interferometer are shown schematically in Fig. 2.23. These are

i.   Two highly polished and optically plane mirrors M_1 and M_2
ii.   Two glass plates G_1 and G_2 with thickness exactly equal and having the same refractive index.
iii. Convex lens L.
iv.   Supports and screws for mirrors and glass plates.

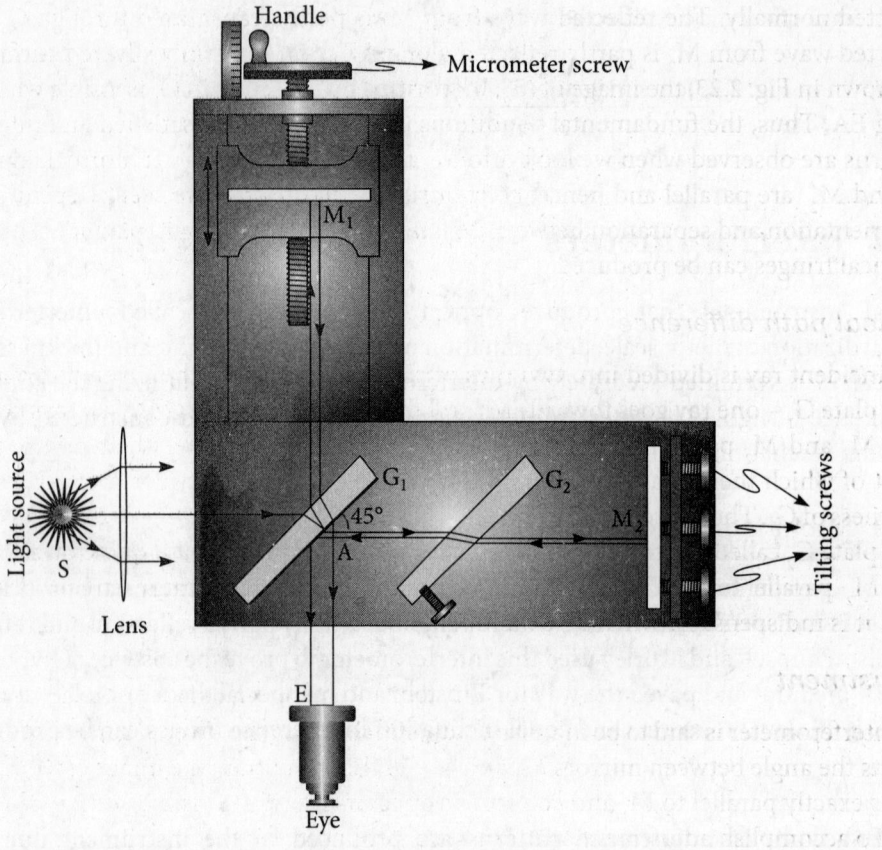

Figure 2.23 | Schematic diagram of a Michelson interferometer

The two glass plates G_1 and G_2 are mounted vertically and exactly parallel to each other on a frame at 45° angle to the interferometer arm as shown. The plate G_1 is so silvered that the light ray coming from the source S is divided by it into reflected and transmitted beams of equal intensity. Mirror M_2 is unmovable whereas mirror M_1 is mounted on a carriage capable of moving along well-machined tracks with the help of a very fine micrometer screw in the double arrow direction. This micrometer screw with very a small least count is calibrated to show the exact distance the mirror M_1 has been moved. The mirror M_2 is capable of slight tilting about the vertical as well as horizontal axis with the help of three screws attached to its back side and thus, mirrors M_1 and M_2 can be made exactly perpendicular to each other so that fringes can be obtained. The interferometer is said to be in normal adjustment when mirrors M_1 and M_2 are exactly perpendicular to each other.

Action of the apparatus

Monochromatic light from an extended source rendered parallel by a lens is divided into two parts of equal intensity by partial reflection from the rear side of G_1. The reflected wave and the transmitted wave proceeds towards mirrors M_1 and M_2 respectively and get reflected normally. The reflected wave from M_1 is partly transmitted through G_1 and the reflected wave from M_2 is partly reflected along AE from the thinly silvered surface of G_1. As shown in Fig. 2.23, the image of M_2, M_2' formed by reflection in G_1 is visible when we see along EA. Thus, the fundamental conditions of interference are satisfied and interference patterns are observed when we look into M_1 along EA through G_1. In normal adjustment, M_1 and M_2' are parallel and hence, concentric circular fringes are seen. Depending upon the orientation and separation between M_1 and M_2, circular, straight, parabolic, hyperbolic elliptical fringes can be produced.

Optical path difference

The incident ray is divided into two rays of equal intensities by the silvered surface of the glass plate G_1 – one ray goes towards M_1 and the other ray towards M_2. The rays concerned with M_1 and M_2 pass through M_2 twice and nil respectively after their origination as a result of which there is an optical path difference of $2(\mu-1)t$ between them. Here t is the thickness of G_1. Therefore, to compensate this path difference of $2(\mu-1)t$, another identical glass plate G_2 called the compensating plate is introduced in the path of the rays concerned with M_2 parallel to G_1. Though it is not essential for producing fringes in monochromatic light, it is indispensable when white light is used.

Adjustment

An interferometer is said to be in normal adjustment when the silvered surface of G_1 exactly bisects the angle between mirrors M_1 and M_2. In this condition, the image of M_2 in M_1, i.e., M_2' is exactly parallel to M_1 and concentric circular fringes are visible when we look along EA. To accomplish adjustment, the following steps are followed.

i.   An extended source is suitable for Michelson interferometer adjustments. If a point source is placed at the focal length of a convex lens, it will behave like an extended source. The total field of view is filled with light when we look into M_1 through G_1.

ii.   Position M_1 and M_2 so that their distances from G_1 are nearly equal, not exactly. If exactly equal, then the field of view would be completely dark.

iii.   Place a pin between the lens and G_1. Two pairs of images of the pin are visible when we look into M_1 along EA – one pair is produced due to reflection from the unsilvered surface and the other pair is produced due to reflection from the silvered surface of G_1.

iv.   Adjust the tilting screws attached to the back side of M_2 so that one pair of images coincides exactly on the other pair and interference fringes appear. When they first appear, the fringes may not be clear unless the eye is focused on or near M_1, so that the observer should look constantly at this mirror while searching for fringes. When they have been found, the adjusting screws should be turned in such a way as to continually increase the width of the fringes and finally, a set of concentric circular fringes will be obtained. M_1 and M_2 are now perpendicular to each other if the angle between M_1 and G_1 is exactly 45°!

Forms of fringes

In a Michelson interferometer, different types of fringes such as straight, circular, elliptical, parabolic, hyperbolic, etc., can be obtained. The shape of the fringe depends upon the angle φ and the separation e between M_1 and M_2'.

i.   **Circular fringes**

These are produced with monochromatic light when the interferometer is in normal adjustments and are used in many kinds of measurements with interferometers. Their origin may be explained with reference to Fig. 2.24.

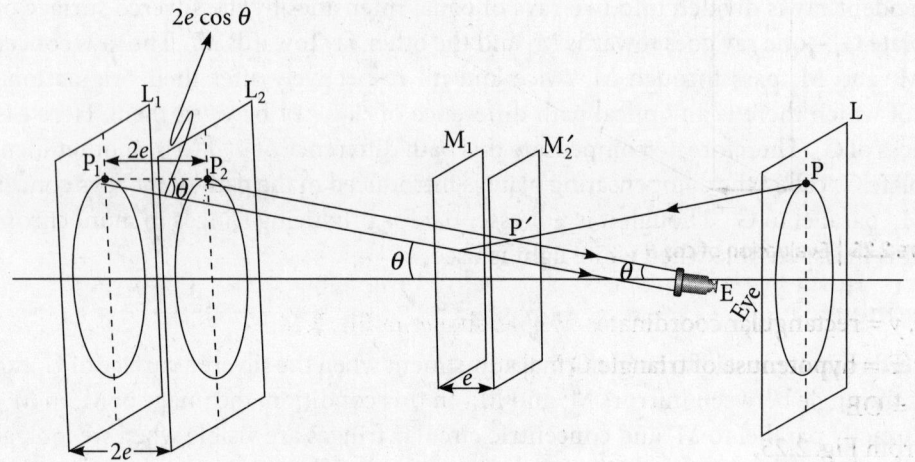

Figure 2.24 | Explanation of the origin of circular fringes

Due to several reflections in the real interferometer, real source L can be thought of as being behind the observer. L_1 and L_2 are its images in M_1 and M_2 respectively. If e is the separation between M_1 and M_2', then $2e$ will be the separation between L_1 and L_2. If P is a point on the real source L, then P_1 and P_2 are its images on L_1 and L_2 respectively. All the light waves originating from P, P_1, and P_2 are in phase all the time and interfere constructively or destructively depending upon their path difference. As shown in Fig. 2.24, $2e\cos\theta$ is the path difference Δ between the waves reaching the eye from P_1 and P_2, i.e.,

$$\Delta = 2e\cos\theta = 2n(\lambda/2) \quad \text{(Bright circular fringe)} \tag{2.86}$$

$$= (2n+1)(\lambda/2) \quad \text{(Dark circular fringe)} \tag{2.87}$$

These equations shows that $e \geq \dfrac{\lambda}{2}$ for the fringe pattern to appear.

Let   O = foot of the perpendicular from eye E on M_2'

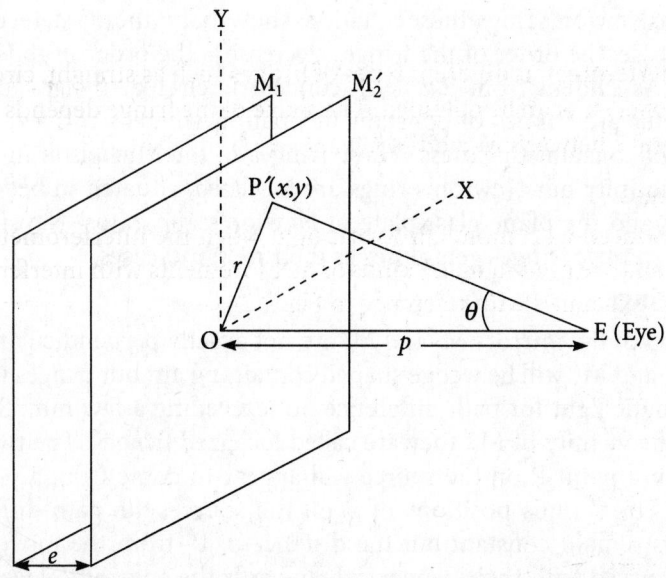

Figure 2.25 | Evaluation of cos θ

x, y = rectangular coordinates of P′ as shown in Fig. 2.24

P′E = hypotenuse of triangle O P′E.

p = OE

From Fig. 2.25,

we have $r^2 = x^2 + y^2$

$$P'E = \sqrt{p^2 + x^2 + y^2} = \sqrt{p^2 + r^2} \quad \text{and hence,}$$

$$\cos\theta = \frac{OE}{P'E} = \frac{p}{\sqrt{p^2 + r^2}}$$

Putting this value of cos θ in Eqs (2.86) and (2.87), we get the radius of the nth bright fringe and dark fringe as

$$r_n = p \times \sqrt{\left(\frac{4e^2}{n^2\lambda^2} - 1\right)} \quad \text{(Bright circular fringe)} \tag{2.88}$$

$$r_n = p \times \sqrt{\left(\frac{16e^2}{(2n+1)^2\lambda^2} - 1\right)} \quad \text{(Dark circular fringe)} \tag{2.89}$$

In contrast to Newton's rings, these equations show that as the diameter of the concentric fringes increase, the order of the fringes decrease – the order of the central fringe is maximum. As evident from Fig. 2.24, concentric circular fringes in the Michelson interferometer are fringes of constant inclination whereas in Newton's rings, they are fringes of constant thickness. These fringes in the Michelson interferometer are situated at infinity but Newton's rings are localized, situated in between the plano-convex lens and the plane glass plate of Newton's apparatus. Nevertheless, in both systems, concentric fringes gets closer as their radii increases.

ii. **Localized fringe**

If orientation of the mirrors M_1 and M_2 are not exactly perpendicular, then the space between M_1 and M_2' will be wedge shaped containing air; but fringes will be seen with monochromatic light for path difference not exceeding a few mm. Since fringes are formed in the vicinity of M_1, they are called localized fringes. The two rays reaching the eye from a point P on the source will appear to come from a point P' near the mirror M_1. For various positions of P on the source, the path difference between the two rays remain constant but the distance of P' from the mirrors change. The fringes are curved with their convex side towards the edge of the wedge as long as e has an appreciable value because there is a variation of the path difference with angle. If the separations of M_1 and M_2' is decreased, the fringes will move towards the left across the field. A new fringe will cross the center each time e changes by $\lambda/2$. As we approach zero path difference, the fringes become straighter. The fringes become perfectly straight when M_1 actually intersects M_2'. Afterwards, they begin to curve in the opposite direction.

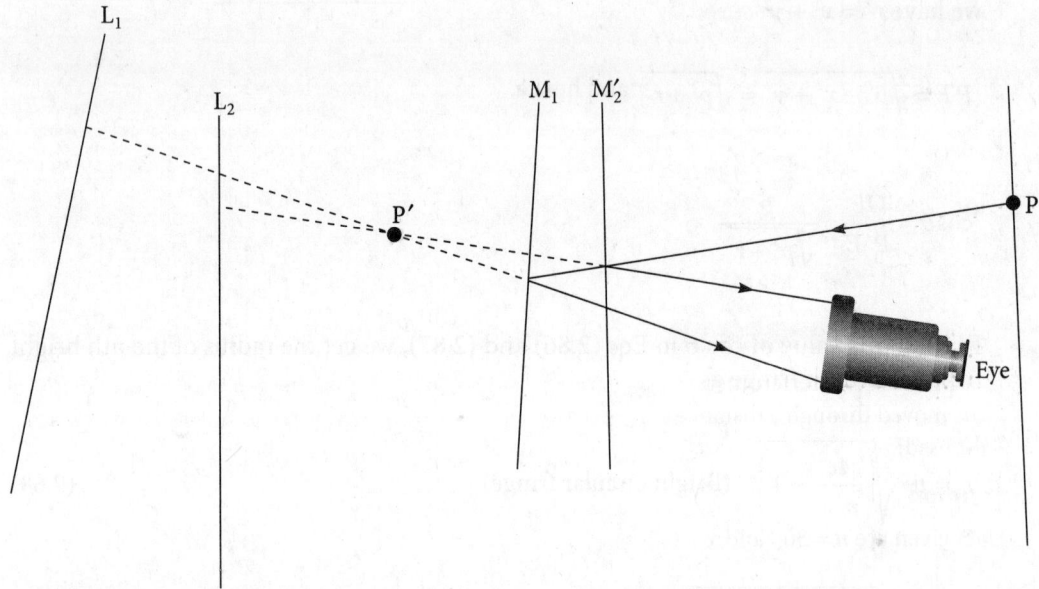

Figure 2.26 | Explanation of origin of fringes when mirrors M_1 and M_2 are not parallel

The field of view becomes blank for large path difference. These fringes are called fringes of equal thickness because they are actually the loci of equal thickness in the wedge-shaped air film and hence, more or less parallel to the edge of the wedge.

Applications

The Michelson interferometer has a wide range of applications in science and technology. We will discuss a few of them like (i) determination of wavelength of monochromatic light, (ii) resolution of spectral lines, (iii) determination of refractive index of materials, and (iv) determination of thickness of a thin sheet.

i. **Determination of wavelength of monochromatic light**

If the Michelson interferometer is in normal adjustment, concentric circular fringes appear in the center of the field of view. As we move M_1 with the help of a micrometer screw, fringes cross the field of view. For the central fringe, since $\cos \theta = 1$, we have

$$de = \frac{\lambda}{2}dn,$$

i.e., one fringe crosses the field of view ($dn = 1$) when M_1 is moved by

$$de = \frac{\lambda}{2} \text{ units.}$$

Thus, when n number of fringes cross the field of view, M_1 is moved by $\dfrac{\lambda n}{2}$ units which can be measured with the help of a micrometer screw. If x is the distance M_1 has been moved as measured by the micrometer screw, we have

$$x = n\frac{\lambda}{2}$$

$$\text{or} \quad \lambda = \frac{2x}{n} \tag{2.90}$$

Example 2.21

In a Michelson interferometer, it is found that 420 fringes crossed field of view when mirror M_1 is moved through a distance of 0.1414 mm. What is the wavelength of the monochromatic light used?

Solution

Data given are $n = 300$ and $x = 0.01414$ cm. $\lambda = ?$

$$\lambda = \frac{2x}{n} = 6733\text{Å}$$

ii. **Resolution of spectral lines**

There are two spectral lines D_1 ($\lambda_1 = 5890$ Å) and D_2 ($\lambda_2 = 5896$ Å) in sodium light. The Michelson interferometer is in normal adjustment and the mirror M_1 is so adjusted that the brightest fringe pattern is obtained. In this case, the bright fringe and dark fringe due to λ_1 coincides with the bright fringe and dark fringe due to λ_2 respectively. When M_1 is moved, two sets of fringes due to λ_1 and λ_2 get out of step and for a certain position of M_1, bright fringes and dark fringes due to λ_1 will coincide with dark fringes and bright fringes due to λ_2 respectively and as a result, no fringe will be visible. As M_1 is moved in the same direction for a certain position of M_1, fringes will be again distinct. This happens when the nth order of the longer wavelength λ_1 coincides with the $(n + 1)$th order of shorter wavelength λ_2. Let n_1 and n_2 be the number of fringes due to wavelengths λ_1 and λ_2 respectively that cross the center of the field of view when M_1 is moved through a distance of x between two consecutive positions of maximum distinctiveness of the fringes. Thus, we have

$$2x = n_1 \lambda_1 = n_2 \lambda_2$$

or $\qquad 2x = n_1 \lambda_1 = (n_1 + 1)\lambda_2$

since $\qquad \lambda_1 > \lambda_2$

or $\qquad n_1 = \dfrac{\lambda_1}{\lambda_1 - \lambda_2}$

Hence, $\quad \Delta\lambda = \lambda_1 - \lambda_2 = \dfrac{\lambda_1 \lambda_2}{2x}$ $\hfill$ (2.91)

Since $\lambda_1 \approx \lambda_2 = \lambda$ this equation becomes

$$\Delta\lambda = \dfrac{\lambda^2}{2x} \hfill (2.92)$$

By using Eq. (2.92), the difference in wavelength of closely spaced spectral lines ($\lambda_1 \approx \lambda_2$) can be determined. The difference in the wave number of two closely spaced spectral lines can be calculated using the formula

$$\dfrac{2\pi}{\lambda_2} - \dfrac{2\pi}{\lambda_1} = \dfrac{2\pi}{2x} \hfill (2.93)$$

Example 2.22

In a Michelson interferometer, two distinct lines 5890 Å and 5896 Å of a sodium vapour lamp were observed. Calculate the distance through which mirror M_1 has moved through.

Solution

Data given are $\lambda_1 = 5890 \text{Å}$, $\lambda_2 = 5896 \text{Å}$, $x = ?$

$$x = \frac{\lambda_1 \lambda_2}{2(\lambda_1 - \lambda_2)} = \frac{5890 \times 5896}{2(5896 - 5890)} \text{Å} = 0.02894 \text{ cm}$$

iii. **Determination of refractive index of materials**

The refractive index of material available in the form of a thin plate can be determined with the help of a Michelson interferometer. The extra optical path difference of $2(\mu - 1)t$ is introduced between two interfering light beams when a thin plate is introduced between glass plate G_1 and M_1. Due to this, a large number of fringes crosses the field of view it is not possible to count each of them in a normal way. The number of fringes that cross the field of view can be counted when the thin sheet is rotated slowly. Suppose n number of fringes cross the field of view when the thin plate is rotated through an angle φ.

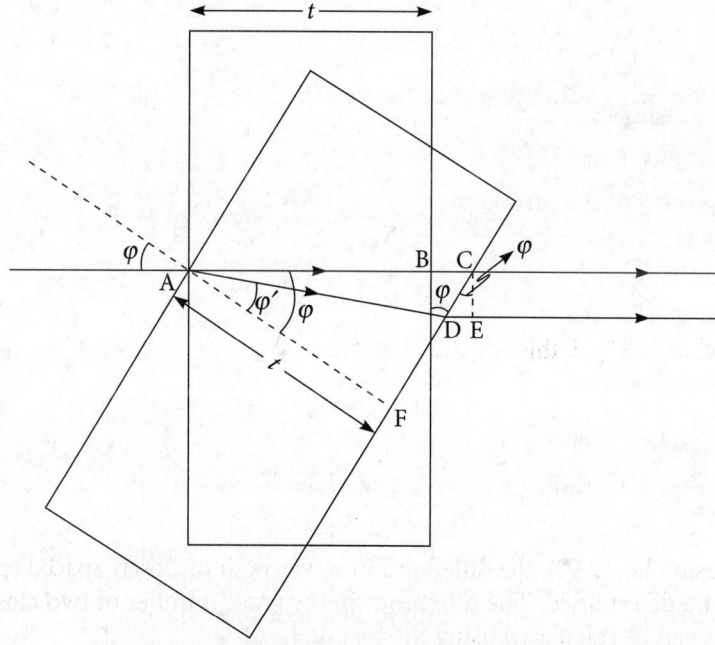

Figure 2.27 | Calculation of optical path difference due to rotation of the plate by φ

The optical path for ABC is $\mu t + BC$ and that for ADE is $\mu AD + DE$ as shown in Fig. 2.27 and hence, the optical path difference is $(\mu AD + DE) - (\mu t + BC)$ due to rotation of the plate by φ. Thus, we have

$$(\mu AD + DE) - (\mu t + BC) = n\frac{\lambda}{2}$$

From Fig. 2.27

i. $\quad AD = \dfrac{t}{\sin\varphi'}$

ii. $\quad BC = \dfrac{t}{\cos\varphi} - t$

iii. $\quad DE = \sin\varphi\left(t\tan\varphi - t\tan\varphi'\right)$ Since $DC = FC - FD = t\left(\tan\varphi - \tan\varphi'\right)$

Putting all these values in aforementioned equation, we get

$$\frac{\mu t}{\cos\varphi'} + t\sin\varphi\left(\tan\varphi - \tan\varphi'\right) - \mu t - t\left(\frac{1}{\cos\varphi} - 1\right) = \frac{n\lambda}{2}$$

Putting $\mu = \dfrac{\sin\varphi}{\sin\varphi'}$ into this equation, we get

$$\frac{\mu}{\cos\varphi'} + \frac{\sin^2\varphi}{\cos\varphi} - \frac{\sin\varphi\sin\varphi'}{\cos\varphi'} - \frac{1}{\cos\varphi} = \frac{n\lambda}{2t} - 1 + \mu$$

or $\quad \dfrac{\mu}{\cos\varphi'} - \dfrac{\sin^2\varphi}{\mu\cos\varphi'} + \dfrac{\sin^2\varphi}{\cos\varphi} - \dfrac{1}{\cos\varphi} = \dfrac{n\lambda}{2t} - 1 + \mu$

or $\quad \dfrac{1}{\mu\cos\varphi'}\left(\mu^2 - \sin^2\varphi\right) - \dfrac{1}{\cos\varphi}\left(1 - \sin^2\varphi\right) = \dfrac{n\lambda}{2t} - 1 + \mu$

Putting $\mu\cos\varphi' = \sqrt{\mu^2 - \sin^2\varphi}$ into this equation, we have

$$\sqrt{\mu^2 - \sin^2\varphi} - \cos\varphi = \frac{n\lambda}{2t} - 1 + \mu$$

or $\mu = \dfrac{(2t-n\lambda)-(1-\cos\varphi)+\dfrac{n^2\lambda^2}{4t}}{2t(1-\cos\varphi)-n\lambda}$

or $\mu \approx \dfrac{(2t-n\lambda)-(1-\cos\varphi)}{2t(1-\cos\varphi)-n\lambda}$ (2.94)

This expression for refractive index contains only measurable quantities. By using the aforementioned formula, the refractive index of material available in the form of a thin plate can be calculated. Also by using the same formula, refractive indices of different gases and liquids contained in thin transparent rectangular containers under different pressures can be studied. In addition to this, $2(\mu-1)t = n\lambda$ can also be used to determine the refractive index of gases as explained in the following example.

Example 2.23

Refractive index of a gas is to be found out by using the Michelson interferometer. A transparent tube of length 20 cm containing gas is introduced between G_1 and M_1. A shift of 200 fringes is observed in the field of view when gas was slowly taken out. Calculate the refractive index of the gas if light of wavelength 5890 Å is used.

Solution

The data given are

$N = 200$

$t = 20$ cm

$\lambda = 5890\text{Å}$

$2(\mu-1)t = n\lambda$

$\mu = \dfrac{n\lambda}{2t} + 1 = 1.0003$

Example 2.24

For accuracy purpose, the refractive index of a material in the form of sheet of a thickness 0.5 cm is to be found out by using the Michelson interferometer. There is a shift of 300 fringes when it is rotated through 53° angle. Sodium vapour lamp is the source of monochromatic light.

Solution

Data given are $t = 0.5$ cm, $\varphi = 50°$, $n = 300$, $\lambda = 5890$ Å.

$$\mu = \frac{(2t - n\lambda) - (1 - \cos\varphi) + \dfrac{n^2\lambda^2}{4t}}{2t(1 - \cos\varphi) - n\lambda}$$

$$\mu = \frac{\left(2 \times 0.5 - 300 \times 5890 \times 10^{-8}\right) - \left(1 - \cos 53\right) + \left(300 \times 5890 \times 10^{-8}\right)^2 \big/ 2}{2 \times 0.5\left(1 - \cos 53\right) - 300 \times 5890 \times 10^{-8}}$$

$$\mu = 1.5354$$

Questions

2.1 What is Huygens' principle with regard to light waves?

2.2 What is a wavefront? How it is produced according to Huygens' principle?

2.3 What type of sources produces a spherical wavefront?

2.4 What type of sources produces a cylindrical wavefront?

2.5 What type of sources produces a plane wavefront?

2.6 At a certain instant, the position of a plane wavefront is given. Applying Huygens' principle, find its position and shape at a later instant.

2.7 At a certain instant, the position of a spherical wavefront is given. Applying Huygens' principle, find its position and shape at a later instant.

2.8 How can you get spherical, cylindrical and plane wavefronts by a point source?

2.9 How can you explain the absence of backward wave propagation?

2.10 What is the basis of Huygens' principle?

2.11 What are secondary wavelets?

2.12 Give few examples of applications of Huygens' principle.

2.13 What is optical path? What is geometrical path?

2.14 What is interference of light?

2.15 Why are interference patterns not visible to the naked eye?

2.16 Does the law of conservation of energy hold good in case of interference phenomenon? If yes, explain

2.17 What are coherent sources of light?

2.18 How are coherent sources practically realized?

2.19   Why was corpuscular theory discarded in favour of wave theory?

2.20   Describe Young's double slit experiment with the necessary theory.

2.21   Classify interference phenomenon. Give few examples of each category.

2.22   Give the mathematical theory of the interference phenomenon.

2.23   Discuss the conditions for interference.

2.24   What is the condition for constructive interference in terms of phase difference between the two interfering waves?

2.25   What is the condition for destructive interference in terms of phase difference between the two interfering waves?

2.26   What is the condition for constructive interference in terms of path difference between the two interfering waves?

2.27   What is the shape of fringes theoretically obtained in Young's double slit experiment?

2.28   Why do we not observe interference effects of light from two candles?

2.29   What do you mean by constructive interference?

2.30   What do you mean by destructive interference?

2.31   Discuss why two independent sources of the same wave length and same amplitude can not produce a visible interference pattern.

2.32   What do you mean by fringe spacing? Give the basic concept about it.

2.33   Derive an expression for fringe spacing of a bright fringe in the fringe pattern produced in Young's double slit experiment.

2.34   Derive an expression for fringe spacing of a dark fringe in the fringe pattern produced in Young's double slit experiment.

2.35   In Young's double slit experiment, what should be the phase difference between the two interfering waves reaching a bright fringe?

2.36   In Young's double slit experiment, what should be the phase difference between the two interfering waves reaching a dark fringe?

2.37   How does fringe spacing vary with the distance of the screen from two coherent sources?

2.38   Derive an expression for the resultant intensity at any point on the screen in case of Young's double slit interference pattern.

2.39   Draw the intensity distribution curve in case of Young's double slit interference pattern.

2.40   Give the analytical treatment of the intensity distribution in case of interference in Young's double slit experiment.

2.41   Draw the diagram of Young's double slit experiment.

2.42   Explain with a diagram how coherent sources are realized in Young's double slit experiment.

2.43   In Young's double slit experiment, would you like the two sources to be closer or farther apart? Give reason for your choice.

2.44 How do you obtain coherent sources in Young's double slit experiment?

2.45 Why is it not possible to obtain a visible interference pattern without the sources being coherent?

2.46 Prove that the law of conservation of energy holds good in case of Young's double slit interference pattern.

2.47 What are the conditions for observation of interference pattern?

2.48 What are the conditions for observation of sustained interference pattern?

2.49 What are the conditions for observation of a good contrast interference pattern?

2.50 Prove analytically that the interference fringes obtained in Young's double slit experiment are hyperbolae.

2.51 Explain how the interference fringes obtained in Young's double slit experiment appear straight although they are actually hyperbolae.

2.52 What is the shape of the interference pattern in space? Explain

2.53 What is the shape of the interference pattern in the *XY*-plane? Explain

2.54 What is the shape of the interference pattern in the *YZ*-plane? Explain

2.55 What is the shape of the interference pattern in the *ZX*-plane? Explain

2.56 What is a Newton's ring?

2.57 Why are Newton's rings circular?

2.58 Describe the experimental setup to obtain Newton's rings.

2.59 Derive the expression for the total phase difference between the two interfering waves to form Newton's rings by reflected light.

2.60 Derive the expression for the total phase difference between the two interfering waves to form Newton's rings by refracted light.

2.61 Under what conditions will the Newton's rings seen by reflected light be bright?

2.62 Under what conditions will the Newton's rings seen by refracted light be bright?

2.63 Under what conditions will the Newton's rings seen by reflected light be dark?

2.64 Under what conditions will the Newton's rings seen by refracted light be dark?

2.65 Explain with a diagram how coherent sources are realized in Newton's ring arrangements.

2.66 Derive the expression for the fringe spacing of the *n*th order bright fringe produced by reflected light.

2.67 Derive the expression for the fringe spacing of the *n*th order dark fringe of Newton's rings produced by reflected light.

2.68 Derive the expression for the fringe spacing of the *n*th order bright fringe of Newton's rings produced by transmitted light.

2.69 Derive the expression for the fringe spacing of the *n*th order dark fringe produced by transmitted light in Newton's rings.

2.70 Prove that fringe spacing decreases with increase of radius of the Newton's rings formed by transmitted light.

2.71 Explain the variation of the radius of the bright fringe of Newton's rings with the refractive index of the thin film enclosed between the plano-convex lens and the glass plate when observed by reflected light.

2.72 Explain the variation of the radius of the dark fringe of Newton's rings with the refractive index of the thin film enclosed between the plano-convex lens and the glass plate when observed by reflected light.

2.73 Explain the variation the radius of the dark fringe of Newton's rings with the refractive index of the thin film enclosed between the plano-convex lens and the glass plate when observed by transmitted light.

2.74 Explain the variation of the radius of the bright fringe of Newton's rings with the refractive index of the thin film enclosed between the plano-convex lens and the glass plate when observed by transmitted light.

2.75 Explain the variation of the diameter of the bright fringe of Newton's rings with the radius of curvature of the plano-convex lens when observed by reflected light.

2.76 Explain the variation of the diameter of the dark fringe of Newton's rings with the radius of curvature of the plano-convex lens when observed by reflected light.

2.77 Explain the variation of the diameter of the dark fringe of Newton's rings with the radius of curvature of the plano-convex lens when observed by transmitted light.

2.78 Explain the variation of the radius of the bright fringe of Newton's rings with the radius of curvature of the plano-convex lens when observed by transmitted light.

2.79 Explain why the central fringe of Newton's rings appears dark as seen by reflected light when a thin air film is enclosed between the glass plate and the spherical surface of the plano-convex lens.

2.80 Explain why the central fringe of Newton's rings appears bright as seen by transmitted light when a thin air film is enclosed between the glass plate and the spherical surface of the plano-convex lens.

2.81 Prove that the central fringe of Newton's rings as seen by reflected light and by transmitted light is complementary to each other.

2.82 Account for the blackness and brightness of the central fringe of Newton's rings.

2.83 How are Newton's rings affected if we use polychromatic light instead of monochromatic light?

2.84 How are Newton's rings affected if we change the light source from red to green?

2.85 Prove that the law of conservation of energy holds good for Newton's rings.

2.86 How do we determine the wavelength of sodium light using Newton's rings?

2.87 Stating the working formula, describe an experiment to determine the wavelength of sodium light using Newton's rings.

2.88 Derive the necessary theory to determine the wavelength of sodium light using Newton's rings.

2.89 How do we determine the refractive index of transparent liquid using Newton's rings?

2.90   Stating the working formula, describe an experiment to determine the refractive index of transparent liquid using Newton's rings.

2.91   Derive the necessary theory to determine the refractive index of transparent liquid using Newton's rings.

2.92   Describe in detail an experiment to determine the wavelength of sodium light using Newton's rings.

2.93   Describe in detail an experiment to determine the refractive index of transparent liquid using Newton's rings.

2.94   Explain the formation of Newton's rings. How can these be used to determine the refractive index of a liquid?

2.95   Explain how Newton's rings are used to determine the wavelength of sodium light.

2.96   Give the theory of Newton's rings and describe an experiment to determine the wavelength of sodium light using these rings.

2.97   Explain, giving necessary theory, how the radius of curvature of a plano-convex lens can be determined using Newton's rings.

2.98   Explain the principle and construction of a Michelson interferometer.

2.99   Explain how we can determine the wavelength of monochromatic light using a Michelson interferometer.

2.100   Derive the working formula for the determination of refractive index of solid materials using a Michelson interferometer.

2.101   Distinguish between circular fringes produced in Newton's ring apparatuses and the Michelson interferometer.

2.102   Explain with theory the determination of refractive index of gas using a Michelson interferometer.

Problems

2.1   The wavelength of light in vacuum is 6500 Å. Calculate the wavelength of the same light in a medium having absolute refractive index 1.5.   [Ans   4333 Å]

In Young's double slit experiment, the separation between the slits is 0.19 cm and the fringe spacing is 0.031 cm at a distance of 1 meter from the slits. Calculate the wavelength of the light.   [Ans   5890 Å]

2.2   In Young's double slit experiment, 1 cm on the screen placed at 200 cm contains 20 fringes. Find the slit separation if the wavelength of light used is 5100 Å.

[Ans   0.051 cm]

2.3   In Young's double slit experiment, two coherent sources are 0.02 cm apart and the fringes are observed on a screen 80 cm away. It is found that the 4th bright fringe is situated at a distance of 1.2 cm from the central fringe. Calculate the wavelength of the monochromatic light used.   [Ans   7500 Å]

2.4   Two coherent sources are 0.07 cm apart and the fringes are observed on a screen 120 cm away. It is found that the 8th dark fringe is situated at a distance of 0.8 cm from the central fringe. Calculate the wavelength of the monochromatic light used.

[Ans   5490 Å]

2.5   A plano-convex lens of radius of curvature 3 meter is placed on an optically plane glass plate in an air medium. It is illuminated by a parallel beam of monochromatic light. The diameter of the 8th bright ring as seen by the reflected light is 0.72 cm. Calculate the wavelength of the light.   [Ans   5082 Å]

2.6   A plano-convex lens of radius of curvature 2.7 m is placed on an optically plane glass plate in an air medium. It is illuminated by a parallel beam of monochromatic light. The diameter of the 10th dark ring as seen by the reflected light is 0.8 cm. Calculate the wavelength of the light.   [Ans   5925 Å]

2.7   A plano-convex lens of radius of curvature 160 cm is placed on an optically plane glass plate. The space between the lens and the glass plate is filled with water having refractive index 1.333 and is illuminated by a parallel beam of monochromatic light. The diameter of the 8th dark ring as seen by the reflected light is 0.4 cm. Calculate the wavelength of the light.   [Ans   4166 Å]

2.8   A plano-convex lens of radius of curvature 3 m is placed on an optically plane glass plate in an air medium and is illuminated by a parallel beam of monochromatic light. The diameter of the 9th bright ring as seen by the transmitted light is 0.74 cm. Calculate the wavelength of the light.   [Ans   5070 Å]

2.9   A plano-convex lens of radius of curvature 2.7 m is placed on an optically plane glass plate in an air medium and is illuminated by a parallel beam of monochromatic light. The diameter of the 10th dark ring as seen by the transmitted light is 0.8 cm. Calculate the wavelength of the light.   [Ans   5644 Å]

2.10   A plano-convex lens of radius of curvature 250 cm is placed on an optically plane glass plate. The space between the lens and the glass plate is filled with ethyl alcohol having refractive index 1.354 and is illuminated by a parallel beam of monochromatic light. The diameter of the 11th dark ring as seen by the transmitted light is 0.72 cm. Calculate the wavelength of the light.   [Ans   6104 Å]

2.11   A plano-convex lens of radius of curvature 290 m is placed on an optically plane glass plate in an air medium and is illuminated by a parallel beam of monochromatic light of wavelength 5893 Å. Calculate the change in diameter of the 7th bright ring when the air medium is replaced by water of refractive index 1.33 and is observed by the transmitted light.   [Ans   0.092 cm]

In a Newton's ring experiment in air, the diameter of the 5th bright ring was 0.336 cm and that of the 15th bright ring was 0.590 cm. If the radius of curvature of the plano-convex lens is 100 cm, calculate the wavelength of the monochromatic light used.

[Ans   5880 Å]

2.12   In a Newton's ring experiment in air, the diameter of the 10th bright ring was 0.372 cm and that of the 15th bright ring was 0.555 cm. If the wavelength of the monochromatic light used is 5893 Å, calculate the radius of curvature of the plano-convex lens used.

[Ans   143.9 cm]

2.13   In a Newton's ring experiment, when certain liquid is inserted in between the plano-convex and the glass plate radius, the radius of the 10th dark ring is measured to be 6 cm. If the wavelength of the monochromatic light used is 5893 Å and the radius of curvature of the plano-convex lens is 203 cm calculate the refractive index of the liquid inserted.   [Ans   1.33]

2.14   In Newton's ring experiment in a laboratory, a source of light having two wavelengths 6000 Å and 4500 Å is used. It is found that the nth dark ring due to 6000 Å coincides with the $(n + 1)$th dark ring due to 4500 Å. Calculate the radii of the nth dark rings due to 6000 Å and 4500 Å if the radius of curvature of the plano-convex lens is 100 cm.   [Ans   0.1342 cm, 0.1162 cm]

In Newton's ring experiment in a laboratory, a sodium vapor lamp having two wavelengths 5890 Å and 5896 Å is used. It is found that the nth dark ring due to 5896 Å coincides with the $(n + 2)$nd dark ring due to 5890 Å. Calculate n. Also calculate the radii of the nth dark rings due to 5896 Å and 5890 Å if the radius of curvature of the plano-convex lens is 200 cm.   [Ans   1963, 4.81 cm, 4.81 cm]

2.15   In Newton's ring experiment in a laboratory, a source of light having two wavelengths 5600 Å and 4800 Å is used. It is found that the nth dark ring due to 5600 Å coincides with the $(n + 2)$nd dark ring due to 4800 Å. Calculate n. Also calculate the radii of the nth dark rings due to 5600 Å and 4800 Å if the radius of curvature of the plano-convex lens is 90 cm.   [Ans   12, 0.2459 cm, 0.2277 cm]

2.16   In the Newton's ring experiment by a light of wavelength 5896 Å, the diameter of the 20th dark ring is 0.73 cm as seen by the reflected light in the air medium. Calculate the radius of curvature of the plano-convex lens and the thickness of the film at this point.   [Ans   112.98 cm, 5.896×10^{-4} cm]

2.17   In the Newton's ring experiment by a light of wavelength 7000 Å, the diameter of the 15th dark ring is 0.565 cm when a thin film of refractive index 1.354 [ethyl alcohol at 20°C] is formed between the lens and the glass plate. Calculate the radius of curvature of the plano-convex lens and the thickness of the film at this point.   [Ans   102.91 cm, 3.877×10^{-4} cm]

2.18   In the Newton's ring experiment by a light of wavelength 6500 Å, the diameter of the 12th dark ring is 0.565 cm when a thin liquid film is formed between the lens and glass plate. If the radius of curvature of the plano-convex lens is 65.5 cm, calculate the refractive index of the liquid.   [Ans   1.461]

2.19   In a biprism experiment, the distance between the source and the eye-piece is 100 cm and that between two virtual sources is 0.075 cm. Find the wavelength of the light used if the fringe spacing is 0.0845 cm.   [Ans 6338×10^{-8} cm]

2.20   In a biprism experiment, the distance between the biprism and the eye-piece is 175 cm and that between the biprism and the monochromatic source ($\lambda = 5 \times 10^{-5}$ cm) is 25 cm. The biprism is made of glass of refractive index 1.5. Calculate the refracting angle.   [Ans   1.14°]

2.21   Fringes of equal inclination are observed in a Michelson interferometer. When mirror M_1 is moved through a distance of 0.1732 mm, 500 number of fringes crossed the field of view. Calculate the wavelength of monochromatic light used.   [Ans   6928 Å]

2.22   Mercury has a distinctive yellow doublet between approximately 575 nm and 580 nm. Calculate the distance through which mirror M_1 has to be moved through so that two distinct lines would be observed.   [Ans   0.03335 mm]

2.23   A material in the form of a sheet of thickness 0.5 cm is introduced between glass plate G_1 and M_1 of the Michelson interferometer. It is found that there is a shift of 500 fringes when it is rotated through a 45° angle. Sodium vapor lamp is the source of monochromatic light. Find its refractive index.   [Ans   2.571]

Multiple Choice Questions

1.   Huygens' idea of secondary waves gives a method to find
      (i)   the focus of a thin lens
      (ii)   the resolving power of a microscope
      (iii) the position and shape of a wavefront
      (iv)   the speed of a light wave

2.   According to Huygens' wave theory of light, the locus of all the points in the same state of vibration is called
      (i)   wavefront   (ii)   zone plate
      (iii) half period zone   (iv)   vibration contour

3.   What is the phase difference between two points situated on a wavefront?
      (i) $\pi/2$ (ii) π
      (iii) 2π (iv)   0

4.   What type of source produces spherical wavefronts?
      (i)   linear source   (ii)   plane source
      (iii) point source   (iv)   none of the above

5.   The rays from two coherent sources reach a certain point with a path difference of $\frac{1}{6}\lambda$. What is the phase difference between them?
      (i) $\frac{\pi}{4}$ (ii) $\frac{\pi}{3}$
      (iii) $\frac{\pi}{2}$ (iv) π

6.   The rays from two coherent sources reach a certain point with a phase difference of $\frac{\pi}{3}$. What is the path difference between them?
      (i) $\frac{\lambda}{6}$ (ii) $\frac{\lambda}{4}$
      (iii) $\frac{\lambda}{2}$ (iv) λ

7. What is the angle between the normal to a wavefront and the direction of propagation of the wave in a homogeneous isotropic medium?

    (i) $\dfrac{\pi}{2}$

    (ii) π

    (iii) 2π

    (iii) 0

8. What is the relation between intensity I and amplitude a of electromagnetic waves?

    (i) $I \propto a^2$

    (ii) $I \propto a$

    (iii) $I \propto \dfrac{1}{a}$

    (iv) $I \propto \dfrac{1}{a^2}$

9. What is the relation between optical path and geometrical path in a medium of refractive index μ?

    (i) optical path = geometrical path

    (ii) optical path = $\mu \times$ geometrical path

    (iii) geometrical path = $\mu \times$ optical path

    (iv) geometrical path $\times$ optical path = μ

10. Who demonstrated the interference phenomenon for the first time?

    (i) Newton

    (ii) Huygens

    (iii) Young

    (iv) Fresnell

11. Can corpuscular theory of light explain the interference phenomenon?

    (i) No

    (ii) Yes

    (iii) Yes or no dependsing on the size of the light corpuscles

    (iv) Yes or no depending on the shape of the light corpuscles

12. What do you mean by the statement, 'Two waves are in the same phase'?

    (i) The phase difference between two waves is only 0

    (ii) The phase difference between two waves is only 2π

    (iii) The phase difference between two waves is either 0 or 2π

    (iv) The phase difference between two waves is π

13. What is the shape of fringes practically obtained on the screen in Young's double slit experiment in the laboratory?

    (i) circular

    (ii) elliptical

    (iii) parabolic

    (iv) straight

14. What is the theoretical shape of fringes obtained on the screen in Young's double slit experiment?

    (i) circular

    (ii) hyperbolic

    (iii) parabolic

    (iv) straight

15. What is the theoretical shape of fringes obtained on the horizontal plane in Young's double slit experiment?

    (i) circular

    (ii) elliptical

    (iii) parabolic

    (iv) straight

16. Which of the following phenomena produces colors in soap bubbles?
    (i)   interference   (ii) diffraction
    (iii) polarization   (iv) dispersion

17. What is the relation between fringe spacing of a bright fringe and a dark fringe?
    (i)   fringe spacing of bright fringe = fringe spacing of dark fringe
    (ii) fringe spacing of bright fringe > fringe spacing of dark fringe
    (iii) fringe spacing of bright fringe < fringe spacing of dark fringe
    (iv) There is no fixed relation between them

18. Young's double slit experiment was performed by taking monochromatic blue, (B), orange (O) and red (R) lights. Which of the following relations for their fringe spacings β is correct?
    (i) $\beta_B > \beta_O > \beta_R$ (ii) $\beta_B < \beta_O < \beta_R$
    (iii) $\beta_O < \beta_B < \beta_R$ (iv) $\beta_B > \beta_R > \beta_O$

19. In Young's double slit experiment, slit spacing are in the ratio 4:9. What will be the ratio of intensity of maxima to minima?
    (i)   5:1   (ii) 3:2
    (iii) 25:1   (iv) 81:16

20. In Young's double slit experiment, the slit central fringe is
    (i)   dark   (ii) bright
    (iii) semi-black   (iv) semi-bright

21. In Young's double slit experiment, slit separation is made half and the separation between the slits and the screen is doubled. What will happen to the fringe spacing?
    (i)   It does not change   (ii) It becomes two times
    (iii) It becomes four times   (iv) It becomes half

22. Does the law of conservation of energy hold good in interference phenomenon?
    (i)   Yes   (ii) No
    (iii) depends upon the type of interference
    (iv) depends upon the experimental arrangements

23. What happens to the fringe spacing of a bright fringe in Young's double slit experiment when the distance between the slits is halved and the perpendicular distance between the screen and plane slits is doubled?
    (i)   unchanged   (ii) doubled
    (iii) halved   (iv) quadrupled

24. What happens to the fringe spacing of a bright fringe in Young's double slit experiment when the distance between the slits is doubled and the perpendicular distance between the screen and plane slits is halved?
    (i)   doubled   (ii) halved
    (iii) one-fourth   (iv) quadrupled

25. Newton's ring illustrates the phenomenon of
    (i) interference
    (ii) diffraction
    (iii) polarization
    (iv) dispersion

26. What is the shape of Newton's rings when you look through the microscope perpendicularly?
    (i) circular
    (ii) elliptical
    (iii) parabolic
    (iv) straight

27. What is the shape of Newton's rings when you look through the microscope making a certain angle with the vertical?
    (i) circular
    (ii) elliptical
    (iii) parabolic
    (iv) straight

28. What happens to the fringe spacings of Newton's rings with increase of radius?
    (i) fringe spacing increases
    (ii) fringe spacing first increases then decreases
    (iii) fringe spacing decreases
    (iv) fringe spacing first decreases then increases

29. The central fringe of Newton's rings as seen by monochromatic reflected light when a thin air film is enclosed between the glass plate and the spherical surface of the plano-convex lens is
    (i) black
    (ii) bright
    (iii) semi-black
    (iv) semi-bright

30. The central fringe of Newton's rings as seen by sunlight when a thin air film is enclosed between the glass plate and the spherical surface of the plano-convex lens is
    (i) black
    (ii) bright
    (iii) colorful
    (iv) central fringe cannot be seen

31. The central fringe of Newton's rings as seen by monochromatic transmitted light when a thin air film is enclosed between the glass plate and the spherical surface of the plano-convex lens is
    (i) black
    (ii) bright
    (iii) semi-black
    (iv) semi-bright

32. Newton's ring experiment was conducted first in air medium then in some liquid medium [transparent liquid is inserted in between the plano-convex lens and glass plate]. What happens to the diameter of a particular ring?
    (i) diameter increases
    (ii) diameter decreases
    (iii) diameter remains unchanged

33. The radius of dark Newton's ring is
    (i) directly proportional to the square root of natural numbers
    (ii) directly proportional to the square root of odd numbers

(iii) directly proportional to the square root of even numbers

(iv) inversely proportional to the square root of natural numbers

34. The radius of bright Newton's ring is

(i) directly proportional to the square root of natural numbers

(ii) directly proportional to the square root of odd numbers

(iii) directly proportional to the square root of even numbers

(iv) inversely proportional to the square root of natural numbers

35. Newton's rings are fringes of

(i) constant inclination   (ii) equal fringe spacing

(iii) constant path difference   (iv) equal thickness

Answers

1 (iii)	2 (i)	3 (iv)	4 (iii)	5 (ii)	6 (ii)	7 (iv)	8 (i)
9 (ii)	10 (iii)	11 (i)	12 (iii)	13 (iv)	14 (ii)	15 (i)	16 (i)
17 (i)	18 (ii)	19 (iii)	20 (ii)	21 (iii)	22 (i)	23 (iv)	24 (iii)
25 (i)	26 (i)	27 (ii)	28 (iii)	29 (i)	30 (i)	31 (ii)	32 (ii)
33 (iii)	34 (ii)	35 (iv)					

3 Diffraction

3.1 Introduction

In a homogeneous medium, light travels in a straight line – if an object is placed in its path, it should cast a sharp shadow of the object. However, the shadow of a fine wire caused by sunlight is not observable on a screen and the shadow of a straight edge as shown in Fig. 3.1 is not sharp when observed minutely. Some light bends into the shadow of the straight edge with the intensity of the light decreasing rapidly as we move into the shadow. The amount of bending depends upon the wavelength of the light and the size of the obstacle. In the region of the fuzzy boundary of the shadow of the straight edge, alternate dark and bright fringes of unequal spacings are seen when viewed by a well-focused microscope against a good contrast background. This optical phenomenon of bending of light around the obstacle is called diffraction and the fringes are called diffraction bands.

The diffraction phenomenon was first discovered by Italian scientist Grimaldi in the year 1665 and was correctly interpreted by Fresnel. He combined the Huygens' principle of secondary wavelets with the principle of interference and concluded that diffraction occurs due to the interference of secondary wavelets originating from various points of the wavefront which are not obstructed by the obstacle. The diffraction phenomenon occurs only when a part of the advancing wavefront is obstructed by some sharp obstacle.

3.2 Classification of Diffraction

The diffraction phenomenon is broadly classified into the following two general classes.

i.   **Fresnel's diffraction**

In this class of diffraction, the light source and the obstacle are separated by finite distance. Therefore, the wavefront incident on the obstacle is either cylindrical or spherical.

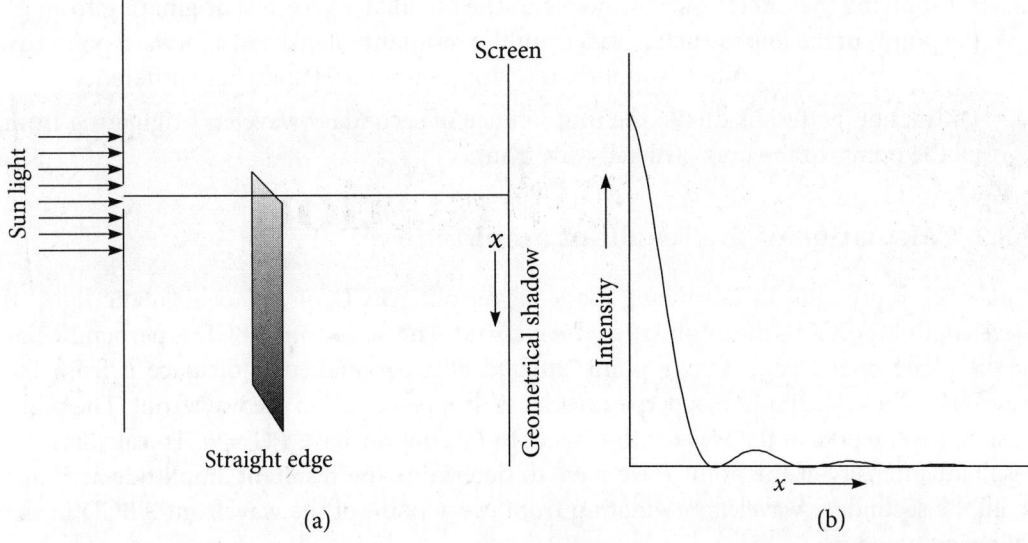

Figure 3.1 | (a) The shadow of a straight edge formed by a parallel beam of sunlight. The shadow is not very sharp. The encroachment of shadow by light is called diffraction. (b) A rough sketch of intensity distribution on the screen

ii. **Fraunhofer's diffraction**

In this class of diffraction, the light source and the obstacle are separated by infinite distance. Therefore, the wavefront incident on the obstacle is plane. In the laboratory, infinite distance is created by placing a lens in between the light source and the obstacle so that the light source is at the focus of the lens.

3.3 Fresnel's Explanation of Rectilinear Propagation of Light

The greatest difficulty encountered by the supporters of the wave theory of light was how to explain the observed fact that light propagates in a straight line. In the year 1815 French physicist Augustin-Jean Fresnel gave the correct interpretation of the rectilinear propagation of light on the basis of wave theory of light by combining Huygens' principle of secondary wavelets with the principle of interference. In addition to this, he concluded that rectilinear propagation of light is only approximate.

3.3.1 Fresnel's assumptions

i. Observable diffraction pattern is produced only when the wavefront is incident on a sharp obstacle or passes through slits.

ii. The principle of interference holds good also for secondary wavelets originating from all the points of the unobstructed wavefront.

iii. By applying the interference principle to the secondary wavelets originating from all the points of the unobstructed wavefront the resultant intensity at a forward point can be calculated taking into account their relative amplitudes and phase differences.

iv. Diffraction pattern is due to the interference of secondary wavelets originating from all the points of the unobstructed wavefront.

3.3.2 Calculation of the resultant amplitude

Figure 3.2 represents an advancing plane wavefront ABCD of monochromatic light of wavelength λ. ABCD is the unobstructed wavefront. The wavefront ABCD is perpendicular to the plane of the page. P is a point situated at a perpendicular distance b from the wavefront. From the point P, a perpendicular is dropped at C_0 on the wavefront. The point C_0 is called the pole of the wave with respect to P. Thus we have $PC_0 = b$. To calculate the resultant intensity at the point P we have to determine the resultant amplitude at P due to all the secondary wavelets originating from every point of the wavefront ABCD in the following manner.

The concentric spheres of radii

$$b+\frac{\lambda}{2}, b+\frac{2\lambda}{2}, b+\frac{3\lambda}{2}, b+\frac{4\lambda}{2}..., \text{ etc.,}$$

are drawn with P as the centre which intersect the wavefront ABCD in circles $C_1, C_2, C_3 \ldots C_n$ etc., respectively. Thus, from Fig. 3.2, we can have

$$PC_1 = b+\frac{\lambda}{2} = PC_0 +\frac{\lambda}{2}$$

or $\quad PC_1 - PC_0 = \frac{\lambda}{2}$

Similarly, $\quad PC_2 - PC_1 = b+\frac{2\lambda}{2}-\left(b+\frac{\lambda}{2}\right)=\frac{\lambda}{2}$

$$PC_3 - PC_2 = b+\frac{3\lambda}{2}-\left(b+\frac{2\lambda}{2}\right)=\frac{\lambda}{2}$$

$$PC_n - PC_{n-1} = b+\frac{n\lambda}{2}-\left(b+\frac{(n-1)\lambda}{2}\right)=\frac{\lambda}{2}$$

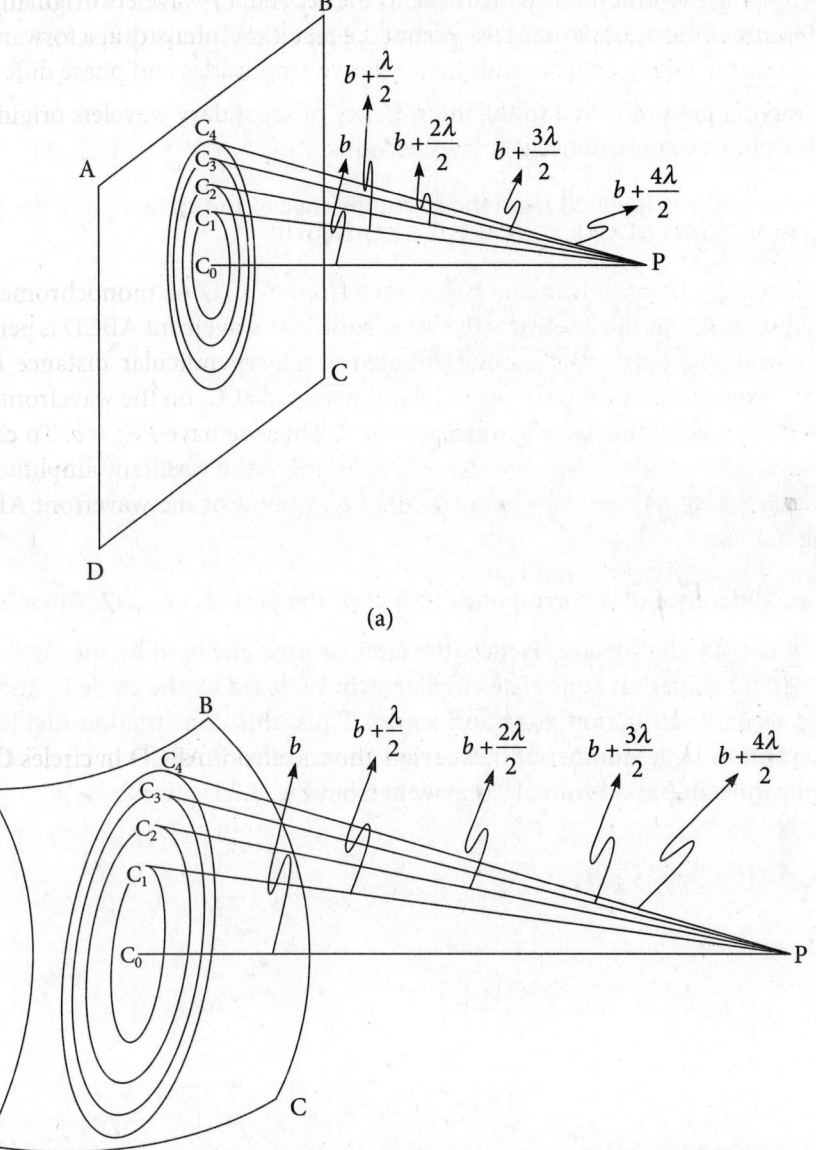

(a)

(b)

Figure 3.2 | (a) Construction of Fresnel's half period zones in a plane wavefront. The plane wavefront ABCD is perpendicular to the plane of the paper. The concentric spheres of radii $b + \dfrac{n\lambda}{2}$ with $n = 1, 2, 3, \ldots$ are drawn with P as the centre which intersect the wavefront ABCD in circles $C_1, C_2, C_3 \ldots, C_n$ respectively. The area of nth half period zone is A_n = (area of the circle C_n) – (area of the circle C_{n-1}). (b) Construction of Fresnel's half period zones in a spherical wavefront

$PC_1 - PC_0$ is the geometrical path difference between the waves originating from the circumference of the circle C_1 and the point C_0. Hence, the phase difference

$$\frac{2\pi}{\lambda}\left(PC_1 - PC_0\right)$$

between the waves originated from the circumference of the circle C_1 and the point C_0 will be given by

$$\frac{2\pi}{\lambda}\left(PC_1 - PC_0\right) = \frac{2\pi}{\lambda}\frac{\lambda}{2} = \pi$$

In general,

$$\frac{2\pi}{\lambda}\left(PC_n - PC_{n-1}\right) = \pi$$

The phase difference of π corresponds to half of the period, i.e., $T/2$, where $T = \dfrac{\lambda}{c}$ is the time period of the light wave. Hence, the circular area enclosed by the circle C_1 is called Fresnel's first half period zone. The circular strip enclosed by the circle C_1 and C_2 is called Fresnel's second half period zone and so on. Thus, this construction divides the entire wavefront into a large number of half period zones called Fresnel's half period zones. The area of the nth half period zone A_n as evident from Fig. 3.2 is given by

$$A_n = \pi\left(C_0C_n^2 - C_0C_{n-1}^2\right)$$

or $$A_n = \pi\left[\left(b + \frac{n\lambda}{2}\right)^2 - \left(b + \frac{(n-1)\lambda}{2}\right)^2\right]$$

or $$A_n = \pi b\lambda + \pi\left(2n-1\right)\frac{\lambda^2}{4} \tag{3.1}$$

Equation shows that higher half period zones have greater area than that of lower half period zones. $\dfrac{\lambda^2}{4}$ is a very small quantity due to which we can neglect the $\pi(2n-1)\dfrac{\lambda^2}{4}$ term in comparison to the term $\pi b\lambda$. Thus, Eq. (3.1) becomes

$$A_n \approx \pi b\lambda \tag{3.2}$$

Equation (3.2) shows that each half period zone has approximately equal area.

3.3.3 Average distance of the *n*th Fresnel's half period zone from the pole

The *n*th half period zone is a circular strip. The distance of the inner circumference from the point P is $b + \dfrac{(n-1)\lambda}{2}$ and that of the outer circumference is $b + \dfrac{n\lambda}{2}$. Therefore, the average distance of the *n*th Fresnel's half period zone d_n from P will be

$$d_n = \frac{b + \dfrac{(n-1)\lambda}{2} + b + \dfrac{n\lambda}{2}}{2}$$

or $\quad d_n = b + (2n-1)\dfrac{\lambda}{4}$ (3.3)

From Eqs (3.1) and (3.3) we have

$$\frac{A_n}{d_n} = \frac{\pi b\lambda + \pi(2n-1)\dfrac{\lambda^2}{4}}{b + (2n-1)\dfrac{\lambda}{4}} = \frac{\pi\lambda\left[b + (2n-1)\dfrac{\lambda}{4}\right]}{b + (2n-1)\dfrac{\lambda}{4}}$$

or $\quad \dfrac{A_n}{d_n} = \pi\lambda$ (3.4)

Let R_n be the amplitude of the secondary wavelets at P due to the *n*th Fresnel's half period zone. The magnitude of R_n may depend upon the following three factors.

i. The number of secondary wavelets originating from Fresnel's half period zone is proportional to the area of the zone. Again, more is the number of secondary wavelets, more is the amplitude at P. Therefore, we conclude

$R_n \propto A_n$

ii. The optical intensity $\left(\propto \text{amplitude}^2\right)$ decreases inversely with the square of the distance. Hence, we can conclude that the amplitude of the *n*th Fresnel's half period zone at P is inversely proportional to the average distance of the *n*th Fresnel's half period zone from P, i.e.,

$R_n \propto \dfrac{1}{d_n}$

iii. $R_n \propto (1 + \cos\theta_n)$, where $(1 + \cos\theta_n)$ is called the obliquity factor or inclination factor. Over a single Fresnel's half period zone, $(1 + \cos\theta_n)$ remains nearly constant since variation of θ_n is very small. As shown in Fig. 3.3, as θ increases from zero to $\frac{\pi}{2}$, $\cos\theta$ decreases slowly at first and then rapidly for larger values of θ. The obliquity factor $(1 + \cos\theta_n)$ varies in the same manner.

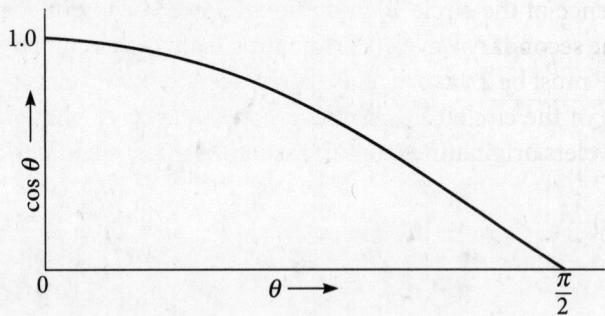

Figure 3.3   Variation of $\cos\theta$ with θ for $0 \leq \theta \leq \frac{\pi}{2}$

By combining (i), (ii), and (iii), we have

$$R_n = K\frac{A_n}{d_n}\left(1 + \cos\theta_n\right)$$

Putting the value of $\frac{A_n}{d_n}$ from Eq. (3.3) into equation, we have

$$R_n = K\pi\lambda(1 + \cos\theta_n)$$

or $R_n = k\left(1 + \cos\theta_n\right)$ (3.5)

Thus, for a particular light, the magnitude of R_n depends only upon the obliquity factors $(1 + \cos\theta_n)$. As θ increases from zero to $\frac{\pi}{2}$, $\cos\theta$ decreases slowly at first and then rapidly for larger values of θ and so do the obliquity factors $(1 + \cos\theta_n)$. R_n varies exactly in the same manner. Therefore, the magnitudes of the amplitudes R_1, R_2, R_3 ..., R_n decrease continuously.

3.3.4 Phase difference among half period zones

Suppose at any particular instant, the phase of the secondary wavelets originating from C_0 and reaching P is zero.

i. At this instant, the phase difference at P between the wavelets originating from the circumference of the circle C_1 and the point C_0 is π. Therefore the average phase difference of all the secondary wavelets originating from the first Fresnel's half period zone reaching P will be

$$\frac{0+\pi}{2} = \frac{\pi}{2}.$$

ii. At the same instant, the phase difference at P between the wavelets originating from the circumference of the circle C_1 and the circumference of the circle C_2 is π. Hence, the phase of the secondary wavelets originating from the circumference of the circle C_2 and reaching P must be 2π as the phase of the secondary wavelets originating from the circumference of the circle C_1 is π. Therefore, the average phase difference of all the secondary wavelets originating from the second Fresnel's half period zone reaching P will be

$$\frac{\pi+2\pi}{2} = \frac{3\pi}{2}.$$

iii. At the same instant, the phase difference at P between the wavelets originating from the circumference of the circle C_2 and the circumference of the circle C_3 is π. Hence, the phase of the secondary wavelets originating from the circumference of the circle C_3 and reaching P must be 3π as the phase of the secondary wavelets originating from the circumference of the circle C_2 is 2π. Therefore, the average phase difference of all the secondary wavelets originating from the third Fresnel's half period zone reaching P will be

$$\frac{2\pi+3\pi}{2} = \frac{5\pi}{2}.$$

iv. Similarly, proceeding further, the average phase difference of all the secondary wavelets originating from the nth Fresnel's half period zone reaching P will be

$$\frac{(2n-1)\pi}{2}.$$

Let R_1, R_2, R_3 ..., R_n be the amplitudes of the secondary wavelets at P due to the first, second, third Fresnel's half period zones respectively. From the earlier discussions, the odd numbered Fresnel's half period zones are in phase and the even numbered Fresnel's half period zones are out of phase. Therefore, the resultant amplitude R of the secondary wavelets originating from all the Fresnel's half period zones at P will be given by

$$R = R_1 - R_2 + R_3 - R_4 + \ldots + (-1)^{n-1} R_n \qquad (3.6)$$

3.3.5 Schuster's method of summing the series

According to Eq. (3.5), the magnitude of each term of the RHS of the Eq. (3.6) is less than the previous one.

Case A: n is odd

If n is odd, Eq. (3.6) becomes

$$R = R_1 - R_2 + R_3 - R_4 + \ldots + R_n$$

In this case, RHS of the aforementioned equation can be grouped in the following two alternative ways; either

$$R = \frac{1}{2}R_1 + \left(\frac{1}{2}R_1 - R_2 + \frac{1}{2}R_3\right) + \left(\frac{1}{2}R_3 - R_4 + \frac{1}{2}R_5\right) + \ldots + \frac{1}{2}R_n \tag{3.7}$$

or $$R = R_1 - \frac{1}{2}R_2 - \left(\frac{1}{2}R_2 - R_3 + \frac{1}{2}R_4\right) - \left(\frac{1}{2}R_4 - R_5 + \frac{1}{2}R_6\right) - \ldots - \frac{1}{2}R_{n-1} + R_n \tag{3.8}$$

The results of Eqs (3.7) and (3.8) should be the same simultaneously. Let us assume that

$$R_i > \frac{R_{i-1} + R_{i+1}}{2}, \; i = 2, 3, 4, \ldots$$

Under this assumption, the bracketed terms in Eqs (3.7) and (3.8) are negative. So Eqs (3.7) and (3.8) can be written respectively as

$$R = \frac{1}{2}R_1 + \frac{1}{2}R_n - \alpha \tag{3.9}$$

$$R = R_1 - \frac{1}{2}R_2 - \frac{1}{2}R_{n-1} + R_n + \beta \tag{3.10}$$

where
$$\alpha = \text{sum of all the bracketed terms in Eq. (3.7)}$$
$$\beta = \text{sum of all the bracketed terms in Eq. (3.8)}$$

From Eqs (3.9) and (3.10), we have respectively

$$\frac{1}{2}R_1 + \frac{1}{2}R_n > R \tag{3.11}$$

and $\quad R > R_1 - \frac{1}{2}R_2 - \frac{1}{2}R_{n-1} + R_n \tag{3.12}$

Experimentally, $R_1 \approx R_2$ and $R_{n-1} \approx R_n$, $n = 2, 3, 4, \ldots$. Under this condition Eq. (3.12) becomes

$$R > \frac{1}{2}R_1 + \frac{1}{2}R_n \tag{3.13}$$

The Eqs (3.11) and (3.13) are simultaneously true only if

$$R = \frac{1}{2}R_1 + \frac{1}{2}R_n \tag{3.14}$$

Case B: n is even

If n is even, Eq. (3.6) becomes

$$R = R_1 - R_2 + R_3 - R_4 + \ldots + R_{n-1} - R_n$$

In this case, RHS of the aforementioned equation can be grouped in the following two alternative ways; either

$$R = \frac{1}{2}R_1 + \left(\frac{1}{2}R_1 - R_2 + \frac{1}{2}R_3\right) + \left(\frac{1}{2}R_3 - R_4 + \frac{1}{2}R_5\right) + \ldots - \frac{1}{2}R_n \tag{3.15}$$

or $\quad R = R_1 - \frac{1}{2}R_2 - \left(\frac{1}{2}R_2 - R_3 + \frac{1}{2}R_4\right) - \left(\frac{1}{2}R_4 - R_5 + \frac{1}{2}R_6\right) - \ldots + \frac{1}{2}R_{n-1} - R_n \tag{3.16}$

The results of Eqs (3.15) and (3.16) should be the same simultaneously. Let us assume that

$$R_i > \frac{R_{i-1} + R_{i+1}}{2}, \quad i = 2, 3, 4, \ldots$$

Under this assumption the bracketed terms in Eqs (3.15) and (3.16) are negative. Hence, Eqs (3.15) and (3.16) can be written respectively as

$$R = \frac{1}{2}R_1 - \frac{1}{2}R_n - \alpha \tag{3.17}$$

$$R = R_1 - \frac{1}{2}R_2 + \frac{1}{2}R_{n-1} - R_n + \beta \tag{3.18}$$

α and β have been defined earlier. From Eqs (3.17) and (3.18), we have respectively

$$\frac{1}{2}R_1 - \frac{1}{2}R_n > R \tag{3.19}$$

and $R > R_1 - \frac{1}{2}R_2 + \frac{1}{2}R_{n-1} - R_n \tag{3.20}$

Experimentally, $R_1 \approx R_2$ and $R_{n-1} \approx R_n$, $n = 2, 3, 4, \ldots$. Under this condition Eq. (3.20) becomes

$$R > \frac{1}{2}R_1 - \frac{1}{2}R_n \tag{3.21}$$

Equations (3.19) and (3.21) are simultaneously true only if

$$R = \frac{1}{2}R_1 - \frac{1}{2}R_n \tag{3.22}$$

According to Eq. (3.5), the amplitude due to the nth half period zone at P R_n is negligibly small for large values of n. So we can neglect R_n in comparison to R_1 in Eqs (3.14) and (3.22). Therefore, whether n is odd or even, the resultant amplitude R at P due to the entire wavefront ABCD is given by

$$R = \frac{1}{2}R_1 \tag{3.23}$$

The resultant amplitude at P due to the entire wavefront ABCD is half of the amplitude at P due to the first Fresnel's half period zone.

In a similar manner, we can construct Fresnel's half period zones in case of cylindrical wavefronts and spherical wavefronts. We can then determine the resultant amplitude at a point due to the entire wavefront.

3.4 Zone Plate

The half period zone theory of Fresnel can be experimentally confirmed with help of an optical device called the zone plate. In reality, either all the odd numbered terms or all the even numbered terms of the RHS of Eq. (3.6) can be removed increasing the resultant amplitude to many times. This concept has been applied practically in constructing a zone plate. A zone plate is simply a thin parallel glass plate containing concentric circles of radii accurately proportional to the square root of natural numbers (recollect Newton's rings). The annular space so created is deeply blackened alternately. Practically, a highly reduced photograph of the Newton's rings is taken on a thin parallel glass plate.

3.4.1 Types of zone plates

Either the even numbered annular spaces are blacked leaving the odd numbered annular space transparent or the odd numbered annular spaces are blacked leaving the even numbered annular space transparent. Depending upon this, the zone plates are of two types.

i. **Positive zone plate**

    When the central circular zone, i.e., the first half period zone on the thin glass plate is transparent, the zone plate is called a positive zone plate. In a positive zone plate, even numbered half period zones, i.e., 2nd, 4th, 6th, …, etc., are blackened to make them opaque for light, leaving odd numbered half period zones, i.e., 1st, 3rd, 5th, … etc., transparent. A typical positive zone plate is shown in Fig. 3.4(a).

ii. **Negative zone plate**

    When the central circular zone, i.e., the first half period zone on the thin glass plate is opaque, the zone plate is called a negative zone plate. In a negative zone plate, odd numbered half period zones, i.e., 1st, 3rd, 5th, …, etc., are blackened to make them opaque for light, leaving even numbered half period zones, i.e., 2nd, 4th, 6th, …, etc., transparent. A typical negative zone plate is shown in Fig. 3.4(b).

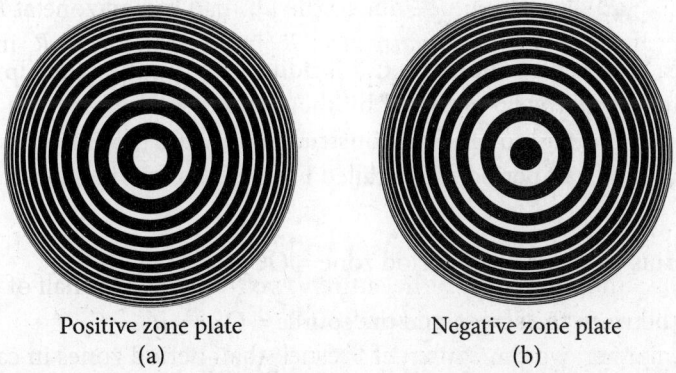

Positive zone plate
(a)

Negative zone plate
(b)

Figure 3.4 | (a) Positive zone plate where the central circular zone is transparent. (b) Negative zone plate where the central circular zone is opaque. Both figures are drawn not to scale

3.4.2 Action of the zone plate

YY' in Fig. 3.5 is a thin transparent screen placed perpendicular to the plane of the page and S is a point source emitting spherical wavefronts of monochromatic light of wavelength λ. P is a point situated on the other side of the screen so that SP passing through the point O on the screen is perpendicular to the screen. We can find the resultant amplitude at P due to the secondary wavelets diverging from various points of the screen by dividing the transparent screen into a large number of half period zones though the screen itself is not a wavefront. We can divide the transparent screen into a large number of half period zones by the following process.

We mark points C_1, C_2, C_3 ..., C_n along YY' on the transparent screen so that

$$SC_1P - SOP = \frac{\lambda}{2}$$

$$SC_2P - SOP = \frac{2\lambda}{2}$$

$$SC_3P - SOP = \frac{3\lambda}{2}$$

$$\dots\dots\dots\dots\dots = \dots$$

$$\dots\dots\dots\dots\dots = \dots$$

$$\dots\dots\dots\dots\dots = \dots$$

$$SC_nP - SOP = \frac{n\lambda}{2}. \tag{3.24}$$

The circular area enclosed by the circle C_1 of radius OC_1 may be called Fresnel's first half period zone. The circular strip enclosed by the circle C_1 and C_2 may be called Fresnel's second half period zone and so on. This construction divides the entire transparent screen into a large number of half period zones called Fresnel's half period zones.

Let

The outer radius of the nth half period zone = $OC_n = r_n$

The distance between the screen and the source = $OS = -u$

The distance between the screen and the point $P = OP = v$

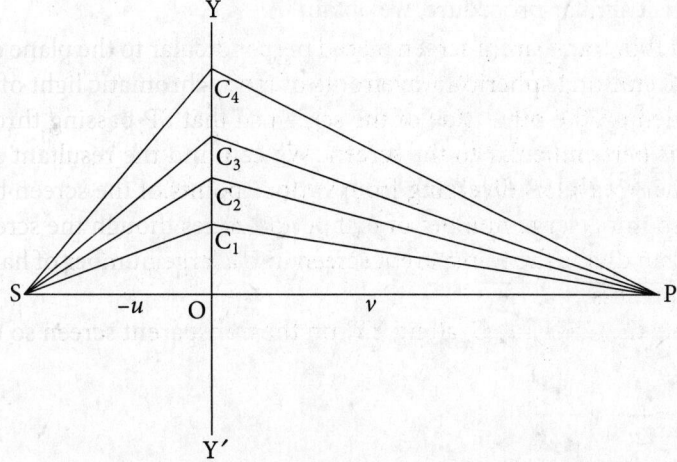

Figure 3.5 | Theory of zone plate. Division of a transparent screen into a large number of half period zones. Either all the odd numbered terms or all the even numbered terms of the RHS of Eq. (3.6) can be removed increasing the resultant amplitude to many times by blackening either all the odd numbered or all the even numbered half period zones

Now applying the Pythagoras theorem to the right angled triangle SOC_n, we get

$$SC_n = \sqrt{OS^2 + OC_n^2} = \sqrt{OS^2 + r_n^2}$$

or $\quad SC_n = OS\left(1 + \dfrac{r_n^2}{OS^2}\right)^{\frac{1}{2}}$

Since in general, $OS \gg r_n$, we can apply the binomial theorem to the this equation to get

$$SC_n = OS\left(1 + \dfrac{1}{2}\dfrac{r_n^2}{OS^2}\right)$$

or $\quad SC_n = OS + \dfrac{1}{2}\dfrac{r_n^2}{OS}$

According to our sign convention, $OS = -u$. Putting this value of OS into the this equation, we get

$$SC_n = -u + \frac{r_n^2}{2(-u)} = -u - \frac{r_n^2}{2u} \qquad (3.25)$$

Following the exact similar procedure, we obtain

$$PC_n = v + \frac{r_n^2}{2v}$$

(3.26)

Combining Eqs (3.25) and (3.26), we get

$$SC_n + PC_n = -u - \frac{r_n^2}{2u} + v + \frac{r_n^2}{2v}$$

or $$SC_n P = -u - \frac{r_n^2}{2u} + v + \frac{r_n^2}{2v}$$

(3.27)

Putting the value of $SC_n P$ from Eq. (3.27) into Eq. (3.24), we obtain

$$-u - \frac{r_n^2}{2u} + v + \frac{r_n^2}{2v} - SOP = \frac{n\lambda}{2}$$

or $$-u - \frac{r_n^2}{2u} + v + \frac{r_n^2}{2v} + u - v = \frac{n\lambda}{2}$$

or $$\frac{r_n^2}{2}\left(\frac{1}{v} - \frac{1}{u}\right) = \frac{n\lambda}{2}$$

(3.28)

or $$r_n^2 = \frac{nuv\lambda}{u - v}$$

(3.29)

or $$r_1^2 = \frac{uv\lambda}{u - v}$$

Putting this equation into Eq. (3.29), we get

$$r_n^2 = n r_1^2$$

(3.30)

or $$r_n = r_1 \sqrt{n}$$

(3.31)

Putting $n = 1, 2, 3, 4, \ldots$, into Eq. (3.31), we can get the radii of the first, second and third Fresnel's half period zone. Thus, the radii of half period zones are directly proportional to the square root of natural numbers.

Area of the nth half period zone

The square of the outer radius of the nth half period zone according to Eq. (3.29) is

$$r_n^2 = \frac{nuv\lambda}{u-v}$$

The inner radius of the nth half period zone is equal to the outer radius of $(n - 1)$th half period zone. So we have

$$r_{n-1}^2 = \frac{(n-1)uv\lambda}{u-v}$$

Therefore, area of the nth half period zone A_n will be equal to

$$A_n = \pi\left(r_n^2 - r_{n-1}^2\right)$$

or $\quad A_n = \pi\left(\dfrac{nuv\lambda}{u-v} - \dfrac{(n-1)uv\lambda}{u-v}\right)$

or $\quad A_n = \dfrac{\pi uv\lambda}{u-v}$ $\hfill (3.32)$

or $\quad A_n = \dfrac{\pi v\lambda}{1 - \dfrac{v}{u}}$ $\hfill (3.33)$

This expression for area of the nth half period zone is independent of n. Therefore, each half period zone has equal area. If the monochromatic light source is at infinity, i.e., $u \to \infty$, the wavefront will be a plane. Putting $u \to \infty$ in Eq. (3.33), the area of the nth half period zone in case of a plane wavefront will be given by

$$A_n = \pi v\lambda$$ $\hfill (3.34)$

3.4.2 B Intensity at P

Let R_1, R_2, R_3, ..., R_n be the amplitudes of the secondary wavelets at P due to the first, second, and third Fresnel's half period zones respectively. The resultant amplitude R of the secondary wavelets originating from all the Fresnel's half period zones at P will be given according to Eq. (3.6) as

$$R = R_1 - R_2 + R_3 - R_4 + ... + (-1)^{n-1} R_n$$

$$R = \frac{1}{2} R_1 \quad (\text{as } n \to \infty)$$

Therefore, the intensity I at P due to the entire wavefront will be given by

$$I = \frac{k}{4} R_1^2 \tag{3.35}$$

where k is the proportionality constant.

The resultant amplitude at P due to only the odd numbered half period zone is

$$R_{odd} = R_1 + R_3 + R_5 + R_7 ... \approx \frac{1}{2} N R_1 \tag{3.36}$$

The resultant amplitude at P due to only the even numbered half period zone is

$$R_{even} = R_2 + R_4 + R_6 + R_8 ... \approx \frac{1}{2} N R_1 \tag{3.37}$$

where N is the number of half period zones on the transparent screen and $N/2$ the number of zones that are exposed. Therefore, the intensity at P due to only odd numbered half period zones I_{odd} will be given by

$$I_{odd} \approx \frac{k}{4} N^2 R_1^2 \tag{3.38}$$

Thus, the intensity at P due to only odd numbered half period zones I_{odd} is N^2 times more than that of the intensity I at P due to the entire wavefront. Similarly, the intensity at P due to only even numbered half period zones I_{even} will be given by

$$I_{even} \approx \frac{k}{4} N^2 R_1^2 \tag{3.39}$$

Thus, the intensity at P due to only even numbered half period zones I_{even} is N^2 times more than that of the intensity I at P due to the entire wavefront.

3.4.3 Principle behind zone plates

From the previous discussions, we conclude that the intensity at point due to only either even numbered half period zones or due to only odd numbered half period zones is many times more than that of the intensity at the same point due to the entire wavefront. If somehow either all the odd numbered half period zones or all the even numbered half period zones on the screen were blocked off, the intensity at a point will be many times more than that of the intensity at the same point due to the entire wavefront. In a zone plate, either all the odd numbered half period zones or all the even numbered half period zones are blackened or made opaque. Hence, when light passes through a zone plate, the intensity at a forward point will be maximum and is the image of the light source. The position of the image should satisfy the Eq. (3.28), i.e.,

$$\frac{r_n^2}{2}\left(\frac{1}{v} - \frac{1}{u}\right) = \frac{n\lambda}{2}$$

or $\quad \dfrac{1}{v} - \dfrac{1}{u} = \dfrac{n\lambda}{r_n^2}$ $\hfill (3.40)$

Equation (3.40) is in the form of the equation of a thin lens. Hence, the zone plate behaves like a lens. Here u is the distance of the source from the zone plate and v is the distance of the image from the zone plate.

3.4.4 Multiple foci of a zone plate

The focal length of a lens or a zone plate is defined as the image distance when the object distance is infinity. Therefore, by putting $u \to \infty$ and $v = f_n$ into Eq. (3.40), we will get the expression for focal length of a zone plate. Hence, we have

$$\frac{1}{f_n} - \frac{1}{\infty} = \frac{n\lambda}{r_n^2}$$

or $\quad f_n = \dfrac{r_n^2}{n\lambda}$ $\hfill (3.41)$

Equation (3.41) shows that a zone plate has a number of foci and a number of focal lengths for different values of n. By putting $n = 1$ into this equation, we get

$$f_1 = \frac{r_1^2}{\lambda}$$ $\hfill (3.42)$

The focal length f_1 obtained by putting $n = 1$ into Eq. (3.41) is called first order focal length or primary focal length of the zone plate and the corresponding focus is called first order focus or primary focus of the zone plate. The intensity of the primary focus is maximum as compared to the other foci. The intensity of the other foci goes on decreasing as we move towards the zone plate along the axis. The intensity of the primary focus I_1 is given, according to Eq. (3.36), by

$$I_1 = k\left(R_1 + R_3 + R_5 + R_7 \ldots\right)^2 \approx \frac{k}{4}N^2 R_1^2 \tag{3.43}$$

3.4.5 Presence of odd numbered foci

The presence of only odd numbered foci both in case of positive and negative zone plates can be explained in the following way. The areas of all the Fresnel's half period zones are approximately equal. The area of a Fresnel's half period zone of a plane wavefront incident on a zone plate is approximately given by Eq. (3.2) as

$$A = \pi \lambda v$$

If v in this equation is replaced by focal length f_1 which is true for a plane incident wavefront, we can have

$$A = \pi \lambda f_1$$

or $\quad \pi\lambda \dfrac{f_1}{2m+1} = \dfrac{A}{2m+1}$ $\hfill (3.44)$

where $m = 0, 1, 2, 3, \ldots$, etc.

Putting $m = 1$ into Eq. (3.44), we get

$$\pi\lambda \frac{f_1}{3} = \frac{A}{3} \tag{3.45}$$

Equation (3.45) shows that when v becomes $\dfrac{f_1}{3}$, area of a Fresnel's zone becomes $\dfrac{A}{3}$, i.e., each Fresnel's half period zone is divided into three half period zones each having area $\dfrac{A}{3}$.

Let us consider the case of a positive zone plate (central zone is transparent). The transparent central zone, i.e., the first zone of the positive zone plate contains the 1st, 2nd, and 3rd half period zone and its second opaque zone contains the 4th, 5th, and 6th half period zone of the incident plane wavefront. Similarly, its transparent third zone contains the 7th, 8th, and 9th half period zone and the fourth opaque zone contains the 10th, 11th, and 12th half period zone of the incident plane wavefront. The odd numbered transparent half period zones of a positive zone plate allows the secondary wavelets to pass and its even numbered opaque half period zones does not allow the secondary wavelets to pass. Hence, the secondary wavelets originating from the 1st, 2nd, 3rd; 7th, 8th, and 9th; 13th, 14th, 15th, and etc., of the Fresnel's half period zones of the incident plane wavefront can pass through a positive zone plate. The resultants of secondary wavelets originating from these Fresnel's half period zones will have equal amplitudes and alternately, they have opposite phases. Therefore, the resultant amplitude A_3 at a distance of $\dfrac{f_1}{3}$ from the zone plate will be given by

$$A_3 = \left(R_1 - R_2 + R_3\right) + \left(R_7 - R_8 + R_9\right) + \left(R_{13} - R_{14} + R_{15}\right) + \left(R_{19} - R_{20} + R_{21}\right) + \dots$$

or $\quad A_3 \approx \left\{R_1 - \dfrac{1}{2}\left(R_1 + R_3\right) + R_3\right\} + \left\{R_7 - \dfrac{1}{2}\left(R_7 + R_9\right) + R_9\right\}$

$$+ \left\{R_{13} - \dfrac{1}{2}\left(R_{13} + R_{15}\right) + R_{15}\right\} + \left\{R_{19} - \dfrac{1}{2}\left(R_{19} + R_{21}\right) + R_{21}\right\} + \dots$$

$$= \left\{\dfrac{1}{2}R_1 + \dfrac{1}{2}R_3\right\} + \left\{\dfrac{1}{2}R_7 + \dfrac{1}{2}R_9\right\} + \left\{\dfrac{1}{2}R_{13} + \dfrac{1}{2}R_{15}\right\} + \left\{\dfrac{1}{2}R_{19} + \dfrac{1}{2}R_{21}\right\} + \dots$$

or $\quad A_3 \approx \dfrac{1}{2}\left(R_1 + R_3 + R_7 + R_9 + R_{13} + R_{15} + R_{19} + R_{21} + \dots\right)$

The intensity of the third order focus I_3 is given by

$$I_3 = \dfrac{k}{4}\left(R_1 + R_3 + R_7 + R_9 + R_{13} + R_{15} + R_{19} + R_{21} + \dots\right)^2 \tag{3.46}$$

Equations (3.43) and (3.46) proves that

$$I_3 < I_1$$

i.e., the intensity of the third order focus is less than the intensity of the primary/first order focus. The position of the third order focus from the zone plate is obviously

$$f_3 = \dfrac{f_1}{3} \tag{3.47}$$

3.4.6 Intensity of fifth order focus

Putting $m = 2$ into Eq. (3.44), we get

$$\pi\lambda\frac{f_1}{5} = \frac{A}{5}$$ (3.48)

Equation (3.48) shows that when v becomes $\frac{f_1}{5}$, area of a Fresnel's zone becomes $\frac{A}{5}$, i.e., each Fresnel's half period zone is divided into five half period zones each having area $\frac{A}{5}$.

As earlier, let us consider the case of a positive zone plate. The first zone of the positive zone plate contains the 1st, 2nd, 3rd, 4th, 5th half period zone and its second opaque zone contains the 6th, 7th, and 8th, 9th, 10th half period zone of the incident plane wavefront. Similarly its transparent third zone contains the 11th, 12th, 13th, 14th, 15th half period zone and the fourth opaque zone contains the 16th, 17th, 18th, 19th, 20th half period zone of the incident plane wavefront. As before the secondary wavelets originating from the 1st, 2nd, 3rd; 4th, 5th, 11th; 12th, 13th, 14th, 15th, and etc., of the Fresnel's half period zones of the incident plane wavefront can pass through a positive zone plate. The resultants of secondary wavelets originating from these Fresnel's half period zones will have equal amplitudes and alternately, opposite phases. Therefore, the resultant amplitude A_5 at a distance of $\frac{f_1}{5}$ from the zone plate will be given by

$$A_3 = \left(R_1 - R_2 + R_3 - R_4 + R_5\right) + \left(R_{11} - R_{12} + R_{13} - R_{14} + R_{15}\right) + \ldots$$

or $\quad A_5 \approx \left\{R_1 - \frac{1}{2}\left(R_1 + R_3\right) + R_3 - \frac{1}{2}\left(R_3 + R_5\right) + R_5\right\} +$

$$\left\{R_{11} - \frac{1}{2}\left(R_{11} + R_{13}\right) + R_{13} - \frac{1}{2}\left(R_{13} + R_{15}\right) + R_{15}\right\} + \ldots$$

or $\quad A_5 \approx \left\{R_1 - \frac{1}{2}R_1 - \frac{1}{2}R_3 + R_3 - \frac{1}{2}R_3 - \frac{1}{2}R_5 + R_5\right\} +$

$$\left\{R_{11} - \frac{1}{2}R_{11} - \frac{1}{2}R_{13} + R_{13} - \frac{1}{2}R_{13} - \frac{1}{2}R_{15} + R_{15}\right\} + \ldots$$

or $\quad A_5 \approx \left\{ \dfrac{1}{2} R_1 + \dfrac{1}{2} R_5 \right\} + \left\{ \dfrac{1}{2} R_{11} + \dfrac{1}{2} R_{15} \right\} + \dots$

or $\quad A_5 \approx \dfrac{1}{2} \left(R_1 + R_5 + R_{11} + R_{15} + R_{21} + R_{25} \right) \dots$

The intensity of the fifth order focus I_5 is given by

$$I_5 = \frac{k}{4} \left(R_1 + R_5 + R_{11} + R_{15} + \dots \right)^2 \tag{3.49}$$

Equations (3.43), (3.46), and (3.49) proves that

$$I_5 < I_3 < I_1$$

i.e., intensity of the fifth order focus is less than the intensity of the third order focus. In general, intensity of the higher order foci is less than the intensity of the lower order foci. Closer is the foci to the zone plate, lesser is the intensity of the foci. Though in the earlier discussions, we have taken the zone plate as positive, similar type of logic is applicable to a negative zone plate. The position of the fifth order focus from the zone plate is obviously

$$f_5 = \frac{f_1}{5} \tag{3.50}$$

Taking into account Eqs (3.42), (3.47), and (3.50), the general expression for odd numbered focal lengths is given by

$$f_{2m-1} = \frac{f_1}{2m-1} \quad \text{with} \quad f_1 = \frac{r_1^2}{\lambda} \tag{3.51}$$

or $\quad f_{2m-1} = \dfrac{r_1^2}{(2m-1)\lambda} \tag{3.52}$

where r_1 is the radius of the Fresnel's first half period zone of the incident plane wavefront on the transparent screen and $m = 1, 2, 3 \dots$. Equations (3.51) or (3.52) give the position of foci for a zone plate.

3.4.7 Absence of even numbered foci

The absence of even numbered foci both in case of positive and negative zone plates can be explained in the following way. The areas of all the Fresnel's half period zones are approximately equal. The area of a Fresnel's half period zone of a plane wavefront incident on a zone plate is approximately given by Eq. (3.2) as

$$A = \pi \lambda v$$

If v in this equation is replaced by focal length f which is true for a plane incident wavefront, we can have

$$A = \pi \lambda f_1$$

or $\quad \pi \lambda \dfrac{f_1}{2m} = \dfrac{A}{2m}$ $\hfill$ (3.53)

where $m = 1, 2, 3, \ldots$, etc.,
   Putting $m = 1$ into Eq. (3.53), we get

$$\pi \lambda \dfrac{f_1}{2} = \dfrac{A}{2}$$ $\hfill$ (3.54)

Equation (3.54) shows that when v becomes $\dfrac{f_1}{2}$, area of a Fresnel's zone becomes $\dfrac{A}{2}$, i.e., each Fresnel's half period zone is divided into two half period zones each having area $\dfrac{A}{2}$.

   Let us consider the case of a positive zone plate again. The transparent central zone, i.e., the first zone contains the 1st, and 2nd half period zone and the second opaque zone contains the 3rd, and 4th half period zone of the incident plane wavefront. Similarly, the transparent third zone contains the 5th and 6th half period zone and the fourth opaque zone contains the 7th and 8th half period zone of the incident plane wavefront. The odd numbered transparent half period zones of a positive zone plate allow the secondary wavelets to pass and its even numbered opaque half period zones do not allow the secondary wavelets to pass. Hence, the secondary wavelets originating from the 1st, 2nd and 5th; 6th, …, etc., of the Fresnel's half period zones of the incident plane wavefront can pass through a positive zone plate. The resultants of secondary wavelets originating from these Fresnel's half period zones will have equal amplitudes and alternately, opposite phases. Therefore, the resultant amplitude A_2 at a distance of $\dfrac{f_1}{2}$ from the zone plate will be given by

$$A_2 = (R_1 - R_2) + (R_5 - R_6) + (R_9 - R_{10}) + (R_{13} - R_{14}) + \ldots$$ $\hfill$ (3.55)

The magnitudes of the bracketed terms in Eq. (3.55) are very small because $R_1 \approx R_2$, $R_5 \approx R_6$, $R_9 \approx R_{10}$, $R_{13} \approx R_{14}$, and etc. Hence, the resultant amplitude A_2 will be very small as a result of which intensity $I_2 \left(= A_2^2 \right)$ at a distance of $\dfrac{f_1}{2}$ from the zone plate will be negligibly low.

3.4.8 Intensity of the fourth order focus

Putting $m = 2$ into Eq. (3.53), we get

$$\pi\lambda\frac{f_1}{4} = \frac{A}{4} \tag{3.56}$$

Equation (3.56) shows that when v becomes $\dfrac{f_1}{4}$, area of a Fresnel's zone becomes $\dfrac{A}{4}$, i.e., each Fresnel's half period zone is divided into four half period zones each having area $\dfrac{A}{4}$.

As earlier, let us consider the case of a positive zone plate. The first zone contains the 1st, 2nd, 3rd, 4th half period zone and the second opaque zone contains the 5th, 6th, 7th, 8th half period zone of the incident plane wavefront. Similarly, the transparent third zone contains the 9th, 10th, 11th, 13th half period zone and the fourth opaque zone contains the 14th, 15th, 16th, 17th half period zone of the incident plane wavefront. As before, the secondary wavelets originating from the 1st, 2nd, 3rd; 4th, 9th; 10th, 11th, 12th, and etc., of Fresnel's half period zones of the incident plane wavefront can pass through a positive zone plate. The resultants of secondary wavelets originating from these Fresnel's half period zones will have equal amplitudes and alternately opposite phases. Therefore, the resultant amplitude A_4 at a distance of $\dfrac{f_1}{4}$ from the zone plate will given by

$$A_4 = \left(R_1 - R_2 + R_3 - R_4\right) + \left(R_9 - R_{10} + R_{11} - R_{12}\right) + \left(R_{17} - R_{18} + R_{19} - R_{20}\right) + \ldots$$

The magnitudes of the bracketed terms in this equation are negligibly very small because $R_1 \approx R_2 \approx R_3 \approx R_4$, $R_9 \approx R_{10} \approx R_{11} \approx R_{12}$, $R_{17} \approx R_{18} \approx R_{19} \approx R_{20}$ … etc. Hence, the resultant amplitude A_4 will be very small as a result of which intensity I_4 at a distance of $\dfrac{f_1}{4}$ from the zone plate will be negligibly low.

Similar logic holds for all the other even numbered foci. Therefore, practically, even numbered foci do not exist for positive zone plates. Though in these discussions we have only considered positive zone plates, a similar logic is applicable to negative zone plates.

3.4.9 Comparison of a zone plate with a convex lens

The similarities and dissimilarities between a convex lens and a zone plate are elaborated here.

Similarities

	Convex lens	Zone plate
1	Convex lens has focusing action.	Zone plate also has focusing action.
2	The real image is formed on the other side of the source.	The real image is also formed on the other side of the source.
3	Focal length of a convex lens depends upon the wavelength.	Focal length of a convex lens also depends upon the wavelength.
4	Convex lens suffers chromatic aberrations.	Zone plate also suffers chromatic aberrations.
5	In a convex lens, the image distance and object distance are connected by $\dfrac{1}{v}-\dfrac{1}{u}=\dfrac{1}{f}$	In a zone plate, the image distance and object distance are also connected by a similar equation $\dfrac{1}{v}-\dfrac{1}{u}=\dfrac{n\lambda}{r_n^2}$

Dissimilarities

	Convex lens	Zone plate
1	Focusing action is due to refraction of light through the lens	Focusing action is due to diffraction of light through the zone plate.
2	All the light waves refracting through a lens are in phase.	The secondary wavelets diffracted through a zone plate are not in phase.
3	All the monochromatic light waves refracting through a lens travel the same optical paths.	The monochromatic secondary wavelets diffracting through a zone plate travel different optical paths.
4	For monochromatic light, a convex lens has a single focus.	For monochromatic light a zone plate has multiple foci.
5	The intensity of the focus is maximum in case of convex lens.	The intensity of the primary focus of a zone plate is less than the focus of a lens.
6	In case of convex lens, the focal length of violet light is less than that of red light.	In case of a zone plate, the focal length of violet light is more than that of red light.
7	For a given wavelength of light, a convex lens has a single focal length given by $\dfrac{1}{f}=(\mu-1)\left(\dfrac{1}{R_1}-\dfrac{1}{R_2}\right)$	For a given wavelength of light, a zone plate has multiple focal lengths given by $f_{2m+1}=\dfrac{r_1^2}{(2m+1)\lambda}$

8	A convex lens has no virtual focus on the side of the source.	A zone plate has virtual foci on the side of the source.
9	A convex lens cannot act like a concave lens in a single medium.	A zone plate can act like a concave lens.
10	A concave lens only acts like a converging lens.	A zone plate can act like a converging as well as a diverging lens.

Example 3.1

The principal focal length of a positive zone plate is 1.5 m for a light of wavelength 6500 Å. Determine the radii of the 1st, 2nd, and 3rd transparent zones.

Solution

The radius of the nth half period zones of a zone plate is given by

$$r_n = \sqrt{nf\lambda}$$

The radius of the first half period zone will be

$$r_1 = \sqrt{f\lambda} = \sqrt{150 \times 6500 \times 10^{-8}} \text{ cm} = 0.10 \text{ cm}.$$

In a zone plate, alternate zones are transparent. Hence, the radius of the 2nd and 3rd half period zones are obtained by putting $n = 3$ and 5 respectively. Hence, we have

$$r_2 = r_1\sqrt{3} = 0.17 \text{ cm} \quad \text{and} \quad r_3 = r_1\sqrt{5} = 0.22 \text{ cm}$$

Example 3.2

What is the radius of the first zone of a zone plate with focal length 30 cm for a light of wavelength 6000 Å?

Solution

The data given are

$$n = 1$$

$$f = 30 \text{ cm}$$

$$\lambda = 6000 \text{ Å} = 6 \times 10^{-5} \text{ cm}$$

$$r_1 = ?$$

The required formula is (3.42), $f_1 = \dfrac{r_1^2}{\lambda}$

or $r_1 = \sqrt{f_1 \lambda} = \sqrt{30 \times 6 \times 10^{-5}} \text{ cm} = 0.042 \text{ cm}$

The radii of the other zones of the zone plate for the same light will be $r_2 = r_1 \sqrt{n}$, $n = 2, 3, \ldots$

Example 3.3

A point source of light of wavelength 5000 Å A is placed on the axis of a zone plate. The strongest and next strongest images of the source are formed on the other side of the zone plate at distances of 30 cm and 10 cm respectively. Calculate (a) the distance of the point source from the zone plate, (b) the radius of the first zone and (c) the principal focal length of the zone plate.

Solution

The data given are $\lambda = 6000$ Å $= 6 \times 10^{-5}$ cm, $v_1 = +30$ cm, and $v_2 = +5$ cm. We know that

$$\frac{1}{v} - \frac{1}{u} = \frac{(2m-1)\lambda}{r_1^2} = \frac{1}{f_{2m-1}}$$

Hence, we can have

$$\frac{1}{v_1} - \frac{1}{u} = \frac{\lambda}{r_1^2}$$

and

$$\frac{1}{v_2} - \frac{1}{u} = \frac{3\lambda}{r_1^2}$$

Putting the values of v_1 and v_2 into these equations, we get

$$\frac{1}{30} - \frac{1}{u} = \frac{\lambda}{r_1^2}$$

$$\frac{1}{5} - \frac{1}{u} = \frac{3\lambda}{r_1^2}$$

Solving the aforementioned two equation for u and r_1, we get

$$u = -20 \text{ cm} \Rightarrow$$

that point source is placed at a distance of 20 cm to the left side of the zone plate.

and $r_1 = 0.027$ cm

The principal focal length of the zone plate ($m = 1$) is given by

$$f_1 = \frac{r_1^2}{\lambda} = 12 \, \text{cm}$$

Example 3.4

A zone plate is constructed by taking the print of Newton's ring formed by a plano-convex lens of radius of curvature 1 m on a thin glass plate. Find the principal focal length of the zone plate constructed in this way.

Solution

The square of the radii of the nth order dark ring of Newton's rings formed in an air medium is given by

$$r_n^2 = n\lambda R$$

or $r_1^2 = \lambda R$

where R = radius of curvature of the curved surface of the plano-convex lens.
The principal focal length of a zone plate is given by

$$f_1 = \frac{r_1^2}{\lambda}$$

or $f_1 = \dfrac{\lambda R}{\lambda} = R = 1 \, \text{m}$

Example 3.5

The radius of the first half period zone of a zone plate is 0.05 cm. What should be the position of a screen so that the brightest spot is formed on the screen when a plane monochromatic light of wavelength 5890 Å is incident normally on the zone plate.

Solution

The data given are

$r_1 = 0.05$ cm

$$\lambda = 5890 \text{ Å} = 5890 \times 10^{-8} \text{ cm}$$

The screen should be at the principal focal plane of the zone plate so that the brightest spot will be formed. The principal focal length of a zone plate is given by

$$f_1 = \frac{r_1^2}{\lambda} = \frac{0.05^2}{5890 \times 10^{-8}} \text{ cm} = 42.4 \text{ cm}$$

Example 3.6

The diameter of the central zone of a zone plate is 0.20 cm. If a point source of light of wavelength 5500 Å is placed at a distance of 50 cm from the zone plate, find the positions of the strongest image and other weaker images.

Solution

The data given are

$$r_1 = \frac{0.2 \text{ cm}}{2} = 0.1 \text{ cm} \quad \text{and} \quad \lambda = 5500 \text{ Å} = 5500 \times 10^{-8} \text{ cm}.$$

The positions all the point images are given by

$$f_{2m-1} = \frac{r_1^2}{(2m-1)\lambda}, \quad m = 1, 2, 3, \ldots.$$

The position of the strongest point image ($m = 1$) will be

$$f_1 = \frac{r_1^2}{\lambda} = \frac{0.1^2}{5500 \times 10^{-8}} \text{ cm} = 182 \text{ cm}$$

The position of the other point images ($m = 2, 3, 4, \ldots$) will be

$$f_3 = \frac{r_1^2}{3\lambda} = 60.6 \text{ cm}$$

$$f_5 = \frac{r_1^2}{5\lambda} = 36.4 \text{ cm} \quad \text{and etc.}$$

3.5 Fraunhofer Diffraction

In the previous sections, we have discussed the Fresnel class of diffraction in which separation between source and diffracting aperture is finite. Now we shall discuss the Fraunhofer class of diffraction in which separation between the source and diffracting aperture is infinite. Accordingly, the incident wavefront is plane. Secondary wavelets originating from the plane of the aperture are in phase and the diffracted wavefront is in plane.

3.5.1 Fraunhofer diffraction due to a single slit

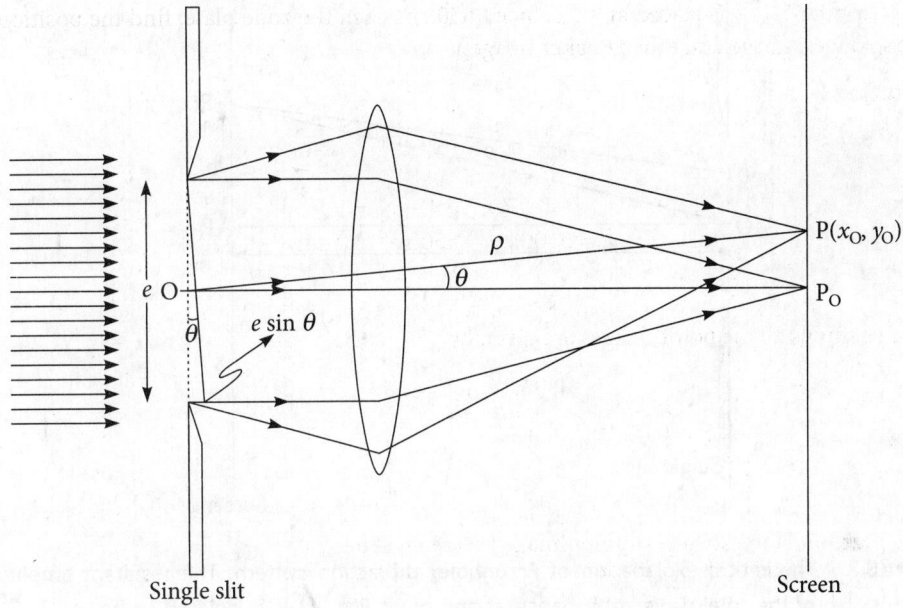

Single slit   Screen

Figure 3.6 | Fraunhofer diffraction of a plane wavefront due to a single slit consists of a central bright band and fainter side bands on a screen. The path difference between the extreme diffracted rays, $e \sin \theta$, has been shown

In Fig. 3.6, a plane wavefront of monochromatic light of wavelength λ is incident on a slit of width e so that the plane of the slit and the plane wavefront is parallel to each other. When the diffracted beam is focused on a screen with the help of a converging lens, a Fraunhofer diffraction pattern consisting of a central bright band and fainter side bands is obtained on the screen. The theoretical explanation of this observation is as follows.

In the Fig. 3.7, the point O, the origin of our coordinate system is the centre of the slit of width e. The plane of the slit coincides with the Y-axis and the central line OP_0 is along the X-axis. $P(x_0, y_0)$ is a point on the screen. OP makes an angle θ with OP_0. Let dy having coordinates $(0, y)$ be an elemental portion of the incident wavefront on the plane of the slit.

r = The distance between dy and $P(x_0, y_0)$.

x_0 = The distance between the plane of the slit and the screen.

ρ = $OP(x_0, y_0)$ = The distance between O and $P(x_0, y_0)$

θ = The angle between OP_0 and $OP(x_0, y_0)$.

$(0, y)$ = Coordinate of dy

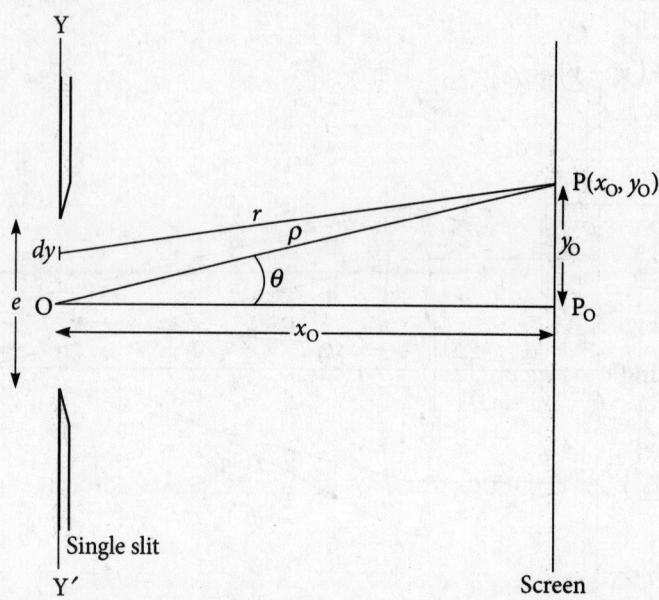

Figure 3.7 | Theoretical explanation of Fraunhofer diffraction pattern. The resultant amplitude of the total wave disturbance at any point $P(x_0, y_0)$ is calculated to be $A = ke\dfrac{\sin \alpha}{\alpha}$ with $\alpha = \dfrac{\pi e \sin \theta}{\lambda}$

The wave disturbance $d\psi$ at the point $P(x_0, y_0)$ due to dy at any given instant is given by

$$d\psi = kdy \sin 2\pi \left(\frac{t}{T} - \frac{r}{\lambda} \right)$$

The total wave disturbance ψ at the point $P(x_0, y_0)$ due to the entire wavefront of width e at any given instant is given as

$$\psi = \int_{-\frac{e}{2}}^{+\frac{e}{2}} k \sin 2\pi \left(\frac{t}{T} - \frac{r}{\lambda} \right) dy \tag{3.57}$$

From the Fig. 3.7, we get

$$r^2 = x_0^2 + (y_0 - y)^2 \tag{3.58}$$

$$\rho^2 = x_0^2 + y_0^2 \tag{3.59}$$

Putting the value of x_0^2 from Eq. (3.59) into Eq. (3.58), we get

$$r^2 = \rho^2 - y_0^2 + (y_0 - y)^2 = \rho^2 + y^2 - 2yy_0$$

or $\quad$ $$r^2 = \rho^2 \left(1 + \frac{y^2}{\rho^2} - \frac{2yy_0}{\rho^2} \right) \tag{3.60}$$

As shown in the Fig. 3.7, $\rho \gg y$ as a result of which $\dfrac{y^2}{\rho^2}$ is negligibly small. Hence, in Eq. (3.60), neglecting $\dfrac{y^2}{\rho^2}$, we obtain

$$r = \rho \left(1 - \frac{2yy_0}{\rho^2} \right)^{\frac{1}{2}} \tag{3.61}$$

As $\dfrac{2yy_0}{\rho^2}$ is much less than 1, we can expand the RHS of Eq. (3.61) up to the second term to have

$$r = \rho - \frac{yy_0}{\rho} \tag{3.62}$$

From the Fig. 3.7, we can have

$$\frac{y_0}{\rho} = \sin\theta$$

Putting this value of $\dfrac{y_0}{\rho}$ into Eq. (3.62), we get

$$r = \rho - y\sin\theta \tag{3.63}$$

Substituting this value of r into Eq. (3.57), we get

$$\psi = \int_{-\frac{e}{2}}^{+\frac{e}{2}} k \sin 2\pi \left(\frac{t}{T} - \frac{\rho - y \sin \theta}{\lambda} \right) dy$$

or $\quad \psi = \int_{-\frac{e}{2}}^{+\frac{e}{2}} k \sin 2\pi \left(\frac{t}{T} - \frac{\rho}{\lambda} + \frac{y \sin \theta}{\lambda} \right) dy$ $\hfill$ (3.64)

For a given point and given wavelength at any particular instant, T, t, ρ λ, and θ do not change with respect to y. The integration (3.64) can be evaluated by the method of substitution.

Let

$$u = 2\pi \left(\frac{t}{T} - \frac{\rho}{\lambda} + \frac{y \sin \theta}{\lambda} \right) \hfill (3.65)$$

or $\quad du = 2\pi \dfrac{\sin \theta}{\lambda} dy$

or $\quad dy = \dfrac{\lambda}{2\pi \sin \theta} du$ $\hfill$ (3.66)

Putting these values of dy and u into Eq. (3.64), we have

$$\psi = \int_{-\frac{e}{2}}^{+\frac{e}{2}} k \sin u \times \frac{\lambda}{2\pi \sin \theta} du$$

or $\quad \psi = k \dfrac{\lambda}{2\pi \sin \theta} \displaystyle\int_{-\frac{e}{2}}^{+\frac{e}{2}} \sin u\, du$

or $\quad \psi = k \dfrac{\lambda}{2\pi \sin \theta} \left(-\cos u \Big|_{-\frac{e}{2}}^{+\frac{e}{2}} \right)$

Re-substituting the value of u from Eq. (3.65) into this equation, we get

$$\psi = -k\frac{\lambda}{2\pi\sin\theta}\left(\cos 2\pi\left(\frac{t}{T}-\frac{\rho}{\lambda}+\frac{y\sin\theta}{\lambda}\right)\right)\Big|_{-\frac{e}{2}}^{+\frac{e}{2}}$$

$$\psi = -k\frac{\lambda}{2\pi\sin\theta}\left(\cos 2\pi\left(\frac{t}{T}-\frac{\rho}{\lambda}+\frac{e\sin\theta}{2\lambda}\right)-\cos 2\pi\left(\frac{t}{T}-\frac{\rho}{\lambda}-\frac{e\sin\theta}{2\lambda}\right)\right)$$

or $\quad \psi = k\dfrac{\lambda}{\pi\sin\theta}\sin 2\pi\left(\dfrac{t}{T}-\dfrac{\rho}{\lambda}\right)\sin\left(\dfrac{\pi e\sin\theta}{\lambda}\right)$

$$= ke\frac{\lambda}{\pi e\sin\theta}\sin 2\pi\left(\frac{t}{T}-\frac{\rho}{\lambda}\right)\sin\left(\frac{\pi e\sin\theta}{\lambda}\right)$$

$$= \frac{ke}{\alpha}\sin 2\pi\left(\frac{t}{T}-\frac{\rho}{\lambda}\right)\sin\alpha$$

or $\quad \psi = ke\dfrac{\sin\alpha}{\alpha}\sin 2\pi\left(\dfrac{t}{T}-\dfrac{\rho}{\lambda}\right)$ $\hfill$ (3.67)

where $\quad \alpha = \dfrac{\pi e\sin\theta}{\lambda}$ $\hfill$ (3.68)

Equation (3.67) is of the form

$$\psi = A\sin\left(\frac{t}{T}-\frac{\rho}{\lambda}\right) \quad \text{with} \quad A = ke\frac{\sin\alpha}{\alpha}.$$

Hence, Eq. (3.67) shows that the amplitude A of the resultant wave disturbance at any point $P(x_0, y_0)$ due to all the secondary wavelets is

$$A = ke\frac{\sin\alpha}{\alpha} \hfill (3.69)$$

Therefore, the resultant intensity I at $P(x_0, y_0)$ will be

$$I = k^2 e^2 \frac{\sin^2 \alpha}{\alpha^2} \tag{3.70}$$

As shown in Fig. 3.7, the central band corresponds to $\theta \to 0$. When $\theta \to 0$ according to Eq. (3.68), $\alpha \to 0$ and the value of

$$\lim_{\alpha \to 0} \frac{\sin \alpha}{\alpha} = 1$$

as a result of which the intensity of the central band I_0 will be given by

$$I_0 = k^2 e^2 \tag{3.71}$$

Putting this value into Eq. (3.70), the expression for the intensity at any point $P(x_0, y_0)$ on the screen will be given by

$$I = I_0 \frac{\sin^2 \alpha}{\alpha^2} \quad \text{with} \quad \alpha = \frac{\pi e \sin \theta}{\lambda} \tag{3.72}$$

3.5.2 Intensity distribution

Now we shall proceed to analyze the intensity distribution on the screen due to Fraunhofer diffraction by a single slit. The resultant amplitude A of the wave disturbance at any point $P(x_0, y_0)$ due to all the secondary wavelets is given by Eq. (3.69). Expanding the $\sin \alpha$ in the power series, we have from Eq. (3.69)

$$A = \frac{ke}{\alpha} \left(\alpha - \frac{\alpha^3}{3!} + \frac{\alpha^5}{5!} - \frac{\alpha^7}{7!} + \dots \right)$$

or $\quad A = ke \left(1 - \frac{\alpha^2}{3!} + \frac{\alpha^4}{5!} - \frac{\alpha^6}{7!} + \dots \right) \tag{3.73}$

Principal maximum

All the terms after second term in Eq. 3.73 are negligibly small and the amplitude A will be maximum if second term is made zero. This is possible when $\alpha = 0$. For a central band as discussed earlier, $\alpha = 0$. When this value is put into Eq. (3.73), it gives the maximum value of the amplitude as

$$A = ke$$

Hence, at the central band, the resultant amplitude of all secondary wavelets is maximum and the intensity at the central band will be maximum. The central band in case of the Fraunhofer single slit diffraction is the brightest band having intensity

$$I_O = k^2 e^2$$

Here $\alpha = 0$ implies that $\theta = 0$, which corresponds to the central band. This shows that the central band is formed by the secondary wavelets originating from the slit travelling perpendicular ($\theta = 0$) to the plane of the slit. This maximum is called the principal maximum. The principal maximum is surrounded symmetrically by alternate dark and bright bands of decreasing intensity as we move away from the principal maximum.

Positions of dark bands

According to Eqs (3.69) or (3.72), the intensity will be minimum at the points for which

$$\sin \alpha = 0$$

or $\alpha = \pm n\pi$ with $n = 1, 2, 3, 4, \ldots$

or $$\frac{\pi e \sin \theta}{\lambda} = \pm n\pi$$

or $$e \sin \theta = \pm 2n \frac{\lambda}{2} \tag{3.74}$$

As shown in Fig. 3.6, the path difference between extreme diffracted rays is $e \sin \theta$. Then according to Eq. (3.74), the rays diffracted at an angle θ will interfere destructively if the path difference between extreme diffracted rays is an even multiple of $\frac{\lambda}{2}$. The same equation also proves that Fraunhofer diffraction does not occur if slit width is less than the wavelength of the monochromatic light used. $\alpha = \pm n\pi$ gives the positions of the minima or dark bands in case of single slit diffraction pattern and are shown in Fig. 3.8.

Positions of secondary maxima

In the Fraunhofer diffraction pattern in addition to the principal maximum, the secondary maxima of lower intensity are also present on both sides of the principal maximum on the screen. In between the secondary maxima, there are dark bands. The positions of the secondary maxima are found in the following way. The intensity I is maximum with respect to α if $\dfrac{dI}{d\alpha} = 0$ and $\dfrac{d^2 I}{d\alpha^2} < 0$. Applying this concept to Eq. (3.72), we have

$$\frac{d}{d\alpha} I_0 \frac{\sin^2 \alpha}{\alpha^2} = 0$$

or $\quad 2I_0 \dfrac{\sin \alpha}{\alpha} \left(\dfrac{\alpha \cos \alpha - \sin \alpha}{\alpha^2} \right) = 0$

This equation shows that either $\dfrac{\sin \alpha}{\alpha} = 0$ or $\left(\dfrac{\alpha \cos \alpha - \sin \alpha}{\alpha^2} \right) = 0$. As discussed earlier, $\dfrac{\sin \alpha}{\alpha} = 0$ (or $\sin \alpha = 0$) gives the position of dark bands. The positions of the secondary maxima are given by the conditions

$$\left(\frac{\alpha \cos \alpha - \sin \alpha}{\alpha^2} \right) = 0$$

or $\quad \alpha \cos \alpha - \sin \alpha = 0$

or $\quad \alpha \cos \alpha = \sin \alpha$

or $\quad \tan \alpha = \alpha$ $\hspace{4cm}$ (3.75)

From Eq. (3.75), the values of α can be found out graphically. As shown in Fig. 3.8(b), $y = \alpha$ and $y = \tan \alpha$ are plotted on the same graph. From the points of intersection of the two plots, the values of α are determined. The values of α as obtained from Fig. 3.8(b) are given approximately by

$$\alpha = 0 \quad \text{and} \quad \alpha \approx \pm(2n+1)\frac{\pi}{2}$$

As discussed earlier, $\alpha = 0$ gives the position of the principal/primary maximum. $\alpha \approx \pm(2n+1)\dfrac{\pi}{2}$ gives the positions of the secondary maxima. Putting $\alpha \approx \pm(2n+1)\dfrac{\pi}{2}$ into Eq. (3.72), the intensities of the secondary maxima are obtained as

$$I_n \approx \frac{I_0}{\left(\dfrac{2n+1}{2} \right)^2 \pi^2} \left(\sin(2n+1)\frac{\pi}{2} \right)^2$$

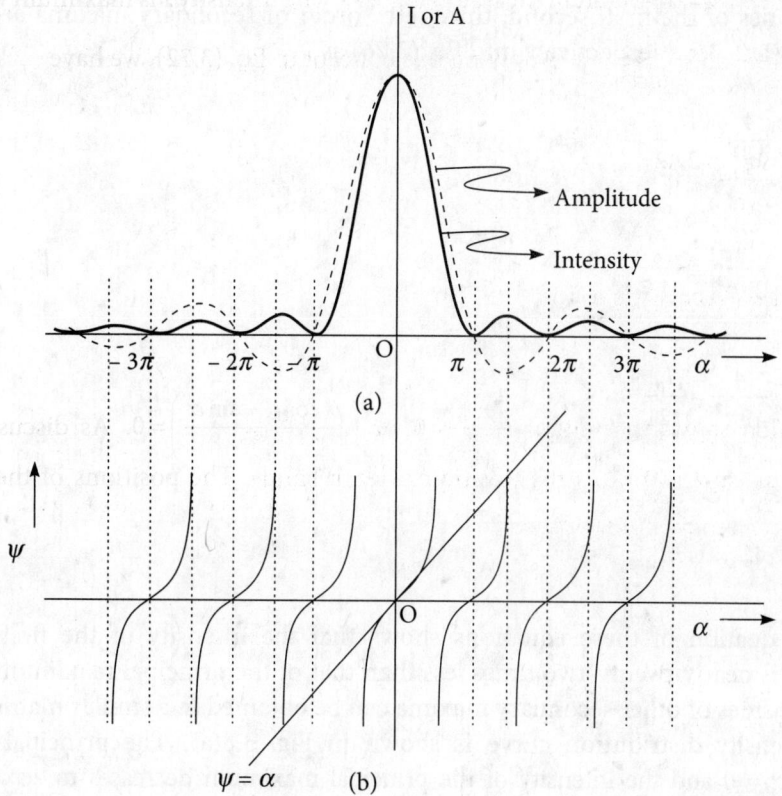

Figure 3.8 (a) shows the graphical solution of $\tan \alpha = \alpha$. $y = \alpha$ and $y = \tan \alpha$ plotted on the same graph. From the points of intersection of the two plots, the values of α are determined. (b) Shows the distribution of intensity with respect to α. $\alpha = 0$ gives the position of the principal/primary maximum, $\alpha \approx \pm(2n+1)\dfrac{\pi}{2}$ give the positions of the secondary maxima and $\alpha = \pm n\pi$ gives the positions of the minima

or
$$I_n \approx \frac{I_0}{\left(\dfrac{2n+1}{2}\right)^2 \pi^2} (\pm 1)^2$$

or
$$I_n \approx \frac{I_0}{\left(\dfrac{2n+1}{2}\right)^2 \pi^2}$$

or
$$I_n \approx I_0 \frac{4}{(2n+1)^2 \pi^2} \tag{3.76}$$

The intensities of the first, second, third, etc., order of secondary maxima are found by putting $n = 1, 2, 3, \ldots$ respectively into Eq. (3.76) as

$$I_1 \approx I_0 \frac{4}{9\pi^2} = \frac{I_0}{22.2}$$

$$I_2 \approx I_0 \frac{4}{25\pi^2} = \frac{I_0}{61.7}$$

$$I_3 \approx I_0 \frac{4}{49\pi^2} = \frac{I_0}{121}$$

....

....

The first equation of these equations shows that the intensity of the first secondary maximum is nearly twenty-two times less than that of the principal maximum. Inference about intensities of other secondary maxima can be obtained in a similar manner.

The intensity distribution curve is shown in Fig. 3.8(a). The principal maximum occurs at $\alpha = 0$ and the intensity of the principal maximum decreases to zero at $\alpha = \pm\pi$. The 1st, 2nd, 3rd, ... order secondary maxima occur at

$$\alpha \approx \frac{3\pi}{2}, \alpha \approx \frac{5\pi}{2}, \alpha \approx \frac{7\pi}{2}\ldots$$

respectively. In general, the $(2n + 1)$th order secondary maxima occur at

$$\alpha_{2n-1} \approx \pm\frac{(2n+1)\pi}{2}, \quad n = 1, 2, 3, \ldots$$

or $$\frac{\pi e \sin\theta_{2n-1}}{\lambda} \approx \pm\frac{(2n+1)\pi}{2}$$

or $$\sin\theta_{2n-1} \approx \pm\frac{(2n+1)\lambda}{2e}$$

or $$\theta_{2n-1} \approx \sin^{-1}\left(\frac{\pm(2n+1)\lambda}{2e}\right)$$

This equation shows that the position of the secondary maxima depends upon the relative values of slit width and wavelength of the light.

3.5.3 Width of the principal maximum

The direction of the first dark band is obtained from Eq. (3.74) as

$$\sin\theta = \frac{\lambda}{e}$$

$$\theta = \sin^{-1}\frac{\lambda}{e} \tag{3.77}$$

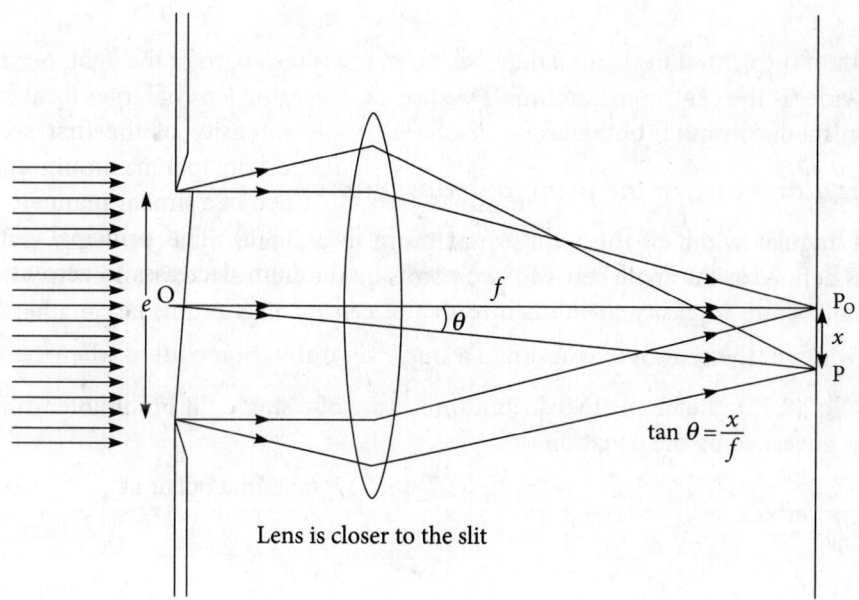

$$\tan\theta = \frac{x}{f}$$

Lens is closer to the slit

Figure 3.9 | Calculation of $\tan\theta = \dfrac{x}{f}$ is depicted. P_0 is the mid-point of the central maximum. P is the centre of the first minimum. x is half the width of the central maximum. When a plane wavefront is incident on the slit, it is diffracted through an angle of θ downward. The diffraction may be upward.

As shown in Fig. 3.9, we have

$$\tan\theta = \frac{x}{f} \tag{3.78}$$

where x is the distance between the centre of the central maximum and centre of the first minimum, i.e., x is half the width of the central maximum. Putting the value of θ from Eq. (3.77) into Eq. (3.78), we obtain

$$\frac{x}{f} = \tan \, \sin^{-1} \frac{\lambda}{e}$$

or $x = f \times \tan \, \sin^{-1} \frac{\lambda}{e} = $ half the width of the central maximum.

Hence, the width of the central maximum $2x$ is

$$2x = 2f \times \tan \, \sin^{-1} \frac{\lambda}{e} \tag{3.79}$$

The width of the central maximum depends upon the wavelength of the light. Narrower is the slit, wider is the central maximum. If we use a converging lens of larger focal length, a wider central maximum is obtained.

Half angular width of the principal maximum

The half angular width of the central maximum in a single slit Fraunhofer diffraction pattern is defined as the angle between two points on the central maximum where intensity is one half of the intensity at the centre of the central maximum, i.e., mathematically $\theta =$ half width of the central maximum in a single slit diffraction pattern when $I = \frac{I_0}{2}$.

From Eq. (3.72), the intensity distribution in case of a single slit Fraunhofer diffraction pattern is governed by the equation

$$I = I_0 \frac{\sin^2 \alpha}{\alpha^2} \quad \text{with} \quad \alpha = \frac{\pi e \sin \theta}{\lambda}$$

or $$\frac{1}{2} = \frac{\sin^2 \alpha}{\alpha^2}$$

or $$\frac{\sin \alpha}{\alpha} = \frac{1}{\sqrt{2}}$$

By solving this equation graphically for α $\left(\text{with } \alpha = \dfrac{\pi e \sin \theta}{\lambda}\right)$, the half width of the central maximum θ can be found. The graphical solution of the aforementioned equation gives

$$\alpha = 1.391$$

$$\frac{\pi e \sin\theta}{\lambda} = 1.391$$

or $\quad \sin\theta = 1.391\dfrac{\lambda}{\pi e}$

or $\quad \theta = \sin^{-1}\left(\dfrac{1\cdot 391\lambda}{\pi e}\right)$

is the half angular width of the central maximum in a single slit diffraction pattern.

There are some clear differences between interference and diffraction phenomena in optics.

	Interference	Diffraction
i	Interference phenomenon occurs between two waves originating from two coherent sources.	Diffraction phenomenon occurs between secondary wavelets originating from the unobstructed portion of the incident wavefront.
ii	The dark fringe of the interference fringe is almost perfectly dark.	The dark band of the diffraction band is not perfectly dark.
iii	In the interference pattern, the spacings of the interference fringes may (double slit) or may not (Newton's rings) be uniform.	In the diffraction pattern, the spacings of the diffraction bands are never equal.
iv	In the interference pattern, the maxima are of equal intensity.	In the diffraction pattern, the maxima are of unequal intensity.

In the following, we mention the basic differences between Fresnel's diffraction and Fraunhofer's diffraction.

	Fresnel's diffraction	Fraunhofer's diffraction
i	In Fresnel's diffraction, the separation between the light source and the obstacle is finite.	In Fraunhofer's diffraction, the separation between the light source and the obstacle is infinite.
ii	The wavefront incident on the obstacle is either spherical or cylindrical.	The wavefront incident on the obstacle is plane.
iii	No lens is used.	A lens is used to produce a plane wavefront from spherical or cylindrical wavefronts.

Example 3.7

A single slit of width 12×10^{-5} cm is illuminated by a parallel beam of monochromatic light of wavelength 6000 Å. Calculate the angular width of the central maximum.

Solution

The data given are $e = 12 \times 10^{-5}$ cm and $\lambda = 6000$ Å $= 6 \times 10^{-5}$ cm.

If θ is the angular width of the central maximum, we have from Eq. (3.77)

$$\sin \theta = \frac{\lambda}{e}$$

or $\theta = \sin^{-1} \dfrac{6 \times 10^{-5}}{12 \times 10^{-5}} = 30°$

Example 3.8

A single slit is illuminated by a parallel beam of monochromatic light of wavelength 6000 Å. Calculate the slit width if the angular width of the central maximum is 20°.

Solution

The angular width of the central maximum is 20°. Hence, the half angular width of the central maximum will be 10°. The slit width will be given by

$$e = \frac{\lambda}{\sin \theta} = \frac{6000 \times 10^{-8}}{\sin 10} = 3.46 \times 10^{-4} \text{ cm}$$

Example 3.9

A slit of width 0.02 cm is illuminated by a parallel beam of monochromatic light. A screen is placed at a distance of 2 m from the converging lens to obtain the diffraction pattern. Calculate the wavelength of the monochromatic light if the first minima lie at 0.5 cm on either side of the central maximum.

Solution

The screen is placed at a distance of 2 m from the converging lens to obtain the diffraction pattern due to a single slit. Hence, the focal length of the converging lens must be 2 m.

The data given are $e = 0.02$ cm, $f = 2$ m $= 200$ cm, $x = 0.5$ cm.

From Eq. (3.79), we have

$$x = \frac{\lambda f}{e}$$

or $\lambda = \dfrac{xe}{f} = \dfrac{0.5 \times 0.02}{200}$ cm $= 5000$ Å

Example 3.10

A slit of width 0.02 cm is illuminated by a parallel beam of monochromatic light of wavelength 5000 Å. The diffracted beam is focused on a screen by a converging lens of focal length 100 cm to obtain the diffraction pattern. Calculate the distance between the centre of the first minima and the central maximum on the screen.

Solution

The data given are $e = 0.02$ cm, $\lambda = 5000$ Å, and $f = 100$ cm
The distance x between the centre of the first minima and the central maximum on the screen is given by

$$x = \frac{\lambda f}{e} = \frac{5000 \times 10^{-8} \times 100}{0.02} \text{ cm} = 0.25 \text{ cm}$$

Example 3.11

For what value of slit width will the first minimum for sodium vapor light of wavelength 5890 Å fall at $\theta = 20°$?

Solution

The data given are $n = 1$, $\lambda = 5890$ Å and $\theta = 20°$.
The direction of the nth minimum of diffraction pattern due to a single slit is given by

$$\sin \theta = \frac{n\lambda}{e}$$

Hence, we have $e = \dfrac{n\lambda}{\sin \theta} = \dfrac{1 \times 5890 \times 10^{-8}}{\sin 20} \text{ cm} = 1.72 \times 10^{-4} \text{ cm}$

Example 3.12

In a single slit diffraction pattern, the distance between the first minimum on left to the first minimum on the right is 0.0546 cm for a light of wavelength 5460 Å. Calculate the slit width if the distance of the screen from the slit is 80 cm.

Solution

The data given are $x = 0.0273$ cm, $\lambda = 5460$ Å $= 5460 \times 10^{-8}$ cm, and $f = 80$ cm.
In a single slit diffraction pattern the distance between the first minimum on left to the first minimum on the right is given by

$$2x = \frac{2\lambda f}{e}$$

or $e = \dfrac{\lambda f}{x} = \dfrac{5460 \times 10^{-8} \times 80}{0.0273}$ cm $= 0.16$ cm

Example 3.13

A single slit of width 0.02 cm is illuminated by a parallel beam of monochromatic light of wavelength 5890 Å. A converging lens is placed behind the slit to focus the diffraction pattern on a screen placed at a distance of 100 cm. What is the distance of the first minimum and the second minimum from the centre of the central maximum?

Solution

The data given are $e = 0.02$ cm, $\lambda = 5890$ Å $= 5890 \times 10^{-8}$ cm, and $f = 100$ cm.

In a single slit diffraction pattern, the distance of the nth minimum to the centre of the central maximum is given by

$$x_n = \frac{n\lambda f}{e}$$

The distance of the first minimum to the centre of the central maximum will be given by

$$x_1 = \frac{1 \times 5890 \times 10^{-8} \times 100}{0.02} \text{ cm} = 0.294 \text{ cm}$$

The distance of the second minimum to the centre of the central maximum will be given by

$$x_2 = \frac{2 \times 5890 \times 10^{-8} \times 100}{0.02} \text{ cm} = 0.589 \text{ cm}$$

Example 3.14

What is the angular width of the central diffraction pattern due to a single slit when the slit width is 2λ?

Solution

The datum given is $e = 2\lambda$.

The angular width of the single slit central diffraction pattern is given by

$$2\theta = 2\sin^{-1}\left(1.391\frac{\lambda}{\pi e}\right) = 2\sin^{-1}\left(1.391\frac{\lambda}{\pi \times 2\lambda}\right) = 25.58°$$

Example 3.15

A single slit of width 5.3×10^{-4}cm is illuminated by a parallel beam of monochromatic light of wavelength 6300 Å. What is the half width of the central maximum?

Solution

The data given are $e = 0.0053$ cm and $\lambda = 5300$ Å.

The angular width of the single slit central diffraction pattern is given by

$$2\theta = 2\sin^{-1}\left(1.391\frac{\lambda}{\pi e}\right) = 2\sin^{-1}\left(1.391\frac{6300 \times 10^{-8}}{\pi \times 5 \cdot 3 \times 10^{-4}}\right) = 6.03°$$

3.6 Plane Diffraction Grating

A plane diffraction grating is an optical device in which a very large number of parallel slits of same widths, separated by equal opaque spaces are arranged within a small linear region on a plane. In a diffraction grating, slit widths are comparable to the wavelength of light.

3.6.1 Theory of plane diffraction grating under normal incidence

When a monochromatic wavefront is incident on the plane diffraction grating normally, the opaque spaces between the slits obstruct the wavefront. Each of the slits becomes the source of secondary wavelets having the same amplitude and the same initial phase. Now we shall see the resultant effect of all the secondary wavelets emitting from N slits at any point $P(x_0, y_0)$ on the screen.

The resultant effect due to a single slit at a point $P(x_0, y_0)$ is, according to Eq. (3.64), given by

$$\psi = \int_{-\frac{e}{2}}^{+\frac{e}{2}} k \sin 2\pi \left(\frac{t}{T} - \frac{\rho}{\lambda} + \frac{y \sin \theta}{\lambda}\right) dy$$

The resultant effect due to two similar slits at a point $P(x_0, y_0)$ will be given by

$$\psi = \int_{-\frac{e}{2}}^{+\frac{e}{2}} k \sin 2\pi \left(\frac{t}{T} - \frac{\rho}{\lambda} + \frac{y \sin \theta}{\lambda}\right) dy + \int_{d-\frac{e}{2}}^{d+\frac{e}{2}} k \sin 2\pi \left(\frac{t}{T} - \frac{\rho}{\lambda} + \frac{y \sin \theta}{\lambda}\right) dy$$

where $d = e + b$ is called the grating element. Thus, the grating element is defined as

$$d = e + b \tag{3.80}$$

Similarly, the resultant effect due to N similar slits at a point $P(x_0, y_0)$ will be given by

$$\psi = \int_{-\frac{e}{2}}^{+\frac{e}{2}} k \sin 2\pi \left(\frac{t}{T} - \frac{\rho}{\lambda} + \frac{y \sin \theta}{\lambda} \right) dy + \int_{d-\frac{e}{2}}^{d+\frac{e}{2}} k \sin 2\pi \left(\frac{t}{T} - \frac{\rho}{\lambda} + \frac{y \sin \theta}{\lambda} \right) dy +$$

$$\int_{2d-\frac{e}{2}}^{2d+\frac{e}{2}} k \sin 2\pi \left(\frac{t}{T} - \frac{\rho}{\lambda} + \frac{y \sin \theta}{\lambda} \right) dy + \ldots +$$

$$\int_{(N-1)d-\frac{e}{2}}^{(N-1)d+\frac{e}{2}} k \sin 2\pi \left(\frac{t}{T} - \frac{\rho}{\lambda} + \frac{y \sin \theta}{\lambda} \right) dy$$

Evaluating all the integrals, we have

$$\psi = ke \frac{\sin \alpha}{\alpha} \sin 2\pi \left(\frac{t}{T} - \frac{\rho}{\lambda} \right) + ke \frac{\sin \alpha}{\alpha} \sin 2\pi \left(\frac{t}{T} - \frac{\rho}{\lambda} + \frac{d \sin \theta}{\lambda} \right) +$$

$$ke \frac{\sin \alpha}{\alpha} \sin 2\pi \left(\frac{t}{T} - \frac{\rho}{\lambda} + \frac{2d \sin \theta}{\lambda} \right) + \ldots + ke \frac{\sin \alpha}{\alpha} \sin 2\pi \left(\frac{t}{T} - \frac{\rho}{\lambda} + \frac{(N-1)d \sin \theta}{\lambda} \right)$$

where $\alpha = \dfrac{\pi e \sin \theta}{\lambda}$

or $\quad \psi = ke \dfrac{\sin \alpha}{\alpha} \displaystyle\sum_{p=0}^{p=N-1} \sin 2\pi \left(\frac{t}{T} - \frac{\rho}{\lambda} + p \frac{d \sin \theta}{\lambda} \right)$

$$= ke \frac{\sin \alpha}{\alpha} \sum_{p=0}^{p=N-1} \sin \left(\frac{2\pi t}{T} - \frac{2\pi \rho}{\lambda} + 2p \frac{\pi d \sin \theta}{\lambda} \right)$$

or $\quad \psi = ke \dfrac{\sin \alpha}{\alpha} \displaystyle\sum_{p=0}^{p=N-1} \sin \left(x + 2p\beta \right)$ $\hfill$ (3.81) where $\beta = \dfrac{\pi d \sin \theta}{\lambda}$ (3.82)

and $x = \dfrac{2\pi t}{T} - \dfrac{2\pi \rho}{\lambda}$ (3.83)

Applying the formula

$$\sum_{p=0}^{p=n} \sin(x + pm) = \frac{\sin\left(x + \dfrac{nm}{2}\right) \sin(n+1)\dfrac{m}{2}}{\sin\dfrac{m}{2}}$$

to the Eq. (3.81), we get

$$\psi = ke\frac{\sin \alpha}{\alpha} \frac{\sin\left[x + (N-1)\beta\right] \sin(N\beta)}{\sin \beta}$$

or $\psi = ke\dfrac{\sin \alpha}{\alpha} \dfrac{\sin(N\beta)}{\sin \beta} \sin\left[x + (N-1)\beta\right]$ (3.84)

This equation shows that $ke\dfrac{\sin \alpha}{\alpha} \dfrac{\sin(N\beta)}{\sin \beta}$ is the amplitude of the resultant wave at point $P(x_0, y_0)$ on the screen. Hence, the resultant intensity at point $P(x_0, y_0)$ will be

$$I = k^2 e^2 \frac{\sin^2 \alpha}{\alpha^2} \frac{\sin^2(N\beta)}{\sin^2 \beta}$$

or $I = I_0 \dfrac{\sin^2 \alpha}{\alpha^2} \dfrac{\sin^2(N\beta)}{\sin^2 \beta}$ (3.85)

The term $\dfrac{\sin^2 \alpha}{\alpha^2}$ represents the intensity distribution of a single slit diffraction pattern and the additional term $\dfrac{\sin^2(N\beta)}{\sin^2 \beta}$ represents the interference effect due to secondary wavelets emitted from N slits. The intensity is maximum, i.e., $I \rightarrow I_0$ at the point where

$\dfrac{\sin \alpha}{\alpha} \to 1$ i.e. $\alpha \to 0$ and $\dfrac{\sin(N\beta)}{\sin \beta} \to N$ i.e. $\beta \to 0$. Putting $N = 1, 2, 3, \ldots,$ etc into the Eq. (3.85), we can obtain the intensity distribution function of the diffraction pattern for single slit, double slit and triple slit, … etc. The intensity distribution function of the diffraction pattern due to a grating containing N slits is governed by Eq. (3.85).

Principal maxima

The intensity is maximum when $\sin \beta = 0$ is put in the expression (3.85). The condition for maxima is therefore

$$\sin \beta = 0$$

or $\quad \beta = \pm n\pi$ (3.86)

When $\beta = \pm n\pi$, the value of $\dfrac{\sin(N\beta)}{\sin \beta} = N$. Equation (3.85) becomes

$$I = I_0 N^2 \frac{\sin^2 \alpha}{\alpha^2} \tag{3.87}$$

Since $\beta = \dfrac{\pi d \sin \theta}{\lambda}$, we get

$$\pm n\pi = \frac{\pi d \sin \theta}{\lambda}$$

Putting the value of $d = e + b$ into this equation, we get

$$(e+b)\sin \theta = \pm n\lambda, \quad n = 0, 1, 2, 3, \ldots \tag{3.88}$$

Equation (3.88) gives the direction of the principal maxima of zero order, 1st order, 2nd order, 3rd order, … etc. for values of $n = 0, 1, 2, 3, \ldots,$ etc., respectively. $(e+b)\sin \theta$ is the path difference between two parallel rays diffracted from any pair of corresponding points on the grating. The whole number n is called the order of the interference maximum. The expression (3.88) is independent of N. Hence, the direction of principal maxima depends only on the grating element $(e + b)$.

Minima

The intensity is minimum when simultaneously $\sin N\beta = 0$ and $\sin \beta \neq 0$ in the expression (3.85). The condition for minima is therefore

$$\sin N\beta = 0 \quad \text{and} \quad \sin \beta \neq 0$$

or $N\beta = \pm m\pi$, with $m \neq nN$; $n = 0, 1, 2, 3, \dots$, etc.   (3.89)

m can take any integral value except 0, N, $2N$, $3N$, ... etc. Putting $m = 0, N, 2N, 3N$, ..., etc. into Eq. (3.89) gives the condition for principal maxima. Putting $m = nN$ into Eq. (3.89), we get $\beta = \pm n\pi$ or $\sin\beta = 0$ which is the condition of principal maxima. So $m \neq nN$.

Putting the value of $\beta = \dfrac{\pi d \sin\theta}{\lambda}$ into Eq. (3.89), we get

$$N\frac{\pi d \sin\theta}{\lambda} = \pm m\pi ; \quad m \neq nN$$

Putting the value of $d = e + b$ into this equation, we get

$$N(e + b)\sin\theta = \pm m\lambda, \quad m \neq nN$$

or $(e + b)\sin\theta = \pm \dfrac{m}{N}\lambda, \quad m \neq nN$ (3.90)

Equation (3.90) gives the direction of minima. The direction of minima are given by

$$(e + b)\sin\theta = \pm\frac{1}{N}\lambda,\ \pm\frac{2}{N}\lambda,\ \pm\frac{3}{N}\lambda,\dots\pm\frac{N-1}{N}\lambda,\ \pm\frac{N+1}{N}\lambda,\ \pm\frac{N+2}{N}\lambda,\ \pm\frac{N+3}{N}\lambda,\ \text{etc.}$$

(3.91)

This Eq. (3.91) is obtained by putting $m = 1, 2, 3, \dots, N - 1, N + 1, N + 2, N + 3, \dots$, etc. in the Eq. (3.90).

The direction of principal maxima are given by

$$(e + b)\sin\theta = 0;\ \pm\frac{N}{N}\lambda,\ \pm\frac{2N}{N}\lambda,\ \pm\frac{3N}{N}\lambda,\ \dots\ \text{etc.}$$ (3.92)

Equation (3.92) derived from Eq. (3.90) by putting $m = 1N, 2N, 3N, 4N, \dots$, etc. The values of m between 1 to N are

$$\frac{1}{N}\lambda,\ \pm\frac{2}{N}\lambda,\ \pm\frac{3}{N}\lambda,\ \dots\ \pm\frac{N-1}{N}\lambda$$

and their total number is $(N - 1)$. $m = 0$ and 1 give the direction of first two principal maxima of the zero order and the first order.

$$m = \frac{1}{N}\lambda, \pm\frac{2}{N}\lambda, \pm\frac{3}{N}\lambda, \ldots \pm\frac{N-1}{N}\lambda$$

give the direction of first $(N - 1)$ minima. Therefore, in between the first two principal maxima of zero order and first order there are $(N - 1)$ minima.

Secondary maxima

In between any two consecutive principal maxima, there are $(N - 1)$ minima. Since there are minima, there must be maxima; otherwise, the minima will not be distinguishable. Since in between two principal maxima there are $(N - 1)$ minima, there must be $(N - 2)$ maxima called secondary maxima. The positions of the secondary maxima are found out

in the following way. The intensity I is maximum with respect to β if $\dfrac{dI}{d\beta} = 0$ and $\dfrac{d^2I}{d\beta^2} < 0$. Applying this concept to Eq. (3.85), we have

$$\frac{d}{d\beta} I_0 \frac{\sin^2\alpha}{\alpha^2} \frac{\sin^2 N\beta}{\sin^2\beta} = 0$$

or $\qquad 2I_0 \dfrac{\sin^2\alpha}{\alpha^2} \dfrac{\sin N\beta}{\sin\beta} \dfrac{d}{d\beta} \dfrac{\sin N\beta}{\sin\beta} = 0$

or $\qquad \dfrac{\sin N\beta}{\sin\beta} \dfrac{d}{d\beta} \dfrac{\sin N\beta}{\sin\beta} = 0$

This equation shows that either

$$\frac{\sin N\beta}{\sin\beta} = 0 \text{ or } \frac{d}{d\beta} \frac{\sin N\beta}{\sin\beta} = 0.$$

As discussed earlier, $\dfrac{\sin N\beta}{\sin\beta} = 0$ with $\sin\beta \neq 0$ gives the position of the minima. The positions of the secondary maxima are given by the conditions

$$\frac{d}{d\beta} \frac{\sin N\beta}{\sin\beta} = 0$$

or $\qquad N\tan\beta = \tan N\beta$ $\hfill$ (3.93)

Equation (3.93) can be solved for the values of β and the values of β except $\beta = \pm n\pi$ gives the direction of the secondary maxima. The secondary maxima are not of equal intensity but decreases gradually as we go away from either side of the principal maxima.

3.6.2 Theory of plane diffraction grating under oblique incidence

Let a parallel beam of light be incident obliquely on a plane diffraction grating at an angle of incident i as shown in Fig. 3.10.

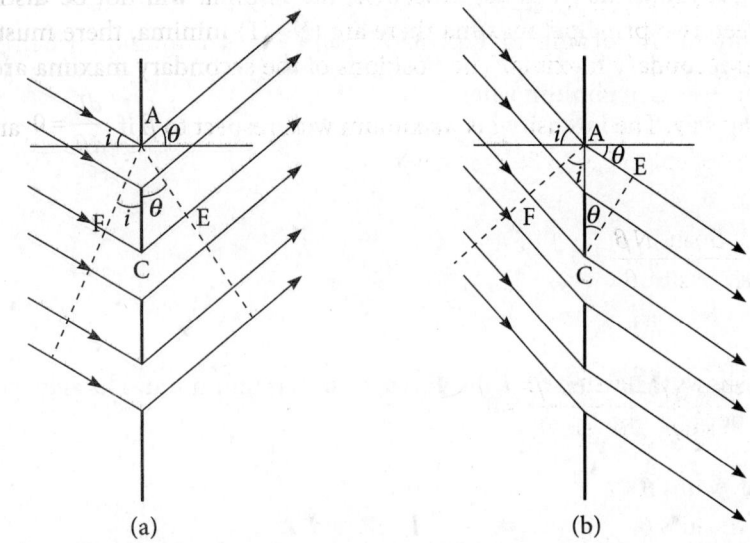

(a)	(b)

Figure 3.10 | Plane diffraction grating under oblique incidence. i = angle of incidence. (a) incident wave is diffracted upward; Here, the path difference is FC + CE; (b) incident wave is diffracted downward. Here, the path difference is AE – FC

The path difference between the secondary wavelets passing through the points A and C is

$$FC + CE = (e+b)\sin i + (e+b)\sin\theta = (e+b)(\sin\theta + \sin i) \tag{3.94}$$

Equation (3.94) is true when incident wave is diffracted upward as shown in Fig. 3.10(a). When incident wave is diffracted downward as shown in Fig. 3.10(b), the path difference between the secondary wavelets passing through the points A and C is

$$FC - AE = (e+b)\sin i - (e+b)\sin\theta = (e+b)(\sin i - \sin\theta) \tag{3.95}$$

The direction of the nth order diffraction maxima in case of upward diffraction is obtained from Eq. (3.94) as

$$(e+b)(\sin\theta + \sin i) = n\lambda$$

or $(e+b)\left(2\sin\dfrac{\theta+i}{2}\cos\dfrac{\theta-i}{2}\right)=n\lambda$ (3.96)

or $\sin\dfrac{\theta+i}{2}=\dfrac{n\lambda}{2(e+b)\cos\dfrac{\theta-i}{2}}$ (3.97)

The deviation of the diffracted beam is $\theta+i$. When deviation $\theta+i$ is minimum, $\sin\dfrac{\theta+i}{2}$ must be minimum. According to Eq. (3.97), $\sin\dfrac{\theta+i}{2}$ is minimum when $\cos\dfrac{\theta-i}{2}^{2}$ is maximum. $\cos\dfrac{\theta-i}{2}$ is maximum when

$$\dfrac{\theta-i}{2}=0$$

or $\theta=i$

This equation shows that when $\theta=i$, the deviation $\theta+i$ is minimum. The angle of minimum deviation will be

$$\delta_m=\theta+i\ \ \text{with}\ \ \theta=i$$

Hence, we have

$$\theta=\dfrac{\delta_m}{2}\ \ \text{and}\ \ i=\dfrac{\delta_m}{2}$$

In terms of angle of minimum deviation, the direction of the nth order diffraction maxima will be obtained from

$$2(e+b)\sin\dfrac{\delta_m}{2}=n\lambda$$ (3.98)

3.6.3 Angular width of the principal maxima

The angular width of a principal maximum is defined as the angle between two straight lines connecting the extreme ends of the principal maximum on the screen to the central point of the grating. The angular width of a principal maximum is a measure of the sharpness of the principal maximum. Lesser is the angular width of a principal maximum, sharper is the principal maximum. The mathematical expression for the angular width of a principal maximum may be derived in the following way.

The directions of the nth order principal maximum when a parallel beam of light is incident normally on a diffraction grating is obtained from Eq. (3.88) as

$$(e+b)\sin\theta_n = n\lambda \tag{3.99}$$

Let $\theta_n + d\theta_n$ and $\theta_n - d\theta_n$ be the directions of first minima on the two sides of the nth order primary maximum. Then directions of these two minima will be given by

$$(e+b)\sin\left(\theta_n \pm d\theta_n\right) = n\lambda \pm \frac{1}{N}\lambda \tag{3.100}$$

where N is the number of lines on the grating.

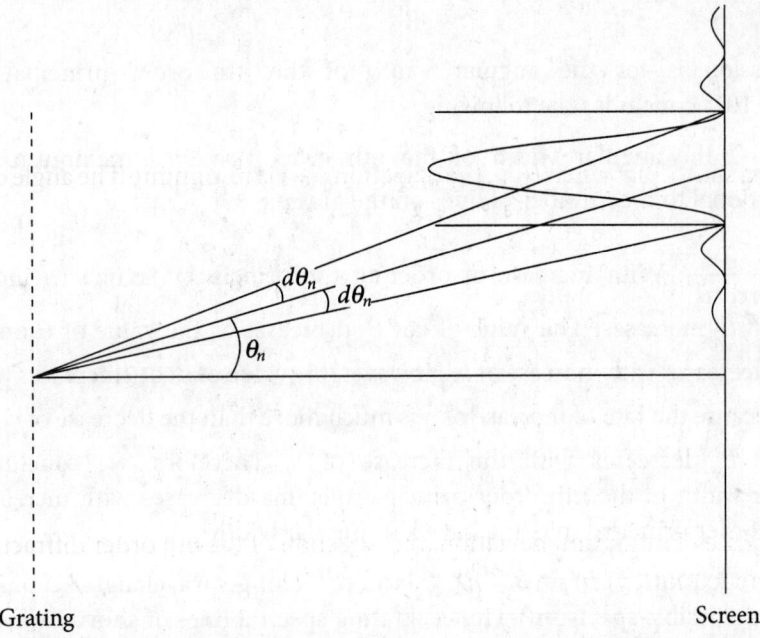

Figure 3.11 | θ_n is the direction of the nth order principal maximum. $\theta_n \pm d\theta_n$ are the directions of two minima on the two sides of the nth order principal maximum. $2d\theta_n$ is the angular width of the nth order principal maximum as is visible from the figure

Dividing Eq. (3.100) by Eq. (3.99), we get

$$\frac{(e+b)\sin\left(\theta_n \pm d\theta_n\right)}{(e+b)\sin\theta_n} = \frac{n\lambda \pm \dfrac{\lambda}{N}}{n\lambda}$$

or $\dfrac{\sin\theta_n\cos d\theta_n \pm \cos\theta_n\sin d\theta_n}{\sin\theta_n} = 1 \pm \dfrac{1}{nN}$

Since $d\theta_n$ is a small quantity, we can set safely $\cos d\theta_n = 1$ and $\sin d\theta_n = d\theta_n$ in the previous equation to obtain

$$\frac{\sin\theta_n \pm d\theta_n\cos\theta_n}{\sin\theta_n} = 1 \pm \frac{1}{nN}$$

or $\dfrac{d\theta_n\cos\theta_n}{\sin\theta_n} = \dfrac{1}{nN}$

or $2d\theta_n = \dfrac{2}{Nn\cot\theta_n}$ (3.101)

This expression is for the angular width of the nth order principal maximum. Equation (3.101) concludes the following.

i. $d\theta_n \propto \dfrac{1}{N}$: the angular width of the nth order principal maximum is inversely proportional to the number of lines on the grating.

ii. $d\theta_n \propto \dfrac{1}{n\cot\theta_n}$: With increase of order of the principal maxima (n increases), the value of θ_n increases. The value of $\cot\theta_n$ decreases as the value of θ_n increases, i.e., $\cot\theta_n$ decreases with increase of n. However, the value of $n\cot\theta$ increases with increase of n, because the rate of increase of n is much more than the decrease of $\cot\theta_n$. Hence, $1/n\cot\theta_n$ decreases with the increase of n. Therefore, we conclude that the angular width of the nth order principal maxima decreases with increase of n, i.e., higher order principal maxima have less angular width.

As seen from equation $(e+b)\sin\theta_n = n\lambda$, θ_n is more for longer wavelength keeping n constant (red end of the visible spectrum). Hence, grating spectral lines of shorter wavelengths are sharper than the longer wavelengths.

Putting the value of n from Eq. (3.99) into Eq. (3.101), we get

$$d\theta_n = \frac{1}{N\cot\theta_n}\frac{\lambda}{(e+b)\sin\theta_n} = \frac{\lambda}{(e+b)N\cos\theta_n}$$ (3.102)

Now if there are N number of lines on the grating surface, the width of the grating will be

$$(N-1)(e+b) \approx N(e+b)$$ since $N \gg 1$

Hence, we have

$$d\theta_n = \frac{1}{\text{width of the grating}} \frac{\lambda}{\cos\theta_n} \qquad (3.103)$$

or $d\theta_n \propto \dfrac{1}{\text{width of the grating}}$ if $\dfrac{\lambda}{\cos\theta_n}$ is constant.

This equation shows that grating spectral lines are more sharper in grating having more width.

3.6.4 Formation of spectra by diffraction grating

The directions or positions of principal maxima of nth order when a parallel beam of light is incident normally on a diffraction grating is obtained from Eq. (3.88) as

$$(e+b)\sin\theta = n\lambda \qquad (3.104)$$

When the number of slits per unit length on the grating is very large of the order of 15000 slits/inch or 6000 slits/cm, i.e., the slit width is comparable to the wave length of the incident light. The slits on the grating, in general, are called lines. When the primary source is narrow, the diffraction maxima are seen as bright sharp lines on the focal plane of the eye-piece of the telescope adjusted for parallel rays. Hence, the principal maxima of a diffraction pattern due to a grating are called spectral lines. It is obvious from Eq. (3.104) that for a given order of principal maxima or spectral lines with $n \neq 0$, the diffraction angle depends upon the wavelength λ. With the increase of wavelength λ, the diffraction angle increases. For different wavelengths, the diffraction angle θ will be different and the position of diffraction maxima of each wavelength will be different. Hence, we shall observe as many spectral lines as there are different wavelengths in the light source. $n = 0 \Rightarrow \theta = 0$; in which case, the principal maximum is situated on the central axis and is called the central principal maximum. Whatever may be the values of λ, $(e + b)\sin\theta = 0$ for the central principal maximum. Therefore, the central maxima of all the wavelengths or colors coincide at a single point giving a central maximum of the same color as that of the source. On either side of the central maximum, there exist spectral lines whose wavelengths increase as we move away from the central maximum. In other words, the spectral lines of shorter wavelengths are closer to the central maximum.

Absence of spectral lines

From Eq. (3.104), we get

$$\sin\theta = \frac{n\lambda}{e+b}$$

Since the value of sin θ cannot be greater than 1, the spectral lines of the wavelengths for which $(e + b) > n\lambda$ are absent.

The condition for absent spectra can be obtained from the following consideration. From Eq. (3.104), we get $e\sin\theta + b\sin\theta = n\lambda$. If the value of the slit width e and diffraction angle θ is such that $e\sin\theta = \lambda$, then the slit has no effect on the diffraction pattern. Dividing $(e+b)\sin\theta = n\lambda$ by $e\sin\theta = \lambda$, we get

$$\frac{(e+b)\sin\theta}{e\sin\theta} = \frac{n\lambda}{\lambda}$$

or $\quad \dfrac{e+b}{e} = n$ $\hfill (3.105)$

In Eq. (3.105), n is the order of the spectra that are absent in the diffraction pattern. Equation (3.105) shows that $n = 1$ when $b = 0$. In this case, grating reduces to a single slit and the first order spectrum is absent. In a single slit, there exists only zero order central maximum. Equation (3.105) shows that $n = 2$ when $b = e$. In this case, the second order spectrum is absent in the diffraction pattern.

Overlapping of spectral lines of a grating

If spectral lines of a grating overlap, it means that spectral lines of different wavelengths or color are formed at the same space. If polychromatic light containing a large number of wavelengths is incident normally on a plane diffraction grating, the spectral lines of higher order and lower wavelengths may overlap with the spectral lines of lower order and higher wavelengths. If first order spectral lines of wavelength λ_1, the second order spectral lines of wavelength λ_2 and the third order spectral lines of wavelength λ_3 overlap each other, we must have

$$(e+b)\sin\theta = 1\times\lambda_1 = 2\times\lambda_2 = 3\times\lambda_3$$

For example, the third order spectral line of wavelength 7000 Å (red light), the fourth order spectral line of wavelength 5250 Å (green light) and the fifth order spectral line of wavelength 4200 Å (violet light) overlap each other or form at the same place because 3×7000 Å $= 4 \times 5250$ Å $= 5 \times 4200$ Å. If the n_1th order of wavelength λ_1 coincides with the n_2th order of the wavelength λ_2, we can have

$$n_1\lambda_1 = n_2\lambda_2$$ $\hfill (3.106)$

Example 3.16

What is the maximum wavelength of the visible spectrum so that the third order spectrum can be observed by a grating having 6000 lines/cm?

Solution

The data given are $n = 3$ and there are 6000 lines/cm on the grating surface.

Hence, the grating element $e + b$ will be $e + b = \dfrac{1}{6000} \text{ cm} = 1.67 \times 10^{-4} \text{ cm}$.

The directions of the nth order spectral line when a parallel beam of light is incident normally on a diffraction grating is obtained from Eq. (3.108) as

$$(e + b)\sin\theta = n\lambda$$

$$\lambda = \frac{(e+b)\sin\theta}{n}$$

The wavelength is maximum when the value of $\sin\theta$ is maximum, i.e., $\sin\theta = 1$. Hence, we have

$$\lambda = \frac{(e+b)}{n}$$

Putting the values of $e + b$ and n into this equation, we get

$$\lambda = \frac{1}{6000 \times 3} \text{ cm} = 5556 \times 10^{-8} \text{ cm} = 5556\,\text{Å}$$

The third order diffraction lines of all the wavelengths less than 5556 Å of the visible spectrum can be observed by this grating.

Example 3.17

What are the angles at which maximum intensity beam occur when a monochromatic light of wavelength 5890 Å is incident normally on a diffraction grating having 3000 lines/cm?

Solution

The data given are $\lambda = 5890$ Å and on the grating there are 3000 lines/cm.

Hence, the grating element $e + b$ will be given by $e + b = \dfrac{1}{3000} \text{ cm}$.

The directions of the nth order spectral line when a parallel beam of light is incident normally on a diffraction grating is obtained from the Eq. (3.108) as

$$(e + b)\sin\theta = n\lambda, \quad \text{with } n = 0, 1, 3, 4, \ldots$$

$$\sin\theta = \frac{n\lambda}{e+b} \quad \text{or} \quad \theta = \sin^{-1}\frac{n\lambda}{e+b}$$

Putting the values of $e + b$, λ into this equation, we get

$$\theta = \sin^{-1}\frac{n \times 5890 \times 10^{-8}}{\dfrac{1}{3000}} = \sin^{-1}(0.1767n)$$

or $\theta = \sin^{-1}(0.1767n)$

Putting $n = 0, 1, 2, 3, \ldots$, we get

$$\theta = \sin^{-1}(0.1767 \times 0) = \sin^{-1}0 = 0° \text{ for } n = 0$$

$$\theta = \sin^{-1}(0.1767 \times 1) = \sin^{-1}(0.1767) = \pm 10.18° \text{ for } n = 1$$

$$\theta = \sin^{-1}(0.1767 \times 2) = \pm 20.70° \text{ for } n = 2$$

$$\theta = \sin^{-1}(0.1767 \times 3) = \pm 32.01° \text{ for } n = 3$$

$$\theta = \sin^{-1}(0.1767 \times 4) = \pm 44.98° \text{ for } n = 4$$

$$\theta = \sin^{-1}(0.1767 \times 5) = \pm 62.07° \text{ for } n = 5$$

Thus, the angles for which diffracted beam has maximum intensity are

$0°, \ \pm 10.18°, \ \pm 20.70°, \ \pm 32.01°, \ \pm 44.98°, \ \pm 62.07°.$

Example 3.18

A sodium vapour light is incident on a grating with 15000 lines/inch. Find the angles of diffraction for the principal maxima of the two lines of the sodium light 5890 Å and 5896 Å in the first order spectrum. Can this grating resolve the sodium vapour light?

Solution

The data given are $n = 1$, $\lambda_1 = 5890$ Å, $\lambda_2 = 5896$ Å and on the grating there are 15000 lines/inch $= \dfrac{15000 \text{ lines}}{2.54 \text{ cm}} = 5906 \text{ lines/cm}$.

Hence, the grating element $= e + b = \dfrac{1}{5906} \text{ cm} = 1.693 \times 10^{-4} \text{ cm}$

The directions of the nth order spectral line when a parallel beam of light is incident normally on a diffraction grating is obtained from Eq. (3.108) as

$$(e + b)\sin\theta = n\lambda$$

or $\theta = \sin^{-1}\left(\dfrac{n\lambda}{e+b}\right)$

The angles of diffraction of the spectral line of the sodium light 5890 Å in the first order spectrum is

$$\theta_1 = \sin^{-1}\left(\dfrac{1 \times 5890 \times 10^{-8}}{1.693 \times 10^{-4}}\right) = 20.36°$$

The angles of diffraction of the spectral line of the sodium light 5896 Å in the first order spectrum is

$$\theta_2 = \sin^{-1}\left(\dfrac{1 \times 5896 \times 10^{-8}}{1.693 \times 10^{-4}}\right) = 20.38°$$

The condition for just resolution is

$$\dfrac{\lambda}{d\lambda} = nN, \quad N = \text{number of lines on the grating surface.}$$

or $N = \dfrac{\lambda}{nd\lambda} = \dfrac{5890 \times 10^{-8}}{1 \times 6 \times 10^{-8}} = 982 \text{ lines}$

Since there are 5906 lines/cm much greater than 982 lines, the two wavelengths of the sodium vapour light can be well resolved by this grating.

Example 3.19

A parallel beam of monochromatic light is incident normally on a grating surface having 8489 lines/cm. The first order spectral line is observed to be diffracted through an angle 30°. Calculate the wavelength of the monochromatic light.

Solution

The data given are $n = 1$ and on the grating there are 8489 lines/cm.

Hence, the grating element $= e + b = \dfrac{1}{8489}$ cm $= 1.178 \times 10^{-4}$ cm

The wavelength of the monochromatic light incident normally on a diffraction grating is obtained from the Eq. (3.104) as

$$\lambda = \frac{(e+b)\sin\theta}{n}$$

$$= \frac{1.178 \times 10^{-4} \sin 30}{1} \text{cm} = 5 \cdot 89 \times 10^{-5} \text{cm} = 5890 \text{ Å}$$

Example 3.20

A parallel beam of monochromatic light of wavelength 6000 Å is incident normally on a grating surface having 6000 lines/cm. What is the highest order spectrum that can be observed?

Solution

The data given are $\lambda = 6000 \text{ Å} = 6000 \times 10^{-8}$ cm and there are 6000 lines/cm on the given grating.

Hence, the grating element of the grating will be $e + b = \dfrac{1}{6000}$ cm $= 1.67 \times 10^{-4}$ cm
The grating equation is

$$(e+b)\sin\theta = n\lambda$$

or　$n = \dfrac{(e+b)\sin\theta}{\lambda}$

n is maximum when $\sin\theta$ is maximum. The maximum value of $\sin\theta$ is 1. Therefore the maximum value of n is given by

$$n_{max} = \frac{(e+b)}{\lambda} = \frac{1.67 \times 10^{-4}}{6000 \times 10^{-8}} = 2.8 \approx 3$$

The highest order spectrum that can be observed is three.

Example 3.21

A plane transmission grating has 5000 lines/cm. Calculate the angular separation between two wavelengths 5890 Å and 5896 Å of a sodium vapour light for a second order spectrum.

Solution

The data given are $n = 2$, $\lambda_1 = 5890$ Å $= 5890 \times 10^{-8}$ cm, $\lambda_2 = 5896$ Å $= 5896 \times 10^{-8}$ cm and on the grating there are 5000 lines/cm.

Hence, the grating element $= e + b = \dfrac{1}{5000}$ cm $= 2.0 \times 10^{-4}$ cm

The directions of the nth order spectral line when a parallel beam of light is incident normally on a diffraction grating is obtained from Eq. (3.104) as

$$\theta = \sin^{-1}\left(\frac{n\lambda}{e+b}\right)$$

The angles of diffraction of the spectral line of the sodium light 5890 Å in the second order spectrum is

$$\theta_1 = \sin^{-1}\left(\frac{2 \times 5890 \times 10^{-8}}{2.0 \times 10^{-4}}\right) = 36.09° = 36°5'24''$$

The angles of diffraction of the spectral line of the sodium light 5896 Å in the second order spectrum is

$$\theta_2 = \sin^{-1}\left(\frac{2 \times 5896 \times 10^{-8}}{2.0 \times 10^{-4}}\right) = 36.13° = 36°7'48''$$

The angle of separation $\theta_2 - \theta_1$ of sodium lines 5890 Å and 5896 Å in the second order spectrum becomes

$$\theta_2 - \theta_1 = 36.13° - 36.09° = 0.04° = 0°2'24''$$

Example 3.22

Calculate the angle of separation of helium lines 5048 Å and 5016 Å in the second order spectrum when a parallel beam of helium light is incident normally on a grating containing 6000 lines/cm.

Solution

The data given are $n = 2$, $\lambda_1 = 5048$ Å $= 5048 \times 10^{-8}$ cm, $\lambda_2 = 5016$ Å $= 5016 \times 10^{-8}$ cm and on the grating there are 6000 lines/cm.

Hence, the grating element $= e + b = \dfrac{1}{6000}$ cm $= 1.67 \times 10^{-4}$ cm

The directions of the nth order spectral line when a parallel beam of light is incident normally on a diffraction grating is obtained from Eq. (3.104) as

$$\theta = \sin^{-1}\left(\frac{n\lambda}{e+b}\right)$$

The angles of diffraction of the spectral line of the helium light 5048 Å in the second order spectrum is

$$\theta_1 = \sin^{-1}\left(\frac{2 \times 5048 \times 10^{-8}}{1.67 \times 10^{-4}}\right) = 37.2° = 37°12'$$

The angles of diffraction of the spectral line of the helium light 5016 Å in the second order spectrum is

$$\theta_2 = \sin^{-1}\left(\frac{2 \times 5016 \times 10^{-8}}{1.67 \times 10^{-4}}\right) = 36.92° = 36°55'$$

The angle of separation of helium lines 5048 Å and 5016 Å in the second order spectrum becomes

$$\theta_1 - \theta_2 = 37.2° - 36.92° = 0.28° = 0°16'48''$$

Example 3.23

A sodium vapour light containing two wavelengths 5890 Å and 5896 Å is incident normally on a grating having 10000 lines/cm. A lens of focal length 100 cm is used to observe the spectrum on a screen. Calculate the separation in cm of the two lines in the first order spectrum.

Solution

The data given are $n = 1$, $\lambda_1 = 5890$ Å $= 5890 \times 10^{-8}$ cm, $\lambda_2 = 5896$ Å $= 5896 \times 10^{-8}$ cm, $f = 100$ cm and on the grating, there are 10000 lines/cm.

Hence, the grating element $= e + b = \dfrac{1}{10000}$ cm $= 1.00 \times 10^{-4}$ cm

The directions of the spectral lines of the nth order spectrum when a parallel beam of light is incident normally on a diffraction grating is obtained from Eq. (3.104) as

$$\theta = \sin^{-1}\left(\frac{n\lambda}{e+b}\right)$$

The positions of the spectral lines of the nth order spectrum from the central line when a parallel beam of light is incident normally on a diffraction grating is obtained from the equation

$$\tan\theta = \frac{x}{f}$$

or $x = f\tan\theta$

or $x = f\tan\sin^{-1}\left(\frac{n\lambda}{e+b}\right)$

The position of the spectral line from the central line ($n = 0$) of the sodium light 5890 Å in the first order spectrum is

$$x_1 = 100 \times \tan\,\sin^{-1}\left(\frac{1 \times 5890 \times 10^{-8}}{1.00 \times 10^{-4}}\right) = 72.88\,\text{cm}$$

The position of the spectral line from the central line ($n = 0$) of the sodium light 5896 Å in the first order spectrum is

$$x_2 = 100 \times \tan\,\sin^{-1}\left(\frac{1 \times 5896 \times 10^{-8}}{1.00 \times 10^{-4}}\right) = 73.00\,\text{cm}$$

The separation in cm of the two lines in the first order spectrum of the sodium light is.

$$x_2 - x_1 = 0.12\text{cm}$$

Example 3.24

A monochromatic light of wavelength 5500 Å is incident normally on a plane transmission grating having 5000 lines/cm. Calculate the difference between the diffraction angles of first order and third order spectra.

Solution

The data given are $n = 1, 2$, $\lambda = 5500$ Å $= 5500 \times 10^{-8}$ cm, and number of lines per cm = 5000.

The grating element ($e + b$) will be obtained as $e + b = \dfrac{1}{5000}\,\text{cm} = 2.0 \times 10^{-4}\,\text{cm}.$

The directions of the nth order spectral line when a parallel beam of light is incident normally on a diffraction grating is obtained from Eq. (3.104) as

$$\theta = \sin^{-1}\left(\frac{n\lambda}{e+b}\right)$$

Hence, we have

$$\theta_1 = \sin^{-1}\left(\frac{1 \times 5500 \times 10^{-8}}{2.0 \times 10^{-4}}\right) = 15°57'43''$$

and $\theta_3 = \sin^{-1}\left(\frac{2 \times 5500 \times 10^{-8}}{2.0 \times 10^{-4}}\right) = 33°22'1''$

Therefore, the difference between the diffraction angles of first order and third order spectra is

$$\theta_3 - \theta_1 = 17°24'2''$$

Example 3.25

In a plane transmission grating, the angle of diffraction for the first order principal maximum is 20° for a wavelength of 6500 Å. Calculate the number of lines in one cm of the grating surface.

Solution

The data given are $n = 1$, $\lambda = 6500$ Å $= 6500 \times 10^{-8}$ cm, $\theta = 20°$.

The directions of the nth order spectral line when a parallel beam of light is incident normally on a diffraction grating is obtained from Eq. (3.104) as

$$(e+b) = \frac{n\lambda}{\sin\theta}$$

Hence, the grating element $e+b = \dfrac{1 \times 6500 \times 10^{-8}}{\sin 20} = 1.9 \times 10^{-4}$

Therefore, the number of lines per cm will be $\dfrac{1}{e+b} = 5263$

Example 3.26

How many orders of diffraction bands will be visible theoretically if the wavelength of incident radiation is 5893 Å and the number of lines per cm on the grating is 6000?

Solution

The data given are $\lambda = 5893$ Å $= 5893 \times 10^{-8}$ cm and the number of lines per cm = 5000.
    The grating element $(e + b)$ will be obtained as

$$e + b = \frac{1}{5000} \text{cm} = 2.0 \times 10^{-4} \text{cm}$$

The directions of the nth order spectral line when a parallel beam of light is incident normally on a diffraction grating is obtained from Eq. (3.104) as

$$n = \frac{(e+b)\sin\theta}{\lambda}$$

or $n_{max} = \dfrac{(e+b)}{\lambda}$ because the maximum value of $\sin\theta = 1$

Putting the values of $e + b$ and λ into this equation, we get

$$n_{max} = \frac{2 \times 10^{-4}}{5893 \times 10^{-8}} = 3.4$$

The maximum orders of diffraction bands that will be visible theoretically is 3

Example 3.27

A parallel beam of monochromatic light is allowed to incident normally on a grating having 8000 lines/cm. The second order spectral line is found to be diffracted through an angle 40°. Calculate the wavelength of the light used.

Solution

The data given are $n = 2$, $\theta = 40°$ and the number of lines per cm = 8000
The grating element $(e + b)$ will be obtained as

$$e + b = \frac{1}{8000} \text{cm} = 1.25 \times 10^{-4} \text{cm}$$

The directions of the nth order spectral line when a parallel beam of light is incident normally on a diffraction grating is obtained from Eq. (3.104) as

$$\lambda = \frac{(e+b)\sin\theta}{n}$$

Putting the values of $e + b$, n and θ into this equation, we obtain the values of the wavelength of the light used as

$$\lambda = \frac{1.25 \times 10^{-4} \sin 40}{2} = 4017\,\text{Å}$$

Example 3.28

A diffraction grating used at normal incidence gives a line with wavelength 6000 Å in a certain order superimposed on another line of wavelength 5000 Å of the next higher order. If the angle of diffraction for the two wavelengths is 25°, how many lines are there in one cm of the grating?

Solution

The data given are $\lambda_1 = 6000$ Å, $\lambda_2 = 5000$ Å, $\theta = 25° = \theta_1 = \theta_2$ and $n_2 = n_1 + 1$

The directions of nth order spectral line when a parallel beam of light is incident normally on a diffraction grating is obtained from Eq. (3.104) as

$$(e + b)\sin\theta = n\lambda$$

For the two wavelengths, we have

$$(e + b)\sin\theta_1 = n_1\lambda_1 \text{ and } (e + b)\sin\theta_2 = n_2\lambda_2 \text{ s}$$

Since $\theta_1 = \theta_2$, we have from these equations

$$n_1\lambda_1 = n_2\lambda_2$$

or $n_1\lambda_1 = (n_1 + 1)\lambda_2$

or $n_1(\lambda_1 - \lambda_2) = \lambda_2$

or $n_1 = \dfrac{\lambda_2}{\lambda_2 - \lambda_2} = \dfrac{5000}{1000} = 5$

$$n_2 = n_1 + 1 = 5 + 1 = 6$$

Putting the values of θ_1, λ_1 and n_1 into equation $(e+b)\sin\theta_1 = n_1\lambda_1$, we get the expression for the grating element

$$e+b = \frac{n_1\lambda_1}{\sin\theta_1} = \frac{5\times6000\times10^{-8}}{\sin 25} \text{ cm} = 7.099\times10^{-4}\text{ cm}$$

The number of lines in one cm of the grating is $\dfrac{1}{e+b} = 1409$

Example 3.29

The limits of the visible spectrum are approximately 4000 Å to 7000 Å. Find the angular width of the first order visible spectrum produced by a plane transmission grating having 8000 lines/cm when light is incident normally on the grating.

Solution

The data given are $\lambda_1 = 4000$ Å, violet light, $\lambda_2 = 7000$ Å, red light, $n_1 = n_2 = 1$, and the number of lines per cm = 8000.

Hence, the grating element

$$e+b = \frac{1}{8000}\text{ cm} = 1.25\times10^{-4}\text{ cm}$$

The directions of the spectral lines of the nth order spectrum when a parallel beam of light is incident normally on a diffraction grating is obtained from Eq. (3.104) as

$$\theta = \sin^{-1}\left(\frac{n\lambda}{e+b}\right)$$

The angular deviation of the violet light in the first order spectrum is

$$\theta_V = \sin^{-1}\left(\frac{1\times4000\times10^{-8}}{1.25\times10^{-4}}\right) = 18.66°$$

The angular deviation of the red light in the first order spectrum is

$$\theta_R = \sin^{-1}\left(\frac{1\times7000\times10^{-8}}{1.25\times10^{-4}}\right) = 34.06°$$

Therefore, the angular width $\theta_R - \theta_V$ of the first order visible spectrum by this grating is

$$\theta_R - \theta_V = 34.06° - 18.66° = 15.4° = 15°24'$$

Example 3.30

A sodium vapour light (5890 Å and 5896 Å) is incident normally on a plane transmission grating and is viewed in third order at an angle of 80° to the central line. The sodium doublets are not resolved. Calculate the grating element.

Solution

The mean wavelength of the sodium vapour light is

$$\frac{1}{2}\left(5890\,\text{Å} + 5896\,\text{Å}\right) = 5893\,\text{Å}.$$

The grating element will be

$$e + b = \frac{n\lambda}{\sin\theta} = \frac{3 \times 5893 \times 10^{-8}}{\sin 80} = 1.8 \times 10^{-4}\,\text{cm}$$

Example 3.31

A grating having 2000 lines/cm is illuminated normally by a polychromatic light and the spectrum is formed on a screen placed 50 cm from the grating. If a square hole of 1 cm side is cut on the screen so that the edge of the square hole is at 5 cm from the central maximum, what is the range of wavelengths that can pass through the hole?

Solution

The data given are $f = 50$ cm and the number of lines per cm of the grating surface is 2000. Hence, the grating element will be

$$e + b = \frac{1}{2000}\,\text{cm}.$$

The positions of two points on the screen are 5 cm and 6 cm from the central maximum. All the wavelengths in between $y = 5$ cm to $y = (5+1)$ cm = 6 cm can pass through the hole.

The directions of the spectral lines of first order spectrum when a parallel beam of light is incident normally on a diffraction grating is obtained from Eq. (3.104) as

$$(e + b)\sin\theta = \lambda$$

The positions of the spectral lines of first order spectrum from the central line when a parallel beam of light is incident normally on a diffraction grating is obtained from the equation

$$\tan\theta = \frac{y}{f}$$

or $\quad \theta = \tan^{-1}\dfrac{y}{f}$

or $\quad \lambda = (e+b)\sin\tan^{-1}\dfrac{y}{f}$

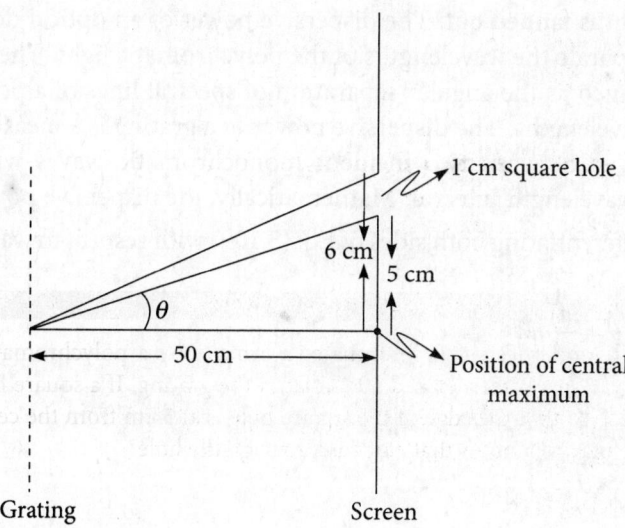

Grating   Screen

Figure 3.12 | Solution of Example 3.31

The wavelength that is incident at $x = 5$ cm is

$$\lambda_1 = (e+b)\sin\tan^{-1}\frac{5}{f}$$

The wavelength that is incident at $x = 6$ cm is

$$\lambda_2 = (e+b)\sin\tan^{-1}\frac{6}{f}$$

Putting the values of $e + b$ and f into these equations, we get

$$\lambda_1 = \frac{1}{2000} \sin \tan^{-1} \frac{5}{50} = 4975\,\text{Å}$$

$$\lambda_2 = \frac{1}{2000} \sin \tan^{-1} \frac{6}{50} = 5957\,\text{Å}$$

The range of wavelengths that pass through the hole is 4975 Å to 5957 Å.

3.7 Dispersion

Dispersion is defined as a phenomenon in which constituent wavelengths, i.e., colors of a polychromatic light is fanned out. The dispersive power of an optical device is the ability of the device to separate the wavelengths of the polychromatic light. The dispersive power of a grating is defined as the angular separation of spectral lines of a polychromatic light with respect to wavelengths. The dispersive power of a grating is a measure of the angular separation produced between two incident monochromatic waves whose wavelengths differ by a small wavelength interval. Mathematically, the dispersive power of a grating is defined as $\frac{d\theta}{d\lambda}$. Differentiating both sides of Eq. (3.109) with respect to wavelength λ, we get

$$\frac{d}{d\lambda}(e+b)\sin\theta = \frac{d}{d\lambda}n\lambda$$

or $$\frac{d\theta}{d\lambda} = \frac{n}{(e+b)\cos\theta}$$

or $$\frac{d\theta}{d\lambda} = \frac{nN'}{\cos\theta} \qquad (3.107)$$

where $N' = \frac{1}{e+b}$ is the number of lines per unit lengths of the grating.

Equation (3.107) shows that the dispersive power of a grating is

i.   Directly proportional to the number of slits per unit length of the grating.
ii.   Directly proportional to the order of the spectral lines.
iii.   Inversely proportional to the cosine of the angle of diffraction.

The value of $\cos\theta$ from equation $(e+b)\sin\theta = n\lambda$ is obtained as

$$\cos\theta = \sqrt{1 - \frac{n^2\lambda^2}{(e+b)^2}} = \frac{\sqrt{(e+b)^2 - n^2\lambda}}{(e+b)}$$

Putting the value of $\cos\theta$ into Eq. (3.107), we get the expression for the dispersive power of a grating as

$$\frac{d\theta}{d\lambda} = \frac{n(e+b)}{(e+b)\sqrt{(e+b)^2 - n^2\lambda^2}}$$

or $\quad \dfrac{d\theta}{d\lambda} = \dfrac{n}{\sqrt{(e+b)^2 - n^2\lambda^2}}$ \hfill (3.108)

If the linear separation between two consecutive spectral lines of wavelengths λ and $\lambda + d\lambda$ is dx on the focal plane of the telescope objective or on the photographic plate, then

$$d\theta = \frac{dx}{f}$$

where f is the focal length of the telescope objective. From this equation, the expression for linear dispersion is obtained as

$$dx = fd\theta$$

or $\quad \dfrac{dx}{d\lambda} = f\dfrac{d\theta}{d\lambda}$

or $\quad \dfrac{dx}{d\lambda} = f\dfrac{nN'}{\cos\theta}$

From this equation, the expression for linear dispersion is obtained as

$$dx = f\frac{nN'}{\cos\theta}d\lambda$$ \hfill (3.109)

Example 3.32

Calculate the dispersive power of a grating in the third order spectrum of a wavelength region 5890 Å having 5000 lines/cm for a normal incidence.

Solution

The data given are $n = 3$, $\lambda = 5890$ Å and $N' = 5000$ lines/cm.
Hence,

$$e + b = \frac{1}{N'} = \frac{1}{5000} \text{cm}$$

The dispersive power of a grating for normal incidence is given from Eq. (3.109) by

$$\frac{d\theta}{d\lambda} = \frac{n}{\sqrt{(e+b)^2 - n^2\lambda^2}}$$

Putting the values of n, $e + b$ and λ into this equation, we get the dispersive power of the grating

$$\frac{d\theta}{d\lambda} = \frac{3}{\sqrt{\left(\dfrac{1}{5000}\right)^2 - 3^2(5890 \times 10^{-8})^2}} = 32021.8 \frac{\text{rad}}{\text{cm}}$$

$$= 32022 \frac{\text{rad}}{\text{cm}}.$$

or $\dfrac{d\theta}{d\lambda} = 3.20 \times 10^{-4} \dfrac{\text{rad}}{\text{Å}}$

Example 3.33

What is the expected dispersive power in the vicinity of a sodium line 5890 Å in the first three orders in a grating having 5250 lines/cm?

Solution

The data given are $N' = \dfrac{1}{e+b} = 5250$ lines/cm, $\lambda = 5890$ Å and $n = 1, 2, 3$.

The directions of the spectral lines of the nth order spectrum when a parallel beam of light is incident normally on a diffraction grating is obtained from Eq. (3.104) as

$$\theta = \sin^{-1}\left(\frac{n\lambda}{e+b}\right)$$

In the first order ($n = 1$), the angle of diffraction is

$$\theta_1 = \sin^{-1}\left(5250 \times 1 \times 5890 \times 10^{-8}\right) = 18°.$$

In the second order ($n = 2$), the angle of diffraction is

$$\theta_2 = \sin^{-1}\left(5250 \times 2 \times 5890 \times 10^{-8}\right) = 38°.$$

In the third order ($n = 3$), the angle of diffraction is

$$\theta_3 = \sin^{-1}\left(5250 \times 3 \times 5890 \times 10^{-8}\right) = 68°$$

The dispersive power is given by

$$\frac{d\theta}{d\lambda} = \frac{nN'}{\cos\theta}$$

The dispersive power in the first order will be

$$\frac{d\theta}{d\lambda} = \frac{1 \times 5250 \times 10^{-8}}{\cos 18} = 5.52 \times 10^{-5} \frac{\text{rad}}{\text{Å}}$$

The dispersive power in the second order will be

$$\frac{d\theta}{d\lambda} = \frac{2 \times 5250 \times 10^{-8}}{\cos 38} = 13.33 \times 10^{-5} \frac{\text{rad}}{\text{Å}}$$

The dispersive power in the third order will be

$$\frac{d\theta}{d\lambda} = \frac{3 \times 5250 \times 10^{-8}}{\cos 68} = 42.04 \times 10^{-5} \frac{\text{rad}}{\text{Å}}$$

Example 3.34

A grating having 3000 lines/cm is illuminated by light from mercury vapour discharges. What is the expected dispersive power in the third order in the vicinity of the intense green line ($\lambda = 5460$ Å)?

Solution

The data given are $n = 3$, $\lambda = 5460$ Å and $N' = 3000$ lines/cm.
Hence,

$$e + b = \frac{1}{N'} = \frac{1}{3000} \text{ cm}$$

The dispersive power of a grating for normal incidence is given from Eq. (3.108) by

$$\frac{d\theta}{d\lambda} = \frac{n}{\sqrt{(e+b)^2 - n^2\lambda^2}}$$

Putting the values of n, $e + b$ and λ into this equation, we get the dispersive power of the grating

$$\frac{d\theta}{d\lambda} = \frac{3}{\sqrt{\left(\dfrac{1}{3000}\right)^2 - 3^2\left(5460 \times 10^{-8}\right)^2}} = 10333.7 \frac{\text{rad}}{\text{cm}} = 10334 \frac{\text{rad}}{\text{cm}}.$$

or $\dfrac{d\theta}{d\lambda} = 1.03 \times 10^{-4} \dfrac{\text{rad}}{\text{Å}}$

3.8 Determination of Wavelength of Light by Grating

Generally, plane transmission grating is used in laboratories to measure the wavelength of monochromatic light.

3.8.1 Theory

The directions or positions of the principal maxima of the nth order when a parallel beam of light is incident normally on a diffraction grating is obtained from Eq. (3.104) as

$$(e + b)\sin\theta = n\lambda$$

Here $(e + b)$ is the grating element, θ is the angle of diffraction corresponding to the wavelength λ of the light. The expression for the wave length is obtained from this equation as

$$\lambda = \frac{(e+b)\sin\theta}{n}$$

The number of lines per inch is written on the grating by the manufacture. Taking the inverse of the number of lines per inch, the grating element $(e + b)$ is calculated in inches. Multiplying it with 2.54, we will get the grating element in cm, i.e.,

$$e+b = \frac{1}{\text{number of lines per inch}} \times 2.54 \text{ cm}$$

The unknown quantity left in the expression for wavelength λ is the angle of diffraction θ for a particular order n.

In the laboratory, the grating spectrum of a source of light is obtained by using a spectrometer. Generally, in physics laboratories, a sodium vapour lamp is used as a source of monochromatic light source. Before performing the experiment to measure the angle of diffraction, the following adjustments has to be carried perfectly.

3.8.2 Adjustments

i.   The spectrometer has to be adjusted for parallel rays by Schuster's method.

ii. The adjustment of the grating for normal incidence of light is done in the following ways.

The slit of the collimator is illuminated by a monochromatic light source and the position of the telescope is adjusted so that the image of the collimator slit is obtained parallel to the vertical cross-wire on the field of view of the telescope. Now the axes of the collimator and telescope are collinear. The position of the telescope is noted with help of the circular vernier scale attached to the spectrometer. From this position, the telescope is rotated through 90° and is fixed with the help of a fixing screw. The axis of the collimator and the axis of the telescope are perpendicular to each other in this position. The transmission grating is placed at the centre of the prism table so that the plane of the grating is perpendicular to the plane of the prism table. Now the prism table is rotated in such a way that the image of the collimator slit formed in the grating surface is visible through the telescope at the centre of the field of view. This is possible when the parallel beam from the sodium light source is incident on the grating surface at 45° as the axes of the telescope and the collimator are perpendicular to each other. The prism table is rotated again through 45° in such a way that the parallel rays of light from the collimator incident normally on the grating surface.

3.8.3 Measurement of θ

The collimator slit is illuminated by a monochromatic light source and is seen through the telescope. The telescope is moved slowly towards the left and is stopped when a vertical bright line is visible. By adjusting more finely, the vertical wire of the cross-wire of the telescope is made to coincide with the bright line. In this position, the reading R_{L1} of the vernier scale is noted down. This reading R_{L1} is the position of the telescope for the first order ($n = 1$) spectrum in the left. In a similar manner, moving further towards the left, the position of the telescope for the second order ($n = 2$) spectrum in the left R_{L2} is obtained. Similarly moving towards the right side, the positions of the first and second order spectrum R_{R1} and R_{R2} are obtained. Practically, in case of the sodium vapour lamp, the spectra up to the third order are visible using a well adjusted spectrometer.

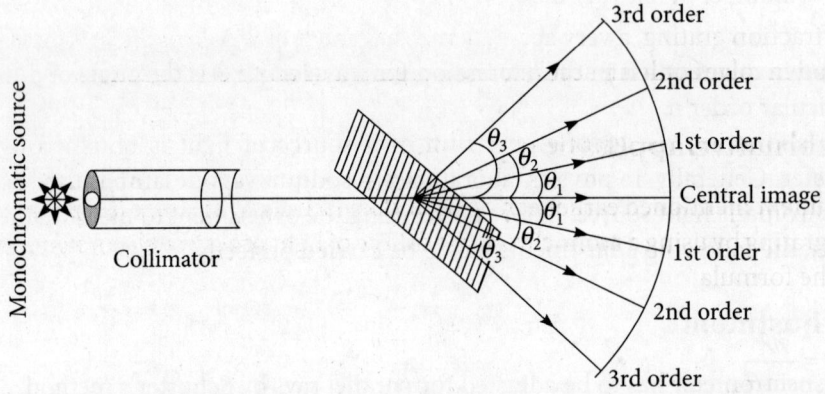

Figure 3.13 | Determination of the wavelength of light using a plane transmission grating. The readings R_{L1}, R_{L2}, R_{L3}, and the readings R_{R1}, R_{R2}, R_{R3} give the positions of the telescope for the 1st , 2nd and 3rd order spectra on the left side and right side respectively. $\theta_1 = \dfrac{R_{L1} \sim R_{R1}}{2}$, $\theta_2 = \dfrac{R_{L2} \sim R_{R2}}{2}$, $\theta_3 = \dfrac{R_{L3} \sim R_{R3}}{2}$.

As can be seen from Fig. 3.13, the angles of diffraction for the first order spectrum θ_1, second order spectrum θ_2 and third order spectrum θ_3 are obtained by

$$\theta_1 = \frac{R_{L1} \sim R_{R1}}{2} = \text{diffraction angle for the first order spectrum}$$

and $\quad \theta_2 = \dfrac{R_{L2} \sim R_{R2}}{2} = \text{diffraction angle for the second order spectrum}$

$$\theta_3 = \frac{R_{L3} \sim R_{R3}}{2} = \text{diffraction angle for the third order spectrum}$$

3.8.4 Calculation of λ

The wavelength of the monochromatic light as obtained for three orders of diffraction will be

$$\lambda_1 = \frac{(e+b)\sin\theta_1}{1}, \ \lambda_2 = \frac{(e+b)\sin\theta_2}{2} \text{ and } \lambda_3 = \frac{(e+b)\sin\theta_3}{3}$$

respectively. The actual wavelength of the monochromatic light λ will be the mean of λ_1, λ_2, and λ_3, i.e.,

$$\lambda = \frac{\lambda_1 + \lambda_2 + \lambda_3}{3}.$$

With a diffraction grating, a very accurate measurement of wavelengths of light is possible. The method involves only an accurate measurement of angle of diffractions.

3.8.5 Alternative application

The experiment mentioned earlier can be used alternatively to measure the grating element of a given grating by using a monochromatic source of light of accurately known wavelength by using the formula

$$e+b = \frac{n\lambda}{\sin\theta}$$

Questions

3.1   What is diffraction?

3.2   Why is the shadow of a fine wire due to sunlight not observable on a screen?

3.3   Differentiate between interference and diffraction.

3.4   How can we remove all the odd terms or even terms from the RHS of Eq. (3.6) practically?

3.5   What is Fresnel diffraction?

3.6   What is Fraunhofer diffraction?

3.7   Fraunhofer diffraction cannot occur if slit width is less than wavelength. why?

3.8   Differentiate between Fresnel and Fraunhofer diffraction.

3.9   What is Fresnel's half period zones?

3.10   Why are Fresnel's half period zones so named?

3.11   Describe the procedure to construct Fresnel's half period zones.

3.12   Explain why higher order Fresnel's half period zones have greater area than that of lower order?

3.13   Prove that the ratio of area of the *n*th half period zone to its average distance from the pole of the wavefront is constant for a particular wavefront.

3.14   The resultant amplitude at pole point due to the entire wavefront is half the amplitude due to the first Fresnel's half period zone. Explain.

3.15   On which factor does the intensity at an axial point due to Fresnel's half period zone depend? Explain.

3.16   How does the radius of Fresnel's half period zone depend on the wavelength of light?

3.17   What is a zone plate?

3.18   Differentiate between positive zone plate and negative zone plate.

3.19   Prove that the radii of half period zones of a zone plate are proportional to the square root of natural numbers.

3.20   Prove that each half period zone of a zone plate have equal area.

3.21   Prove that the intensity at an axial point due to only odd numbered half period zones is many times more than that of the intensity at the point due to the entire wavefront.

3.22   Prove that the intensity at an axial point due to only even numbered half period zones is many times more than that of the intensity at the point due to the entire wavefront.

3.23   Explain how the zone plate behaves like a lens.

3.24   Derive an expression for the focal length of a zone plate.

3.25   How do the focal lengths of a zone plate depend on the wavelength of light?

3.26   Prove that a zone plate has a number of foci.

3.27   Explain the presence of only odd numbered foci in a zone plate.

3.28   Explain the absence of even numbered foci in a zone plate.

3.29   Explain how the intensity of the focus of a zone plate decreases with increase in the order of the focus.

3.30   Explain how the intensity of the focus decreases as it moves towards the zone plate.

3.31   What are the similarities between a zone plate and a lens?

3.32   What are the dissimilarities between a zone plate and a lens?

3.33   Derive an expression for the intensity distribution of Fraunhofer single slit diffraction pattern.

3.34   Derive an approximate expression for the intensity of the secondary maxima in case of Fraunhofer's single slit diffraction.

3.35   Graphically show the intensity distribution due to Fraunhofer's single slit diffraction.

3.36   Franhofer diffraction cannot occur if the slit width is less than the wavelength. Explain.

3.37   What is a plane grating?

3.38   What does inverse of a grating element give?

3.39   How will you know with the naked eye whether the given specimen is a glass piece or a grating?

3.40   Derive an expression for intensity of the diffracted beam diffracted by a plane grating on normal incidence.

3.41   Derive an expression for the principal maxima of the diffracted beam diffracted by a plane grating on normal incidence.

3.42   Derive an expression for the direction of the principal maxima of the diffracted beam diffracted by a plane grating on normal incidence.

3.43   Derive an expression for the direction of minima of the diffracted beam diffracted by a plane grating on normal incidence.

3.44   Derive an expression for direction principal maxima of the diffracted beam diffracted by a plane grating on oblique incidence.

3.45   Explain why higher order principal maxima have lesser angular width.

3.46   The grating element of a grating is 3.5×10^{-5} cm. Can violet light be diffracted by this grating? Justify your answer.

3.47   Explain why grating spectral lines of shorter wavelengths are sharper than the longer wavelengths.

3.48   Show that the third order grating spectrum of violet light ($\lambda = 4000$ Å) always overlaps the second order grating spectrum of orange color light ($\lambda = 6000$ Å).

3.49   Differentiate between grating spectrum and a single-slit spectrum of a monochromatic light.

3.50   What is the maximum value of wavelength to obtain a grating spectra in a grating?

3.51   Explain why in general, x-ray cannot be diffracted by optical grating.

3.52   What is the implication of angular width of the principal maxima in spectroscopy?

3.53   Derive an expression for the angular width of the principal maxima.

3.54   What are the factors on which the angular width of the principal maxima depends?

3.55   What do you mean by dispersive power of a grating?

3.56   Derive an expression for dispersive power of a grating?

3.57   Derive the relation $\dfrac{d\theta}{d\lambda} = \dfrac{\tan \theta}{\lambda}$

3.58   What are the factors on which the dispersive power of a grating depends?

3.59   Prove that for lower order spectral lines the half angular width $d\theta$ in doublet lines is given by $d\theta = \dfrac{nd\lambda}{e+b}$.

3.60   What are the characteristics of grating spectra?

3.61   What is the maximum value of the order of grating spectra that can be obtained by a grating?

3.62   In a grating, opaque space is three times the slit width. What will happen to the spectra?

3.63   Describe an experiment to determine the wavelength of monochromatic light.

3.64   Describe an experiment to determine the grating element of a grating.

3.65   Describe an experiment to determine the number of lines on one cm of a given grating surface.

Problems

3.1   The principal focal length of a positive zone plate is 1.5 m for a light of wavelength 6500 Å. Determine the radii of the 1st, 2nd, and 3rd opaque zones of the zone plate.

[Ans   0.140 cm, 0.198 cm, 0.242 cm]

3.2   What is the radius of the first zone of a zone plate with focal length 20 cm for a light of wavelength 5000 Å?   [Ans   0.0316 cm]

3.3   A point source of light of wavelength 6000 Å is placed on the axis of a zone plate. The strongest and next strongest images of the source are formed on the other side of the zone plate at distances of 30 cm and 6 cm respectively. Calculate (a) the distance of the point source from the zone plate, (b) the radius of the first zone, and (c) the principal focal length of the zone plate.   [Ans   −30 cm, 3×10^{-2} cm, 15 cm]

3.4   A zone plate is constructed by taking the print of Newton's ring formed by a plano-convex lens of radius of curvature 2 m on a thin glass plate. Find the principal focal length of the zone plate constructed in this way.   [Ans   200 cm]

3.5   The radius of the first half period zone of a zone plate is 0.05 cm. What should be the position of a screen so that the brightest spot is formed on the screen when a plane monochromatic light of wavelength 6500 Å is incident normally on the zone plate.

[Ans   38.46 cm]

3.6   The diameter of the central zone of a zone plate is 0.24 cm. If a point source of light of wavelength 6000 Å is placed at a distance of 60cm from the zone plate find the positions of the strongest image and other weaker images.

[Ans   240 cm, 80 cm, 48 cm, 34.29 cm, …, etc.]

3.7   A single slit of width 14×10^{-5} cm is illuminated by a parallel beam of monochromatic light of wavelength 7000 Å. Calculate the angular width of the central maximum.

[Ans   30°]

3.8   A slit of width 0.04 cm is illuminated by a parallel beam of monochromatic light. A screen is placed at a distance of 1m from the converging lens to obtain the diffraction pattern. Calculate the wavelength of the monochromatic light if the first minima lie at 0.15 cm on either side of the central maximum.   [Ans   6000 Å]

3.9   A single slit is illuminated by a parallel beam of monochromatic light of wavelength 5600 Å. Calculate the slit width if the angular width of the central maximum is 10°.

[Ans 6.42×10^{-4} cm]

3.10   A slit of width 0.04 cm is illuminated by a parallel beam of monochromatic light of wavelength 5890 Å. The diffracted beam is focused on a screen by a converging lens of focal length 80 cm to obtain the diffraction pattern. Calculate the distance between the centre of the first minima and the central maximum on the screen.

[Ans   0.118 cm]

3.11 For what value of slit width will the first minimum for red light of wavelength 6500 Å fall at $\theta = 30°$? [Ans 1.3 × 10⁻⁴cm]

3.12 In a single slit diffraction pattern, the distance between the first minimum on the left to the first minimum on the right is 0.052 cm for a light of wavelength 5460 Å. Calculate the slit width if the distance of the screen from the slit is 80 cm.
[Ans 0.0168 cm]

3.13 A single slit of width 0.04 cm is illuminated by a parallel beam of monochromatic light of wavelength 5900 Å. A converging lens is placed behind the slit to focus the diffraction pattern on a screen placed at a distance of 70 cm. What is the distance of the first minimum and second minimum from the centre of the central maximum?
[Ans 0.103 cm, 0.206 cm]

3.14 What is the half width of the central diffraction pattern due to a single slit when slit width is 5λ? [Ans 10.16°]

3.15 A single slit of width 0.04 cm is illuminated by a parallel beam of monochromatic light of wavelength 5900 Å. What is the half width of the central maximum? [Ans 0.075°]

3.16 What is the maximum wavelength of the visible spectrum so that the fifth order spectrum can be observed by a grating having 4000 lines/cm? [Ans 5000 Å]

3.17 What are the angles at which a maximum intensity beam occurs when a monochromatic light of wavelength 5000 Å is incident normally on a diffraction grating having 4000 lines/cm? [Ans 0°, ±11.54°, ±23.58°, ±36.87°, ±53.13°]

3.18 A sodium vapour light is incident on a grating with 10000 lines/inch. Find the angles of diffraction for the principal maxima of the two lines of the sodium light 5890 Å and 5896 Å in the second order spectrum. [Ans 27.63°, 27.66°]

3.19 A parallel beam of light is incident normally on a grating surface having 11800 lines/cm. The first order spectral line is observed to be diffracted through an angle 30°. Calculate the wavelength of the monochromatic light. [Ans 4237 Å]

3.20 A parallel beam of monochromatic light of wavelength 6300 Å is incident normally on a grating surface having 4000 lines/cm. What is the highest order spectrum that can be observed? [Ans fourth]

3.21 Calculate the angle of separation of sodium lines 5896 Å and 5890 Å in the second order spectrum when a parallel beam of helium light is incident normally on a grating containing 5000 lines/cm. [Ans 0.043° = 0°2'35"]

3.22 A plane transmission grating having 6000 lines/cm is used to obtain a spectrum of light from sodium lamp in the second order. Calculate the angle of separation of sodium lines 5896 Å and 5890 Å. [Ans 0.057° – 0°3'25"]

3.23 A helium light containing two wavelengths 5048 Å and 5016 Å is incident normally on a grating having 8000 lines/cm. A lens of focal length 150 cm is used to observe the spectrum on a screen. Calculate the separation in cm of the two lines in the second order spectrum. [Ans 3.68 cm]

3.24 A monochromatic light of wavelength 5000 Å is incident normally on a plane transmission grating having 6000 lines/cm. Calculate the difference between the diffraction angles of the first order and third order spectra. [Ans 46.7° = 46°42'2"]

3.25 In a plane transmission grating, the angle of diffraction for the second order principal maximum is 20° for a wavelength of 6000 Å. Calculate the number of lines in one cm of the grating surface. [Ans 2850 lines/cm]

3.26 How many orders of diffraction bands will be visible theoretically if the wavelength of incident radiation is 5000 Å and the number of lines per cm on the grating surface is 5000? [Ans 4]

3.27 How many orders of diffraction bands will be visible theoretically if the wavelength of incident radiation is 5000 Å and the number of lines per inch on the grating surface is 2620? [Ans 19]

3.28 A parallel beam of monochromatic light is allowed to fall incident normally on a grating having 6000 lines/cm and the first order spectral line is found to be diffracted through an angle 20°. Calculate the wavelength of the light used. [Ans 5700 Å]

3.29 A diffraction grating used at normal incidence gives a line with wavelength 6000 Å in a certain order superimposed on another line of wavelength 4500 Å of the next higher order. If the angle of diffraction for the two wavelengths is 30°, how many lines are there in one cm of the grating? [Ans 2778 lines/cm]

3.30 A plane transmission grating has 6000 lines/cm. Show that the angular separation between two wavelengths 5890 Å and 5896 Å of sodium vapour light for a second order spectrum is approximately 3 minutes of an arc.

3.31 Light, which is a mixture of two wavelengths 5500 Å and 5555 Å, is incident normally on a plane transmission grating having 10000 lines/cm. A lens of focal length 150 cm is used to observe the spectrum on a screen. Calculate the separation in cm of the two lines in the first order spectrum. [Ans 1.43 cm]

3.32 A diffraction grating used at normal incidence gives a green line 5400 Å in a certain order superimposed on the violet line 4050 Å of the next higher order. If the angle of diffraction for the two lines is 30°, how many lines are there on the one centimeter grating surface. [Ans 3086 lines/cm]

3.33 Design a grating that will spread the first order visible spectrum through an angular range of 20° if the range of visible spectrum is 4300 Å to 6800 Å.
[Ans 10900 lines/cm]

3.34 What is the expected dispersion D for a sodium line 5896 Å in the first two orders in a grating having 3000 lines/cm? [Ans 3.05×10^{-5} rad/Å, 6.42×10^{-5} rad/Å]

3.35 Calculate the dispersive power of a grating in the third order spectrum in the wavelength region 5000 Å having 4000 lines/cm for normal incidence.
[Ans 1.5×10^{-4} rad/Å]

Multiple Choice Questions

1. What type of wavefront is incident in the case of Fresnel's class of diffraction?
   (i) plane   (ii) spherical
   (iii) cylindrical   (iv) elliptical

2. What type of wavefront is incident in the case of Fraunhofer's class of diffraction?

   (i)   plane   (ii)   spherical

   (iii) cylindrical   (iv)   elliptical

3. Who gave the correct interpretation of rectilinear propagation of light?

   (i)   Fresnel   (ii)   Fraunhofer

   (iii) Newton   (iv)   Maxwell

4. What is the phase difference between two waves originating from the inner and outer perimeter of Fresnel's half period zones?

   (i) 2π (ii) π

   (iii) $\dfrac{\pi}{2}$ (iv)   0

5. What is the phase difference between two waves originating from two consecutive Fresnel's half period zones

   (i)   0   (ii) $\dfrac{\pi}{2}$

   (iii) π (iv) 2π

6. The amplitude at a pole point due to a half period zone depends only upon

   (i)   area of the half period zone

   (ii) average distance of the half period zones from the pole point

   (iii) obliquity factor/inclination factor

   (iv) all the above

7. What is (are) the factor(s) on which intensity due to any Fresnel's half period zone at an axial point depends?

   (i)   area of the half period zone

   (ii) average distance of the half period zones from the pole point

   (iii) obliquity factor/inclination factor

   (iv) all the above

8. In a positive zone plate, central zone is opaque.

   (i)   True   (ii)   False

9. In a negative zone plate, central zone is opaque.

   (i)   True   (ii)   False

10. The shape of the central zone in a zone plate is

   (i)   square   (ii)   rectangle

   (iii) circular   (iv)   elliptical

11. The intensity of foci decreases as we move towards the zone plate.

   (i)   True   (ii)   False

12. The radii of half period zones of a zone plate are proportional to the square root of
    (i) even natural numbers
    (ii) odd natural numbers
    (iii) integers
    (iv) natural numbers

13. The focus of the first order of a zone plate is situated at a point
    (i) nearest to the zone plate
    (ii) farthest from the zone plate
    (iii) mid-way between nearest and farthest points

14. The principal, i.e., primary focal length of a zone plate does not depend upon
    (i) wavelength of the light
    (ii) radius of the first half period zone
    (iii) frequency of the light
    (iv) none of the above

15. I_1, I_3 and I_5 are the intensities of the first three foci of a zone plate. Which of the following relations is correct?
    (i) $I_5 > I_3 > I_1$
    (ii) $I_1 < I_3 < I_5$
    (iii) $I_5 < I_3 < I_1$
    (iv) $I_3 > I_5 > I_1$

16. In single slit diffraction, the intensity of the first secondary maximum is less than that of the principal maximum by
    (i) 20 times
    (ii) 21 times
    (iii) 22 times
    (iv) 23 times

17. For the 5th order grating spectrum of the following lights which angle of diffraction is more?
    (i) violet
    (ii) green
    (iii) yellow
    (iv) red

18. What is the unit of dispersive power of a grating?
    (i) radian
    (ii) watt
    (iii) radian/cm
    (iv) radian/degree

19. When white light is incident on a grating, which of the following colors deviates most?
    (i) violet
    (ii) blue
    (iii) orange
    (iv) yellow

20. The maximum order spectrum produced in a grating is
    (i) directly proportional to the grating element
    (ii) inversely proportional to the grating element
    (iii) directly proportional to the wavelength
    (iv) directly proportional to the order of the spectrum

21. In single slit diffraction pattern, orange light is replaced by red light without changing the experimental setup. The diffraction pattern will
    (i) disappear
    (ii) be unchanged
    (iii) be wider
    (iv) be narrower

22. A CD gives a sensation of rainbow colors because of
    (i) interference   (ii) polarization
    (iii) diffraction   (iv) scattering

23. The maximum number of orders produced by grating is
    (i) directly proportional to both the grating element and the wavelength
    (ii) inversely proportional to both the grating element and the wavelength
    (iii) directly proportional to the grating element and inversely to the wavelength
    (iv) directly proportional to the wavelength and inversely to the grating element

24. When white light is incident on a grating which of the following colors deviates least?
    (i) violet   (ii) blue
    (iii) orange   (iv) yellow

25. In a grating, the grating element is three times the slit width. Which order of spectrum will be absent
    (i) 1   (ii) 2
    (iii) 3   (iv) 4

Answers

1 (ii)	2 (i)	3 (i)	4 (ii)	5 (iii)	6 (iii)	7 (iii)	8 (ii)
9 (i)	10 (iii)	11 (i)	12 (iv)	13 (ii)	14 (iv)	15 (iii)	16 (iii)
17 (iv)	18 (iii)	19 (iii)	20 (i)	21 (iv)	22 (iii)	23 (iii)	24 (i)
25 (iii)							

4 Polarization

4.1 Introduction

The phenomena of interference and diffraction proved successfully the wave character of light. However, it cannot confirm whether light is a transverse wave or a longitudinal wave. Interference and diffraction occur in transverse as well as longitudinal waves. The transverse nature of light was first confirmed by an optical phenomenon called polarization. Polarization is defined as a process of restricting the vibrations of a transverse wave to one direction or one plane only.

4.2 Polarization of Waves

The appearance of a longitudinal wave is the same when viewed along any direction. It is perfectly symmetrical about the direction of propagation. However, this is not so with transverse waves. All electromagnetic waves are transverse waves. Light is an electromagnetic wave consisting of mutually perpendicular electric vector and magnetic vector. The electric vector is also called light vector. In case of transverse waves, the particles of the medium vibrate at right angles to the direction of propagation. If vibration of electric vectors of the light wave is confined to the XY plane, (i.e., the wave is plane polarized/linearly polarized with the plane of vibration along the XY plane) and the views taken along the X, Y, and Z direction will be different from each other. Ordinary light behaves in such a manner that it appears perfectly symmetrical about the direction of propagation, though there is no doubt about its transverse character – this is due to the fact that millions of light vectors undergo rapid changes in their direction within a metre length; not only does the direction of the transverse vibrations of the light wave change, but the character of the vibration also changes. The transverse vibration of the light wave may undergo changes from linear to circular, circular to elliptical. By using suitable devices, the light vector may be constrained to describe linear, circular, elliptical patterns of fixed orientation transverse to the direction of propagation.

4.2.1 Mechanical demonstration of polarization of waves

A simple mechanical experiment can be designed to demonstrate polarization of mechanical transverse waves. Suppose as shown in Fig. 4.1, a loosely stretched thin string AB passes through two narrow rectangular slits N_1 and N_2, each a little wider than the diameter of the string. Let end B be fixed to a rigid wall and transverse waves be produced in it by shaking the free end A. The direction of shaking, i.e., vibration can be varied in an arbitrary manner. The portion of the string within AN_1 vibrates in an arbitrary direction implying that the wave in this portion of the string is not polarized. The slit N_1 allows those vibrations which are parallel along its length.

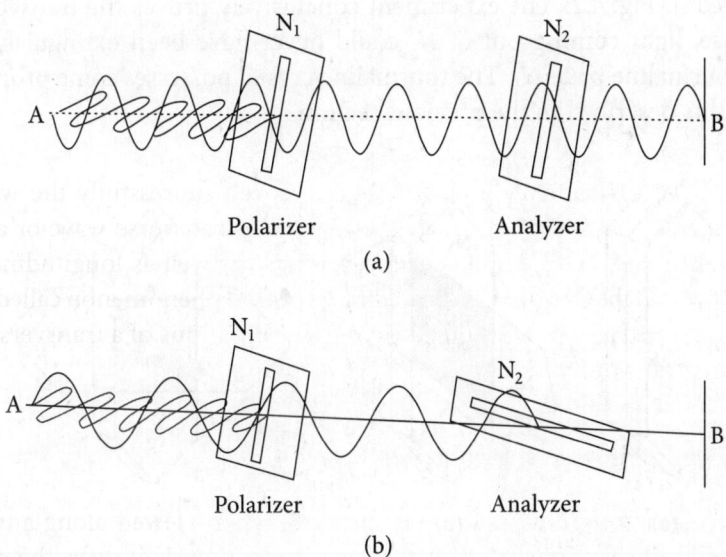

Figure 4.1 | Mechanical experiment to demonstrate the polarization of a wave. In Fig. 4.1(a), the two slits N_1 and N_2 are parallel to each other, whereas in Fig. 4.1(b) they are perpendicular to each other

The wave along the string between the two slits N_1 and N_2 is confined along a plane passing through the two slits as shown in Fig. 4.1(a). Thus, the wave between the two slits N_1 and N_2 is plane polarized or linearly polarized. In this experiment, slit N_1 is called a polarizer. The polarized wave can be completely transmitted through slit N_2 only if it is parallel to N_1. It cannot be transmitted through slit N_2 if N_2 is perpendicular to N_1 as shown in Fig. 4.1(b). In this case, there is no wave along the string beyond N_2. Slit N_2 acts as the analyzer or detector of polarization of incoming waves. The transmission of longitudinal waves remain unaffected either to the presence of the slits or to the relative orientation of the two slits N_1 and N_2.

4.2.2 Demonstration of optical polarization of waves

An optical experiment similar to the mechanical one described earlier can be designed which throws considerable light on the nature of vibration of light vector in a light wave. When a ray of ordinary light is allowed to fall on a tourmaline plate N_1 cut with faces parallel to its vertical axis, only a part of the light is transmitted and is found to be plane polarized. No remarkable changes in the transmitted light are observed when tourmaline plate N_1 is rotated. If this transmitted light from N_1 is further passed through a second similar tourmaline plate N_2 kept with its vertical axis parallel to the vertical axis of N_1, no change in intensity occurs. Moreover, no change is observed in the intensity of light coming out of N_2 when both N_1 and N_2 are rotated together. If N_1 is kept fixed and N_2 is rotated, the light coming out of N_2 becomes dimmer and dimmer and vanishes completely when two vertical axes are perpendicular to each other. With further rotation of N_2, the light intensity increases and becomes maximum when both the vertical axes are parallel to each other. This is depicted in Fig. 4.2. The experiment conclusively proves the transverse nature of light; otherwise, light coming out of N_2 could never have been extinguished simply by rotating the tourmaline plate N_2. The tourmaline crystal possesses some properties similar to that of the slits described in the previous section.

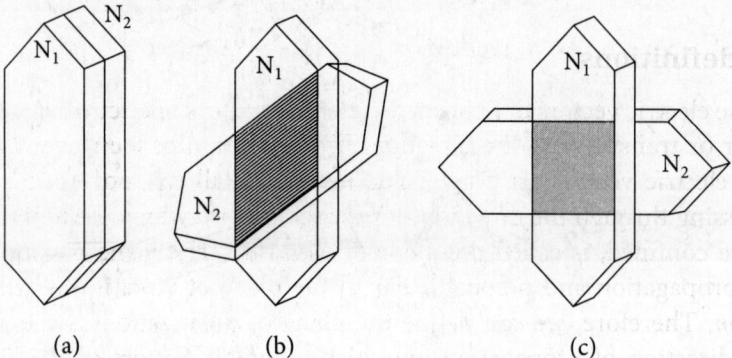

(a)   (b)   (c)

Figure 4.2 | Optical demonstration of a polarization of a light wave in tourmaline crystal. (a) the two vertical axes are parallel to each other and no decrease of light occurs when it comes out of N_2. (b) the two vertical axes are not parallel to each other and some light comes out of N_2. (c) the two vertical axes are perpendicular to each other and no light comes out of N_2

4.2.3 Pictorial representation of light

Light is an electromagnetic wave. It consists of millions of electromagnetic waves of mutually perpendicular vibrating electric and magnetic vectors. The planes of vibration of light vectors have random orientations due to the random emissions of radiations from excited atoms or molecules of the source. The light vectors are therefore arranged symmetrically about the direction of propagation. The end-on view of the unpolarized beam of light is shown in Fig. 4.3. In this figure, a beam of light is coming out of the plane of the paper perpendicularly.

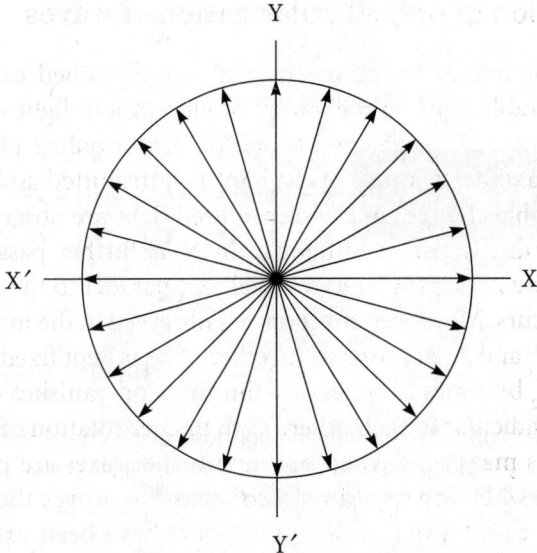

Figure 4.3 | End-on view of an unpolarized ray of light coming out of the plane of the paper perpendicularly

4.2.4 Few definitions

The transverse electric vectors of light waves (electric vectors of electromagnetic waves are perpendicular or transverse to the direction of propagation of the wave) are called *light vectors* since electric vectors are responsible for almost all types of optical phenomena. The plane passing through the direction of propagation, on which the light vectors of the light wave are confined, is called the *plane of vibration*. The plane passing through the direction of propagation and perpendicular to the plane of vibration is called the *plane of polarization*. Therefore, we can define the plane of polarization as the plane passing through the direction of propagation on which magnetic vectors of the light wave are confined. The plane of vibration and the plane of polarization are shown in Fig. 4.4. The plane passing through the incident ray and the the normal drawn at the point of incidence is called the *plane of incidence*. The plane passing through the reflected ray and the normal drawn at the point of incidence is called the *plane of reflection*. The plane passing through the refracted ray and the normal drawn at the point of incidence is called the *plane of refraction*. The electric vectors of the light wave which are vibrating in the plane of incidence are called *parallel vibrations* and the light vectors which are vibrating perpendicular to the plane of incidence are called *perpendicular vibrations*.

4.3 Classification of Polarized Light

Generally, there are three different types of polarization states: linear, circular and elliptical. Each of these commonly encountered states is characterized by differing motion of the electric field vector with respect to the direction of propagation of the light wave.

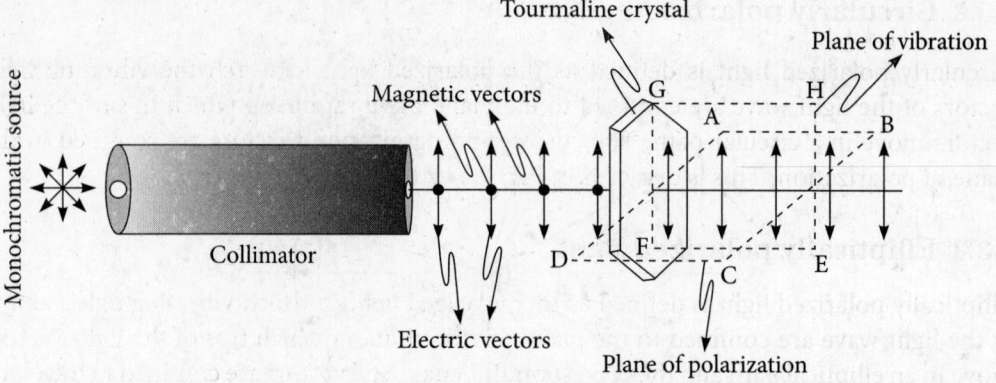

Figure 4.4 | Unpolarized light after passing through tourmaline crystal become plane polarized. The EFGH plane along the vertical axis of the tourmaline crystal is the plane of vibration and the ABCD plane is the plane of polarization

4.3.1 Plane polarized light

Plane polarized light is also called linearly polarized light. In plane polarized light, the vibrating electric vectors of the light wave are confined to a single plane called the plane of vibration and the corresponding magnetic vectors are confined in the plane of polarization. This is depicted in Fig. 4.5.

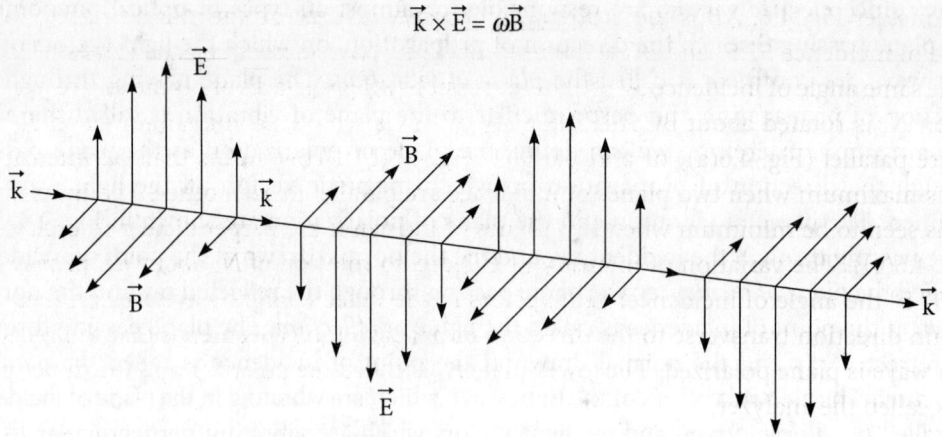

Figure 4.5 | Instantaneous snapshot of a plane polarized light wave showing the electric vectors, magnetic vectors and propagation vector. The plane containing the vibrating electric vector $\vec{E}$ and the propagation vector $\vec{k}$ is called the plane of vibration and the plane containing the vibrating magnetic vector $\vec{B}$ and the propagation vector $\vec{k}$ is called the plane of polarization. The vectors $\vec{E}$, $\vec{B}$, and $\vec{k}$ are mutually perpendicular to each other at any instant of time

4.3.2 Circularly polarized light

Circularly polarized light is defined as the polarized light in which the vibrating light vectors of the light wave are confined to the plane of vibration, on which tips of the light vectors move in a circular path. The corresponding magnetic vectors are confined in the plane of polarization. This is depicted in Fig. 4.27 later in the chapter.

4.3.3 Elliptically polarized light

Elliptically polarized light is defined as the polarized light in which vibrating light vectors of the light wave are confined to the plane of vibration, on which tips of the light vectors move in an elliptical path and the corresponding magnetic vectors are confined in the plane of polarization. This is depicted in Fig. 4.29 later in the chapter.

4.4 Polarization by Reflection

The simplest way to produce a plane polarized light is by the process of reflection. When ordinary light is allowed to fall on the polished surface of transparent dielectric media, for certain angles of incidence, the reflected light wave is plane polarized with the plane of polarization perpendicular to the reflecting surface and passing through the incident direction. The angle of incidence (equal to the angle of reflection) for which the reflected light is most completely plane polarized is called polarizing angle φ_p as shown in Fig. 4.7 later. It depends upon the nature of the interface and the wavelength of the light wave. For air and glass interface, this polarizing angle is 57.5°.

Consider Fig. 4.6. *AB* is the ordinary light wave incident on the glass surface N_1 with angle of incidence 57.5° and *BC* is the reflected light wave incident on the glass surface N_2 at the same angle of incidence 57.5°. This is possible when surfaces of N_1 and N_2 are parallel. When N_2 is rotated about BC, the intensity of *CD* is seen to be maximum when N_1 and N_2 are parallel (Fig. 4.6(a)) or anti-parallel (Fig. 4.6(b)). This means that the intensity of *CD* is maximum when two planes of incidence are parallel to each other. The intensity of *CD* is seen to be minimum when two planes of incidence are perpendicular to each other (Fig. 4.6(c)). The variation of intensity of *CD* due to rotation of N_2 about *BC* proves that for 57.5° the angle of incidence, light vectors in the reflected light wave are confined to a certain direction transverse to the direction of propagation. This means that the reflected light wave is plane polarized. The lower plate N_1 is called the polarizer and the upper plate N_2 is called the analyzer.

4.4.1 Explanation of polarization by reflection

In the following, a most simplified explanation of polarization by reflection is given. The light vectors in the unpolarized light wave can be resolved at the point of incidence into two components, one component perpendicular to the plane of incidence and the other

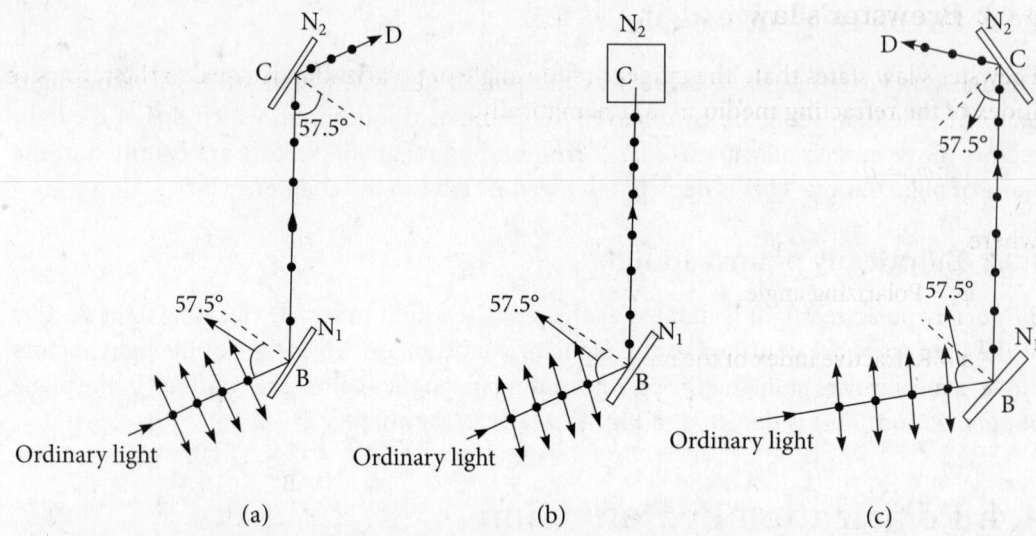

(a)   (b)   (c)

Figure 4.6 | Polarization by reflection from glass surface. (a)The intensity of CD is maximum when N_1 and N_2 are parallel or (b) anti-parallel and (c) the intensity of CD is minimum when N_1 and N_2 are perpendicular to each other implying that reflected light wave is plane polarized

component parallel to the plane of incidence. At the polarizing angle, the angle between the reflected ray and the refracted ray is 90°. The light vectors which are in the plane of incidence will become parallel to the direction of reflected ray to produce longitudinal waves. However, light is a transverse wave having no longitudinal components of light vector. Therefore, we conclude that the light vectors in the plane of incidence of the incident wave are absent in the reflected wave and hence, are 100% present in the refracted wave. The light vectors perpendicular to the plane of incidence are partly transmitted and partly reflected. Therefore, we conclude that the reflected wave is completely plane polarized with low intensity whereas the refracted wave is partially polarized with comparatively high intensity. All this happens only if the angle of incidence is equal to the polarizing angle.

The intensity of the wave CD as said in Section 4.4 depends upon the relative orientation of N_1 and N_2. The reason is as follows. In Fig. 4.6, wave BC contains light vector perpendicular to the plane of incidence containing the incident wave AB. When N_1 and N_2 are parallel or anti-parallel to each other, their planes of incidence are the same. Therefore, light vectors perpendicular to the plane of incidence are partly reflected along CD and partly refracted by N_2 making the intensity of CD maximum.

However, when N_1 and N_2 are perpendicular to each other, their planes of incidence become perpendicular to each other. Therefore, light vectors perpendicular to the plane of incidence in N_1 become parallel to the plane of incidence in N_2. Light vectors parallel to the plane of incidence of N_2 can only be refracted and cannot be reflected, and the intensity of the wave CD becomes minimum. Therefore, we can conclude that the intensity of CD depends on the angle between two planes of reflection.

4.4.2 Brewster's law

Brewster's law states that, 'the tangent of the angle of polarization is equal to the refractive index of the refracting medium'. Mathematically,

$$\tan \varphi_p = \mu$$

where

φ_p = Polarizing angle.

μ = Refractive index of the medium.

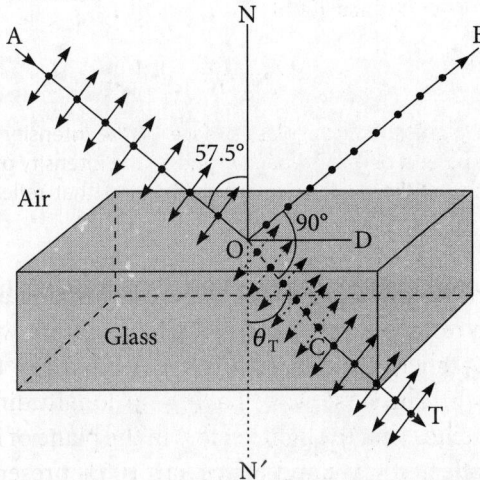

Figure 4.7 | Illustration of Brewster's law. AO, OC and OB are the direction of incident, refracted and reflected wave. AO, OB, OD, OC, and CT all lie in a single plane. NN' is the normal drawn at the point of incidence O. ∠BOC = 90°. The reflected light OB is plane polarized with the plane of vibration passing through the reflected ray and perpendicular to the plane of incidence/plane of reflection. The refracted or transmitted light OC is plane polarized to some extent with the plane of incidence/plane of refraction as the plane of vibration

At the polarizing angle, it is found experimentally that the angle between the reflected ray and the refracted ray is 90°. Thus from Fig. 4.7, we have

$$\angle BOC = 90°$$

or $$\angle BOD + \angle DOC = 90°$$

or $90° - \angle NOB + 90° - \angle NOC = 90°$

or $\angle NOB + \angle NOC = 90°$

or $\angle NOA + \angle NOC = 90°$

or $\varphi_p + \theta_T = 90°$ (4.1)

At the polarizing angle, Snell's law becomes

$$\frac{\sin \varphi_p}{\sin \theta_T} = \mu$$

or $\sin \varphi_p = \mu \sin \theta_T$

Putting the value of θ_T from Eq. (4.1) into the above equation we have

$$\sin \varphi_p = \mu \sin \left(90° - \varphi_p \right)$$

or $\tan \varphi_p = \mu$ (4.2)

Example 4.1

A glass plate of refractive index 1.7 is to be used as a polarizer. Calculate the polarizing angle and angle of refraction.

Solution

From Eq. (4.2), we have

$$\varphi_p = \tan^{-1} 1.7 = 59.54°$$

$$\theta_T = 90° - \varphi_p = 90° - 59.54° = 30.46°$$

Example 4.2

When a light ray is incident on water surface at an angle 53°, it is found that reflected and refracted rays are perpendicular to each other. Find the polarizing angle and the refractive index of water.

Solution

According to the question, the reflected ray is perpendicular to the refracted ray. Therefore,

Angle of polarization = Angle of incidence = angle of polarization = 53°

Applying Brewster's law, we have

$$\mu = \tan 53° = 1.33$$

Hence, the polarizing angle and refractive index of water are 53° and 1.33 respectively.

4.5 Polarization by Refraction

As shown in Fig. 4.7, if ordinary light wave is incident on the surface of a glass of refractive index 1.5 at polarizing angle 57.5°, 100% of light vectors vibrating parallel to the plane of incidence and 85% of light vectors vibrating perpendicular to the plane of incidence are refracted, whereas only 15% light vectors vibrating perpendicular to the plane of incidence is reflected. Therefore, ordinary light wave incident on the surface of the first glass plate in a pile of glass plates at the polarizing angle, 15% vibrations perpendicular to the plane of incidence are reflected and the rest containing 100% parallel vibrations and 85% transverse vibrations are refracted. The refracted wave containing 100% parallel and 85% transverse vibrations incident on the first surface of the second plate of the pile of glass plates at polarizing angle with all parallel vibrations and 85% of 85% perpendicular vibrations are refracted and 15% of 85% are reflected. Thus, by each reflection and refraction phenomena occurring at each glass plate at the polarizing angle, 15% of perpendicular vibrations are sieved out from the refracted light wave. Hence, the perpendicular vibrations contained in the refracted wave are decreased at the rate of 15% per plate. The process continues and when light wave has travelled through a pile of glass plates containing around 35 plates, the transmitted light wave is nearly free of perpendicular vibrations and contains only parallel vibrations. Thus, we get a plane polarized light with the help of a pile of glass plates, vibrations being in the plane of incidence.

The degree of polarization of the refracted wave is small for a single surface. If the number of plates in the pile is more, we can quench out the perpendicular vibrations from the refracted wave, making the refracted wave more completely plane polarized containing light vectors parallel to the plane of incidence. The degree of polarization P is defined as

$$P = \frac{I_{//} - I_\perp}{I_{//} + I_\perp} = \frac{N}{N + \dfrac{2\mu^2}{1-\mu^2}} \tag{4.3}$$

where

$I_{\parallel}$ = sum of intensities of vibrations parallel to the plane of incidence.

$I_{\perp}$ = sum of intensities of vibrations perpendicular to the plane of incidence.

N = number of glass plates in the pile of glass plates.

μ = refractive index of the glass plate.

4.5.1 Malus's law

If a light wave plane polarized by reflection at one plane surface (polarizer) is incident at the polarizing angle on the second plane surface (analyzer), the intensity of the twice-reflected light wave (CD in Fig. 4.6) varies with the angle between the planes of the two surfaces. The same is also true for twice the transmitted wave from polarizer to analyzer. The Malus law states that, 'The intensity of the polarized light transmitted through the analyzer varies as the square of the cosine of the angle between the plane of refraction of the analyzer and the plane of the polarizer'.

Any polarized vibration can be resolved into two rectangular components – one component parallel to the plane of refraction and the other component perpendicular to the plane of refraction as shown in Fig. 4.8.

Let

r = OP be the amplitude of the vibrations transmitted by the polarizer.

θ = angle between the planes of the polarizer and the analyzer.

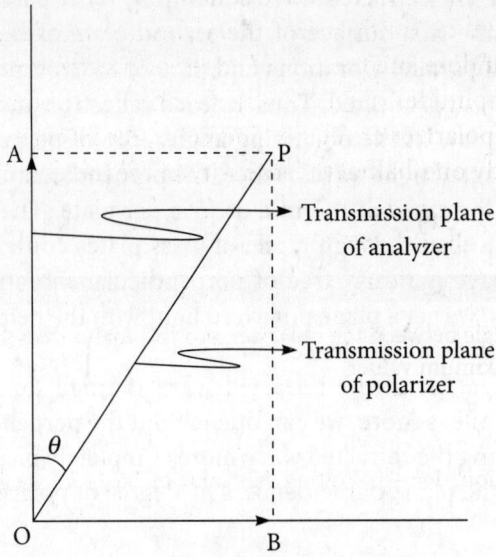

Figure 4.8 | Resolution of amplitude of light vector of plane polarized light when the angle between the polarizer plane and the analyzer plane is θ

The component of OP along OA = OP cos θ = r cos θ

The component of OP along OB = OP sin θ = r sin θ

The component of OP along OA = r cos θ is transmitted/refracted through the analyzer. The intensity of the light transmitted through the analyzer is directly proportional to the square of the amplitude component of OP along OA = r cos θ. Mathematically, we have

$$I \propto r^2 \cos^2 \theta$$

or $$I = cr^2 \cos^2 \theta \qquad (4.4)$$

where c is the proportionality constant.

$$I = cr^2 = I_0 = \text{maximum intensity for } \theta = 0$$

Putting this value of $cr^2 = I_0$ into Eq. (4.4), we get

$$I = I_0 \cos^2 \theta \qquad (4.5)$$

Equation (4.5) is the mathematical form of the Malus law.

In Eq. (4.5), I_0 is the intensity of the plane polarized light incident on the analyzer. The intensity transmitted by the polarizer is found to be exactly one-half of the intensity of the unpolarized light incident on the polarizer. This is explained as follows. The light incident on the polarizer is a random mixture of all states of polarization. The average values of the resultants of the amplitudes parallel to the principal section and perpendicular to the principal section of the polarizer are equal. Therefore, the intensity of the light transmitted by the polarizer is exactly one-half of the intensity of the incident light.

Example 4.3

What should be the angle between the polarizer and the analyzer so that intensity is reduced to one-fourth of its maximum value?

Solution

According to the question, $I = \dfrac{I_0}{4}$. Putting this value in Eq. (4.1), we get

$$\frac{I_0}{4} = I_0 \cos^2 \theta$$

or $\cos^2 \theta = \dfrac{1}{4}$

or $\theta = 60°$

Example 4.4

The plane of vibration of the incident ray makes an angle of 60° with the optic axis. Compare the intensities of the ordinary ray and the extraordinary ray (explained later in this chapter).

Solution

Intensity of ordinary ray is given by

$I_0 = r^2 \sin^2 \theta$

Intensity of extraordinary ray is given by

$I_E = r^2 \cos^2 \theta$

Hence, $\dfrac{I_0}{I_E} = \tan^2 \theta$

According to the question, $\theta = 60°$

$\therefore$ $\dfrac{I_0}{I_E} = \tan^2 60° = 3$

Thus, the intensity of an ordinary ray is three times the intensity of an extraordinary ray.

4.6 Polarization by Scattering

The transverse electric vector in the light wave is responsible for all optical phenomena. Let unpolarized sunlight be allowed to fall on an air molecule present in the atmosphere. An air molecule is neutral, containing equal amounts of opposite charges. The vibrating light vector of the light wave will exert force in opposite directions on the positive and the negative charges of the molecule. The direction of the forces on the positive and negative charges changes rapidly since light vector in the light wave is vibrating. Since the charges in the molecule are not rigidly bound to it, oscillations of charges occur synchronously with the vibrating light vectors. The frequency of the oscillation of charges and the vibration frequency of light vectors of the light wave is the same. Thus, the vibrating light vectors of the light wave produce accelerations of the charges. According to electromagnetic theory, accelerated charges emit radiations which form the scattered light.

Suppose unpolarized sunlight consisting of two polarized components is incident on an air molecule along the X-axis as shown in Fig. 4.9. If the light vector of the incident light wave is vibrating in the XY plane, it will make the charge of the air molecule to vibrate along the Y-axis. The vibrating charge of air molecule will emit radiations in the form of scattered light in all directions except along the Y-axis. To emit the radiation along the Y-axis, light wave would have to be a longitudinal wave.

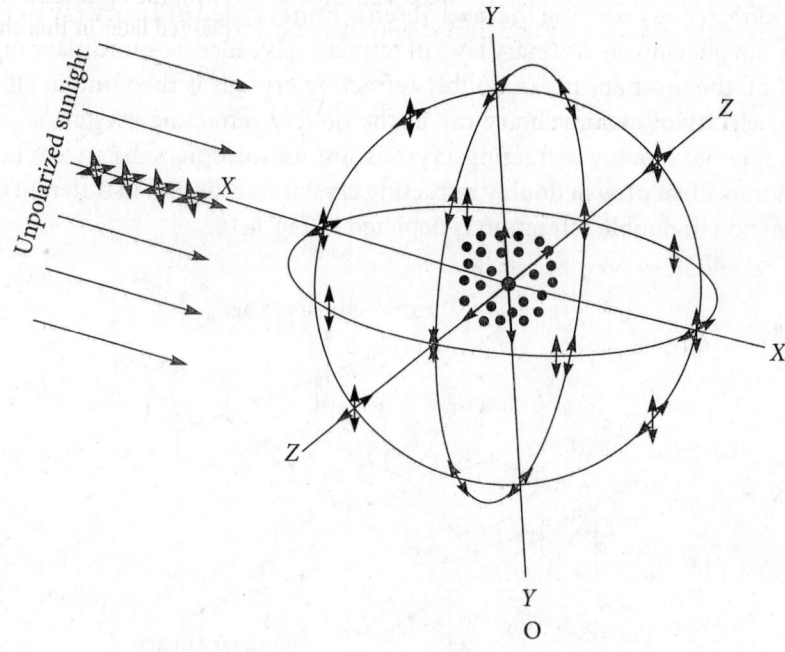

Figure 4.9 | The polarization of sunlight by scattering from the fine particles present in the atmosphere. O is the observer looking upward

Again assume that the incident light vector is vibrating along the Z-axis. The vibrating charge of air molecule will emit radiations in the form of scattered light in all directions except along the Z-axis. To emit the radiation along the Z-axis light, the wave would have to be a longitudinal wave. Therefore, an observer at O looking vertically upward will find the blue light plane polarized with light vectors vibrating parallel to the Z-axis. No air molecule at P can be set to vibrate along the X-axis, since this would contradict the transverse nature of the light wave.

4.7 Double Refraction

If we place a calcite, a crystallized form of calcium carbonate $CaCO_3$, on a point marked on paper and look through the calcite, we will see two images of the point simultaneously.

Again, if we rotate the calcite crystal on the paper over the point, out of the two images of the point, one image rotates about the other. The appearance of two images of a single point is due to the double refraction or bi-refringence phenomenon. The light reflected from the point on the paper while passing through the calcite is divided into two beams and these two beams enter into our eyes as a result of which we are able to see two images of the point.

The phenomenon of splitting of a ray of ordinary light incident on calcite into two beams after refraction, out of which one obeys the laws of refraction whereas the other does not, is called double refraction. The two beams emerge from the calcite parallel to each other, the distance between the two emergent beams being proportional to the thickness of the crystal. The refracted ray obeying the laws of refraction is called ordinary ray or O-ray. The refracted ray not obeying the ordinary laws of refraction is called extraordinary ray or E-ray. The velocity of the ordinary ray in doubly refracting crystals is the same in all directions whereas the velocity of extraordinary ray in the doubly refracting crystals is different in different directions. Doubly refracting crystals are anisotropic substances, because the velocity of extraordinary ray in doubly refracting crystals is different in different directions. The phenomenon of double refraction is depicted in Fig. 4.10.

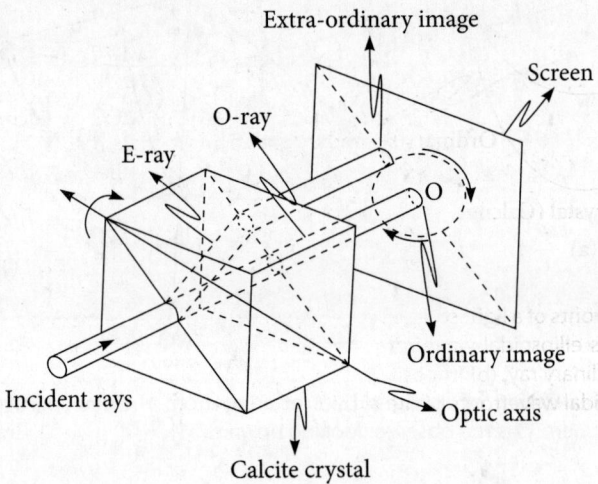

Figure 4.10 | Splitting of ordinary light into two beams of plane polarized light by a doubly refracting crystal

Suppose a point source of light is embedded inside a doubly refracting crystal. We will have two sets of wavefronts – one for extraordinary ray and the other for ordinary ray. Since the speed of ordinary ray is the same in all directions, the wavefront of ordinary ray will be spherical. Again, since the speed of extraordinary ray is different in different directions, the wavefront of the extraordinary ray will be ellipsoidal or an ellipsoid of revolution, the axis of revolution being the optic axis. In negative crystals like calcite, the ellipsoidal wavefront of the extraordinary ray encompasses the spherical wavefront of the ordinary ray, as the speed of the extraordinary ray is more than that of the ordinary ray in negative crystals. In positive crystals like quartz, the ellipsoidal wavefront of the extraordinary ray is encompassed by the spherical wavefront of the ordinary ray, as speed of the ordinary ray is more than that of the extraordinary ray in positive crystals. These observations are depicted in Fig. 4.11. After the discovery of the phenomenon by Dutch philosopher Erasmos Bartholinus in the year 1669, Huygens, in the year 1690, proved that the two beams are plane polarized with the plane of polarization being perpendicular to each other.

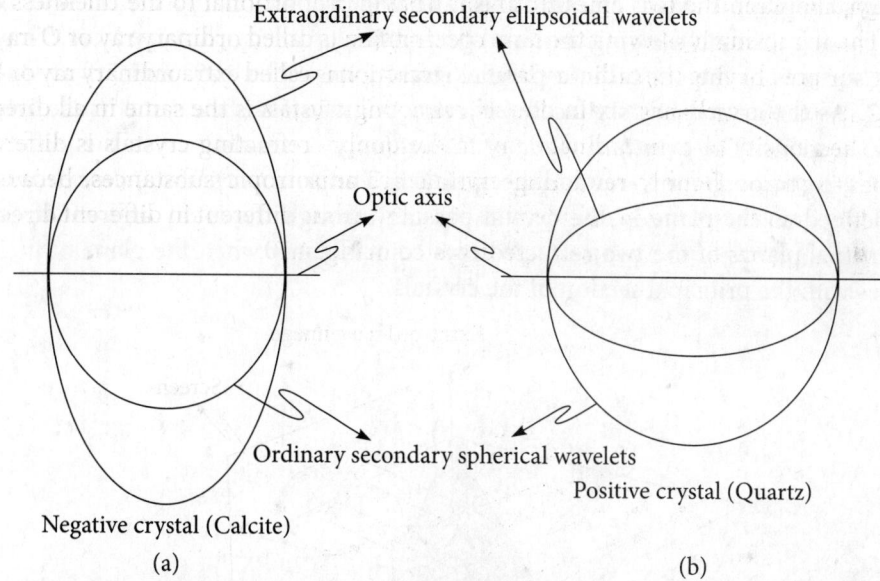

Extraordinary secondary ellipsoidal wavelets

Optic axis

Ordinary secondary spherical wavelets

Positive crystal (Quartz)

Negative crystal (Calcite)

(a)

(b)

Figure 4.11 | Wavefronts of a light source embedded inside a doubly refracting crystal. (a) In negative crystals ellipsoidal wavefronts of the extraordinary ray cover spherical wavefronts of the ordinary ray. (b) In positive crystals, spherical wavefronts of the ordinary ray cover ellipsoidal wavefronts of the extraordinary ray

4.7.1 Few terms connected with the double refraction phenomenon

Crystals which exhibit double refraction phenomenon are called *doubly refracting crystals*. An *anisotropic substance* is defined as a substance which has different properties along different directions. A refracted ray which obeys the laws of refraction is called an *ordinary ray* or O-ray. A refracted ray which does not obey the laws of refraction is called an *extraordinary ray* or E-ray. A *negative crystal* is defined as a doubly refracting crystal in which the speed of the extraordinary ray along the optic axis is greater than or equal to the speed of the ordinary ray. A *positive crystal* is defined as a doubly refracting crystal in which the speed of the ordinary ray along the optic axis is greater than or equal to the speed of the extraordinary ray. The two diagonally opposite solid angles of calcite are formed by the junction of three obtuse angles of three faces. A line into the crystal at one of these corners and equally inclined to these three faces is called the *crystallographic axis*. The *optic axis* of a doubly refracting crystal is defined as the direction along which a light ray falling on the crystal face is not separated out into O-rays and E-rays as they have the same speed. A *uniaxial crystal* is defined as a doubly refracting crystal in which the speed of the extraordinary ray and the ordinary ray is the same along only one direction. A *biaxial crystal* is defined as a doubly refracting crystal in which the speed of the extraordinary ray and the ordinary ray is the same along two directions. Any direction in the crystal parallel to the crystallographic axis is an optic axis. The *principal section of a crystal* is defined as a plane passing through the optic axis and perpendicular to any cleavage faces. This plane cuts the surfaces of the crystal in a parallelogram with angles 109° and 71° as shown in Fig. 4.12. As the crystal has six faces, for every point inside the crystal, there are three principal sections. The *principal plane of the ordinary ray* is defined as the plane in the crystal passing through the O-ray and the optic axis. The *principal plane of the extraordinary ray* is defined as the plane in the crystal passing through the E-ray and the optic axis. The principal planes of the two refracted rays coincide only when the plane of incidence coincides with the principal section of the crystals.

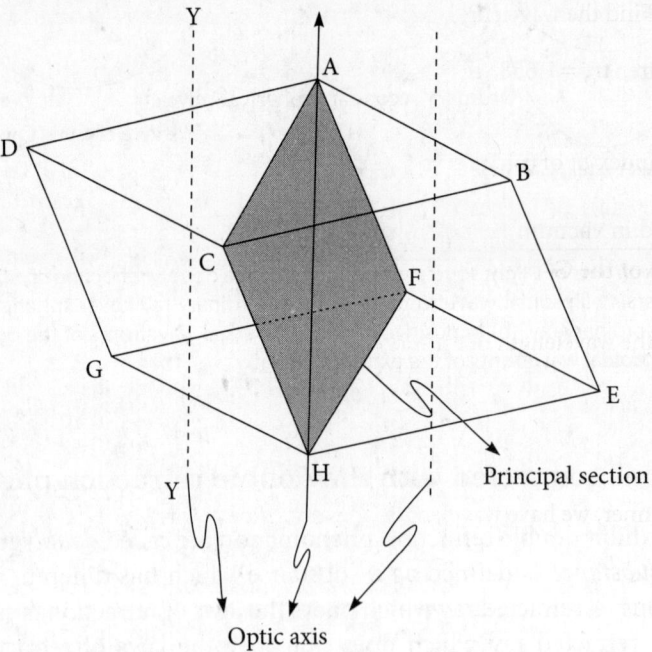

Figure 4.12 | Rhombohendral structure of calcite. ACHF is the principal plane, AH is the crystallo graphic axis and the direction AH and YY are the optic axes

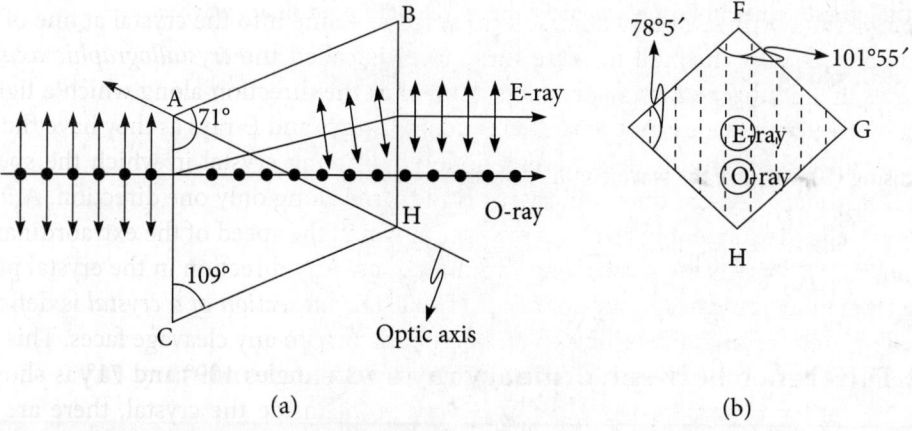

Figure 4.13 | Principal sections in a calcite crystal

Example 4.5

A parallel beam of linearly polarized light of wavelength 5890 Å (in vacuum) is incident on a calcite crystal. Find the wavelengths of ordinary and extraordinary waves in the crystal.

Given: for calcite, $\mu_O = 1.658$, $\mu_E = 1.486$.

Solution

The refractive index μ_0 of calcite crystal for O-ray is defined as

$$\mu_O = \frac{\text{Speed in vacuum}}{\text{Speed of the O-ray}} = \frac{c}{v_0} = \frac{\lambda}{\lambda_0}$$

Thus, we have the wavelength of ordinary wave

$$\lambda_0 = \frac{\lambda}{\mu_0} \tag{A}$$

In a similar manner, we have wavelength of extraordinary wave

$$\lambda_E = \frac{\lambda}{\mu_E} \tag{B}$$

In Eqs (A) and (B), λ is the wavelength of the light in vacuum.
According to the question, for calcite,

$$\mu_O = 1.658, \ \mu_E = 1.486 \ \text{and} \ \lambda = 5890 \,\text{Å}.$$

Putting these values in Eq. (A), we have the wavelength of ordinary wave as

$$\lambda_O = \frac{5890\,\text{Å}}{1.658} = 3552\,\text{Å}$$

By using (B), we have the wavelength of ordinary wave as

$$\lambda_E = \frac{5890\,\text{Å}}{1.486} = 3964\,\text{Å}$$

4.7.2 Difference between ordinary ray and extraordinary ray

	Ordinary ray	Extraordinary ray
1	A refracted ray obeying the ordinary laws of refraction is called ordinary ray or O-ray.	A refracted ray not obeying the ordinary laws of refraction is called extraordinary ray or E-ray.
2	The image produced by an O-ray does not rotate upon rotation of the calcite crystal.	The image produced by an E-ray rotates upon rotation of the calcite crystal.
3	The speed of an ordinary ray is greater than an E-ray in positive crystals like quartz.	The speed of an extraordinary ray is greater than an O-ray in negative crystals like calcite.
4	O-ray is always in the plane of incidence.	E-ray is not necessarily always in the plane of incidence.
5	The speed of O-ray is same in all directions.	The speed of E-ray is not the same in all directions.
6	O-ray is polarized in the principal section, i.e., the plane of polarization lies in the principal section.	E-ray is polarized perpendicular to the principal section, i.e., the plane of polarization is perpendicular to the principal section.
7	The plane of vibration of an O-wave is perpendicular to the principal section.	The plane of vibration of an E-wave lies in the principal section.
8	The ratio $(\sin i / \sin r)$ is the same for any angle of incidence.	The ratio $(\sin i / \sin r)$ varies with angles of incidence.
9	The refractive index μ_0 of the crystal for an O-ray is defined as $$\mu_O = \frac{\text{Speed in vacuum}}{\text{Speed of the O-ray}}$$	The refractive index μ_E of the negative crystal for an E-ray is defined as $$\mu_E = \frac{\text{Speed in vacuum}}{\text{Maximum speed of the E-ray}}$$ and the refractive index μ_E of the positive crystal for an E-ray is defined as $$\mu_E = \frac{\text{Speed in vacuum}}{\text{Minimum speed of the E-ray}}.$$

4.7.3 Polarization by double refraction

When the O-ray and E-ray emerging from the calcite crystal are analyzed by an analyzer, it is found that as the analyzer is rotated in its own plane, the intensity of either one varies in the opposite manner. That is, the intensity of one of the ray increases only when the intensity of the other ray decreases on the rotation of the analyzer. When the longer axis of the analyzer is exactly parallel to the longer diagonal of the end face of the calcite, the brightness of the O-image maximum and the E-image disappears completely. On rotating the analyzer by 90° further, the brightness of the E-image becomes maximum and the O-image disappears completely. In this case, the longer axis of the analyzer is exactly perpendicular to the longer diagonal of the end face of the calcite. This is possible only when the O-wave and the E-wave is polarized with the plane of polarization of the O-wave in the principal section and the plane of polarization of the E-wave perpendicular to it.

4.7.4 Huygens' experiment on polarization by double refraction

Two calcite crystals C_1 (ACHF) and C_2 (A'C'H'F') of equal thickness are placed in such a manner that their principal sections are parallel and symmetrical to each other. Ordinary unpolarized light, after passing through C_1, normally is split into O-ray and E-ray and both rays enter into the calcite crystal C_2 as shown in Fig. 4.14. The O-ray emerges from C_2 without any deviation from the direction of incidence of unpolarized light while the E-ray gets deviated and emerges from C_2 parallel to O-ray. We assign the symbols O_0 and E_E to the images formed by the O-ray and the E-ray emerging from C_2 respectively. Now when C_2 is rotated slightly taking

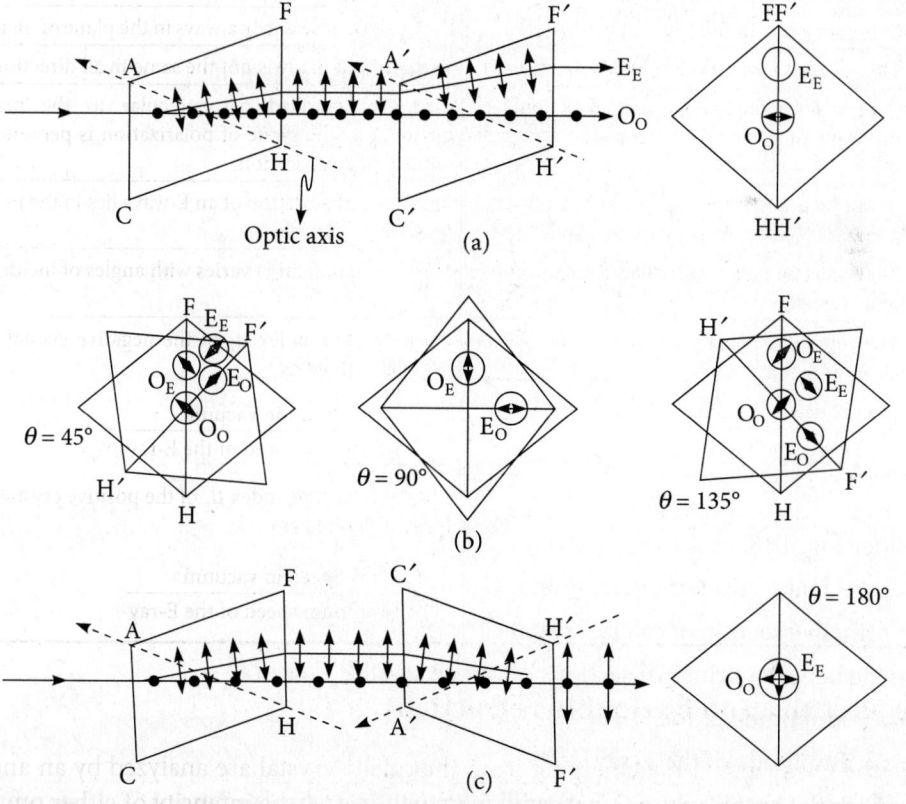

Figure 4.14 Huygens' experiment with two calcite crystals

O-ray as an axis, O-ray and E-ray suffer double refraction in C_2 giving rise to four images with two faint old images O_0 and E_E; two faint new images O_E and E_0. On further rotation of C_2, it is found that images O_0 and O_E remain fixed while image E_0 rotates around O_0 and E_E rotates around O_E. Again, the intensity of O_0 images decreases while the intensity of O_E and E_0 images increases. The following observations were observed.

i.   At 45° rotation of C_2, four images have the same intensity.

ii. At 90° rotation of C_2, two images O_0 and E_0 disappear and images O_E and E_0 have maximum intensity.

iii. On further rotation of C_2, two images O_0 and E_E reappear and the intensity of images O_E and E_0 decreases.

iv. At 135° rotation of C_2, four images have the same intensity again.

v.   At 180° rotation of C_2, the principal sections of the two calcite crystals are parallel while the optic axes have opposite directions (see Fig. 4.14(a)). Therefore, images O_E and E_0 disappear and images O_0 and E_E coalesce.

The same changes take place in the reverse order when rotation continues from 180° to 360°.

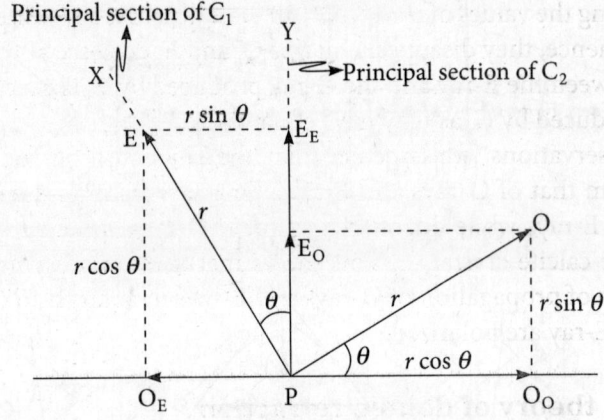

Figure 4.15 | Explanations for Huygens' experiment with two calcite crystals

Consider Fig. 4.15.

PX = principal section of calcite crystal C_1

PY = principal section of calcite crystal C_2

θ = angle between principal sections of calcite crystals C_1 and C_2

PO = r = amplitude of the O-ray

PE = r = amplitude of the E-ray

$PO_0 = r\cos\theta$ = Amplitude of O_0 perpendicular to the principal section of C_2. This component gives rise to the ordinary image O_0.

Intensity of the ordinary image $O_0 = kr^2 \cos^2 \theta = I_0 \cos^2 \theta$ (A)

$PE_0 = r \sin \theta =$ Amplitude of O_0 along the principal section of C_2. This component gives rise to the extraordinary image E_0.

Intensity of the extraordinary image $E_0 = kr^2 \sin^2 \theta = I_0 \sin^2 \theta$ (B)

$PE_E = r \cos \theta =$ Amplitude of E_E along the principal section of C_2. This component gives rise to the extraordinary image E_E.

Intensity of the extraordinary image $E_E = kr^2 \cos^2 \theta = I_0 \cos^2 \theta$ (C)

$PO_E = r \sin \theta =$ Amplitude of E_E perpendicular to the principal section of C_2. This component gives rise to the ordinary image O_E.

Intensity of the ordinary image $O_E = kr^2 \sin^2 \theta = I_0 \sin^2 \theta$ (D)

Our observations discussed from (i) to (v) can be explained with the help of equations from (A) to (D) by putting the values of $\theta = 0°$, $45°$, $90°$, and $135°$. At $180°$, intensities of O_E and E_0 become zero and hence, they disappear. Images O_0 and E_E coalesce at this position because the separation between the E-ray and the O-ray produced by C_1 is exactly compensated by the separation produced by C_2 as they have equal thickness.

From these observations, we conclude that the characteristics of ordinary light are quite different from that of O-rays and E-rays. Ordinary light on double refraction gives rise to O-rays and E-rays while double refraction of O-rays and E-rays depends upon the orientations of the calcite crystal C_2. This shows that there is no symmetry in vibrations about the direction of propagation of O-rays and E-rays in contrast to ordinary light. That means O-ray and E-ray are polarized.

4.7.5 Huygens' theory of double refraction

Huygens gave a satisfactory explanation for the origin of O-waves and E-waves with his secondary wavelets concept of wave theory of light. Accordingly, he put forward the following postulates

i.   Every internal point and surface point of a crystal disturbed by the incident wavefront must become the source of two type of wavelets – ordinary wavelets and extraordinary wavelets. This explains the origin of O-waves and E-waves in the doubly refracting crystals.

ii.   The O-ray wavefront is spherical in the doubly refracting crystal. This postulate explains the constancy of the speed of an O-wave in crystals in all directions and the validity of Snell's law for O-waves. The ratio $\sin i / \sin r_0$ for the O-wave does not vary with direction of its propagation relative to the optic axis. Hence, the refractive index of the doubly refracting crystal is independent of the direction of propagation of the O-wave relative to the optic axis.

iii. The E-ray wavefront is ellipsoid of revolution about its principal axes in the doubly refracting crystal. This postulate explains the directional dependence of the speed of E-waves in the crystals. The speed of propagation of E-waves is different in different directions of incidence in the crystal. Hence, E-waves do not obey Snell's law and the value of the refractive index of the doubly refracting crystal depends upon the angle of incidence and the angle of refraction. The ratio $\sin i / \sin r_E$ for the E-wave varies with the direction of its propagation relative to the optic axis. Hence, the refractive index of the doubly refracting crystal is not independent of the direction of propagation of the E-wave relative to the optic axis.

iv. The properties of the doubly refracting crystals are symmetric about the optic axis. The speed of E-waves must be equal along the directions equally inclined to the optic axis and as a result, the ellipsoid of revolution must be symmetrical about the optic axis. To account for all these observations, Huygens postulated that the axis of revolution of the ellipsoid should be along the optic axis through the point of origin of the wavelets.

v. The last postulate put forward by Huygens is that the spherical wavefront and the ellipsoidal wavefront touch each other along the extremities of the axis of revolution (optic axis) of the ellipse generating the ellipsoidal wavefront. That is, the spherical O-wavefront and the ellipsoidal E-wavefront touches each other at two points on the optic axis. This explains the equality of speed of O-waves and E-waves along the optic axis and hence, the absence of the double refraction phenomenon along the optic axis. Thus, according to Huygens, every point of the doubly refracting crystal surface disturbed by the incident wave is to be considered as the source of two secondary wavelets – spherical for ordinary waves and ellipsoid of revolution about the optic axis for extraordinary waves, the two wave surfaces touching each other along the extremities of the optic axis.

4.7.6 Phenomenon of double refraction at normal incidence

Suppose a uniaxial crystal slab is cut in such a way that its unique optic axis is neither parallel nor perpendicular to the crystal surface and the optic axis is shown by the broken lines in Fig. 4.16(a). Let a beam of parallel light wave be incident normally on the plane surface of the uniaxial crystal slab as shown in the Fig. 4.16. Here the plane of incidence is the plane of the page. The optic axis has the direction shown by the broken lines in Fig. 4.16. WW' is the advancing wavefront and AC is the incident wavefront. Three points A, B, and C on the refracting surface PQ are taken into account as the sources of Huygens' secondary wavelets. After a short time interval, Huygens' wavelets entering the crystal from these points will have the form as shown in Fig. 4.16. In order to construct the ordinary ray wavefront, we draw spheres of certain radii at points A, B, and C. The common tangent plane OO' to these spheres at points A', B', and C' is the ordinary ray wavefront. When the points A', B', and C' are joined to the points A, B, and C, the refracted ordinary ray is obtained.

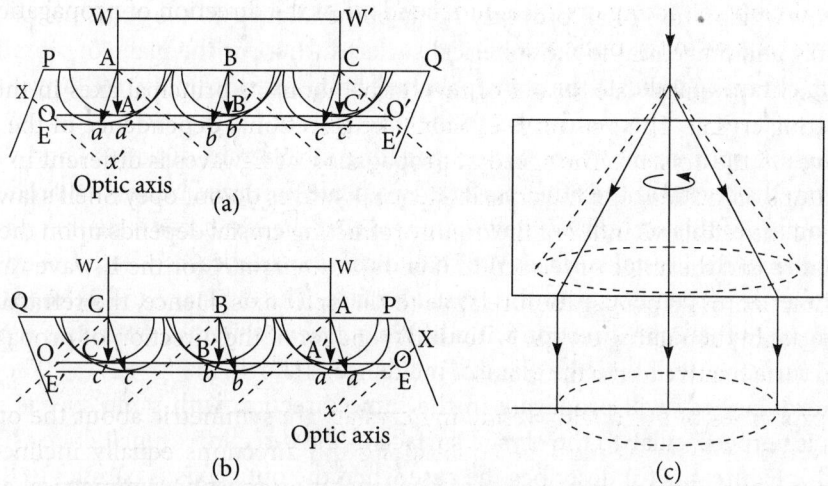

Figure 4.16 Huygens' construction of an ordinary ray wavefront and an extraordinary ray wavefront of a plane wave incident normally on a negative uniaxial crystal. In figures (a) and (b), the optic axis is the direction shown by the broken line parallel to the plane of the page. (c) For a ray incident normally, if the crystal is rotated about a line normal to the surface of the crystal, the extraordinary ray will rotate making a cone inside the crystal and if projected on a screen it will rotate making a circular path on the screen. Here, the vertical section of a negative uniaxial crystal is shown

In order to construct the extraordinary ray wavefront, we draw ellipses centered at A, B, and C with their minor axes along the optic axis. The ellipsoids of revolution are obtained by rotating the ellipses about the minor axis along optic axis. The common tangent plane EE' to these ellipsoids of revolution at points a, b, and c is the extraordinary ray wavefront. When the points a, b, and c are joined to the points A, B, and C, the refracted extraordinary ray is obtained.

Let

t = time taken by the extraordinary wave front EE' to reach the position as shown in the Fig. 4.16(a). The ray velocity of the extraordinary ray is defined as

$$\text{ray velocity} = \frac{Aa}{t} = \frac{Bb}{t} = \frac{Cc}{t}$$

and the normal velocity of the extraordinary ray is defined as

$$\text{normal velocity} = \frac{Aa'}{t} = \frac{Bb'}{t} = \frac{Cc'}{t}$$

The symbols are depicted in Fig. 4.16(a) and (b).

If the uniaxial crystal slab is cut in such a way that its unique optic axis makes a certain angle with the crystal surface as shown by the broken lines in Fig. 4.16(b), then although the direction of the ordinary ray will not change, the direction of the extraordinary ray will change and will propagate in the direction shown in Fig. 4.16(b).

The plane of vibration of an ordinary ray is perpendicular to the principal plane of the ordinary ray and tangential to the spherical surfaces where as the plane of vibration of an extraordinary ray is parallel to the principal plane of the extraordinary ray, tangential to the ellipsoidal surfaces and perpendicular to the direction of propagation. Thus, an ordinary incident light is split up into two rays propagating in different directions and when they emerge from the crystal at the other surface, two plane polarized beams are produced.

At the outset of this section, we have assumed that the optic axis is neither parallel nor perpendicular to the crystal surface. We shall discuss special cases where the optic axis is either parallel or perpendicular to the crystal surface. The ordinary and extraordinary ray will propagate in the same direction with different speeds when the optic axis is parallel to the crystal surface as well as to the plane of incidence as shown in Fig. 4.17(a). The ordinary and extraordinary ray will propagate in the same direction with same speeds when the optic axis is perpendicular to the crystal surface in the direction of incidence as shown in Fig. 4.17(b). Figure 4.17(c) describes the case when the optic axis is parallel to the crystal surface and perpendicular to the plane of the page or plane of the incidence. In this case, the vertical section of the extraordinary ray wavefront will be circular as that of ordinary rays and both the rays propagate along the same direction with different speeds as in the Fig. 4.17(a).

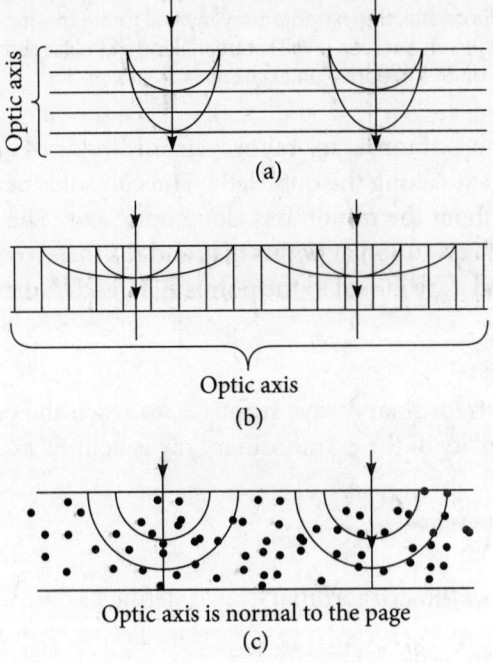

Figure 4.17 | Illustration of double refraction phenomenon due to a plane polarized wave incident normally on a negative uniaxial crystal. In Fig. 4.17(a), the optic axis is parallel to the refracting surface as well as to the plane of incidence. In Fig. 4.17(b), the optic axis is perpendicular to the refracting surface and parallel to the plane of incidence. In Fig. 4.17(c), the optic axis is perpendicular to the plane of incidence and parallel to the refracting surface and is shown as dots. The vertical section of a negative uniaxial crystal is shown

4.7.7 Phenomenon of double refraction at oblique incidence

Suppose a uniaxial crystal slab is cut in such a way that its unique optic axis is neither parallel nor perpendicular to the crystal surface and the optic axis makes an arbitrary angle with the crystal surface as shown by the broken lines in Fig. 4.18. Let a beam of parallel light wave be incident obliquely on the plane surface of the uniaxial crystal slab as shown in the figure. Here the plane of incidence is the plane of the page. Let BD represent the incident wavefront. At the instant when the incident wavefront touches the crystal surface at B, this point on the crystal becomes the source of Huygens' secondary wavelets for ordinary and extraordinary rays. As the incident wavefront advances, every point along the crystal surface PQ between the points B and F successively becomes the source of Huygens' secondary wavelets for ordinary and extraordinary rays. All these points would have advanced different distances till point F on the crystal is just being disturbed by the advancing wavefront.

Let

c = speed of light in air.

v_0 = speed of O-ray in the crystal.

v_E = speed of E-ray in the crystal perpendicular to the optic axis.

t = time taken by the wave to reach F from D.

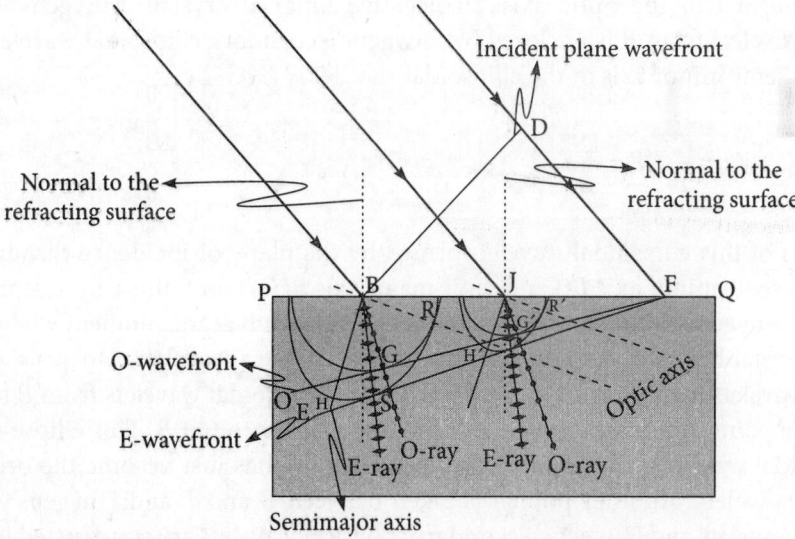

Figure 4.18 | Refraction of a plane wave for oblique incidence on a negative uniaxial crystal. The broken lines represent the optic axes. In this figure, the vertical section of a negative uniaxial crystal is shown

Hence, we have

$$t = \frac{DF}{c}$$

During this time interval, Huygens' secondary spherical wavelets from B would acquire a radius BG within the crystal given by

$$BG = v_0 t = \frac{v_0}{c} DF = \frac{DF}{\mu_0}$$

where μ_0 is the refractive index of the crystal for ordinary rays. The section of this spherical wavelet by the plane of incidence through G is a circle of radius BG. Hence, the circle of radius BG represents the position of Huygens' secondary spherical wavelets from B at the moment when the point F on the crystal becomes a source of Huygens' secondary wavelets.

The semi-major axis a of Huygens' secondary ellipsoidal wavelets from B at the moment when F just becomes the origin of Huygens' secondary wavelets will be given by

$$a = BH = v_E t = \frac{v_E}{c} DF = \frac{DF}{\mu_E}$$

where μ_E is the refractive index of the crystal for extraordinary rays propagating in a direction normal to the optic axis. In negative uniaxial crystals, Huygens' secondary spherical wavelets from B is enclosed by Huygens' secondary ellipsoidal wavelets from B. Hence, the semi-minor axis of the ellipsoidal wavelet is given by

$$b = BG = v_0 t = \frac{v_0}{c} DF = \frac{DF}{\mu_0}$$

The section of this ellipsoidal wavelet formed by the plane of incidence through G is an ellipse with semi-minor axis BG and semi-major axis BH. Hence, this ellipse represents the position of Huygens' secondary ellipsoidal wavelets from B at the moment when the point F on the crystal becomes a source of Huygens' secondary wavelets. Huygens' secondary spherical wavelets from B and Huygens' secondary ellipsoidal wavelets from B touch each other at the point of intersection R with the optic axis through B. This ellipse represents the ellipsoidal wavelets from B at the moment when F has just become the origin of the secondary wavelets. Another point J is taken between B and F and Huygens' secondary ellipsoidal wavelets and Huygens' secondary spherical wavelets are constructed in a similar manner.

The position of secondary wavelets originating from all the points between B and F are obtained in the same manner as described earlier. The common tangential plane FO perpendicular to the plane of the page to these spheres at points G, G', and FO is the ordinary ray wavefront when F has just become the origin of the secondary wavelets. When the points G and G' are joined to the points B and J respectively, the refracted ordinary rays are obtained from the points B and J on the refracting surface.

In order to construct the extraordinary ray wavefront, we draw ellipses centered at B and J with their minor axes along the optic axis. The ellipsoids of revolution are obtained by rotating the ellipses about the optic axis. The common tangential plane FE perpendicular to the plane of the page to these ellipsoids of revolution at points S and T is the extraordinary ray wavefront when F has just become the origin of the secondary wavelets. When the points S and T are joined to the points B and J respectively, the refracted extraordinary rays are obtained from the points B and J on the refracting surface.

The plane of vibration of the ordinary ray is perpendicular to the principal plane of the ordinary ray and is tangential to the spherical surfaces, whereas the plane of vibration of the extraordinary ray is along the principal plane of the extraordinary ray and is tangential to the ellipsoidal surfaces. Thus, an ordinary incident light splits into two rays propagating in different directions and when they emerge from the crystal at the other surface, two plane polarized beams are produced.

4.7.8 Special cases

At the beginning of this section, we have assumed that the optic axis is neither parallel nor perpendicular to the crystal surface. We shall now discuss special cases where the optic axis is either parallel or perpendicular to the crystal surface in different orientation.

Case 1:

Suppose a uniaxial crystal slab is cut in such a way that its unique optic axis is parallel to the refracting surface of the crystal and it lies along the plane of incidence. These conditions are depicted in Fig. 4.19.

Let a beam of parallel light wave be incident obliquely on the plane surface of a uniaxial crystal slab as shown in Fig. 4.19. Here the plane of incidence is the plane of the page. Let BD represent the incident wavefront. At the instant when the incident wavefront touches the crystal surface at B, this point on the crystal becomes the source of Huygens' secondary wavelets for ordinary and extraordinary rays. As the incident wavefront advances, every point along the crystal surface PQ between B and F successively becomes the source of Huygens' secondary wavelets for ordinary and extraordinary rays. All these points would have advanced different distances till the point F on the crystal is just being disturbed by the advancing wavefront.

During this time interval, Huygens' secondary spherical wavelets from B would acquire a radius BG within the crystal. The section of this spherical wavelet formed by the plane of incidence through G is a circle of radius BG. Hence, the circle of radius BG represents the position of Huygens' secondary spherical wavelets from B at the moment when the point F on the crystal becomes the source of Huygens' secondary wavelets. The position of secondary spherical wavelets originating from all the points between B and F are obtained in the same manner as described here. Another point J is taken between B and F and Huygens' secondary spherical wavelets are constructed in a similar manner.

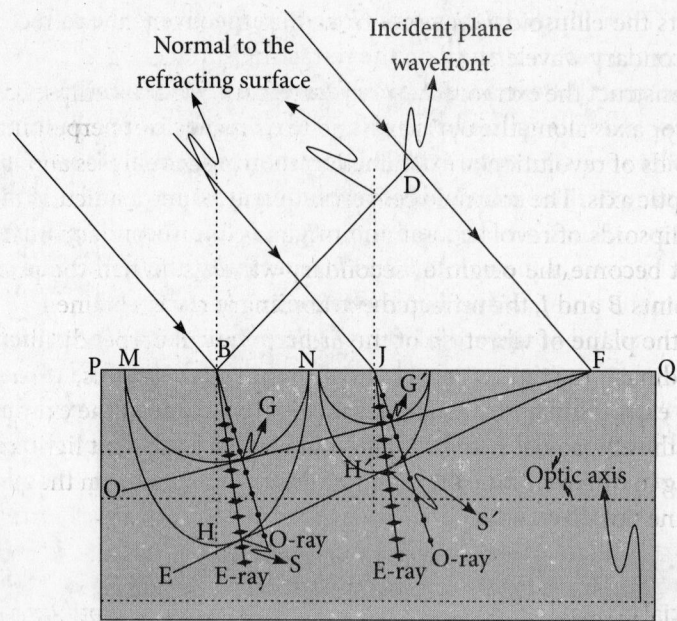

Figure 4.19 | Refraction of a plane wave for oblique incidence on a negative uniaxial crystal. The optic axis is parallel to the refracting surface of the crystal and lies along the plane of incidence. In this figure, the vertical section of a negative uniaxial crystal is shown

The common tangential plane *FO* perpendicular to the plane of the page to these spheres at points *G* and *G'* is the ordinary ray wavefront when *F* has just become the origin of secondary wavelets. When the points *G* and *G'* are joined to the points *B* and *J* respectively, the refracted ordinary ray is obtained from the points *B* and *J* on the refracting surface.

According to the given condition, the optic axis is parallel to the refracting surface of the crystal and lies along the plane of incidence. In this case, the optic axis lies on the surface of the refracting surface. The semi-major axis of Huygens' secondary ellipsoidal wavelets from *B*, at the moment when *F* just becomes the origin of Huygens' secondary wavelets is *BH* perpendicular to the refracting surface along the plane of incidence. In negative uniaxial crystals, Huygens' secondary spherical wavelets from *B* are enclosed by Huygens' secondary ellipsoidal wavelets from *B* touching each other at the points *M* and *N* on the optic axis through the point *B*. Hence, the semi-minor axis of Huygens' secondary ellipsoidal wavelets from *B*, at the moment when *F* just becomes the origin of Huygens' secondary wavelets is *BM* = *BN* (= *BG*) along the optic axis through the point *B* lying along the plane of incidence. The section of this ellipsoidal wavelet formed by the plane of incidence through *G* is an ellipse of semi-minor axis *BN* = *BM* and semi-major axis *BH*. Hence, this ellipse represents the position of Huygens' secondary ellipsoidal wavelets from *B* at the moment when the point F on the crystal becomes a source of Huygens' secondary wavelets. Huygens' secondary spherical wavelets from *B* and Huygens' secondary ellipsoidal wavelets from *B* touch each other at the points *M* and *N* on the optic axis through *B*. This ellipse represents the ellipsoidal wavelets from B at the moment when F has just become the origin of secondary wavelets.

In order to construct the extraordinary ray wavefront, we draw ellipses centered at *B* and *J* with their minor axes along the optic axis and their major axes perpendicular to the optic axis. The ellipsoids of revolution are obtained by rotating the ellipses about the semi-minor axis along the optic axis. The common tangential plane *FE* perpendicular to the plane of the page to these ellipsoids of revolution at points *S* and *S'* is the extraordinary ray wavefront when *F* has just become the origin of secondary wavelets. When the points *S* and *S'* are joined to the points *B* and *J*, the refracted extraordinary ray is obtained.

In this case, the plane of vibration of the ordinary ray is perpendicular to the principal plane of the ordinary ray and tangential to the spherical surfaces, whereas the plane of vibration of the extraordinary ray is along the principal plane of the extraordinary ray and is tangential to the ellipsoidal surfaces. Thus, an ordinary incident light split ups into two rays propagating in different directions and when they emerge from the crystal at the other surface, two plane polarized beams are produced.

Case 2:

Suppose a uniaxial crystal slab is cut in such a way that its unique optic axis is perpendicular to the refracting surface of the crystal and lies along the plane of incidence. These conditions are depicted in Fig. 4.20.

Let a beam of parallel light wave be incident obliquely on the plane surface of a uniaxial crystal slab as shown in Fig. 4.20. Here the plane of incidence is the plane of the page. Let *BD* represent the incident wavefront. At the instant when incident wavefront touches the crystal surface at *B*, this point on the crystal becomes the source of Huygens' secondary wavelets for ordinary and extraordinary rays. As the incident wavefront advances, every point along the crystal surface *PQ* between *B* and *F* successively becomes the sources of Huygens' secondary wavelets for ordinary and extraordinary rays. All these points would have advanced different distances till the point *F* on the crystal is just being disturbed by the advancing wavefront.

During this time interval, Huygens' secondary spherical wavelets from *B* would acquire a radius *BG* within the crystal. The section of this spherical wavelet formed by the plane of incidence through *G* is a circle of radius *BG*. Hence, the circle of radius *BG* represents the position of Huygens' secondary spherical wavelets from *B* at the moment when the point *F* on the crystal becomes Huygens' secondary wavelets. The position of secondary spherical wavelets originating from all the points between *B* and *F* are obtained in the same manner as described here. Another point *J* is taken between *B* and *F* and Huygens' secondary spherical wavelets are constructed in a similar manner. The common tangential plane *FO* perpendicular to the plane of the page to these spheres at points *G* and *G'* is the ordinary ray wavefront when *F* has just become the origin of secondary wavelets. When the points *G* and *G'* are joined to the points *B* and *J* respectively, the refracted ordinary ray is obtained.

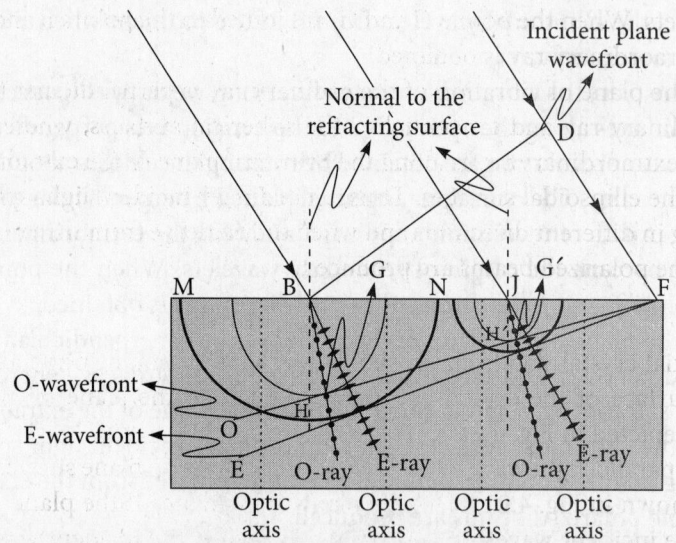

Figure 4.20 | Refraction of a plane wave for oblique incidence on a negative uniaxial crystal. The optic axis is perpendicular to the refracting surface of the crystal and lies along the plane of incidence. In this figure, the vertical section of a negative uniaxial crystal is shown

According to the given condition, the optic axis is perpendicular to the refracting surface of the crystal and lies on the plane of incidence. The semi-major axis of Huygens' secondary ellipsoidal wavelets from B, at the moment when F just becomes the origin of Huygens' secondary wavelets is $BM = BN$ perpendicular to the optic axis on the refracting surface along the plane of incidence. In negative uniaxial crystals, Huygens' secondary spherical wavelets from B are enclosed by Huygens' secondary ellipsoidal wavelets from B touching each other at the extremities of the optic axis through the point B. Hence, the semi-minor axis of Huygens' secondary ellipsoidal wavelets from B, at the moment when F just becomes the origin of Huygens' secondary wavelets is BH along the optic axis through the point B lying along the plane of incidence. The section of this ellipsoidal wavelet by the plane of incidence through G is an ellipse with semi-minor axis BH and semi-major axis $BM = BN$. Hence, this ellipse represents the position of Huygens' secondary ellipsoidal wavelets from B at the moment when the point F on the crystal becomes a source of Huygens' secondary wavelets. This ellipse represents the ellipsoidal wavelets from B at the moment when F has just become the origin of secondary wavelets. Another point J is taken between B and F and Huygens' secondary ellipsoidal wavelet is constructed in a similar manner.

In order to construct the extraordinary ray wavefront, we draw ellipses centered at B and J with their minor axes inside the crystal along the optic axis and their major axes on the refracting surface perpendicular to the optic axis. The ellipsoids of revolution are obtained by rotating the ellipses about the semi-minor axis along the optic axis. The common tangent plane FE perpendicular to the plane of the page to these ellipsoids of revolution at points G and G' is the extraordinary ray wavefront when F has just become the origin of secondary wavelets. When the points G and G' are joined to the points B and J respectively, the refracted extraordinary ray is obtained.

In this case, the plane of vibration of the ordinary ray is perpendicular to the principal plane of the ordinary ray and tangential to the spherical surfaces, whereas the plane of vibration of the extraordinary ray is along the principal plane of the extraordinary ray and is tangential to the ellipsoidal surfaces. Thus, an ordinary incident light splits up into two rays propagating in different directions and when they emerge from the crystal at the other surface, two plane polarized beams are produced.

Case 3:

Suppose a uniaxial crystal slab is cut in such a way that its unique optic axis is parallel to the refracting surface of the crystal and perpendicular to the plane of incidence. These conditions are depicted in Fig. 4.21.

Let a beam of parallel light wave be incident obliquely on the plane surface of the uniaxial crystal slab as shown in Fig. 4.21. Here the plane of incidence is the plane of the page. Let BD represent the incident wavefront. At the instant when the incident wavefront touches the crystal surface at B, this point on the crystal becomes the source of Huygens' secondary wavelets for ordinary and extraordinary rays. As the incident wavefront advances, every point along the crystal surface PQ between B and F successively becomes the source of Huygens' secondary wavelets for ordinary and extraordinary rays. All these points would have advanced different distances till the point F on the crystal is just being disturbed by the advancing wavefront.

Rotating the ellipse about the minor axis, which in this case is along the optic axis, forms the ellipsoidal wavelets in the negative uniaxial crystal slab. Therefore, the axis of revolution in this case is perpendicular to the plane of the page and the other two axes lie on the plane of the page as shown in Fig. 4.21. Hence, the section of the ellipsoidal wavelets of the extraordinary wave from the point B are semicircles having radius equal to the semi-major axis of the ellipsoidal wavelet and the section of the spherical wavelets of ordinary waves from the point B by the plane of incidence are semicircles having radius equal to the semi-minor axis of the ellipsoidal wavelet when F has just become the origin of secondary wavelets. Another point J is taken between B and F and Huygens' secondary spherical wavelets and ellipsoidal wavelets are constructed in a similar manner. The position of secondary spherical wavelets and secondary ellipsoidal wavelets originating from all the points between B and F are obtained in the same manner as described earlier. The common tangential plane FO perpendicular to the plane of the page to these spheres at points G and G' is the ordinary ray wavefront when F has just become the origin of secondary wavelets. When the points G and G' are joined to the points B and J respectively, the refracted ordinary ray is obtained. The common tangential plane FE perpendicular to the plane of the page to these ellipsoids of revolution at points S and S' is the extraordinary ray wavefront when F has just become the origin of secondary wavelets. When the points S and S' are joined to the points B and, J respectively, the refracted extraordinary ray is obtained.

Thus, we conclude that when the optic axis is parallel to the refracting surface of the crystal and perpendicular to the plane of incidence, the speed of the extraordinary wave is the same along any direction on the plane of incidence and the extraordinary wave travels perpendicularly to the extraordinary wavefront. Thus, when the optic the axis is parallel to the refracting surface of the crystal and perpendicular to the plane of incidence, the refractive index of the crystal for the extraordinary wave is constant. Under this condition, the extraordinary wave as well as the ordinary wave obeys the two laws of refraction. In this case, also as evident from Fig. 4.19, the plane of incidence, the principal plane for the

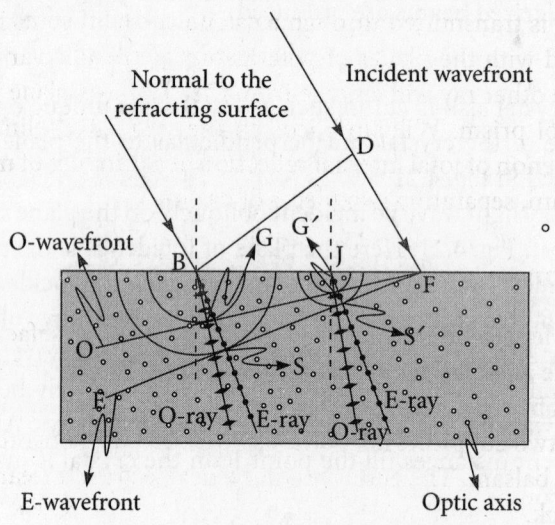

° Represents optic axis perpendicular
to the plane of the paper

Figure 4.21 | Refraction of a plane wave for oblique incidence on a negative uniaxial crystal. The optic axis is parallel to the refracting surface of the crystal and perpendicular to the plane of incidence. In this figure, O represents the optic axis perpendicular to the plane of the paper. Here the vertical section of a negative uniaxial crystal is shown

ordinary wave and the principal plane for the extraordinary wave do not coincide with each other. The principal plane for the extraordinary wave is perpendicular to the plane of the page as it is the plane containing the optic axis and the extraordinary refracted ray *BS*. The vibrations of the extraordinary wave are tangential to the ellipsoidal wave surfaces along the principal plane for the extraordinary wave. Hence, the direction of vibrations of the extraordinary wave is perpendicular to the plane of the page or plane of incidence. The principal plane for the ordinary wave is perpendicular to the plane of the page as it is the plane containing the optic axis and the ordinary refracted ray *BG*. The vibrations of the ordinary wave are tangential to the spherical wave surfaces and perpendicular to the principal plane for the ordinary wave. Hence, the direction of vibrations of the extraordinary wave is parallel to the plane of the page or plane of incidence.

4.8 Nicol Prism

The Nicol prism is an optical device designed from a calcite crystal for producing and analyzing plane polarized light.

4.8.1 Principle

When ordinary light is transmitted through a calcite crystal it splits into an O-ray and an E-ray plane polarized with the planes of polarization perpendicular to each other. If one ray is eliminated, the other ray will emerge from the crystal as plane polarized. This is the principle of the Nicol prism. William Nicol, in the year 1828, eliminated the O-ray by utilizing the phenomenon of total internal reflection at a thin film of non-doubly refracting plaster, Canada balsam, separating two pieces of calcite.

4.8.2 Construction

A calcite crystal with length three times its width is taken. The end faces of this crystal is cut in such a way that the angle in the principal section become 68° and 112°. Now the crystal is cut into two pieces by a plane perpendicular to the principal section as well as to PR and $P'R'$ (Fig. 4.22). The two cut pieces are optically polished and cemented together again by a thin layer of Canada balsam. The end faces of the device are left clear while the remaining faces are painted black.

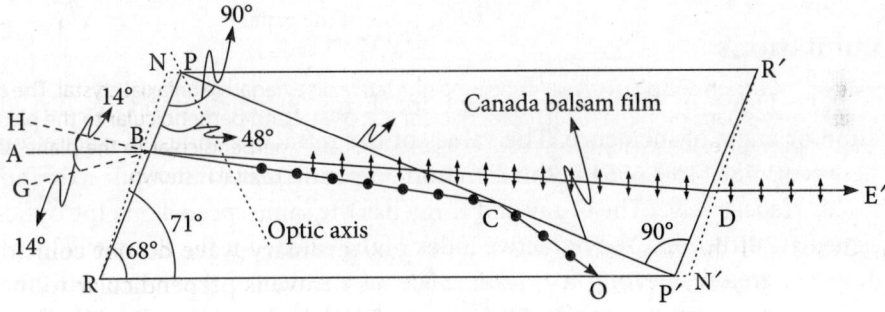

Figure 4.22 | Illustration of action of a Nicol prism

4.8.3 Action of a Nicol prism

The refractive index of Canada balsam μ_B is intermediate between the refractive indices of calcite crystal for O-ray μ_0 and E-ray μ_E. For sodium light, having mean wavelength 5893 Å, $\mu_E(1.486) < \mu_B(1.550) < \mu_0(1.658)$. That means for this light

i.   Canada balsam is optically more dense than calcite for E-ray.

ii.   Canada balsam is optically less dense than calcite for O-ray.

iii.   The calcite is less dense for E-ray than O-ray.

Canada balsam behaves as a rarer medium for O-ray and as a denser medium for E-ray with respect to calcite.

When monochromatic unpolarized sodium light ray AB enters the device through the face PR at the point B in a direction parallel to the longer side, it splits into polarized O-ray along BO and E-ray along BE, the latter making the greater angle with the optic axis (see Fig. 4.22). The plane of vibration of the O-ray is perpendicular to the principal section and the plane of vibration of E-ray is along the principal section. Passing through a denser medium (calcite), the O-ray is incident at an angle more than the critical angle on the rarer medium (Canada balsam) at C. The critical angle of incidence is

$$\theta_C = \sin^{-1} \frac{\mu_B}{\mu_0} = 69.2°$$

for the aforementioned two media, namely Canada balsam and calcite for O-ray. This is accomplished simply by the way of constructing and dimension of the Nicol prism. After reflection of the O-ray by the Canada balsam, it follows the path CO to be absorbed by the blackened sidewall. On the other hand, the E-ray passing through the rarer (calcite) medium is incident on a denser medium (Canada balsam) and get refracted to emerge from the face $P'R'$ parallel to the incident ray. Thus, finally the Nicol prism transmits only the E-ray plane polarized with vibrations confined to the principal section.

4.8.4 Limitations

The refractive index of the calcite crystal for E-ray depends upon its direction of propagation or angle of incidence. The values of the refractive index of calcite for E-ray will lie between 1.4864 to 1.6554 in the direction perpendicular to the optic axis and along the optic axis respectively. (The O-ray and E-ray has the same speed along the optic axis!). For intermediate directions, the refractive index of the calcite for the E-ray will lie between these limits. Therefore, beyond a certain angle of incidence, the E-ray will be totally reflected. The dimensions of the Nicol prism are such that this angle is at about 14°. Thus, to avoid the total reflection of E-ray and transmission of O-ray by the Canada balsam, the angle between the extreme rays of light incident on the Nicol is limited to about 28°.

4.8.5 Parallel and crossed Nicol prisms

The Nicol prism can be used as a polarizer and an analyzer. When two Nicol prisms are placed coaxially end to end, the Nicol prism on which unpolarized light is incident first is the polarizer N_1 and the second Nicol prism on which polarized E-ray is incident is called the analyzer N_2. The vibrations of the polarized E-ray are along the principal section of the polarizer N_1. If the two principal sections of the two Nicol prisms are parallel to each other, then the vibrations of the polarized E-ray emerging from N_1 are along the principal section of the analyzer N_2 and hence, the E-ray will be transmitted freely through the analyzer N_2. This combination of two Nicol prisms placed coaxially where the principal sections of the polarizer and the analyzer are parallel to each other in a single plane is called parallel Nicols or parallel Nicol prism.

If two Nicol prisms are placed in such a way that the principal sections of the two Nicol prisms are perpendicular to each other, then the vibrations of the polarized E-ray emerging from N_1 are perpendicular to the principal section of the analyzer N_2 and hence, the E-ray will be incident as an O-ray on the Canada balsam of the analyzer N_2 and will be reflected by the analyzer N_2 exactly in the same manner as the O-ray was totally reflected in the polarizer N_1. Thus, no light emerges from the analyzer N_2 in case of crossed Nicols. This combination of two Nicols placed perpendicular to each other so that the principal sections of the polarizer and the analyzer are perpendicular to each other is called crossed Nicols or crossed Nicol prism.

When the analyzer N_2 is rotated 90° from the parallel position, it becomes a crossed Nicol where the principal sections of the two Nicols (polarizer N_1 and analyzer N_2) are perpendicular to each other. With further rotation of N_2 by 90°, the principal sections of polarizer N_1 and analyzer N_2 are parallel to each other and the E-ray is transmitted freely by the analyzer N_2. Again with further rotation of N_2 through 90°, the principal sections of polarizer N_1 and analyzer N_2 are perpendicular to each other and no light is transmitted by this combination.

In the intermediate position between crossed and parallel settings of the combination, some E-ray is transmitted. Just on entering into the analyzer N_2, the amplitudes of the E-ray are resolved into two components namely $r \cos \theta$ parallel to the principal section of N_2 and $r \sin \theta$ perpendicular to the principal section of N_2. See Fig. 4.23.

The component $r \cos \theta$ parallel to the principal section of N_2 is transmitted and $r \sin \theta$ perpendicular to the principal section of N_2 is totally reflected by the analyzer N_2. Thus, the intensity of light coming out for intermediate positions of two Nicols will be given by

$$I = I_0 \cos^2 \theta \qquad (4.6)$$

where

I_0 = Intensity of the transmitted light of parallel Nicols

= Intensity of the transmitted light when $\theta = 0°$

θ = Angle between principal sections of the two Nicols

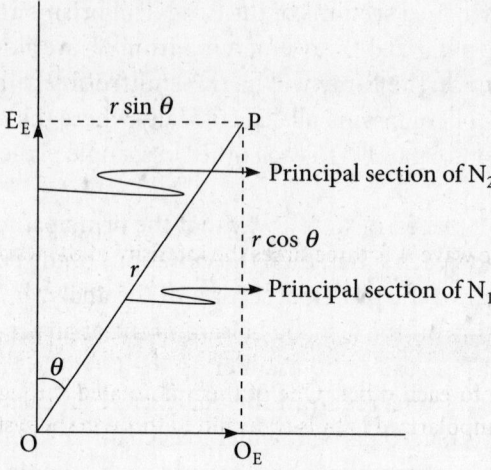

Figure 4.23 | Resolutions for amplitudes of the E-ray for intermediate positions of the analyzer N_2 and the polarizer N_1

Example 4.6

Two adjacent plane polarized waves A and B with planes of polarization perpendicular to each other, are analyzed by an analyzer. In one orientation of the analyzer, the intensity of the wave B is zero. When the analyzer is rotated through an angle 60° from this orientation, the intensities of the two waves are equal. Compare the intensities of the two waves.

Solution

Suppose a plane polarized light is coming from a polarizer and is incident on the analyzer. According to the question, in one orientation of the analyzer, the intensity of the wave B is zero. According to Eq. (4.6) in this orientation for the wave B, $\theta = 90°$ and this implies that $\theta = 0°$ for the wave A. Now the analyzer is rotated through an angle 60° as a result of which for the wave A, $\theta = 60°$ and for the wave B, $\theta = 30°$. The intensities of the two waves are equal. Hence, from Eq. (4.6), we have

$$I_A = I_{A0} \cos^2 60$$

$$I_B = I_{B0} \cos^2 30$$

Since $I_A = I_B$, we write

$$I_{AO} \cos^2 60 = I_{BO} \cos^2 30$$

or $I_{A0} \times \dfrac{1}{4} = I_{B0} \times \dfrac{3}{4}$

or $I_{A0} = 3 \times I_{B0}$

Thus, the intensity of the wave A is three times the intensity of the wave B.

Example 4.7

Two Nicols are crossed to each other. One of them is rotated through an angle 30°. What percentage of incident unpolarized light is transmitted through the system?

Solution

We know from Section 4.11 that if I_0 is the intensity of the unpolarized light incident on the polarizer, then only $\dfrac{I_0}{2}$ is transmitted through the polarizer. According to the question, one of Nicols is rotated through an angle 30° from the crossed position. Therefore, the angle between the two planes of the Nicols is 60°. Hence, the intensity of the light transmitted by the analyzer I is obtained from Eq. (4.6) as

$$I = \dfrac{I_0}{2} \cos^2 60$$

or $\dfrac{I}{I_0} = \dfrac{1}{2} \times \dfrac{3}{4}$

or $\dfrac{I}{I_0} = 0.375$

Hence, the percentage of the incident unpolarized light transmitted through the system is 37.5%.

4.9 Retardation Plates

The speed of the O-ray v_0 and speed of the E-ray v_E inside a doubly refracting crystal is in general not the same and $v_E \geq v_0$. Therefore, a path difference or phase difference is introduced between the O-ray and the E-ray during their propagation inside the crystal. The magnitude of the path difference or phase difference introduced between the E-ray or the O-ray during their propagation inside the doubly refracting crystal depends upon the thickness of the doubly refracting crystal in the direction of propagation and on its refractive index.

A plate cut from a doubly refracting crystal by a plane parallel to the optic axis to introduce a given phase difference/path difference between the O-ray and the E-ray the during the normal transmission of light through it is called a retardation plate.

Let

t = thickness of the retardation plate in the direction of propagation of light.

μ_0 = refractive index of the retardation plate for the O-ray in the direction of propagation of the O-ray.

μ_E = refractive index of the retardation plate for the E-ray in the direction of propagation of the E-ray.

Hence, we have

$\mu_0 t$ = optical path of the O-ray in the direction of propagation of the O-ray inside the retardation plate.

$\mu_E t$ = optical path of the E-ray in the direction of propagation of the E-ray inside the retardation plate.

The optical path difference Δ between the O-ray and the E-ray inside the retardation plate is therefore given by

$$\Delta = \mu_0 t - \mu_E t$$

or $\qquad \Delta = \left(\mu_0 - \mu_E\right)t$ $\hfill$ (4.7)

The corresponding phase difference between the O-ray and the E-ray is therefore given by

$$\delta = \frac{2\pi}{\lambda}\left(\mu_0 - \mu_E\right)t \hfill (4.8)$$

The retardation plates are of two types depending upon the phase or path difference between the O-ray and the E-ray. The retardation plate is said to be a half-wave plate if

$$\Delta = (2n+1)\frac{\lambda}{2} \quad \text{or} \quad \delta = (2n+1)\pi$$

and a quarter-wave plate if

$$\Delta = (2n+1)\frac{\lambda}{4} \quad \text{or} \quad \delta = (2n+1)\frac{\pi}{2}$$

4.9.1 Half-wave plate

A plate cut from a doubly refracting crystal by a plane parallel to the optic axis of such a thickness that a phase difference of $(2n+1)\pi$ or path difference of $(2n+1)\dfrac{\lambda}{2}$ is introduced between the O-ray and the E-ray during the normal transmission of light through it is called a half-wave plate. In a half-wave plate, lagging of the O-ray from the E-ray starts from the point of incidence and goes on increasing continuously throughout their propagation inside the crystal and finally, when they emerge from the crystal, they have a phase difference of π or path difference of $\dfrac{\lambda}{2}$ if the half-wave plate has minimum thickness. See Fig. 4.24.

The thickness of the half-wave plate can be calculated from Eqs (4.7) or (4.8) so that a path difference of $(2n+1)\dfrac{\lambda}{2}$ or a phase difference of $(2n+1)\pi$ is introduced between

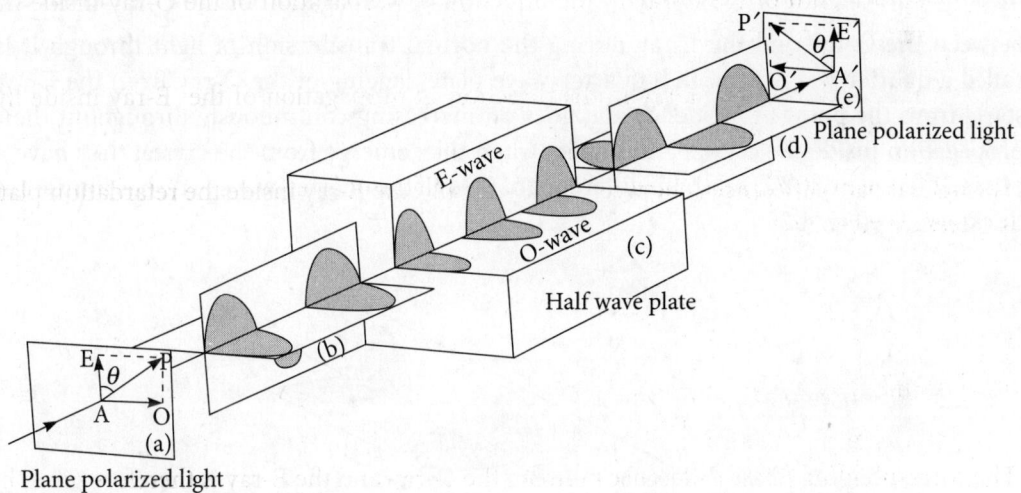

Figure 4.24 | Illustration of action of a half-wave plate. A half-wave plate rotates the plane of polarization by 2θ when light is incident on the plate at an angle θ with the optic axis

the O-wave and the E-wave while they are coming out of the half-wave plate. From the definition of a half-wave plate, we have

$$\Delta = (2n+1)\frac{\lambda}{2}, \quad n = 0, 1, 2, 3, \ldots$$

Putting this value of Δ into the LHS of Eq. (4.7), we get

$$(2n+1)\frac{\lambda}{2} = \left(\mu_0 - \mu_E\right)t$$

or $\quad t = \dfrac{1}{\mu_0 - \mu_E}\left(\dfrac{\lambda}{2}\right)(2n+1)$ $\qquad\qquad$ (4.9)

Equation (4.9) gives the thickness of the half-wave plate. The minimum thickness of the half-wave plate is obtained by putting $n = 0$ in Eq. (4.9) and is given by

$$t_{\min} = \dfrac{1}{\mu_0 - \mu_E}\left(\dfrac{\lambda}{2}\right)$$ $\qquad\qquad$ (4.10)

4.9.2 Quarter-wave plate

A plate cut from a doubly refracting crystal by a plane parallel to the optic axis of such a thickness that a phase difference of $(2n+1)\dfrac{\pi}{2}$ or path difference of $(2n+1)\dfrac{\lambda}{4}$ is introduced between the O-ray and the E-ray during the normal transmission of light through it is called a quarter-wave plate. In a quarter-wave plate, lagging of the O-ray from the E-ray starts from the point of incidence and goes on increasing continuously throughout their propagation inside the crystal and finally, when they emerge from the crystal they have a phase difference of $(\pi/2)$ or a path difference of $(\lambda/4)$ if the quarter-wave plate has minimum thickness. See Fig. 4.25.

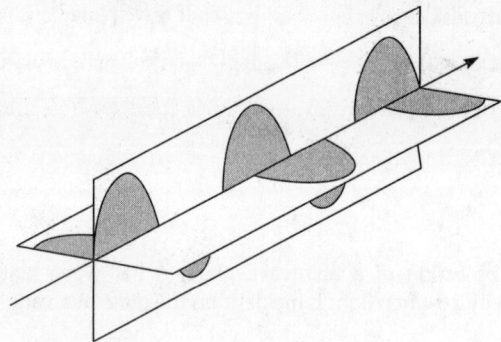

Figure 4.25 | Illustration of action of a quarter-wave plate

The thickness of the quarter-wave plate can be calculated from Eqs (4.7) or (4.8) so that a path difference of $(2n+1)\dfrac{\lambda}{4}$ or phase difference of $(2n+1)\dfrac{\pi}{2}$ is introduced between the O-wave and the E-wave while they are coming out of the quarter-wave plate. From the definition of a quarter-wave plate, we have

$$\Delta = (2n+1)\dfrac{\lambda}{4}, \quad n = 0, 1, 2, 3, \ldots$$

Putting this value of Δ into the LHS of Eq. (4.7), we get

$$(2n+1)\frac{\lambda}{4} = (\mu_0 - \mu_E)t$$

or $\qquad t = \dfrac{1}{\mu_0 - \mu_E}\left(\dfrac{\lambda}{4}\right)(2n+1)$ $\qquad\qquad\qquad\qquad$ (4.11)

This equation gives the thickness of the quarter-wave plate. The minimum thickness of a quarter-wave plate is obtained by putting $n = 0$ in the Eq. (4.11) and is given by

$$t_{min} = \dfrac{1}{\mu_0 - \mu_E}\left(\dfrac{\lambda}{4}\right)$$ $\qquad\qquad\qquad\qquad$ (4.12)

Example 4.8

Plane polarized light is incident on a piece of quartz cut parallel to the axis. Find the least thickness for which the ordinary ray and the extraordinary ray combine to form the plane polarized light. Given: $\mu_0 = 1.5442$, $\mu_E = 1.5533$, and $\lambda = 5 \times 10^{-5}$ cm.

Solution

The ordinary ray and the extraordinary ray combine to form the plane polarized light on emergence if the plate introduces a phase difference of π or a path difference of $\dfrac{\lambda}{2}$ between the ordinary ray and the extraordinary ray. The plate which introduces a phase difference of π or path difference of $\dfrac{\lambda}{2}$ between the ordinary ray and the extraordinary ray is the half-wave plate.

The data given are

$\mu_0 = 1.5442$

$\mu_E = 1.5533$

$\lambda = 5 \times 10^{-5}$ cm.

Here,

$$t = \dfrac{1}{\mu_E - \mu_0}\left(\dfrac{\lambda}{2}\right)$$

is the least thickness of a half-wave plate in positive crystals

$$\therefore \qquad t = \dfrac{1}{1.5442 - 1.5533} \times \dfrac{5 \times 10^{-5}}{2}\ \text{cm} = 2.75 \times 10^{-3}\ \text{cm}$$

Example 4.9

Calculate the thickness of the mica sheet required to make a quarter-wave plate and a half-wave plate for $\lambda = 5460$ Å. The indices of refraction for the ordinary and the extraordinary waves in mica are 1.586 and 1.592 respectively.

Solution

The data given are

$\mu_0 = 1.586$

$\mu_E = 1.592$

$\lambda = 5460 \times 10^{-8}$ cm.

Thickness for the quarter-wave plate is $t_4 = \dfrac{1}{\mu_E - \mu_0}\left(\dfrac{\lambda}{4}\right)$ in positive crystals

or $t_4 = \dfrac{1}{1.592 - 1.586}\left(\dfrac{5460 \times 10^{-8}}{4}\right)$ cms $= 2.275 \times 10^{-3}$ cms

Thickness for the half-wave plate is $t_2 = \dfrac{1}{\mu_E - \mu_0}\left(\dfrac{\lambda}{2}\right)$ in positive crystals

$t_2 = \dfrac{1}{1.592 - 1.586}\left(\dfrac{5460 \times 10^{-8}}{2}\right)$ cms $= 4.55 \times 10^{-3}$ cms

4.10 Production of Circularly Polarized Light

4.10.1 Principle

Circular motion of light results when two mutually perpendicular coherent linear vibrations of equal amplitudes and frequency but differing in phase by $\dfrac{\pi}{2}$, are compounded together. The two mutually perpendicular coherent linear vibrations of equal amplitudes and frequency but differing in phase of $\dfrac{\pi}{2}$ can be represented as

$$x = \frac{a}{\sqrt{2}}\sin\left(\omega t + \frac{\pi}{2}\right)$$

(4.13)

$$y = \frac{a}{\sqrt{2}} \sin \omega t \qquad (4.14)$$

Equation (4.13) can be re-written as

$$x = \frac{a}{\sqrt{2}} \cos \omega t \qquad (4.15)$$

From Eqs (4.14) and (4.15), we have

$$y^2 + x^2 = \frac{a^2}{2} \sin^2 \omega t + \frac{a^2}{2} \cos^2 \omega t$$

or $\quad x^2 + y^2 = \dfrac{a^2}{2} \qquad (4.16)$

This is the equation of a circle representing the circular motion of radius $\dfrac{a}{\sqrt{2}}$.

4.10.2 Production

The experimental arrangement for the production of circularly polarized light is shown in Fig. 4.26. The plane polarized light from Nicol prism N_1 is allowed to fall on another Nicol prism N_2 placed at a certain distance in the crossed position. Since two Nicol prisms N_1 and N_2 are in the crossed position, the field of view is dark. Now in this position, a quarter-wave plate mounted in the tube T_1 is introduced between the two Nicols and held normal to the ray of light coming from Nicol N_1.

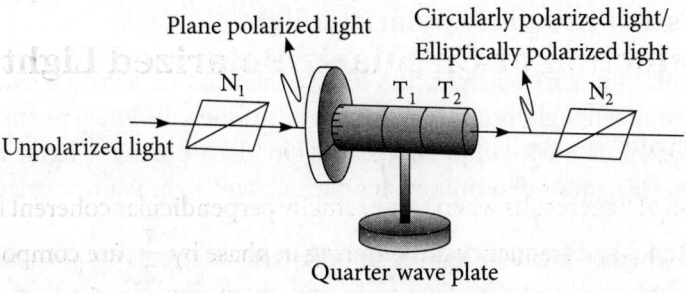

Figure 4.26 | Experimental arrangement for the production of circularly and elliptically polarized light

Tube T_1 can be rotated about the outer fixed tube T_2. Thus, the quarter-wave plate mounted in tube T_1 can be rotated about a horizontal axis through any angle we desire. After the introduction of a quarter-wave plate between two Nicol prisms N_1 and N_2, the field of view beyond N_2 may not be dark. If the field of view is not dark, the tube T_1 containing the quarter-wave plate is rotated so as to make the field of view dark. In this position, the vibrations of the plane polarized light incident normally on the quarter-wave plate are along the optic axis of the quarter wave plate and hence, perpendicular to Nicol prism N_2. Now the tube T_1 containing the quarter-wave plate is rotated through 45° so that the vibrations of the plane polarized light incident normally on the quarter-wave plate make an angle of 45° with the optic axis of the quarter-wave plate. At this position, the incident polarized light is split up into two rectangular components – the O-ray and the E-ray having equal amplitudes ($\because \sin 45° = \cos 45°$) and frequency. According to the properties of the quarter-wave plate, there is a phase difference of ($\pi/2$) between the emergent O-wave and the E-wave so that the resultant beam after the quarter-wave plate is circularly polarized. See Fig. 4.27.

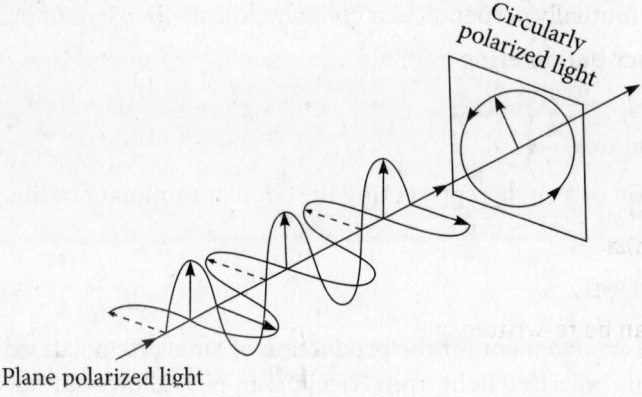

Figure 4.27 Conversion of plane polarized light into circularly polarized light by a quarter-wave plate when the plane polarized light is incident on the quarter-wave plate at an angle of 45° with the optic axis

4.10.3 Analysis of circularly polarized light

To analyze circularly polarized light, the light is allowed to fall on a Nicol prism. The intensity of the emergent light beam remains uniform when the Nicol prism is rotated. This observation is common to both unpolarized and circularly polarized light. Therefore, from this observation, we cannot determine whether incident light is unpolarized or circularly polarized.

To conclude whether incident light is unpolarized or circularly polarized, the beam is allowed to fall first on a quarter-wave plate and then on a Nicol prism. If the beam is circularly polarized after passing through a quarter-wave plate, it becomes plane polarized. When this emergent beam from the quarter-wave plate is incident on a rotating Nicol prism, the field of view becomes dark twice during each complete rotation of the Nicol prism, which implies that the incident original light is circularly polarized. On the other hand, if the emergent beam from the quarter-wave plate is incident on a rotating Nicol prism, the field of view does not become dark twice during each complete rotation of the Nicol prism, which implies that the incident original light is unpolarized. Hence, we can conclude that if the original beam after passing through the quarter-wave plate is extinguished twice in each rotation while passing through the rotating Nicol, it is circularly polarized.

4.11 Production of Elliptically Polarized Light

4.11.1 Principle

Elliptical motion of light results when two mutually perpendicular coherent linear vibrations of unequal amplitudes, equal frequency but differing in phase by $\dfrac{\pi}{2}$, are compounded together. The two mutually perpendicular coherent linear vibrations of unequal amplitudes and equal frequency but differing in phase by $\dfrac{\pi}{2}$ can be represented as

$$x = a \cos \theta \sin\left(\omega t + \frac{\pi}{2} \right) \tag{4.17}$$

$$y = a \sin \theta \sin \omega t \tag{4.18}$$

Equation (4.17) can be re-written as

$$\frac{x}{a \cos \theta} = \cos \omega t \tag{4.19}$$

Equation (4.18) can be re-written as

$$\frac{y}{a \sin \theta} = \sin \omega t \tag{4.20}$$

From Eqs (4.19) and (4.20) (squaring and adding both sides), we have

$$\frac{x^2}{a^2 \cos^2 \theta} + \frac{y^2}{a^2 \sin^2 \theta} = \sin^2 \omega t + \cos^2 \omega t$$

or $\quad \dfrac{x^2}{a^2 \cos^2 \theta} + \dfrac{y^2}{a^2 \sin^2 \theta} = 1 \tag{4.21}$

This is the equation of an ellipse representing elliptical motion.

4.11.2 Production

The production of elliptically polarized light can be explained with the help of the experimental arrangement shown in Fig. 4.26. The plane polarized light from the Nicol prism N_1 is allowed to fall on another Nicol prism N_2 placed at a certain distance in a crossed position. Since two Nicol prisms N_1 and N_2 are in the crossed position, the field of view is dark. Now in this position, a quarter-wave plate mounted in tube T_1 is introduced between the two Nicols and held normal to the ray of light coming from Nicol N_1. Tube T_1 can be rotated about the outer fixed tube T_2. Thus, the quarter-wave plate mounted in tube T_1 can be rotated about a horizontal axis through any angle we desire. After the introduction of a quarter-wave plate between two Nicol prisms N_1 and N_2, the field of view beyond N_2 may not be dark. If the field of view is not dark, tube T_1 containing the quarter-wave plate is rotated so as to make the field of view dark. In this position, the vibrations of the plane polarized light incident normally on the quarter-wave plate are along the optic axis of the quarter wave plate and hence, perpendicular to Nicol prism N_2. Now tube T_1 containing the quarter-wave plate is rotated through 30° (any angle not equal to 45°) so that the vibrations of the plane polarized light incident normally on the quarter-wave plate make an angle of 30° with the optic axis of the quarter-wave plate. At this position, the incident polarized light is split up into two rectangular components, the O-ray and the E-ray having unequal amplitudes ($\because \sin 30° \neq \cos 30°$) and equal frequency. According to the properties of the quarter-wave plate, there is a phase difference of $\dfrac{\pi}{2}$ between the emergent O-wave and the E-wave so that the resultant beam after the quarter-wave plate is elliptically polarized. See Fig. 4.28.

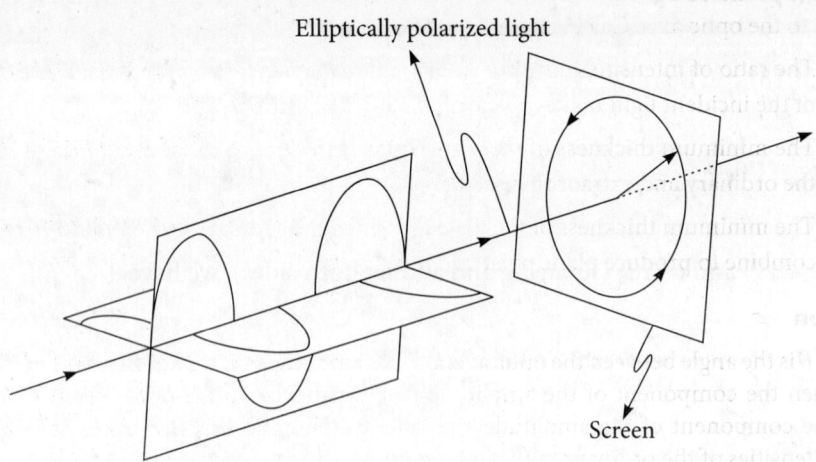

Elliptically polarized light

Screen

Figure 4.28 | Conversion of plane polarized light into an elliptically polarized light by a quarter-wave plate when plane polarized light is incident on the quarter-wave plate at an angle excluding 0°, 45° and 90° with the optic axis

4.11.3 Analysis of elliptically polarized light

To analyze the elliptically polarized light, the light is allowed to fall on a Nicol prism. The intensity of the emergent light beam varies between minimum to maximum when the Nicol prism is rotated. This observation is common to both unpolarized and elliptically polarized light. Therefore, from this observation we cannot determine whether the incident light is unpolarized or elliptically polarized. The maximum or minimum intensity depends upon the condition whether the principal plane of the Nicol prism is parallel to the major axis or minor axis of the ellipse.

To conclude whether incident light is a mixture of unpolarized or polarized light, the beam is allowed to fall first on a quarter-wave plate and then on a Nicol prism. If the beam is elliptically polarized after passing through a quarter-wave plate, it becomes plane polarized. When this emergent beam from the quarter-wave plate is incident on a rotational Nicol prism, the field of view become dark twice during each complete rotation of the Nicol prism, which implies that the incident original light is elliptically polarized. On the other hand, if the emergent beam from the quarter-wave plate is incident on a rotational Nicol prism, the field of view does not become dark twice during each complete rotation of the Nicol prism, which implies that the incident original light is a mixture of unpolarized and polarized light. Hence, we can conclude that if the original beam after passing through the quarter-wave plate is extinguished twice in each rotation, when studied by a rotational Nicol, it is elliptically polarized.

Example 4.10

The plane polarized light of wavelength 5890 Å is incident on a thin quartz plate cut with faces parallel to the optic axis. Calculate

i. The ratio of intensities of the ordinary and the extraordinary ray if the plane of vibration of the incident light makes an angle of 30° with the optic axis.

ii. The minimum thickness of the plate which introduces a phase difference of 30° between the ordinary and extraordinary rays.

iii. The minimum thickness of the plate for which the ordinary and extraordinary waves will combine to produce plane polarized light. Given $\mu_0 = 1.586$ and $\mu_E = 1.592$

Solution

i. If θ is the angle between the optic axis and the amplitude vector r of the plane of vibration then the component of the amplitude r perpendicular to the optic axis is $r \sin \theta$ and the component of the amplitude r parallel to the optic axis is $r \cos \theta$. Therefore, the intensities of the ordinary ray I_0 and the extraordinary I_E ray will be given by

$$I_0 = cr^2 \sin^2 \theta$$

$$I_E = cr^2 \cos^2 \theta$$

where c is the constant of proportionality.
Hence, we have

$$\frac{I_0}{I_E} = \frac{\sin^2 \theta}{\cos^2 \theta} = \frac{\sin^2 30}{\cos^2 30} = \frac{1}{3}$$

ii. From Eq. (4.8), we have

$$\delta = \frac{2\pi}{\lambda}\left(\mu_E - \mu_0\right)t$$

According to the question, phase difference of

$$30° \left(= \frac{\pi}{6}\right)$$

exists between the ordinary and extraordinary rays. Hence, putting this in the aforementioned equation, we get

$$\frac{\pi}{6} = \frac{2\pi}{\lambda}\left(\mu_E - \mu_0\right)t$$

$$t = \frac{\lambda}{12\left(\mu_E - \mu_0\right)}$$

Putting the values of λ, μ_E, and μ_0 into this equation, we get

$$t = \frac{5890 \times 10^{-8}}{12(1.592 - 1.586)} \text{cm} = 8.18 \times 10^{-4} \text{ cm.}$$ (Ans.)

iii. The ordinary and extraordinary waves will combine to produce a plane polarized light only if they have a phase difference of π or path difference of $\frac{\lambda}{2}$. Therefore, the minimum thickness of the plate for which the ordinary and extraordinary waves will combine to produce plane polarized light is obtained from Eq. (4.8) as

$$\pi = \frac{2\pi}{\lambda}\left(\mu_E - \mu_0\right)t$$

or $t = \dfrac{\lambda}{2\left(\mu_E - \mu_0\right)} = \dfrac{5890 \times 10^{-8}}{2(1.592 - 1.586)} \text{cms} = 4.908 \times 10^{-3} \text{cms.}$

4.12 Analysis of Light

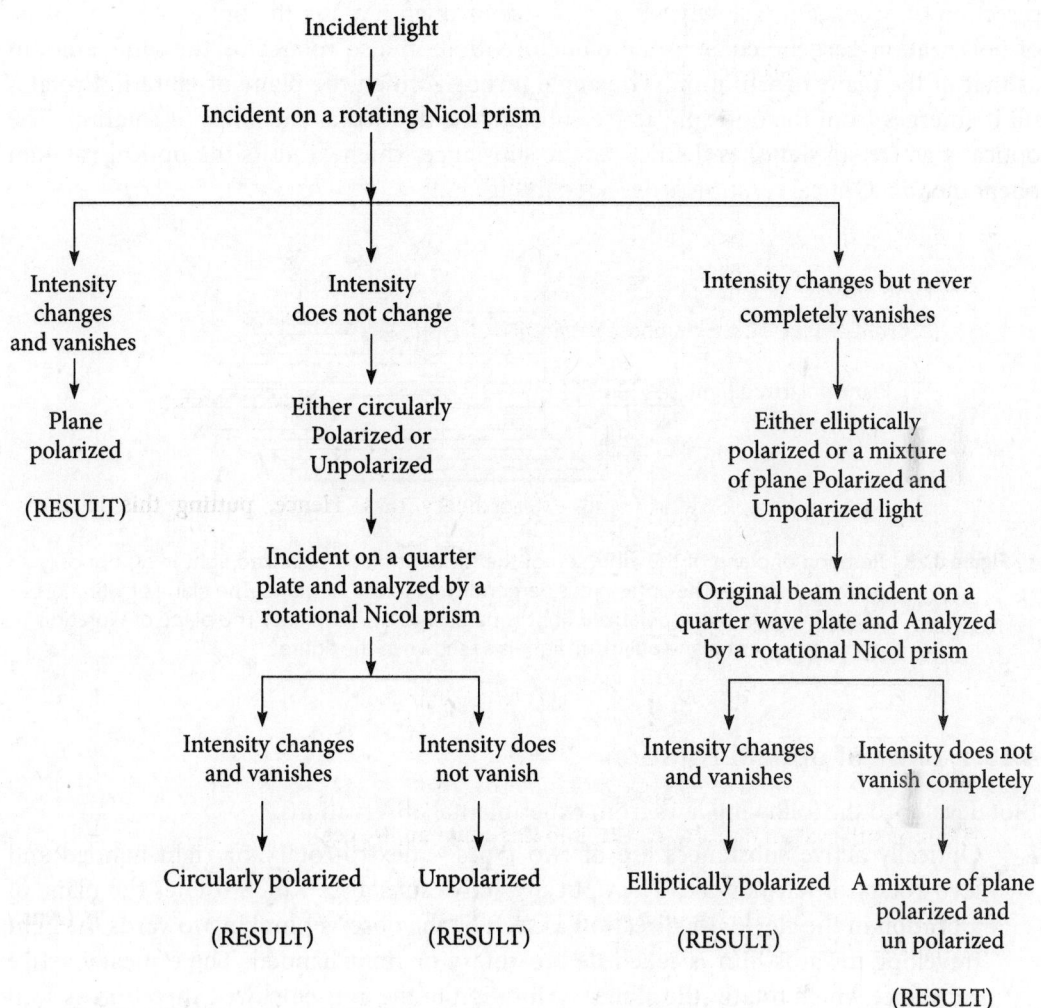

Incident light

↓

Incident on a rotating Nicol prism

- **Intensity changes and vanishes** → **Plane polarized** (RESULT)

- **Intensity does not change** → **Either circularly Polarized or Unpolarized** → Incident on a quarter plate and analyzed by a rotational Nicol prism
  - **Intensity changes and vanishes** → **Circularly polarized** (RESULT)
  - **Intensity does not vanish** → **Unpolarized** (RESULT)

- **Intensity changes but never completely vanishes** → **Either elliptically polarized or a mixture of plane Polarized and Unpolarized light** → Original beam incident on a quarter wave plate and Analyzed by a rotational Nicol prism
  - **Intensity changes and vanishes** → **Elliptically polarized** (RESULT)
  - **Intensity does not vanish completely** → **A mixture of plane polarized and un polarized** (RESULT)

4.13 Optical Rotation

Optical rotation or optical activity was first observed in a quartz crystal in 1811 by a French physicist Francois Arago. Another French physicist, Jean-Baptiste Biot, found in 1815, that liquid solutions of tartaric acid or of sugar are optically active, as are liquid or vaporous turpentine. Louis Pasteur was the first to recognize that optical activity arises from the dissymmetric arrangement of atoms in crystalline structures or in individual molecules of certain compounds.

Optical rotation is defined as the phenomenon by virtue of which substances like quartz, calcite, tourmaline, cinnabar, sodium chlorate, sugar solution, tartaric acid solution, turpentine, etc., rotate the plane of vibration of the plane polarized light about the direction of propagation slowly when it passes through it along the optic axis. The plane of polarization perpendicular to the plane of vibration also rotates by the same amount as that of the plane of vibration. The angle through which the plane of vibration rotates till it emerges from the optically active substance is defined as the angle of rotation. The optically active substance is defined as the substance which exhibits the optical rotation phenomenon. Optical rotation is depicted in Fig. 4.29.

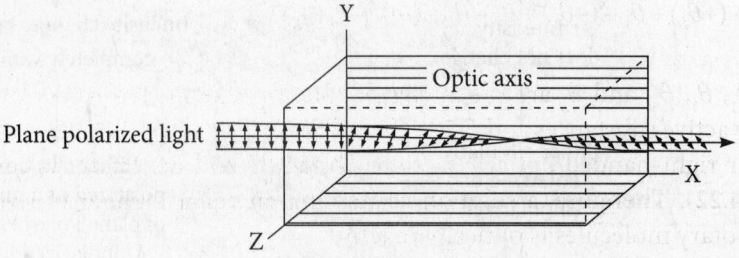

Figure 4.29 | Rotation of plane of the vibration of the incident plane polarized light in an optically active substance. The optic axis is perpendicular to the *YZ* plane. The plane of vibration of the incident plane polarized light is parallel to the *XY* plane. The plane of vibration of the emergent plane polarized light is as shown in the figure

4.13.1 Laws of optical rotation

Biot developed the following laws from experimental observations.

i.   Optically active substances are of two types – dextro-rotary or right-handed and levo-rotary or left-handed. The optically active substance which rotates the plane of vibration in the clockwise direction as seen by the observer looking towards the light travelling towards him is called dextro-rotary or right-handed. The optically active substance which rotates the plane of vibration in the anti-clockwise direction as seen by the observer looking towards the light travelling towards him is called levo-rotary or left-handed.

ii. For levo-rotary or left-handed optically active substances, the angle of rotation is taken as positive while for dextro-rotary or right-handed optically active substances, the angle of rotation is taken as negative.

iii. The angle of rotation θ of the plane of vibration produced by the optically active solid substance for a given wavelength is directly proportional to the length travelled by the light inside the substance. Mathematically, it is translated as $\theta \propto \ell$.

iv. The angle of rotation θ of the plane of vibration produced by the optically active fluid for a given wavelength and path length is directly proportional to the concentration C of the solution or vapour. Mathematically, it is translated as $\theta \propto C$.

v. The angle of rotation θ of the plane of vibration produced by the optically active substance for a given wavelength and path length is approximately inversely proportional to the square of the wavelength λ of the light employed. Mathematically, it is given as $\theta \propto \dfrac{1}{\lambda^2}$. Thus, in the visible spectrum, the angle of rotation θ is least for red and greatest for violet.

vi. The total angle of rotation θ produced by a number of optically active chemically non-reactive substances in a mixture is equal to the algebraic sum of all the angles of rotation produced by individual specimens separately. Mathematically, it is translated as

$$\theta = \theta_1 + (-\theta_2) + \theta_3 + (-\theta_4) + \theta_5 + \theta_6 + (-\theta_7) + (-\theta_8) + \dots \tag{4.22}$$

where θ_1, θ_3, θ_5, and θ_6 are the angles of rotation due to levo-rotary or left-handed optically active substances and θ_2, θ_4, θ_7, and θ_8 are the angles of rotation due to dextro-rotary or right-handed optically active substances and hence, are taken as negative in Eq. (4.22). Therefore, a solution containing an equal number of levo-rotary and dextro-rotary molecules is optically inactive.

4.13.2 Fresnel's theory of optical rotation

Augustin-Jean Fresnel explained the phenomenon of optical rotation assuming the following facts

i. A beam of plane polarized light entering into an optically active crystal is broken up into two circularly polarized vibrations – one right handed and other left handed.

ii. The two circularly polarized vibrations rotate with the same frequency in an optically inactive crystal.

iii. The two circularly polarized vibrations travel with different velocities in an optically active crystal. In dextro-rotary substances, the velocity of the right-handed circularly polarized vibration is greater than the velocity of the left-handed circularly polarized vibration while in levo-rotary substances, the velocity of the left-handed circularly polarized vibration is greater than the velocity of the right-handed circularly polarized vibration.

iv. After emerging from the substance, the two opposite circularly polarized vibrations recombine to produce a plane polarized wave. In this emerging plane polarized wave, the plane of vibration is rotated with respect to the plane of the vibration of incident plane polarized light.

Case 1: Optically inactive substances

Let us consider the case of an optically inactive crystal like calcite, in which according to Fresnel, plane polarized vibration on entering into the crystal along the optic axis, is resolved into two circularly polarized vibrations rotating with the same angular frequency. In Fig. 4.30, *OL* is the circularly polarized vibration rotating in the anti-clockwise direction

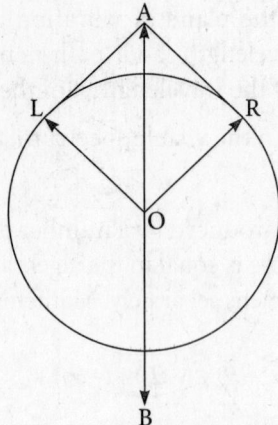

and *OR* is the circularly polarized vibration rotating in the clockwise direction. These two vectors *OL* and *OR* rotate with the same frequency. *OA* is the resultant of *OL* and *OR* and does not vary with respect to time. This resultant *OA* is always in the same plane as that of the incident vibration. Hence, coming out of the crystal, the two circular vibrations will again combine to give a plane polarized light having the same plane of vibration as that of the incident plane polarized light along *AB*. Therefore, an optically inactive crystal like calcite does not rotate the plane of vibration.

Case 2: Optically active substances

Let us consider the case of an optically active crystal like a right-handed quartz crystal, in which according to Fresnel, plane polarized vibration on entering into the crystal along the optic axis is resolved into two circularly polarized vibrations rotating with different angular frequencies. In a dextro-rotary crystal like quartz, the angular velocity of the right-handed circularly polarized vibration is greater than the angular velocity of the left-handed circularly polarized vibration. In Fig. 4.31, *OL* is the circularly polarized vibration rotating in the anti-clockwise direction and *OR* is the circularly polarized vibration rotating in the clockwise direction. These two vectors *OL* and *OR* rotate with different frequency. As a result on emergence, clock-wise vibrations make a greater angle than the angle made by anticlock-wise vibrations.

The resultant of *OL* and *OR* is *OA′* and varies in direction with respect to time inside the crystal. Hence, coming out of the crystal, the two circular vibrations will again combine to give a plane polarized light with the plane of vibration *A′B′* different from that of the incident plane polarized light *AB*. Therefore, an optically active crystal like quartz rotates the plane of vibration.

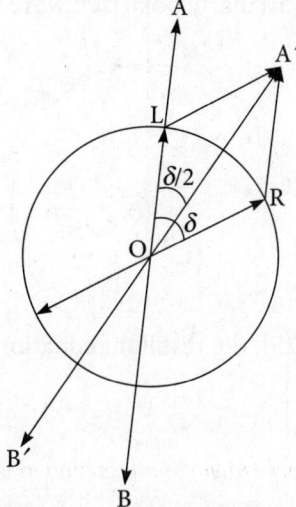

Figure 4.31 | *OL* is the circularly polarized vibration rotating in the anti-clockwise direction and *OR* is the circularly polarized vibration rotating in the clockwise direction. These two vectors *OL* and *OR* rotate with different frequencies in opposite direction in the quartz crystal. *OA* is the resultant of *OL* and *OR* and its direction varies with respect to time. This resultant *OA* is not in the same plane as that of the incident vibration

4.13.3 Mathematical analysis of Fresnel's theory of optical rotation

Suppose a plane polarized light is travelling inside an optically active crystal along the positive Z direction. According to Fresnel, it is broken up into two left-handed and right-handed circularly polarized waves. The plane of vibration of the two circularly polarized waves will be parallel to the XY plane.

Let

v_1 = velocity of the right-handed circularly polarized wave.

v_2 = velocity of the left-handed circularly polarized wave.

The equation of the right-handed circularly polarized wave will be

$$x_1 = r \, \sin\frac{2\pi}{T}\left(t - \frac{z}{v_1}\right) \tag{4.23}$$

$$y_1 = r \, \cos\frac{2\pi}{T}\left(t - \frac{z}{v_1}\right) \tag{4.24}$$

The equation of the left-handed circularly polarized wave will be

$$x_2 = r \ \sin\frac{2\pi}{T}\left(t - \frac{z}{v_2}\right) \tag{4.25}$$

$$y_2 = -r \ \cos\frac{2\pi}{T}\left(t - \frac{z}{v_2}\right) \tag{4.26}$$

By combining Eqs (4.23) and (4.25), the resultant equation of motion along the X-axis will be given by

$$x = x_1 + x_2$$

or $\quad x = r \ \sin\frac{2\pi}{T}\left(t - \frac{z}{v_1}\right) + r \ \sin\frac{2\pi}{T}\left(t - \frac{z}{v_2}\right)$

or $\quad x = r\left[\sin\frac{2\pi}{T}\left(t - \frac{z}{v_1}\right) + \sin\frac{2\pi}{T}\left(t - \frac{z}{v_2}\right)\right]$

or $\quad x = 2r \ \sin\frac{2\pi}{T}\left[t - \frac{z}{2}\left(\frac{1}{v_1} + \frac{1}{v_2}\right)\right] \ \cos\frac{\pi}{T}\left(\frac{1}{v_2} - \frac{1}{v_1}\right) \tag{4.27}$

By combining Eqs (4.24) and (4.26), the resultant equation of motion along the Y-axis will be given by

$$y = y_1 + y_2$$

or $\quad y = r \ \cos\frac{2\pi}{T}\left(t - \frac{z}{v_1}\right) - r\cos\frac{2\pi}{T}\left(t - \frac{z}{v_2}\right)$

or $\quad y = r\left[\cos\frac{2\pi}{T}\left(t - \frac{z}{v_1}\right) - \cos\frac{2\pi}{T}\left(t - \frac{z}{v_2}\right)\right]$

or $\quad y = -2r \, \sin\dfrac{2\pi}{T}\left[t - \dfrac{z}{2}\left(\dfrac{1}{v_1} + \dfrac{1}{v_2}\right)\right] \sin\dfrac{\pi}{T}\left(\dfrac{1}{v_2} - \dfrac{1}{v_1}\right)$ $\hspace{2cm}$ (4.28)

Taking the ratio of Eqs (4.26) and (4.25), we have

$$\dfrac{y}{x} = -\tan\dfrac{\pi \, z}{T}\left(\dfrac{1}{v_2} - \dfrac{1}{v_1}\right)$$

or $\quad \dfrac{y}{x} = \tan\left[-\dfrac{\pi \, z}{T}\left(\dfrac{1}{v_2} - \dfrac{1}{v_1}\right)\right]$

or $\quad \dfrac{y}{x} = \tan\theta$

or $\quad y = x \, \tan\theta + 0 \quad (y = mx + c)$ $\hspace{2cm}$ (4.29)

where

$$\theta = -\dfrac{\pi z}{T}\left(\dfrac{1}{v_2} - \dfrac{1}{v_1}\right)$$ $\hspace{2cm}$ (4.30)

Equation (4.29) represents a straight line making an angle θ with the X-axis. Therefore, the emergent light is a plane polarized light with the plane of vibration making an angle θ with the X-axis as shown in Fig. 4.32.

Interpretation of Eq. (4.30)

i. $v_1 > v_2$, θ is negative; rotation of the plane of vibration is clockwise as in the case of right-handed crystals like quartz.

ii. $v_1 < v_2$, θ is positive; rotation of the plane of vibration is anti-clockwise as in the case of left-handed crystals like turpentine.

iii. $v_1 = v_2$, θ is zero; there is no rotation of the plane of vibration as in the case of optically inactive crystals like calcite.

iv. The angle of rotation θ is directly proportional to the length z of the material travelled by the light.

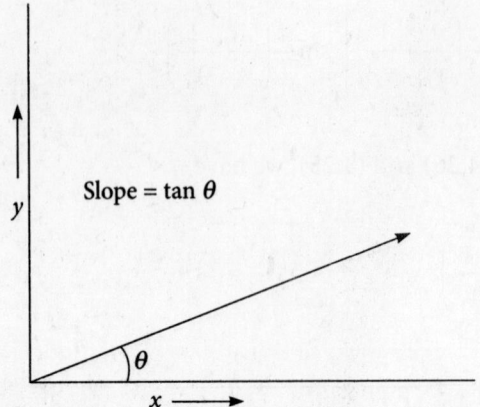

Figure 4.32 | The emergent light is a plane polarized light with the plane of vibration making an angle θ with the X-axis

4.13.4 Calculation of the angle of optical rotation

With the help of Eq. (4.30), we can express the angle of optical rotation in terms of the refractive index of the crystal for the right-handed circularly polarized light μ_R and the left-handed circularly polarized light μ_L. From Eq. (4.30), we have

$$\theta = \frac{\pi\, l}{T}\left(\frac{1}{v_R} - \frac{1}{v_L}\right) \tag{4.31}$$

where

v_R = velocity of the right-handed circularly polarized wave

v_L = velocity of the left-handed circularly polarized wave

ℓ = length of the path travelled by the plane polarized light inside the optically active crystal.

Let

μ_R = refractive index of the crystal for the right-handed circularly polarized light.

μ_L = refractive index of the crystal for the left-handed circularly polarized light.

c = speed of light in vacuum.

From the definition of refractive index of a material, we have

$$\mu_R = \frac{c}{v_R}$$

or $\dfrac{1}{v_R} = \dfrac{\mu_R}{c}$ (4.32)

and $\mu_L = \dfrac{c}{v_L}$

or $\dfrac{1}{v_L} = \dfrac{\mu_L}{c}$ (4.33)

Putting the values of $\dfrac{1}{v_R}$ and $\dfrac{1}{v_L}$ from Eq. (4.32) and (4.33) into the Eq. (4.31), we get

$$\theta = \dfrac{\pi \ell}{T}\left(\dfrac{\mu_R}{c} - \dfrac{\mu_L}{c}\right)$$

or $\theta = \dfrac{\pi \ell}{T \times c}(\mu_R - \mu_L)$

or $\theta = \dfrac{\pi \ell}{\lambda}(\mu_R - \mu_L)$ $(\because \lambda = T \times c)$ (4.34)

4.13.5 Specific rotation

The specific rotation or rotary power ω of an optically active substance at a particular temperature and at a particular wavelength is defined as the rotation produced by a 10 cm column of liquid containing 1 gm of active substance for every cubic centimetre of solution. This statement can be stated mathematically as

$$\omega = \dfrac{10\theta}{\ell \times C}$$ (4.35)

where

ω = specific rotation in units of $\dfrac{degree \times cm^2}{gm}$

θ = angle of rotation in degree

ℓ = length of the light path in centimetres

C = number of grammes of optically active substance per cubic centimetre of solution, i.e., concentration in $\dfrac{gm}{cm^3}$.

The specific rotation or rotary power ω of optically active substances varies not only with the concentration of the optically active substances but also with solvents. Optical rotation is assigned a negative value if it is clockwise with respect to an observer facing the light source, positive if counter-clockwise. A substance with a negative specific rotation is described as dextro-rotatory and denoted by the prefix d or ($-$); one with a positive specific rotation is levo-rotatory, designated by the prefix l or ($+$).

Example 4.11

Calculate the specific rotation of a sugar solution if the plane of polarization is rotated through $28°$ passing through a length of 0.2 m of 30% sugar solution.

Solution

The data given are

$$\theta = 28^O$$

$$\ell = 0.2m = 20cm$$

$$C = 0.3 \text{ gm/cm}^3$$

The specific rotation ω is given by

$$\omega = \frac{10\theta}{\ell \times C} = \frac{10 \times 28° cm^2}{20 \times 0.3gm} = 46.67° \frac{cm^2}{gm}$$

Example 4.12

The plane of polarization of a plane polarized light is turned through an angle $12.6°$ passing through a 10% sugar solution of length 20 cm. Calculate the specific rotation.

Solution

The data given are

$$\theta = 12.6°$$

$$\ell = 20cm$$

$$C = 10\% = 0.1 \text{gm/cm}^3$$

The specific rotation ω is given by

$$\omega = \frac{10\theta}{\ell \times C} = \frac{10 \times 12.6}{20 \times 0.1} = 63° \text{cm}^2/\text{gm}$$

Example 4.13

An optically active right-handed solution of 20 cm length rotates the plane of polarization through 40° and a 30 cm length of optically active left-handed solution rotates the plane of polarization through 24°. These two solutions in the ratio 1 : 2 are contained in a 30 cm tube. Calculate the angle of optical rotation.

Solution

The angle of optical rotation is directly proportional to the length of the solution. The two solutions are chemically non-reactive. The right- and left-handed molecules rotate the plane of polarization separately and their addition gives the resultant rotation. According to the question, two solutions are mixed in a 1 : 2 volume ratio and contained in a 30 cm tube. Hence, we can divide the tube of 30 cm in the ratio 1 : 2. The first solution is assumed to be in the first portion of the tube of length 10 cm and the second solution is assumed to be in the second portion of the tube of length 20 cm.

According to the question, 20 cm length of the first (right-handed) solution rotates the plane of polarization through 40°. Hence, 10 cm length of the first (right-handed) solution rotates the plane of polarization through

$$\theta_1 = 40° \times \frac{10 \text{ cms}}{20 \text{ cms}} = -20° \quad \text{(−ve since it is right-handed)}$$

Again according to the question, 30 cm length of the second (left-handed) solution rotates the plane of polarization through 24°. Hence, 20 cm length of the second (left-handed) solution rotates the plane of polarization through

$$\theta_2 = 24° \times \frac{20 \text{ cms}}{30 \text{ cms}} = +16° \quad \text{(+ve since it is right-handed)}$$

Therefore, the total optical rotation θ according to Eq. (4.22) is given by

$$\theta = \theta_1 + \theta_2$$

$$\theta = -20° + 16° = -4°$$

Thus, the mixture behaves as a right-handed solution.

Example 4.14

The refractive indices for right- and left-handed vibrations are 1.5580 and 1.55821 respectively for quartz for sodium light of wavelength 5890 Å. Find the optical rotation for the same light by a plate of quartz of thickness 1.00 mm when its faces are cut perpendicular to the optic axis.

Solution

The angle of rotation of the plane of vibration is given by Eq. (4.34) as

$$\theta = \frac{\pi \ell}{\lambda} \left(\mu_R - \mu_L \right)$$

The data given are

$\ell = 1.00 \text{ mm} = 0.100 \text{cm}$

$\lambda = 5890 \text{Å} = 5890 \times 10^{-8} \text{cm}$

$\mu_R = 1.55810$

$\mu_L = 1.55821$

Putting these data into the aforementioned equation, we get the angle of rotation as

$$\theta = \frac{\pi \times 0.1}{5890 \times 10^{-8}} (1.55821 - 1.55810) = 0.5870 \text{ rad} = 33.63°$$

4.14 Polarimeter

French physicist Jean-Baptiste Biot discovered that in a mixture of optically active substances and optically inactive substances, the angle of rotation of the mixture is directly proportional to the concentration of the optically active substance. This discovery has been exploited extensively in commerce and industry to estimate the percentage of sugar present in an optically inactive impurity. A number of devices known as polarimeters have been designed for the purpose.

Polarimeters are devices by virtue of which the angle of rotation of optically active substances in pure form or in impure form can be measured. The polarimeter calibrated to read directly the percentage of cane sugar in a solution is called sacharimeter. Polarimeters can be used to estimate the specific rotation or rotary power of a substance or of a mixture of optically active and optically inactive substances.

The first version of polarimeters as used by Mitcherlich were not very accurate in determining the angle of rotation as the field of view remained completely dark not for a single position but for a large range of rotation of the analyzer. This draw-back was removed by using the half shadow principle in which the field of view was divided into two halves situated side by side – this was done by inserting a circular disc in a certain orientation between the polarizer and the analyzer. Half of the circular disc is made of glass and the other half is made of a specially cut quartz and is known as Laurent half shade. As a result of this insertion, the analyzer can be accurately set for equal brightness of the two halves of the field of view. This highly sensitive polarimeter based on the half shadow principle is called Laurent's half-shade polarimeter.

4.14.1 Laurent's half-shade polarimeter

Principle of the polarimeter

When the plane polarized light is allowed to pass along the optic axis of an optically active substance, its plane of vibration rotates about the direction of propagation and emerges from the optically active substance rotating through a certain angle known as the angle of rotation. The angle of rotation is directly proportional to the length travelled by the light inside the substance, concentration C of the solution, inversely proportional to the square of the wavelength λ of the light employed (approximately). The total angle of rotation θ produced by a number of optically active chemically non-reactive substances in a mixture is equal to the algebraic sum of all the angles of rotation produced by individual specimens separately.

Construction

The essential components of Laurent's half-shade polarimeter are depicted in Fig. 4.33. It consists of two Nicol prisms N_1 and N_2. N_1 is the polarizer and N_2 is the analyzer. The two Nicol prisms N_1 and N_2 can be rotated about a common axis. The rotation of N_1 and N_2 can be read with help of a circular vernier scale (not shown) having least count of a fraction of a degree. Behind the polarizer, there is Laurent's half-shade disc. As mentioned earlier, half of Laurent's half-shade disc is made of a semi-circular half-wave plate made of quartz and the other half is made of semicircular glass. The thickness of the semicircular glass plate is such that it transmits and absorbs the same amount of light as that of the semicircular quartz half-wave plate. Behind the analyzer, there is a glass tube having a larger diameter at its middle portion. The solution containing optically active substance is filled in this glass tube so that there are no air bubbles and the two ends of the tube are closed by transparent cover-slips. The air bubble, if any is arrested, is moved into the middle portion having larger diameter. A convex lens is placed in front of the polarizer N_1 so that a monochromatic source of light is at the focus of the lens. By this arrangement, the monochromatic light incident on the polarizer N_1 is made parallel. The emergent light from the analyzer N_2 is viewed through a short focus telescope.

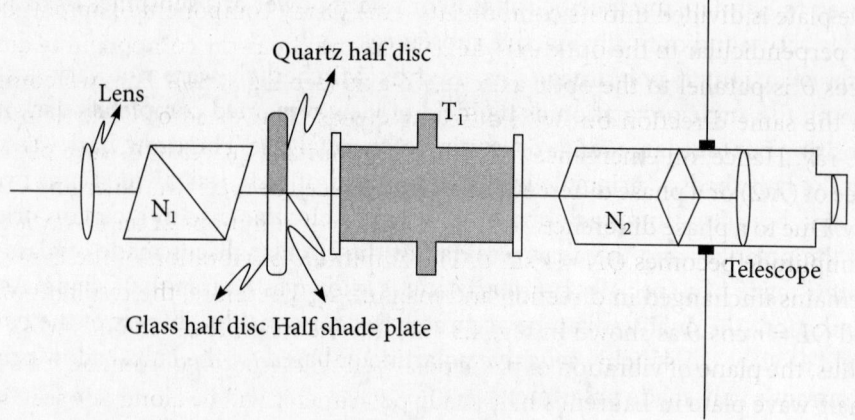

Figure 4.33 | Laurent's half-shade polarimeter/sacharimeter

Action of Laurent's half-shade disc

Let the plane of vibration of the plane polarized light incident normally on the half-shade disc be along PQ as shown in Fig. 4.34 making an angle θ with AC parallel to the optic axis.

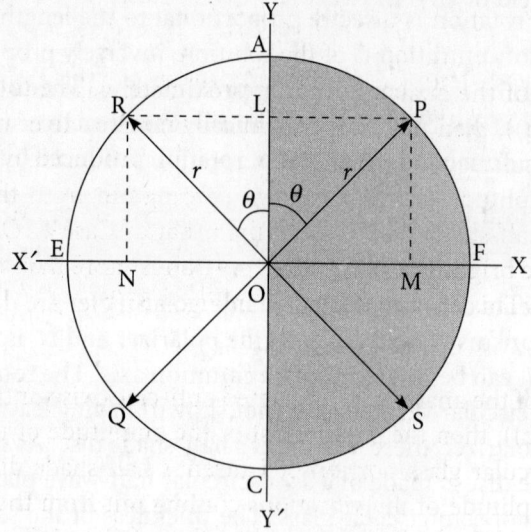

Figure 4.34 | Action of Laurent's half-shade disc

AC is the interface between the semi-circular quartz half-wave plate and the semicircular glass plate. The plane of vibrations of the incident plane polarized light passing through the glass plate remains unchanged. Hence, the vibrations of the plane polarized light emerging from the semi-circular glass disc are along the plane PQ as that of the incident light. However, the incident plane polarized light passing through the semi-circular quartz half-wave plate is divided into its components – the O-ray component of amplitude $OM = r \sin \theta$ is perpendicular to the optic axis, i.e., X-axis and the E-ray component of amplitude $OL = r \cos \theta$ is parallel to the optic axis, i.e., Y-axis (see Fig. 4.34). The two components travel in the same direction but with different speeds. The speed of O-ray is more than that of E-ray. Hence, on emergence from the semi-circular quartz half-wave plate, a path difference of $(\lambda/2)$ or a phase difference of π is introduced between the emerging O-ray and the E-ray. Due to a phase difference of π, the direction of vibrations of the O-ray is reversed and its amplitude becomes $ON = r \sin \theta$. The amplitude of vibration of the E-ray, $OL = r \cos \theta$, remains unchanged in direction and magnitude. Therefore, the resultant of $ON = r \sin \theta$ and $OL = r \cos \theta$ as shown in Fig. 4.34 is OR. The magnitudes of OR and OP are the same. Thus, the plane of vibration of plane polarized light emerging from the semi-circular quartz half-wave plate in Laurent's half-shade polarimeter will be along RS (see Fig. 4.34).

Case 1:

Suppose the principal plane of the analyzer N_2 is parallel to EOF of Laurent's half-shade disc. See Fig. 4.35(a). The component of the amplitude of the vibrations coming out from the semi-circular glass portion of Laurent's half-shade disc is $ON = r \sin \theta$ and the component of the amplitude of the vibrations coming out from the semi-circular quartz portion of Laurent's half-shade disc is $OM = r \sin \theta$. Thus, both these amplitudes are equal. Therefore, the two halves of the field of view have the same brightness.

Case 2:

If the principal plane of the analyzer N_2 is rotated clock-wise until OP is perpendicular to $E'F'$ (see Fig. 4.35(b)), then the component of the amplitude of the vibrations coming out from the semi-circular quartz portion of Laurent's half-shade disc is $OM = 0$ and the component of the amplitude of the vibrations coming out from the semi-circular glass portion of Laurent's half-shade disc is more than that of Case 1 ($ON' > ON$). Therefore, in the field of view, the brightness of the glass portion is more than that of Case 1 and the quartz portion is dark. This has been depicted in Fig. 4.35(b).

Case 3:

If the principal plane of the analyzer N_2 is rotated anti-clockwise until OR is perpendicular to $E'F'$ (see Fig. 4.35(c)), then the component of the amplitude of the vibrations coming out from the semi-circular glass portion of Laurent's half-shade disc is $ON = 0$ and the component of the amplitude of the vibrations coming out from the semi-circular quartz portion of Laurent's half-shade disc is more than that of Case 1. Therefore, in the field of view, the brightness of the quartz portion is more than that of Case 1 and the glass portion is dark. This has been depicted in Fig. 4.35(c).

If the principal plane of the analyzer N_2 is parallel to AOC of Laurent's half-shade disc, then Case 1 is repeated qualitatively and quantitatively. Case 2 is repeated qualitatively if the principal plane of the analyzer N_2 is parallel to ROS of Laurent's half-shade disc. Case 3 is repeated qualitatively if the principal plane of the analyzer N_2 is parallel to POQ of Laurent's half-shade disc (see Fig. 4.34).

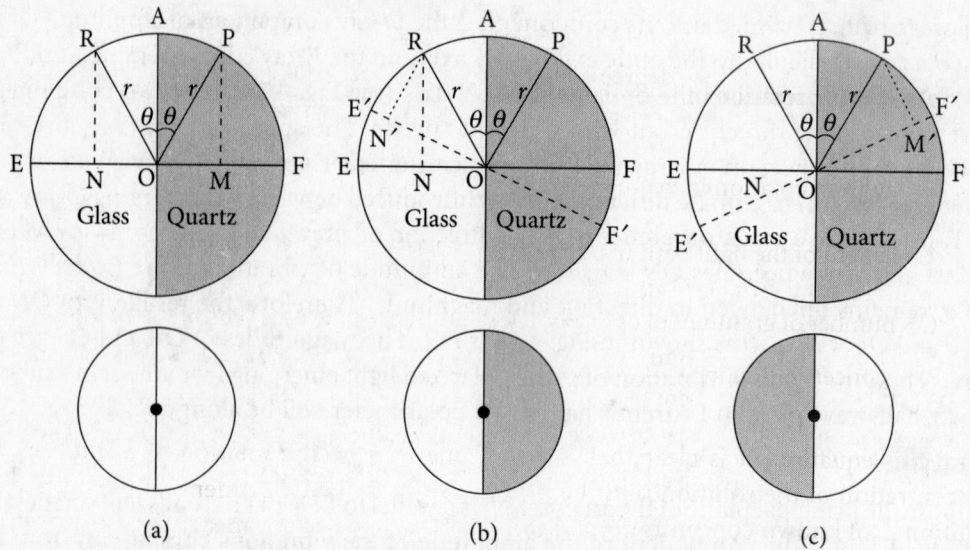

Figure 4.35 | Principle of Laurent's half-shade polarimeter/sacharimeter

Thus, the Laurent's half-shade device described here divides the field of view into two halves.

Accuracy

To increase the accuracy of the polarimeter, the half-shadow angle θ is fixed at the proper angle by rotating the polarizer N_1. The components of vibrations emerging from the semi-circular quartz half-wave plate and the semi-circular glass plate transmitted through the analyzer N_2 depend upon the angle θ on which the accuracy of the polarimeter depends on. The device is so accurate that if the principal plane of the analyzer is rotated through a fraction of a degree with respect to *EOF*, a remarkable change in intensity of the two halves of the field of view is noticed.

Limitation

Laurent's half-shade polarimeter or sacharimeter can be used only for that light source having wavelength for which the quartz half-shade disc behaves as a half-wave plate.

Determination of specific rotation

We know that the specific rotation ω from Eq. (4.35) is given by

$$\omega = \frac{10}{\ell} \times \frac{\theta}{C}$$

where

ω = specific rotation in $\dfrac{\text{degree} \times \text{cm}^2}{\text{gm}}$

θ = angle of rotation in degree

ℓ = length of the light path in centimetre

C = number of grammes of optically active substance per cubic centimetre of solution, i.e.,

concentration in $\dfrac{\text{gm}}{\text{cm}^3}$.

From this equation, it is clear that to determine the specific rotation of a solution, the concentration of the solution must be known priory to the experiment. The length of the solution ℓ of known concentration filled in the glass tube can be measured directly or can be taken from the working manual of the polarimeter. In the aforementioned equation, the only measurable quantity left unmeasured is θ, the angle of rotation. The angle of rotation θ can be measured in the following way.

The experimental setup is depicted in Fig. 4.33. First, the experimental tube is filled with clean water and placed in its proper position. The short focus telescope is focused on Laurent's half-shade disc. The analyzer N_2 is rotated so that the brightness of the two halves of the field of view due to the semi-circular glass disc and the semi-circular quartz disc become equal. At this position, the readings of the two circular verniers attached to the polarizer and the analyzer are noted. Now the glass tube containing pure water is taken out and water is drained out. Then the glass tube is dried up and filled with a optically active solution of known concentration and placed in its proper position without disturbing the polarizer and the analyzer. When the tube containing optically active solution of known concentration is placed, the planes of vibrations of the quartz half and the glass half are rotated. In dextro-rotary substances like cane sugar solution, PQ and RS are rotated clockwise and in levo-rotary substances like turpentine, PQ and RS are rotated anti-clockwise. Therefore, the two halves of the field of view now are not equally bright. To make the two halves of the field of view equally bright, the analyzer is rotated clockwise if the optically active solution is dextro-rotary; otherwise, it is rotated anti-clock wise. When the two halves of the field of view are equally bright, the readings of the two circular verniers attached to the polarizer and the analyzer are noted for the second time. The difference in the two readings of the same vernier gives the angle of rotation of the optically active solution of known concentration. The experiment is repeated by taking solutions of the same solute but of different concentrations. The angle of rotation is measured for each concentration.

As the rotary power of an optically active substance depends upon its properties, from Eq. (4.34), it is clear that the angle of rotation of an optically active substance is directly proportional to its concentration. Therefore, if we plot the angle of rotation versus concentration, we shall obtain a straight line passing through the origin as shown in Fig. 4.36. From the graph, taking the slope of the straight line, the value of the ratio θ/c is determined.

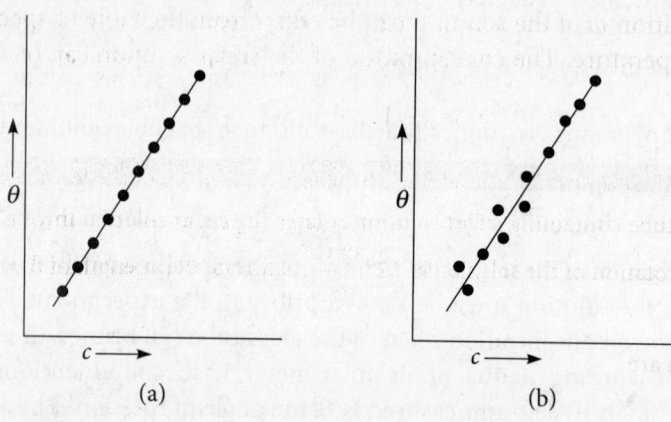

$$\text{(a)} \qquad\qquad \text{(b)}$$

Figure 4.36 | Calculation of $\dfrac{\theta}{c}$ to be put in Eq. (4.35) to determine the rotary power ω of a substance using Laurent's half-shade polarimeter. (a) is the theoretical plot and (b) is the experimental curve. Due to various reasons, all the experimental points may not lie in a straight line. The straight line has been drawn by the least-square method

Putting this value of the ratio θ/c in Eq. (4.35), the value of specific rotation or rotary power ω of an optically active substance is estimated.

Determination of concentration of sugar solution

From Eq. (4.35), we have

$$C = \frac{10\theta}{\ell\omega} \tag{4.36}$$

where

C = number of grammes of optically active substance per cubic centimetre of solution, i.e., concentration in $\dfrac{\text{gm}}{\text{cm}^3}$

θ = angle of rotation in degree

ℓ = length of the light path in centimetre

ω = rotary power or specific rotation in $\dfrac{\text{degree} \times \text{cm}^2}{\text{gm}}$.

Equation (4.36) is the working formula for the determination of strength or concentration of sugar solution.

The sugar solution is optically active with a dextro-rotary nature. To determine the concentration of a given sugar solution, the experiment is conducted in the same manner as described earlier. The length of the solution ℓ and the angle of rotation θ are measured. The specific rotation ω of the solution can be taken from the table of specific rotations at a particular temperature. The concentration of the sugar solution can be calculated using Eq. (4.36).

Example 4.15

A 20 cm long tube containing sugar solution rotates the sugar solution through an angle 21°. If the specific rotation of the solution is $62° \dfrac{cm^2}{gm}$, calculate the strength of the sugar solution.

Solution

The data given are

$$\theta = 21°$$

$$\ell = 20 \text{ cms}$$

$$\omega = 62° \frac{cm^2}{gm}$$

From Eq. (4.36), we get the expression for concentration C as

$$C = \frac{10\theta}{\ell\omega} = \frac{10 \times 21}{20 \times 62} = 0.17 \text{ gms/cm}^3 = 17\%$$

Example 4.16

A sugar solution of certain strength is filled in a tube of length 20 cm. Find the strength of the solution if the rotary power of the solution is $60° \dfrac{cm^2}{gm}$ and the optical rotation produced is 12°.

Solution

The data given are

$$\theta = 12°$$

$$\ell = 20 \text{cm}$$

$$\omega = 60° \frac{cm^2}{gm}$$

From Eq. (4.36), we get the expression for concentration C as

$$C = \frac{10\theta}{\ell\omega} = \frac{10 \times 12}{20 \times 60} = 0.10 \text{ gms/cm}^3 = 10\%$$

Example 4.17

A sugar solution was prepared by adding 100 grammes of cane sugar to one litre of water. By observation, it is found that this gave the solution an optical rotation of 12° when filled in a 20 cm tube. If the specific rotation of pure sugar is $66° \frac{cm^2}{gm}$, find the percentage of purity of the sugar.

Solution

The data given are

$$\theta = 12°$$

$$\ell = 20 \text{ cms}$$

$$\omega = 66° \frac{cm^2}{gm}$$

From Eq. (4.36) we get the expression for concentration C as

$$C = \frac{10\theta}{\ell\omega} = \frac{10 \times 12}{20 \times 66} = 0.091 \text{ gms/cm}^3 = 91 \text{gms/liter}$$

According to the question, the concentration of the sugar solution is 100 gm/litre. Hence, the sugar sample (100 gm) dissolved in one litre of water contains 91 gm of pure sugar. That is, 100 gm of sugar contains 91 gm of pure sugar. Hence, the percentage of purity of the sugar sample is 91%.

Questions

4.1   What is polarization?

4.2   Can radio waves be polarized? *X*-rays? Ultrasonic?

4.3   Explain the statement, 'Only transverse waves can be polarized'.

4.4   Describe a mechanical experiment to explain polarization.

4.5   Why does light appear perfectly symmetrical about the direction of propagation?

4.6   Distinguish between polarized and ordinary light.

4.7   What is light vector?

4.8   Design an optical experiment to demonstrate polarization of light wave.

4.9   Explain why it is not possible to polarize a longitudinal wave.

4.10 Distinguish between plane of polarization and plane of vibration.

4.11 What is the difference between linearly polarized light and plane polarized light?

4.12 Explain how light is polarized by reflection.

4.13 Can sound waves be polarized?

4.14 What is the plane of transmission?

4.15 What is the plain of incidence?

4.16 State and explain Brewster's law.

4.17 What is the angle of polarization and on what factors does the angle of polarization depend upon?

4.18 State and explain Malus' law.

4.19 Explain how light is polarized by refraction.

4.20 Mention various methods for production of polarized light.

4.21 What do you mean by double refraction?

4.22 Explain what happens when light passes through doubly refracting material.

4.23 Distinguish between positive doubly refracting crystal and negative doubly refracting crystal.

4.24 What is optic axis? What are its properties?

4.25 Show that when light is incident on a plane parallel plate of glass at the polarizing angle for the upper surface, the refracted ray strikes the lower surface at an angle of polarization for that surface.

4.26 What do you mean by the term degree of polarization?

4.27 Explain how light is polarized by scattering.

4.28 What are ordinary and extraordinary rays? Do they obey Snell's law?

4.29 What is the principal section of a crystal? What is its relation to the optic axis?

4.30 Distinguish between ordinary and extraordinary rays.

4.31 Explain how light is polarized by double refraction.

4.32 What are the postulates in Huygens' theory of double refraction?

4.33 Describe Huygens' experiment on polarization by double refraction.

4.34 Prove that planes of polarization of the O-ray and the E-ray are perpendicular to each other.

4.35 Under what condition(s) are the O-ray and the E-ray not separated out?

4.36 What are the three ways that polarized light can be produced other than by using a polarizing filter?

4.37 How can the optic axis of a calcite block be determined?

4.38 What should be the orientation of the optic axis in a negative uniaxial crystal so that the E-ray and the O-ray both obey both laws of refraction?

4.39 Explain with a diagram the nature of refracted O-ray and E-ray when the optic axis is inclined to the refracting surface of the crystal and lying in the plane of incidence.

4.40 Explain with a diagram the nature of refracted O-ray and E-ray when the optic axis is parallel to the refracting surface of the crystal and lying in the plane of incidence.

4.41 Explain with a diagram the nature of refracted O-ray and E-ray when the optic axis is perpendicular to the refracting surface of the crystal and lying in the plane of incidence.

4.42 Explain with a diagram the nature of refracted O-ray and E-ray when the optic axis is perpendicular to the refracting surface as well as to the plane of incidence.

4.43 Describe the construction and working of a Nicol prism.

4.44 Why should the angle of incidence on a Nicol prism be around 14°?

4.45 Explain the action of a Nicol prism.

4.46 Explain how Nicol prism produces plain polarized light.

4.47 Explain how Nicol prism can be used as a polarizer.

4.48 Explain how Nicol prism can be used as an analyzer.

4.49 What do you mean by parallel Nicol prisms?

4.50 Explain the action of parallel Nicols.

4.51 What do you mean by crossed Nicol prisms?

4.52 A student argued that if a polarizer is inserted in between two crossed Nicols, some transmission of light may occur. Explain how far the student is correct. How can adding a polarizer in between two crossed Nicols increase the transmission?

4.53 What is the function of Canada balsam in a Nicol prism?

4.54 Explain the action of a retardation plate.

4.55 Describe the action of a half-wave plate.

4.56 Derive an expression for the thickness of a half-wave plate.

4.57 Describe the action of a quarter-wave plate.

4.58 Derive an expression for the thickness of a quarter-wave plate.

4.59 Why is the half-wave plate so called?

4.60 What is a quarter-wave plate? Derive an expression for its thickness for a given wavelength in terms of its refractive index.

4.61   How many times is the length of a half-wave plate more than that of a quarter-wave plate for a particular monochromatic light.

4.62   Is it necessary to make a full-wave plate? Derive an expression for the thickness of a full-wave plate.

4.63   What are the functions of a half-wave plate and a quarter-wave plate?

4.64   Distinguish between a half-wave plate and a quarter-wave plate.

4.65   If you are given a half-wave plate, a quarter-wave plate and a glass plate, how will you distinguish between them?

4.66   What is a half-wave plate? Derive an expression for its thickness for a given wavelength in terms of its refractive index.

4.67   What do you mean by a half-wave plate and a quarter-wave plate? How will you distinguish between the two?

4.68   What is elliptically polarized light? How it is produced?

4.69   How you will distinguish between elliptically polarized light and a mixture of plain polarized light and unpolarized light?

4.70   What are(is) the condition(s) for production of elliptically polarized light?

4.71   What is circularly polarized light? How it is produced?

4.72   How can you know whether a given light is plain polarized light or circularly polarized light?

4.73   How can you know whether a given light is elliptically polarized or not?

4.74   A beam of right circularly polarized light is incident normally and get reflected. Is the reflected beam right or left polarized? Explain.

4.75   How you will distinguish between circularly polarized light and a mixture of plain polarized light and unpolarized light?

4.76   What are(is) the condition(s) for production of circularly polarized light?

4.77   How you will distinguish between circularly polarized light and elliptically polarized light?

4.78   How can you know whether a given light is partially plain polarized or not?

4.79   How can you know whether a given light is unpolarized or not?

4.80   How can you obtain circularly polarized light from unpolarized light?

4.81   How would you change a right-handed circularly polarized light into a left-handed circularly polarized light?

4.82   How would you change a left-handed circularly polarized light into a right-handed circularly polarized light?

4.83   How would you change a plane polarized light into a circularly polarized light?

4.84   How would you change a plane polarized light into an elliptically polarized light?

4.85   How would you change a circularly polarized light into a plane polarized light?

4.86   How would you change an elliptically polarized light into a plane polarized light?

4.87   What is optical rotation or optical activity?

4.88   What do you mean by optically active substance? Give examples

4.89   Distinguish between levo-rotary and dextro-rotary substances. Give at least two examples for each.

4.90   How does optical rotation affect the plane polarized light?

4.91   What is angle of rotation? On what factors does the angle of rotation of the plane of vibration depend?

4.92   What is rotary power or specific rotation of an optically active substance? On what factors does the specific rotation of an optically active substance depend?

4.93   What are the laws of optical rotation?

4.94   Apply Fresnel's theory of optical rotation to explain the behavior of optically active substances.

4.95   Prove mathematically that the speed of the right-handed circularly polarized wave and the speed of the left-handed circularly polarized wave are not the same in optically active substances.

4.96   What is specific rotation? Describe in detail the construction and working of Laurent's half-shade polarimeter. Explain how you would use it to determine the specific rotation of a sugar solution.

4.97   Explain why and how Laurent's half-shade polarimeter can be used to show that nature of a sugar solution. On what factors does the angle of rotation depend on?

4.98   What is/are limitation(s) of Laurent's half-shade polarimeter?

4.99   Explain how Laurent's half-shade polarimeter can be used to verify the purity of a liquid or of an optically active dissolvable substance.

Problems

4.1   What is the polarizing angle for a glass surface having refractive index 1.54. [Ans 57°]

4.2   The angle of polarization for diamond is 67.58°. What is the refractive index of diamond?
[Ans   2.424]

4.3   What is the Brewster's angle of incidence for a light on water with refractive index 1.33?
[Ans   53.1°]

4.4   The plane of vibration of an incident ray makes an angle of 30° with the optic axis. Compare the intensities of the ordinary ray and the extraordinary ray.   [Ans $I_E = 3I_0$]

4.5   Critical angle in a certain substance is 45°. What is the polarizing angle?   [Ans   54.74]

4.6   Calculate the range of the polarizing angles for white light incident on crown glass. Assume that for white light, the wavelength limits are 4000 Å to 7000 Å and the corresponding refractive indices are 1.5233 (blue) and 1.5146 (red) respectively.
[Ans   56.71° (blue) to 56.57° (red)]

4.7   A parallel beam of light is incident at an angle of 68° on a plane glass plate. The reflected beam is completely plain polarized. Determine the refractive index of the glass and the angle of refraction for this angle of incidence.   [Ans   2.48, 22°]

4.8   A parallel beam of plane polarized light strikes a calcite crystal in such a way that light vectors makes an angle 60° with the optic axis of the calcite crystal. What is the ratio of the amplitudes and intensities of the two refracted ray?   [Ans   1.732:1, 3:1]

4.9   A parallel beam of linearly polarized light of wavelength 5890 Å (in vacuum) is incident on a quartz crystal. Find the wavelengths of ordinary and extraordinary waves in the crystal. Given: for quartz $\mu_0 = 1.544$, $\mu_E = 1.553$.   [Ans   3815 Å, 3793 Å]

4.10   A parallel beam of linearly polarized light of wavelength 5890 Å (in vacuum) is incident on a tourmaline crystal. Find the wavelengths of ordinary and extraordinary waves in the crystal. Given: for tourmaline, $\mu_0 = 1.640$, $\mu_E = 1.620$.   [Ans   3591 Å, 3636 Å]

4.11   A parallel beam of linearly polarized light of wavelength 5890 Å (in vacuum) is incident on ice. Find the wavelengths of ordinary and extraordinary waves in the crystal. Given: for ice $\mu_0 = 1.360$, $\mu_E = 1.307$.   [Ans   4510 Å, 4506 Å]

4.12   The refractive index of a certain glass is 1.65. For what angle of incidence is light reflected from the surface of this glass completely polarized if the glass is immersed in water of refractive index 1.33?   [Ans   51.1°]

4.13   A polarizer and an analyzer are so oriented that the maximum amount of light is transmitted. To what fraction of its maximum value is the intensity of the transmitted light reduced when the analyzer is rotated through 35°, 45°, 60° and 90°.
          [Ans   0.75, 0.50, 0.25, 0]

4.14   What minimum thickness of calcite is needed to introduce a phase difference of 45°, 90°, and 180° between the emergent O-ray and E-ray when plane polarized light is incident normally on it?   [Ans 4.28×10^{-5} cm, 8.55×10^{-5} cm, 1.71×10^{-4} cm]

4.15   The refractive indices of quartz for sodium light are 1.544 and 1.533. Calculate the thickness of the quartz which will give the O-ray and the E-ray a path difference of 1.25 wavelengths.   [Ans 8.2×10^{-3} cm]

4.16   The refractive indices of quartz for a light of wavelength 5000 Å are 1.544 and 1.533. Calculate the thickness of a half-wave plate.   [Ans 2.78×10^{-3} cm]

4.17   The refractive indices of quartz for a light of wavelength 5890 Å are 1.55 and 1.54. Calculate the thickness of a quarter-wave plate.   [Ans 1.4725×10^{-3} cm]

4.18   A sugar solution was prepared by adding 80 grammes of cane sugar to one litre of water. By observation, it is found that it gave an optical rotation of 9.9° when filled into a 20 cm tube. If the specific rotation of pure sugar is $66° \dfrac{\text{cm}^2}{\text{gm}}$, find the percentage of purity of the sugar.   [Ans   ≈ 94%]

4.19   An optically active solution of rotary power $55° \dfrac{\text{cm}^2}{\text{gm}}$ causes an optical rotation of 12° in a tube of 11 cm long. What is the concentration of the solution? [Ans   0.198 gm/cc]

4.20   How will you orient the polarizer and analyzer so that the intensity of a ray of sunlight is reduced to (i) 0.5, (ii) 0.25, (iii) 0.75, (iv) 0.125 of its original intensity.

[Ans   45°, 60°, 30°, and 69°]

4.21   Calculate the specific rotation/rotary power of a 20% sugar solution contained in a 20 cm length tube if the plane of polarization of plane polarized light is rotated by an angle of 26.4°.

[Ans $66° \dfrac{cm^2}{gm}$]

Multiple Choice Questions

1.   Which of the following phenomena proves the transverse nature of light?

(i)   Interference   (ii)   Refraction

(iii)   Diffraction   (iv)   Polarization

2.   What is plane of incidence?

(i)   The plane containing the incidence ray and the normal to the surface at the point of incidence

(ii)   The plane containing the light vector and the normal to the plane of incidence at the point of incidence

(iii)   The plane containing the light vector and passing through the direction of propagation

(iv)   The plane on which incidence occurs

3.   What is plane of reflection?

(i)   The plane containing the incident ray and the refracted ray

(ii)   The plane containing the incidence ray and the reflected ray

(iii)   The plane containing the reflected ray and the normal to the surface at the point of incidence

(iv)   The plane on which reflection occurs

4.   What is plane of refraction?

(i)   The plane containing the incident ray and the refracted ray

(ii)   The plane containing the incidence ray and the reflected ray

(iii)   The plane containing the refracted ray and the normal to the plane of incidence at the point of incidence

(iv)   The plane on which refraction occurs

5.   What is light vectors?

(i)   The vector which have small values

(ii)   The vector which represent light

(iii)   The magnetic vector of the light wave transverse to the direction of propagation

(iv)   The electric vector of the light wave transverse to the direction of propagation

6. What is plane of vibration?

    (i)   The plane passing through the direction of propagation containing light vectors

    (ii) The plane passing through the direction of propagation containing electric vectors

    (iii) The plane passing through the direction of propagation containing magnetic vectors

    (iv) The plain on which vibration occurs

7. The orientation of the plane of vibration of polarized light does not change.

    (i)   True   (ii) False

8. What is polarization of a wave?

    (i)   Confinement of vibrations of wave into a single plane

    (ii) Confinement of vibrations of wave into a small space

    (iii) Confinement of vibrations of wave into a single line

    (iv) Combination of vibrations of wave into one vibration

9. Polarization is the characteristic of

    (i)   Only longitudinal wave

    (ii) Only transverse wave

    (iii) Both transverse and longitudinal waves

    (iv) Only light wave

10. What is plane of polarization?

    (i)   The plane passing through the direction of propagation containing light vectors

    (ii) The plane passing through the direction of propagation containing electric vectors

    (iii) The plane passing through the direction of propagation containing magnetic vectors

    (iv) The plain on which vibration occurs

11. The plane of polarization is perpendicular to

    (i)   The plane of vibration

    (ii) Direction of propagation

    (iii) Electric vectors of the light wave

    (iv) Magnetic vectors of the light wave

12. The plane of vibration and the plane of polarization are perpendicular to each other

    (i)   True   (ii) False

13. What is angle of polarization?

    (i)   The angle of incidence at which light is slightly polarized

    (ii) The angle of incidence at which light is partially polarized

(iii) The angle of incidence at which light is mostly polarized

(iv) The angle of incidence at which light cannot be polarized

14. The angle of polarization depends upon

(i) angle of incidence   (ii) angle of reflection

(iii) nature of the interface   (iv) wavelength of the light

15. At angle of polarization, the angle of reflection is not equal to the angle of incidence.

(i) True   (ii) False

16. What is the value of the sum of the angle of polarization and the angle of refraction

(i) $\dfrac{\pi}{4}$ (ii) $\dfrac{\pi}{3}$

(iii) $\dfrac{\pi}{2}$ (iv) $\dfrac{\pi}{1}$ or π

17. The refractive index of a certain glass is 1.5. What is the polarizing angle for this glass surface?

(i) 55°   (ii) 56°

(iii) 57°   (iv) 58°

18. What is the chemical formula of calcite crystal?

(i) $CuCO_3$ (ii) $CaCO_3$

(iii) $CoCO_3$ (iv) $CoCO_3$

19. A calcite crystal cube has how many principal sections?

(i) 1   (ii) 2

(iii) 3   (iv) 4

20. The two beams of light produced due to double refraction do not obey laws of refraction

(i) True   (ii) False

21. Which of the following is incorrect

(i) O-ray obeys the laws of refraction

(ii) E-ray does not obey the laws of refraction

(iii) Separation between the O-ray and the E-ray does not depend upon the thickness of the doubly refracting crystal

(iv) O-ray and E-ray emerges from the calcite crystal parallel to each other

(v) Speed of O-ray inside a doubly refracting crystal is same in all directions

(vi) Speed of E-ray inside a doubly refracting crystal is same in all directions

(vii) In a doubly refracting negative crystal, the speed of E-ray is more than that of O-ray

(viii) In a doubly refracting positive crystal, the speed of E-ray is less than that of O-ray

(ix) Wavefront of E-ray is spherical

(x) Wavefront of O-ray is spherical

22. The optic axis is a line
    (i)   True   (ii) False

23. An isotropic substance is defined as a substance having
    (i)   the same properties at every point in the substance
    (ii) the same properties along every direction in the substance
    (iii) different properties at every point in the substance
    (iv) different properties along every direction in the substance

24. In a positive crystal, the speed of O-ray is greater than E-ray
    (i)   True   (ii) False

25. In a negative crystal, the speed of O-ray is greater than E-ray
    (i)   True   (ii) False

26. The speed of O-ray and E-ray are the same along the optic axis in positive and negative crystals
    (i)   True   (ii) False

27. The orientation between the optic axis and the crystallographic axis is
    (i)   Optic axis and crystallographic axis are parallel to each other
    (ii) Optic axis and crystallographic axis are parallel to each other
    (iii) Optic axis and crystallographic axis makes 71° with each other
    (iv) They can have any orientation with each other

28. The orientation between the plane of polarization of O-ray and the plane of polarization of E ray is
    (i)   Plane of polarization of O-ray and plane of polarization of E-ray is parallel to each other
    (ii) Plane of polarization of O-ray and plane of polarization of E-ray is perpendicular to each other
    (iii) Plane of polarization of O-ray makes an angle 109° with plane of polarization of E-ray
    (iv) They can have any orientation with each other

29. Which of the following is correct?
    (i)   Characteristics of O-ray and E-ray are the same
    (ii) Characteristics of ordinary light and O-ray are the same
    (iii) Characteristics of ordinary light and E-ray are the same
    (iv) None of the above

30. Which of the following is correct?
    (i)   Refractive indices of a doubly refracting crystal for O-ray and E-ray are the same in all directions inside the crystal
    (ii) Refractive indices of a doubly refracting crystal for O-ray and E-ray are the same in a particular direction inside the crystal

(iii) Refractive index of a doubly refracting crystal for E-ray is the same in all direction but is not the same for O-ray in all directions inside the crystal

(iv) None of the above

31. In a Nicol prism, μ_E and μ_0 are the refractive indices for E-ray and O-ray respectively. If μ_B is the refractive index of the Canada balsam which of the following is correct

(i) $\mu_0 < \mu_B < \mu_E$ (ii) $\mu_B < \mu_0 < \mu_E$

(iii) $\mu_0 < \mu_E < \mu_B$ (iv) $\mu_E < \mu_B < \mu_0$

32. Which of the following is correct

(i) Canada balsam is the rarer medium for E-ray and denser medium for O-ray

(ii) Canada balsam is the rarer medium for E-ray and rarer medium for O-ray

(iii) Canada balsam is the denser medium for E-ray and rarer medium for O-ray

(iv) Canada balsam is the denser medium for E-ray and denser medium for O-ray

33. Which ray emerges out from a Nicol prism?

(i) O-ray   (ii) E-ray

(iii) Unpolarized light ray   (iv) Both O-ray and E-ray

34. Which ray is absorbed by the blackened side wall in a Nicol prism?

(i) O-ray   (ii) E-ray

(iii) Unpolarized light ray   (iv) Both O-ray and E-ray

35. No light emerges from crossed Nicol prisms because

(i) In crossed Nicols, the principal sections of the polarizer and the analyzer are parallel to each other

(ii) In crossed Nicols, the principal sections of the polarizer and the analyzer are perpendicular to each other

(iii) In crossed Nicols, the polarizer does not allow any light to reach the analyzer

(iv) In crossed Nicols, light cannot enter into the polarizer

36. Which of the following equations is/are correct for a half-wave plate?

(i) $\delta = \dfrac{\pi}{2}$ (ii) $\delta = \pi$

(iii) $t = \dfrac{\lambda}{2(\mu_0 - \mu_E)}$ (iv) $t = \dfrac{\lambda}{4(\mu_0 - \mu_E)}$

37. Which of the following equations is/are correct for a quarter-wave plate?

(i) $\delta = \dfrac{\pi}{2}$ (ii) $\delta = \pi$

(iii) $t = \dfrac{\lambda}{2(\mu_0 - \mu_E)}$ (iv) $t = \dfrac{\lambda}{4(\mu_0 - \mu_E)}$

38. Light is incident on a rotational Nicol prism and the intensity of the emergent light beam becomes maximum and zero periodically. We can conclude that the incident light is

(i) circularly polarized   (ii) elliptically polarized

(iii) unpolarized   (iv) linearly/plane polarized

39. Light is incident on a rotating Nicol prism and the intensity of the emergent light beam remains unchanged. We can conclude that the incident light is

(i) Either circularly polarized or unpolarized

(ii) Either elliptically polarized or unpolarized

(iii) Unpolarized

(iv) Linearly polarized or un polarized

40. Light is incident on a rotating Nicol prism and the intensity of the emergent light beam becomes maximum and minimum but not zero. We can conclude that the incident light is

(i) Either circularly polarized or partially polarized

(ii) Either elliptically polarized or partially polarized

(iii) Unpolarized

(iv) Linearly polarized or un polarized

41. The light passing through a quarter-wave plate is analyzed by a rotating Nicol prism. It is found that the intensity of the emergent light beam from the rotating Nicol prism becomes maximum and minimum but not zero. We can conclude that the incident light is

(i) circularly polarized   (ii) elliptically polarized

(iii) unpolarized   (iv) linearly polarized

42. The light passing through a quarter-wave plate is analyzed by a rotating Nicol prism. It is found that the intensity of the emergent light beam from the rotating Nicol prism does not vary. We can conclude that incident light is

(i) circularly polarized   (ii) elliptically polarized

(iii) unpolarized   (iv) linearly polarized

43. The light passing through a quarter-wave plate is analyzed by a rotating Nicol prism. It is found that the intensity of the emergent light beam from the rotating Nicol prism becomes maximum and minimum equal to zero. We can conclude that the incident light is

(i) circularly polarized   (ii) elliptically polarized

(iii) unpolarized   (iv) linearly polarized

44. The light passing through a quarter-wave plate is analyzed by a rotating Nicol prism. It is found that the intensity of the emergent light beam from the rotating Nicol prism becomes maximum and minimum not equal to zero. We can conclude that the incident light is

    (i) circularly polarized
    (ii) elliptically polarized
    (iii) partially polarized
    (iv) linearly polarized

45. Optically active substances are those substances which produce

    (i) double refraction phenomenon
    (ii) optical rotation phenomenon
    (iii) double refraction and optical rotation phenomena
    (iv) optical communication action

46. An optically active substance is dextro-rotary if

    (i) as seen by an observer looking in the direction of propagation of light, the plane of vibration is rotated clockwise
    (ii) as seen by an observer looking against the direction of propagation of light, the plane of vibration is rotated clockwise
    (iii) as seen by an observer looking in the direction of propagation of light, the plane of vibration is rotated anti-clockwise
    (iv) as seen by an observer looking against the direction of propagation of light, the plane of vibration is rotated anti-clockwise

47. An optically active substance is levo-rotary if

    (i) as seen by an observer looking in the direction of propagation of light, the plane of vibration is rotated clock wise
    (ii) as seen by an observer looking against the direction of propagation of light, the plane of vibration is rotated clockwise
    (iii) as seen by an observer looking in the direction of propagation of light, the plane of vibration is rotated anti-clockwise
    (iv) as seen by an observer looking against the direction of propagation of light, the plane of vibration is rotated anti clockwise

48. The angle of rotation inside an optically active substance is least for

    (i) Blue light
    (ii) Red light
    (iii) Green light
    (iv) Violet light

49. The angle of rotation inside an optically active substance is greatest for

    (i) Red light
    (ii) Blue light
    (iii) Green light
    (iv) Violet light

50. If θ is the angle of rotation of the plane of vibration, t is the length of the optically active solution, C is the concentration of an optically active substance in the solution and λ is the wave length of the light employed, then which of the following is/are incorrect

(i) $\theta \propto t$

(ii) $\theta \propto C$

(iii) $\theta \propto \dfrac{1}{\lambda}$

(iv) None of the above

51. The solution containing equal number of levo-rotary and dextro-rotary molecules is

(i) levo-rotary

(ii) dextro-rotary

(iii) optically active

(iv) optically inactive

52. The specific rotation ω is defined as $\omega = \dfrac{10\theta}{\ell C}$. The unit of specific rotation is

(i) $\dfrac{\text{degree} \times \text{meter}^2}{\text{kg}}$

(ii) $\dfrac{\text{radian} \times \text{meter}^2}{\text{kg}}$

(iii) $\dfrac{\text{degree} \times \text{cm}^2}{\text{gram}}$

(iv) $\dfrac{\text{radian} \times \text{cm}^2}{\text{gram}}$

53. The graph plotted between the angle of the rotation of plane of vibration and concentration of sugar solution is

(i) Linear

(ii) Elliptical

(iii) Hyperbolic

(iv) Parabolic

54. The device which measures the percentage of cane sugar in a solution is called

(i) Polarimeter

(ii) Sacharimeter

(iii) Barometer

(iv) Manometer

55. Polarimeter consists of

(i) One Nicols prism

(ii) Two Nicols prisms

(iii) Three Nicols prisms

(iv) Four Nicols prisms

Answers

1 (iv)	2 (i)	3 (iii)	4 (iii)	5 (iv)	6 (ii)	7 (ii)	8 (i)
9 (ii)	10 (iii)	11 (i & iii)	12 (i)	13 (iii)	14 (iii & iv)	15 (ii)	16 (iii)
17 (ii)	18 (ii)	19 (iii)	20 (i)	21 (iii, vi & ix)	22 (ii)	23 (ii)	24 (i)
25 (ii)	26 (i)	27 (i)	28 (ii)	29 (iv)	30 (ii)	31 (iv)	32 (iii)
33 (ii)	34 (i)	35 (ii)	36 (ii & iii)	37 (i & iv)	38 (iv)	39 (ii)	40 (ii)
41 (i)	42 (iii)	43 (ii)	44 (iii)	45 (ii)	46 (ii)	47 (iv)	48 (ii)
49 (iv)	50 (iii)	51 (iv)	52 (iii)	53 (i)	54 (ii)	55 (ii)	

5 Electromagnetism

5.1 Introduction

The phenomenon of electromagnetism in any medium is completely described by a set of four first order partial differential equations called Maxwell's equations. Maxwell's equations are the relationships between electric and magnetic fields in the presence of electric charges and currents, whether steady or rapidly fluctuating, in vacuum or in matter. The equations represent one of the most elegant and concise way to describe the fundamentals of electromagnetism. Maxwell's equations are a combination of the works of Gauss, Faraday, Ampère, Biot, Savart, and others. Remarkably, Maxwell's equations are perfectly consistent with the transformation equations of the special theory of relativity. To be more exact, these equations constitute a complete description of the behavior of electric and magnetic fields separately or jointly in any medium.

5.2 Vector Calculus

In vector calculus, the spatial derivatives of one types of vector and scalar fields give other types of vector or scalar fields. Depending upon the requirement, the first order differential operator (see Eq. 5.2) may be applied to a scalar function to obtain a vector function or vice versa. It may also give rise to one type of vector field from another type of vector field! The concept of vector calculus was fully exploited by James Clerk Maxwell in a simple way in discovering the missing link between the electric field and the magnetic field, thus establishing the electromagnetic nature of light. Therefore, for complete appreciation of complexities of electromagnetism, at least a brief explanation of vector calculus would be highly beneficial.

The *field*, in vector calculus, is defined as a region within which every physical quantity can be expressed as a continuous function of the position of a point in the region. The corresponding function is called a point function. Broadly speaking, fields are of two types – *scalar fields* and *vector fields*. All the quantities in a scalar field are scalars and all the quantities in a vector field are vectors. All quantities in both scalar and vector fields are functions of positions and times. The vectors in vector fields and the scalars in scalar fields may change with respect to positions and times. First, we shall discuss what are line integrals, surface integrals and volume integrals.

5.2.1 Line integrals

The integration of a vector along a curve in a vector field is called a *line integral*. Let C be a curve in a vector field connecting any two points situated inside the vector field and $\vec{F}(x, y, z)$ a vector function in the vector field. The line integral of $\vec{F}(x, y, z)$ along the path C is given by $\int_C \vec{F}.\overline{d\ell}$. If the path is a closed path, then the line integral of $\vec{F}(x, y, z)$ along the closed path is written as $\oint_C \vec{F}.\overline{d\ell}$. Here, $\overline{d\ell}$ is an elemental vector length of the path. The integral $\oint_C \vec{F}.\overline{d\ell}$ is also called the *circulation* of $\vec{F}(x, y, z)$ around the closed path C.

Physical significance of a line integral

If $\vec{F}(x, y, z)$ is a force and $\overline{d\ell}$ is the elemental vector length or small displacement along the path of a particle, then line integral $\int_A^B \vec{F}.\overline{d\ell}$ is the work done in displacing the particle from A to B. Again if $\vec{F}(x, y, z)$ is the electric field and $\overline{d\ell}$ is the elemental vector length or small displacement along the path connecting two points A and B, then the line integral $\int_A^B \vec{F}.\overline{d\ell}$ is the electric potential difference between A and B.

Example 5.1

Calculate the circulation of $\vec{F} = \hat{x}zy + \hat{y}x^2 - \hat{z}z^2$ around the closed path shown in the unit cube illustrated in Fig. 5.1.

Solution

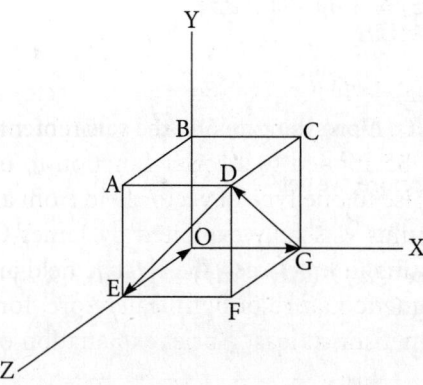

Figure 5.1 | Evaluation of line integral of a vector function as defined in Example 1.1

The line integral of $\vec{F}$ around the closed path EOGDE is given by

$$\oint_{EOGDE} \vec{F} \cdot \vec{d\ell} = \int_{EO} \vec{F} \cdot \vec{d\ell} + \int_{OG} \vec{F} \cdot \vec{d\ell} + \int_{GD} \vec{F} \cdot \vec{d\ell} + \int_{DE} \vec{F} \cdot \vec{d\ell} \qquad \text{(A)}$$

In general, $\vec{F} = \hat{x}zy + \hat{y}x^2 - \hat{z}z^2$ and $\vec{d\ell} = \hat{x}dx + \hat{y}dy + \hat{z}dz$

i.   Evaluation of $\int_{EO} \vec{F} \cdot \vec{d\ell}$: Along the path EO, the values of $x = y = 0$ and hence, $\vec{d\ell} = \hat{z}dz$. Therefore, we get

$$\int_{EO} \vec{F} \cdot \vec{d\ell} = \int_{EO} (\hat{x}zy + \hat{y}x^2 - \hat{z}z^2) \cdot (\hat{z}dz) = \int_{1}^{0} (-z^2)dz = \frac{1}{3}$$

ii.   Evaluation of $\int_{OG} \vec{F} \cdot \vec{d\ell}$: Along the path OG, the values of $y = z = 0$ and hence $\vec{d\ell} = \hat{x}dx$. Therefore, we get

$$\int_{OG} \vec{F} \cdot \vec{d\ell} = \int_{OG} (\hat{x}zy + \hat{y}x^2 - \hat{z}z^2) \cdot (\hat{x}dx) = \int_{OG} yzdx = \int_{0}^{1} 0 \times 0 \, dx = 0$$

iii. Evaluation of $\int_{GD} \vec{F} \cdot \vec{d\ell}$: Along the path GD, the values of $y = z$ and $x = 1$; hence, $dy = dz$ and $\vec{d\ell} = \hat{y}dy + \hat{z}dz$. Therefore, we get

$$\int_{GD} \vec{F} \cdot \vec{d\ell} = \int_{GD} (\hat{x}zy + \hat{y}x^2 - \hat{z}z^2) \cdot (\hat{y}dy + \hat{z}dz) = \int_{GD} (x^2 dy - z^2 dz)$$

$$= \int_{GD} (dy - z^2 dz) = \int_{0}^{1} dy - \int_{0}^{1} z^2 dz = \frac{2}{3}$$

iv.   Evaluation of $\int_{DE} \vec{F} \cdot \vec{d\ell}$: Along the path DE, the values of $x = y$ and $z = 1$; $dx = dy$ and $\vec{d\ell} = \hat{x}dx + \hat{y}dy$. Therefore, we get

$$\int_{DE} \vec{F} \cdot \vec{d\ell} = \int_{DE} (\hat{x}zy + \hat{y}x^2 - \hat{z}z^2) \cdot (\hat{x}dx + \hat{y}dy) = \int_{DE} (zydx + x^2 dy)$$

$$= \int_{DE} (ydx + x^2 dy) = \int_{DE} (xdx + x^2 dx) = \int_{1}^{0} (xdx + x^2 dx) = -\frac{5}{6}$$

Putting the values of $\int_{EO} \vec{F} \cdot \overline{d\ell}$, $\int_{OG} \vec{F} \cdot \overline{d\ell}$, $\int_{GD} \vec{F} \cdot \overline{d\ell}$ and $\int_{DE} \vec{F} \cdot \overline{d\ell}$ into Eq. (A) we get

$$\oint_{EOGDE} \vec{F} \cdot \overline{d\ell} = \frac{1}{3} + 0 + \frac{2}{3} + \left(-\frac{5}{6}\right) = \frac{1}{6}$$

5.2.2 Surface integrals

Let us consider the vector field $\vec{F}$ defined by $\vec{F} = \vec{F}(x, y, z)$. Draw a surface S in this field. Let $\overrightarrow{ds}$ be an elemental vector area on this surface. $\hat{n}$ is a unit vector perpendicular to $\overrightarrow{ds}$ in the outward direction (this is the direction of the elemental vector area $\overrightarrow{ds}$). The surface integral of $\vec{F}$ over the surface S is $\int_S \vec{F} \cdot \hat{n} ds$. If the surface is a closed surface, then the surface integral of $\vec{F}$ over the surface S is written as $\oint_S \vec{F} \cdot \hat{n} ds$. The surface integral of a vector field gives the total amount of flux passing through the surface S.

Physical significance of surface integral

Let $\vec{F}$ be the velocity vector of a moving fluid at any point in the fluid, and S a fixed surface imagined inside the fluid. The surface integral $\oint_S \vec{F} \cdot \hat{n} ds$ is the total amount of fluid or fluid flux passing through the surface S in unit time. If $\vec{F}$ is the magnetic induction vector and S the surface through which magnetic flux passes, then the surface integral $\oint_S \vec{F} \cdot \hat{n} ds$ is the total amount of magnetic flux passing through the surface S.

Example 5.2

Evaluate $\oint_S \vec{F} \cdot \overrightarrow{ds}$, where $\vec{F} = \hat{x} 4xz - \hat{y} y^2 + \hat{z} yz$, S is the surface of the cube bounded by $x = 0$, $x = 1$; $y = 0$, $y = 1$; $z = 0$, $z = 1$ planes.

Solution

For the surfaces of the cube, we have

$$\oint_S \vec{F} \cdot \overrightarrow{ds} = \oint_S \vec{F} \cdot \hat{n} ds = \iint_{ABCD} \vec{F} \cdot \hat{z} dxdy + \iint_{DEOA} \vec{F} \cdot \hat{y} ds + \iint_{AOGB} \vec{F} \cdot \hat{x} ds$$

$$+ \iint_{BCFG} \vec{F} \cdot \hat{y} ds + \iint_{GOEF} \vec{F} \cdot \hat{z} ds + \iint_{FEDC} \vec{F} \cdot \hat{n} ds \tag{A}$$

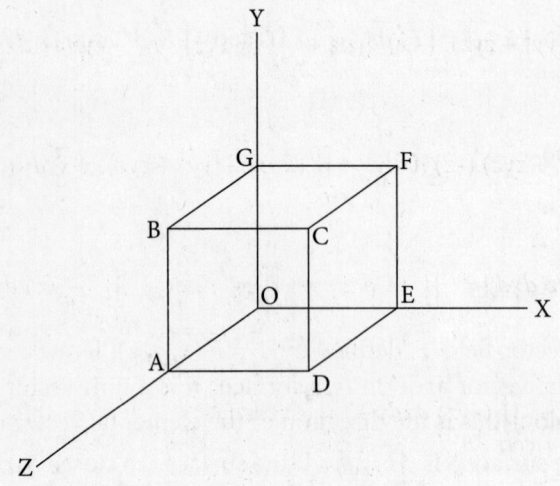

Figure 5.2 | Evaluation of surface integral of a vector function over the surface of a unit cube

As shown in the figure, we have

i.   For the ABCD surface, $ds = dxdy$, $\hat{n} = \hat{z}$ and $z = 1$

ii. For the DEOA surface $ds = dzdx$, $\hat{n} = -\hat{y}$ and $y = 0$

iii. For the AOGB surface $ds = dydz$, $\hat{n} = -\hat{x}$ and $x = 0$

iv. For the BCFG surface $ds = dzdx$, $\hat{n} = \hat{y}$ and $y = 1$

v.   For the GOEF surface $ds = dxdy$, $\hat{n} = -\hat{z}$ and $z = 0$

vi. For the FEDC surface $ds = dydz$, $\hat{n} = \hat{x}$ and $x = 1$

Putting all these values into Eq. (A), we have

$$\vec{F} = \hat{x}4xz - \hat{y}y^2 + \hat{z}yz$$

$$\oint_S \vec{F} \cdot \vec{ds} = \oint_S \vec{F} . \hat{n} ds = \iint_{ABCD} \vec{F} \cdot \hat{n} ds + \iint_{DEOA} \vec{F} \cdot \hat{n} ds + \iint_{AOGB} \vec{F} \cdot \hat{n} ds + \iint_{BCFG} \vec{F} \cdot \hat{n} ds$$

$$+ \iint_{GOEF} \vec{F} \cdot \hat{n} ds + \iint_{FEDC} \vec{F} \cdot \hat{n} ds$$

or $$\oint_S \vec{F} . \hat{n} ds = \iint_{ABCD} (\hat{x}4xz - \hat{y}y^2 + \hat{z}yz) . \hat{z} \, dxdy + \iint_{DEOA} (\hat{x}4xz - \hat{y}y^2 + \hat{z}yz) . (-\hat{y}) \, dxdz$$

$$+ \iint\limits_{AOGB} (\hat{x}4xz - \hat{y}y^2 + \hat{z}yz).\,(-\hat{x})\,dydz + \iint\limits_{BCFG} (\hat{x}4xz - \hat{y}y^2 + \hat{z}yz).\hat{y}\,dxdz +$$

$$\iint\limits_{GOEF} (\hat{x}4xz - \hat{y}y^2 + \hat{z}yz).(-\hat{z})\,dxdy + \iint\limits_{FEDC} (\hat{x}4xz - \hat{y}y^2 + \hat{z}yz).\,\hat{x}\,dzdy$$

or $\oint\limits_{S} \vec{F}.\hat{n}ds = \iint\limits_{ABCD} yz\,dxdy = \iint\limits_{ABCD} yz\,dxdy + \iint\limits_{DEOA} y^2\,dxdz + \iint\limits_{AOGB} (-4xz)\,dydz +$

$$\iint\limits_{BCFG} (-y^2)\,dxdz + \iint\limits_{GOEF} (-yz)\,dxdy + \iint\limits_{FEDC} 4xzdzdy$$

Putting the values of x, y and z for all the six surfaces into this equation, we have

$$\oint\limits_{S} \vec{F}.\hat{n}ds = \iint\limits_{ABCD} y\times 1\,dxdy + \iint\limits_{DEOA} 0^2\,dxdz + \iint\limits_{AOGB} (-4\times 0\times z)\,dydz + \iint\limits_{BCFG} (-1^2)\,dxdz +$$

$$\iint\limits_{GOEF} (-y\times 0)\,dxdy + \iint\limits_{FEDC} 4\times 1\times zdzdy$$

$$= \iint\limits_{ABCD} ydxdy + 0 + 0 - \iint\limits_{BCFG} dxdz + 0 + 4\iint\limits_{FEDC} zdzdy$$

$$= \frac{1}{2} + 0 + 0 - 1 + 0 + 4\times\frac{1}{2} = \frac{3}{2}$$

or $\oint\limits_{S} \vec{F}.\hat{n}ds = \dfrac{3}{2}$

5.2.3 Volume integral

If we consider a closed surface in space enclosing a volume V, then $\oint\limits_{V} \vec{F}dv$ is defined as the volume integration. In the Cartesian coordinate system,

$$\oint\limits_{V} \vec{F}dv = \int\limits_{X}\int\limits_{Y}\int\limits_{Z} \vec{F}dxdydz = \int\limits_{X}\int\limits_{Y}\int\limits_{Z} (\hat{x}\,F_X + \hat{y}\,F_Y + \hat{z}\,F_Z)\,dx\,dy\,dz$$

or $\quad \displaystyle\oint_V \vec{F}dv = \hat{x}\iiint_{XYZ}F_X dxdydz + \hat{y}\iiint_{XYZ}F_Y dxdydz + \hat{z}\iiint_{XYZ}F_Z dxdydz$

Here $\hat{x}$, $\hat{y}$, and $\hat{z}$ are the unit vectors along the +X, +Y, and +Z axes, $\vec{F} = \hat{x}F_X + \hat{y}F_Y + \hat{z}F_Z$ and F_X, F_Y, and F_Z are the X, Y, and Z components of $\vec{F}$.

Example 5.3

Evaluate $\displaystyle\int_V \vec{F}dv$, where $\vec{F} = \hat{x}2zy - \hat{y}z + \hat{z}x^2$, V is the volume bounded by the surfaces $z = 6$, $x = 6$, $y = z^2$, and $y = 4$.

Solution

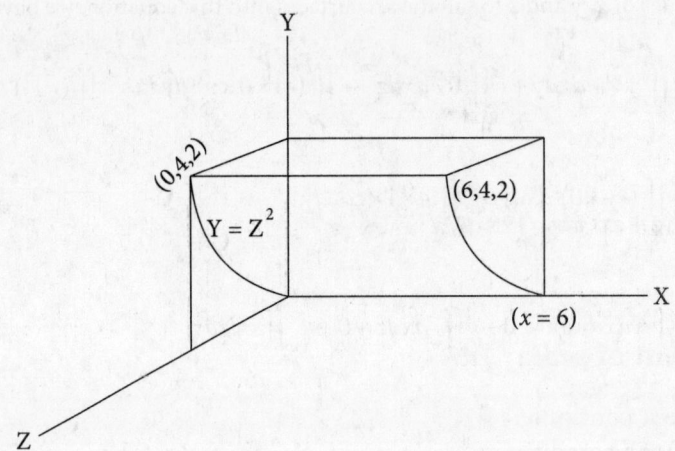

Figure 5.3 | Evaluation of volume integral of a vector function over the given volume

$$\int_V \vec{F}dv = \iiint_{xyz}(\hat{x}2yz - \hat{y}z + \hat{z}x^2)dxdydz$$

$$= \hat{x}\int_0^6\int_{z^2}^4\int_0^2 2yzdxdydz - \hat{y}\int_0^6\int_{z^2}^4\int_0^2 zdxdydz + \hat{z}\int_0^6\int_{z^2}^4\int_0^2 x^2dxdydz$$

$$= \hat{x}2\int_0^6\int_0^2 \left.\frac{y^2}{2}\right|_{z^2}^4 zdxdz - \hat{y}\int_0^6\int_0^2 \left.y\right|_{z^2}^4 zdxdz + \hat{z}\int_0^6\int_0^2 \left.y\right|_{z^2}^4 x^2dxdz$$

$$= \hat{x}\int_0^6\int_0^2(16-z^4)zdxdz - \hat{y}\int_0^6\int_0^2(4-z^2)zdxdz + \hat{z}\int_0^6\int_0^2(4-z^2)x^2dxdz$$

$$= \hat{x}\int_0^6\int_0^2(16z-z^5)dxdz - \hat{y}\int_0^6\int_0^2(4-z^3)dxdz + \hat{z}\int_0^6\int_0^2(4-z^2)x^2dxdz$$

$$= \hat{x}\int_0^6 dx\int_0^2(16z-z^5)dz - \hat{y}\int_0^6 dx\int_0^2(4-z^3)dz + \hat{z}\int_0^6 x^2dx\int_0^2(4-z^3)dz$$

$$= \hat{x}|x|_0^6 \times \left|8z^2 - \frac{z^6}{6}\right|_0^2 - \hat{z}|x|_0^6 \times \left|4z - \frac{z^4}{4}\right|_0^2 + \hat{z}\left|\frac{x^3}{3}\right|_0^6 \times \left|4z - \frac{z^4}{4}\right|_0^2$$

$$= \hat{x}6 \times \left(32 - \frac{64}{6}\right) - \hat{z}6 \times \left(8 - \frac{16}{4}\right) + \hat{z}\frac{216}{3} \times \left(8 - \frac{16}{4}\right)$$

or $\int_V (\hat{x}2yz - \hat{y}z + \hat{z}x^2)dv = 128\hat{x} - 24\hat{y} + 288\hat{z}$

5.2.4 Gradient of scalar function

Let $\varphi(x,y,z)$ be a continuous differentiable function in a scalar field (i.e., $\varphi(x,y,z)$ is a scalar function). The gradient of $\varphi(x,y,z)$, i.e., grad φ in the rectangular coordinate system is defined as

$$\text{grad}\varphi = \vec{\nabla}\varphi = \left(\hat{x}\frac{\partial}{\partial x} + \hat{y}\frac{\partial}{\partial y} + \hat{z}\frac{\partial}{\partial z}\right)\varphi = \hat{x}\frac{\partial\varphi}{\partial x} + \hat{y}\frac{\partial\varphi}{\partial y} + \hat{z}\frac{\partial\varphi}{\partial z} \tag{5.1}$$

where $\hat{x}$, $\hat{y}$, and $\hat{z}$ are the unit vectors along the $+X$, $+Y$, and $+Z$ axes. From Eq. (5.1), we may have

$$\vec{\nabla} = \hat{x}\frac{\partial}{\partial x} + \hat{y}\frac{\partial}{\partial y} + \hat{z}\frac{\partial}{\partial z} \tag{5.2}$$

and $\vec{\nabla}.\vec{\nabla} = \nabla^2 = \dfrac{\partial^2}{\partial x^2} + \dfrac{\partial^2}{\partial y^2} + \dfrac{\partial^2}{\partial z} \tag{5.3}$

where ∇^2 is called a Laplacian operator.

The gradient of a scalar function in integral form can also be defined as

$$\vec{\nabla}\varphi = \lim_{V \to 0} \frac{\int_s \varphi ds}{V} \tag{5.4}$$

where $\vec{ds}$ is the vector area element on the volume $V = \int_V dxdydz$. The mathematical steps connecting Eq. (5.1) with (5.4) are beyond the scope of the book. The gradient of $\varphi(r, \theta, \phi)$, i.e., grad φ in the spherical polar coordinates system is defined as

$$\text{grad}\,\varphi = \vec{\nabla}\varphi = \hat{r}\frac{\partial\varphi}{\partial r} + \hat{\theta}\frac{1}{r}\frac{\partial\varphi}{\partial\theta} + \hat{\phi}\frac{1}{r\sin\theta}\frac{\partial\varphi}{\partial\phi} \tag{5.5}$$

where $\hat{r}$, $\hat{\theta}$, and $\hat{\phi}$ are unit vectors in the direction of $\vec{r}$, $\vec{\theta}$ and $\vec{\phi}$ respectively. The gradient of a scalar function is a vector function. A vector field derived from a scalar field (by taking the gradient of the field) is called a scalar potential field or lamellar vector field.

Physical significance of gradient of scalar field

$\vec{\nabla}\varphi$ is a vector. This vector tells us how $\varphi(x, y, z)$ varies around a point with the change of x, y, and z. In a scalar field of φ, we can connect all the points in the region having the same values of φ. These points will lie on a surface (recollect equi-potential surface in electrostatics). Since φ has the same values at every point on the surface, φ does not change on the surface when we move from point to point. That means $d\varphi$ or the magnitude of $\vec{\nabla}\varphi$, $|\vec{\nabla}\varphi| = 0$ along the surface. The function φ changes most rapidly in a direction perpendicular to the surface. The value $|\vec{\nabla}\varphi|$ is the maximum space rate of change of φ. This means that 'most rapid rate' at which φ changes is the magnitude of the vector $\vec{\nabla}\varphi$. The direction of $\vec{\nabla}\varphi$ is the direction along which φ increases most rapidly.

Example 5.4

Find the gradient of the scalar function $\varphi(x, y, z) = xy + yz + zx$ at the point (1, 2, 3).

Solution

$$\vec{\nabla}\varphi(x, y, z) = \left(\hat{x}\frac{\partial}{\partial x} + \hat{y}\frac{\partial}{\partial y} + \hat{z}\frac{\partial}{\partial z}\right)(xy + yz + zx)$$

$$= \left(\hat{x}\frac{\partial}{\partial x}(xy + yz + zx)\right) + \hat{y}\frac{\partial}{\partial y}(xy + yz + zx) + \hat{z}\frac{\partial}{\partial z}(xy + yz + zx)$$

or $\vec{\nabla}\varphi(x, y, z) = \hat{x}(y + z) + \hat{y}(z + x) + \hat{z}(x + y)$

So $\vec{\nabla}\varphi(1, 2, 3) = \hat{x}(2 + 3) + \hat{y}(3 + 1) + \hat{z}(1 + 2) = 5\hat{x} + 4\hat{y} + 3\hat{z}$

Example 5.5

If a vector function $\vec{F}$ is a function of space coordinates x, y, z, and time coordinates t, then prove that $d\vec{F} = \left(\vec{dr} \cdot \nabla\right)\vec{F} + \dfrac{\partial \vec{F}}{\partial t}dt$

Solution

According to the question, $\vec{F} = \vec{F}(x, y, z, t)$

$$\therefore \quad d\vec{F} = \frac{\partial \vec{F}}{\partial x}dx + \frac{\partial \vec{F}}{\partial y}dy + \frac{\partial \vec{F}}{\partial z}dz + \frac{\partial \vec{F}}{\partial t}dt$$

$$= \left(\frac{\partial}{\partial x}dx + \frac{\partial}{\partial y}dy + \frac{\partial}{\partial z}dz\right)\vec{F} + \frac{\partial \vec{F}}{\partial t}dt$$

$$= \left[\left(\hat{x}\frac{\partial}{\partial x} + \hat{y}\frac{\partial}{\partial y} + \hat{z}\frac{\partial}{\partial z}\right) \cdot \left(\hat{x}dx + \hat{y}dy + \hat{z}dz\right)\right]\vec{F} + \frac{\partial \vec{F}}{\partial t}dt$$

$$= \left(\vec{\nabla} \cdot \vec{dr}\right)\vec{F} + \frac{\partial \vec{F}}{\partial t}dt$$

or $d\vec{F} = \left(\vec{dr} \cdot \vec{\nabla}\right)\vec{F} + \dfrac{\partial \vec{F}}{\partial t}dt$

Example 5.6

If $\vec{a}$ is a constant vector and $\vec{r}$ is the position vector, then prove that $\vec{\nabla}\left(\vec{a} \cdot \vec{r}\right) = \vec{a}$.

Solution

LHS

$$\vec{\nabla}\left(\vec{a}\cdot\vec{r}\right) = \nabla\left[\left(\hat{x}a_x + \hat{y}a_y + za_z\right)\cdot\left(\hat{x}x + \hat{y}y + \hat{z}z\right)\right]$$

$$= \nabla\left[\left(\hat{x}a_x + \hat{y}a_y + za_z\right)\cdot\left(\hat{x}x + \hat{y}y + \hat{z}z\right)\right]$$

$$= \left(\hat{x}\frac{\partial}{\partial x} + \hat{y}\frac{\partial}{\partial y} + \hat{z}\frac{\partial}{\partial z}\right)\left(a_x x + a_y y + a_z z\right)$$

$$= \hat{x}\frac{\partial}{\partial x}\left(a_x x + a_y y + a_z z\right) + \hat{y}\frac{\partial}{\partial y}\left(a_x x + a_y y + a_z z\right) + \hat{z}\frac{\partial}{\partial z}\left(a_x x + a_y y + a_z z\right)$$

$$= \left(\hat{x}a_x + 0 + 0\right) + \left(0 + \hat{y}a_y + 0\right) + \left(\hat{z}a_z + 0 + 0\right)$$

$$= \hat{x}a_x + \hat{y}a_y + \hat{z}a_z$$

or $\vec{\nabla}\left(\vec{a}\cdot\vec{r}\right) = \vec{a}.$

5.2.5 Divergence of a vector function

The divergence of a vector field is defined as the net amount of flux of the vector field diverging or converging per unit volume. The flux of the vector field $\vec{F}$ is defined as $\oint_S \vec{F}.\hat{n}ds$. Therefore, we can define the divergence of a vector $\vec{F}$, i.e., $\operatorname{div}\vec{F}$ is the limit of its surface integral per unit volume as the volume enclosed by the surface goes to zero. Mathematically,

$$\operatorname{div}\vec{F} = \vec{\nabla}\cdot\vec{F} = \lim_{V\to 0}\frac{\oint_S \vec{F}.\hat{n}ds}{V} \tag{5.6}$$

where $\hat{n}$ is a unit vector perpendicular to the elemental surface area ds in the outward direction. $\vec{F}.\hat{n}$ is the component of $\vec{F}$ along $\hat{n}$, i.e., normal to ds in an outward direction. V is the volume enclosed by the surface S. The integration $\oint_S$ is taken over the closed surface S.

If $\vec{F}(x,y,z)$ is a continuous differentiable function in a vector field (i.e. $\vec{F}(x,y,z)$ is a vector function), divergence of $\vec{F}$, i.e., div $\vec{F}$ in the rectangular coordinate system is

$$\text{div}\,\vec{F} = \vec{\nabla}\cdot F = \left(\hat{x}\frac{\partial}{\partial x} + \hat{y}\frac{\partial}{\partial y} + \hat{z}\frac{\partial}{\partial z}\right)\cdot\left(\hat{x}F_X + \hat{y}F_Y + \hat{z}F_Z\right)$$

or $\quad$ $\text{div}\,\vec{F} = \dfrac{\partial F_X}{\partial x} + \dfrac{\partial F_Y}{\partial y} + \dfrac{\partial F_Z}{\partial z}$ $\hfill$ (5.7)

where $\hat{x}$, $\hat{y}$, and $\hat{z}$ are the unit vectors along the $+X$, $+Y$, and $+Z$ axes. Here, $\vec{F} = \hat{x}F_X + \hat{y}F_Y + \hat{z}F_Z$ and F_X, F_Y, and F_Z are the X, Y, and Z components of $\vec{F}$. The divergence of $\vec{F}(r,\theta,\phi)$, i.e., div $\vec{F}$ in the spherical polar coordinates system is given as

$$\text{div}\,F = \vec{\nabla}.\varphi = \frac{1}{r^2}\frac{\partial}{\partial r}(r^2 F_r) + \frac{1}{r\sin\theta}\frac{\partial}{\partial\theta}(\sin\theta\,F_\theta) + \frac{1}{r\sin\theta}\frac{\partial F_\phi}{\partial\phi}$$

$\hfill$ (5.8)

where F_r, F_θ, and F_ϕ are the r, θ, and ϕ components of the vector $\vec{F}$ respectively.

Physical significance of divergence of a vector field

The divergence of a vector field is a scalar field because it gives the net amount of flux coming out of the volume or the net amount of flux going into the volume. Quantitatively, the divergence of a vector field is the net amount of flux of the vector field diverging or converging per unit volume. If the divergence of a vector function is positive at a point in the flux, then either the flux is expanding and its density is falling with time or the flux is produced at the point. If the divergence of a vector function is negative, then either the flux is contracting and its density at the point is increasing or the flux is swallowed up at the point. If the net flux entering the volume is exactly balanced by the amount of flux leaving the volume, then div $\vec{F} = 0$. A vector field which satisfies this condition is called a solenoidal vector field.

Example 5.7

If $\vec{a}$ is a constant vector and $\vec{r}$ is the position vector, then prove that $\vec{\nabla}\cdot\left(\vec{a}\times\vec{r}\right) = 0$

Solution

$$\vec{\nabla}\cdot\left(\vec{a}\times\vec{r}\right) = \left(\hat{x}\frac{\partial}{\partial x} + \hat{y}\frac{\partial}{\partial y} + \hat{z}\frac{\partial}{\partial z}\right)\cdot\left[\left(a_x x + a_y y + a_z z\right)\times\left(\hat{x}x + \hat{y}y + \hat{z}z\right)\right]$$

$$= \left(\hat{x}\frac{\partial}{\partial x} + \hat{y}\frac{\partial}{\partial y} + \hat{z}\frac{\partial}{\partial z}\right)\cdot\left[\hat{x}\left(a_y z - a_z y\right) + \hat{y}\left(a_z x - a_x z\right) + \hat{z}\left(a_x y - a_y x\right)\right]$$

$$= \frac{\partial}{\partial x}\left(a_y z - a_z y\right) + \frac{\partial}{\partial y}\left(a_z x - a_x z\right) + \frac{\partial}{\partial z}\left(a_x y - a_y x\right)] = 0.$$

Example 5.8

If $\hat{u}$ is a unit vector and $\vec{r}$ is a position vector, prove that $\vec{\nabla} \cdot \left[\left(\hat{u} \cdot \vec{r}\right)\hat{u}\right] = 1$

Solution

According to the question, $\hat{u}$ is a unit vector. Hence, we have

$$\hat{u} = \hat{x}u_x + \hat{y}u_y + \hat{z}u_z$$

or $u_x^2 + u_y^2 + u_z^2 = u^2 = 1$

where u_x, u_y, and u_z are the X, Y, and Z components of the unit vector $\hat{u}$.
As $\hat{u}$ is a unit vector, we have

$$\hat{u} \cdot \vec{r} = \left(\hat{x}u_x + \hat{y}u_y + \hat{z}u_z\right) \cdot \left(\hat{x}x + \hat{y}y + \hat{z}z\right) = xu_x + yu_y + zu_z$$

Now

$$\text{LHS} = \vec{\nabla} \cdot \left[\left(\hat{u} \cdot \vec{r}\right)\hat{u}\right] = \hat{u} \cdot \vec{\nabla}\left(\hat{u} \cdot \vec{r}\right) + \left(\hat{u} \cdot \vec{r}\right)\vec{\nabla} \cdot \hat{u} = u \cdot \nabla\left(u \cdot r\right)$$

$$= \hat{u} \cdot \vec{\nabla}\left(xu_x + yu_y + zu_z\right)$$

$$= \left(\hat{x}u_x + \hat{y}u_y + \hat{z}u_z\right) \cdot \left[\left(\hat{x}\frac{\partial}{\partial x} + \hat{y}\frac{\partial}{\partial y} + \hat{z}\frac{\partial}{\partial z}\right)\left(xu_x + yu_y + zu_z\right)\right]$$

$$= \left(\hat{x}u_x + \hat{y}u_y + \hat{z}u_z\right) \cdot \left[\left(\hat{x}\frac{\partial}{\partial x}\left(xu_x + yu_y + zu_z\right)\right) + \hat{y}\frac{\partial}{\partial y}\left(xu_x + yu_y + zu_z\right)\right.$$

$$\left. + \hat{z}\frac{\partial}{\partial z}\left(xu_x + yu_y + zu_z\right)\right]$$

$$= \left(\hat{x}u_x + \hat{y}u_y + \hat{z}u_z \right) \cdot \left[\left(\hat{x}u_x + 0 + 0 \right) + \left(0 + \hat{y}u_y + 0 \right) + \left(0 + 0 + \hat{z}u_z \right) \right]$$

$$= \left(\hat{x}u_x + \hat{y}u_y + \hat{z}u_z \right) \cdot \left(\hat{x}u_x + \hat{y}u_y + \hat{z}u_z \right) = \hat{u} \cdot \hat{u} = u^2 = 1.$$

Example 5.9

If $\hat{u}$ is a unit vector and $\vec{r}$ is a position vector, prove that $\vec{\nabla} \cdot \left[\left(\hat{u} \times \vec{r} \right) \times \hat{u} \right] = 2$

Solution

According to the question, $\hat{u}$ is a unit vector. Hence, we have $\left| \hat{u} \right| = 1$.

$$\left(\hat{u} \times \vec{r} \right) \times \hat{u} = \left(\hat{u} \cdot \hat{u} \right)\vec{r} - \left(\hat{u} \cdot \vec{r} \right)\hat{u} = \vec{r} - \left(\hat{u} \cdot \vec{r} \right)\hat{u}$$

LHS

$$\vec{\nabla} \cdot \left[\left(\hat{u} \times \vec{r} \right) \times \hat{u} \right] = \vec{\nabla} \cdot \left[\vec{r} - \left(\hat{u} \cdot \vec{r} \right)\hat{u} \right] = \vec{\nabla} \cdot \vec{r} - \vec{\nabla} \cdot \left[\left(\hat{u} \cdot \vec{r} \right)\hat{u} \right] = 3 - \vec{\nabla} \cdot \left[\left(\hat{u} \cdot \vec{r} \right)\hat{u} \right] = 2.$$

Example 5.10

Prove that $\vec{F} = \dfrac{\vec{r}}{r^2}$ is non-solenoidal, where $\vec{r}$ is the position vector.

Solution

$$\vec{\nabla} \cdot \vec{F} = \vec{\nabla} \cdot \frac{\vec{r}}{r^2} = \vec{\nabla} \cdot \frac{\hat{x}x + \hat{y}y + \hat{z}z}{x^2 + y^2 + z^2} = \nabla \cdot \left[\frac{\hat{x}x}{x^2 + y^2 + z^2} + \frac{\hat{y}y}{x^2 + y^2 + z^2} + \frac{\hat{z}z}{x^2 + y^2 + z^2} \right]$$

$$= \left(\hat{x}\frac{\partial}{\partial x} + \hat{y}\frac{\partial}{\partial y} + \hat{z}\frac{\partial}{\partial z} \right) \cdot \left(\frac{\hat{x}x}{x^2 + y^2 + z^2} + \frac{\hat{y}y}{x^2 + y^2 + z^2} + \frac{\hat{z}z}{x^2 + y^2 + z^2} \right)$$

$$= \frac{\partial}{\partial x}\frac{x}{x^2 + y^2 + z^2} + \frac{\partial}{\partial y}\frac{y}{x^2 + y^2 + z^2} + \frac{\partial}{\partial z}\frac{z}{x^2 + y^2 + z^2}$$

$$= \frac{1}{x^2 + y^2 + z^2} - \frac{2x^2}{\left(x^2 + y^2 + z^2 \right)^2} + \frac{1}{x^2 + y^2 + z^2} -$$

$$\frac{2y^2}{\left(x^2+y^2+z^2\right)^2}+\frac{1}{x^2+y^2+z^2}-\frac{2z^2}{\left(x^2+y^2+z^2\right)^2}$$

$$=\frac{3}{x^2+y^2+z^2}-\frac{2\left(x^2+y^2+z^2\right)}{\left(x^2+y^2+z^2\right)^2}=\frac{3}{x^2+y^2+z^2}-\frac{2}{x^2+y^2+z^2}=\frac{1}{r^2}$$

Thus, we have $\vec{\nabla}\cdot\vec{F}=\dfrac{1}{r^2}$ which is a non-zero quantity, implying that the vector function is non-solenoidal.

5.2.6 Curl of a vector function

The curl of a vector $\operatorname{curl}\vec{F}$, is itself a vector quantity. To find the component of $\operatorname{curl}\vec{F}$ along any chosen direction, draw a small closed path of area S lying in the plane normal to that direction, and evaluate the line integral $\oint\vec{F}.\vec{d\ell}$ around the closed path. As the closed path is shrunk in size, (i.e., $S\to0$) the integral diminishes with the area, and the value of $\dfrac{\oint_C\vec{F}.\vec{d\ell}}{S}$ as $S\to0$ is the component of $\operatorname{curl}\vec{F}$ in the chosen direction. The direction in which the vector $\operatorname{curl}\vec{F}$ points is the direction in which $\oint_C\vec{F}.\vec{d\ell}$ is largest. Thus, we can define the curl of a vector function in the following way.

The component of $\operatorname{curl}\vec{F}$ in the direction of the unit vector $\hat{n}$ (i.e., $\hat{n}\cdot\operatorname{curl}\vec{F}$) is the limit of a line integral per unit area as the enclosed area goes to zero, this area being perpendicular to $\hat{n}$. Mathematically, the statement translates

$$\hat{n}\cdot\operatorname{curl}\vec{F}=\vec{\nabla}\times\vec{F}=\lim_{S\to0}\frac{\oint_C\vec{F}.\,\hat{r}d\ell}{S}\tag{5.9}$$

Here $\hat{r}$ is a unit vector in the direction of $\vec{d\ell}$ (i.e., direction in which $\vec{d\ell}$ is travelling). The curve C encloses the surface S. We can also define the same function in another equivalent way.

The curl of a vector is the limit of the ratio of the integral of its cross product with the outward drawn normal, over a closed surface, to the volume enclosed by the surface as the volume goes to zero. Mathematically, the statement is translated as

$$\operatorname{curl}\vec{F}=\vec{\nabla}\times\vec{F}=\lim_{V\to0}\frac{\oint_S\hat{n}\times\vec{F}ds}{V}\tag{5.10}$$

Equations (5.9) and (5.10) are absolutely equivalent.

If $\vec{F}(x,y,z)$ is a continuous differentiable function in a vector field. (i.e. $\vec{F}(x,y,z)$ is a vector function), the curl of $\vec{F}$, i.e., curl$\vec{F}$ in the rectangular coordinate system is

$$\text{curl}\vec{F} = \vec{\nabla} \times F = \hat{x}\left(\frac{\partial F_Z}{\partial y} - \frac{\partial F_Y}{\partial z}\right) + \hat{y}\left(\frac{\partial F_X}{\partial z} - \frac{\partial F_Z}{\partial x}\right) + \hat{z}\left(\frac{\partial F_Y}{\partial x} - \frac{\partial F_X}{\partial y}\right) \qquad (5.11)$$

or $\quad \text{curl}\vec{F} = \vec{\nabla} \times F = \begin{vmatrix} \hat{x} & \hat{y} & \hat{z} \\ \dfrac{\partial}{\partial x} & \dfrac{\partial}{\partial y} & \dfrac{\partial}{\partial z} \\ F_X & F_Y & F_Z \end{vmatrix}$ $\qquad\qquad\qquad (5.12)$

Here $\hat{x}$, $\hat{y}$, and $\hat{z}$ are the unit vectors along the +X, +Y, and +Z axes, $\vec{F} = \hat{x}F_X + \hat{y}F_Y + \hat{z}F_Z$ and F_X, F_Y and F_Z are the X, Y, and Z components of $\vec{F}$. The curl of $\vec{F}(r,\theta,\phi)$ i.e., curl$\vec{F}$ (i.e., $\vec{\nabla} \times F$) in the spherical polar coordinates system is given by

$$\text{curl}\vec{F} = \vec{\nabla} \times \vec{F} = \frac{1}{r\sin\theta}\left[\frac{\partial}{\partial\theta}(\sin\theta\ F_\theta) - \frac{\partial F}{\partial\phi}\right]\hat{r} +$$

$$\frac{1}{r}\left[\frac{1}{\sin\theta}\frac{\partial F_\phi}{\partial\phi} - \frac{\partial(rF_\phi)}{\partial r}\right]\hat{\theta} + \frac{1}{r}\left[\frac{\partial(rF_\theta)}{\partial r} - \frac{\partial F_r}{\partial\theta}\right]\hat{\phi} \qquad (5.13)$$

where $\hat{r}$, $\hat{\theta}$, and $\hat{\phi}$ are the unit vectors along $\hat{r}$, $\hat{\theta}$, and $\vec{\phi}$ axes. F_r, F_θ, and F_ϕ are the $\hat{r}$, $\hat{\theta}$, and $\vec{\phi}$ components of the vector $\vec{F}(r,\theta,\phi)$ respectively.

Physical significance of curl of a vector field

The value of a line integral around a closed path in a vector field is not zero if the vector field is not the gradient of a scalar field. The vector fields which are not the gradients of scalar fields are called non-lamellar vector fields. The value of the line integral around a closed path in non-lamellar vector fields depends upon the orientation of the small vector area enclosed by the path. There is a certain orientation of the elemental vector area for which the value of the line integral around the path enclosing the elemental vector area is maximum. This greatest line integral when computed for unit area is the curl of the vector field. The curl of a vector field is a vector quantity directed along the normal to the elemental vector area which is in a position giving this line integral the greatest value.

The curl is associated with rotation of flux at a point. The curl of the velocity vector provides a measure of the angular velocity of the fluid at any point in the flow field. The curl of linear velocity of a particle in the flow field is equal, both in magnitude and direction, to two times the angular velocity of the particle (i.e., $\vec{\nabla} \times \vec{v} = 2\vec{\omega}$). The physical quantity obtained by multiplying 2 with the angular velocity is defined as vorticity. When the curl of the velocity vector of the particles in the flow field is zero $\left(\vec{\nabla}\times\vec{v}=0\right)$ at every point of the region, the flow is irrotational i.e., the flow has no vertex or turbulence at any point in the flow. In physics of fluid flow, hydrodynamics, and aerodynamics the concept of curl is of central importance. Finally, we conclude that curl and rot (short form of rotation) are synonymous with each other. Even now in German, the word 'rot' is used for curl.

Example 5.11

If vector $\vec{A}$ and $\vec{B}$ are irrotational, prove that $\vec{A}\times\vec{B}$ is solenoidal.

Solution

According to the question, vector $\vec{A}$ and $\vec{B}$ are irrotational. Hence, we have $\vec{\nabla}\times\vec{A}=0$ and $\vec{\nabla}\times\vec{B}=0$.

As vector $\vec{A}\times\vec{B}$ is solenoidal, we have to prove that $\vec{\nabla}\cdot\left(\vec{A}\times\vec{B}\right)=0$

Now

$$\vec{\nabla}\cdot\left(\vec{A}\times\vec{B}\right)=\vec{B}\cdot\left(\nabla\times\vec{A}\right)-\vec{A}\cdot\left(\nabla\times\vec{B}\right)=\vec{B}\cdot0-\vec{A}\cdot0=0$$

Since divergence of $\vec{A}\times\vec{B}$ vector is zero, $\vec{A}\times\vec{B}$ is solenoidal under the given condition.

Example 5.12

If $\vec{a}$ is a constant vector and $\vec{r}$ is the position vector, then prove that $\vec{\nabla}\times\left(\vec{a}\times\vec{r}\right)=2\vec{a}$.

Solution

$$\vec{\nabla}\times\left(\vec{a}\times\vec{r}\right)=\left(\hat{x}\frac{\partial}{\partial x}+\hat{y}\frac{\partial}{\partial y}+\hat{z}\frac{\partial}{\partial z}\right)\times\left[\left(a_x x+a_y y+a_z z\right)\times\left(\hat{x}x+\hat{y}y+\hat{z}z\right)\right]$$

$$=\left(\hat{x}\frac{\partial}{\partial x}+\hat{y}\frac{\partial}{\partial y}+\hat{z}\frac{\partial}{\partial z}\right)\times\left[\hat{x}\left(a_y z-a_z y\right)+\hat{y}\left(a_z x-a_x z\right)+\hat{z}\left(a_x y-a_y x\right)\right]$$

$$=\hat{x}\left[\frac{\partial}{\partial y}\left(a_x y-a_y x\right)-\frac{\partial}{\partial z}\left(a_z x-a_x z\right)\right]+\hat{y}\left[\frac{\partial}{\partial z}\left(a_y z-a_z y\right)-\frac{\partial}{\partial x}\left(a_x y-a_y x\right)\right]$$

$$+\hat{z}\left[\frac{\partial}{\partial x}\left(a_z x-a_x z\right)-\frac{\partial}{\partial y}\left(a_y z-a_z y\right)\right]$$

$$=\hat{x}\left[a_x+a_x\right]+\hat{y}\left[a_y+a_y\right]+\hat{z}\left[a_z+a_z\right]$$

or $\nabla \times \left(\vec{a} \times \vec{r} \right) = 2 \left(\hat{x} a_x + \hat{y} a_y + \hat{z} a_z \right) = 2 \left(\hat{x} a_x + \hat{y} a_y + \hat{z} a_z \right) = 2\vec{a}.$

Example 5.13

If $\hat{u}$ is a unit vector and $\vec{r}$ is a position vector, prove that $\vec{\nabla} \times \left[\left(\hat{u} \cdot \vec{r} \right) \hat{u} \right] = 0.$

Solution

$$\vec{\nabla} \times \left[\left(\hat{u} \cdot \vec{r} \right) \hat{u} \right] = \left(\hat{u} \cdot \vec{r} \right) \vec{\nabla} \times \hat{u} - \hat{u} \times \vec{\nabla} \left(\hat{u} \cdot \vec{r} \right) = \left(\hat{u} \cdot \vec{r} \right) \times 0 - \hat{u} \times \hat{u} = 0.$$

Example 5.14

If $\hat{u}$ is a unit vector and $\vec{r}$ is a position vector, prove that $\vec{\nabla} \times \left[\left(\hat{u} \times \vec{r} \right) \times \hat{u} \right] = 0.$

Solution

$$\left(\hat{u} \times \vec{r} \right) \times \hat{u} = \left(\hat{u} \cdot \hat{u} \right) \vec{r} - \left(\hat{u} \cdot \vec{r} \right) \hat{u} = \vec{r} - \left(\hat{u} \cdot \vec{r} \right) \hat{u}$$

Hence, we have

$$\vec{\nabla} \times \left[\left(\hat{u} \times \vec{r} \right) \times \hat{u} \right] = \vec{\nabla} \times \left[\vec{r} - \left(\hat{u} \cdot \vec{r} \right) \hat{u} \right] = \vec{\nabla} \times \vec{r} - \vec{\nabla} \times \left(\hat{u} \cdot \vec{r} \right) \hat{u} = 0$$

Example 5.15

Show that $\vec{\nabla} \times \vec{v} = 2\vec{\omega}$, where $\vec{v}$ is the linear velocity and $\vec{\omega}$ is the angular velocity.

Solution

We know that

$$\vec{v} = \vec{\omega} \times \vec{r} = \hat{x} \left(\omega_y z - \omega_z y \right) + \hat{y} \left(\omega_z x - \omega_x z \right) + \hat{z} \left(\omega_x y - \omega_y x \right)$$

or $\vec{v} = \hat{x} \omega_1 + \hat{y} \omega_2 + \hat{z} \omega_3$ where

$$\omega_1 = \omega_y z - \omega_z y$$

$$\omega_2 = \omega_z x - \omega_x z$$

$$\omega_3 = \omega_x y - \omega_y x$$

$$\vec{\nabla} \times \vec{v} = \begin{vmatrix} x & y & z \\ \dfrac{\partial}{\partial x} & \dfrac{\partial}{\partial y} & \dfrac{\partial}{\partial z} \\ \omega_1 & \omega_2 & \omega_3 \end{vmatrix}$$

or　$\vec{\nabla} \times \vec{v} = \hat{x}\left(\omega_x + \omega_x\right) + \hat{y}\left(\omega_y + \omega_y\right) + \hat{z}\left(\omega_z + \omega_z\right)$

$$= \hat{x}\left(2\omega_x\right) + \hat{y}\left(2\omega_y\right) + \hat{z}\left(2\omega_z\right)$$

or　$\vec{\nabla} \times \vec{v} = 2\left(\hat{x}\omega_x + \hat{y}\omega_y + \hat{z}\omega_z\right) = 2\vec{\omega}$

$2\vec{\omega}$ is called vorticity in fluid dynamics. Since the curl of the velocity vector is non-zero, the velocity function is rotational in this case.

Example 5.16

Prove that $\vec{F}(r) = \dfrac{\vec{r}}{r^2}$ is irrotational where r is the position vector.

Solution

$$\vec{\nabla} \times \vec{F}(r) = \vec{\nabla} \times \frac{\vec{r}}{r^2} = \vec{\nabla} \times \frac{\hat{x}x + \hat{y}y + \hat{z}z}{x^2 + y^2 + z^2}$$

$$= \vec{\nabla} \times \left[\frac{\hat{x}x}{x^2 + y^2 + z^2} + \frac{\hat{y}y}{x^2 + y^2 + z^2} + \frac{\hat{z}z}{x^2 + y^2 + z^2} \right]$$

$$= \left(\hat{x}\frac{\partial}{\partial x} + \hat{y}\frac{\partial}{\partial y} + \hat{z}\frac{\partial}{\partial z} \right) \vec{\nabla} \times \left[\frac{\hat{x}x}{x^2 + y^2 + z^2} + \frac{\hat{y}y}{x^2 + y^2 + z^2} + \frac{\hat{z}z}{x^2 + y^2 + z^2} \right]$$

$$= \hat{x}\left(\frac{\partial}{\partial y}\frac{z}{x^2 + y^2 + z^2} - \frac{\partial}{\partial z}\frac{y}{x^2 + y^2 + z^2} \right) + \hat{y}\left(\frac{\partial}{\partial z}\frac{x}{x^2 + y^2 + z^2} - \frac{\partial}{\partial x}\frac{z}{x^2 + y^2 + z^2} \right)$$

$$+ \hat{z}\left(\frac{\partial}{\partial x}\frac{y}{x^2 + y^2 + z^2} - \frac{\partial}{\partial y}\frac{x}{x^2 + y^2 + z^2} \right)$$

$$= \hat{x}\left(-\frac{2yz}{x^2+y^2+z^2}+\frac{2yz}{x^2+y^2+z^2}\right)+\hat{y}\left(-\frac{2zx}{x^2+y^2+z^2}+\frac{2xz}{x^2+y^2+z^2}\right)$$

$$+\hat{z}\left(-\frac{2xy}{x^2+y^2+z^2}+\frac{2yx}{x^2+y^2+z^2}\right)=0$$

Thus, we have $\vec{\nabla}\times\vec{F}(r)=0$ implying that the vector function defined by the equation $\vec{F}(r)=\dfrac{\vec{r}}{r^2}$ is irrotational

Example 5.17

Prove that $F(r)\vec{r}$ is irrotational, where $\vec{r}$ is the position vector and $F(r)$ is a scalar differentiable function.

Solution

$$\vec{\nabla}\times F(r)\vec{r}=F(r)\vec{\nabla}\times\vec{r}-\vec{r}\times\vec{\nabla}F(r)=F(r)0-\vec{r}\times\nabla F(r)=0-\vec{r}\times\frac{dF(r)}{dr}\hat{r},\ \ \hat{r}$$

is a unit vector in the direction of the position vector $\vec{r}$. Thus we have

$$\vec{\nabla}\times F(r)\vec{r}=0-r\frac{dF(r)}{dr}\sin 0=0$$

implying that $F(r)\vec{r}$ is irrotational.

5.2.7 Gauss's divergence theorem

Gauss's divergence theorem enables us to go from volume integral to surface integral and vice versa. The theorem states that volume integral of divergence of a vector field F taken over any volume V is equal to the surface integral of F taken over the closed surface enclosing the volume V. Mathematically, this statement translates as

$$\int_V \text{div}\vec{F}dv=\oint_S \vec{F}.\hat{n}ds \tag{5.14}$$

Here $\hat{n}$ is a unit vector normal to $\vec{ds}$ in an outward direction. $\vec{F}.\hat{n}$ is the component of $\vec{F}$ along $\hat{n}$, i.e., normal to $\vec{ds}$ in an outward direction.

Example 5.18

Verify Gauss' divergence theorem for the vector $\vec{F} = \hat{x}x^2 + \hat{y}y^2 + \hat{z}z^2$ taken over the cube $0 \le x,\ y,\ z \le 1$.

Solution

Consider the cube as shown in Fig. 5.4.

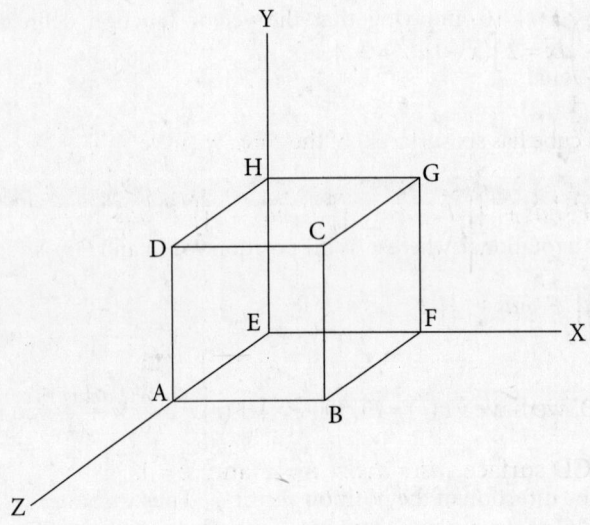

Figure 5.4 | Verification of Gauss's divergence theorem

Given $\vec{F} = \hat{x}x^2 + \hat{y}y^2 + \hat{z}z^2$

Hence, $\vec{\nabla} \cdot \vec{F} = \left(\hat{x}\dfrac{\partial}{\partial x} + \hat{y}\dfrac{\partial}{\partial y} + \hat{z}\dfrac{\partial}{\partial z} \right) \cdot \left(\hat{x}x^2 + \hat{y}y^2 + \hat{z}z^2 \right) = 2(x + y + z)$

Now, the values of x, y, and z vary from 0 to 1 and therefore, the volume of the cube,

$$\int_V \operatorname{div} \vec{F} dv = \int_0^1 \int_0^1 \int_0^1 2(x + y + z) dx dy dz$$

$$= 2 \int_0^1 \int_0^1 \left[xz + yz + \frac{z^2}{2} \right]_0^1 dx dy$$

$$= 2 \int_0^1 \int_0^1 \left[x + y + \frac{1}{2} \right] dx dy$$

$$= 2 \int_0^1 \left[xy + \frac{y^2}{2} + \frac{1}{2} y \right]_0^1 dx$$

$$= 2 \int_0^1 \left(x + \frac{1}{2} + \frac{1}{2} \right) dx = 2 \int_0^1 (x+1) dx = 3 \tag{A}$$

For the surfaces (a cube has six surfaces) of the cube, we have

$$\oint_S \vec{F} . \hat{n} ds = \iint_{ABCD} \vec{F} \cdot \hat{n} ds + \iint_{BFGC} \vec{F} \cdot \hat{n} ds + \iint_{FGHE} \vec{F} \cdot \hat{n} ds + \iint_{EHDA} \vec{F} \cdot \hat{n} ds$$

$$+ \iint_{CDHG} \vec{F} \cdot \hat{n} ds + \iint_{ABFE} \vec{F} \cdot \hat{n} ds \tag{B}$$

As shown in Fig. 5.3, we have

i. For the ABCD surface, $ds = dydz$, $\hat{n} = \hat{x}$ and $x = 1$

ii. For the BFGC surface, $ds = dzdx$, $\hat{n} = \hat{y}$ and $y = 1$

iii. For the FGHE surface, $ds = dydz$, $\hat{n} = -\hat{x}$ and $x = 0$

iv. For the EHDA surface, $ds = dzdx$, $\hat{n} = -\hat{y}$ and $y = 0$

v. For the CDHG surface, $ds = dxdy$, $\hat{n} = \hat{z}$ and $z = 1$

vi. For the ABFE surface, $ds = dxdy$, $\hat{n} = -\hat{z}$ and $z = 0$

Putting all these values into Eq. (B), we have

$$\oint_S \vec{F} . \hat{n} \; ds = \iint_{ABCD} \vec{F} . \hat{x} dydz + \iint_{BFGC} \vec{F} . \hat{y} dxdz + \iint_{FGHE} \vec{F} . (-\hat{x}) dzdy +$$

$$\iint_{EHDA} \vec{F} . (-\hat{y}) dxdz + \iint_{CDHG} \vec{F} . \hat{z} dxdz + \iint_{ABFE} \vec{F} . (-\hat{z}) dxdy$$

$$= \iint_{ABCD} \left(\hat{x} x^2 + \hat{y} y^2 + \hat{z} z^2 \right) . \hat{x} dydz + \iint_{BFGC} \left(\hat{x} x^2 + \hat{y} y^2 + \hat{z} z^2 \right) . \hat{y} dxdz +$$

$$\iint\limits_{FGHE} \left(\hat{x}x^2 + \hat{y}y^2 + \hat{z}z^2\right).\left(-\hat{x}\right)dydz + \iint\limits_{EHDA} \left(\hat{x}x^2 + \hat{y}y^2 + \hat{z}z^2\right).\left(-\hat{y}\right)dxdz$$

$$+ \iint\limits_{CDHG} \left(\hat{x}x^2 + \hat{y}y^2 + \hat{z}z^2\right).\hat{z}dxdz + \iint\limits_{ABFE} \left(\hat{x}x^2 + \hat{y}y^2 + \hat{z}z^2\right).\left(-\hat{z}\right)dxdy$$

$$= \iint\limits_{ABCD} x^2 dydz + \iint\limits_{BFGC} y^2 dxdz + \iint\limits_{FGHE} \left(-x^2\right)dydz + \iint\limits_{EHDA} \left(-y^2\right)dxdz$$

$$+ \iint\limits_{CDHG} z^2 dxdz + \iint\limits_{ABFE} \left(-z^2\right)dxdy$$

Putting the values of x, y, and z for all the six surfaces into this equation, we have

$$\oint\limits_{S} \vec{F}.\hat{n}ds = \iint\limits_{ABCD} 1\times dydz + \iint\limits_{BFGC} 1\times dxdz + \iint\limits_{FGHE} 0\times dydz + \iint\limits_{EHDA} 0\times dxdz$$

$$+ \iint\limits_{CDHG} 1\times dxdz + \iint\limits_{ABFE} 0\times dxdy$$

$$= \iint\limits_{ABCD} dydz + \iint\limits_{BFGC} dxdz + 0 + 0 + \iint\limits_{CDHG} dxdz + 0$$

Since each side of the cube is of unit length, the area of each face is 1, i.e., the area of ABCD = 1, area of BFGH = 1, area of CDHG = 1. Hence, we have

$$\oint\limits_{S} \vec{F}.\hat{n}ds = 3 \tag{C}$$

Therefore, (A) = (C)

Thus, LHS = RHS.

or $\displaystyle\int\limits_{V} \text{div}\vec{F}dv = \oint\limits_{S} \vec{F}.\hat{n}ds$

5.2.8 Stokes' theorem

This theorem states that the tangential line integral of a vector function F around any closed path is equal to the normal surface integral of the curl of that function over an enclosed surface which has the curve for its bounding edge. Mathematically, this statement translates as

$$\oint_C \vec{F}.\hat{r}d\ell = \oint_S (\vec{\nabla}\times\vec{F}).\hat{n}ds \tag{5.15}$$

Here $\hat{n}$ is a unit vector normal to $\vec{ds}$ in an outward direction and $\hat{r}$ is a unit vector in the direction of $\vec{d\ell}$ (i.e., direction in which $\vec{d\ell}$ is travelling). $\vec{F}\cdot\hat{r}$ is the component of $\vec{F}$ along the vector length $\vec{d\ell}$ or $\vec{F}\cdot\hat{r}$ is the component of $\vec{F}$ tangential to $\vec{d\ell}$ where as $(\vec{\nabla}\times\vec{F}).\hat{n}$ is the component of $\vec{\nabla}\times\vec{F}$ along $\hat{n}$, i.e., normal to $\vec{ds}$ in an outward direction.

Example 5.19

Verify Stokes' theorem for the vector $\vec{F} = \hat{x}z^2 + \hat{y}5x + \hat{z}0$ taken over a square defined by $0 \leq x \leq 1$ and $0 \leq y \leq 1;\ \ z = 1$.

Solution

According to Stoke's theorem

$$\oint_C \vec{F}\cdot\vec{d\ell} = \oint_S \vec{\nabla}\times\vec{F}\cdot\hat{n}ds$$

The LHS of the mathematical expression of Stokes's theorem is $\oint_C \vec{F}\cdot\vec{d\ell}$

The circulation of the given vector function $\vec{F}$ around the closed path OABCO is given by

$$\oint_C \vec{F}\cdot\vec{d\ell} = \oint_{OABCO} \vec{F}\cdot\vec{d\ell} = \int_{OA}\vec{F}\cdot\vec{d\ell} + \int_{AB}\vec{F}\cdot\vec{d\ell} + \int_{BC}\vec{F}\cdot\vec{d\ell} + \int_{CO}\vec{F}\cdot\vec{d\ell} \tag{A}$$

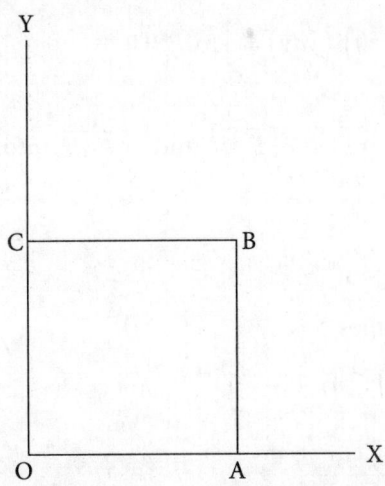

Figure 5.5 | Verification of Stokes' theorem

Given that $\vec{F} = \hat{x}z^2 + \hat{y}5x + \hat{z}0$ and $\vec{d\ell} = \hat{x}dx + \hat{y}dy$

i.   Evaluation of $\int_{OA} \vec{F} \cdot \vec{d\ell}$. Along the path OA, the value of $y = 0$ and hence, $\vec{d\ell} = \hat{x}dx$. Hence, we get

$$\int_{OA} \vec{F} \cdot \vec{d\ell} = \int_{OA} \left(\hat{x}z^2 + \hat{y}5x + \hat{z}0 \right) \cdot \left(\hat{x}dx \right) = \int_0^1 z^2 dx = z^2 \int_0^1 dx = z^2 = 1$$

ii. Evaluation of $\int_{AB} \vec{F} \cdot \vec{dl}$. Along the path AB, the value of $x = 1$ and hence, $\vec{d\ell} = \hat{y}dy$. Hence, we get

$$\int_{AB} \vec{F} \cdot \vec{d\ell} = \int_{AB} \left(\hat{x}z^2 + \hat{y}5x - \hat{z}0 \right) \cdot \left(\hat{y}dy \right) = \int_{AB} 5x dy = 5x \int_0^1 dy = 5$$

iii. Evaluation of $\int_{BC} \vec{F} \cdot \vec{d\ell}$. Along the path GD, the value of $y = 1$ and hence, $\vec{d\ell} = \hat{x}dx$. Therefore, we get

$$\int_{BC} \vec{F} \cdot \vec{d\ell} = \int_{BC} \left(\hat{x}z^2 + \hat{y}5x + \hat{z}0 \right) \cdot \left(\hat{x}dx \right) = \int_{BC} z^2 dx = z^2 \int_1^0 dx = -z^2 = -1$$

iv. Evaluation of $\int_{DE} \vec{F} \cdot \vec{d\ell}$. Along the path CO, the value of $x = 0$ and $\vec{d\ell} = \hat{y}dy$. Hence, we get

$$\int_{CO} \vec{F} \cdot \vec{d\ell} = \int_{CO} \left(\hat{x}z^2 + \hat{y}5x + \hat{z}0 \right) \cdot \left(\hat{y}dy \right) = \int_{CO} 5x dy = 0$$

Putting the values of $\int_{OA} \vec{F} \cdot \vec{d\ell}$, $\int_{AB} \vec{F} \cdot \vec{d\ell}$, $\int_{BC} \vec{F} \cdot \vec{d\ell}$ and $\int_{CO} \vec{F} \cdot \vec{d\ell}$ into Eq. (A), we get

$$\oint_C \vec{F} \cdot \vec{d\ell} = \oint_{OABCO} \vec{F} \cdot \vec{d\ell} = 5$$

LHS of Stokes' theorem becomes

$$\oint_C \vec{F} \cdot \vec{d\ell} = 5.$$

(A)

RHS of Stokes' theorem is $\oint_S \vec{\nabla} \times \vec{F} \cdot \hat{n} ds$

Now $\vec{\nabla} \times \vec{F} = \vec{\nabla} \times (\hat{x}z^2 + \hat{y}5x + \hat{z}0) = \begin{vmatrix} \hat{x} & \hat{y} & \hat{z} \\ \dfrac{\partial}{\partial x} & \dfrac{\partial}{\partial y} & \dfrac{\partial}{\partial z} \\ z^2 & 5x & 0 \end{vmatrix}$

or $\quad \vec{\nabla} \times \vec{F} = \hat{y}2z + \hat{z}5$

Putting this value of $\vec{\nabla} \times \vec{F}$ into the RHS of Stokes' theorem, we get

$$\oint_S \vec{\nabla} \times \vec{F} \cdot \hat{n} ds = \oint_S (\hat{y}2z + \hat{z}5) \cdot \hat{z} dx dy$$

or $\quad \oint_S \vec{\nabla} \times \vec{F} \cdot \hat{n} ds = \oint_S 5 dx dy = 5 \times \oint_S dx dy = 5$ \hfill (B)

Thus, (A) = (B)

    LHS of Stokes' theorem = RHS of Stokes' theorem

    Thus, Stokes' law is verified.

5.2.9 Green's theorem

This theorem enables us to express an integral taken over the surfaces of a number of bodies as an integral taken through the space between them. If there are two scalar functions of space f and g, then Green's theorem is used to change the volume integral into a surface integral. Thus, the theorem is expressed analytically as

$$\int_V \left(f \nabla^2 g - g \nabla^2 f \right) dv = \int_S \left(f \vec{\nabla} g - g \vec{\nabla} f \right) \cdot \vec{ds} \hfill (5.16)$$

where volume V is enclosed by the surface S.

Green's theorem in the plane

The theorem states that, if S is a closed region in the x–y plane bounded by a closed curve C and if M and N are continuous functions of x and y having a continuous derivative in S, then

$$\oint_C \left(M dx + N dy \right) = \oint_S \left(\frac{\partial N}{\partial x} - \frac{\partial M}{\partial y} \right) dx dy \hfill (5.17)$$

where C is travelled in the positive (counter-clockwise or outward) direction.

Example 5.20

Verify Green's theorem in the plane for $\oint_C\left[\left(xy+y^2\right)dx+x^2dy\right)\right]$ where C is the closed curve

of the region bounded by $y = x$ and $y = x^2$.

Solution

$y = x$ and $y = x^2$ intersect at $(0,0)$ and $(1,1)$. The positive direction in traveling C is shown in Fig. 5.6.

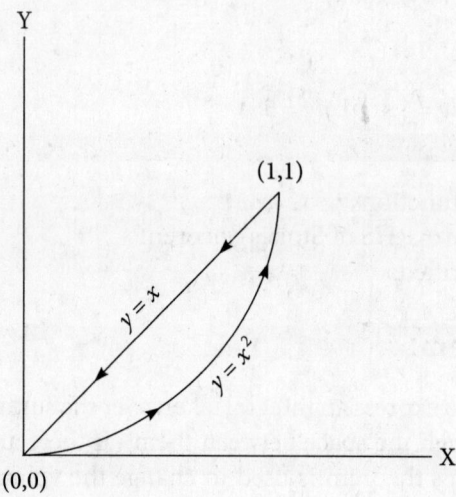

Figure 5.6 | Verification of Green's theorem

LHS :

Along $y = x^2$

$$\oint_C\left[\left(xy+y^2\right)dx+x^2dy\right)\right]=\int_0^1\left(xx^2+x^4\right)dx+\int_0^1x^2(2x)dx=\frac{19}{20} \tag{A}$$

Along $y = x$ from $(1,1)$ to $(0,0)$ (see Fig. 5.6)

$$\oint_C\left[\left(xy+y^2\right)dx+x^2dy\right)\right]=\int_1^0\left(x^2+x^2\right)dx+\int_1^0x^2dx=-1$$

The line integral around the closed path

$$C=\frac{19}{20}+\left(-1\right)=\frac{1}{20}$$

RHS :

$$\oint_S \left(\frac{\partial N}{\partial x} - \frac{\partial M}{\partial y} \right) dxdy = \oint_S \left[\left(\frac{\partial}{\partial x} (x^2) \right) - \frac{\partial}{\partial y} (xy + y^2) \right] dxdy$$

$$= \oint_S (x - 2y) dxdy = \int_0^1 \int_{x^2}^X (x - 2y) dydx = \int_0^1 (xy - y^2) \Big|_{x^2}^x dx = -\frac{1}{20} \qquad \text{(B)}$$

(A) = (B)

Thus, LHS = RHS.

Example 5.21

If f and g are the scalar functions of x, y, and z, prove that $\vec{\nabla} \cdot (\vec{\nabla} f \times \vec{\nabla} g) = 0$.

Solution

$$\vec{\nabla} \cdot (\vec{\nabla} f \times \vec{\nabla} g) = \vec{\nabla} g \cdot (\vec{\nabla} \times \vec{\nabla} f) - \vec{\nabla} f \cdot (\vec{\nabla} \times \vec{\nabla} g) = \nabla g \cdot 0 - \nabla f \cdot 0 = 0.$$

Example 5.22

Prove that $\int_V \vec{\nabla} f dv = \int_S f \vec{ds}$ if f is a scalar function of x, y, and z.

Solution

From Gauss's law, we have

$$\int_V \vec{\nabla} \cdot \vec{F} \, dv = \int_S \vec{F} \cdot \vec{ds}$$

Let $\vec{F} = \vec{A} f$, where $\vec{A}$ is a constant vector.
Hence, Gauss's law becomes

$$\int_V \vec{\nabla} \cdot \vec{A} f \, dv = \int_S \vec{A} f \cdot \vec{ds}$$

or $\int_V (\vec{A} \cdot \vec{\nabla} f - f \vec{\nabla} \cdot \vec{A}) dv = \vec{A} \cdot \int_S f \, \vec{ds}$

or $\int_V (\vec{A} \cdot \vec{\nabla} f - 0) dv = \vec{A} \cdot \int_S f \, \vec{ds}$

or $\vec{A} \cdot \int_V \vec{\nabla} f \, dv - \vec{A} \cdot \int_S f \, \vec{ds} = 0$

or $\vec{A} \cdot \left[\int_V \vec{\nabla} f \, dv - \int_S f \, \vec{ds} \right] = 0$

Since the constant vector $\vec{A}$ is an arbitrary constant, we can have from this equation

$$\int_V \vec{\nabla} f dv = \int_S f \vec{ds}$$

Example 5.23

Prove that $\int_C f dr = \int_S \hat{n} \times \vec{\nabla} f ds$ if f is a differentiable scalar function of x, y, and z and $\vec{r}$ is the position vector.

Solution

Let us define a vector $\vec{F} = \vec{k} f$, where $\vec{k}$ is a non-zero constant vector and f is a differentiable scalar function of x, y, and z and $\vec{r}$ is the position vector. Let us consider the integral

$$\int_C \left(\vec{k} f \right) \cdot \vec{dr}.$$

Applying Stokes' theorem to this integral, we get

$$\int_C \left(\vec{k} f \right) \cdot \vec{dr} = \int_S \vec{\nabla} \times \left(\vec{k} f \right) \cdot \hat{n} ds = \int_S \left(f \vec{\nabla} \times \vec{k} - \vec{k} \times \vec{\nabla} f \right) \cdot \hat{n} ds$$

$$= \int_S \left(-\vec{k} \times \vec{\nabla} f \right) \cdot \hat{n} ds = \int_S \left(\vec{\nabla} f \times \vec{k} \right) \cdot \hat{n} ds = \int_S \vec{k} \cdot \left(\hat{n} \times \vec{\nabla} f \right) ds$$

or $\int_C \left(\vec{k} f \right) \cdot \vec{dr} - \vec{k} \cdot \int_S \hat{n} \times \vec{\nabla} f ds = 0$

or $\vec{k} \cdot \left[\int_C f \vec{dr} - \int_S \hat{n} \times \vec{\nabla} f ds \right] = 0$

or $\int_C f \vec{dr} = \int_S \hat{n} \times \vec{\nabla} f ds$

5.2.10 Useful vector relations

General relations

1. $\vec{A} \times \vec{B} = \hat{x}(A_Y B_Z - A_Z B_Y) + \hat{y}(A_Z B_X - A_X B_Z) + \hat{z}(A_X B_Y - A_Y B_X)$

2. $\vec{A} \cdot \vec{B} = A_X B_X + A_Y B_Y + A_Z B_Z$

3. $\vec{A} \cdot \vec{B} \times \vec{C} = \vec{A} \times \vec{B} \cdot \vec{C} = \vec{C} \times \vec{A} \cdot \vec{B}$

4. $\vec{A} \times (\vec{B} \times \vec{C}) = (\vec{A} \cdot \vec{C})\vec{B} - (\vec{A} \cdot \vec{B})\vec{C}$

5. $\vec{A} \times (\vec{B} \times \vec{C}) + \vec{B} \times (\vec{C} \times \vec{A}) + \vec{C} \times (\vec{A} \times \vec{B}) = 0$

6. $(\vec{A} \times \vec{B}) \cdot (\vec{C} \times \vec{D}) = \vec{A} \cdot [\vec{B} \times (\vec{C} \times \vec{D}) = (\vec{A} \cdot \vec{C})(\vec{B} \cdot \vec{D}) - (\vec{A}.\vec{D})(\vec{B}.\vec{C})$

7. $(\vec{A} \times \vec{B}) \times (\vec{C} \times \vec{D}) = (\vec{A} \times \vec{B} \cdot \vec{D})\vec{C} - (\vec{A} \times \vec{B}.\vec{C})\vec{D}$

Operator $\vec{\nabla}$ in rectangular coordinates (x, y, z)

8. $\vec{\nabla} = \hat{x}\dfrac{\partial}{\partial x} + \hat{y}\dfrac{\partial}{\partial y} + \hat{z}\dfrac{\partial}{\partial z}$

9. $\vec{\nabla} \cdot \vec{\nabla} = \nabla^2 = \dfrac{\partial^2}{\partial x^2} + \dfrac{\partial^2}{\partial y} + \dfrac{\partial^2}{\partial z}$

Operator $\vec{\nabla}$ in circular cylindrical coordinates (r, θ, z)

10. $\vec{\nabla} = \hat{r}\dfrac{\partial}{\partial r} + \hat{\theta}\dfrac{1}{r}\dfrac{\partial}{\partial \theta} + \hat{z}\dfrac{\partial}{\partial z}$

11. $\nabla^2 = \dfrac{1}{r}\dfrac{\partial}{\partial r}\left(r\dfrac{\partial}{\partial r}\right) + \dfrac{1}{r}\dfrac{\partial}{\partial r} + \dfrac{1}{r^2}\dfrac{\partial^2}{\partial \theta^2} + \dfrac{\partial^2}{\partial z^2}$

Operator $\vec{\nabla}$ in spherical polar coordinates (r, θ, φ)

12. $\vec{\nabla} = \hat{r}\dfrac{\partial}{\partial r} + \hat{\theta}\dfrac{1}{r}\dfrac{\partial}{\partial \theta} + \hat{\phi}\dfrac{1}{r\sin\theta}\dfrac{\partial}{\partial \phi}$

13. $\nabla^2 = \dfrac{1}{r^2}\dfrac{\partial}{\partial r}\left(r^2\dfrac{\partial}{\partial r}\right) + \dfrac{1}{r^2\sin\theta}\dfrac{\partial}{\partial \theta}\left(\sin\theta\dfrac{\partial}{\partial \theta}\right) + \dfrac{1}{r^2\sin\theta}\dfrac{\partial^2}{\partial \phi^2}$

14. $\vec{\nabla}(\varphi\Psi) = \varphi\vec{\nabla}\Psi + \Psi\vec{\nabla}\varphi$

15. $\vec{\nabla}\cdot\left(\varphi\vec{A}\right) = \vec{A}\cdot\vec{\nabla}\varphi + \varphi\vec{\nabla}\cdot\vec{A}$

16. $\vec{\nabla}\times\left(\varphi\vec{A}\right) = \varphi\vec{\nabla}\times\vec{A} - \vec{A}\times\vec{\nabla}\varphi$

17. $\vec{\nabla}\cdot\left(\vec{A}\times\vec{B}\right) = \vec{B}\cdot\left(\vec{\nabla}\times\vec{A}\right) - \vec{A}\cdot\left(\vec{\nabla}\times\vec{B}\right)$

18. $\vec{\nabla}\times\left(\vec{A}\times\vec{B}\right) = \vec{A}\left(\vec{\nabla}\cdot\vec{B}\right) - \vec{B}\left(\vec{\nabla}\cdot\vec{A}\right) + \left(\vec{B}\cdot\vec{\nabla}\right)\vec{A} - \left(\vec{A}\cdot\vec{\nabla}\right)\vec{B}$

19. $\vec{\nabla}\times\left(\vec{A}\cdot\vec{B}\right) = \vec{A}\times\left(\vec{\nabla}\times\vec{B}\right) + \vec{B}\times\left(\vec{\nabla}\times\vec{A}\right) + \left(\vec{B}\cdot\vec{\nabla}\right)\vec{A} + \left(\vec{A}\cdot\vec{\nabla}\right)\vec{B}$

20. $\nabla^2\varphi = \vec{\nabla}\cdot\vec{\nabla}\varphi$

21. $\vec{\nabla}\times\left(\vec{\nabla}\times\vec{A}\right) = \vec{\nabla}\left(\vec{\nabla}\cdot\vec{A}\right) - \nabla^2\vec{A}$

22. $\vec{\nabla}\times\left(\vec{\nabla}\varphi\right) = 0$

23. $\vec{\nabla}\cdot\left(\vec{\nabla}\times\vec{A}\right) = 0$

Special Relations

If $\vec{r} = \hat{x}x + \hat{y}y + \hat{z}z$ $\vec{r}$ and $\vec{k}$ = constant vector, then

24. $\vec{\nabla}\cdot\vec{r} = 3$

25. $\vec{\nabla} \times \vec{r} = 0$

26. $\vec{\nabla} |\vec{r}| = \dfrac{\vec{r}}{|\vec{r}|}$

27. $\vec{\nabla}\left(\dfrac{1}{|\vec{r}|}\right) = -\dfrac{\vec{r}}{|\vec{r}|^3}$

28. $\vec{\nabla}\dfrac{\vec{r}}{|\vec{r}|^3} = -\nabla^2\left(\dfrac{1}{|\vec{r}|}\right) = 0$ if $|\vec{r}| \neq 0$

29. $\vec{\nabla}\cdot\left(\dfrac{\vec{k}}{|\vec{r}|}\right) = k\cdot\left[\vec{\nabla}\left(\dfrac{1}{|\vec{r}|}\right)\right] = -\left(\dfrac{\vec{k}\cdot\vec{r}}{|\vec{r}|^3}\right)$

30. $\vec{\nabla}\times\left[\vec{k}\times\left(\dfrac{\vec{r}}{|\vec{r}|^3}\right)\right] = -\vec{\nabla}\left(\dfrac{\vec{k}\cdot\vec{r}}{|\vec{r}|^3}\right)$ if $|\vec{r}| \neq 0$

31. $\nabla^2\left(\dfrac{\vec{k}}{|\vec{r}|}\right) = \vec{k}\nabla^2\left(\dfrac{1}{|\vec{r}|}\right) = 0$ if $|\vec{r}| \neq 0$

32. $\vec{\nabla}\times\left(\vec{k}\times\vec{B}\right) = \vec{k}\left(\vec{\nabla}\cdot\vec{B}\right) + \vec{k}\times\left(\vec{\nabla}\times\vec{B}\right) - \vec{\nabla}\left(\vec{k}\cdot\vec{B}\right)$

Integral relations

33. $\oint_S \vec{F}\cdot\vec{ds} = \int_V (\vec{\nabla}\cdot\vec{F})dv$ (Gauss's divergence theorem)

34. $\oint_C \vec{F}\cdot\vec{dr} = \int_S (\vec{\nabla}\times\vec{F})\cdot\vec{ds}$ (Stokes' theorem)

35. $\oint_S \varphi\, ds = \int_V (\vec{\nabla}\varphi)dv$

36. $\oint\limits_{S}\overrightarrow{ds}\times\overrightarrow{F} = \int\limits_{V}\left(\overrightarrow{\nabla}\times\overrightarrow{F}\right)dv$

If S is an open surface bounded by the curve C of which ds is the line element, then

37. $\oint\limits_{C}\varphi\overrightarrow{dr} = \int\limits_{S}\overrightarrow{ds}\times\overrightarrow{\nabla}\varphi$

38. $\varphi_2 - \varphi_1 = \int\limits_{C}\overrightarrow{\nabla}\varphi\cdot dr$

In rectangular coordinates (x, y, z)

39. $\overrightarrow{\nabla}\varphi = \hat{x}\dfrac{\partial\varphi}{\partial x} + \hat{y}\dfrac{\partial\varphi}{\partial y} + \hat{z}\dfrac{\partial\varphi}{\partial z}$

40. $\overrightarrow{\nabla}\cdot\overrightarrow{F} = \dfrac{\partial\overrightarrow{F}}{\partial x} + \dfrac{\partial\overrightarrow{F}}{\partial y} + \dfrac{\partial\overrightarrow{F}}{\partial z}$

41. $\overrightarrow{\nabla}\times\overrightarrow{F} = \hat{x}\left(\dfrac{\partial F_Z}{\partial y} - \dfrac{\partial F_Y}{\partial z}\right) + \hat{y}\left(\dfrac{\partial F_X}{\partial z} - \dfrac{\partial F_Z}{\partial x}\right) + \hat{z}\left(\dfrac{\partial F_Y}{\partial x} - \dfrac{\partial F_X}{\partial y}\right)$

In circular cylindrical coordinates (r, θ, z)

42. $\overrightarrow{\nabla}\varphi = \hat{r}\dfrac{\partial\varphi}{\partial r} + \hat{\theta}\dfrac{1}{r}\dfrac{\partial\varphi}{\partial\theta} + \hat{z}\dfrac{\partial\varphi}{\partial z}$

43. $\overrightarrow{\nabla}\cdot\overrightarrow{F} = \hat{r}\dfrac{\partial F_r}{\partial r} + \hat{\theta}\dfrac{1}{r}\dfrac{\partial F_\theta}{\partial\theta} + \hat{z}\dfrac{\partial F_z}{\partial z}$

44. $\overrightarrow{\nabla}\times\overrightarrow{F} = \hat{r}\left(\dfrac{1}{r}\dfrac{\partial F_Z}{\partial\theta} - \dfrac{\partial F_\theta}{\partial z}\right) + \hat{\theta}\left(\dfrac{\partial F_r}{\partial\theta} - \dfrac{\partial F_Z}{\partial r}\right) + \hat{z}\left(\dfrac{\partial F_\theta}{\partial z} + \dfrac{F_\theta}{r} - \dfrac{1}{r}\dfrac{\partial F_r}{\partial\theta}\right)$

In spherical polar coordinates (r, θ, ϕ)

45. $\overrightarrow{\nabla}\varphi = \hat{r}\dfrac{\partial\varphi}{\partial r} + \hat{\theta}\dfrac{1}{r}\dfrac{\partial\varphi}{\partial\theta} + \hat{\phi}\dfrac{1}{r\sin\theta}\dfrac{\partial\varphi}{\partial\phi}$

46. $\vec{\nabla}\cdot\vec{F} = \dfrac{1}{r^2}\dfrac{\partial}{\partial r}\left(r^2 F_r\right) + \dfrac{1}{r\sin\theta}\dfrac{\partial}{\partial\theta}\left(\sin\theta\, F_\theta\right) + \dfrac{1}{r\sin\theta}\dfrac{\partial F_\phi}{\partial\phi}$

47. $\vec{\nabla}\times\vec{F} = \hat{r}\dfrac{1}{r\sin\theta}\left[\dfrac{\partial}{\partial\theta}\left(\sin\theta\, F_\phi\right) - \dfrac{\partial F_\theta}{\partial\theta}\right] + \hat{\theta}\left[\dfrac{1}{\sin\theta}\dfrac{\partial F_r}{\partial\phi^2} - \dfrac{\partial\left(rF_\phi\right)}{\partial r}\right] + \dfrac{\hat{\phi}}{r}\left[\dfrac{\partial\left(rF_\theta\right)}{\partial r} - \dfrac{\partial F_r}{\partial\theta}\right]$

5.3 Gauss's Law

5.3.1 Gauss's law of electrostatics in free space

Gauss's law states that the surface integral of electric field intensity $\vec{E}$ at a point P on the closed surface S is equal to $\dfrac{1}{\varepsilon_0}$ times the total charge q enclosed by the surface S (This hypothetical surface S passes through the point P at which the electric field intensity $\vec{E}$ is found out and is called a Gaussian surface). Mathematically, Gauss's law is given by

$$\text{Electric flux} = \varphi_E = \oint_S \vec{E}\cdot\hat{n}\, ds = \frac{q}{\varepsilon_0} \tag{5.18}$$

where $\hat{n}$ is the unit vector normal to $\vec{ds}$ in an outward direction and ε_0 is the permittivity of free space. In terms of the volume charge density function ρ, the total charge contained in the volume is given by

$$q = \oint_V dq = \int_V \rho\, dv \quad \left(\because\ \rho = \frac{dq}{dv}\right) \tag{5.19}$$

Hence, the mathematical formulation of Gauss's law in terms of charge density function ρ becomes

$$\oint_S \vec{E}\cdot\hat{n}\, ds = \frac{1}{\varepsilon_0}\oint_V \rho\, dv \tag{5.20}$$

Here, $\oint \rho\, dv$ is the total charge enclosed by the Gaussian surface S.

The differential form of Gauss's law, as obtained by applying the divergence theorem to Eq. (5.20), is

$$\nabla\cdot\vec{E} = \frac{\rho}{\varepsilon_0} \tag{5.21}$$

Equation (5.21) is the differential form of Gauss's law of electrostatics in free space.

5.3.2 Gauss's law of electrostatics in a dielectric medium

Gauss's law in a dielectric medium can be expressed as

$$\oint_S \vec{E} \cdot \hat{n} ds = \frac{q'}{\varepsilon}$$

or $\quad \oint_S \varepsilon \vec{E} \cdot \hat{n} ds = q'$

Here q' is not the total charge but the free charge only. The details have been discussed in Chapter 9, 'Dielectric Materials' in Part II (Principles of Engineering Physics II).

Since $\varepsilon \vec{E} = \vec{D}$, the previous equation becomes

$$\oint_S \vec{D} \cdot \hat{n} ds = q'$$

or $\quad \oint_S \vec{D} \cdot \hat{n} ds = \oint_V \rho' dv$ (5.22)

Here ρ' is the free charge density only and polarization charges or induced surface charges are excluded. ε is the permittivity of the dielectric medium. Equation (5.22) is the integral form of Gauss's law of electrostatics in the dielectric medium.

The differential form of Gauss's law in a linear dielectric medium, as obtained by applying the divergence theorem to Eq. (5.22), is

$$\nabla \cdot \vec{D} = \rho'$$ (5.23)

where ρ' is the free charge density.

5.3.3 Applications of Gauss's law

Gauss's law can be used to find out the electric field intensity if the charge distribution is so symmetrical that we can easily evaluate the integral in Eq. (5.20) by a proper choice of the hypothetical surface or Gaussian surface S. Now we shall consider some electrostatic problems where Gauss's law is applicable. First of all, we shall calculate the total charge enclosed by the Gaussian surface and shall put this value in the place of

$$\oint_V \rho dv$$

to calculate the electric field intensity. One example of a spherically symmetric charge distribution will make it clear.

A spherically symmetric charge distribution is defined as the distribution of charge where volume charge density at any point depends only on the radial distance – not upon the direction. In this case, we can calculate the total charge enclosed by the Gaussian surface by evaluating the integral

$$\oint_V \rho(r)dv = \oint \rho(r)4\pi r^2 dr.$$

As an example to evaluate this integral (i.e., to calculate the total charge enclosed by the spherical Gaussian surface), let us consider the case where a spherically symmetric charge distribution of radius R is characterized by the following volume charge density function

$$\rho(r) = \rho_0\left(1 - \frac{r^2}{R^2}\right) \text{ for } r \leq R \text{ (i.e., inside the sphere)}$$

and

$$\rho(r) = 0 \text{ for } r \geq R \text{ (i.e., outside the sphere, there is no charge)}$$

Consider a thin spherical shell of radius x (the value of x varies from 0 to R) and thickness dx (just like coconut shell) concentric with the spherically symmetric charge distribution. The volume of this shell (elemental shell) will be the surface area of the spherical shell × thickness of the shell (this is the definition of volume!), i.e.,

$$dv = 4\pi x^2 \times dx$$

The charge on this shell (elemental charge) will be given by

$$dQ = \rho(x) \times dv = \rho(x) \times 4\pi x^2 dx$$

The total charge enclosed by the Gaussian surface will be given by

$$Q = \oint_V \rho(x)dv$$

$$= \int_0^R \rho_0\left(1 - \frac{x^2}{R^2}\right) \times 4\pi x^2 dx$$

$$= 4\pi\rho_0\left[\int_0^R x^2 dx - \int_0^R \frac{x^2}{R^2}x^2 dx\right]$$

or $\quad Q = \dfrac{8}{15} \pi \rho_0 R^3$

If charge is uniformly distributed throughout the sphere, then the volume charge density function ρ will be independent of the radial distance and will be a constant which can be easily taken outside the integral sign. The calculation will be easier and the total charge enclosed by the Gaussian surface will simply be equal to ρV, where V is the volume enclosed by the Gaussian surface.

Thus, what we have described here is the method of calculating total charge enclosed by a Gaussian surface. In all our discussions, P will be the point where electric field intensity is to be found out.

Electric field due to a point charge

Consider a point charge Q is placed at the point O. We have to determine the electric field intensity E at a point P situated at a distance r from the point charge Q.

Imagine a spherical Gaussian surface through the point P with the centre at the charge Q. By symmetry, E has same magnitude at any point on this Gaussian surface and its direction is always perpendicular to the surface, i.e., the angle between E and $\hat{n}$ is $0°$ at any point on the Gaussian surface. See Fig. 5.7. In a number of cases, we shall use Eq. (5.20) and rewriting it, we have

$$\oint_S \vec{E} \cdot \hat{n} ds = \frac{1}{\varepsilon_0} \oint_V \rho dv \qquad (A)$$

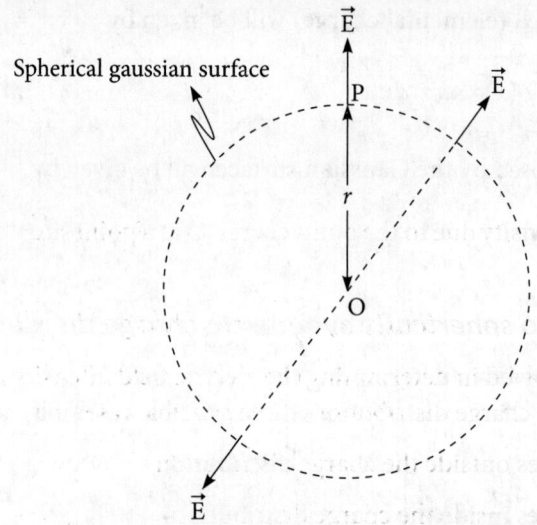

Figure 5.7 | Application of Gauss's law to a point charge

Here $\oint\limits_{V} \rho dv = \oint\limits_{V} dQ$ is the total charge enclosed by the Gaussian surface. In this case, $\oint \rho dv$ is equal to Q. Putting this value in Eq. (A), we have

$$\oint\limits_{S} E|\hat{n}|\cos 0 ds = \frac{Q}{\varepsilon_0}$$

or $$\oint\limits_{S} E ds = \frac{Q}{\varepsilon_0}.$$

The magnitude of $\vec{E}$ is constant everywhere on the Gaussian surface. Hence, we can take E outside the integral sign. We have

$$E \oint\limits_{S} ds = \frac{Q}{\varepsilon_0} \qquad\qquad (B)$$

Here ds is the magnitude of the elemental vector surface area of the spherical Gaussian surface. Hence, $\oint\limits_{S} ds$ = total surface area of the spherical Gaussian surface = $4\pi r^2$ (surface area of a sphere). Putting this value in (B), we have

$$E \, 4\pi r^2 = \frac{Q}{\varepsilon_0}$$

or $$E = \frac{1}{4\pi\varepsilon_0} \frac{Q}{r^2}$$

$$\vec{E} = \frac{1}{4\pi\varepsilon_0} \frac{Q}{r^3} \vec{r}$$

is the electric field intensity due to the point charge Q at a point situated at a distance r from the charge Q.

Electric field due to spherically symmetric charge distribution

In the calculation involved in determining the electric field intensity at any point in case of spherically symmetric charge distribution, three possible cases may arise.

Case 1: The point P lies outside the charge distribution.

Case 2: The point P lies inside the charge distribution.

Case 3: The point P lies on the surface of the charge distribution

We shall discuss Case 2 in great detail and the other two cases are left as exercises to the readers. In Case 2, the point P lies inside the charge distribution and we are to determine the electric field intensity at an internal point.

Let P be an internal point at a distance r from the centre of the spherically symmetric charge distribution of charge Q. R is the radius of the spherically symmetric charge distribution. Through the point imagine a spherical Gaussian surface of radius r with the centre at the centre of the spherically symmetric charge distribution. This has been depicted in Fig. 5.8. By symmetry, $\vec{E}$ has the same magnitude at any point on this spherical Gaussian surface of radius r and its direction is always perpendicular to the surface S, i.e., the angle between $\vec{E}$ and $\hat{n}$ is 0° at any point on this Gaussian surface.

Consider a thin spherical shell of radius x (the value of x varies from 0 to r) and thickness dx concentric with the spherically symmetric charge distribution. The volume of this shell will be

$$dv = 4\pi x^2 \times dx$$

The charge on this shell (elemental charge) will be given by

$$dq = \rho \times dv = \rho \times 4\pi x^2 dx \quad \text{(See Fig. 5.8(b))}$$

The volume of the spherically symmetric charge distribution of radius $R = \dfrac{4\pi R^3}{3}$. This spherically symmetric charge distribution of radius R contains Q charge. Therefore, volume charge density ρ will be given by

$$\rho = \frac{Q}{\dfrac{4\pi R^3}{3}} = \frac{3Q}{4\pi R^3}$$

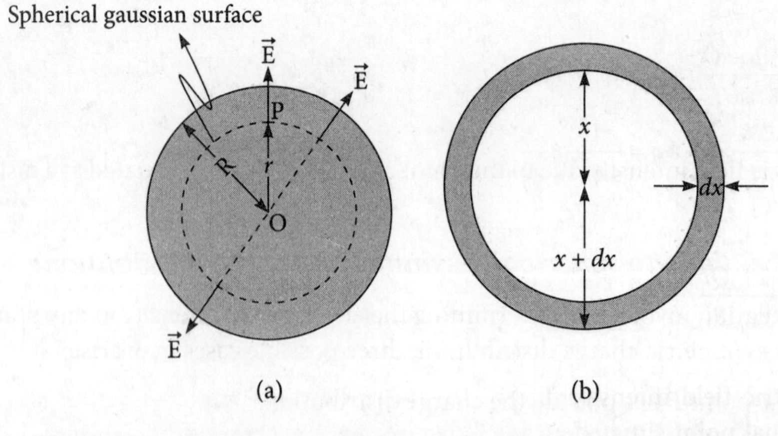

(a)   (b)

Figure 5.8 | (a) Electric field intensity due to spherically symmetric charge distribution when point P is situated inside the charge distribution. The broken line circle represents the spherical Gaussian surface. (b) Calculation of $dv = 4\pi x^2 dx$

Hence, the total charge enclosed by the Gaussian surface will be given by

$$q = \oint_V \rho \, dv = \rho \int_0^r 4\pi x^2 dx = \frac{4\pi r^3}{3}\rho = \frac{4\pi \, r^3}{3} \times \frac{3Q}{4\pi \, R^3} = \frac{Q \, r^3}{R^3}$$

In this case, $\oint \rho \, dv$, is the total charge enclosed by the Gaussian surface S is equal to q. Putting this value in Eq. (5.20), we have

$$\oint_S E \, |\hat{n}| \cos 0 \, ds = \frac{q}{\varepsilon_0}$$

or $$\oint_S E \, ds = \frac{q}{\varepsilon_0}.$$

The magnitude of $\vec{E} = E$ is constant everywhere on the Gaussian surface. Hence, we can take E outside the integral sign. We have

$$E \oint_S ds = \frac{q}{\varepsilon_0} \tag{A}$$

Here ds is the magnitude of the elemental vector surface area of the spherical Gaussian surface. Hence, $\oint_S ds =$ total surface area of the spherical Gaussian surface $= 4\pi r^2$. Putting this value in Eq. (A), we have

$$E \, 4\pi r^2 = \frac{q}{\varepsilon_0}$$

or $$E = \frac{1}{4\pi r^2} \times \frac{1}{\varepsilon_0} \times q = \frac{1}{4\pi r^2} \times \frac{1}{\varepsilon_0} \times \frac{Q \, r^3}{R^3}$$

or $$\vec{E} = \frac{1}{4\pi \varepsilon_0} \frac{Q \vec{r}}{R^3}$$

is the electric field intensity due to the spherically symmetric distribution of charge Q at an internal point situated at a distance r from the centre of the spherically symmetric distribution charge $(r < R)$.

All the three cases are regarding the location of P represented graphically in Fig. 5.9.

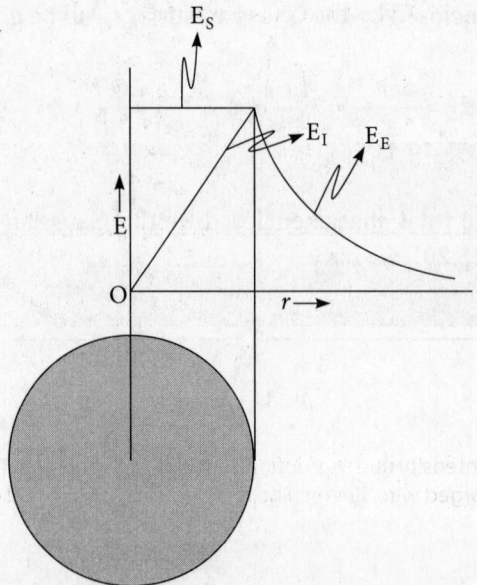

Figure 5.9 The variation of electric field intensity in case of a spherically symmetric distribution of charge. The curves marked E_E, E_I, and E_S represents Case 1, Case 2, and Case 3 respectively

Electric field due to a linear charge of uniform distribution

Let us consider a wire of infinite length placed horizontally to be charged uniformly. Hence, the wire has uniform linear charge density λ (charge per unit length measured in C/m). P is a point situated above the wire at a distance r from the wire where electric field intensity is to be found out. The Gaussian surface passing through this point P will be a right cylinder of radius r. Let its length be L. See Fig. 5.10. The electric flux will come out radially from the charged wire. Hence, no flux passes through the two circular faces of the cylindrical Gaussian surface. Because of axial symmetry of the cylindrical Gaussian surface, the magnitude of the electric field at any point on the curved surface will be the same and will be directed radially outward. For the curved surface S_1, the angle between E and the unit vector $\hat{n}_1$ is zero and for circular faces S_2 and S_3 of the cylindrical Gaussian surface, the angle is 90°. By putting these conditions into Eq. (5.20), we have

$$\oint_{S1} \vec{E} \cdot \hat{n} ds + \oint_{S2} \vec{E} \cdot \hat{n} ds + \oint_{S3} \vec{E} \cdot \hat{n} ds = \frac{1}{\varepsilon_0} \oint_C \lambda dx$$

Since the wire is uniformly charged, λ is constant and we can take it outside the integral sign. The previous equation becomes

$$\oint_{S1} \vec{E} \cdot \hat{n} ds + \oint_{S2} \vec{E} \cdot \hat{n} ds + \oint_{S3} \vec{E} \cdot \hat{n} ds = \frac{1}{\varepsilon_0} \times \lambda \int_0^L dx$$

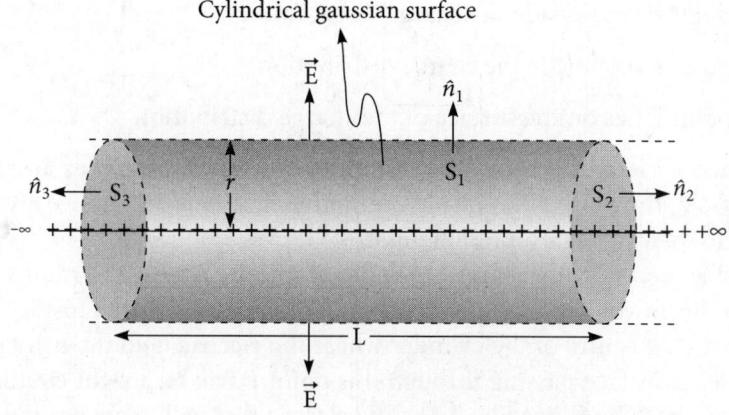

Figure 5.10 | Electric field intensity due to an infinite line of positive charge. The central thick line is the positively charged wire. Broken line represents the cylindrical Gaussian surface

or $\oint\limits_{S1} E|\hat{n}|\cos 0 ds + \oint\limits_{S2} E|\hat{n}|\cos 90° ds + \oint\limits_{S3} E|\hat{n}|\cos 90° ds = \dfrac{1}{\varepsilon_0} \times \lambda L$

or $E\oint\limits_{S1} ds = \dfrac{\lambda L}{\varepsilon_0}$ (B)

Now $\oint\limits_{S1} ds$ is the total curved surface area of the cylindrical Gaussian surface which is equal to $2\pi r L$ (the curved surface area of a cylinder). Putting this value in Eq. (B), we have

$$E \times 2\pi\, rL = \dfrac{\lambda L}{\varepsilon_0}$$

or $\vec{E} = \dfrac{\lambda}{2\pi\varepsilon_0\, r^2}\vec{r}$

is the electric field intensity due to the charged wire with linear charge density λ of infinite length at any point situated above the wire at a distance r from the wire. The direction of vector $\vec{E}$ is radially outward for a line of positive charge and radially inward for a line of negative charge.

Electric field due to a cylindrically symmetric charge distribution

In the calculation involving the determination of the electric field intensity at any point in case of a cylindrically symmetric charge distribution, three possible cases may arise.

Case 1: The point P lies outside the charge distribution.

Case 2: The point P lies inside the charge distribution.

Case 3: The point P lies on the surface of the charge distribution

We shall discuss Case 2 in great detail and other two cases are left as an exercise to the readers. In Case 2, the point I lies inside the charge distribution and we are to determine the electric field intensity at an internal point.

Let us consider a uniformly charged cylinder of infinite length and radius R. Hence, the cylinder has uniform volume charge density θ. P is the point situated inside the cylinder at a distance r from the centre of the cylinder where the electric field intensity is to be found out. The Gaussian surface passing through this point P will be a right circular cylinder of radius r. Let its length be L. See Fig. 5.11. The electric flux will come out radially from the uniformly charged cylinder. Hence, no flux passes through the two circular faces of the cylindrical Gaussian surface. Because of the axial symmetry of the cylindrical Gaussian surface, the magnitude of the electric field at any point on the curved surface will be the same and will be directed radially outward. For curved surface S_1, the angle between $\vec{E}$ and unit vector $\hat{n}_1$ is zero and for circular faces S_2 and S_3 of the cylindrical Gaussian surface, the angle is 90°. Under these conditions, we have

$$\oint_S \vec{E} \cdot \hat{n} ds = \frac{1}{\varepsilon_0} \oint_V \rho dv \tag{A}$$

Here, $\oint_V dQ = \oint \rho dv$ is the total charge enclosed by the cylindrical Gaussian surface of length L and radius r. In this case, $\oint_V \rho dv = Q$. Putting this value in Eq. (A), we have

$$\oint_{S1} \vec{E} \cdot \hat{n} ds + \oint_{S2} \vec{E} \cdot \hat{n} ds + \oint_{S3} \vec{E} \cdot \hat{n} ds = \frac{1}{\varepsilon_0} \times Q$$

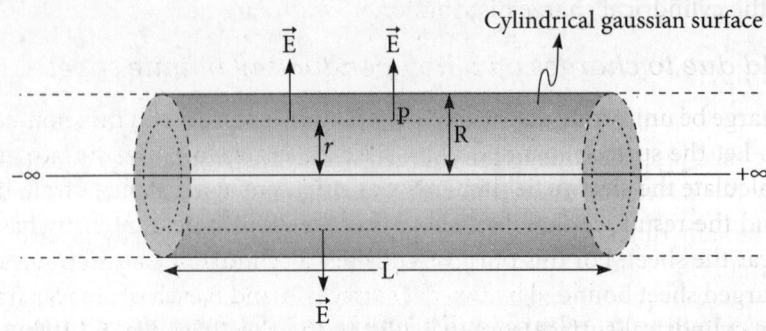

Figure 5.11 Electric field intensity due to the cylindrical symmetric distribution of charges when the point lies inside the charge distribution. The broken line represents the cylindrical Gaussian surface

or $\displaystyle\oint_{S1}\vec{E}\cdot\hat{n}ds+\oint_{S2}\vec{E}\cdot\hat{n}ds+\oint_{S3}\vec{E}\cdot\hat{n}ds=\frac{Q}{\varepsilon_0}$

Since the angle between $S1$ and the unit vector $\hat{n}$ is $0°$ and that between $S2$ and $\hat{n}$, $S3$ and $\hat{n}$ are $90°$ each, this equation becomes

$$\oint_{S1}E|\hat{n}|\cos 0ds+\oint_{S2}E|\hat{n}|\cos 90°ds+\oint_{S3}E|\hat{n}|\cos 90°ds=\frac{Q}{\varepsilon_0}$$

or $\displaystyle E\oint_{S1}ds=\frac{Q}{\varepsilon_0}$ (B)

Now $\displaystyle\oint_{S1}ds$ is the total curved surface area of the cylindrical Gaussian surface which is equal to $2\pi rL$. Q is the total charge enclosed by the cylindrical Gaussian surface of radius r and length L. Since $\pi r^2 L$ is the volume of the cylinder of length L and radius r and ρ is the uniform charge density, we have $Q=\rho\times\pi\, r^2 L$. Putting these two values in Eq. (B), we have

$$E\times 2\pi\, rL=\frac{1}{\varepsilon_O}\rho\times\pi r^2 L$$

or $\displaystyle\vec{E}=\frac{\rho\vec{r}}{2\varepsilon_O}$

is the electric field intensity due to the charged cylinder of infinite length with volume charge density ρ at any point situated inside the cylinder at a distance r from the centre of the cylinder. The direction of vector $\vec{E}$ is radially outward for a cylinder of positive charge and radially inward for a cylinder of negative charge. The result is independent of the radius of the charged cylinder and depends on the distance r between the point P and the centre of the cylindrical charge distribution.

Electric field due to charges on a non-conducting infinite sheet

Let electric charge be uniformly distributed over the plane surface of a thin non-conducting infinite sheet. Let the surface charge density (i.e., the charge per unit surface area) be σ. We have to calculate the electric field intensity at any point at a distance r from the surface charge. To find the result we shall first show that the electric field intensity has the same magnitude near the sheet. For this purpose consider a cylindrical Gaussian surface on one side of the charged sheet bounded by two flat surfaces A and B each of area S parallel to the sheet with the cylindrical surface perpendicular to the sheet. See Fig. 5.12. By symmetry, the electric field strength everywhere is normal to the flat bases and parallel to the curved surface. Let $\vec{E}_1$ and $\vec{E}_2$ be the electric field strength at the flat bases A and B respectively.

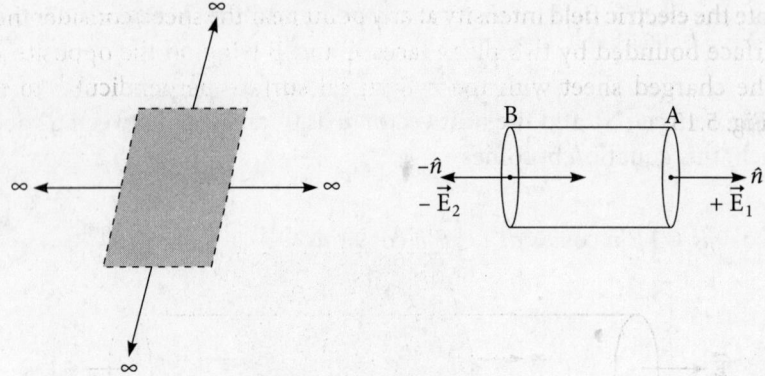

Figure 5.12 | Electric field intensity has the same magnitude on both sides of a non-conducting infinite plane sheet of charges

As the electric field strength is parallel to the curved surface, the contribution to electric flux due to the curved surface is zero.

The contribution to electric flux due to flat surface A

$$= \oint_S \vec{E}_1 \cdot \hat{n} ds = +E_1 S$$

The contribution to electric flux due to flat surface B

$$= \oint_S \vec{E}_2 \cdot \hat{n} ds = -E_2 S$$

From Gauss's law Eq. (5.20), we have

$$\oint_S \vec{E} \cdot \hat{n} ds = \frac{1}{\varepsilon_0} \oint_V \rho dv$$

or $\quad \oint_S \vec{E} \cdot \hat{n} ds = \frac{1}{\varepsilon_0} \times (\text{charge enclosed by the Gaussian surface})$

In this case, we have

$$-E_2 S + E_1 S = \frac{1}{\varepsilon_0} \times 0 = 0$$

or $\quad E_1 = E_2$

That means the electric field intensity is the same at all points near the sheet.

To calculate the electric field intensity at any point near the sheet, consider the cylindrical Gaussian surface bounded by two plane faces A and B lying on the opposite sides of and parallel to the charged sheet with the cylindrical surface perpendicular to the sheet as depicted in Fig. 5.13.

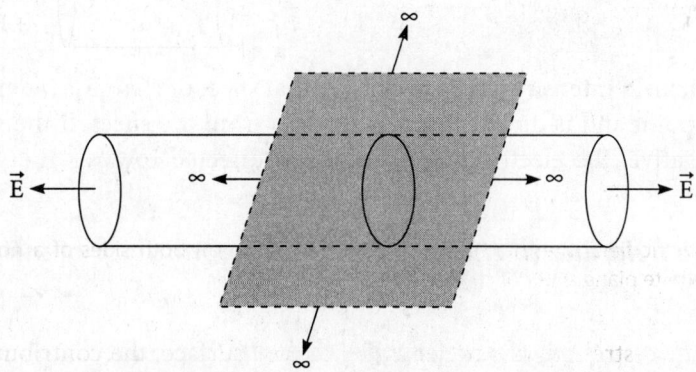

Figure 5.13 | Electric field intensity calculation due to an infinite non-conducting plane sheet of charges

As already explained the electric field intensity at every point on the flat surfaces is the same. By symmetry, the electric field intensity E is normal outward at the points on the two plane surfaces and parallel to the curved surface. Thus, the contribution to the total electric flux due to, curved surface is zero.

The contribution to the electric flux due to the flat surface A is equal to

$$\oint_S \vec{E} \cdot \hat{n}ds = ES$$

The contribution to the electric flux due to the flat surface B is equal to

$$\oint_S \vec{E} \cdot \hat{n}ds = ES$$

Thus, the total electric flux = $ES + ES = 2ES$

According to Gauss's law,

$$\oint_S \vec{E} \cdot \hat{n}ds = \frac{1}{\varepsilon_0} \times (\text{charge enclosed by the Gaussian surface})$$

or   Total electric flux =

$$\frac{1}{\varepsilon_0} \times (\text{Total charge enclosed by the Gaussian surface})$$

or $\quad 2ES = 2ES = \dfrac{1}{\varepsilon_0} \times (\sigma S)$

or $\quad E = \dfrac{\sigma}{2\varepsilon_0}$

Thus, the electric field intensity due to an infinite flat sheet of charge is independent of the distance of the point and is directed normally away from the sheet. If the surface charge density ρ is negative, the electric field intensity is directed towards the surface charge normally.

5.3.3F *Electric field due to charges on a plane conductor of infinite extent*

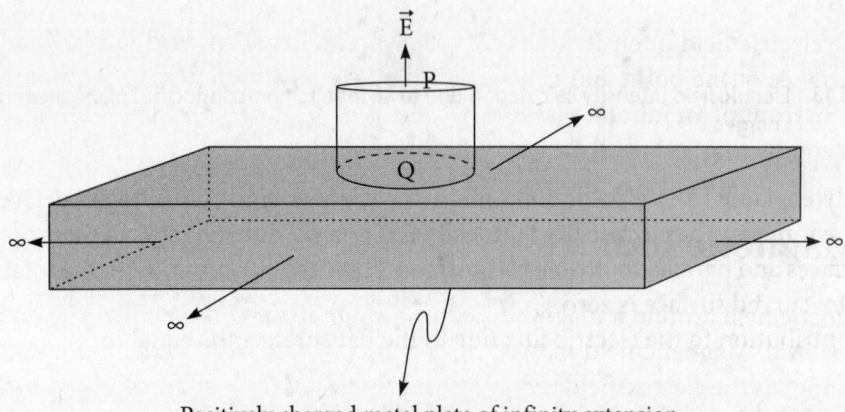

Positively charged metal plate of infinity extension

Figure 5.14 Electric field intensity calculation due to an infinite conducting plane sheet of charges

Figure 5.14 represents a portion of a charged infinite plane conductor sheet. By symmetry, the electric field strength everywhere is directed normally away from the plane of the sheet. To find the electric field intensity at any point near the sheet, consider a cylindrical Gaussian surface bounded by two plane bases P and Q each of area S parallel to the plane of the conductor with the cylindrical surface normal to the conductor. As the electric field intensity at every point inside the conductor is zero, the electric flux due to the base Q is zero. Moreover, as the curved surface is parallel to the electric flux, its contribution is zero. Therefore, it is only the plane base P that contributes to the total electric flux. Thus, we have

$$\text{Total electric flux} = \text{contribution to electric flux due to base } P = \oint_S \vec{E} \cdot \hat{n}\, ds = ES$$

According to Gauss's law

$$\oint_S \vec{E} \cdot \hat{n}\, ds = \frac{1}{\varepsilon_0} \times (\text{charge enclosed by the Gaussian surface})$$

or   Total electric flux $= \dfrac{1}{\varepsilon_0} \times (\text{Total charge enclosed by the Gaussian surface})$

or $ES = \dfrac{1}{\varepsilon_0} \times (\sigma\, S)$

or $E = \dfrac{\sigma}{\varepsilon_0}$

Thus, the electric field intensity due to a charged infinite plane conductor is independent of the distance of the point and is directed normally away from the charge. Electric field intensity in front of an infinite plane conductor is twice the electric field intensity in front of a non-conducting infinite thin sheet of charge.

5.4 Magnetic Induction

The magnetic field around a magnet is defined as the space within which magnetic effects can be realized. Magnetic field may be represented by magnetic lines of induction, in the same manner as the electric field is represented by electric lines of force. We can define the magnetic lines of induction as the imaginary lines drawn in a magnetic field in such a way that tangent to it at any point gives the direction of magnetic induction $\vec{B}$ at that point. The magnitude of magnetic induction is more at the places where concentration of magnetic lines of induction is more. With this background information, we shall discuss magnetic induction quantitatively in the following way.

When a charge moves in a magnetic field, in addition to the Coulomb force between charges, a new force will act on the moving the charge. This force $\vec{F}$ depends upon the charge q, velocity of the charge $\vec{v}$ and most importantly on a property of the magnetic field called magnetic induction $\vec{B}$ given by

$$\vec{F} = q\left(\vec{v} \times \vec{B}\right) \tag{5.24}$$

If in the concerned space, there is an electric field of intensity $\vec{E}$ as well as a magnetic field of induction $\vec{B}$, then the total force acting on the moving charge will be

$$q\vec{E} + q\left(\vec{v} \times \vec{B}\right)$$

The magnetic induction $\vec{B}$ can also be defined in the following way. When charges move in a conductor placed in a magnetic field, a new force will act on the conductor. This force $\vec{F}$ depends upon the electric current i, length of the conductor and most importantly on a property of the magnetic field called magnetic induction $\vec{B}$ given by

$$\vec{F} = \int_0^\ell i\,\vec{d\ell} \times \vec{B} \tag{5.25}$$

where $\vec{d\ell}$ is the elemental vector length of the conductor, its direction being the direction of current. $i\,d\ell$ is called a current element.

The magnetic induction $\vec{B}$ can also be defined as the magnetic flux or magnetic lines of induction ϕ per unit area of the surface perpendicular to $\vec{B}$. This has been depicted in Fig. 5.15. Mathematically, we can write the relation as

$$B = \frac{d\varphi}{dS} \tag{5.26}$$

or $\quad \varphi = \int_S \vec{B} \cdot \vec{dS}. \tag{5.27}$

For uniform distribution of magnetic lines of induction or magnetic flux and for plane area Eq. (5.27) becomes

$$\varphi = \vec{B} \cdot \vec{S} = BS \cos\theta$$

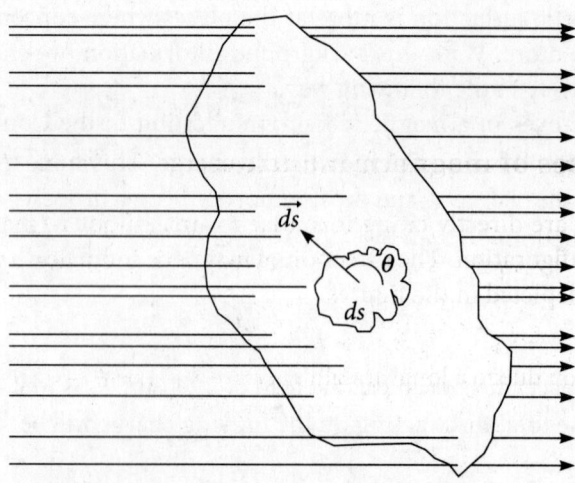

Figure 5.15 | Magnetic flux through a plane area making an angle θ with the flux

5.4.1 Units of magnetic induction

The SI unit of magnetic induction is found out to be $\dfrac{N}{Am}$ from Eq. (5.25) and $\dfrac{Wb}{m^2}$ from Eq. (5.26), Weber being the unit of magnetic flux. In cgs emu (electromagnetic unit), the unit of magnetic flux is Maxwell and so the unit of magnetic induction becomes $\dfrac{Maxwell}{cm^2}$ or $\dfrac{dyne}{abAcm}$. Hence, we have

$$1 \text{ Weber} = \frac{1Nm}{A}$$

$$1 \text{ Maxwell} = \frac{1\,dyne\,cm}{abA}$$

$$1 \text{ Weber} = 10^8 \text{ Maxwell}$$

$$1 \text{ abA} = 10A$$

$$\frac{1N}{Am} = \frac{1Wb}{m^2} = 1 Tesla \quad \text{(SI units)}$$

$$\frac{1dyne}{abAcm} = \frac{1Maxwell}{cm^2} = 1 gauss \quad \text{(cgs emu)}$$

$$1T = 10^4 \text{ gauss}$$

5.4.2 Special cases of magnetic induction

In the following, we are directly citing formulae for magnitude of magnetic induction in different current configuration. The direction of magnetic induction and the symbols used in the formulae are depicted in the figures.

i. Magnetic induction due to a long straight wire $= \dfrac{\mu_0}{4\pi}\dfrac{I_0}{r}\left(\sin\theta_2 - \sin\theta_1\right)$ (5.28)

ii. Magnetic induction due to an infinitely long straight wire $= \dfrac{\mu_0}{2\pi}\dfrac{I_0}{r}$ (5.29)

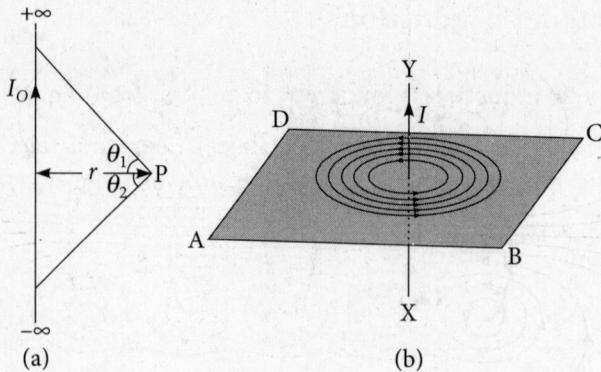

(a)   (b)

Figure 5.16 (a) Magnetic induction due to a long straight wire carrying a steady current I_0. If the wire is infinitely long then $\theta_1 = \dfrac{\pi}{2}$ and $\theta_2 = -\dfrac{\pi}{2}$. The direction of $\vec{B}$ at P is into the paper perpendicularly. (b) Shape and direction of the magnetic field due to a long current carrying wire

iii. Magnetic induction at any point on the axis of a circular coil $= \dfrac{\mu_0 n I_0 r^2}{2(r^2 + x^2)^{3/2}}$ $\qquad$ (5.30)

iv. Magnetic induction at the centre of a circular coil $= \dfrac{\mu_0 n I_0}{2r}$ $\qquad$ (5.31)

v. Magnetic induction at any point on the axis of a circular coil if $x >>> . r = \dfrac{\mu_0}{2} \dfrac{n I_0}{x^3} r^2$ $\qquad$ (5.32)

vi. Magnetic induction at any point on the axis of a solenoid $= \mu_0 n I_0$ $\qquad$ (5.33)

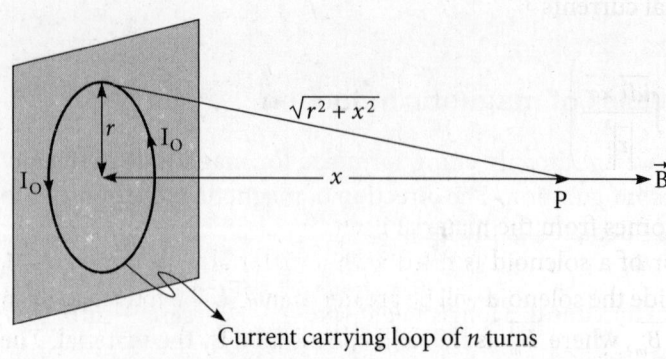

Current carrying loop of n turns

Figure 5.17 Magnetic induction at any point on the axis of a circular coil of n turns carrying current I_0. At the centre of the coil $x = 0$. If the point is very far away from the centre of the coil, then we can neglect r in comparison to x. The direction of $\vec{B}$ is along the axis away from the centre of the coil

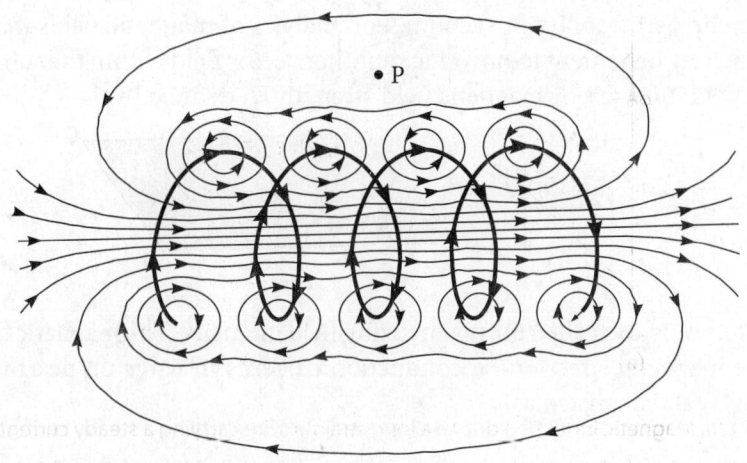

Figure 5.18 | Magnetic induction at any point on the axis of a solenoid having n turns per unit length carrying current I_0. The direction of B is along the axis of the solenoid

5.5 Magnetic Field Strength (Intensity)

The magnetic field generated by currents

$$\overrightarrow{B_O} = \frac{\mu_O}{4\pi} \int \frac{i \overrightarrow{d\ell} \times \overrightarrow{r}}{\left|\overrightarrow{r}\right|^3}$$

passing through wires wound over magnetic materials, result in the materials becoming magnetized. Thus, it contributes to the total magnetic field and the total magnetic field $\overrightarrow{B}$ inside the material will be different from $\overrightarrow{B_0}$. We shall now discuss what part of $\overrightarrow{B}$ comes from the external currents

$$\left(\overrightarrow{B_O} = \frac{\mu_0}{4\pi} \int \frac{i \overrightarrow{d\ell} \times \overrightarrow{r}}{\left|\overrightarrow{r}\right|^3} \right)$$

and what part comes from the material itself.

If the interior of a solenoid is filled with a material with non-zero magnetization, the total field $\overrightarrow{B}$ inside the solenoid will be greater than $\overrightarrow{B_0}$. The total field $\overrightarrow{B}$ inside the material will be $\overrightarrow{B} = \overrightarrow{B_0} + \overrightarrow{B_m}$, where $\overrightarrow{B_m}$ is the field contributed by the material. The component $\overrightarrow{B_m}$ is proportional to the magnetization vector $\overrightarrow{M}$ (total magnetic dipole moments per unit volume). Thus, we have

$$\overrightarrow{B} = \overrightarrow{B_0} + \mu_0 \overrightarrow{M} \tag{5.34}$$

μ_0 is the magnetic permeability of vacuum. For analysis of magnetic fields that arise from magnetization, it is convenient to introduce another vector field within the substance called the magnetic field intensity or magnetic field strength $\overline{H}$, defined by

$$\overline{H} = \frac{\overline{B}_0}{\mu_0} - \overline{M} \left(= \frac{1}{4\pi} \int \frac{i\overline{d\ell} \times \vec{r}}{\left|\vec{r}\right|^3} \right) \tag{5.35}$$

The same vector $\overline{H}$ is also called the magnetizing field intensity. The magnetic field strength represents the magnetic effect of the conduction currents in wires on nearby substances. Thus, Eq. (5.34) can be written as

$$\vec{B} = \mu_0 \left(\overline{H} + \overline{M} \right) \tag{5.36}$$

$\overline{H}$ and $\overline{M}$ will have the same units Am^{-1}. According to Eq. (5.36) the total magnetic field $\vec{B}$ $(= \mu \overline{H}$, μ being the magnetic permeability of the material) inside the material can be devided into two parts $\mu_0 \overline{H}$ is due to external factors such as the current and $\mu_0 \overline{M}$ is due to the specific nature of the magnetic material.

5.6 Ampere's Circuital Law

Ampere's circuital law states that the line integral of magnetic induction $\vec{B}$ around a closed path C is equal to μ_0 times the total electric current I enclosed by the path C when there is no magnetic materials present. Mathematically, Ampere's circuital law is given by

$$\oint_C \vec{B}.\overline{d\ell} = \mu_0 I \tag{5.37}$$

where $\overline{d\ell}$ is an elemental vector length of the path C. The direction of $\overline{d\ell}$ is the direction in which the magnetic flux will encircle the current carrying wire. If $\hat{r}$ is a unit vector in the direction of $\overline{d\ell}$, Eq. (5.37) becomes

$$\oint_C \vec{B} \cdot \hat{r} d\ell = \mu_0 I \tag{5.38}$$

Taking the magnitude of the scalar product, Eq. (5.38) becomes

$$\oint_C B d\ell \cos\theta = \mu_0 I \tag{5.39}$$

where θ is the angle between the direction of magnetic induction $\vec{B}$ and the unit vector $\hat{r}$. The magnitude of $\hat{r}$ is 1 since it is a unit vector in the direction of $\overline{d\ell}$.

The direction of magnetic induction $\vec{B}$ near a wire carrying current is given by the right-hand rule. According to the rule, if the wire is held with the right, thumb pointing in the direction of the current, then the direction of the magnetic intensity $\vec{B}$ will be the direction in which the other fingers are curled towards. This is depicted in the Fig. 5.19.

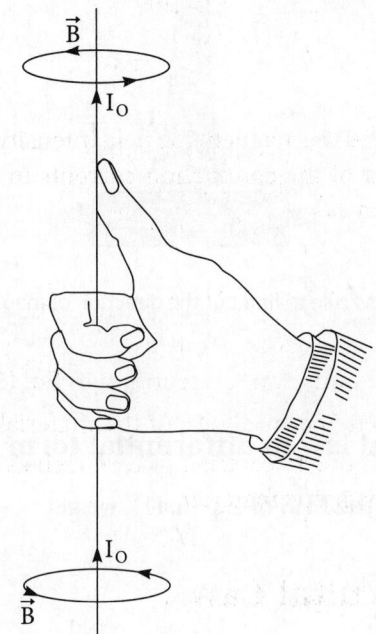

Figure 5.19 | Right-hand rule to find out the direction of magnetic induction $\vec{B}$ produced by electric current

We can also find out the direction of magnetic induction by the rotation of a right-hand screw. Place the screw perpendicularly on the cross-section of the current carrying wire. The direction of magnetic induction will be the direction of rotation of the right-hand screw so that the screw will move in the direction of the current. This is shown in Fig. 5.20. In terms of current density, J (i.e., $J = \dfrac{dI}{ds}$ current per unit cross-sectional area), total electric current is given by

$$I = \int_S \vec{J} \cdot \vec{ds} \quad \left(\because dI = \vec{J} \cdot \vec{ds} \right)$$ (5.40)

Hence, mathematical formulation of Ampere's circuital law becomes

$$\oint_C \vec{B} \cdot \vec{d\ell} = \mu_0 \int_S \vec{J} \cdot \vec{ds}$$ (5.41)

where $\hat{n}$ is the unit vector normal to the cross-sectional area in the outward direction.

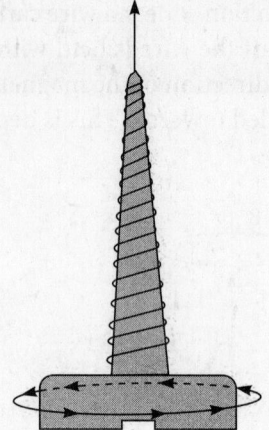

Figure 5.20 Maxwell's corkscrew rule to find out the direction of magnetic induction $\vec{B}$ produced by electric current

5.6.1 Ampere's circuital law in differential form

Applying Stokes' theorem to the LHS of Eq. (5.41), we get

$$\int_C \vec{\nabla} \times \vec{B} \cdot \vec{ds} = \mu_0 \int_S \vec{J} \cdot \vec{ds}$$

or $\quad \vec{\nabla} \times \vec{B} = \mu_0 \vec{J}$ (5.42)

5.6.2 Applications of Ampere's circuital law

Magnetic induction due to an infinitely long straight current carrying wire

Let us consider an infinite length of straight wire carrying a steady current I_0. Let P be the point at which magnetic induction $\vec{B}$ is to be found out. The perpendicular distance between the wire and the point P is r. Now imagine a circular path C through the point P so that the wire is normal to the plane of the circular path C at the centre of the circular path C. Thus, the circular path C encloses the current carrying wire with the centre at the wire. This is shown in Fig. 5.21.

Then from Ampere's law, we have

$$\oint_C \vec{B} \cdot \hat{r} \, d\ell = \mu_0 I_0$$

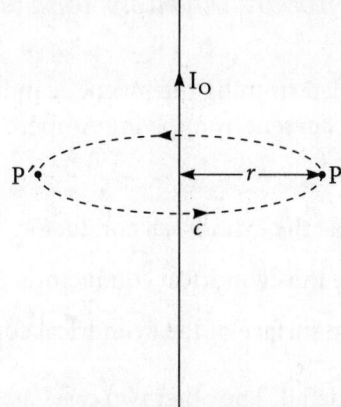

Figure 5.21 | Magnetic induction due to an infinitely long straight wire carrying current I_0. The direction of magnetic induction $\vec{B}$ at P is normal into the plane of the paper and at P' it is normal out of the paper

Here the path C encloses the current I_0 flowing through the wire. In this case, the direction of $\vec{B}$ and dl (or unit vector $\hat{r}$) are the same, i.e., angle between the magnetic induction $\vec{B}$ and the unit vector $\hat{r}$ is 0 at any point on the path C. In this case, the shapes of the magnetic fluxes are concentric circles. Figure 5.16(b) may be consulted. The previous equation becomes

$$\oint_C B d\ell = \mu_0 I_0$$

Now the magnitude of the magnetic induction $\vec{B}$ is the same, i.e., it is constant at any point on the circular path C. Hence, we can take it outside the integration sign. The previous equation becomes

$$B \oint_C d\ell = \mu_0 I_0 \tag{A}$$

Here $\oint_C d\ell$ is the total length of the path C. The radius of the circular path C is r (equal to the perpendicular distance between the wire and the point P). Hence, the total length of the circular path C will be equal to $2\pi r$, i.e.,

$$\oint_C d\ell = 2\pi r$$

Hence, from (A) we have

$$B = \frac{\mu_0 I_0}{2\pi r}$$

is the magnitude of magnetic induction $\vec{B}$ at a point situated at a perpendicular distance r from the wire carrying a steady current I_0.

Magnetic induction due to an infinitely long straight current carrying cylinder

In the calculation involved in determining the magnetic induction at any point in case of a cylindrical conductor carrying current by applying Ampere's circuital, three possible cases may arise.

Case 1: The point P lies outside the cylindrical conductor.

Case 2: The point P lies inside the cylindrical conductor.

Case 3: The point P lies on the surface of the cylindrical conductor.

We shall discuss the Case 2 in detail. The other two cases are left as exercises to the readers. In Case 2, the point P lies inside the charge distribution and we are to find out electric field intensity at an internal point.

The point P at which magnetic induction is to be found out lies inside at a perpendicular distance of r from the central axis of the cylindrical conductor of radius R. This is shown in Fig. 5.22.

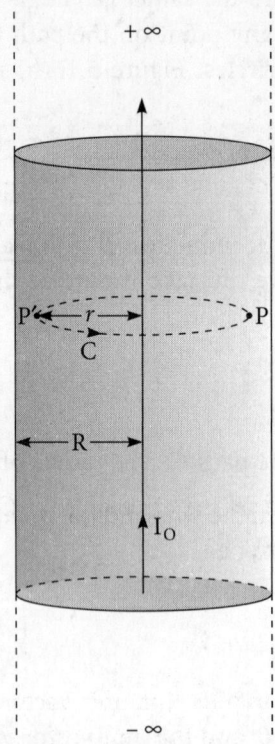

Figure 5.22 | Magnetic induction due to an infinitely long straight wire carrying current I_0 when point P is situated inside the cylinder. The direction of magnetic induction $\vec{B}$ at P is normal into the plane of the paper and at P' it is normal out of the paper

Let us consider an infinite length straight cylindrical conductor carrying a steady current I_0 distributed uniformly throughout the cross-sectional area. Let P the point at which the magnetic induction $\vec{B}$ is to be found out be situated inside the cylindrical conductor. The perpendicular distance between the central axis of the cylindrical conductor and the point P is r. Now imagine a circular path C through the point P so that the cylindrical conductor is normal to the plane of the circular path C at the centre of the circular path C. See Fig. 5.22. Then according to Ampere's circuital law, we have

$$\oint_C \vec{B} \cdot \hat{r} d\ell = \mu_0 I_0 \tag{A}$$

Here I_0 is the current flowing through the cross-sectional area of the cylindrical conductor enclosed by the circular path C. However, in our present case, the amount of current flowing through the cross-sectional area of the cylindrical conductor is not the same as the amount of current flowing through the cross-sectional area enclosed by the circular path C. Therefore, we shall first calculate the amount of current flowing through the cross-sectional area enclosed by the circular path C. The cross-sectional area of the cylindrical conductor is πR^2 since the radius of the cylindrical conductor is R. Through the cross-sectional area πR^2, the amount of current flowing is I_0. Therefore, the amount of current flowing per unit cross-sectional area will be given by

$$\frac{I_0}{\pi R^2}$$

The cross-sectional area enclosed by the circular path C is πr^2 since r is the radius of the path C. Therefore, the amount of current flowing through the cross-sectional area πr^2 will be given by

$$\frac{I_0}{\pi R^2} \times \pi \, r^2 = I_0 \times \frac{r^2}{R^2}$$

or $\quad I = I_0 \times \dfrac{r^2}{R^2}$

is the amount of current enclosed by the circular path C. Putting this value of current in Eq. (A), we have

$$\oint_C \vec{B} \cdot \hat{r} d\ell = \mu_0 \left(I_0 \times \frac{r^2}{R^2} \right)$$

In this case, the direction of $\vec{B}$ and $\overline{d\ell}$ (or unit vector $\hat{r}$) are the same, i.e., the angle between the magnetic induction $\vec{B}$ and the unit vector $\hat{r}$ is 0 at any point on the path C. The previous equation becomes

$$\oint_C B d\ell = \mu_0 \left(I_0 \times \frac{r^2}{R^2} \right)$$

Now the magnitude of the magnetic induction $\vec{B}$ is the same, i.e., is constant at any point on the circular path C. Hence, we can take it outside the integration sign. The previous equation becomes

$$B\oint_C d\ell = \mu_0 I_0 \times \frac{r^2}{R^2} \tag{B}$$

Here $\oint_C d\ell$ is the total length of the path C. The radius of the circular path C is r. Hence, the total length of the circular path C will be equal to $2\pi r$, i.e.,

$$\oint_C dl = 2\pi r$$

Hence, from (B), we get

$$B = \frac{\mu_0 I_0}{2\pi \, R^2} \times r$$

is the magnitude of the magnetic induction $\vec{B}$ at a point situated inside the cylindrical conductor at a perpendicular distance r from the central axis of the cylindrical conductor carrying a steady current I_0. In this case, the variation of the magnetic induction $\vec{B}$ with r is not absurd. The magnetic induction at any point on the central axis of a cylindrical conductor carrying a current I_0 is zero.

All the previously mentioned three cases are represented graphically in Fig. 5.23.

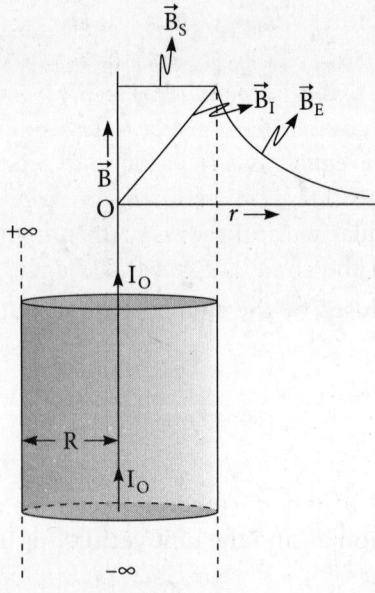

Figure 5.23 | The variation of magnetic induction in case of a cylindrical current carrying conductor. The curves marked B_E, B_I, and B_S represents Case 1, Case 2, and Case 3 respectively

5.7 Faraday's Law of Electromagnetic Induction

Faraday's law of electromagnetic induction states that the induced emf in a circuit is equal to the negative rate at which the magnetic flux φ through the circuit is changing. The mathematical statement of the law is

$$\text{Induced emf} = -\frac{d\varphi}{dt} \tag{5.43}$$

The negative sign appears in Eq. (5.43) due to the fact that the direction of induced emf is such that it opposes the very cause which causes it. This statement for the negative sign is called Lenz's law. The magnetic flux linked with a coil can be changed either by moving the coil through the magnetic filed or by changing the magnetic field it self. Whenever magnetic flux is linked with the coil changes, induced emf is produced in the coil. The induced emf exists in the coil as long as the magnetic flux is linked with the coil changes. More rapid is the change of magnetic flux more is the magnitude of induced emf.

The time-varying magnetic flux

$$\left(\frac{d\varphi}{dt} \neq 0 \right)$$

sets up induced electric field E at various points around the coil. The induced electric field is non-conservative whereas the electric field produced by static charge is conservative. This induced electric field is as real as the electric field produced by static charges. The electric force (i.e., Coulomb force) exerted on a test charge q_0 due to this induced electric field will be $q_0 E$. Thus, we can say that a changing magnetic field produces an electric field. If the time-varying magnetic flux is perpendicular to the plane of the paper, the electric flux produced will be concentric circles. The concept is depicted in the Fig. 5.24.

The induced emf will be equal to the amount of work done (work done = force × length of the path in the direction of force) in circulating a unit positive charge around the path. For a circular path of radius r, the amount of work done in moving a unit positive charge around the circle is $E2\pi r$ and is equal to the induced emf. For any other path the induced emf is given by line integral of the induced emf. Thus, we have

$$\text{Induced emf} = \oint_C \vec{E} \cdot \vec{dr} \tag{5.44}$$

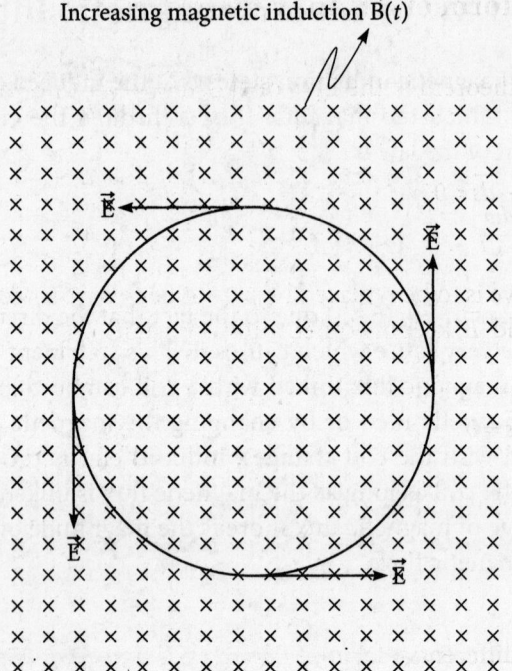

Figure 5.24 | Induced electric fields are produced by time-varying increasing magnetic fields. The time-varying magnetic flux are into the plane of the paper in a perpendicular direction

or $\qquad \oint_C \vec{E} \cdot \vec{dr} = -\dfrac{d\varphi}{dt}$ $\hfill$ (5.45)

5.7.1 Integral form of Faraday's law

From Eq. (5.45), we have

$$\oint_C \vec{E} \cdot \vec{dr} = -\frac{d\varphi}{dt} = -\frac{d}{dt} \oint_S \vec{B} \cdot \vec{ds}$$

Since surface S does not change its position or shape with time, we have

$$\oint_C \vec{E} \cdot \vec{dr} = -\oint_S \frac{d\vec{B}}{dt} \cdot \vec{ds}. \hfill (5.46)$$

Equation (5.46) is the integral form of Faraday's law of electromagnetic induction.

5.7.2 Differential form of Faraday's law

Now applying Stokes' theorem to the LHS of $\oint_C \vec{E} \cdot \vec{dr} = -\oint_S \frac{d\vec{B}}{dt} \cdot \vec{ds}$, we have

$$\oint_S \vec{\nabla} \times \vec{E} \cdot \vec{ds} + \oint_S \frac{d\vec{B}}{dt} \cdot \vec{ds} = 0$$

The total time derivative is to be replaced by partial derivative since we are only concerned with changes in the field $\vec{B}$, keeping other things constant.

$$\oint_S \left(\vec{\nabla} \times \vec{E} + \frac{\partial \vec{B}}{\partial t} \right) \cdot \vec{ds} = 0$$

or $\qquad \vec{\nabla} \times \vec{E} = -\frac{\partial \vec{B}}{\partial t}$ $\hfill (5.47)$

Equation (5.47) is the differential form of Faraday's law of electromagnetic induction.

The vector field is defined to be conservative (work done is independent of path; electric field is produced by static charges; gravitational field of earth) if curl of the vector field is zero. Otherwise, it is non-conservative. Equation (5.47) shows that the electric field has a non-conservative part due to changing magnetic flux density and a conservative part due to electric charge density. Faraday's law of induction shows how electric and magnetic fields are interconnected. Their independent nature vanishes when they become time dependent. Thus, these two fields are combined as a single field known as electromagnetic field.

Suppose in a region of space, magnetic induction (hence magnetic flux) B is changing at the constant rate $\frac{dB}{dt}$. Imagine a circular path of radius r, whose plane is perpendicular to magnetic flux. An electric field of constant magnitude will be produced at all the points on this path with its direction tangent to the path. In this case, the angle between $\vec{B}$ and $\vec{ds}$ is zero and that between $\vec{E}$ and $\vec{dr}$ is zero. Under these conditions, Eq. (5.46) becomes

$$E \oint_C dr = -\frac{dB}{dt} \oint_S ds$$

or $E \times$ length of the circular path $= -\frac{dB}{dt} \times$ cross-sectional area of the circular path.

$$E = -\frac{1}{2} r \frac{dB}{dt}.$$

Here, the minus sign indicates the direction of electric field intensity. The direction of electric field intensity is such that it opposes the change of magnetic intensity or change of magnetic flux (Lenz's law).

Example 5.24

A wire of length 2 m perpendicular to X-Y plane is moving with a constant velocity $\vec{v} = 2\hat{x} + 3\hat{y} + \hat{z}$ m/s through a region of uniform magnetic field of induction $\vec{B} = \hat{x} + 2\hat{y} \dfrac{Wb}{m^2}$. Calculate the potential difference between the two ends of the wire.

Solution

When a rod moves in a magnetic field, the magnetic field exerts $q(v \times B)$force on the moving rod. For the constancy of the velocity of the rod, this magnetic force has to be balanced. The magnetic force $q\left(\vec{v} \times \vec{B}\right)$ is balanced by the electric force $q\vec{E}$ developed by the internally produced electric field. Thus, we have

$$q\vec{E} + q\left(\vec{v} \times \vec{B}\right) = 0$$

or $\vec{E} = -\left(\vec{v} \times \vec{B}\right) = \left(\hat{x} + 2\hat{y}\right) \times \left(2\hat{x} + 3\hat{y} + \hat{z}\right) = \left(2\hat{x} - \hat{y} + \hat{z}\right)$

If L is the length of the rod, then the potential difference developed between the two ends of the rod will be

$$e = E \cdot L = \left(2\hat{x} - \hat{y} + \hat{z}\right) \cdot 2\hat{z} = 2V$$

($L = 2\hat{z}$ because its length is along the Z-axis)

Example 5.25

The magnetic flux φ through a coil perpendicular to its plane is varying with time t obeying the equation $\varphi = 5t^2 + 7t + 9$ Wb. Calculate the emf induced in the coil at time = 2 seconds.

Solution

The induced emf produced in the coil, according to Faraday's law of induction, is given by

$$e = -\frac{d\varphi}{dt}$$

$$e = \frac{d}{dt}(5t^2 + 7t + 9)\frac{Wb}{s} = (10t + 7)V$$

At $t = 2$ seconds   induced emf = 27 V.

5.8 Displacement Current

According to Ampere's circuital law, $\oint_C \vec{B}.\overline{d\ell} = \mu_0 I$ if there are no magnetic materials present in the vicinity of the current carrying wire. In the presence of magnetic materials, Ampere's law is modified and given by

$$\oint_C \vec{B}.\overline{d\ell} = \mu_0 I_C + \mu_0 I_M \tag{5.48}$$

where I_C is the conduction current and I_M is the magnetizing current.

According to Faraday's law of electromagnetic induction, a changing magnetic field produces electric field ($-\dfrac{d\varphi}{dt} = \oint_C \vec{E}.\overline{d\ell}$, φ is the magnetic flux). Nature loves symmetry. Therefore, we may expect the reverse concept to be true. That means a changing electric field produces magnetic field, i.e.,

$$\mu_0 \varepsilon_0 \frac{d\varphi_E}{dt} = \oint_C \vec{B}.\overline{d\ell} \tag{5.49}$$

where φ_E is the electric flux or electric line of force. Students should verify that

$$\varepsilon_0 \frac{d\varphi_E}{dt}$$

has the dimension of current. Though there is no motion of charge, we name

$$\varepsilon_0 \frac{d\varphi_E}{dt}$$

as displacement current I_d. Thus, we have

$$I_d = \varepsilon_0 \frac{d\varphi_E}{dt} \tag{5.50}$$

Equation (5.50) can also be derived by differentiating Eq. (5.18) with respect to time.

The magnetic field can be set up by (a) magnetizing bodies, (b) electric current and (c) time-varying electric field. Ampere's circuital law, which gives the quantitative determination of magnetic induction, can be expressed with full generality as

$$\oint_C \vec{B}.\overline{d\ell} = \mu_0 I_C + \mu_0 I_M + \mu_0 I_d \tag{5.51}$$

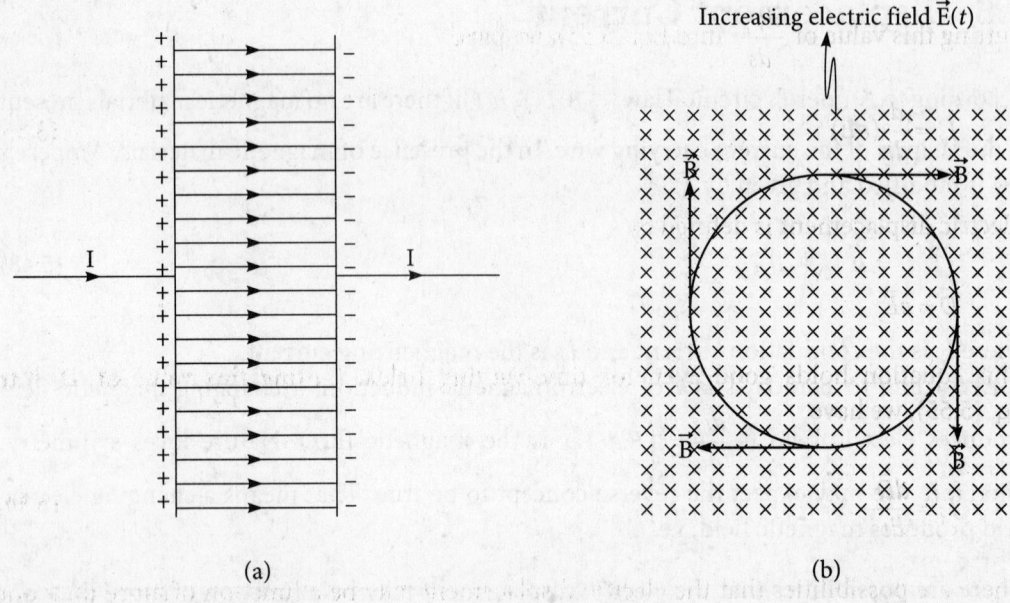

(a)　　　　　　　　　　　　　　　　(b)

Figure 5.25 | (a) The production of induced magnetic field in the gap between two plates of a parallel plate capacitor due to a time-varying electric field. The time-varying electric field is established between the two plates of the capacitor during its charging and discharging. In both the cases, the variation of the electric flux is exponential in nature. (b) Here, the induced magnetic field is produced due to increasing electric field The induced magnetic flux is circular with its plane perpendicular to the electric flux

The displacement current density J_d is defined as the displacement current per unit cross-sectional area. Hence, mathematically, the displacement current density is given as

$$J_d = \frac{dI_d}{ds} \qquad (5.52)$$

Putting the value of displacement current I_d from Eq. (5.50) into Eq. (5.52), we get

$$J_d = \frac{d}{ds}\left(\varepsilon \frac{d\varphi_E}{dt}\right) = \frac{d}{dt}\left(\varepsilon \frac{d\varphi_E}{ds}\right) \qquad (5.53)$$

Now Gauss's law can be re-written here as $\varphi_E = \oint_S \vec{E} \cdot \overline{ds}$

or $\qquad \dfrac{d\varphi_E}{ds} = E \qquad (5.54)$

Putting this value of $\dfrac{d\varphi_E}{ds}$ into Eq. (5.53), we have

$$J_d = \frac{d}{dt}(\varepsilon E) \tag{5.55}$$

Electric displacement is defined by

$$\vec{D} = \varepsilon \vec{E}$$

This equation holds good even for time-varying fields. Putting this value of $\vec{D}$ into Eq. (5.55), we have

$$\vec{J}_d = \frac{d\vec{D}}{dt} \tag{5.56}$$

There are possibilities that the electric displacement may be a function of more than one variable. Therefore, we modify Eq. (5.56) and obtain

$$\vec{J}_d = \frac{\partial \vec{D}}{\partial t}$$

5.8.1 Physical significance of displacement current

Equation (5.50) gives the magnitude of the displacement current. Unlike conduction current, which is produced due to the motion of the charges, displacement current is produced due to the time variation of the electric flux. Like conduction current, the displacement current is continuous. However, conduction current is not continuous across the capacitor gap whereas displacement current is continuous across the capacitor gap, i.e., displacement current flows through the capacitor gap. Otherwise, the capacitor plates cannot be charged by steady conduction current. The magnitude of the displacement current flowing through the capacitor gap is equal to the conduction current flowing in the wires connecting the plates of the capacitor. This is depicted pictorially in Fig. 5.25.

5.8.2 Distinction between conduction current and displacement current

	Conduction current	Displacement current
1	Conduction current is produced due to the actual flow charge carriers in conductors.	Displacement current is produced due to time-varying electric flux.
2	The flow of conduction current is possible only in a conducting material medium.	The flow of displacement current is possible in any medium, even in vacuum.
3	The magnitude of conduction current depends upon the applied potential difference and resistivity of the conductors.	The magnitude of displacement current depends upon the rate of change of electric flux with respect to time and permittivity of the medium.
4	Conduction current obeys Ohm's law.	Displacement current does not obey Ohm's law.
5	In almost all conductors, conduction current dominates displacement current normally.	In almost all conductors, displacement current is dominated by conduction current normally.
6	Thus, conduction current lags behind displacement current by a phase difference of $\frac{\pi}{2}$ when an alternating electric field is applied.	Thus, displacement current leads the conduction current by a phase difference of $\frac{\pi}{2}$ when an alternating electric field is applied.
7	$I_C = \dfrac{\text{Potential difference}}{\text{Resistance}}$	$I_d = \varepsilon \dfrac{d\varphi_E}{dt}$

Example 5.26

The electric field intensity between two circular plates each of radius R of a parallel plate capacitor is increasing at the rate of $\dfrac{dE}{dt}$. Calculate the induced magnetic field B at various points x on the vertical section. Figure 5.25 may be referred.

Solution

According to Eq. (5.49), we have

$$\oint_C \vec{B} \cdot \vec{d\ell} = \varepsilon_0 \mu_0 \frac{d\varphi_E}{dt}.$$

where $\vec{d\ell}$ is an elemental length on the circular path C. C is a circular path imagined on the vertical section passing between the two plates of the capacitor so that the electric flux are perpendicular to the plane of this circular path and the centre of the circular plate lie on a straight line perpendicular to the plates of the capacitor. Since the path is circular, angle between magnetic induction $\vec{B}$ and $\vec{d\ell}$ is 0°. Hence, the previous equation becomes

$$\oint_C B d\ell = \varepsilon_0 \mu_0 \frac{d\varphi_E}{dt} \tag{A}$$

Case 1:   Suppose $x \leq R$ (i.e., the point lies inside the two circular plates). Here, x is the distance of the point at which the induced magnetic induction is to be found out. Through this point, the circular path C is imagined. Obviously, the radius of this circular path will be x. The magnitude of the induced magnetic induction at any point on the circular path is constant. Hence, Eq. (A) becomes

$$B \int_0^{2\pi x} d\ell = \varepsilon_0 \mu_0 \frac{d\varphi_E}{dt}$$

Since the electric field intensity E is uniform between two plates of the capacitor, the amount of flux passing through a circular path of radius x is $\varphi_E = \pi x^2 E$ and putting this vale of φ_E in the previous equation, we have

or $$B \times 2\pi x = \varepsilon_0 \mu_0 \frac{d \, \pi x^2 E}{dt}$$

or $$B = \frac{\varepsilon_0 \mu_0 x}{2} \times \frac{dE}{dt}.$$

Case 2:   Suppose $x \geq R$ (i.e., the point lies outside the two circular plates). Here, x is the distance of the point at which the induced magnetic induction is to be found out. Through this point, the circular path C is imagined. Obviously, the radius of this circular path will be x. The magnitude of the induced magnetic induction at any point on the circular path is constant. Hence, Eq. (A) becomes

$$B \int_0^{2\pi x} d\ell = \varepsilon_0 \mu_0 \frac{d\varphi_E}{dt}.$$

Since the electric field intensity E is uniform between two plates of the capacitor, the amount of flux passing through a circular path of radius R is $\varphi_E = \pi R^2 E$ and putting this vale of φ_E in the previous equation, we have

$$B \times 2\pi x = \varepsilon_0 \mu_0 \frac{d \, \pi R^2 E}{dt}$$

or $$B = \frac{\varepsilon_0 \mu_0 R^2}{2x} \frac{dE}{dt}$$

Example 5.27

The electric field intensity E between two circular plates each of radius R of a parallel plate capacitor is increasing at the rate of $\dfrac{dE}{dt}$. Calculate the displacement current. Figure 5.25 may be seen.

Solution

According to Eq. (5.50), we have

$$I_d = \varepsilon_0 \frac{d\varphi_E}{dt}$$

Since the electric field intensity E is uniform between the two circular plates of the capacitor, the amount of flux passing between the two circular plates, each of radius R, will be $\varphi_E = \pi R^2 E$. Putting this vale of φ_E in the previous equation, we have

$$I_d = \varepsilon_0 \frac{d\,\pi R^2 E}{dt}$$

or $I_d = \varepsilon_0 \pi R^2 \dfrac{dE}{dt}$

The displacement current can be expressed in another form in the following way for any capacitor.

Let A be the area of each plate. Hence, from the previous equation, we have

$$I_d = \varepsilon_0 A \frac{dE}{dt} \qquad (A)$$

The uniform electric field intensity between the two plates of the capacitor is given by

$$E = \frac{\sigma}{\varepsilon_0},$$

where σ is the surface charge density $= \dfrac{q}{A}$.
Hence, we have

$$E = \frac{q}{A\varepsilon_0} = \frac{CV}{A\varepsilon_0}$$

where C is the capacitance of the capacitor and V is the potential difference between the two plates of the capacitor at any time t.

Putting this value of E into Eq. (A), we have

$$I_d = \varepsilon_0 A \frac{d}{dt}\left(\frac{CV}{A\varepsilon_0}\right)$$

or $I_d = C\dfrac{dV}{dt}$.

Example 5.28

Prove that the displacement current flowing between the two plates of the capacitor is equal to the conduction current utilized in charging the capacitor.

Solution

Figure 5.25(a) may be consulted. According to Eq. (5.50), the displacement current I_d is given by

$$I_d = \varepsilon_0 \frac{d\varphi_E}{dt}$$

Since the electric field intensity E is uniform between the two plates of the capacitor, the amount of flux passing between the two plates each of area A, will be $\varphi_E = AE$ and putting this vale of φ_E in the previous equation, we have

$$I_d = \varepsilon_0 \frac{dAE}{dt}$$

or $I_d = \varepsilon_0 A \dfrac{dE}{dt}$ (A)

The uniform electric field intensity between the two plates of the capacitor is given by

$$E = \frac{\sigma}{\varepsilon_0},$$

where $\sigma = \dfrac{q}{A}$ = surface charge density

Hence, we have

$$E = \frac{q}{A\varepsilon_0}$$

Now putting this value of E into Eq. (A), we have

$$I_d = I_d = \varepsilon_0 A \frac{d}{dt} \left(\frac{q}{A\varepsilon_0} \right)$$

or $I_d = \frac{dq}{dt} = I_C = $ conduction current.

Example 5.29

In a certain silicon rod of resistivity 2000 Ωm carrying a current, the electric field varies with time according to the equation $E = E_0 \sin \omega t$. Calculate (a) the maximum conduction current density and (b) the maximum displacement current density if $\varepsilon = \varepsilon_0$. Given, $E_0 = 0.1$ V/m, $\omega = 2\pi \times 60$ Hz.

Solution

The conduction current I_C, is given by Ohm's law as

$$I_C = \frac{V}{R} = \frac{VA}{\rho\ell} = \frac{EA}{\rho}$$

where

$V = $ Potential difference between the two ends of the conductor.

$R = $ Resistance of the conductor.

$A = $ Cross-sectional area of the conductor.

$E = $ Electric field intensity applied to the conductor.

$\ell = $ Length of the conductor.

$\rho = $ Resistivity of the material of the conductor.

From this equation, the conduction current density J_C is given by

$$J_C = \frac{E}{\rho}$$

According to the question, this equation becomes

$$J_C = \frac{E_0 \sin \omega t}{\rho}.$$

Hence, the maximum value of the conduction current density from this equation will be

$$J_{CMAX} = \frac{E_0}{\rho}$$

Since maximum value of $\sin \omega t = 1$

Putting the numerical values of different quantities from the question into this equation, we get

$$J_{CMAX} = \frac{0.1\,\text{Vm}^{-1}}{2000\,\Omega\text{m}} = 5 \times 10^{-5}\,\text{Am}^{-2}$$

According to Eq. (5.50), displacement current density is given by

$$J_d = \frac{dD}{dt} = \frac{d\varepsilon_0 E}{dt}$$

$$J_d = \varepsilon_0 \frac{dE}{dt}$$

According to the question, this equation becomes

$$J_d = \varepsilon_0 \frac{dE_0 \sin \omega t}{dt} = \varepsilon_0 \omega E_0 \cos \omega t$$

Hence, the maximum value of displacement current density from this equation will be

$$J_{dMAX} = \varepsilon_0 \omega E_0$$

Putting the numerical values of different quantities from the question into this equation, we get

$$J_{dMAX} = 8.85 \times 10^{-12}\,\text{C}^2\text{N}^{-1}\text{m}^{-2} \times 2\pi \times 60\,\text{Hz} \times 0.1\,\text{Vm}^{-1} = 3.34 \times 10^{-10}\,\text{Am}^{-2}.$$

Example 5.30

In Example 5.29, (a) what is the phase difference between the conduction current and the displacement current? (b) At what frequency would the maximum conduction current density and the maximum displacement current density becomes equal.

Solution

(a) In Example 5.29, we have obtained

$$J_C = \frac{E_0 \sin \omega t}{\rho} \quad \text{and} \quad J_d = \varepsilon_0 \omega E_0 \cos \omega t$$

for the same specimen having constant area of cross-section A. Hence, the above two equations can be written as

$$I_C = \frac{A E_0 \sin \omega t}{\rho} \quad \text{and} \quad I_d = A\varepsilon_0 \omega E_0 \cos \omega t = A\varepsilon_0 \omega E_0 \sin\left(\omega t + \frac{\pi}{2}\right)$$

Thus, displacement current leads the conduction current by a phase difference of $\frac{\pi}{2}$.

(b) From Example 5.29, the conduction current density is 5×10^{-5} Am^{-2} and the displacement current density is obtained as $\varepsilon_0 \omega E_0$. According to the question, these two current densities are to be equal. Hence, we have

$$\varepsilon_0 \omega E_0 = 5 \times 10^{-5} \, \text{Am}^{-2}$$

or $2\pi v = \dfrac{5 \times 10^{-5} \, \text{Am}^{-2}}{\varepsilon_0 E_0}$

or $v = \dfrac{5 \times 10^{-5} \, \text{Am}^{-2}}{2\pi \times 8.85 \times 10^{-12} \, \text{Co}^2\text{N}^{-1}\text{m}^{-2} \times 0.1 \text{Vm}^{-1}} = 9.0 \, \text{MHz}$

Example 5.31

Find the displacement current density next to your radio set in air where the local radio station provides a carrier wave having $\vec{H} = 0.2 \cos(6 \times 10^8 t - 2x)\hat{z}$ A/m.

Solution

The magnetic field component is given as

$$\vec{H} = 0.2 \cos(6\pi \times 10^8 t - 2x)\hat{z} \, \text{A/m}$$

We know that $\varepsilon \omega \vec{E} = \vec{H} \times \vec{k}$

or $E = \dfrac{Hk}{\varepsilon \omega} = \dfrac{H}{\varepsilon c} \qquad \left(\because \quad \dfrac{\omega}{k} = v \right)$

We know that the displacement current density is given by

$$J_d = \varepsilon \frac{dE}{dt} = \frac{1}{c}\frac{dH}{dt} = \frac{1}{c}\frac{d}{dt}(0.2\cos(6\pi \times 10^8 t - 2x)$$

$$= \frac{1}{3 \times 10^8}(0.2\sin(6\pi \times 10^8 t - 2x) \times 6\pi \times 10^8 \, \hat{y}\text{A/m}^2$$

or $J_d = (0.4\pi \sin(6\pi \times 10^8 t - 2x)\hat{y}\text{A/m}^2$.

5.9 Maxwell's Electromagnetic Equations

Maxwell's equations represent mathematical expressions of certain experimental results. They are applicable to all macroscopic situations. Like conservation principles (conservation of energy, conservation of linear momentum and conservation of angular momentum are three pillars of physics) they are the fundamental principles of electromagnetic waves.

5.9.1 Maxwell's electromagnetic equations in differential form

The four equations which the field vectors $\vec{E}, \vec{D}, \vec{B}$, and $\vec{H}$ satisfy everywhere are:

i. $\vec{\nabla} \cdot \vec{D} = \rho$ (5.57)

ii. $\vec{\nabla} \cdot \vec{B} = 0$ (5.58)

iii. $\vec{\nabla} \times \vec{E} = -\frac{\partial \vec{B}}{\partial t}$ (5.59)

iv. $\vec{\nabla} \times \vec{H} = \vec{J} + \frac{\partial \vec{D}}{\partial t}$ (5.60)

These four equations are the fundamental equations in the electromagnetic field and are well known as Maxwell's equations in differential form.

Example 5.32

For a dielectric medium $\sigma = 0$, $\mu_r = 10$, and $\varepsilon_r = 2.5$. Examine whether the pair of fields $E = 3y\hat{y}$ and $H = 7x\hat{x}$ satisfy Maxwell's equation.

Solution

For a perfect dielectric ($\rho = 0$), Maxwell's first equation becomes

$$\nabla \cdot \vec{E} = 0$$

Now LHS of this equation is

$$\vec{\nabla} \cdot \vec{E} = \frac{\partial E_x}{\partial x} + \frac{\partial E_y}{\partial y} + \frac{\partial E_z}{\partial z} = 0 + \frac{\partial 3y}{\partial y} + 0 = 3$$

Thus, we get $\vec{\nabla} \cdot \vec{E} \neq 0$ implying that the given electric field does not satisfy Maxwell's first equation.

From Maxwell's second equation, we get

$$\vec{\nabla} \cdot \vec{H} = 0$$

Now LHS of this equation is

$$\vec{\nabla} \cdot \vec{H} = \frac{\partial H_x}{\partial x} + \frac{\partial H_y}{\partial y} + \frac{\partial H_z}{\partial z} = \frac{\partial 7x}{\partial x} + 0 + 0 = 7$$

Thus, we get $\vec{\nabla} \cdot \vec{H} \neq 0$ implying that the given magnetic field does not satisfy Maxwell's second equation.

Maxwell's third equation is

$$\vec{\nabla} \times \vec{E} = -\frac{\partial \vec{B}}{\partial t}.$$

The LHS of this equation is

$$\vec{\nabla} \times \vec{E} = \hat{x}\left(\frac{\partial E_z}{\partial y} - \frac{\partial E_Y}{\partial z}\right) + \hat{y}\left(\frac{\partial E_X}{\partial z} - \frac{\partial E_z}{\partial x}\right) + \hat{z}\left(\frac{\partial E_Y}{\partial x} - \frac{\partial E_X}{\partial y}\right) = 0$$

The RHS of Eq. (A) is

$$-\frac{\partial B}{\partial t} = -\mu\frac{\partial H}{\partial t} = -\mu\frac{\partial 7x}{\partial t} = 0.$$

Thus, LHS = RHS

Maxwell's third equation is satisfied by the given fields.

Maxwell's fourth equation is

$$\vec{\nabla} \times \vec{H} = \vec{J} + \frac{\partial \vec{D}}{\partial t} = \sigma \vec{E} + \varepsilon \frac{\partial \vec{E}}{\partial t}$$

For a perfect dielectric, $\sigma = 0$. Hence, this equation becomes

$$\vec{\nabla} \times \vec{H} = \varepsilon \frac{\partial \vec{E}}{\partial t} \tag{A}$$

The LHS of this equation is

$$\vec{\nabla} \times \vec{H} = \hat{x}\left(\frac{\partial H_Z}{\partial y} - \frac{\partial H_Y}{\partial z}\right) + \hat{y}\left(\frac{\partial H_X}{\partial z} - \frac{\partial H_Z}{\partial x}\right) + \hat{z}\left(\frac{\partial H_Y}{\partial x} - \frac{\partial H_X}{\partial y}\right) = 0$$

The RHS of Eq. (A) is

$$\varepsilon \frac{\partial \vec{E}}{\partial t} = \varepsilon \frac{\partial 3y}{\partial t} = 0$$

Thus, LHS = RHS

Therefore, we conclude that the given fields do not satisfy all the four Maxwell's equations.

5.9.2 Special cases

Maxwell's electromagnetic equation in free space

In free space, the volume charge density $\rho = 0$ and current density $J = 0$. Therefore Maxwell's electromagnetic equation in free space are given by

i. $\vec{\nabla} \cdot \vec{D} = 0$ (5.61)

ii. $\vec{\nabla} \cdot \vec{B} = 0$ (5.62)

iii. $\vec{\nabla} \times \vec{E} = -\dfrac{\partial \vec{B}}{\partial t}$ (5.63)

iv. $\vec{\nabla} \times \vec{H} = \dfrac{\partial \vec{D}}{\partial t}$ (5.64)

Maxwell's electromagnetic equations in linear isotropic media

In linear isotropic media, electric displacement $\vec{D} = \varepsilon \vec{E}$ and magnetic field intensity $\vec{H} = \dfrac{\vec{B}}{\mu}$. Therefore, Maxwell's electromagnetic equation in linear isotropic media are given by

i. $\vec{\nabla} \cdot \vec{E} = \dfrac{\rho}{\varepsilon}$ (5.65)

ii. $\vec{\nabla} \cdot \vec{H} = 0$ (5.66)

iii. $\vec{\nabla} \times \vec{E} = -\mu \dfrac{\partial \vec{H}}{\partial t}$ (5.67)

iv. $\vec{\nabla} \times \vec{H} = \vec{J} + \varepsilon \dfrac{\partial \vec{E}}{\partial t}$ (5.68)

Maxwell's electromagnetic equations for a harmonically varying field

If the magnetic field and electric field vary harmonically obeying the equations $\vec{D} = \vec{D}_0 e^{i\omega t}$ and $\vec{B} = \vec{B}_0 e^{i\omega t}$, which is in complex form, then Maxwell's equations become

i. $\vec{\nabla} \cdot \vec{D} = \rho$ (5.69)

ii. $\vec{\nabla} \cdot \vec{B} = 0$ (5.70)

iii. $\vec{\nabla} \times \vec{E} + i\omega \vec{B} = 0$ $\left(\because \dfrac{\partial \vec{B}}{\partial t} = i\omega \vec{B} \right)$ (5.71)

iv. $\vec{\nabla} \times \vec{H} - i\omega \vec{D} = \vec{J}$ $\left(\because \dfrac{\partial \vec{D}}{\partial t} = i\omega \vec{D} \right)$ (5.72)

Example 5.33

Prove $\vec{\nabla} \cdot \vec{D} = \rho$ and $\vec{\nabla} \cdot \vec{B} = 0$ from equations $\vec{\nabla} \times \vec{E} = -\dfrac{\partial \vec{B}}{\partial t}$ and $\vec{\nabla} \times \vec{H} = \vec{J} + \dfrac{\partial \vec{D}}{\partial t}$.

Solution

Now taking the divergence of both sides of the equation $\vec{\nabla} \times \vec{E} = -\dfrac{\partial \vec{B}}{\partial t}$, we get

$$\vec{\nabla} \cdot \vec{\nabla} \times \vec{E} = -\vec{\nabla} \cdot \dfrac{\partial \vec{B}}{\partial t}$$

However, we know that the divergence of curl of any vector function is always zero. Hence, this equation becomes

$$-\dfrac{\partial}{\partial t}\left(\vec{\nabla} \cdot \vec{B}\right) = 0$$

If at each point in space in the past or in the future $\vec{\nabla} \cdot B = 0$, Eq. (A) will give

$$\vec{\nabla} \cdot \vec{B} = 0 \qquad \text{(Proved)}$$

Now taking the divergence of both sides of the equation $\vec{\nabla} \times \vec{H} = \vec{J} + \dfrac{\partial \vec{D}}{\partial t}$, we get

$$\vec{\nabla} \cdot \left(\vec{\nabla} \times \vec{H}\right) = \vec{\nabla} \cdot \left(\vec{J} + \dfrac{\partial \vec{D}}{\partial t}\right)$$

However, we know that the divergence of curl of any vector function is always zero. Hence, this equation becomes

$$\vec{\nabla} \cdot \vec{J} + \vec{\nabla} \cdot \dfrac{\partial \vec{D}}{\partial t} = 0 \qquad \text{(B)}$$

Putting the value of $\vec{\nabla} \cdot \vec{J}$ from the equation of continuity $\left(\vec{\nabla} \cdot \vec{J} + \dfrac{\partial \rho}{\partial t} = 0\right)$ into Eq. (B), we have

$$\dfrac{\partial}{\partial t}\left(-\rho + \vec{\nabla} \cdot \vec{D}\right) = 0 \qquad \text{(C)}$$

If at each point in space in the past or in the future $\vec{\nabla} \cdot \vec{D} - \rho = 0$, equation (C) will give

$$\vec{\nabla} \cdot \vec{D} = \rho \qquad \text{(Proved)}$$

Example 5.34

Derive Coulomb's law of electrostatic force from Maxwell's electromagnetic equation $\vec{\nabla} \cdot \vec{D} = \rho$.

Solution

We know $\vec{\nabla} \cdot \vec{D} = \rho$ (Maxwell's first equation)

Integrating both sides of this equation over the volume V, we have

$$\int_V \nabla \cdot \vec{D} dv = \int \rho dv$$

or $$\int_V \nabla \cdot \vec{D} dv = q$$

Applying Gauss's divergence theorem to the LHS of this equation, we have

$$q = \int_S \vec{D} \cdot \vec{ds} = D \int_S ds = D 4\pi r^2 = \varepsilon E 4\pi r^2$$

or $$E = \frac{q}{4\pi \varepsilon \; r^2}$$

The force on a test charge q_0 will be

$$F = E q_0 = \frac{1}{4\pi\varepsilon} \frac{q_0 q}{r^2}, \quad \text{Coulomb's law.}$$

5.9.3 Maxwell's electromagnetic equations in integral form

Case 1: The integral form of the first equation $\vec{\nabla} \cdot D = \rho$ is given by

$$\oint_S \vec{D} \cdot \vec{ds} = \int_V \rho dv = q \tag{5.73}$$

where q is the net charge contained in the volume V, S is the surface bounding the volume V. The integral form of Maxwell's first equation states that the total electric displacement through the surface enclosing a volume is equal to the total charge within the volume.

Case 2: The integral form of the second equation $\vec{\nabla} \cdot \vec{B} = 0$ is given by

$$\oint_S \vec{B} \cdot \vec{ds} = 0 \tag{5.74}$$

which signifies that, the total outward flux of magnetic induction B through any closed surface S is equal to zero.'

Case 3: The integral form of the third equation $\vec{\nabla} \times \vec{E} = -\dfrac{\partial \vec{B}}{\partial t}$ is given by

$$\oint_C \vec{E}.\overrightarrow{d\ell} = -\frac{\partial}{\partial t} \oint_S \vec{B}.\overrightarrow{ds} \tag{5.75}$$

which signifies that the electromotive force around a closed path is equal to the time derivative of the magnetic displacement through any surface bounded by the path.

Case 4: The integral form of fourth equation $\vec{\nabla} \times \vec{H} = \vec{J} + \dfrac{\partial \vec{D}}{\partial t}$ is re-written in integral form as

$$\oint_C \vec{H} \cdot \overrightarrow{d\ell} = \oint_S \left(\vec{J} + \frac{\partial \vec{D}}{\partial t} \right) \cdot \overrightarrow{ds} \tag{5.76}$$

which signifies that the magnetomotive force around a closed path is equal to the conduction current plus the time derivative of the electric displacement through any surface bounded by the path.

Questions

5.1   What is the characteristic of a vector in a vector field?

5.2   What is the characteristic of a scalar in a scalar field?

5.3   What is a line integral? Give a few physical examples of line integrals.

5.4   What is circulation of a vector field?

5.5   What is a surface integral? Give a few physical examples of surface integrals.

5.6   What is the gradient of a scalar function? What is its direction?

5.7   What is a lamellar vector field? What is a non-lamellar vector field?

5.8   What is the physical significance of divergence of a vector function?

5.9   What is flux of a vector function?

5.10 Define gradient of a scalar function in terms of integrals.

5.11 Define divergence of a vector function. Is the resultant a vector or a scalar?

5.12 Define irrotational vector field. Give examples.

5.13 Which theorem transforms surface integrals to volume integrals? State it.

5.14 State Gauss's law in the electrostatic field in integral form.

5.15 Write down Gauss's law in the gravitational field of a body.

5.16   How do you define magnetic flux linked with a surface in case of non-uniform magnetic field?

5.17   Write down the conditions for the validity of Ampere's circuital law.

5.18   Derive the relation between displacement current and electric displacement.

5.19   State Maxwell's electromagnetic equation, which is a consequence of Gauss's law in electrostatics.

5.20   Explain why the electric field due to static charges is a lamellar vector field.

5.21   Evaluate the gradient of the magnitude of a position vector.   [Ans. Unit vector]

5.22   What is the physical significance of a gradient of a scalar function?

5.23   Define divergence of a vector function in terms of integrals.

5.24   Define curl of a vector function. Is the resultant a vector or a scalar?

5.25   What is the physical significance of curl of a vector function?

5.26   Define curl of a vector function in terms of integrals.

5.27   How can you get a scalar function from a vector field and how can you get a vector function from a scalar field? Give physical examples of both the cases.

5.28   Define solenoidal vector field. Give examples.

5.29   Define conservative vector field. Give examples.

5.30   Define rotational vector field. Give examples.

5.31   Define non-conservative vector field. Give examples.

5.32   Which theorem transforms line integrals to surface integrals? State it.

5.33   State Green's theorem in a plane.

5.34   Evaluate the divergence of a position vector.

5.35   Evaluate the gradient of a position vector.

5.36   What is the gradient of a scalar function in terms of rectangular coordinates?

5.37   What is the divergence of a vector function in terms of rectangular coordinates?

5.38   What is the curl of a vector function in terms of rectangular coordinates?

5.39   Define electric field intensity in terms of electric flux.

5.40   Can you define electric field intensity in terms of electric potential?

5.41   What are polar and non-polar molecules of a dielectric? Give some examples.

5.42   What do you mean by electric displacement?

5.43   Distinguish between electric field intensity vector and electric displacement vector.

5.44   With what charges are electric field intensity vector, electric displacement vector and electric polarization vector related to?

5.45   What is the inherent meaning of Gauss's law?

5.46   What is a Gaussian surface?

5.47   Transform Gauss's law in electrostatic field in the integral form into its corresponding differential form.

5.48 State Gauss's law in electrostatic field in differential form.

5.49 State Gauss's law in magnetic field in differential form.

5.50 State Gauss's law in magnetic field in integral form.

5.51 Use Gauss's divergence theorem to transform the integral form of Gauss's law in magnetic field into its differential form.

5.52 State Gauss's law in electrostatic field in the integral form for a dielectric medium.

5.53 State Gauss's law in electrostatic field in differential form for a dielectric medium.

5.54 Using Gauss's law, calculate the field intensity due to a point charge.

5.55 Using Gauss's law, calculate the field intensity due a spherically symmetric charge distribution of uniform density when the point lies inside the charge distribution.

5.56 Using Gauss's law, calculate the field intensity due a spherically symmetric charge distribution of uniform density when the point lies outside the charge distribution.

5.57 Using Gauss's law, calculate the field intensity due a spherically symmetric charge distribution of uniform density when the point lies on the surface of the charge distribution.

5.58 Using Gauss's law, calculate the field intensity due a cylindrically symmetric charge distribution of uniform density of infinite length when the point lies inside the charge distribution.

5.59 Using Gauss's law, calculate the field intensity due a cylindrically symmetric charge distribution of uniform density of infinite length when the point lies outside the charge distribution.

5.60 Using Gauss's law, calculate the field intensity due a cylindrically symmetric charge distribution of uniform density of infinite length when point lies on the surface of the charge distribution.

5.61 Using Gauss's law, calculate the field intensity due a linear distribution of charge of infinite length.

5.62 Plot the variation of electric field intensities in case of a cylindrically symmetric charge distribution of charge of uniform density.

5.63 Using Gauss's law, calculate the field intensity at nearby points of both sides of a uniformly charged non-conducting sheet of infinite extension.

5.64 Using Gauss's law, calculate the field intensity at a nearby point over a uniformly charged conducting body of infinite extension.

5.65 What will be Gauss's law for magnetic field?

5.66 Prove Coulomb's law in electrostatics from Gauss's law.

5.67 A certain region of space bounded by an imaginary closed surface (Gaussian surface) contains no charge. Is the electric field intensity always zero everywhere on the surface? If not, under what circumstances is it zero on the surface?

5.68 Would Gauss's law hold true if the exponent in Coulomb's law were not exactly 2?

5.69 Would Gauss's law hold true if the surface area of a sphere is not $4\pi r^2$?

5.70 Plot the variation of electric field intensities with respect to radial distance in case of a spherically symmetric charge distribution of uniform density.

5.71 What is magnetic induction? What are its units?

5.72 What are the dimensions of magnetic induction in the SI system?

5.73 What is magnetic field intensity vector? What are its units?

5.74 What are the dimensions of the magnetic intensity vector in SI system?

5.75 Is there any relation between magnetic induction vector and magnetic intensity vector? If yes, write it for different media.

5.76 What is Maxwell's corkscrew rule? Explain it.

5.77 What is the right-hand rule for finding the direction of magnetic induction near a wire carrying current?

5.78 What is the effect of magnetic field on a moving charge?

5.79 Discuss the motion of a charge under the combined effects of electric and magnetic fields.

5.80 Is magnetic induction constant for points that lie on a given line of induction? Explain.

5.81 Current is flowing from down to up in a vertical wire. Depict the direction of the magnetic induction and magnetic intensity at different points on a fixed horizontal plane.

5.82 What is the inherent meaning of Ampere's circuital law?

5.83 Write down Ampere's circuital law in integral and differential form due to a current carrying wire.

5.84 Compare Gauss's law and Ampere's circuital law.

5.85 A certain closed path encloses no electric current. Is the magnetic induction always zero everywhere on the closed path? If not, under what circumstances is it zero on the path?

5.86 Would Ampere's law hold true if the exponent in Biot–Savart's law were not exactly 2?

5.87 Would Ampere's law hold true if the perimeter of a circular path is not $2\pi r$?

5.88 In a current carrying circular loop of wire is the magnetic induction uniform for all the points inside the loop?

5.89 Using Ampere's circuital law, calculate the magnetic induction at a nearby point due to a current carrying wire of infinite extension.

5.90 Using Ampere's circuital law, calculate the magnetic induction due a cylindrical current carrying conductor of infinite length when the point lies inside the conductor.

5.91 Using Ampere's circuital law ,calculate the magnetic induction due a cylindrical current carrying conductor of infinite length when the point lies outside the conductor.

5.92 Using Ampere's circuital law, calculate the magnetic induction due a cylindrical current carrying conductor of infinite length when the point lies on the surface of the conductor.

5.93   Plot the variation of magnetic induction in case of a cylindrical current carrying conductor with respect to radial distance.

5.94   What is displacement current?

5.95   How can displacement current be produced? Explain.

5.96   Why is the quantity $\varepsilon_0 \dfrac{d\varphi_E}{dt}$ referred to as (displacement) current?

5.97   State and prove the relation between displacement current density and electric displacement.

5.98   What are the different ways of setting up of magnetic fields?

5.99   Generalize Ampere's circuital law taking into consideration the conduction current, magnetizing current and displacement current.

5.100 State Faraday's law of electromagnetic induction.

5.101 State Lenz's law.

5.102 Is induced electric field conservative or non-conservative?

5.103 Is static electric field conservative or non-conservative?

5.104 Compare static electric field and induced electric field.

5.105 Can induced magnetic field like induced electric field be produced? If yes explain how?

5.106 Can electric field for which $\vec{\nabla} \times E \neq 0$ be produced? If yes how can it be produced?

5.107 State Faraday's law of electromagnetic induction in integral form.

5.108 State Faraday's law of electromagnetic induction in differential form.

5.109 What does equation of continuity in electromagnetism represent?

5.110 State Maxwell's electromagnetic equations in differential form.

5.111 State Maxwell's electromagnetic equations in integral form.

5.112 State Maxwell's electromagnetic equations in free space/vacuum.

5.113 State Maxwell's electromagnetic equations in a medium containing charges and currents.

5.114 State Maxwell's electromagnetic equations in a medium containing no charges and currents.

5.115 State Maxwell's electromagnetic equations in a linear and isotropic medium containing charges and currents.

5.116 State Maxwell's electromagnetic equations in a linear and isotropic medium containing no charges and currents.

5.117 Write the Maxwell's electromagnetic equations, which are not changed due to the presence of currents and charges.

5.118 Write Maxwell's electromagnetic equations, which are changed due to the presence of currents and charges.

5.119 State Maxwell's electromagnetic equation which follows from the generalized Ampere's circuital law.

5.120 State Maxwell's equation which supports the concept of non-existence of magnetic monopole.

5.121 State Maxwell's electromagnetic equation which is a consequence of Faraday's law of electromagnetic induction.

5.122 Write down Maxwell's electromagnetic equations for a harmonically varying field defined by $\vec{D} = \vec{D}_0 e^{i\omega t}$ and $\vec{B} = \vec{B}_0 e^{i\omega t}$.

Problems

5.1 Evaluate $\vec{\nabla} \times (\vec{\nabla}\varphi)$ where φ is a scalar function. [Ans   Zero]

5.2 What is the value of $\vec{\nabla} \times (\vec{\nabla} \times \vec{A})$ if $\vec{\nabla} \cdot \vec{A} = 0$. [Ans $-\nabla^2 \vec{A}$]

5.3 If $\vec{r}$ is the position vector, then prove that $\vec{\nabla} r = \hat{r}$.

5.4 If $\vec{r}$ is the position vector, then prove that $\vec{\nabla} \cdot \vec{r} = 3$.

5.5 If $\vec{r}$ is the position vector, then prove that $\vec{\nabla} \times \vec{r} = 0$.

5.6 Evaluate $\vec{\nabla}(\vec{a} \cdot \vec{r})$, where $\vec{a}$ and $\vec{r}$ are the constant vector and position vector respectively.

5.7 Evaluate the line integral for the vector function $F(x) = \dfrac{kq}{x^2}\hat{x}$ from infinity to x_0 if k and q are constants. [Ans $\dfrac{kq}{x_0}$]

5.8 Evaluate the line integral for the vector function $\vec{F} = \hat{x}(3x + 6y) - \hat{y}(14yz) + \hat{z}(20xz^2)$ along the straight line from (0,0,0) to (1,1,1). [Ans   13/3]

5.9 Evaluate the surface integral for the vector function $\vec{F} = 4xz\hat{x} - y^2\hat{y} + yz\hat{z}$ over the surface of the cube bounded by $x = 0, x = 1, y = 0, y = 1, z = 0, z = 1$ planes. [Ans   3/2]

5.10 Electric potential φ on the X–Y plane is given by $\varphi = ax^2 - by^2$, where a and b are constants. Find the components of the electric field intensity. [Ans $-2ax, -2ay$]

5.11 Electric potential φ in a region is represented by $\varphi = 2x + 3y - z$. Obtain the expression for gradient and electric field intensity. [Ans $2\hat{x} + 3\hat{y} - \hat{z}, -2\hat{x} - 3\hat{y} + \hat{z}$]

5.12 If $\varphi(x, y, z) = 3x^2 y - 2y^2 z^3$, find $\vec{\nabla}\varphi$ at the point (1, –2, –1). [Ans $-(12\hat{x} + 5\hat{y} + 24\hat{z})$]

5.13 If $\vec{k}$ is a constant scalar, prove that $\vec{\nabla}(\vec{k}) = 0$.

5.14 If $\vec{k}$ is a constant vector, prove that $\vec{\nabla} \cdot \vec{k} = 0$.

5.15 If $\vec{k}$ is a constant vector, prove that $\vec{\nabla} \times \vec{k} = 0$.

5.16 If $\vec{u}$ is a unit vector and $\vec{r}$ is a position vector, prove that $\vec{\nabla}(\vec{u} \cdot \vec{r}) = \vec{u}$.

5.17 If $\vec{u}$ is a unit vector and $\vec{r}$ is a position vector, prove that $\vec{\nabla} \cdot [(\vec{u} \times \vec{r}) \times \vec{u}] = 2$.

5.18 If $\vec{u}$ is a unit vector and $\vec{r}$ is a position vector, prove that $\vec{\nabla} \times [(\vec{u} \times \vec{r}) \times \vec{u}] = 0$.

5.19 If $\vec{u}$ is a unit vector and $\vec{r}$ is a position vector, prove that $\vec{\nabla} \times [(\vec{u} \cdot \vec{r})\vec{u}] = 0$.

5.20 If $\vec{u}$ is a constant scalar, prove that $\vec{\nabla}u = 0$.

5.21 If $\vec{u}$ is a constant vector, prove that $\vec{\nabla} \cdot \vec{u} = 0$.

5.22 If $\vec{u}$ is a constant vector, prove that $\vec{\nabla} \times \vec{u} = 0$.

5.23 If $\vec{r}$ is a position vector, prove that $\nabla^2 \dfrac{1}{r} = 0$.

5.24 Prove that when $\vec{B}$ is solenoidal (i.e., $\vec{\nabla} \cdot \vec{B} = 0$), then there exists a vector such that $\vec{B} = \vec{\nabla} \times \vec{A}$.

5.25 If $\vec{r}$ is a position vector, prove that $r^n \vec{r}$ is an irrotational vector for any value of n whereas it is solenoidal only when $n = -3$.

5.26 A parallel plate capacitor having plate area $100\,cm^2$ and plate separation 1.4 cm is charged to a potential difference of 100 volts. Calculate the electric field intensity, electric displacement and electric polarization vector in the empty space between the two plates of the capacitor. [Ans 1.000×10^4 V/m, 8.85×10^{-8} C/m^2, 0]

5.27 In Problem 26, a dielectric slab of thickness 0.5 cm and dielectric constant 7.0 is introduced between the two plates of the capacitor. Calculate the electric field intensity, electric displacement and electric polarization vector in the dielectric slab. [Ans 1.43×10^3 V/m, 8.85×10^{-8} C/m^2, 7.5×10^{-3} C/m^2]

5.28 Current is flowing in an infinity length wire placed on the plane of the page. Calculate the total magnetic flux passing through a strip of paper of length b and width w with the closer side at r distance from the wire.

[Hints and Ans $\quad \varphi_M = \int\limits_{r}^{r+w} \vec{B} \cdot \vec{ds} = \int\limits_{r}^{r+w} \vec{B}(x) \cdot \overrightarrow{bdx} = \dfrac{\mu_0 i}{2\pi} \ln\left(1 + \dfrac{w}{r}\right)$]

5.29 A current is flowing in a long circular conductor of radius a. The current is distributed in the wire in such a way that the current density at a distance r from the axis is given by $J = J_0\left(1 + \dfrac{r}{a}\right)$. Find the total current in the wire and the magnetic induction at both inside and out-side points of the conductor. [Ans $\dfrac{5}{3} J_0 \pi a^2$, $\dfrac{5\mu_0 J_0 a^2}{6b}$, $\dfrac{\mu_0 J_0 b}{2}\left(1 + \dfrac{2b}{3a}\right)$]

5.30 A current is flowing in a long circular conductor of radius 2 cm. The current is distributed in the wire in such a way that the current density at a distance r from the axis is given by $J = 0.8\left(1 + \dfrac{r^2}{4}\right)$. Find the total current in the wire and the magnetic induction at points 3 cm and 1 cm from the axis of the conductor.

[Ans 15.1 Amp., 1.01×10^{-8} T, 5.65×10^{-9} T]

5.31   Calculate the magnetic induction at a perpendicular distance a from the centre of a strip of width w and infinite length, carrying a current i.   [Ans $\dfrac{\mu_0 i}{\pi w}\tan^{-1}\dfrac{w}{2a}$]

5.32   Calculate the magnetic induction at a perpendicularly large distance a from the centre of a strip of width w and infinite length, carrying a current i.   [Ans $\dfrac{\mu_0 i}{2\pi a}$]

5.33   A metal rod of length L moves in a magnetic field of magnetic induction B with velocity v. Show that the ends of the rod are at potential difference $\vec{B}\cdot\vec{L}\times\vec{v}$.

5.34   A copper rod of length L rotates about an axis passing through one end of the rod with an angular frequency ω in a uniform magnetic field of induction B. Calculate the potential difference (emf) developed between the two ends of the rod.

$$\text{[Ans} \quad \frac{1}{2}B\omega L^2 \text{]}$$

5.35   The magnetic flux linked through a loop perpendicular to the plane of the coil and directed out of the paper is varying according to equation $\varphi = 5t^2 + 10t + 17$. What is the magnitude and direction of the induced emf at time $t = 3$ second.

[Ans   40V, anticlockwise]

5.36   A plane circular disc of radius r rotates at n revolutions per second about an axis perpendicular to it at its centre. A uniform magnetic field B exists parallel to the axis. Prove that there is an emf between the centre of the disc and its rim of magnitude $n\pi B r^2$.

5.37   A spatially uniform magnetic field $B = B_0 \sin \omega t$ is directed at an angle θ to the normal of the plane of a circular loop. Calculate the emf induced in the field.

[Ans $B_0 \pi r^2 \omega \cos\theta \cos\omega t$]

5.38   Electric field of strength 10^{13} V/m normal to an area 0.01 m^2 is doubled in one second. Find the displacement current.   [Ans   0.89 Ampere]

5.39   The parallel plate capacitor has plates each of area 2 m^2 and the plates are separated by a dielectric of thickness 1 mm, dielectric constant 3. The potential difference and conduction current at certain instant of time is 100 V and 2 mA, respectively. Find the displacement current flowing between the two plates of the capacitor.

[Ans   2 mA]

5.40   In a certain copper conductor of resistivity 2×10^{-8} Ωm carrying a current, the electric field varies with time according to the equation $E = E_0 \sin \omega t$. Calculate (a) the maximum conduction current density and (b) the maximum displacement current density if $\varepsilon = \varepsilon_0$. Given, $E_0 = 0.1$ volt/meter, $\omega = 2\pi \times 60$ Hz.

[Ans   (a) 5.0×10^6 Am^{-2} (b) 3.34×10^{-10} Am^{-2}]

5.41   Two metallic rods of same metal and same radii were joined by a special type of gum of dielectric constant k resulting in a parallel plate capacitor. The thickness of the gum t is comparatively small in comparison to the radius r of the rod. There is a current I in each rod. Calculate (a) the conduction current density, (b) the displacement current density and (c) the electric field intensity in the gum.

[Ans (a) $\dfrac{I}{\pi r^2}$, (b) $\dfrac{I}{\pi r^2}$, (c) $\dfrac{q}{\varepsilon_0 k\pi r^2}$]

5.42 A voltage $50 \sin 10^3\, t$ is applied to the plates of a parallel plate capacitor with a plate area of 5 cm² and plate separation 3 mm. Calculate the displacement current and conduction current assuming the dielectric to have permittivity two times the permittivity of free space. [Ans $147.4 \times 10^{-9} \cos 10^3\, t\, A$]

5.43 In free space $E = \hat{y}20 \cos(\omega t - 50x)$V/m. Calculate the displacement current density, magnetic field intensity vector and angular frequency of the wave.

[Ans $-\hat{y}20\omega\varepsilon_0 \sin(\omega t - 50x)$A/m², $\hat{z}\, 0.4\omega\varepsilon_0 \cos(\omega t - 50x)$A/m, 1.5×10^{10} rad/s]

5.44 Consider a charge moving with a uniform velocity along the axis of a circle and calculate the displacement current that flows through the circle. Next, calculate the conduction current through the circle and show that the two results are identical.

Multiple Choice Questions

1. In Fig. 5.26 which shows a differential volume, match the items in the left side to those in the right side.

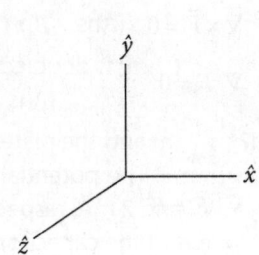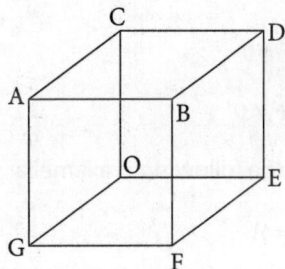

Figure 5.26

(i) $\overrightarrow{d\ell}$ from A to B

(ii) $\overrightarrow{d\ell}$ from A to G

(iii) $\overrightarrow{d\ell}$ from A to C

(iv) $\overrightarrow{ds}$ for face ABDC

(v) $\overrightarrow{ds}$ for face AGOC

(a) $-\hat{x}dydz$

(b) $\hat{y}dxdz$

(c) $\hat{z}dxdy$

(d) $-\hat{y}dxdz$

(e) $\hat{x}dx$

(vi) $\overrightarrow{ds}$ for face OEFG

(f) $-\hat{y}dy$

(vii) $\overrightarrow{ds}$ for face ABFG

(g) $-\hat{z}dz$

2. If $\vec{r}$ is the position vector of the point (x, y, z), then which of the following is incorrect?

(i) $\vec{\nabla}r = \dfrac{\vec{r}}{|\vec{r}|}$

(ii) $\vec{\nabla}\cdot\vec{r} = 1$

(iii) $\vec{\nabla}|\vec{r}| = \dfrac{\vec{r}}{|\vec{r}|}$

(iv) $\nabla^2(\vec{r}\cdot\vec{r}) = 6$

3. Which of the following is a meaningless combination?

(i) grad div

(ii) div curl

(iii) curl grad

(iv) div grad

(v) div curl

(vi) grad curl

(vii) curl div

4. Which of the following is zero?

(i) grad div

(ii) div curl

(iii) curl grad

(iv) div grad

(v) div curl

5. Which of the following is solenoidal?

(i) $\oint_C \vec{F}\cdot\overrightarrow{d\ell} = 0$

(ii) $\oint_S \vec{F}\cdot\overrightarrow{ds} = 0$

(iii) $\vec{\nabla}\cdot\vec{F} = 0$

(iv) $\vec{\nabla}\times\vec{F} = 0$

(v) $\vec{\nabla}\times\vec{F} \neq 0$

(vi) $\nabla^2\vec{F} = 0$

6. Which of the following is a lamellar vector field?

(i) $\vec{\nabla}\varphi = \vec{A}$

(ii) $\vec{\nabla}\cdot\vec{F} = \vec{B}$

(iii) $\vec{\nabla}\times\vec{F} = \vec{A}$

(iv) $\vec{\nabla}\times\vec{F} \neq 0$

(v) $\nabla^2\vec{F} = \vec{A}$

7. Which of the following is a conservative field?

(i) $\oint_C \vec{F}\cdot\overrightarrow{d\ell} = 0$

(ii) $\oint_S \vec{F}\cdot\overrightarrow{ds} = 0$

(iii) $\vec{\nabla}\cdot\vec{F} = 0$

(iv) $\vec{\nabla}\times\vec{F} = 0$

(v) $\vec{\nabla}\times\vec{F} \neq 0$

(vi) $\oint_S \vec{F}dv \neq 0$

8. Which of the following is an irrotational vector?

(i) $\oint_C \vec{F} \cdot \vec{d\ell} = 0$ (ii) $\oint_S \vec{F} \cdot \vec{ds} = 0$

(iii) $\vec{\nabla} \cdot \vec{F} = 0$ (iv) $\vec{\nabla} \times \vec{F} = 0$

(v) $\vec{\nabla} \times \vec{F} \neq 0$ (vi) $\oint_S \vec{F} dv \neq 0$

9. The electric flux density on the surface of a sphere is the same for a point charge q located at the centre and for the charge q uniformly distributed throughout the sphere.

(i)   Yes   (ii)   No

(iii)   Not necessarily

10. By saying that electrostatic field is conservative, we do not mean that

(i)   It is the gradient of a scalar potential

(ii)   Its circulation is identically zero

(iii)   Its curl is identically zero

(iv)   The work done in a closed path inside the field is zero

(v)   The potential difference between any two points is zero

11. Sea water has the relative permittivity 80. Its permittivity is

(i)   81   (ii)   79

(iii) 7.08×10^{-10} Coul^2 / Newton · meter^2 (iv) 70.08×10^{-10} Coul^2 / Newton · meter^2

12. Both permittivity and electric susceptibility are dimensionless

(i)   True   (ii)   False

(iii)   Cannot be said

13. If $\vec{\nabla} \cdot \vec{D} = \varepsilon \vec{\nabla} \cdot \vec{E}$ and $\vec{\nabla} \cdot \vec{J} = \sigma \vec{\nabla} \cdot \vec{E}$ in a given material, the material is said to be

(i)   Linear   (ii)   Isotropic

(iii)   Homogeneous   (iv)   Linear and homogeneous

(v)   Linear isotropic   (vi)   Isotropic and homogeneous

14. If $\vec{D} = \varepsilon \vec{E}$ and $\vec{B} = \mu \vec{H}$ in a given material, the material is said to be

(i)   Linear   (ii)   Isotropic

(iii)   Homogeneous   (iv)   Linear and homogeneous

(v)   Linear isotropic   (vi)   Isotropic and homogeneous

15. One of the following is not a source of magnetostatic fields

(i)   A dc in a wire   (ii)   A permanent magnet

(iii)   An accelerated charge

(iv)   An electric field linearly changing with time

(v)   A charged disc rotating at uniform speed

16. Identify the configuration in Fig. 5.27 that is not a correct representation of current I and magnetic induction $\vec{B}$.

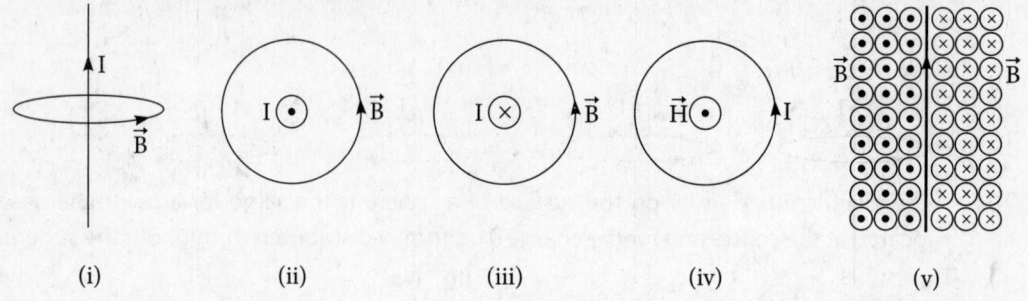

    (i)   (ii)   (iii)   (iv)   (v)

Figure 5.27

17. Which of the following statements is not characteristic of a magnetostatic field?
    (i) It is solenoidal
    (ii) It is conservative
    (iii) It has no sinks or sources
    (iv) Magnetic flux lines are always closed
    (v) Total number of flux lines entering a given region is equal to the total number of flux lines leaving the region

18. One of the following equations is not Maxwell's equation for a static electromagnetic field in a linear homogeneous medium.
    (i) $\vec{\nabla} \cdot \vec{B} = 0$ (ii) $\vec{\nabla} \times \vec{D} = 0$

    (iii) $\oint_C \vec{B} \cdot \vec{d\ell} = \mu_0 \vec{J}$ (iv) $\oint_S \vec{D} \cdot \vec{ds} = q$

    (v) $\nabla^2 \vec{A} = \mu_0 \vec{J}$

19. The flux through each turn of a 100-turn coil is $(t^3 - 2t) \times 10^{-3}$ Wb, where t is in seconds. The induced emf at $t = 2$ seconds is
    (i) 1 volt   (ii) –1 volt
    (iii) 4 milli volt   (iv) 0.4 volt
    (v) –0.4 volt

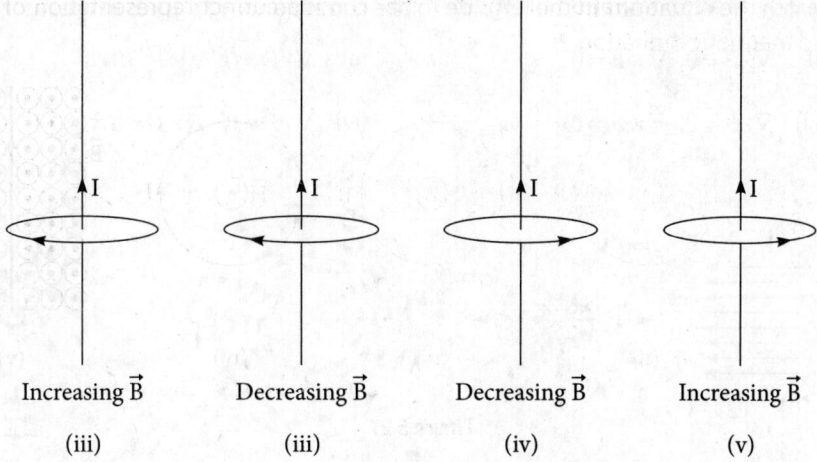

Increasing $\vec{B}$ Decreasing $\vec{B}$ Decreasing $\vec{B}$ Increasing $\vec{B}$

(iii)   (iii)   (iv)   (v)

Figure 5.28

20. Assuming that each loop is stationary and the time-varying magnetic field B induces current I, which of the configurations in Fig. 5.28 are incorrect?

21. The concept of displacement current was a major contribution attributed to

(i) Faraday   (ii) Lenz

(iii) Maxwell   (iv) Lorentz

22. Identify which of the following equations are not Maxwell's equations for time-varying fields

(i) $\vec{\nabla}\cdot\vec{J}+\dfrac{\partial\rho}{\partial t}=0$ (ii) $\vec{\nabla}\cdot\vec{D}=\rho$

(iii) $\vec{\nabla}\cdot\vec{E}=-\dfrac{\partial\vec{B}}{\partial t}$ (iv) $\oint_{C}\vec{H}\cdot\vec{d\ell}=\int_{S}\left(\sigma\vec{E}+\varepsilon\dfrac{\partial\vec{E}}{\partial t}\right)\cdot\vec{ds}$

(v) $\oint_{S}\vec{B}\cdot\vec{ds}=0$

23. An electromagnetic field is said to be non-existent or not Maxwellian if it fails to satisfy Maxwell's equations and the wave equations derived from them. Which of the following fields in free space are not Maxwellian?

(i) $\vec{H}=\hat{x}\cos x\,\cos10^{6}t$ (ii) $\vec{H}=\hat{x}10\cos\left(10^{5}t-\dfrac{z}{10}\right)$

(iii) $\vec{E}=\hat{x}100\cos\omega t$ (iv) $\vec{D}=\hat{z}\,e^{-10y}\sin(10^{5}-10y)$

(v) $\vec{B}=\hat{z}\,0.4\sin10^{4}t$ (vi) $\vec{B}=\hat{z}(1-\rho^{2})\sin\omega t$

24. Match the equations in the left side to the corresponding figures in the right side.

   (i) $\vec{\nabla} \cdot \vec{F} = 0,\ \vec{\nabla} \times \vec{F} = 0$ (ii) $\vec{\nabla} \cdot \vec{F} \neq 0,\ \vec{\nabla} \times \vec{F} = 0$

   (iii) $\vec{\nabla} \cdot \vec{F} = 0,\ \vec{\nabla} \times \vec{F} \neq 0$ (iv) $\vec{\nabla} \cdot \vec{F} \neq 0,\ \vec{\nabla} \times \vec{F} \neq 0$

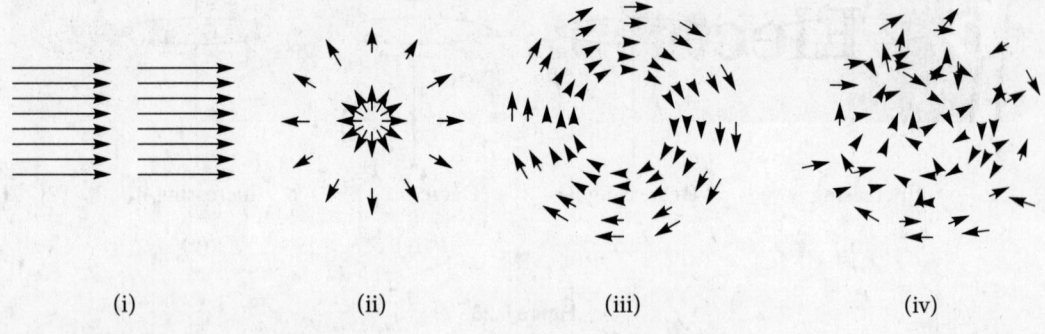

     (i)   (ii)   (iii)   (iv)

Figure 5.29

25. Poynting's vector physically denotes the power density leaving or entering a given volume in a time-varying field.

   (i) True   (ii) False

26. For electrostatic field, $\nabla \times \vec{E} = 0$

   (i) True   (ii) False

27. In an anisotropic dielectric, like crystalline quartz, the direction of $\varepsilon_0 \vec{E}$ and the polarization vector are the same.

   (i) True   (ii) False

Answers

1 (i – e, ii – f, iii – g, iv – b, v – a, vi – d, vii – c)   2 (ii)   3 (vi, vii)   4 (iii, v)   5 (ii)

6 (i)   7 (i)   8 (iv)   9 (i)   10 (v)   11 (iii)   12 (ii)   13 (iv)

14 (i)   15 (v)   16 (iii)   17 (ii)   18 (v)   19 (ii)   20 (ii)   21 (iii)

22 (i)   23 (iii)   24 (i – a, ii – b, iii – c, iv – d)   25 (i)   26 (i)   27 (ii)

6 Electromagnetic Waves

6.1 Introduction

In 1865, James Clark Maxwell laid the foundation stone for the grand unification theory (GUT) by combining electric and magnetic phenomena. He showed, mathematically, that electric and magnetic fields fluctuating together could form a propagating wave, appropriately called an electromagnetic wave. In the previous chapter, we have established Maxwell's electromagnetic equations. Sunlight is a mixture of electromagnetic waves having different wavelengths. Each wavelength corresponds to a unique color in the visible range. An electromagnetic wave in a vacuum consists of mutually perpendicular and oscillating electric and magnetic fields. The wave is a transverse wave since the fields are mutually perpendicular to the direction along which the wave travels. In general, waves are the means of transporting energy or information. A wave is a function of both space and time. This chapter examines the general properties of electromagnetic waves.

6.2 Electromagnetic Energy Density

Energy is transported through space by means of electromagnetic waves obeying Maxwell's electromagnetic equations. We assume that space is linear, isotropic, homogeneous characterized by constant permeability μ, constant permittivity ε, and constant conductivity σ.

Maxwell's equations taken from Chapter 5 are

$$\vec{\nabla} \cdot \vec{B} = 0 \tag{6.1}$$

$$\vec{\nabla} \cdot \vec{D} = \rho \tag{6.2}$$

$$\vec{\nabla} \times \vec{E} = -\frac{\partial \vec{B}}{\partial t} \tag{6.3}$$

$$\vec{\nabla} \times \vec{H} = \vec{J} + \frac{\partial \vec{D}}{\partial t} \tag{6.4}$$

Taking into consideration these equations, we can have

$$\vec{\nabla} \cdot (\vec{E} \times \vec{H}) = \vec{H} \cdot \vec{\nabla} \times \vec{E} - \vec{E} \cdot \vec{\nabla} \times \vec{H} =$$

$$-\vec{H} \cdot \frac{\partial \vec{B}}{\partial t} - \vec{E} \cdot \left(\vec{J} + \frac{\partial \vec{D}}{\partial t} \right) = -\vec{H} \cdot \frac{\partial \vec{B}}{\partial t} - \vec{E} \cdot \vec{J} - \vec{E} \cdot \frac{\partial \vec{D}}{\partial t} \tag{6.5}$$

For a linear isotropic medium,

$$\vec{B} = \mu \vec{H} \quad \text{and} \quad \vec{D} = \varepsilon \vec{E}. \tag{6.6}$$

Therefore, Eq. (6.5) for a linear isotropic medium becomes

$$\vec{\nabla} \cdot (\vec{E} \times \vec{H}) = -\vec{H} \cdot \frac{\partial}{\partial t}(\mu \vec{H}) - \vec{E} \cdot \frac{\partial}{\partial t}(\varepsilon \vec{E}) - \vec{J} \cdot \vec{E}$$

$$= -\mu \vec{H} \cdot \frac{\partial}{\partial t}(\vec{H}) - \varepsilon \vec{E} \cdot \frac{\partial}{\partial t}(\vec{E}) - \vec{J} \cdot \vec{E}$$

$$= -\frac{1}{2}\frac{\partial}{\partial t}(\mu H^2) - \frac{1}{2}\frac{\partial}{\partial t}(\varepsilon E^2) - J \cdot E$$

$$= -\frac{1}{2}\frac{\partial}{\partial t}(\mu \vec{H} \cdot \vec{H}) - \frac{1}{2}\frac{\partial}{\partial t}(\varepsilon \vec{E} \cdot \vec{E}) - \vec{J} \cdot \vec{E}$$

$$= -\frac{\partial}{\partial t}\left[\frac{1}{2}\vec{B} \cdot \vec{H} + \frac{1}{2}\vec{D} \cdot \vec{E} \right] - \vec{J} \cdot \vec{E}$$

or   we have $\vec{J}.\vec{E} + \dfrac{\partial}{\partial t}\left[\dfrac{1}{2}\vec{B} \cdot \vec{H} + \dfrac{1}{2}\vec{D} \cdot \vec{E} \right] = -\vec{\nabla} \cdot (\vec{E} \times \vec{H})$ \hfill (6.7)

Now integrating both sides over the volume V enclosed by the surface S, we have

$$\oint_V \vec{J}.\vec{E}dv + \oint_V \frac{\partial}{\partial t}\left[\frac{1}{2}\vec{B}\cdot\vec{H} + \frac{1}{2}\vec{D}\cdot\vec{E}\right]dv = -\oint_V \vec{\nabla}\cdot(\vec{E}\times\vec{H})dv$$

Applying Gauss's divergence theorem to the RHS of this equation, we have

$$\oint_V \vec{J}.\vec{E}dv + \oint_V \frac{\partial}{\partial t}\left[\frac{1}{2}\vec{B}\cdot\vec{H} + \frac{1}{2}\vec{D}\cdot\vec{E}\right]dv = -\oint_S (\vec{E}\times\vec{H})ds \qquad (6.8)$$

6.2.1 Interpretation of the left-hand side of Eq. (6.8)

The LHS of Eq. (6.8) consists of two terms. The first term in the LHS of Eq. (6.8) is $\oint_V \vec{J}.\vec{E}dv$.

Suppose a charged particle q moves with constant velocity under the combined influence of the mechanical $\vec{F}_M$, electrical $q\vec{E}$ and magnetic forces $q(\vec{v}\times\vec{B})$. Since the charged particle q moves with constant velocity $\vec{v}$, we have from Newton's laws of motion

$$\vec{F}_M + q\vec{E} + q(\vec{v}\times\vec{B}) = 0$$

or $\quad \vec{F}_M = -q\left[\vec{E} - (\vec{v}\times\vec{B})\right] \qquad (6.9)$

Now $\vec{F}_M \cdot \vec{v}$ is the rate at which mechanical force $\vec{F}_M$ does work on the charged particle and it is equal to the rate at which the electromagnetic force $q\left[\vec{E} + (\vec{v}\times\vec{B})\right]$ does work on it. From Eq. (6.9), we have

$$\vec{F}_M \cdot \vec{v} = -q\left[\vec{E}\ (\vec{v}\times\vec{B})\right]\cdot\vec{v} = -q\vec{E}\cdot\vec{v} - q(\vec{v}\times\vec{B})\cdot\vec{v} = -q\vec{E}\cdot\vec{v} \qquad (6.10)$$

Let ρ be the charge density of the medium. If ρ charges move with velocity $\vec{v}$, then the rate at which the mechanical force $\vec{F}_M$ does work on the unit volume (containing ρ charges moving with velocity $\vec{v}$) will be given by, from Eq. (6.10),

$$\vec{F}_M \cdot v = -E\cdot\rho v \qquad (6.11)$$

The volume passing through the cross-sectional area ds with velocity v in unit time is vds. See Fig. 6.1. This volume vds contains charge $\rho\vec{v}\cdot\vec{ds}$. The amount of charge passing through the cross-sectional area ds per unit time is $\rho\vec{v}\cdot\vec{ds}$. By definition of current, we have

$$I = \rho\vec{v}\cdot\vec{ds}$$

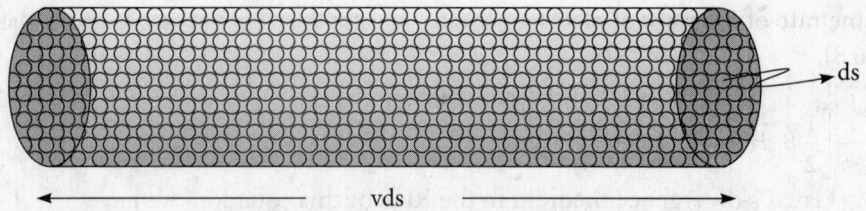

Figure 6.1 | The calculation of the amount of charge passing through the cross-sectional area *ds* per unit time

or $\quad \dfrac{I}{ds} = \rho \vec{v}$

or $\quad \vec{J} = \rho \vec{v}$ is the electric current density. $\hfill (6.12)$

Putting this value $\vec{J}$ in Eq. (6.11), we get

$$\vec{F}_M \cdot \vec{v} = -\vec{E} \cdot \vec{J} = -\vec{J} \cdot \vec{E} \hfill (6.13)$$

This is the rate at which mechanical force does work on unit volume and is called power density. The amount of energy transferred to unit volume of the electromagnetic field per unit time is $\vec{J} \cdot \vec{E}$. The amount of energy transferred to dv volume of electromagnetic field per unit time is $\vec{J} \cdot \vec{E}\, dv$. The total amount of energy transferred per unit time to the entire electromagnetic field of volume V is equal $\oint_V \vec{J} . \vec{E} dv$.

The second term in the LHS of Eq. (6.8) is

$$\oint_V \frac{\partial}{\partial t} \left[\frac{1}{2} \vec{B} \cdot \vec{H} + \frac{1}{2} \vec{D} \cdot \vec{E} \right] dv. \quad \frac{1}{2} \vec{B} \cdot \vec{H}$$

is the magnetic energy density (magnetic energy per unit volume) and $\dfrac{1}{2} \vec{D} \cdot \vec{E}$ is the electric energy density (electric energy per unit volume). Therefore, the term

$$\frac{1}{2} \vec{B} \cdot \vec{H} + \frac{1}{2} \vec{D} \cdot \vec{E}$$

is the electromagnetic energy density in space. The term

$$\frac{\partial}{\partial t} \left[\frac{1}{2} \vec{B} \cdot \vec{H} + \frac{1}{2} \vec{D} \cdot \vec{E} \right]$$

is the time rate of increase of electromagnetic energy density. Therefore, the second term of Eq. (6.8),

$$\oint_V \frac{\partial}{\partial t}\left[\frac{1}{2}\vec{B}\cdot\vec{H}+\frac{1}{2}\vec{D}\cdot\vec{E}\right]dv,$$

will be the time rate of increase of electromagnetic energy in the volume V.

Combining all the discussions here, we arrive at the conclusion that left side of Eq. (6.8) represents the sum of the power expended by the field due to motion of charges and the time rate of increase of stored electromagnetic energy in fields.

6.2.2 Interpretation of the right-hand side of Eq. (6.8)

The term in the RHS of Eq. (6.8) is $-\oint_S (\vec{E}\times\vec{H})\cdot\vec{ds}$

Since the left side of Eq. (6.8) represents the sum of the power expended by the field due to motion of charges and the time rate of increase of stored electromagnetic energy in fields, the right side of the same equation must represent the power flow into the volume V through the surface S or power flow out of the volume V through the surface S. The term $\vec{E}\times\vec{H}$ must represent the flow of energy density flux per unit time associated with the electromagnetic field.

6.3 Poynting's Vector

One of the important equations of propagation of electromagnetic energy is the Eq. (6.8), re-written here as

$$\oint_V \vec{J}.\vec{E}dv+\oint_V \frac{\partial}{\partial t}\left[\frac{1}{2}\vec{B}\cdot\vec{H}+\frac{1}{2}\vec{D}\cdot\vec{E}\right]dv=-\oint_S (\vec{E}\times\vec{H})\cdot\vec{ds} \tag{6.14}$$

The vector $\vec{E}\times\vec{H}$ is called Poynting's vector $\vec{P}$. It is the power density associated with electromagnetic field. Poynting's vector $\vec{P}$ must be the energy that flows out of or into the boundary through unit surface area per unit time. This fact can be checked by dimensional analysis of Poynting's vector. The dimensional formula of Poynting's vector is given by

$$[P]=[E][H]$$

We know that the dimensional formula for electric field intensity E is given by

$$[E]=\frac{[F]}{[q]}=\frac{[F]}{[It]}=\frac{[MLT^{-2}]}{[AT]}$$

or $\quad [E] = \left[MLT^{-3} A^{-1} \right]$

We know that the dimensional formula for magnetic field intensity H is given by

$$[H] = \frac{[Id\ell \sin \theta]}{[r^2]} = \frac{[I][d\ell]}{[r^2]} \qquad \left(\because H = \int_C \frac{Idl \sin \theta}{r^2} \right)$$

or $\quad [H] = \frac{[A][L]}{[L^2]} = [AL^{-1}]$

Therefore, dimensional formula of Poynting's vector becomes

$$[P] = [E][H] = [MLT^{-3} A^{-1}][AL^{-1}] = [MT^{-3}] = \frac{ML^2 T^{-2}}{L^2 T^1} \quad \left(= \frac{energy}{time \times area} \right)$$

or $\quad [P] = \frac{watt}{meter^2}$

Thus, the dimension of Poynting's vector is same as that of power per unit area or power density. Hence, we can say that Poynting's vector is the power density associated with electromag.netic field.

As discussed earlier, the vector $\vec{E} \times \vec{H}$ is called Poynting's vector $\vec{P}$. Hence, we have

$$\vec{P} = \vec{E} \times \vec{H} \tag{6.15}$$

Its direction can be found out by the rule of cross-product of two vectors. For a plane wave, the direction of propagation of the wave is the direction of Poynting's vector. The magnitude of Poynting's vector for electromagnetic wave is

$$P = EH \tag{6.16}$$

since the angle between the electric vector $\vec{E}$ and the magnetic vector $\vec{H}$ is 90°.

Example 6.1

Calculate the value of Poynting's vector on the surface of the sun if the power radiated by the sun is 4×10^{26} W while its radius is 7×10^8 m.

Solution

Poynting's vector is the power passing through the unit surface area. The surface area of the sun is

$$4\pi r^2 = 4 \times \frac{22}{7} \times 49 \times 10^{16}\,\mathrm{m}^2 = 616 \times 10^{16}\,\mathrm{m}^2$$

Hence, Poynting's vector $= \dfrac{4 \times 10^{26}\,\mathrm{W}}{616 \times 10^{16}\,\mathrm{m}^2} = 6.5 \times 10^7\,\dfrac{\mathrm{W}}{\mathrm{m}^2}.$

Example 6.2

An electromagnetic wave is propagating in free space with electric vector $E(z,t) = 50\cos(\omega t - kz)\hat{x}$. How much average energy is crossing a circular area of radius 2 m on the XY plane in unit time.

Solution

The electromagnetic wave is propagating along the $+Z$ direction. The average energy passing perpendicularly through the unit area in unit time (i.e., average value of Poynting's vector) is given by

$$<P> = \frac{1}{2}\sqrt{\frac{\varepsilon}{\mu}} \times E_0^2 = \frac{1}{2} \times 2.654 \times 10^{-3} \times 2500\,\mathrm{W/m}^2 = 3.318\,\mathrm{W/m}^2$$

Therefore, the amount of energy passing perpendicularly through $\pi r^2 = 3.14 \times 4 = 12.56\,\mathrm{m}^2$ area per unit time will be

$$= 3.318\,\mathrm{W/m}^2 \times 12.56\,\mathrm{m}^2 = 41.67\,\mathrm{W}$$

Example 6.3

An electromagnetic wave is propagating in free space with electric vector $E(z,t) = 150\cos(\omega t - kz)\hat{x}$. How much average energy passes through a rectangular hole of length 3 cm and width 1.5 cm on the XY-plane in one minute time.

Solution

The electromagnetic wave is propagating along the $+Z$ direction. The average energy passing perpendicularly through unit area in unit time (i.e., average value of Poynting's vector) is given by

$$<P> = \frac{1}{2}\sqrt{\frac{\varepsilon}{\mu}} \times E_0^2 = \frac{1}{240\pi} \times E_0^2 = \frac{1}{240\pi} \times 150^2\,\mathrm{W/m}^2 = 29.84\,\mathrm{W/m}^2$$

Therefore, the amount of energy passing perpendicularly through $3 \times 10^{-2} \times 1.5 \times 10^{-2} = 4.5 \times 10^{-4}\,\mathrm{m}^2$ area in unit time will be

$$= 29.84\,\mathrm{W/m^2} \times 4.5 \times 10^{-4}\,\mathrm{m}^2 = 1.34 \times 10^{-2}\,\mathrm{W}$$

Therefore, the amount of energy passing perpendicularly through $3 \times 10^{-2} \times 1.5 \times 10^{-2} = 4.5 \times 10^{-4}\,\mathrm{m}^2$ area in 60 seconds will be

$$= 1.34 \times 10{-2} \times 60\,\mathrm{J} = 0.81\,\mathrm{J}$$

6.4 Poynting's Theorem

One of the important equations of propagation of electromagnetic energy is the Eq. (6.8), re-written here as

$$-\oint_V \frac{\partial}{\partial t}\left[\frac{1}{2}\vec{B}\cdot\vec{H} + \frac{1}{2}\vec{D}\cdot\vec{E}\right]dv = \oint_V \vec{J}.\vec{E}dv + \oint_S (\vec{E}\times\vec{H})\cdot\vec{ds} \qquad (6.17)$$

The left-hand side of Eq. (6.17) is the decrease (due to negative sign) in electromagnetic energy per unit time in a certain volume V. The first term in the right-hand side of Eq. (6.17) is the work done by the field forces (Lorentz force = electromagnetic force = $q[\vec{E} + (\vec{v}\times\vec{B})]$, (see Eqs (6.9), (6.10)) per unit time on the charges contained in volume V. The second term in the RHS is the outward (due to positive sign before the term) energy flux flow per unit time. This is the interpretation of Eq. (6.17) and is called Poynting's theorem. Thus, we can express Poynting's theorem in the following way.

Poynting's theorem states that decrease of electromagnetic energy per unit time in a certain volume V is equal to the sum of the work done by the field forces per unit time and the outward energy flux per unit time. Poynting's theorem is the law of conservation of energy in electromagnetic fields. Equation (6.17) is the mathematical formulation of Poynting's theorem.

6.5 Vector Potential and Scalar Potential

6.5.1 Magnetic scalar potential

In electrostatics, there is a physical quantity called electrostatic potential φ_E gradient which with a negative sign gives a vector quantity called electric field intensity, i.e.,

$$\vec{E} = -\vec{\nabla}\varphi_E \qquad (6.18)$$

If $\vec{\nabla} \times \vec{E} = 0$, E can be expressed as a gradient of a scalar function. We shall search and try to find out if there is a similar expression in magnetostatics. We know that $\vec{\nabla} \times \vec{B} = 0$ when current density $\vec{J}$ is zero. With the analogy from electrostatics, we chose a scalar function φ_M to express $\vec{B}$ in the following form

$$\vec{B} = -\vec{\nabla}\varphi_M \qquad (6.19)$$

and call φ_M as the magnetic scalar potential or scalar potential. Taking the divergence of both sides of Eq. (6.19), we have

$$-\vec{\nabla} \cdot \vec{\nabla}\varphi_M = \vec{\nabla} \cdot \vec{B} = 0$$

or $\quad \nabla^2 \varphi_M = 0 \qquad (6.20)$

Thus, φ_M satisfies Laplace's equation. The magnetic scalar potential is only meaningful in the region where there is no current.

6.5.2 Magnetic vector potential

We know that divergence of any curl is zero, i.e.,

$$\vec{\nabla} \cdot \vec{\nabla} \times \vec{A} = 0 \qquad (6.21)$$

Maxwell's equation for magnetic induction vector $\vec{B}$ is

$$\vec{\nabla} \cdot \vec{B} = 0 \qquad (6.22)$$

Comparing Eq. (6.22) with Eq. (6.21), we have

$$\vec{B} = \vec{\nabla} \times \vec{A} \qquad (6.23)$$

The vector $\vec{A}$ satisfying Eq. (6.23) is called a magnetic vector potential. To make the calculation easier, the vector potential $\vec{A}$ has to satisfy the gauge

$$\vec{\nabla} \cdot \vec{A} = -\varepsilon\mu \frac{\partial \varphi_E}{\partial t} \qquad (6.24)$$

Equations (6.23) and (6.24) completely define the vector potential $\vec{A}$. Equation (6.24) is called the Lorentz gauge. Taking the curl of both sides of Eq. (6.23), we have

$$\vec{\nabla} \times \vec{B} = \vec{\nabla} \times \vec{\nabla} \times \vec{A} = \vec{\nabla}(\vec{\nabla} \cdot \vec{A}) - \nabla^2 \vec{A}$$

In magnetostatics, a convenient condition that $\vec{A}$ has to satisfy is the Coulomb gauge condition, i.e., $\vec{\nabla} \cdot \vec{A} = 0$. Under this condition, the previous equation becomes

$$\nabla^2 \vec{A} = -\mu_0 \vec{J} \qquad \text{since } \nabla \times B = \mu_0 J \qquad (6.25)$$

Equation (6.25) is Poisson's equation for the magnetic vector potential $\vec{A}$. Poisson's equation for electrostatic potential V is

$$\nabla^2 \varphi_E = -\frac{\rho}{\varepsilon_0}. \qquad (6.26)$$

Example 6.4

A current distribution gives rise to magnetic vector potential $\vec{A} = x^2 y\hat{x} + y^2 x\hat{y} - 4xyz\hat{z}$. Calculate the magnetic induction at the point $(-1, 2, 3)$.

Solution

$$\vec{\nabla} \times \vec{A} = \begin{vmatrix} \hat{x} & \hat{y} & \hat{z} \\ \dfrac{\partial}{\partial x} & \dfrac{\partial}{\partial y} & \dfrac{\partial}{\partial z} \\ x^2 y & y^2 x & -4xyz \end{vmatrix} = -4xz\hat{x} + 4yz\hat{y} + (y^2 - x^2)\hat{z}$$

or $\quad \vec{B} = 12\hat{x} + 24\hat{y} + 3\hat{z}$

Example 6.5

Obtain an expression for magnetic vector potential $\vec{A}$ at a point due to an infinite length straight current carrying straight conductor.

Solution

Method 1:

Outside the conductor

We know that in circular cylindrical coordinates (r, θ, z), the magnetic induction vector $\vec{B}$ due to an infinite length straight current carrying straight conductor is given by

$$\vec{B} = \frac{\mu_0 I}{2\pi r} \hat{\theta}$$

Here the length of the straight conductor is along the z-axis.

Using the definition of the magnetic vector potential $\vec{A}$, we have

$$\vec{\nabla} \times \vec{A} = \frac{\mu_0 I}{2\pi r} \hat{\theta}$$

Since $\vec{B}$ has only the θ component $\nabla \times \vec{A}$ has only the θ component. Hence, this equation can be expressed as

$$\frac{\partial A_r}{\partial z} - \frac{\partial A_z}{\partial r} = \frac{\mu_0 I}{2\pi r}$$

Since the length of the wire is along the z-axis, $\vec{A}$ cannot be a function of z. Hence, the aforementioned equation becomes

$$-\frac{dA_z}{dr} = \frac{\mu_0 I}{2\pi r}$$

Integrating this equation, we get

$$A_z = -\frac{\mu_0 I}{2\pi} \ell_n r + C$$

If the magnetic vector potential $\vec{A}$ is zero on the surface of the wire, the value of C will be

$$C = \frac{\mu_0 I}{2\pi} \ell_n a$$

where a is the radius of the wire. Putting this value of C in the previous expression for A_z, we get

$$A_z = \frac{\mu_0 I}{2\pi} \ell_n a - \frac{\mu_0 I}{2\pi r} \ell_n r$$

or $\quad \vec{A} = \frac{\mu_0 I}{2\pi} \ell_n \left(\frac{a}{r}\right) \hat{z}$

Thus, $\vec{A} = \frac{\mu_0 I}{2\pi} \ell_n \left(\frac{a}{r}\right) \hat{z}$ is the magnetic vector potential at a point 'r' due to an infinite length straight current carrying wire.

Method 2:

(A) *Inside the straight conductor*

The magnetic vector potential can also be found out with help of Poisson's Eq. (6.24). Each component of vector $\vec{A}$ satisfies Eq. (6.24). In our case, vector $\vec{A}$ has only the z-component. Therefore, Eq. (6.24) becomes

$$\nabla^2 A_z = -\mu J_z$$

where J_z is the z-component of the current density vector. Hence, putting $J_z = \dfrac{I}{\pi a^2}$ into this equation, we get

$$\nabla^2 A_z = -\frac{\mu I}{\pi a^2} \qquad\qquad (A)$$

In circular cylindrical coordinates, Eq. (A) becomes

$$\frac{1}{r}\frac{\partial}{\partial r}\left(r\frac{\partial A_z}{\partial r}\right) + \frac{1}{r}\frac{\partial A_z}{\partial r} + \frac{1}{r^2}\frac{\partial^2 A_z}{\partial \theta^2} + \frac{\partial^2 A_z}{\partial z^2} = -\frac{\mu I}{\pi a^2}$$

From physical intuition of the problem, we know that J_z is independent of z and θ and is a function of only r. Hence, the aforementioned equation reduces to

$$\frac{1}{r}\frac{d}{dr}\left(r\frac{dA_z}{dr}\right) = -\frac{\mu I}{\pi a^2}$$

Integrating this equation, we get

$$r\frac{dA_z}{dr} = -\frac{\mu I r^2}{2\pi a^2} + C_1$$

This equation is correct for all values of r. Putting $r = 0$ into this equation, we get $C_1 = 0$. Hence, the equation becomes

$$r\frac{dA_z}{dr} = -\frac{\mu I r^2}{2\pi a^2}$$

or $\dfrac{dA_z}{dr} = -\dfrac{\mu I r}{2\pi a^2}$

Integrating this equation, we get

$$A_z = -\frac{\mu I r^2}{4\pi a^2} + C_2 \qquad\qquad (B)$$

We know that $A_z = 0$ at $r = a$, i.e., on the surface of the current carrying straight conductor. Under this condition, Eq. (B) gives

$$C_2 = \frac{\mu I a^2}{4\pi a^2} = \frac{\mu I}{4\pi}$$

Putting this value of C_2 into Eq. (B), we get

$$A_z = -\frac{\mu I r^2}{4\pi a^2} + \frac{\mu I}{4\pi}$$

or $A_z = \frac{\mu I}{4\pi}\left(1 - \frac{r^2}{a^2}\right)$ at an inside point.

(B)   *Outside the conductor*

Poisson's equation outside the conductor is $\nabla^2 A_z = 0.$ Hence, we have

$$\frac{1}{r}\frac{d}{dr}\left(r\frac{dA_z}{dr}\right) = 0$$

$$A_z = C_1 \ell_n r + C_2 \tag{C}$$

At $r = a$, i.e., on the surface of the conductor, $A_z = 0.$ Hence, Eq. (C) under this condition gives

$$C_2 = -C_1 \ell_n a$$

Putting this value of C_2 into Eq. (C), we have

$$A_z = C_1 \ell_n r - C_1 \ell_n a$$

or $\vec{A} = C_1 \ell_n\left(\frac{r}{a}\right)\hat{z}$

Putting $C_1 = \frac{\mu_0 I}{2\pi}$ into this equation, we get

$\vec{A} = \frac{\mu_0 I}{2\pi} \ell_n\left(\frac{a}{r}\right)\hat{z}$ at an out-side point.

6.6 Electromagnetic Wave Equations for $\vec{E}$ and $\vec{B}$

The most important consequence of Maxwell's equations is the equation for electromagnetic wave propagation. We shall now derive the electromagnetic wave equation for linear, isotropic, homogeneous space characterized by constant permeability μ, constant permittivity ε and constant conductivity σ.

6.6.1 Electromagnetic wave equations for $\vec{E}$

The wave equation for the electric field vector $\vec{E}$, is derived by taking the curl of both sides of Maxwell's Eq. (6.3). Thus, we have

$$\vec{\nabla} \times (\vec{\nabla} \times \vec{E}) = -\vec{\nabla} \times \frac{\partial \vec{B}}{\partial t} = -\mu \frac{\partial}{\partial t} (\vec{\nabla} \times \vec{H}) = -\mu \frac{\partial}{\partial t} \left(\vec{J} + \frac{\partial \vec{D}}{\partial t} \right)$$

or $\quad \vec{\nabla}(\vec{\nabla} \cdot \vec{E}) - \nabla^2 \vec{E} = -\sigma\mu \frac{\partial \vec{E}}{\partial t} + \mu\varepsilon \frac{\partial^2 \vec{E}}{\partial t^2}$

or $\quad \vec{\nabla} \left(\frac{\rho}{\varepsilon} \right) - \nabla^2 \vec{E} = -\sigma\mu \frac{\partial \vec{E}}{\partial t} + \mu\varepsilon \frac{\partial^2 \vec{E}}{\partial t^2}$

Re-arranging this equation, we get

$$\nabla^2 \vec{E} - \mu\varepsilon \frac{\partial^2 \vec{E}}{\partial t^2} - \mu\sigma \frac{\partial \vec{E}}{\partial t} = \nabla \left(\frac{\rho}{\varepsilon} \right) \tag{6.27}$$

For a charge-free space, charge density $\rho = 0$. Under this condition, Eq. (6.27) becomes

$$\nabla^2 \vec{E} - \mu\varepsilon \frac{\partial^2 \vec{E}}{\partial t^2} - \mu\sigma \frac{\partial \vec{E}}{\partial t} = 0 \tag{6.28}$$

6.6.2 Electromagnetic wave equations for $\vec{H}$

The wave equation for the magnetic field vector $\vec{H}$ is derived by taking the curl of both sides of Maxwell's Eq. (6.4). Thus, we have

$$\vec{\nabla} \times (\vec{\nabla} \times \vec{H}) = \vec{\nabla} \times \left(\vec{J} + \frac{\partial \vec{D}}{\partial t} \right) = \sigma(\vec{\nabla} \times \vec{E}) + \varepsilon \frac{\partial}{\partial t}(\vec{\nabla} \times \vec{E}) = \sigma \left(-\frac{\partial \vec{B}}{\partial t} \right) + \varepsilon \frac{\partial}{\partial t} \left(-\frac{\partial \vec{B}}{\partial t} \right) \tag{6.29}$$

or $\quad \vec{\nabla}(\vec{\nabla} \cdot \vec{H}) - \nabla^2 \vec{H} = -\mu\sigma \dfrac{\partial \vec{H}}{\partial t} - \mu\varepsilon \dfrac{\partial^2 \vec{H}}{\partial t^2}$

or $\quad -\nabla^2 \vec{H} = -\mu\sigma \dfrac{\partial \vec{H}}{\partial t} - \mu\varepsilon \dfrac{\partial^2 \vec{H}}{\partial t^2}$ \hfill (6.30)

Re-arranging Eq. (6.30), we get

$$\nabla^2 \vec{H} - \mu\varepsilon \frac{\partial^2 \vec{H}}{\partial t^2} - \mu\sigma \frac{\partial \vec{H}}{\partial t} = 0 \tag{6.31}$$

Equations (6.28) and (6.31) are the electromagnetic wave equations in terms of vector $\vec{E}$ and vector $\vec{H}$ governing the propagation electromagnetic waves in homogeneous, linear, isotropic, charge-free space whether the space is conducting or non-conducting. The important point to note here is that any function satisfying Eqs (6.28) and (6.31) may not satisfy Maxwell's equations.

6.6.3 Electromagnetic wave equations for $\vec{B}$

Now multiplying both sides of Eq. (6.31) with μ, we have

$$\nabla^2 \mu\vec{H} - \mu\varepsilon \frac{\partial^2 \mu\vec{H}}{\partial t^2} - \mu\sigma \frac{\partial \mu\vec{H}}{\partial t} = \mu \times 0$$

or $\quad \nabla^2 \vec{B} - \mu\varepsilon \dfrac{\partial^2 \vec{B}}{\partial t^2} - \mu\sigma \dfrac{\partial \vec{B}}{\partial t} = 0$

This equation is the electromagnetic wave equation in terms of the magnetic induction vector $\vec{B}$.

6.7 Wave Equation in Terms of Scalar and Vector Potentials

The wave Eqs (6.28) and (6.31) are derived on the assumption that space is free of sources, i.e., $\rho = 0$ and $\vec{J} = 0$. However, electromagnetic waves are produced by accelerated sources and propagation of waves occuring in the presence of sources. The equations governing the propagation of electromagnetic waves are expressed in terms of scalar potential φ_M and vector potential $\vec{A}$ defined by Eqs (6.19) and (6.23). Re-writing Eq. (6.23), we have

$$\vec{B} = \vec{\nabla} \times \vec{A}. \tag{6.32}$$

Equation (6.32) gives the magnetic fields in terms of vector potential $\vec{A}$. Now putting this value of $\vec{B}$ from Eq. (6.32) into Maxwell's Eq. (6.3), we have

$$\vec{\nabla} \times \vec{E} = -\frac{\partial}{\partial t}(\vec{\nabla} \times \vec{A})$$

or

$$\nabla \times \left(\vec{E} + \frac{\partial \vec{A}}{\partial t} \right) = 0 \tag{6.33}$$

We know that if the curl of a vector function is zero, then the vector function can be expressed as a gradient of a scalar function. Thus, we can express the bracketed term of Eq. (6.33) as the gradient of a scalar function. In our case, let that scalar function be φ_E. Thus, we have

$$\vec{E} + \frac{\partial \vec{A}}{\partial t} = -\vec{\nabla}\varphi_E$$

or

$$\vec{E} = -\frac{\partial \vec{A}}{\partial t} - \vec{\nabla}\varphi_E \tag{6.34}$$

Equation (6.34) gives the electric fields in terms of scalar potential φ_E.

6.7.1 Wave equation in terms of vector potential $\vec{A}$

From Maxwell's Eq. (6.4), we have

$$\nabla \times \vec{H} = \vec{J} + \frac{\partial \vec{D}}{\partial t}$$

or

$$\vec{\nabla} \times \vec{B} = \mu \left(\sigma \vec{E} + \frac{\partial \varepsilon \vec{E}}{\partial t} \right)$$

Putting the value of $\vec{B}$ from Eq. (6.32) into this equation, we have

$$\vec{\nabla} \times (\vec{\nabla} \times \vec{A}) = \mu\sigma\vec{E} + \mu\varepsilon\frac{\partial \vec{E}}{\partial t}$$

Putting the value of $\vec{E}$ from Eq. (6.34) into this equation, we have

$$\vec{\nabla} \times (\vec{\nabla} \times \vec{A}) = \mu\sigma\left(-\vec{\nabla}\varphi_E - \frac{\partial \vec{A}}{\partial t} \right) + \mu\varepsilon\frac{\partial}{\partial t}\left(-\vec{\nabla}\varphi_E - \frac{\partial \vec{A}}{\partial t} \right)$$

or $\quad \vec{\nabla}(\vec{\nabla} \cdot \vec{A}) - \nabla^2 \vec{A} = -\mu\sigma\vec{\nabla}\varphi_E - \mu\sigma\dfrac{\partial \vec{A}}{\partial t} - \mu\varepsilon\vec{\nabla}\dfrac{\partial \varphi_E}{\partial t} - \mu\varepsilon\dfrac{\partial^2 \vec{A}}{\partial t^2}$

Re-arranging this equation, we get

$$-\nabla^2 \vec{A} + \mu\varepsilon\dfrac{\partial^2 \vec{A}}{\partial t^2} + \vec{\nabla}(\vec{\nabla} \cdot \vec{A}) + \mu\varepsilon\vec{\nabla}\dfrac{\partial \varphi_E}{\partial t} = \mu\sigma\left(-\vec{\nabla}\varphi - \dfrac{\partial \vec{A}}{\partial t} \right)$$

or $\quad -\nabla^2 \vec{A} + \mu\varepsilon\dfrac{\partial^2 \vec{A}}{\partial t^2} + \vec{\nabla}(\vec{\nabla} \cdot \vec{A}) + \mu\varepsilon\vec{\nabla}\dfrac{\partial \varphi_E}{\partial t} = \mu\sigma\vec{E}$

or $\quad -\nabla^2 \vec{A} + \mu\varepsilon\dfrac{\partial^2 \vec{A}}{\partial t^2} + \vec{\nabla}(\vec{\nabla} \cdot \vec{A}) + \mu\varepsilon\vec{\nabla}\dfrac{\partial \varphi_E}{\partial t} = \mu\vec{J}$

or $\quad -\nabla^2 \vec{A} + \mu\varepsilon\dfrac{\partial^2 \vec{A}}{\partial t^2} + \vec{\nabla}\left(\vec{\nabla} \cdot \vec{A} + \mu\varepsilon\dfrac{\partial \varphi_E}{\partial t} \right) = \mu\vec{J}$ $\qquad (6.35)$

Now the Lorentz gauge condition for electromagnetic field is

$$\vec{\nabla} \cdot \vec{A} + \mu\varepsilon\dfrac{\partial \varphi_E}{\partial t} = 0. \qquad (6.36)$$

Therefore, under the Lorentz gauge condition, Eq. (6.35) becomes

$$\nabla^2 \vec{A} - \mu\varepsilon\dfrac{\partial^2 \vec{A}}{\partial t^2} = -\mu\vec{J} \qquad (6.37)$$

Equation (6.37) is the wave equation in terms of vector potential A.

6.7.2 Wave equation in terms of scalar potential φ_E

From Maxwell's Eq. (6.2), we have

$$\rho = \vec{\nabla} \cdot \varepsilon\left(-\vec{\nabla}\varphi_E - \dfrac{\partial \vec{A}}{\partial t} \right) \qquad (6.38)$$

o.r $\quad \vec{\nabla} \cdot \vec{\nabla}\varphi_E + \dfrac{\partial}{\partial t}(\vec{\nabla} \cdot \vec{A}) = -\dfrac{\rho}{\varepsilon} \qquad (6.39)$

Now putting Lorentz gauge condition $\vec{\nabla} \cdot \vec{A} + \mu \varepsilon \dfrac{\partial \varphi_E}{\partial t} = 0$ for electromagnetic field into this equation, we get

$$\nabla^2 \varphi_E - \mu \varepsilon \frac{\partial^2 \varphi_E}{\partial t^2} = -\frac{\rho}{\varepsilon} \tag{6.40}$$

Equation (6.40) is the wave equation in terms of scalar potential φ_E.

6.8 Plane Electromagnetic Waves

The fields produced by accelerating charges leave the sources and travel through space in the form of electromagnetic waves. A plane wave is defined as a wave whose phase is the same at a given instant at all the points on each plane perpendicular to some specified direction. A monochromatic wave is defined as a wave characterized by a single frequency. For a plane monochromatic wave propagating in the $+Z$ direction, $\omega t - kz$ is constant. The surface passing through the points having the same phase is called a wavefront. In case of a plane wave, the wavefront is a plane perpendicular to the direction of propagation. In the plane wave, the field vector component is constant (of course, in magnitude) over all the points on the wavefront. These field vector components on a wavefront are the functions of the perpendicular distance of the wavefront from the origin and are also functions of time. In most cases, we choose our coordinate system in such a manner that wave propagates in the $+Z$ direction. Therefore, in most cases,

$$\frac{\partial}{\partial x} = \frac{\partial}{\partial y} = 0 \; ; \; \frac{\partial}{\partial z} \neq 0 \tag{6.41}$$

and

$$\vec{E}(z,t) = \vec{E}_0 e^{i(\omega t - kz)} \tag{6.42}$$

$$\vec{H}(z,t) = \vec{H}_0 e^{i(\omega t - kz)} \tag{6.43}$$

where $(\omega t - kz)$ is the phase angle. the symbols have their usual meanings.

The electric fields $\vec{E}(z,t)$ and magnetic field $\vec{H}(z,t)$ have the same phase. If possible, let α be the phase difference between the electric field and the magnetic field. Under this condition, Eq. (6.42) and Eq. (6.43) become

$$\vec{E}(z,t) = \vec{E}_0 e^{i(\omega t - kz)} \tag{6.44}$$

$$\vec{H}(z,t) = \vec{H}_0 e^{i(\omega t - kz + \alpha)} \tag{6.45}$$

The fields $\vec{E}(z,t)$ and $\vec{H}(z,t)$ have to satisfy Maxwell's equation

$$\vec{\nabla} \times \vec{E} = -\frac{\partial \vec{B}}{\partial t}. \tag{6.46}$$

Taking the y-component of both sides of Eq. (6.46), we have

$$(\vec{\nabla} \times \vec{E})_Y = -\frac{\partial B_Y}{\partial t}$$

or $\quad \dfrac{\partial E_X}{\partial z} - \dfrac{\partial E_Z}{\partial x} = -\mu \dfrac{\partial H_Y}{\partial t}$

Taking the help of Eq. (6.41), from this equation, we have

$$\frac{\partial E_X}{\partial z} = -\mu \frac{\partial H_Y}{\partial t}$$

or $\quad \dfrac{\partial}{\partial z} E_{OX} e^{i(\omega t - kz)} = -\mu \dfrac{\partial}{\partial t} H_{OY} e^{i(\omega t - kz + \alpha)}$

or $\quad ikE_{OX} e^{i(\omega t - kz)} = i\omega\mu H_{OY} e^{i(\omega t - kz + \alpha)}$

We know that

$\qquad$ Real part of LHS = Real part of RHS

or $\quad kE_{OX} \cos(\omega t - kz) = \omega\mu H_{OY} \cos(\omega t - kz + \alpha) \tag{6.47}$

(Imaginary part of LHS = Imaginary part of RHS; imaginary parts may be taken)

$\quad$ Equation (6.47) is true for all values of z and t which is possible only when α is equal to zero. Therefore, electric field $\vec{E}$ and magnetic field $\vec{H}$ have the same phase ($\alpha = 0$). Hence, Eq. (6.47) becomes

$$kE_{OX} \cos(\omega t - kz) = \omega\mu H_{OY} \cos(\omega t - kz)$$

or $\quad kE_{OX} = \omega\mu H_{OY} \tag{6.48}$

Since plane electromagnetic waves are a good approximation of the actual waves generated by moving charges, most of our discussions will be limited to plane monochromatic waves.

6.9 Transverse Nature of Electromagnetic Waves

We shall determine the relative orientation of vector $\vec{E}$, vector $\vec{H}$ and the propagation vector $\vec{k}$. Suppose, electromagnetic wave is propagating in any arbitrary direction defined by the propagation vector $\vec{k}$. We can represent the electromagnetic wave by the complex equations

$$\vec{E}(\vec{r},t) = \vec{E}_0 e^{i(\omega t - \vec{k}\cdot\vec{r})} \tag{6.49}$$

$$\vec{H}(\vec{r},t) = \vec{H}_0 e^{i(\omega t - \vec{k}\cdot\vec{r})} \tag{6.50}$$

Here, the vectors $\vec{E}_0$ and $\vec{H}_0$ are constants in time, $\vec{k}$ is the propagation vector whose magnitude is equal to $\dfrac{2\pi}{\lambda}$ and its direction is the direction of propagation of wave and $i = \sqrt{-1}$. Equations (6.49) and (6.50) consisting of two parts, real part and imaginary part can be expanded using the formula $e^{i\theta} = \cos\theta + i\sin\theta$. The real parts of Eqs (6.49) and (6.50) are taken to represent electromagnetic waves. The imaginary parts may be used to represent electromagnetic waves without any difficulties. However, care must be taken not to intermix the real part and imaginary part during calculation.

6.9.1 Transverse nature of vector $\vec{E}$

The field vector $\vec{E}(r,t) = \vec{E}_0 e^{i(\omega t - \vec{k}\cdot\vec{r})}$ of Eq. (6.49) must satisfy Maxwell's Eq. (6.2). Thus, we have for charge free regions

$$\vec{\nabla}\cdot\vec{E}(\vec{r},t) = 0$$

or $\left(\hat{x}\dfrac{\partial}{\partial x} + \hat{y}\dfrac{\partial}{\partial y} + \hat{z}\dfrac{\partial}{\partial z}\right)\cdot\left(\hat{x}E_{OX}e^{i(\omega t - \vec{k}\cdot\vec{r})} + \hat{y}E_{OY}e^{i(\omega t - \vec{k}\cdot\vec{r})} + \hat{z}E_{OZ}e^{i(\omega t - \vec{k}\cdot\vec{r})}\right) = 0$

or $iE_{OX}e^{i(\omega t - \vec{k}\cdot\vec{r})}\dfrac{\partial}{\partial x}(\omega t - \vec{k}\cdot\vec{r}) + iE_{OY}e^{i(\omega t - \vec{k}\cdot\vec{r})}\dfrac{\partial}{\partial y}(\omega t - \vec{k}\cdot\vec{r}) +$

$$iE_{OZ}e^{i(\omega t - \vec{k}\cdot\vec{r})}\dfrac{\partial}{\partial z}(\omega t - \vec{k}\cdot\vec{r}) = 0$$

or $iE_{OX}e^{i(\omega t - \vec{k}\cdot\vec{r})}k_X + iE_{OY}e^{i(\omega t - \vec{k}\cdot\vec{r})}k_Y + iE_{OZ}e^{i(\omega t - \vec{k}\cdot\vec{r})}k_Z = 0$

or $i(E_X k_X + E_Y k_Y + E_Z k_Z) = 0$

or $\vec{E} \cdot \vec{k} = 0$ (6.51)

Hence, the dot product of two vectors $\vec{E}$ and $\vec{k}$ is zero. Therefore, they are perpendicular to each other. That means the field vector $\vec{E}$ is perpendicular to the direction of propagation.

6.9.2 Transverse nature of vector $\vec{H}$

The field vector $\vec{H}(\vec{r},t) = \vec{H}_0 e^{i(\omega t - \vec{k}\cdot\vec{r})}$ of Eq. (6.50) must satisfy Maxwell's Eq. (6.1). Thus, we have

$$\vec{\nabla} \cdot \vec{H}(\vec{r},t) = 0$$

or $\left(\hat{x}\dfrac{\partial}{\partial x} + \hat{y}\dfrac{\partial}{\partial y} + \hat{z}\dfrac{\partial}{\partial z} \right) \cdot \left(\hat{x}H_{OX}e^{i(\omega t - \vec{k}\cdot\vec{r})} + \hat{y}H_{OY}e^{i(\omega t - \vec{k}\cdot\vec{r})} + \hat{z}H_{OZ}e^{i(\omega t - \vec{k}\cdot\vec{r})} \right) = 0$

or $iH_{OX}e^{i(\omega t - \vec{k}\cdot\vec{r})}\dfrac{\partial}{\partial x}(\omega t - \vec{k}\cdot\vec{r}) + iH_{OY}e^{i(\omega t - \vec{k}\cdot\vec{r})}\dfrac{\partial}{\partial y}(\omega t - \vec{k}\cdot\vec{r}) +$

$$iH_{OZ}e^{i(\omega t - \vec{k}\cdot\vec{r})}\dfrac{\partial}{\partial z}(\omega t - \vec{k}\cdot\vec{r}) = 0$$

or $iH_{OX}e^{i(\omega t - \vec{k}\cdot\vec{r})}k_X + iH_{OY}e^{i(\omega t - \vec{k}\cdot\vec{r})}k_Y + iH_{OZ}e^{i(\omega t - \vec{k}\cdot\vec{r})}k_Z = 0$

or $\vec{H} \cdot \vec{k} = 0$ (6.52)

Hence, the dot product of two vectors $\vec{H}$ and $\vec{k}$ is zero. Therefore, they are perpendicular to each other. That means the field vector $\vec{H}$ is perpendicular to the direction of propagation.

By combining Eqs (6.51) and (6.52), we conclude that field vectors $\vec{E}$ and $\vec{H}$ are perpendicular to the direction of propagation. Such types of waves are called transverse waves. Therefore, electromagnetic waves are transverse waves.

6.9.3 Relative orientation of $\vec{E}$ and $\vec{H}$

The field vectors $\vec{E}(\vec{r},t) = \vec{E}_0 e^{i(\omega t - \vec{k}\cdot\vec{r})}$ and $\vec{H}(\vec{r},t) = \vec{H}_0 e^{i(\omega t - \vec{k}\cdot\vec{r})}$ must satisfy Maxwell's Eq. (6.3). Thus, we have

$$\vec{\nabla} \times \vec{E}(\vec{r},t) = -\dfrac{\partial \mu \vec{H}(\vec{r},t)}{\partial t}$$

or $\quad \hat{x}\left[\dfrac{\partial}{\partial y}\left(E_{OZ}e^{i(\omega t - \vec{k}\cdot\vec{r})}\right) - \dfrac{\partial}{\partial z}\left(E_{OY}e^{i(\omega t - \vec{k}\cdot\vec{r})}\right)\right] +$

$\hat{y}\left[\dfrac{\partial}{\partial z}\left(E_{OX}e^{i(\omega t - \vec{k}\cdot\vec{r})}\right) - \dfrac{\partial}{\partial x}\left(E_{OZ}e^{i(\omega t - \vec{k}\cdot\vec{r})}\right)\right] +$

$\hat{z}\left[\dfrac{\partial}{\partial x}\left(E_{OY}e^{i(\omega t - \vec{k}\cdot\vec{r})}\right) - \dfrac{\partial}{\partial y}\left(E_{OX}e^{i(\omega t - \vec{k}\cdot\vec{r})}\right)\right] = -\mu\dfrac{\partial \vec{H}(\vec{r},t)}{\partial t}$

or $\quad i\left[\hat{x}\left(E_Z k_Y - E_Y k_Z\right) + \hat{y}\left(E_X k_Z - E_Z k_X\right) + \hat{z}\left(E_Y k_X - E_X k_Y\right)\right]$

$= -\mu i(-\omega)H_0 e^{i(\omega t - \vec{k}\cdot\vec{r})}$

or $\quad \vec{k} \times \vec{E} = \mu\omega\vec{H}$ $\hfill (6.53)$

Equation (6.53) implies that the vector $\vec{H}$ is perpendicular to both the vectors $\vec{k}$ and $\vec{E}$.

Again, the field vectors $\vec{E}$ and $\vec{H}$ of Eqs (6.49) and (6.50) must satisfy Maxwell's Eq. (6.4) in a source-free medium ($\vec{J}=0$). Thus, we have

$$\vec{\nabla} \times \vec{H}(\vec{r},t) = \dfrac{\partial \varepsilon \vec{E}(\vec{r},t)}{\partial t}$$

or $\quad \hat{x}\left[\dfrac{\partial}{\partial y}\left(H_{OZ}e^{i(\omega t - \vec{k}\cdot\vec{r})}\right) - \dfrac{\partial}{\partial z}\left(H_{OY}e^{i(\omega t - \vec{k}\cdot\vec{r})}\right)\right] +$

$\hat{y}\left[\dfrac{\partial}{\partial z}\left(H_{OX}e^{i(\omega t - \vec{k}\cdot\vec{r})}\right) - \dfrac{\partial}{\partial x}\left(H_{OZ}e^{i(\omega t - \vec{k}\cdot\vec{r})}\right)\right] +$

$\hat{z}\left[\dfrac{\partial}{\partial x}\left(H_{OY}e^{i(\omega t - \vec{k}\cdot\vec{r})}\right) - \dfrac{\partial}{\partial y}\left(H_{OX}e^{i(\omega t - \vec{k}\cdot\vec{r})}\right)\right] = \varepsilon\dfrac{\partial \vec{E}(\vec{r},t)}{\partial t}$

or $\quad i\left[\hat{x}\left(H_Z k_Y - H_Y k_Z\right) + \hat{y}\left(H_X k_Z - H_Z k_X\right) + \hat{z}\left(H_Y k_X - H_X k_Y\right)\right]$

$$= \varepsilon\, i(-\omega)E_0 e^{i(\omega t - \vec{k}\cdot\vec{r})}$$

$$i(\vec{k}\times\vec{H}) = -i\varepsilon\omega\vec{E}$$

or $\quad \vec{H}\times\vec{k} = \varepsilon\omega\vec{E}$ $\hspace{6cm}$ (6.54)

Equation (6.54) implies that the vector $\vec{E}$ is perpendicular to both the vectors $\vec{k}$ and $\vec{H}$.

Therefore, by combining Eqs (6.53) and (6.54), we conclude that field vectors $\vec{E}$ and $\vec{H}$ are perpendicular to each other and to the propagation vector $\vec{k}$. Equation (6.51) shows that vector $\vec{k}$ and $\vec{E}$ are perpendicular to each other and Eq. (6.52) shows that vector $\vec{k}$ and $\vec{H}$ are perpendicular to each other. Poynting's vector $\vec{E}\times\vec{H}$ points along the direction of propagation (i.e., $\vec{E}\times\vec{H}$ and $\vec{k}$ have the same direction). Therefore, we conclude that the vectors $\vec{E}$, $\vec{H}$, and $\vec{k}$ form a right-hand orthogonal set. This is depicted in Fig. 6.2.

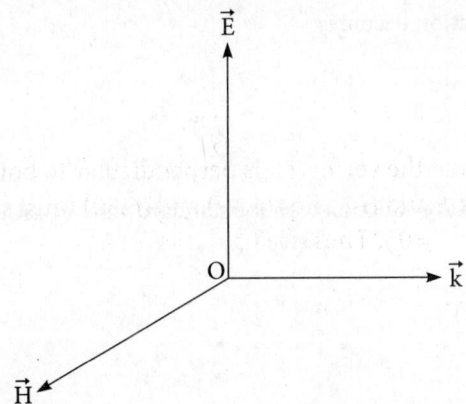

Figure 6.2 | Relative orientation of the magnetic field vector $\vec{H}$, the electric field vector $\vec{E}$, and the propagation vector $\vec{k}$ – together they three form a right-hand orthogonal set

Taking the magnitude of both sides of Eq. (6.53), we get

$$kE \sin 90° = \mu\omega H$$

or $\quad \dfrac{E}{H} = \mu v = \sqrt{\dfrac{\mu}{\varepsilon}} = \eta$ $\qquad \left(\because v = \dfrac{1}{\sqrt{\mu\varepsilon}}\right)$ $\hspace{3cm}$ (6.55)

In Eq. (6.55), E/H has the dimension of impedance and is called intrinsic impedance η of the medium. Here, v is the speed of the electromagnetic wave in the medium. For vacuum,

$$\frac{E}{H} = \mu_0 c = 376.7\Omega.$$

See what happens if you take Eq. (6.54) instead of Eq. (6.53) in this calculation. Interesting!

Example 6.6

If the electric field in free space is given by $\vec{E} = \hat{y}80\cos(10^8 t + \beta x)$V/m, then calculate the value of β and the time it takes to travel a distance of $\lambda/2$. Determine the direction of wave propagation. Also calculate the wavelength.

Solution

a) The value of β is given by

$$\beta = \frac{\omega}{v}.$$

In free space, this equation becomes

$$\beta = \frac{\omega}{c}$$

Comparing the given wave equation with the standard wave equation, we get $\omega = 10^8$ rad/s Therefore,

$$\beta = \frac{10^8}{3 \times 10^8} = \frac{1}{3} \text{rad/s}$$

b) From the definition of time period, we have

$$\omega = \frac{2\pi}{T}$$

or $T = \dfrac{2\pi}{\omega} = \dfrac{2\pi}{10^8} s = 62.83 \times 10^{-9} s$

In one time period, wave travels a distance of one wavelength. Hence, in 62.83×10^{-9}s, wave travels a distance of λ. Therefore, the time taken to travel a distance of $\lambda/2$ is 31.42×10^{-9} s

c) The given electric field has no components other than the Y-component. Due to the positive sign in the expression $(10^8 t + \beta x)$ and due to the equations $\vec{k} \times \vec{E} = \mu\omega\vec{H}$ and $\vec{H} \times \vec{k} = \mu\varepsilon\vec{E}$, we conclude that the electromagnetic wave is propagating along the negative X-axis. The equation can now be written as

$$E_Y = 80\cos\left(10^8 t + \frac{x}{3}\right) \text{V/m}.$$

d) The wavelength of the wave is given by

$$\lambda = cT = 3\times10^8 \text{ m/s} \times 62.83\times10^{-9}\text{ s} = 18.85\text{m}$$

Example 6.7

The electric field intensity of a uniform plane wave in air is given by $E = 800\cos(\omega t - kz)\hat{x}$. If the wavelength of the wave is 2.0 feet, find (a) frequency, (b) time period, (c), value of k, and (d) amplitude of magnetic vector H.

Solution

It is given that wavelength λ = 2vt = 2 × 0.3048 m = 0.61 m and speed c of the wave is 3×10^8 m/s

a. We know that

$$c = \lambda v$$

or $v = \dfrac{c}{\lambda} = \dfrac{3\times10^8}{0.61}\text{Hz} = 4.92\times10^8\text{ Hz}$

b. The time period is given by $\dfrac{1}{v} = 2.03\times10^{-9}$ s

c. The value of k is given by $\dfrac{2\pi}{\lambda} = \dfrac{2\pi}{0.61} = 10.31\,\text{m/s}$

d. We know that

$$\vec{H}\times\vec{k} = \varepsilon\omega\vec{E}$$

or $Hk\sin90° = \varepsilon\omega E$

or $Hk = \varepsilon\omega E$

or $H_0 k = \varepsilon\omega E_0$

or $H_0 = \dfrac{\varepsilon \omega E_0}{k} = \varepsilon c E_0$ $\qquad \left(\because \quad \dfrac{\omega}{k} = v \right)$

Hence, we have $H_0 = 8.85 \times 10^{-12} \times 3 \times 10^8 \times 800\ \text{A/m} = 2.12\ \text{A/m}.$

Example 6.8

The amount of electromagnetic energy received by earth in the form of light from the sun is 1300 W/m². Calculate the root mean square value of the electric vector and the magnetic vector of the light wave on the earth surface.

Solution

The amount of electromagnetic energy received by earth from the sun is 1300 Watt/m². This is nothing but the magnitude of Poynting's vector. We know that the average magnitude of Poynting's vector is given by

$$< P > = \dfrac{1}{2} \sqrt{\dfrac{\varepsilon}{\mu}} \times E_0^2$$

or $< P > = \dfrac{E_0^2}{2\eta}$ $\qquad \left(\because \eta = \sqrt{\dfrac{\mu}{\varepsilon}} \right)$

where η is the intrinsic impedance of the medium. For vacuum or air medium, if η_0 is the intrinsic impedance then this equation becomes

$$< P > = \dfrac{E_0^2}{2\eta_0}$$

or $E_0 = \sqrt{2 < P > \eta_0}$

Putting the values of $<P>$ and η_0 into this equation, we get

$$E_0 = \sqrt{2 \times 1300 \times 377}\ \text{V/m} = 990\ \text{V/m}$$

Hence, the root mean square value of the electric vector of the light wave will be

$$E_{\text{rms}} = \dfrac{E_0}{\sqrt{2}} = 700\ \text{V/m}$$

The root mean square value of the magnetic vector of the light wave will be obtained from

$$H_{rms} = \varepsilon c E_{rms} = 1.858 \, \text{A/m} \, .$$

6.10 Speed of Electromagnetic Waves

For charge-free non-conducting medium ($\rho = 0$ and $\sigma = 0$), Eqs (6.28) and (6.31) become, respectively

$$\nabla^2 E - \mu\varepsilon \frac{\partial^2 E}{\partial t^2} = 0 \qquad (6.56)$$

$$\nabla^2 H - \mu\varepsilon \frac{\partial^2 H}{\partial t^2} = 0 \qquad (6.57)$$

These equations are of the same form as that of the wave equations we are familiar with. The general wave equation in a lossless medium is given by

$$\nabla^2 \Psi - \frac{1}{v^2} \frac{\partial^2 \Psi}{\partial t^2} = 0 \qquad (6.58)$$

where v is the phase speed or the speed of the wave.

Now comparing electromagnetic wave Eqs (6.56) and (6.57) for a charge-free non-conducting medium with the general wave Eq. (6.58), we get

$$\mu\varepsilon = \frac{1}{v^2}$$

or $\quad v = \dfrac{1}{\sqrt{\mu\varepsilon}}$ $\qquad (6.59)$

which is the phase speed or speed of electromagnetic waves in a linear, isotropic, homogeneous, charge-free non-conducting medium characterized by constant permeability μ and constant permittivity ε. For vacuum, Eq. (6.59) becomes

$$c = \frac{1}{\sqrt{\mu_0 \varepsilon_0}} \qquad (6.60)$$

where μ_0 and ε_0 are the permeability and permittivity of the vacuum respectively and c is the phase speed or simply speed of electromagnetic wave in vacuum. The values of μ_0 and ε_0 are given by

$$\mu_0 = 4\pi \times 10^{-7} \frac{Wb}{Am} = 4\pi \times 10^{-7} \frac{N}{A^2}$$

$$\varepsilon_0 = 8\cdot 8547 \times 10^{-12} \frac{C^2}{Nm^2} = 8\cdot 8547 \times 10^{-12} \frac{A^2 s^2}{Nm^2}$$

Putting these values of μ_0 and ε_0 in Eq. (6.60), the phase speed of electromagnetic wave is found to be

$$c = \frac{1}{\sqrt{\mu_0 \varepsilon_0}} = 2.99784 \times 10^8 \, m/s \approx 3 \times 10^8 \, m/s.$$

This is the speed of light in vacuum. Therefore, we conclude that light is a form of electromagnetic radiation or wave. γ-rays, x-rays, ultraviolet, infrared, radio waves, T.V. waves, microwaves and so on are all electromagnetic waves or radiations differing only in the order of magnitude of their wavelengths.

The refractive index n of the medium with respect to vacuum is defined by

$$n = \frac{c}{v} \qquad (6.61)$$

where c is the speed of light in vacuum and v is the speed of light in the medium whose refractive index is n.

Putting the values of c and v in this equation, we get the expression for refractive index of the medium as

$$n = \sqrt{\frac{\mu \varepsilon}{\mu_0 \varepsilon_0}} = \sqrt{\frac{\mu}{\mu_0} \times \frac{\varepsilon}{\varepsilon_0}} = \sqrt{\mu_r \times \varepsilon_r}$$

where $\mu_r = \dfrac{\mu}{\mu_0}$ is called the relative permeability and $\varepsilon_r = \dfrac{\varepsilon}{\varepsilon_0}$ is called the relative permittivity.

The synonyms of relative permittivity are specific inductive capacity, dielectric constant and dielectric coefficient. Here it is important to mention that n, μ_r, and ε_r are determined at the same frequency.

Example 6.9

A non-magnetic medium is characterized by relative permitivitty $\varepsilon_r = 80$. Calculate the speed of an electromagnetic wave in the medium and the refractive index of the medium.

Solution

Since the medium is non-magnetic, $\mu_r = 1$. The speed of the electromagnetic wave is given by the expression

$$v = \frac{1}{\sqrt{\mu\varepsilon}} = \frac{c}{\sqrt{\mu_r \varepsilon_r}}$$

Putting the values of $\varepsilon_r = 80$, $\mu_r = 1$, and $c = 3\times10^8$ m/s into this equation, we get

$$v = \frac{3\times10^8}{\sqrt{1\times80}} \text{m/s} = 3.35\times10^7 \text{m/s}$$

The refractive index n of the medium is given by

$$n = \sqrt{\mu_r \varepsilon_r} = \sqrt{1\times80} = 8.94.$$

6.11 Average Value of Poynting's Vector

Poynting's vector $\vec{P} = \vec{E}\times\vec{H}$ represents the instantaneous rate at which energy radiates across unit area. We shall calculate Poynting's vector in linear, isotropic, homogeneous space characterized by constant permeability μ, constant permittivity ε and constant conductivity σ. Re-writing Eq. (6.15), we have

$$\vec{P} = \vec{E}\times\vec{H} \tag{6.62}$$

The vectors $\vec{E}$, $\vec{H}$, and $\vec{k}$ are mutually perpendicular to each other. Hence, Eq. (6.62) becomes

$$\vec{P} = \hat{k}EH \tag{6.63}$$

where $\hat{k}$ is a unit vector in the direction of the vector $\vec{k}$. From Eq. (6.55), we get

$$H = \frac{E}{\mu v}$$

where v is the speed or phase speed of the wave. By putting this value of H in Eq. (6.63), we get

$$\vec{P} = \hat{k} E \frac{E}{\mu v} = \hat{k} \sqrt{\frac{\varepsilon}{\mu}} \; E^2 \qquad \left(\because v = \frac{1}{\sqrt{\mu \varepsilon}} \Rightarrow \frac{1}{\mu v} = \sqrt{\frac{\varepsilon}{\mu}} \right) \tag{6.64}$$

Equation (6.64) is Poynting's vector in terms of magnitude of electric vector $\vec{E}$. In a similar way, we can derive the expression for Poynting's vector in terms of magnitude of magnetic vector $\vec{H}$. It is left as an exercise to the readers. If the real part of Eq. (6.49) represents the electric component of the electromagnetic waves, then we have

$$\vec{E} = \vec{E}_0 \cos(\omega t - \vec{k} \cdot \vec{r})$$

Putting this value of $\vec{E}$ in Eq. (6.64), we get

$$\vec{P} = \hat{k} \sqrt{\frac{\varepsilon}{\mu}} \; E_0^2 \cos^2(\omega t - \vec{k} \cdot \vec{r}) \tag{6.65}$$

This is the instantaneous flow of energy per unit area per unit time (i.e., Poynting's vector) at a point r and at any time t. The average value of $\vec{P}$ at a point r, over a complete period of oscillation will be given by

$$\left\langle \vec{P} \right\rangle = \left\langle \hat{k} \sqrt{\frac{\varepsilon}{\mu}} \; E_0^{\,2} \cos^2(\omega t - \vec{k} \cdot \vec{r}) \right\rangle$$

Since $\hat{k}$, ε, μ, and E_0 are constants in time, this equation is boiled down to

$$\left\langle \vec{P} \right\rangle = \hat{k} \sqrt{\frac{\varepsilon}{\mu}} \; E_0^{\,2} \left\langle \cos^2(\omega t - \vec{k} \cdot \vec{r}) \right\rangle \tag{6.66}$$

The average value of any function $f(x)$ over the interval (a, b) is given by the expression

$$\left\langle f(x) \right\rangle = \frac{\displaystyle\int_a^b f(x)dx}{\displaystyle\int_a^b dx}$$

Hence, the average value of $\cos^2(\omega t - \vec{k} \cdot \vec{r})$ at a point '$\vec{r}$', over a complete period of oscillation will be given by

$$\langle \cos^2(\omega t - \vec{k} \cdot \vec{r}) \rangle = \frac{\int_0^{2\pi} \cos^2(\omega t - \vec{k} \cdot \vec{r}) d(\omega t)}{\int_0^{2\pi} d(\omega t)}$$

$$= \frac{\int_0^{2\pi} d(\omega t) + \int_0^{2\pi} \cos 2(\omega t - \vec{k} \cdot \vec{r})] d(\omega t)}{4\pi} = \frac{1}{2}$$

Now putting the value of $\langle \cos^2(\omega t - \vec{k} \cdot \vec{r}) \rangle$ in Eq. (6.66), we have

$$\langle \vec{P} \rangle = \frac{1}{2} \sqrt{\frac{\varepsilon}{\mu}} E_0^2 \hat{k} \tag{6.67}$$

The total electromagnetic energy U from the second term of Eq. (6.8) is given by

$$U = \oint_V \left[\frac{1}{2} \vec{B} \cdot \vec{H} + \frac{1}{2} \vec{D} \cdot \vec{E} \right] dv$$

The time-averaged energy density over a complete period of oscillation $\langle U_d \rangle$ associated with electromagnetic wave will be given by

$$\langle U_d \rangle = \frac{\varepsilon}{2} E_0^2 \tag{6.68}$$

or $\quad E_0^2 = \frac{2}{\varepsilon} \langle U_d \rangle$

Putting this value of E_0^2 in Eq. (6.67), we get

$$\langle P \rangle = \sqrt{\frac{1}{\mu\varepsilon}} \langle U_d \rangle \hat{k} = v \langle U_d \rangle \hat{k} \tag{6.69}$$

where v is the phase speed of the wave and $\hat{k}$ is a unit vector in the direction of the vector $\vec{k}$ or in the direction of propagation of electromagnetic waves.

This shows that the time-averaged energy flow per unit area is in the direction of propagation of the wave and flows with the speed or the phase speed of propagation of the wave. Thus, we conclude that energy flows in the electromagnetic field with the same speed as the wave itself.

6.12 Propagation of Electromagnetic Waves in Plasma Medium

The fourth state of matter is the plasma state which is defined as the state in which matter is ionized. Hence, plasma consists of ions and free electrons distributed over a region in space. One example of plasma is the ionosphere of upper atmosphere. In the undisturbed condition, plasma is neutral. If some how it is disturbed, the density of electrons increases at some point, as a result of which they will repel each other by electro-static forces and move to the equilibrium position. As the electrons move to their original position, they gain kinetic energy and overshoot the equilibrium position. Again, they move backward to reach the equilibrium position, but overshoot. Thus, they will oscillate back and forth. The angular frequency with which they make this oscillation is called plasma angular frequency ω_p. Consider a hypothetical rectangular volume in a plasma medium of infinite extension. Suppose the neutral plasma is disturbed due to the propagation of electromagnetic waves and the positive and negative charges are displaced through a distance x from each other as shown in Fig. 6.3. Due to electromagnetic disturbance, at a certain instant, positive charges are present inside a width x on the left side of the rectangular volume, whereas negative charges are present inside a width x on the right side of the rectangular volume. Thus, neutral plasma is confined between two charged layers of width d.

Let N be the number of charged particles per unit volume and σ', the surface charge density of the two sides of the rectangular volume containing positive charges (left side) and negative charges (right side). The volume of unit surface area [unit depth × unit length] and x width of the rectangular volume is $1 \times x = x$. This volume will contain Nx number of charged particles as N is the number of charged particles per unit volume. The charge of a single particle is e in magnitude. The charge contained in the volume $1 \times x = x$ is therefore, Nex. This is also the charge contained in the aforesaid unit surface area. Therefore, we have

$$\sigma' = Nex \qquad (6.70)$$

The neutral plasma is confined between two charged layers of width d. If the width of the neutral plasma is small, then the electric field intensity E between the two charged layers of width d will be given by

$$E = \frac{\sigma'}{\varepsilon_0} = \frac{Nex}{\varepsilon_0} \qquad (6.71)$$

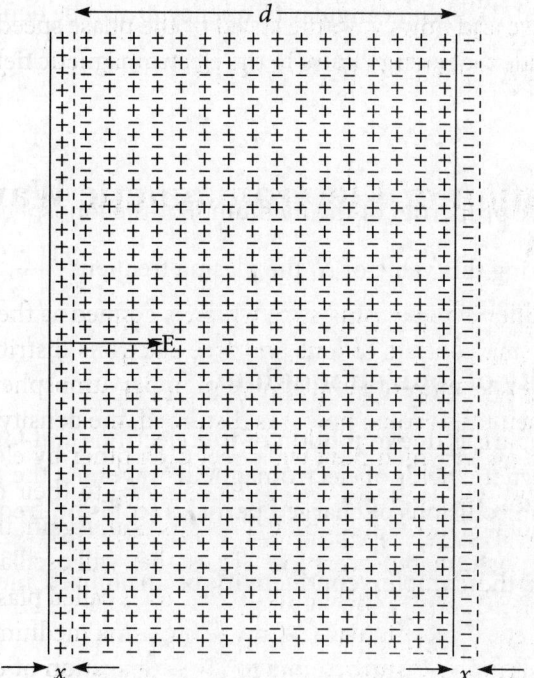

Figure 6.3 | Visualization of idealized plasma medium. Due to ideal electromagnetic disturbance, positive and negative particles are separately displaced through a distance x. Neutral plasma is confined between two charged layers of width d.

The restoring force F on the electrons will be

$$F = -eE = -\frac{Ne^2 x}{\varepsilon_0} \tag{6.72}$$

Equation (6.72) shows that the restoring force is proportional to x, the displacement of the positive or negative charges in opposite direction. This implies that displaced charges execute simple harmonic oscillation. Applying Newton's law of motion to Eq. (6.72), we have

$$m_e \frac{d^2 x}{dt^2} = -\frac{Ne^2}{\varepsilon_0} x$$

or $\quad \dfrac{d^2 x}{dt^2} + \dfrac{Ne^2}{m_e \varepsilon_0} x = 0$

or $\quad \dfrac{d^2 x}{dt^2} + \omega_p^2 x = 0 \tag{6.73}$

Therefore, from Eq. (6.73), plasma angular frequency ω_p can be defined as

$$\omega_p = \left(\frac{Ne^2}{\varepsilon_0 m_e} \right)^{\frac{1}{2}}$$

(6.74)

and is a function of the properties of the medium. For the ionosphere, N is of the order of 10^{11} electrons/m³. Taking this value of N, the plasma frequency $\left(= \dfrac{\omega_p}{2\pi} \right)$ is calculated to be 2.84×10^6 Hz approximately.

6.12.1 Conductivity of ionized medium

In ionized gas, current arises due to motion of free electrons and ions under the influence of the electric and magnetic fields of electromagnetic waves. As the pressure is low, we can assume that there is no collisions of charged particles (collision frequency = 0) and hence, no energy loss.

The expression for the complex conductivity of an ionized medium σ is given here without proof as

$$\sigma = -i \frac{Ne^2}{\omega \, m_e}$$

(6.75)

where

N = no. of electrons per unit volume

E = charge of an electron

ω = angular frequency of the electromagnetic wave

m_e = mass of a electron

Since $\vec{J} = \sigma \vec{E}$, we can write

$$\vec{J} = -i \frac{Ne^2}{\omega m_e} \vec{E} = \frac{Ne^2}{\omega m_e} \vec{E} \, e^{-i \pi/2} \qquad \left(\because \; e^{i\theta} = \cos\theta + i\sin\theta \right)$$

(6.76)

This equation shows that current density vector or simply current lags behind the electric field vector $\vec{E}$ by a phase angle of $\pi/2$ in the ionized medium. Equation (6.76) also shows that $\vec{J} \cdot \vec{E}$ will be imaginary implying that there is no energy loss in the ionized medium.

6.12.2 Wave equation in ionized medium

The equation of propagation for the electric vector $\vec{E}$ is obtained from Eq. (6.27) as

$$\nabla^2 \vec{E} - \mu\varepsilon \frac{\partial^2 \vec{E}}{\partial t^2} - \mu\sigma \frac{\partial \vec{E}}{\partial t} = \vec{\nabla}\left(\frac{\rho}{\varepsilon}\right) \tag{6.77}$$

For a charge-free region, the space charge density $\rho = 0$. In the undisturbed condition, plasma is neutral. Under this condition, Eq. (6.77) becomes

$$\nabla^2 \vec{E} - \mu\varepsilon \frac{\partial^2 \vec{E}}{\partial t^2} - \mu\sigma \frac{\partial \vec{E}}{\partial t} = 0 \tag{6.78}$$

Suppose a plane electromagnetic wave is travelling in the $+Z$ direction so that its electric vector $\vec{E}$ is parallel to the X-axis and magnetic vector $\vec{H}$ parallel to the Y-axis (i.e., electromagnetic wave is plane polarized). Under this condition, Eq. (6.78) becomes

$$\frac{\partial^2 E_X}{\partial z^2} - \mu\varepsilon \frac{\partial^2 E_X}{\partial t^2} - \mu\sigma \frac{\partial E_X}{\partial t} = 0 \tag{6.79}$$

We assume that E_X varies sinusoidally with time so that we may take $E_X = E_0 e^{i\omega t}$. Putting this value of E_X into Eq. (6.79), we get

$$\frac{\partial^2 E_X}{\partial z^2} - (i\omega\mu\sigma - \omega^2 \mu\varepsilon) E_X = 0$$

or
$$\frac{\partial^2 E_X}{\partial z^2} - \gamma^2 E_X = 0 \tag{6.80}$$

where $\gamma^2 = i\omega\mu\sigma - \omega^2 \mu\varepsilon$. $\tag{6.81}$

Here γ is called propagation constant. Equation (6.80) is the wave equation for the electric vector $\vec{E}$ in one dimension though time t does not appear explicitly. We have already assumed time variation of E_X.

The solution of Eq. (6.80) is given by

$$E_X = E_0 e^{-\gamma z} \tag{6.82}$$

Since the propagation constant γ is complex, we can express it in the form of a complex number as

$$\gamma = \alpha + i\beta \tag{6.83}$$

Here α is called the attenuation constant and β is called the phase constant. In a lossless medium, the wave does not attenuate and so $\alpha = 0$ and β is a measure of the radian phase shift per metre. That is, for unit length, phase shift is β and for λ length, the phase shift will be $\beta\lambda$; however, by definition, for λ length, the phase shift (or change) will be 2π. Therefore, we have

$$\beta\lambda = 2\pi$$

or $\quad \lambda = \dfrac{2\pi}{\beta}$ \hfill (6.84)

The phase speed v by definition is given by

$$v = \frac{\lambda}{T} = \frac{\lambda\omega}{2\pi}$$ \hfill (6.85)

Here T is the time period of the electromagnetic wave. Putting the value of λ from Eq. (6.84) into Eq. (6.85), we get

$$v = \frac{\omega}{\beta}$$ \hfill (6.86)

Now putting the value of γ from Eq. (6.83) into Eq. (6.82), we have

$$E_X = E_0 e^{-\alpha z}\, e^{-i\beta z}$$

In this equation, $e^{-\alpha z}$ is called the attenuation factor and $e^{-i\beta z}$ is called the phase factor. It is important to know that the real part of propagation constant γ appears in the attenuation factor and the imaginary part of γ appears in the phase factor of the solution.

Similarly, the equation of propagation for the magnetic field vector $\overrightarrow{H}$ may be obtained from Eq. (6.31).

6.12.3 Propagation constant in an ionized medium

The propagation constant γ is defined from Eq. (6.81) as

$$\gamma = \sqrt{i\omega\mu\sigma - \omega^2\mu\varepsilon}$$ \hfill (6.87)

and in ionized medium this relation becomes

$$\gamma = \sqrt{i\omega\mu_0\sigma - \omega^2\mu_0\varepsilon_0}$$ \hfill (6.88)

Putting the value of σ from Eq. (6.75) into Eq. (6.88), we have

$$\gamma = \sqrt{\frac{Ne^2\mu}{m_e} - \omega^2\mu_0\varepsilon_0} \qquad (6.89)$$

From Eq. (6.74), we have

$$\omega_p^2\varepsilon_0 = \frac{Ne^2}{m_e}$$

Putting this value in Eq. (6.84), we get

$$\gamma = \sqrt{\omega_p^2\mu_0\varepsilon_0 - \omega^2\mu_0\varepsilon_0}$$

$$\text{or} \quad \gamma = \frac{1}{c}\sqrt{\omega_p^2 - \omega^2} \qquad \left(\because \quad c = \frac{1}{\sqrt{\mu_0\varepsilon_0}}\right) \qquad (6.90)$$

which is the propagation constant in an ionized medium. It's value depends upon the values of ω_p and ω.

Case 1: Wave propagation in an ionized medium at high frequency $(\omega > \omega_p)$

If the angular frequency of the electromagnetic wave ω is higher than the plasma angular frequency ω_p, then the propagation constant γ will be purely imaginary. The value of α in Eq. (6.83) will be zero and the value of β can be determined from Eq. (6.90) in the following manner. From Eq. (6.90), we have

$$\gamma = \frac{1}{c}\sqrt{\omega_p^2 - \omega^2}$$

$$\text{or} \quad \gamma = 0 + i\frac{1}{c}\sqrt{\omega^2 - \omega_p^2}$$

This is in the form of $\gamma = \alpha + i\beta$ with

$$\alpha = 0$$

and $\quad \beta = \frac{1}{c}\sqrt{\omega^2 - \omega_p^2} \neq 0 \qquad (6.91)$

Non-zero value of β indicates that there is a radian phase shift per metre of the electromagnetic wave. The electromagnetic wave motion exists in the ionized medium. That is, there is a possibility that the electromagnetic wave can propagate through an ionized media without any attenuation since $\alpha = 0$.

Putting the value of β from Eq. (6.91) into Eq. (6.86), we have

$$v = \frac{\omega}{\frac{1}{c}\sqrt{\omega^2 - \omega_p^2}}$$

or $$v = \frac{c}{\sqrt{1 - \frac{\omega_p^2}{\omega^2}}}$$ (6.92)

which is the speed at which the electromagnetic wave propagates in the ionized medium when its angular frequency is greater than the plasma angular frequency of the medium.

The refractive index of the medium is defined as the ratio of speed of the electromagnetic wave in vacuum to the speed of the electromagnetic wave in the medium. The refractive index of the ionized medium is found out from the Eq. (6.92) to be

$$\frac{c}{v} = \sqrt{1 - \frac{\omega_p^2}{\omega^2}}$$

Putting the value of ω_p^2 from Eq. (6.74) into equation, we get

$$\frac{c}{v} = \sqrt{1 - \frac{Ne^2}{\varepsilon_0 m_e} \times \frac{1}{4\pi^2 v^2}}   \qquad (\because \ \omega = 2\pi v)$$

or $$\frac{c}{v} = \sqrt{1 - \frac{1}{4\pi\varepsilon_0} \frac{Ne^2}{m_e} \times \frac{1}{\pi v^2}}$$

Putting the values of known constants into this equation, we get the expression for refractive index of the ionized medium when the electromagnetic wave propagates at a higher frequency than the plasma frequency as

$$\frac{c}{v} = \sqrt{1 - 80.54 \times \frac{N}{v^2}}$$ (6.93)

From this equation, we come to know that with increase of ionization density N, the refractive index of the ionized medium decreases and vice versa. That is, as ionization density N increases continuously, the medium becomes more and more rarer for the electromagnetic wave. Therefore, when an electromagnetic wave propagates in an ionized medium with continuous increase in ionization density, the wave will deviate away from the normal to the layer as if the electromagnetic wave is passing from a denser medium to a rarer one. This shows the possibility of reflection of the electromagnetic wave with appropriate frequency from a particular layer inside the ionized medium.

Case 2: Wave propagation in an ionized medium at very high frequency $(\omega \gg \omega_p)$

When an electromagnetic wave propagates in an ionized medium at very high frequency, i.e., $\omega \gg \omega_p$, we can neglect the term $\dfrac{\omega_p^2}{\omega^2}$ in Eq. (6.92) to obtain

$$v = \frac{c}{\sqrt{1-0}} = c.$$

This equation shows that the refractive index of the ionized medium for very high frequency electromagnetic wave is unity. Such a type of electromagnetic wave propagates undeviated in the ionized medium. That is why laser frequencies penetrate through the ionosphere without being deviated.

Case 3: Wave propagation in an ionized medium at cut-off frequency $(\omega = \omega_p)$

The cut-off frequency ω_c is defined as the frequency of the electromagnetic wave when the electromagnetic wave is propagating in an ionized medium with plasma frequency ω_p of the ionized medium. If the angular frequency of the electromagnetic wave is equal to the plasma angular frequency ω_p, then the propagation constant γ will be zero. The values of α and β will both be zero in Eq. (6.83). Since the phase factor β is zero, the electromagnetic wave is not propagating in the ionized medium at this particular frequency called the cut-off frequency ω_c.

   The cut-off frequency of an ionized medium is the characteristic of the ionized medium and depends only upon the ionization density N of the medium. The expression for the cut-off frequency ω_c is obtained in the following way.

$$\omega_c = \omega_p = \left(\frac{Ne^2}{\varepsilon_0 m_e} \right)^{\frac{1}{2}}$$

or $\quad v_c = \dfrac{1}{2\pi} \left(\dfrac{Ne^2}{\varepsilon_0 m_e} \right)^{\frac{1}{2}}$ (6.94)

v_c is the cut-off frequency of the ionized medium.

   Therefore, for a particular ionization density, electromagnetic waves with a frequency more than the cut-off frequency can be propagated through the ionized medium, whereas if a wave has a frequency less than the cut-off frequency as we shall see in the next case, it cannot be propagated in that particular ionized medium.

Case 4: Wave propagation in an ionized medium at low frequency $(\omega < \omega_p)$

If the angular frequency of the electromagnetic wave ω is less than the plasma angular frequency ω_p, then the propagation constant γ will be purely real. The value of β in Eq. (6.83) will be zero and the value of α will be found out from Eq. (6.90) in the following manner. From Eq. (6.90), we have

$$\gamma = \frac{1}{c}\sqrt{\omega_p^2 - \omega^2} + i0$$

This is in the form of $\gamma = \alpha + i\beta$ with

$$\alpha = \frac{1}{c}\sqrt{\omega_p^2 - \omega^2}$$

and

$$\beta = 0$$

Zero value of β indicates that there is no radian phase shift as the electromagnetic wave progresses through the medium. This is impossible. That means no electromagnetic wave motion exists in the medium. In other words, there is no possibility that an electromagnetic wave can propagate through an ionized media at frequency less than the plasma frequency.

Example 6.10

In the photosphere of the sun, if there are 6×10^{16} electrons per unit volume, then calculate the minimum frequency of the electromagnetic wave to pass through the photosphere.

Solution

The minimum frequency of the electromagnetic wave to pass through an ionized medium, as explained earlier, is called the cut-off frequency. The cut-off frequency v_C is given by

$$v_C = \frac{1}{2\pi}\left(\frac{Ne^2}{\varepsilon_0 m_e}\right)^{\frac{1}{2}} = \frac{1}{2\pi}\left(\frac{6\times10^{16}\times1.6^2\times10^{-38}}{8.85\times10^{-12}\times9.11\times10^{-31}}\right)^{\frac{1}{2}} \text{Hz} = 2.2\times10^9\,\text{Hz}$$

Example 6.11

In ionosphere there are approximately 10^{11} electrons per unit volume. Calculate the plasma frequency of the medium. Also calculate the speed of the electromagnetic wave having 300 MHz.

Solution

The plasma frequency of the ionosphere is

$$v_p = \frac{1}{2\pi}\left(\frac{Ne^2}{\varepsilon_0 m_e}\right)^{\frac{1}{2}} = 2.84\,\text{MHz}$$

The speed of electromagnetic wave in the ionosphere is given by

$$v = \frac{c}{\sqrt{1 - \dfrac{\omega_p^2}{\omega^2}}} = 3\times10^8\,\text{m/s}$$

6.13 Reflection and Refraction of Electromagnetic Waves at Non-conducting and Conducting Boundaries

In order to discuss the behavior of electromagnetic waves at the boundary during the phenomena of reflection, and refraction, we shall first determine the boundary conditions which the field vectors must satisfy at the surface of discontinuity between the two media. Without giving the proofs, we shall write down the boundary conditions which time-dependent electromagnetic field vectors $\vec{B}$, $\vec{E}$, $\vec{D}$, and $\vec{H}$ satisfy at the interface between two different media.

i.   Boundary condition for $\vec{B}$

The normal component of the magnetic induction vector $\vec{B}$ is continuous across the boundary. If B_{1n} and B_{2n} are the normal components of $\vec{B}$ on the interface in the first medium and the second medium respectively, then

$$B_{1n} = B_{2n} \tag{6.95}$$

ii. Boundary condition for $\vec{E}$

The tangential component of the electric vector $\vec{E}$ is continuous across the interface. If E_{1t} and E_{2t} are the tangential components of $\vec{E}$ on the interface in the first medium and the second medium respectively, then

$$E_{1t} = E_{2t} \tag{6.96}$$

iii. Boundary condition for $\vec{D}$

The normal component of the electric displacement vector $\vec{D}$ is not continuous across the interface and changes by an amount equal to the free surface charge density σ at the interface. If D_{1n} and D_{2n} are the normal components of $\vec{D}$ on the interface in the first medium and the second medium respectively, then

$$D_{1n} - D_{2n} = \sigma \tag{6.97}$$

iv. Boundary condition for $\vec{H}$:

The tangential component of the magnetic field vector $\vec{H}$ is continuous across the interface separating the two dielectric media. If H_{1t} and H_{2t} are the tangential components of $\vec{H}$ on the interface in the first medium and the second medium respectively, then

$$H_{1t} = H_{2t} \tag{6.98}$$

6.13.1 Reflection and refraction of electromagnetic waves at a non-conducting surface

Consider the case of a plane interface separating two different isotropic, homogeneous, charge-free ($\rho = 0$), linear ($D = \varepsilon E$ and $B = \mu H$) non-conducting ($\sigma = 0$) media of infinite extension. Let the first side of the interface be characterized by permittivity ε_1 and permeability μ_1; the second side by ε_2 and μ_2.

Consider that an electromagnetic wave propagating in the first medium is incident obliquely on the interface separating two non-conducting media. In general, a reflected and transmitted wave will be produced. The incident and reflected wave will lie in the first medium and the transmitted wave will lie in the second medium. See Fig. 6.4.

Let

E_I = Electric field component of the incident electromagnetic wave in the first medium.

E_R = Electric field component of the reflected electromagnetic wave in the first medium.

E_T = Electric field component of the transmitted electromagnetic wave in the second medium.

The electric field vectors of the incident, reflected and transmitted electromagnetic waves at the point of incidence are given by

$$E_I = E_{OI}e^{i(\omega_I t - \vec{k}_I \cdot \vec{r})} \tag{6.99}$$

$$E_R = E_{OR}e^{i(\omega_R t - \vec{k}_R \cdot \vec{r})} \tag{6.100}$$

$$E_T = E_{OT}e^{i(\omega_T t - \vec{k}_T \cdot \vec{r})} \tag{6.101}$$

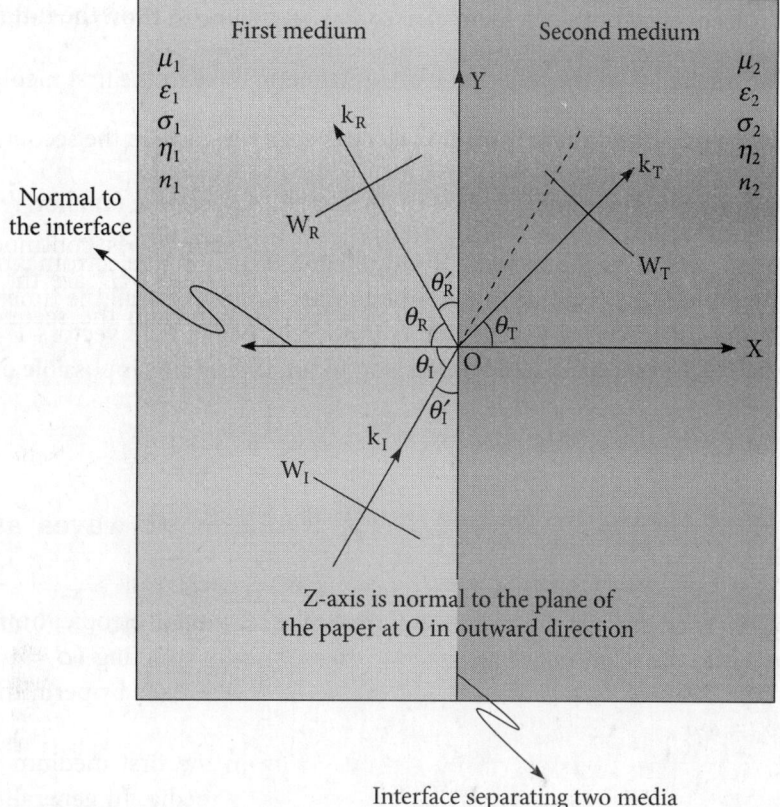

First medium

μ_1
ε_1
σ_1
η_1
n_1

Second medium

μ_2
ε_2
σ_2
η_2
n_2

Normal to the interface

Z-axis is normal to the plane of the paper at O in outward direction

Interface separating two media

Figure 6.4 | Reflection and refraction/transmission of plane electromagnetic waves at a plane interface. The Y-Z plane is the plane interface. The X-Y plane is the plane of incidence. W_I, W_R and W_T are the incident, reflected and refracted plane wavefronts

Here,

E_{OI} = Time-independent amplitude of the electric field component of the incident electromagnetic wave.

E_{OR} = Time-independent amplitude of the electric field component of the reflected electromagnetic wave.

E_{OT} = Time-independent amplitude of the electric field component of the transmitted electromagnetic wave.

$\vec{k}_I$ = Propagation vector of the incident electromagnetic wave in the first medium.

$\vec{k}_R$ = Propagation vector of the reflected electromagnetic wave in the first medium.

$\vec{k}_T$ = Propagation vector of the transmitted electromagnetic wave in the second medium.

ω_I = Angular frequency of the incident electromagnetic wave in the first medium.

ω_R = Angular frequency of the reflected electromagnetic wave in the first medium.

ω_T = Angular frequency of the transmitted electromagnetic wave in the second medium.

$\vec{r}$ and t are the position vector of the point of incidence and the time function respectively.

The electric field vectors of the incident, reflected and transmitted electromagnetic waves must satisfy the boundary conditions described in Section 6.13 for all the time and for all the points on the interface. The tangential components of the field vectors $\vec{E}$ and $\vec{H}$ are continuous across the boundary at all points and for all times. This is possible only if

$$e^{i(\omega_I t - \vec{k}_I \cdot \vec{r})} = e^{i(\omega_R t - \vec{k}_R \cdot \vec{r})} = e^{i(\omega_T t - \vec{k}_T \cdot \vec{r})}$$

or $\omega_I t - \vec{k}_I \cdot \vec{r} = \omega_R t - \vec{k}_R \cdot \vec{r} = \omega_T t - \vec{k}_T \cdot \vec{r}$ (6.102)

This relation is valid for all the points on the interface including $r = 0$ at any time t. Putting $r = 0$ into the relation, we have

$$\omega_I t = \omega_R t = \omega_T t$$

or $v_I = v_R = v_T$

This shows that reflected and transmitted electromagnetic waves have the same frequency as that of the incident electromagnetic wave when incident on the dielectric surface.

Equation (6.102) is also true for $t = 0$. Hence, we have putting $t = 0$ into Eq. (6.102),

$$\vec{k}_I \cdot \vec{r} = \vec{k}_R \cdot \vec{r} = \vec{k}_T \cdot \vec{r}$$ (6.103)

This equation shows that all the three vectors $\vec{k}_I$, $\vec{k}_R$, and $\vec{k}_T$ lie in a single plane. This proves the first law of reflection and refraction.

Now suppose the position vector $\vec{r}$ lies on the interface along the positive y-axis.

Let

θ_I' = Angle between the position vector $\vec{r}$ and the propagation vector $\vec{k}_I$.

θ_R' = Angle between the position vector $\vec{r}$ and the propagation vector $\vec{k}_R$.

θ_T' = Angle between the position vector $\vec{r}$ and the propagation vector $\vec{k}_T$.

$\vec{k}$'s are the propagation vectors whose magnitude is by definition given as

$$k = \frac{2\pi}{\lambda} = \frac{2\pi v}{v} \qquad (\because \; v = \lambda v)$$

or $\quad k = \omega\sqrt{\mu\varepsilon} = \dfrac{\omega}{v}$

Hence, we have

$\vec{k}_I$ = Propagation vector of the incident electromagnetic wave in the first medium

$\quad = \omega\sqrt{\mu_1\varepsilon_1} = \dfrac{\omega}{v_1}$ in magnitude. $\hspace{3cm}$ (6.104)

$\vec{k}_R$ = Propagation vector of the reflected electromagnetic wave in the first medium

$\quad = \omega\sqrt{\mu_1\varepsilon_1} = \dfrac{\omega}{v_1}$ in magnitude. $\hspace{3cm}$ (6.105)

$\vec{k}_T$ = Propagation vector of the transmitted electromagnetic wave in the second medium

$\quad = \omega\sqrt{\mu_2\varepsilon_2} = \dfrac{\omega}{v_2}$ in magnitude. $\hspace{3cm}$ (6.106)

(since $\omega_I = \omega_R = \omega_T = \omega$)

Taking the magnitude of Eq. (6.103), we have [See Fig. 6.4]

$$k_I r \, \cos\theta_I' = k_R r \, \cos\theta_R' = k_T r \, \cos\theta_T'$$

or $\quad k_I \cos\left(\dfrac{\pi}{2} - \theta_I\right) = k_R \cos\left(\dfrac{\pi}{2} - \theta_R\right) = k_T \cos\left(\dfrac{\pi}{2} - \theta_T\right)$

or $\quad k_I \sin\theta_I = k_R \sin\theta_R = k_T \sin\theta_T \hspace{3cm}$ (6.107)

From Eq. (6.107), we get

$$k_I \sin\theta_I = k_R \sin\theta_R$$

However, according to Eq. (6.104) and Eq. (6.105), $k_I = k_R$. This equation becomes

$$\sin\theta_I = \sin\theta_R$$

or $\theta_I = \theta_R$

or   Angle of incidence = angle of reflection.

This proves the second law of reflection.
    Also from Eq. (6.107), we get

$$k_I \sin \theta_I = k_T \sin \theta_T$$

or $\dfrac{\sin \theta_I}{\sin \theta_T} = \dfrac{k_T}{k_I} = \dfrac{v_1}{v_2} = \dfrac{\dfrac{c}{v_2}}{\dfrac{c}{v_1}} = \dfrac{n_2}{n_1} = {}_1^2 n$

In this equation

    c = speed of electromagnetic wave in free space.

    v_1 = speed of electromagnetic wave in the first medium.

    v_2 = speed of electromagnetic wave in the second medium.

    n_1 = refractive index of the first medium.

    n_2 = refractive index of the second medium.

    ${}_1^2 n$ = refractive index of the second medium with respect to the first medium

This proves the second law of refraction and is the well-known Snell's law.

Case 1: Electric vector normal to the plane of incidence

The plane of incidence is defined as the plane containing the propagation vector k_I and normal to the interface at the point of incidence (In ray optics, the plane of incidence is defined as the plane passing through the incident ray and the normal). In our case, the X-Y plane will be the plane of incidence and the Y-Z plane is the interface separating the two non-conducting media. See Fig. 6.5. All the field vectors of the electromagnetic wave obey the boundary conditions described earlier. The tangential components of the electric vector are continuous along the interface. In our case, since electric vectors are perpendicular to the plane of incidence in the inward direction (i.e., along the $-Z$-axis), they are at the same time tangential to the interface. Hence, we have

$$\vec{E}_I + \vec{E}_R = \vec{E}_T$$

Since the exponential parts of Eqs (6.99) to (6.101) are the same, we can write the aforementioned equation as

$$E_{OI} + E_{OR} = E_{OT} \tag{6.108}$$

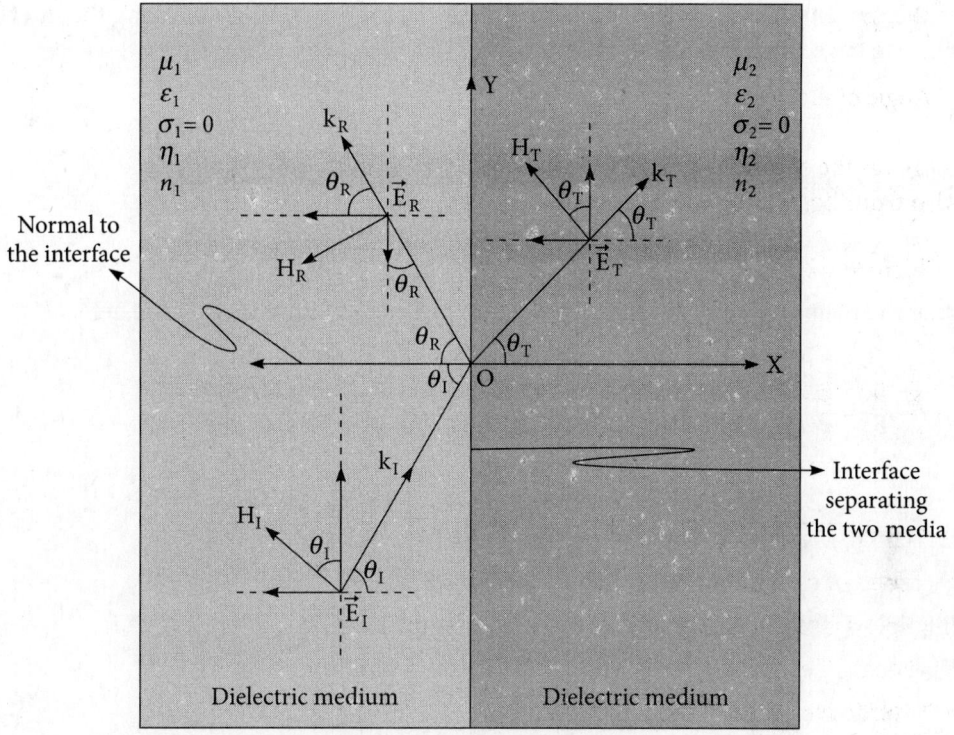

Figure 6.5 | Reflection and refraction of an electromagnetic wave on a dielectric interface with the electric vector normal to the plane of incidence

Since electric vectors are perpendicular to the plane of incidence along the $-Z$-axis, magnetic vectors are parallel to the plane of incidence so that the following vector equations, obtained from Eq. (6.53), are not violated.

$$\vec{H}_I = \frac{\vec{k}_I \times \vec{E}_I}{\omega \mu_1} \tag{6.109}$$

$$\vec{H}_R = \frac{\vec{k}_R \times \vec{E}_R}{\omega \mu_1} \tag{6.110}$$

$$\vec{H}_T = \frac{\vec{k}_T \times \vec{E}_T}{\omega \mu_2} \tag{6.111}$$

(since $\omega_I = \omega_R = \omega_T = \omega$)

Now the tangential components of the magnetic vector are continuous along the interface. Hence, we have from Fig. 6.5,

$$H_I \cos\theta_I - H_R \cos\theta_R = H_T \cos\theta_T$$

Since the exponential parts are the same, we can write this equation as

$$H_{OI} \cos\theta_I - H_{OR} \cos\theta_R = H_{OT} \cos\theta_T \qquad (6.112)$$

Taking the magnitude of the magnetic vectors in Eqs (6.109) to (6.111), we get

$$H_I = \frac{k_I E_I \sin 90°}{\omega\mu_1}$$

or $$H_{OI} = \frac{k_I}{\omega}\frac{E_{OI}}{\mu_1} = \sqrt{\frac{\varepsilon_1}{\mu_1}}\, E_{OI} \qquad \left(\because v = \frac{\omega}{k} = \frac{1}{\sqrt{\mu\varepsilon}} \right) \qquad (6.113)$$

Using the definition of η, we have

$$H_{OI} = \frac{E_{OI}}{\eta_1} \qquad (6.114)$$

Similarly, we have

$$H_{OR} = \frac{E_{OR}}{\eta_1} \qquad (6.115)$$

and $$H_{OT} = \frac{E_{OT}}{\eta_2} \qquad (6.116)$$

Now putting the values of H_{OI}, H_{OR}, and H_{OT} from Eqs (6.114) to (6.116) into Eq. (6.112), we have

$$\frac{E_{OI}}{\eta_1}\cos\theta_I - \frac{E_{OR}}{\eta_1}\cos\theta_R = \frac{E_{OT}}{\eta_2}\cos\theta_T \qquad (6.117)$$

Putting the value of E_{OT} from Eq. (6.108) into Eq. (6.117), we get

$$\frac{E_{OI}}{\eta_1}\cos\theta_I - \frac{E_{OR}}{\eta_1}\cos\theta_R = \frac{\cos\theta_T}{\eta_2}(E_{OI} + E_{OR})$$

or $$\frac{E_{OR}}{\eta_1}\cos\theta_R + E_{OR}\frac{\cos\theta_T}{\eta_2} = E_{OI}\frac{\cos\theta_1}{\eta_1} - E_{OI}\frac{\cos\theta_T}{\eta_2}$$

or $\quad \left(\dfrac{E_{OR}}{E_{OI}}\right)_N = \dfrac{\dfrac{\cos\theta_I}{\eta_1} - \dfrac{\cos\theta_T}{\eta_2}}{\dfrac{\cos\theta_I}{\eta_1} + \dfrac{\cos\theta_T}{\eta_2}}$ $\qquad\qquad$ (6.118)

The subscript N in this equation or in the following equations refers to the fact that the electric field vector is normal to the plane of incidence.

Again putting the value of E_{OR} from Eq. (6.108) into Eq. (6.117), we get

$$\frac{E_{OI}}{\eta_1}\cos\theta_I - \frac{E_{OT} - E_{OI}}{\eta_1}\cos\theta_R = \frac{E_{OT}}{\eta_2}\cos\theta_T$$

or $\quad \dfrac{E_{OI}}{\eta_1}\cos\theta_I - \dfrac{E_{OT}}{\eta_1}\cos\theta_I + \dfrac{E_{OI}}{\eta_1}\cos\theta_I = \dfrac{E_{OT}}{\eta_2}\cos\theta_T$

or $\quad \left(\dfrac{E_{OT}}{E_{OI}}\right)_N = \dfrac{2\dfrac{\cos\theta_I}{\eta_1}}{\dfrac{\cos\theta_I}{\eta_I} + \dfrac{\cos\theta_T}{\eta_2}}$ $\qquad\qquad$ (6.119)

Equations (6.118) and (6.119) are called Fresnel's equations.

If the two media are dielectrics, then, $\mu_1 = \mu_2 = \mu_0$. The intrinsic impedance η defined by $\eta = \sqrt{\dfrac{\mu}{\varepsilon}}$ becomes

$$\eta = \frac{\mu}{\sqrt{\mu\varepsilon}} = \mu v = \frac{\mu}{n}c$$

Here c is the speed of light in vacuum and n is the refractive index of the medium.

Thus, we have for two dielectric media

$$\eta_1 = \frac{\mu_0}{n_1}c$$

and

$$\eta_2 = \frac{\mu_0}{n_2}c$$

By taking the ratio of η_1 and η_2, we get

$$\frac{\eta_1}{\eta_2} = \frac{n_2}{n_1}$$ $\qquad\qquad$ (6.120)

Putting the value of $\dfrac{\eta_1}{\eta_2}$ from Eq. (6.120) into Eq. (6.118), we get one of Fresnel's equations in the form $\left(\dfrac{E_{OR}}{E_{OI}}\right)_N$

$$\left(\frac{E_{OR}}{E_{OI}}\right)_N = \frac{\cos\theta_I - \dfrac{\eta_1}{\eta_2}\cos\theta_T}{\cos\theta_I + \dfrac{\eta_1}{\eta_2}\cos\theta_T} = \frac{\cos\theta_I - \dfrac{n_2}{n_1}\cos\theta_T}{\cos\theta_I + \dfrac{n_2}{n_1}\cos\theta_T} \tag{6.121}$$

Again putting the value of $\dfrac{\eta_1}{\eta_2}$ into Eq. (6.119), we get the second Fresnel's equation in the form

$$\left(\frac{E_{OT}}{E_{OI}}\right)_N = \frac{2\cos\theta_I}{\cos\theta_I + \dfrac{\eta_1}{\eta_2}\cos\theta_T} = \frac{2\cos\theta_I}{\cos\theta_I + \dfrac{n_2}{n_1}\cos\theta_T} \tag{6.122}$$

Applying Snell's law to Eq. (6.121), we have

$$\left(\frac{E_{OR}}{E_{OI}}\right)_N = \frac{\sin\theta_T\,\cos\theta_I - \sin\theta_I\,\cos\theta_T}{\sin\theta_T\,\cos\theta_I + \sin\theta_I\,\cos\theta_T} = \frac{\sin(\theta_I - \theta_T)}{\sin(\theta_I + \theta_T)} \tag{6.123}$$

Again applying Snell's law to Eq. (6.122), we have

$$\left(\frac{E_{OT}}{E_{OI}}\right)_N = \frac{2\,\sin\theta_T\,\cos\theta_I}{\sin\theta_T\,\cos\theta_I + \sin\theta_I\,\cos\theta_T} = \frac{2\,\sin\theta_T\,\cos\theta_I}{\sin(\theta_I + \theta_T)} \tag{6.124}$$

Interpretation of Eq. (6.121)

If an electromagnetic wave is passing from a denser medium (first medium) to a rarer medium (second medium) with $n_1 > n_2$ and $\theta_I < \theta_T$ or $\cos\theta_I > \cos\theta_T$ then the numerator of Eq. (6.121) is always positive, thereby implying that

$$\left(\frac{E_{OR}}{E_{OI}}\right)_N$$

is to be positive. The positive values of

$$\left(\frac{E_{OR}}{E_{OI}}\right)_N$$

indicate that reflected wave is in phase with the incident wave.

If an electromagnetic wave is passing from a rarer medium (first medium) to a denser medium (second medium) with $n_1 < n_2$ and $\theta_I > \theta_T$ or $\cos\theta_I < \cos\theta_T$, then the numerator of Eq. (6.121) is always negative, thereby implying that

$$\frac{E_{OR}}{E_{OI}}$$

is to be negative. The negative values of

$$\left(\frac{E_{OR}}{E_{OI}}\right)_N$$

indicate that the reflected wave is out phase with the incident wave. That means the phase difference between the incident wave and the reflected wave is π.

Interpretation of Eq. (6.122)

Whether the electromagnetic wave is passing from a rarer medium to a denser medium or vice versa, the ratio

$$\left(\frac{E_{OT}}{E_{OI}}\right)_N$$

is always positive for all possible values of θ_I and θ_T $(0 \le \theta_I, \ \theta_T \le 90)$. The positive values of

$$\left(\frac{E_{OT}}{E_{OI}}\right)_N$$

indicate that refracted or transmitted wave is in phase with the incident wave.

If the plane polarized electromagnetic wave with its electric vector perpendicular to the plane of incidence is incident normally ($\theta_I = 0$) on the interface separating two different dielectric media, we have from Eqs (6.121) and (6.123)

$$\left(\frac{E_{OR}}{E_{OI}}\right)_N = \frac{1 - \dfrac{n_2}{n_1}}{1 + \dfrac{n_2}{n_1}} \quad \text{and} \quad \left(\frac{E_{OT}}{E_{OI}}\right)_N = \frac{2}{1 + \dfrac{n_2}{n_1}} \quad \text{respectively.}$$

Case 2: Electric vector parallel to the plane of incidence

In our case, the X-Y plane will be the plane of incidence and the Y-Z plane is the interface separating the two non-conducting media. See Fig. 6.6. The tangential components of the electric vector are continuous along the interface. In our case, since electric vectors are parallel to the plane of incidence, we have

$$E_I \cos\theta_I - E_R \cos\theta_I = E_T \cos\theta_T \qquad (\because \quad \theta_R = \theta_I)$$

Since exponential parts of Eqs (6.99) to (6.101) are the same, we can write the aforementioned equation as

$$E_{OI} \cos\theta_I - E_{OR} \cos\theta_I = E_{OT} \cos\theta_T \tag{6.125}$$

Electric vectors are parallel to the plane of incidence and magnetic vectors are perpendicular to the plane of incidence so that the following vector equations, obtained from Eq. (6.53), are not violated.

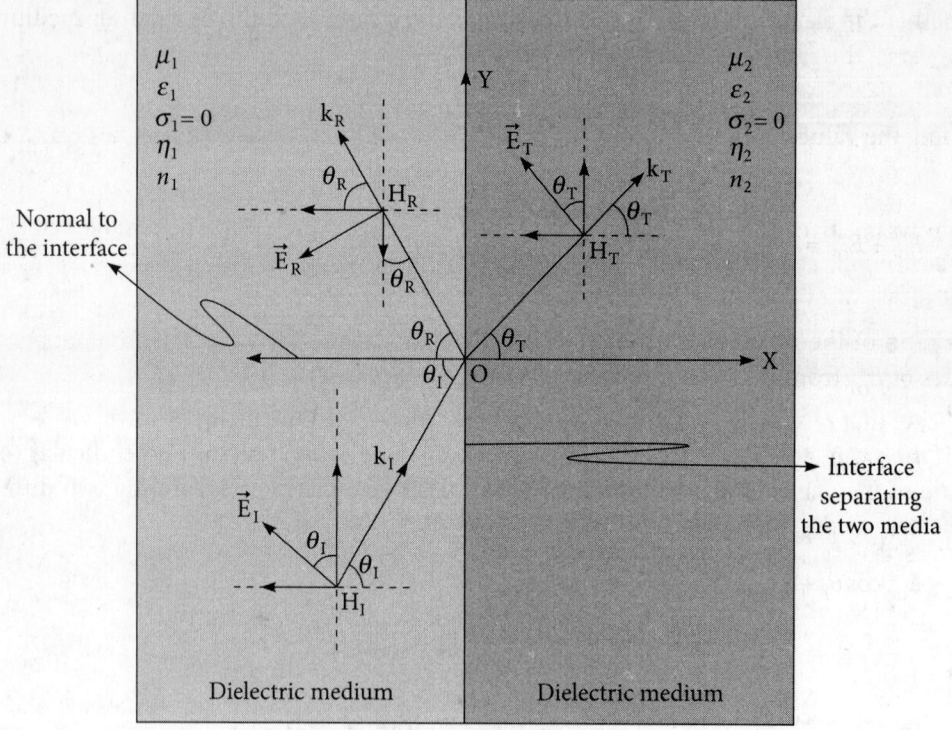

Z-axis is normal to the plane of
the paper at O in outward direction

Figure 6.6 | Reflection and refraction of electromagnetic wave on a dielectric interface with the electric vector parallel to the plane of incidence

The boundary condition for the magnetic field vector is that the tangential component of the magnetic field vector $\vec{H}$ is continuous across the interface separating the two dielectric media. Making use of this boundary condition, we have

$$\vec{H}_I + \vec{H}_R = \vec{H}_T$$

or $\quad H_{OI} + H_{OR} = H_{OT}$

Putting the values of H_{OI}, H_{OR}, and H_{OT} from Eqs (6.114) to (6.116) into this equation, we have

$$\frac{E_{OI}}{\eta_1} + \frac{E_{OR}}{\eta_1} = \frac{E_{OT}}{\eta_2}$$

or $\quad E_{OI} + E_{OR} = \dfrac{\eta_1}{\eta_2} E_{OT}$

Putting the value of $\dfrac{\eta_1}{\eta_2}$ from Eq. (6.120) into the aforementioned equation, we get

$$E_{OI} + E_{OR} = \frac{n_2}{n_1} E_{OT} \tag{6.126}$$

The ratio of the reflected amplitude to the incident amplitude is obtained by putting the values of E_{OT} from Eq. (6.126) into Eq. (6.125). Thus, we have

$$E_{OI} \cos\theta_I - E_{OR} \cos\theta_I = \cos\theta_T \frac{n_1}{n_2}(E_{OI} + E_{OR})$$

or $\quad E_{OR} \cos\theta_I + E_{OR} \cos\theta_T \dfrac{n_1}{n_2} = -E_{OI} \cos\theta_T \dfrac{n_1}{n_2} + E_I \cos\theta_{OI}$

or $\quad \left(\dfrac{\vec{E}_{OR}}{E_{OI}}\right)_P = \dfrac{\cos\theta_I - \dfrac{n_1}{n_2}\cos\theta_T}{\cos\theta_I + \dfrac{n_1}{n_2}\cos\theta_T} \tag{6.127}$

The subscript P in Eq. (6.127) or in the following equations refers to the fact that the electric field vector is parallel to the plane of incidence.

Putting the value of $\dfrac{n_1}{n_2}$ from Snell's law into this equation, we get

$$\left(\frac{E_{OR}}{E_{OI}}\right)_P = \frac{\sin\theta_I \cos\theta_I - \sin\theta_T \cos\theta_T}{\sin\theta_I \cos\theta_I + \sin\theta_T \cos\theta_T} = \frac{\sin 2\theta_I - \sin 2\theta_T}{\sin 2\theta_I + \sin 2\theta_T}$$

$$= \frac{2\cos(\theta_I + \theta_T)\sin(\theta_I - \theta_T)}{2\sin(\theta_I + \theta_T)\cos(\theta_I - \theta_T)}$$

or $\qquad \left(\dfrac{E_{OR}}{E_{OI}}\right)_P = \dfrac{\tan(\theta_I - \theta_T)}{\tan(\theta_I + \theta_T)}$ $\qquad\qquad\qquad\qquad\qquad$ (6.128)

The ratio of the transmitted amplitude to the incident amplitude is obtained by putting the values of E_{OR} from Eq. (6.126) into Eq. (6.125). Thus, we have

$$E_{OI}\cos\theta_I - \cos\theta_I\left(\frac{n_2}{n_1}E_{OT} - E_{OI}\right) = E_{OT}\cos\theta_T$$

or $\qquad \left(\dfrac{E_{OT}}{E_{OI}}\right)_P = \dfrac{2\cos\theta_I}{\cos\theta_T + \dfrac{n_2}{n_1}\cos\theta_I}$ $\qquad\qquad\qquad\qquad$ (6.129)

Putting the value of $\dfrac{n_1}{n_2}$ from Snell's law into this equation, we get

$$\left(\frac{E_{OT}}{E_{OI}}\right)_P = \frac{2\cos\theta_I}{\cos\theta_T + \dfrac{\sin\theta_I}{\sin\theta_T}\cos\theta_I}$$

or $\qquad \left(\dfrac{E_{OT}}{E_{OI}}\right)_P = \dfrac{2\cos\theta_I \sin\theta_T}{\sin(\theta_I + \theta_T)\cos(\theta_I - \theta_T)}$ $\qquad\qquad\qquad$ (6.130)

Equations (6.128) and (6.130) are called Fresnel's equations.

Interpretation of Eq. (6.127)

If an electromagnetic wave passes from a denser medium (first medium) to a rarer medium (second medium) with $n_1 > n_2$ or $\dfrac{n_1}{n_2} > 1$ and $\theta_I > \theta_T$ or $\cos\theta_I < \cos\theta_T$, then the numerator of Eq. (6.127) is always negative thereby implying that

$$\left(\frac{E_{OR}}{E_{OI}}\right)_P$$

is to be negative. The negative values of

$$\left(\frac{E_{OR}}{E_{OI}}\right)_P$$

indicate that the reflected wave is out phase with the incident wave. That means the phase difference between the incident wave and the reflected wave is π.

If an electromagnetic wave passes from a rarer medium (first medium) to a denser medium (second medium) with $n_1 < n_2$ or $\dfrac{n_1}{n_2} < 1$ and $\theta_I > \theta_T$ or $\cos\theta_I < \cos\theta_T$, then the numerator of Eq. (6.125) is always positive thereby implying that

$$\left(\frac{E_{OR}}{E_{OI}}\right)_P$$

is to be positive. The positive values of

$$\left(\frac{E_{OR}}{E_{OI}}\right)_P$$

indicate that the reflected wave is in phase with the incident wave.

Interpretation of Eq. (6.129)

Whether the electromagnetic wave is passing from a rarer medium to a denser medium or vice versa, the ratio $\left(\dfrac{E_{OT}}{E_{OI}}\right)_P$ is always positive for all possible values of θ_I and θ_T $(0 \le \theta_I, \theta_T \le 90)$. The positive values of $\left(\dfrac{E_{OT}}{E_{OI}}\right)_P$ indicate that the refracted or the transmitted wave is in phase with the incident wave.

If the plane polarized electromagnetic wave with its electric vector parallel to the plane of incidence is incident normally ($\theta_I = 0$) on the interface separating two different dielectric media, we have from Eqs (6.127) and (6.129)

$$\left(\frac{E_{OR}}{E_{OI}}\right)_P = \frac{1 - \dfrac{n_2}{n_1}}{1 + \dfrac{n_2}{n_1}} \quad \text{and} \quad \left(\frac{E_{OT}}{E_{OI}}\right)_P = \frac{2}{1 + \dfrac{n_2}{n_1}} \quad \text{respectively.}$$

The phenomena in Case 1 and in Case 2 are the same for normal incidence.

Example 6.12

An electromagnetic wave is refracted from a dielectric medium to free space. Calculate the dielectric constant of the dielectric medium if the angle of incidence is the critical angle 18°.

Solution

The refractive index n of the medium from the given data is

$$n = \frac{1}{\sin\theta_C} = \frac{1}{\sin 18} = 3.24$$

However, we know that refractive index n is

$$n = \frac{c}{v} = \frac{\sqrt{\dfrac{1}{\mu_0\varepsilon_0}}}{\sqrt{\dfrac{1}{\mu\varepsilon}}} = \sqrt{\frac{\mu\varepsilon}{\mu_0\varepsilon_0}} = \sqrt{\frac{\mu_0\varepsilon}{\mu_0\varepsilon_0}}$$

($\because \mu = \mu_0$ for dielectric medium)

or $n = \sqrt{\dfrac{\varepsilon}{\varepsilon_0}} = \sqrt{\varepsilon_r}$

Hence, the dielectric constant of the dielectric medium will be

$$\varepsilon_r = n^2 = 10.5$$

Example 6.13

A plane electromagnetic wave with electric vector $E_I = 10\cos(\omega t - z)\hat{y}\text{V/m}$ propagating in air medium is incident normally on the surface of a lossless dielectric medium characterized by $\varepsilon_r = 3$ and $\mu_r = 1$. Find (a) λ, (b) ω, (c) $\vec{E}_T$ (transmitted electric vector), and (d) $\vec{H}_I$ (incident magnetic vector).

Solution

a.   The wavelength λ of the electromagnetic wave is $\lambda = \dfrac{2\pi}{1}m = 6.28m$

b.   The speed of the electromagnetic wave in air medium is

$$c = \lambda v = \frac{\lambda}{2\pi}2\pi v = \frac{\lambda}{2\pi}\omega$$

or $\omega = c\dfrac{2\pi}{\lambda} = 3 \times 10^8 \, \text{rad/s}$

c. The field of the electromagnetic wave we can have

$$\frac{E_{OT}}{E_{OI}} = \frac{2}{1 + \dfrac{n_2}{n_1}}$$

Putting the values of n_1 and n_2 into this equation, we get the amplitude of transmitted electric vector as

$$E_{OT} = 10 \times \frac{2}{1 + \dfrac{\sqrt{3}}{\sqrt{1}}} = 7.32 \, \text{V/m} \qquad \left(\because n = \sqrt{\varepsilon_r} \right)$$

Hence, the electric vector of the transmitted electromagnetic wave is given by

$$E_T = 7.32 \cos(\omega t - z)\hat{y} \, \text{V/m}.$$

d. The transmitted electric vector can be obtained from

$$\frac{E_T}{H_T} = \eta^2 = \frac{\mu}{\varepsilon} =$$

or $H_T = E_T \dfrac{\varepsilon_r \varepsilon_0}{\mu_r \mu_0}$

Putting the values of $\varepsilon_r \, \varepsilon_0 \, \mu_r \, \mu_0$, and $\vec{E}_T$ into this equation, we get the amplitude of the transmitted electric vector as

$$H_T = 7.32 \times \frac{3 \times 8.85 \times 10^{-12}}{4\pi \times 10^{-7}} \cos(\omega t - z)\hat{x} A/m H_T$$

$$= 1.49 \times 10^6 \cos(\omega t - z)\hat{x} \, \text{A/m}$$

Example 6.14

An electromagnetic wave with the amplitude of electric vector as 1.5×10^{-3} V/m propagates in a dielectric medium having $\varepsilon_r = 8.5$ and $\mu_r = 1$. It is incident normally on the interface between the dielectric medium and the air medium and simultaneously gets reflected and refracted. Calculate the amplitudes of the electric and magnetic vectors of the reflected electromagnetic wave and refracted electromagnetic wave.

Solution

For the normal incidence of the electromagnetic wave on the dielectric interface inside the denser side, we get from Eq. (6.127)

$$\frac{E_{OR}}{E_{OI}} = \frac{1 - \dfrac{\eta_1}{\eta_2}}{1 + \dfrac{\eta_1}{\eta_2}} \qquad \text{(A)}$$

Similarly, from Eq. (6.129), we get

$$\frac{E_{OT}}{E_{OI}} = \frac{2}{1 + \dfrac{\eta_1}{\eta_2}} \qquad \text{(B)}$$

As for the given information, we have

$$E_{OI} = 1.5 \times 10^{-3} \, \text{V/m}, \quad \mu_{r1} = 1, \quad \varepsilon_{r1} = 8.5, \quad \mu_{r2} = 1, \quad \varepsilon_{r2} = 1$$

Now we can calculate η_1 and η_2 as

$$\eta_1 = \sqrt{\frac{\mu_0 \mu_{r1}}{\varepsilon_0 \varepsilon_{r1}}} = 129 \, \Omega$$

and $\eta_2 = \sqrt{\dfrac{\mu_0 \mu_{r2}}{\varepsilon_0 \varepsilon_{r2}}} = 377 \, \Omega$

Putting the values of η_1, η_2, and E_{OI} into the Eq. (A), we get

$$\frac{E_{OR}}{E_{OI}} = \frac{1 - 0.3422}{1 + 0.3422} = 0.4901$$

or $E_{OR} = 0.4901 \times 1.5 \times 10^{-3} \, \text{V/m} = 7.35 \times 10^{-4} \, \text{V/m}.$

Putting the values of η_1, η_2, and E_{OI} into Eq. (B), we get

$$\frac{E_{OT}}{E_{OI}} = \frac{2}{1 + \dfrac{129}{377}} = 1.490$$

or $E_{OT} = 1.4901 \times 1.5 \times 10^{-3} \, \text{V/m} = 2.24 \times 10^{-3} \, \text{V/m}$

From Eq. (6.55), we have

$$H_{OI} = \frac{E_{OI}}{\eta_1} \qquad \text{(C)}$$

Putting the values of η_1 and E_{OI} into Eq. (C), we get

$$H_{OI} = \frac{1.5 \times 10^{-3}\,\text{V/m}}{129\,\Omega} = 1.163 \times 10^{-5}\,\text{A/m}$$

For the amplitude of the magnetic field of the electromagnetic wave, we can have

$$\frac{H_{OR}}{H_{OI}} = \frac{\dfrac{\eta_1}{\eta_2} - 1}{\dfrac{\eta_1}{\eta_2} + 1} \qquad \text{(D)}$$

and

$$\frac{H_{OT}}{H_{OI}} = \frac{2}{1 + \dfrac{\eta_2}{\eta_1}} \qquad \text{(E)}$$

Putting the values of η_1, η_2, and E_{OI} into Eq. (D), we get

$$\frac{H_{OR}}{H_{OI}} = \frac{0.3422 - 1}{0.3422 + 1}$$

or $H_{OR} = -5.70 \times 10^{-6}\,\text{A/m}$

Putting the values of η_1, η_2, and E_{OI} into the Eq. (E), we get

$$\frac{H_{OT}}{H_{OI}} = \frac{2}{1 + 2.92} = 0.51$$

or $H_{OT} = 0.51 \times 1.163 \times 10^{-5}\,\text{A/m} = 5.93 \times 10^{-6}\,\text{A/m}$

Example 6.15

An electromagnetic wave, with the amplitude of electric vector as 1.0 V/m, propagating in free space is incident normally on the surface of a dielectric medium characterized by $\varepsilon_r = 18.5$ and $\mu_r = 198.4$. It simultaneously gets reflected and refracted. Calculate the amplitudes of the electric and magnetic vectors of the reflected electromagnetic wave and the refracted electromagnetic wave.

Solution

For the normal incidence of the electromagnetic wave on the dielectric surface, we get from Eq. (6.127)

$$\frac{E_{OR}}{E_{OI}} = \frac{1 - \frac{\eta_1}{\eta_2}}{1 + \frac{\eta_1}{\eta_2}} \tag{A}$$

Similarly, from Eq. (6.129), we get

$$\frac{E_{OT}}{E_{OI}} = \frac{2}{1 + \frac{\eta_1}{\eta_2}} \tag{B}$$

As for the given information, we have

$$E_{OI} = 1.0 \text{V/m}, \ \mu_{r1} = 1, \ \varepsilon_{r1} = 1, \ \mu_{r2} = 198.4, \ \varepsilon_{r2} = 18.5$$

Now we can calculate η_1 and η_2 as

$$\eta_1 = \sqrt{\frac{\mu_0 \mu_{r1}}{\varepsilon_0 \varepsilon_{r1}}} = 377\Omega$$

and $\eta_2 = \sqrt{\frac{\mu_0 \mu_{r2}}{\varepsilon_0 \varepsilon_{r2}}} = 1234.6\Omega$

Putting the values of η_1, η_2, and E_{OI} into the Eq. (A), we get

$$\frac{E_{OR}}{E_{OI}} = \frac{1 - 0.3055}{1 + 0.3055} = 0.5320$$

or $E_{OR} = 0.5320 \times 1.0 \text{V/m} = 0.5320 \text{V/m}$

Putting the values of η_1, η_2, and E_{OI} into the Eq. (B), we get

$$\frac{E_{OT}}{E_{OI}} = \frac{2}{1 + \frac{377}{1234}} = 1.532$$

or $E_{OT} = 1.532 \times 1.0 \, \text{V/m} = 1.532 \, \text{V/m}$

From Eq. (6.55), we have

$$H_{OI} = \frac{E_{OI}}{\eta_1} \qquad \text{(C)}$$

Putting the values of η_1 and E_{OI} into Eq. (C), we get

$$H_{OI} = \frac{1.0 \, \text{V/m}}{377 \, \Omega} = 2.65 \times 10^{-3} \, \text{A/m}$$

For the amplitude of the magnetic field of the electromagnetic wave, we can have

$$\frac{H_{OR}}{H_{OI}} = \frac{\dfrac{\eta_1}{\eta_2} - 1}{\dfrac{\eta_1}{\eta_2} + 1} \qquad \text{(D)}$$

and
$$\frac{H_{OT}}{H_{OI}} = \frac{2}{1 + \dfrac{\eta_2}{\eta_1}} \qquad \text{(E)}$$

Putting the values of η_1, η_2, and E_{OI} into Eq. (D), we get

$$\frac{H_{OR}}{H_{OI}} = \frac{0.3055 - 1}{0.3055 + 1}$$

or $H_{OR} = -0.53 \times 2.65 \times 10^{-3} \, \text{A/m} = -1.41 \times 10^{-3} \, \text{A/m}$

Putting the values of η_1, η_2, and E_{OI} into the Eq. (E), we get

$$\frac{H_{OT}}{H_{OI}} = \frac{2}{1 + 3.27} = 0.47$$

or $H_{OT} = 0.47 \times 2.65 \times 10^{-3} \, \text{A/m} = 1.25 \times 10^{-3} \, \text{A/m}$

Reflection and transmission coefficient

The reflection coefficient R_R is defined as the square of the ratio of the amplitude of the reflected wave to the amplitude of the incident wave. Mathematically,

$$R_R = \left(\frac{E_{OR}}{E_{OI}} \right)^2 \tag{6.131}$$

The transmission coefficient or refraction coefficient R_T is defined as the square of the ratio of the amplitude of transmitted wave to the amplitude of the incident wave. Mathematically,

$$R_T = \left(\frac{E_{OT}}{E_{OI}} \right)^2 \tag{6.132}$$

It can be checked for surety that the sum of the reflection coefficient and the transmission coefficient is unity both in Case 1 (Electric vector normal to the plane of incidence) and Case 2 (Electric vector parallel to the plane of incidence). R_{RN} and R_{TN} are the reflection and transmission coefficients respectively in the case where the electric vector of the electromagnetic wave is normal to the plane of incidence. R_{RP} and R_{TP} are the reflection and transmission coefficients respectively in the case where the electric vector of the electromagnetic wave is parallel to the plane of incidence.

Example 6.16

The phase constant of a certain good conducting material at a certain frequency is 3.75×10^5 rad/m. Calculate the frequency of the electromagnetic wave. Given $\sigma = 58 \times 10^6$ Siemen/m and $\mu_r = 1$.

Solution

The phase constant β of a good conductor is given by

$$\beta = \sqrt{\frac{\omega \mu \sigma}{2}}$$

Hence, the expression for angular frequency comes out to be

$$\omega = \frac{2\beta^2}{\mu \sigma}$$

or the linear frequency will be $v = \dfrac{2\beta^2}{2\pi \mu \sigma}$

Putting the given values of β, μ, and σ into this equation, we get the value of frequency v of the electromagnetic wave as

$$v = \frac{2(3.75 \times 10^5)^2}{2\pi \times 4\pi \times 10^{-7} \times 58 \times 10^6} Hz = 614.2 \, \text{MHz} .$$

Example 6.17

The fraction of energy transmitted through the interface of two dielectric media in case of normal incidence is half of the fraction of the energy reflected. Calculate the ratio of refractive indices of the two media.

Solution

According to the condition given in the problem,

$$R_T = \frac{1}{2} R_R$$

$$\frac{4 \dfrac{n_1}{n_2}}{\left[\dfrac{n_1}{n_2} + 1\right]^2} = \frac{1}{2}\left(\frac{1 - \dfrac{n_1}{n_2}}{1 + \dfrac{n_1}{n_2}}\right)^2$$

With $x = \dfrac{n_1}{n_2}$, we have

$$4x = \frac{1}{2}(1-x)^2$$

or $x = 5 + 2\sqrt{6}$

or $\dfrac{n_1}{n_2} = 5 + 2\sqrt{6}$

The ratio of refractive indices of the two media is found to be $5 + 2\sqrt{6}$.

Brewster's angle

Brewster's angle or the polarizing angle θ_B is defined as the angle of incidence for which the angle of transmission θ_T is equal to $\dfrac{\pi}{2} - \theta_B$. That means, $\theta_I = \theta_B =$ if $\theta_T = \dfrac{\pi}{2} - \theta_B$. Taking the square of Eq. (6.128), we have

$$R_{RP} = \frac{\tan^2(\theta_I - \theta_T)}{\tan^2(\theta_I + \theta_T)} = \frac{\tan^2(\theta_B - \theta_T)}{\infty} = 0$$

or $\left(\dfrac{E_{OR}}{E_{OI}}\right)_P = 0$

or $E_{OR} = 0$

Thus, the electromagnetic wave with the electric vector parallel to the plane of incidence is made to be incident on the surface of the dielectric medium at Brewster's angle, the electric vector parallel to the plane of incidence is not present in the reflected ray. The reflected electromagnetic wave contains the electric vector perpendicular to the plane of incidence.

Now putting this condition into Snell's law, we have

$$\frac{n_2}{n_1} = \frac{\sin\theta_B}{\sin\left(\frac{\pi}{2} - \theta_B\right)} = \frac{\sin\theta_B}{\cos\theta_B}$$

or $\tan\theta_B = \dfrac{n_2}{n_1}$

or $\theta_B = \tan^{-1}\left(\dfrac{n_2}{n_1}\right)$ (6.133)

Brewster's angle depends of the only on the refractive indices of the first and second media. Knowing the values of refractive indices of the first and second media, we can calculate Brewster's angle. Therefore, we conclude that the unpolarized light (an electromagnetic wave) is made to be incident on an interface with the angle of incidence equal to Brewster's angle, the reflected light will contain the electric vector perpendicular to the plane of incidence, whereas the electric vector parallel to the plane of incidence will be absent and the transmitted light is also plane polarized with the electric vector parallel to the plane of incidence. Thus, reflected light is plane polarized with the electric vector perpendicular to the plane of incidence. The transmitted/refracted light is also plane polarized with the electric vector parallel to the plane of incidence. The plane of polarization is defined as the plane containing the propagation vector $\vec{k}$ and the magnetic vector $\overline{H}$. The plane of vibration is defined as the plane containing the propagation vector $\vec{k}$ and the electric vector $\vec{E}$. The plane of incidence is defined as the plane containing the propagation vector $\vec{k}$ and the normal to the surface at the point of incidence. The plane polarized light with its electric vector parallel to the plane of incidence is incident at Brewster's angle and will be refracted without reflection, whereas plane polarized light with its electric vector perpendicular to the plane of incidence is incident at Brewster's angle and will be reflected without transmission. This is depicted in Fig. 6.7.

It is an experimental fact that the unpolarized light incident at Brewster's angle is reflected predominately polarized with electric vectors perpendicular to the plane of incidence. The degree of polarization δ_p is defined as

$$\delta_p = \frac{R_{RN} - R_{RP}}{R_{RN} + R_{RP}}$$ (6.134)

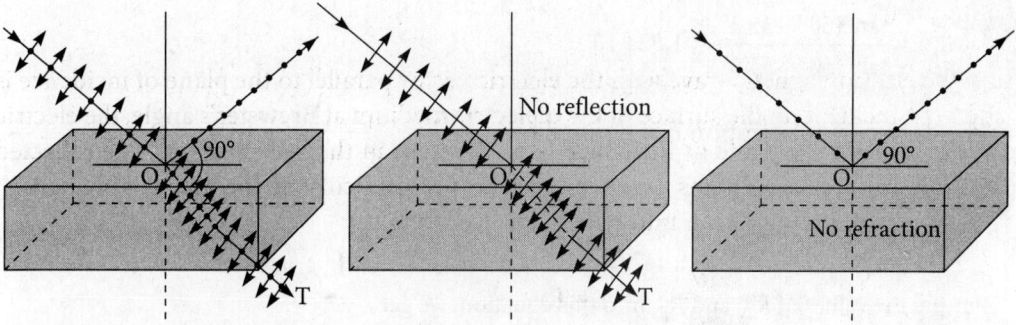

Figure 6.7 | Polarization on reflection and refraction

where

$$R_{RN} = \frac{\sin^2(\theta_I - \theta_T)}{\sin^2(\theta_I + \theta_T)} \qquad\qquad (6.135)$$

$$R_{RP} = \frac{\tan^2(\theta_I - \theta_T)}{\tan^2(\theta_I + \theta_T)} \qquad\qquad (6.136)$$

Example 6.18

Calculate the degree of polarization for ordinary light incident at an angle 60° on a glass having refractive index 1.5

Solution

From Snell's law, from Eq. (6.108), we have

$$\sin\theta_T = \frac{n_1}{n_2}\sin\theta_I$$

or $\theta_T = 35.3°$

$$R_{RN} = \frac{\sin^2(\theta_I - \theta_T)}{\sin^2(\theta_I + \theta_T)}$$

Putting the values of θ_I and θ_T into this equation, we have

$$R_{RN} = 0.176$$

$$R_{RP} = \frac{\tan^2(60° - 35.3°)}{\tan^2(60° + 35.3°)} = 1.82 \times 10^{-3}$$

The degree of polarization δ_p is defined by

$$\delta_p = \frac{R_{RN} - R_{RP}}{R_{RN} + R_{RP}}$$

Putting the values of R_{RN} and R_{RP} into this equation, we get

$$\delta_p = 98\%$$

Example 6.19

Calculate the critical angle and Brewster's angle for an electromagnetic wave passing through non-magnetic quartz of relative permitivitty 5.

Solution

The refractive index of quartz is given by

$$n = \sqrt{\mu_r \varepsilon_r} = 2.24$$

The critical angle for an electromagnetic wave passing through non-magnetic quartz will be obtained from

$$\sin \theta_C = \frac{1}{2.24}$$

or $\theta_C = 26.5°$

Brewster's angle for an electromagnetic wave passing through non-magnetic quartz will be obtained from

$$\tan \theta_B = n$$

or $\theta_B = \tan^{-1} 2.24 = 65.9°$

6.13.2 Reflection and refraction of electromagnetic waves at a conducting surface

Let us consider the case where there is reflection and refraction at the boundary of the dielectric and the conducting media. Let the first medium be a dielectric medium characterized by constant permittivity ε_1, constant permeability μ_1, refractive index n_1, intrinsic impedance η_1 and conductivity $\sigma_1 = 0$; and the second medium be a conducting medium characterized by constant permittivity ε_2, constant permeability μ_2, refractive index n_2, intrinsic impedance η_2 and non-zero constant conductivity σ_2. Both the media are assumed to be isotropic, homogeneous, charge free ($\rho = 0$), linear ($\vec{D} = \varepsilon\vec{E}$ and $\vec{B} = \mu\vec{H}$) of infinite extension.

Consider that an electromagnetic wave propagating in the first medium is incident obliquely on the interface separating the conducting medium from the non-conducting medium. In general, a reflected and a transmitted wave will be produced. The incident and reflected wave will lie in the first medium and the transmitted wave will lie in the second medium. See Fig. 6.8. The laws of reflection and refraction can be proved by the methods discussed earlier.

The propagation vector $\vec{k}_T$ in the second conducting medium is in general found out to be complex. In the charge-free conduction medium, the field vectors E_T and H_T satisfy the following wave equations

$$\nabla^2 E_T - \mu_2\varepsilon_2\frac{\partial^2 E_T}{\partial t^2} - \mu_2\sigma_2\frac{\partial E_T}{\partial t} = 0 \tag{6.137}$$

$$\nabla^2 H_T - \mu_2\varepsilon_2\frac{\partial^2 H_T}{\partial t^2} - \mu_2\sigma_2\frac{\partial H_T}{\partial t} = 0 \tag{6.138}$$

The solution of the wave Eq. (6.137) is given by

$$\vec{E}_T(r,t) = \vec{E}_{OT}e^{i(\omega_T t - \vec{k}_T \cdot \vec{r})}.$$

Substituting this back into Eq. (6.137), we have

$$-k_T^2 E_T(r,t) + \varepsilon_2\mu_2\omega^2 E_T(r,t) + i\sigma_2\mu_2\omega E_T(r,t) = 0$$

or $\quad -k_T^2 + \varepsilon_2\mu_2\omega^2 + i\sigma_2\mu_2\omega = 0$

or $\quad k_T^2 = \varepsilon_2\mu_2\omega^2\left(1 + i\dfrac{\sigma_2}{\varepsilon_2\omega}\right) \tag{6.139}$

The first term corresponds to the displacement current and the second term corresponds to the conduction current. The relation between propagation vector $\vec{k}$ and frequency ω is called the dispersion relation. Equation (6.139) shows that the propagation vector in a conducting medium is complex. Hence, we can express the propagation vector $\vec{k}$ in the form

$$\bar{k}_T = \alpha + i\beta \tag{6.140}$$

Squaring both sides of this equation, we have

$$k_T^2 = \alpha^2 + i^2\beta^2 + i2\alpha\beta = \alpha^2 - \beta^2 + i2\alpha\beta$$

Comparing this equation with Eq. (6.139), we have

$$\alpha^2 - \beta^2 = \varepsilon_2\mu_2\omega^2$$

and $\;\; 2\alpha\beta = \sigma_2\mu_2\omega$

Solving the two equations given here, we have the real part of $\vec{k}_T$ as

$$\alpha = \omega\sqrt{\frac{\mu_2\varepsilon_2}{2}}\left[\left\{1+\left(\frac{\sigma_2}{\varepsilon_2\omega}\right)^2\right\}^{\frac{1}{2}} + 1\right]^{\frac{1}{2}} \tag{6.141}$$

and the imaginary part of $\vec{k}_T$ as

$$\beta = \omega\sqrt{\frac{\mu_2\varepsilon_2}{2}}\left[\left\{1+\left(\frac{\sigma_2}{\varepsilon_2\omega}\right)^2\right\}^{\frac{1}{2}} - 1\right]^{\frac{1}{2}} \tag{6.142}$$

The same value of $\vec{k}_T$ will be obtained if we consider Eq. (6.138) instead of Eq. (6.137).

For a very good conducting medium or metallic medium, $\dfrac{\sigma_2}{\varepsilon_2\omega} \gg 1$. Hence, neglecting 1 in comparison to $\left(\dfrac{\sigma_2}{\varepsilon_2\omega}\right)^2$ in Eqs (6.141) and (6.142), we have respectively

$$\alpha = \omega\sqrt{\frac{\mu_2\varepsilon_2}{2}}\sqrt{\frac{\sigma_2}{\varepsilon_2\omega}} = \sqrt{\frac{\mu_2\sigma_2\omega}{2}} \tag{6.143}$$

$$\beta = \sqrt{\frac{\mu_2\sigma_2\omega}{2}} \tag{6.144}$$

Thus, for a very good conducting medium or metallic medium, $\dfrac{\sigma_2}{\varepsilon_2 \omega} \gg 1$ and the propagation vector $\overrightarrow{k_T}$ in the metallic medium is found out by putting these values of α and β in Eq. (6.140) as

$$k_T = \sqrt{\frac{\mu_2 \sigma_2 \omega}{2}} + i \sqrt{\frac{\mu_2 \sigma_2 \omega}{2}} = \sqrt{\frac{\mu_2 \sigma_2 \omega}{2}}\,(1+i) \tag{6.145}$$

The refractive index n_2 of the second conducting medium is given by

$$n_2 = \frac{c}{\omega} k_T$$

Putting the value of k_T from Eq. (6.145) into this equation, we get the expression for the refractive index of the second conducting medium as

$$n_2 = \frac{c}{\omega} \times \sqrt{\frac{\mu_2 \sigma_2 \omega}{2}}\,(1+i) = c\,\sqrt{\frac{\mu_2 \sigma_2}{2\omega}}(1+i) \tag{6.146}$$

The refractive index of the conducting medium is very large in comparison to that of a non-conducting medium. From Snell's law, we have

$$\sin \theta_T = \left(\frac{n_1}{n_2}\right) \sin \theta_I$$

or
$$\cos \theta_T = \left[1 - \left(\frac{n_1}{n_2}\right)^2 \sin^2 \theta_I \right]^{\frac{1}{2}} \tag{6.147}$$

The refractive index of a conducting medium n_2 is very large in comparison to that of a non-conducting medium n_1. Hence, $\left(\dfrac{n_1}{n_2} \sin \theta_I\right)^2$ becomes negligibly small in comparison to 1 in Eq. (6.147). Therefore, we have from Eq. (6.147),

$$\cos \theta_T \approx 1$$

or
$$\theta_T \approx 0 \tag{6.148}$$

Equation (6.148) shows that the angle of refraction when the electromagnetic wave is incident on the surface of a metallic medium is very close to zero irrespective of the angle of incidence. Thus, in a metallic medium, the electromagnetic wave is transmitted along the normal direction.

Case 1: Electric vector perpendicular to the plane of incidence

The plane of incidence, as mentioned earlier, is defined as the plane containing the propagation vector $\vec{k}_I$ and the normal to the interface at the point of incidence. In our case, the X-Y plane will be the plane of incidence and the Y-Z plane is the interface separating the conducting medium from the non-conducting medium. See Fig. 6.8. All the field vectors of the electromagnetic wave obey the boundary conditions described in Section 6.13. The tangential components of the electric vector are continuous along the interface. In our case, since the electric vectors are perpendicular to the plane of incidence in the inward direction (i.e., along the $-Z$ axis), they are at the same time tangential to the interface. Hence, we have

$$\vec{E}_I + \vec{E}_R = \vec{E}_T$$

or $\quad \vec{E}_{OI} + \vec{E}_{OR} = \vec{E}_{OT}$ (6.149)

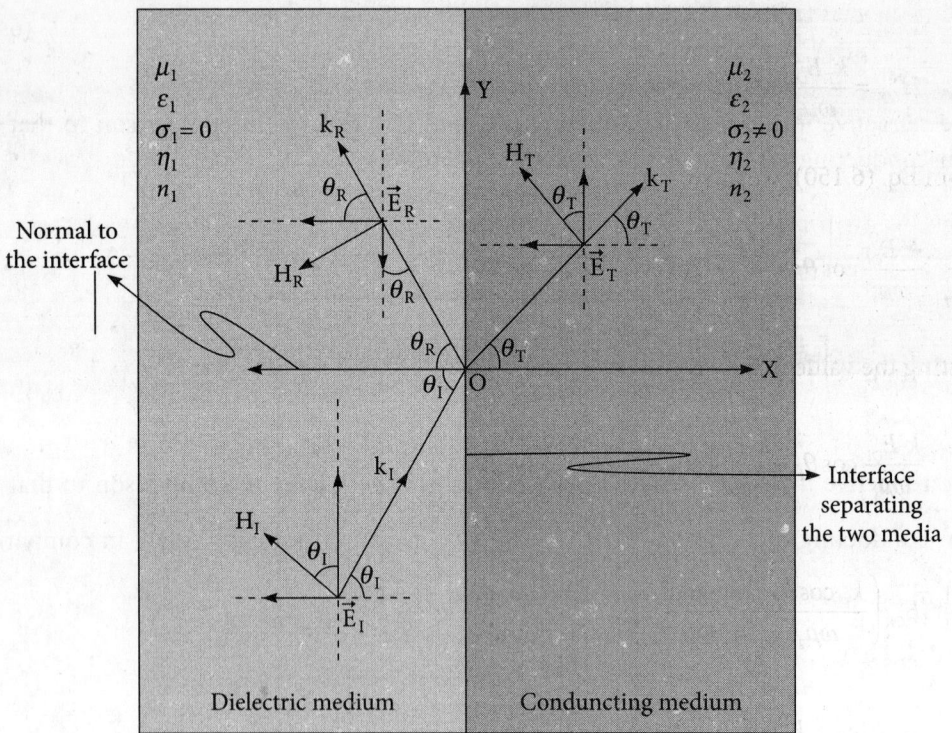

Z-axis is normal to the plane of
the paper at O in outward direction

Figure 6.8 | Reflection and refraction of an electromagnetic wave on the surface of a conducting medium with the electric vector normal to the plane of incidence

Now the tangential components of the magnetic vector are continuous along the interface. Hence, we have from Fig. 6.8,

$$H_I \cos\theta_I - H_R \cos\theta_R = H_T \cos\theta_T$$

or $\quad H_{OI}\cos\theta_I - H_{OR}\cos\theta_R = H_{OT}\cos\theta_T$ (6.150)

Also we know the following relations

$$H_{OI} = \frac{k_I E_{OI}}{\omega\,\mu_1}$$ (6.151)

$$H_{OR} = \frac{k_I E_{OR}}{\omega\mu_1}$$ (6.152)

$$H_{OT} = \frac{k_T E_{OT}}{\omega\mu_2}$$ (6.153)

From Eq. (6.150), we have

$$\frac{k_I E_{OI}}{\omega\mu_1}\cos\theta_I - \frac{k_I E_{OR}}{\omega\mu_1}\cos\theta_R = \frac{k_T E_{OT}}{\omega\mu_2}\cos\theta_T$$ (6.154)

Putting the value of E_{OT} from Eq. (6.149) into Eq. (6.154), we get

$$\frac{k_I E_{OI}}{\omega\mu_1}\cos\theta_I - \frac{k_I E_{OR}}{\omega\mu_1}\cos\theta_R = \frac{k_T \cos\theta_T}{\omega\mu_2}(E_{OI} + E_{OR})$$

or $\quad E_{OR}\left(\dfrac{k_T \cos\theta_T}{\omega\mu_2} + \dfrac{k_I \cos\theta_R}{\omega\mu_1}\right) = E_{OI}\left(\dfrac{k_I \cos\theta_I}{\omega\mu_1} - \dfrac{k_T \cos\theta_T}{\omega\mu_2}\right)$

or $\quad \left(\dfrac{E_{OR}}{E_{OI}}\right)_N = \dfrac{\dfrac{k_I \cos\theta_I}{\mu_1} - \dfrac{k_T \cos\theta_T}{\mu_2}}{\dfrac{k_I \cos\theta_I}{\mu_1} + \dfrac{k_T \cos\theta_T}{\mu_2}} \qquad (\because\ \theta_R = \theta_I)$

Applying the condition for a good conductor, this equation becomes

$$\left(\frac{E_{OR}}{E_{OI}}\right)_N = \frac{\dfrac{k_I \cos\theta_I}{\mu_1} - \dfrac{k_T}{\mu_2}}{\dfrac{k_I \cos\theta_I}{\mu_1} + \dfrac{k_T}{\mu_2}} = \frac{\cos\theta_I - \dfrac{k_T}{k_I}\dfrac{\mu_1}{\mu_2}}{\cos\theta_I + \dfrac{k_T}{k_I}\dfrac{\mu_1}{\mu_2}} \qquad (6.155)$$

Again putting the value of E_{OR} from Eq. (6.149) into Eq. (6.154), we get

$$\frac{k_I E_{OI}}{\omega\mu_1}\cos\theta_I - \frac{k_I \cos\theta_R}{\omega\mu_1}(E_{OT} - E_{OI}) = \frac{k_T E_{OT}}{\omega\mu_2}\cos\theta_T$$

or $\quad \dfrac{k_I E_{OI}}{\mu_1}\cos\theta_I - \dfrac{k_I \cos\theta_I}{\mu_1}E_{OT} + \dfrac{k_I \cos\theta_I}{\mu_1}E_{OI} = \dfrac{k_T E_{OT}}{\mu_2}\cos\theta_T$

or $\quad \left(\dfrac{E_{OT}}{E_{OI}}\right)_N = \dfrac{2\dfrac{k_I \cos\theta_I}{\mu_1}}{\dfrac{k_I \cos\theta_I}{\mu_I} + \dfrac{k_T \cos\theta_T}{\mu_2}}$

Applying the condition for a good conductor, this equation becomes

$$\left(\frac{E_{OT}}{E_{OI}}\right)_N = \frac{2\dfrac{k_I \cos\theta_I}{\mu_1}}{\dfrac{k_I \cos\theta_I}{\mu_I} + \dfrac{k_T}{\mu_2}} = \frac{2\cos\theta_I}{\cos\theta_I + \dfrac{k_T}{k_I}\dfrac{\mu_1}{\mu_2}} \qquad (6.156)$$

Equations (6.150) and (6.156) are called Fresnel's equations in a conducting medium.

Now taking the values of $\bar{k}_I$, $\bar{k}_T$ from Eqs (6.104) and (6.145), we get the expression for $\dfrac{k_T}{k_I}\dfrac{\mu_1}{\mu_2}$ as

$$\frac{k_T}{k_I}\frac{\mu_1}{\mu_2} = \frac{(1+i)}{\sqrt{2}}\frac{\sqrt{\mu_2\sigma_2\omega}}{\omega\sqrt{\mu_1\varepsilon_1}}\frac{\mu_1}{\mu_2}$$

or $\quad \dfrac{k_T}{k_I}\dfrac{\mu_1}{\mu_2} = \dfrac{(1+i)\sqrt{\mu_2\sigma_2\omega\mu_1^2}}{\sqrt{2\mu_1\varepsilon_1\omega^2\mu_2^2}}$

or $\qquad \dfrac{k_T}{k_I}\dfrac{\mu_1}{\mu_2} = (1+i)\sqrt{\dfrac{\mu_1\sigma_2}{2\mu_2\varepsilon_1\omega}}$ $\qquad\qquad\qquad\qquad\qquad\qquad\qquad$ (6.157)

Putting this value of $\dfrac{k_T}{k_I}\dfrac{\mu_1}{\mu_2}$ in Eq. (6.155), we get

$$\left(\frac{E_{OR}}{E_{OI}}\right)_N = \frac{\cos\theta_I - (1+i)\sqrt{\dfrac{\mu_1\sigma_2}{2\omega\mu_2\varepsilon_1}}}{\cos\theta_I + (1+i)\sqrt{\dfrac{\mu_1\sigma_2}{2\omega\mu_2\varepsilon_1}}}$$

The value of $\sqrt{\dfrac{\mu_1\sigma_2}{2\mu_2\varepsilon_1\omega}}$ is very large in comparison to $\cos\theta_I$ for a good conductor. Hence, we can neglect $\cos\theta_I$ in comparison to $\sqrt{\dfrac{\mu_1\sigma_2}{2\mu_2\varepsilon_1\omega}}$. Under this condition, the aforementioned equation becomes

$$\left(\frac{E_{OR}}{E_{OI}}\right)_N \approx \frac{-(1+i)\sqrt{\dfrac{\mu_1\sigma_2}{2\omega\mu_2\varepsilon_1}}}{(1+i)\sqrt{\dfrac{\mu_1\sigma_2}{2\omega\mu_2\varepsilon_1}}}$$

or $\qquad \left(\dfrac{E_{OR}}{E_{OI}}\right)_N \approx -1$ $\qquad\qquad\qquad\qquad\qquad\qquad\qquad\qquad$ (6.158)

This equation shows that the reflected wave is approximately π radians out of phase with the incident wave and reflection is nearly 100%.

Putting the value of $\dfrac{k_T}{k_I}\dfrac{\mu_1}{\mu_2}$ from Eq. (6.157) into Eq. (6.156), we get

$$\left(\frac{E_{OT}}{E_{OI}}\right)_N = \frac{2\cos\theta_I}{\cos\theta_I + (1+i)\sqrt{\dfrac{\mu_1\sigma_2}{2\mu_2\varepsilon_1\omega}}}$$

The value of $\sqrt{\dfrac{\mu_1\sigma_2}{2\mu_2\varepsilon_1\omega}}$ is very large in comparison to $\cos\theta_I$ for a good conductor. Hence, we can neglect $\cos\theta_I$ in comparison to $\sqrt{\dfrac{\mu_1\sigma_2}{2\mu_2\varepsilon_1\omega}}$. Under this condition, the aforementioned equation becomes

$$\left(\frac{E_{OT}}{E_{OI}}\right)_N = \frac{2\cos\theta_I}{(1+i)\sqrt{\dfrac{\mu_1\sigma_2}{2\mu_2\varepsilon_1\omega}}}$$

Since $\cos\theta_I \ll \sqrt{\dfrac{\mu_1\sigma_2}{2\mu_2\varepsilon_1\omega}}$, the LHS of this equation tends to zero. Hence, we can have

$$\left(\frac{E_{OT}}{E_{OI}}\right)_N \approx 0 \tag{6.159}$$

Equation (6.159) shows that the refracted wave is approximately in phase with the incident wave. The attenuation of a transmitted wave is very high.

Case 2: Electric vector parallel to the plane of incidence

In this case the X-Y plane will be the plane of incidence and the Y-Z plane is the interface separating the conducting medium from the non-conducting medium. See Fig. 6.9. The tangential components of the electric vector are continuous along the interface. In our case, since the electric vectors are parallel to the plane of incidence, we have

$$E_I \cos\theta_I - E_R \cos\theta_I = E_T \cos\theta_T$$

Since the exponential parts of Eqs (6.137) to (6.139) are the same, we can write this equation as

$$E_{OI} \cos\theta_I - E_{OR} \cos\theta_I = E_{OT} \cos\theta_T \tag{6.160}$$

Electric vectors are parallel to the plane of incidence and magnetic vectors are perpendicular to the plane of incidence so that the following vector equations, obtained from Eq. (6.53) are not violated. The boundary condition for the magnetic field vector is that the tangential component of the magnetic field vector $\vec{H}$ is continuous across the interface. Making use of this boundary condition, we have

$$\vec{H}_I + \vec{H}_R = \vec{H}_T$$

or $H_{OI} + H_{OR} = H_{OT}$

Putting the values of H_{OI}, H_{OR}, and H_{OT} from Eqs (6.151) to (6.153) into this equation, we have

$$\frac{k_I E_{OI}}{\omega\,\mu_1} + \frac{k_I E_{OR}}{\omega\mu_1} = \frac{k_T E_{OT}}{\omega\mu_2}$$

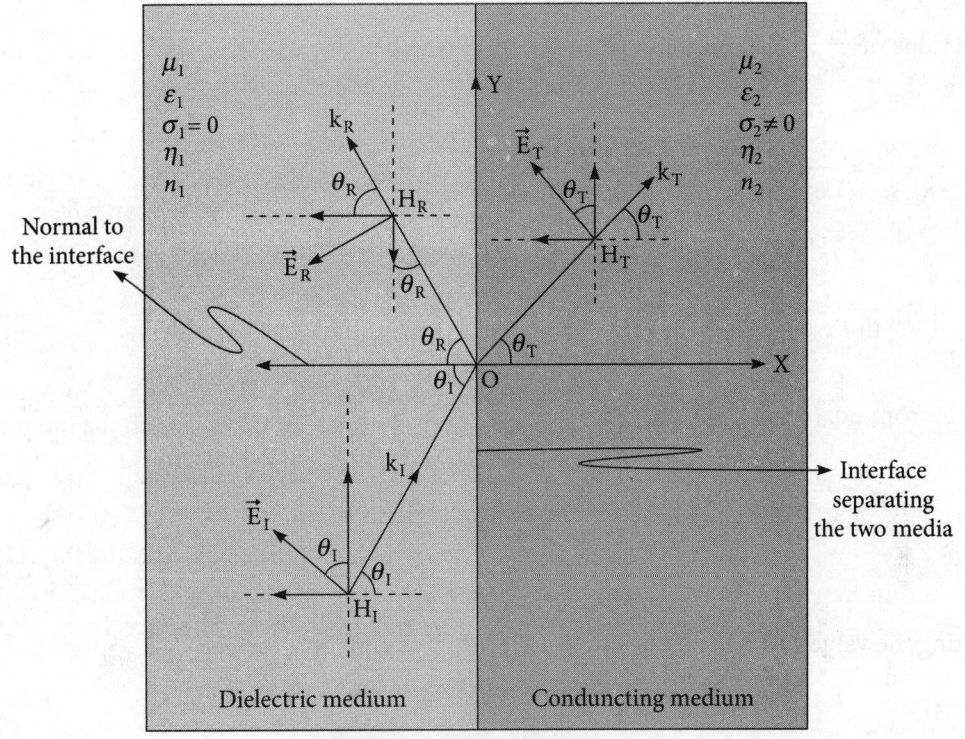

Figure 6.9 Reflection and refraction of an electromagnetic wave on the surface of a conducting medium with the electric vector parallel to the plane of incidence

or $\quad E_{OI} + E_{OR} = \dfrac{k_T}{k_I}\dfrac{\mu_1}{\mu_2} E_{OT}$ (6.161)

or $\quad E_{OR} = \dfrac{k_T}{k_I}\dfrac{\mu_1}{\mu_2} E_{OT} - E_{OI}$

Putting the value of $\dfrac{k_T}{k_I}\dfrac{\mu_1}{\mu_2}$ from Eq. (6.157) into this equation, we get

$$E_{OR} = (1+i)\sqrt{\dfrac{\mu_1\sigma_2}{2\mu_2\varepsilon_1\omega}}\ E_{OT} - E_{OI}$$ (6.162)

The value of $\dfrac{k_I}{k_T}\dfrac{\mu_2}{\mu_1}$ is found out from Eq. (6.157) to be

$$\frac{k_I}{k_T}\frac{\mu_2}{\mu_1} = \frac{1}{1+i}\sqrt{\frac{2\mu_2\varepsilon_1\omega}{\mu_1\sigma_2}} = \frac{1-i}{2}\sqrt{\frac{2\mu_2\varepsilon_1\omega}{\mu_1\sigma_2}} \qquad (6.163)$$

$$\left(\because \quad \frac{1}{1+i} = \frac{1-i}{(1+i)(1-i)} = \frac{1-i}{1-i^2} = \frac{1-i}{2}\right)$$

Again from Eq. (6.161), we have

$$E_{OT} = \frac{k_I}{k_T}\frac{\mu_2}{\mu_1}(E_{OI} + E_{OR})$$

Putting the value of $\dfrac{k_I}{k_T}\dfrac{\mu_2}{\mu_1}$ from Eq. (6.163) into this equation, we get

$$E_{OT} = \frac{1-i}{2}\sqrt{\frac{2\mu_2\varepsilon_1\omega}{\mu_1\sigma_2}}(E_{OI} + E_{OR}) \qquad (6.164)$$

The ratio of the reflected amplitude to the incident amplitude is obtained by putting the values of E_{OT} from Eq. (6.164) into Eq. (6.160). Thus, we have

$$E_{OI}\cos\theta_I - E_{OR}\cos\theta_I = \cos\theta_T\left(\frac{1-i}{2}\sqrt{\frac{2\mu_2\varepsilon_1\omega}{\mu_1\sigma_2}}E_{OI} + \frac{1-i}{2}\sqrt{\frac{2\mu_2\varepsilon_1\omega}{\mu_1\sigma_2}}E_{OR}\right)$$

$$= \cos\theta_T\frac{1-i}{2}\sqrt{\frac{2\mu_2\varepsilon_1\omega}{\mu_1\sigma_2}}E_{OI} + \cos\theta_T\frac{1-i}{2}\sqrt{\frac{2\mu_2\varepsilon_1\omega}{\mu_1\sigma_2}}E_{OR}$$

or $\quad \cos\theta_T\dfrac{1-i}{2}\sqrt{\dfrac{2\mu_2\varepsilon_1\omega}{\mu_1\sigma_2}}E_{OR} + E_{OR}\cos\theta_I = \cos\theta_I E_{OI} - \cos\theta_T\dfrac{1-i}{2}\sqrt{\dfrac{2\mu_2\varepsilon_1\omega}{\mu_1\sigma_2}}E_{OI}$

or $\quad E_{OR}\left(\cos\theta_T\dfrac{1-i}{2}\sqrt{\dfrac{2\mu_2\varepsilon_1\omega}{\mu_1\sigma_2}} + \cos\theta_I\right) = E_{OI}\left(\cos\theta_I - \cos\theta_T\dfrac{1-i}{2}\sqrt{\dfrac{2\mu_2\varepsilon_1\omega}{\mu_1\sigma_2}}\right)$

or $\quad \left(\dfrac{E_{OR}}{E_{OI}}\right)_P = \dfrac{\cos\theta_I - \cos\theta_T \dfrac{1-i}{2}\sqrt{\dfrac{2\mu_2\varepsilon_1\omega}{\mu_1\sigma_2}}}{\cos\theta_I + \cos\theta_T \dfrac{1-i}{2}\sqrt{\dfrac{2\mu_2\varepsilon_1\omega}{\mu_1\sigma_2}}}$

For a good conductor, the value of $\sqrt{\dfrac{\mu_1\sigma_2}{2\mu_2\varepsilon_1\omega}}$ is very large in comparison to $\cos\theta_I$. Hence,

we can neglect $\cos\theta_I$ in comparison to $\sqrt{\dfrac{\mu_1\sigma_2}{2\mu_2\varepsilon_1\omega}}$. Under this condition, the aforementioned equation becomes

$$\left(\dfrac{E_{OR}}{E_{OI}}\right)_P \approx -1 \qquad\qquad (6.165)$$

The subscript P in Eq. (6.165) or in the following equations refers to the fact that the electric field vector is parallel to the plane of incidence. The ratio of the refracted amplitude to the incident amplitude is obtained by putting the values of E_{OR} from Eq. (6.162) into Eq. (6.160). Thus, we have

$$E_{OI}\cos\theta_I - \cos\theta_I(1+i)\sqrt{\dfrac{\mu_1\sigma_2}{2\mu_2\varepsilon_1\omega}}\,E_{OT} - E_{OI}\cos\theta_I = E_{OT}\cos\theta_T$$

or $\quad E_{OI}\cos\theta_I + \cos\theta_I E_{OI} = E_{OT}\cos\theta_T + \cos\theta_I(1+i)\sqrt{\dfrac{\mu_1\sigma_2}{2\mu_2\varepsilon_1\omega}}E_{OT}$

or $\quad \left(\dfrac{E_{OT}}{E_{OI}}\right)_P = \dfrac{2\cos\theta_I}{\cos\theta_T + \cos\theta_I(1+i)\sqrt{\dfrac{\mu_1\sigma_2}{2\mu_2\varepsilon_1\omega}}}$

or $\quad \left(\dfrac{E_{OT}}{E_{OI}}\right)_P = \dfrac{2\cos\theta_I}{1 + \cos\theta_I(1+i)\sqrt{\dfrac{\mu_1\sigma_2}{2\mu_2\varepsilon_1\omega}}}$

For a good conducting medium, $\theta_T \approx 0$ and $\sqrt{\dfrac{\mu_1\sigma_2}{2\mu_2\varepsilon_1\omega}}$ is very large as a result of which the denominator of RHS of the aforementioned equation is very large in comparison to the numerator. Therefore, the equation becomes

$$\left(\dfrac{E_{OT}}{E_{OI}}\right)_P \approx 0 \qquad\qquad (6.166)$$

Equations (6.165) and (6.166) show that the reflected wave is approximately out of phase with the incident wave and the transmitted wave is nearly in phase with the incident wave. The attenuation of the refracted wave is very high, where as reflection is nearly 100%.

Reflection and transmission coefficients at the metallic surface

The reflection coefficient R_R is defined as the square of the ratio of the amplitude of reflected wave to the amplitude of the incident wave. Mathematically,

$$R_R = \left(\frac{E_{OR}}{E_{OI}} \right)^2 \tag{6.167}$$

The transmission coefficient or refraction coefficient R_T is defined as the square of the ratio of the amplitude of the transmitted wave to the amplitude of the incident wave. Mathematically,

$$R_T = \left(\frac{E_{OT}}{E_{OI}} \right)^2 \tag{6.168}$$

It can be checked for surety that the sum of the reflection coefficient and the transmission coefficient is unity both in Case 1 (Electric vector prependicular to the plane of incidence) and Case 2 (Electric vector parallel to the plane of incidence). R_{RN} and R_{TN} are the reflection and transmission coefficients respectively in the case where the electric vector of the electromagnetic wave is normal to the plane of incidence. R_{RP} and R_{TP} are the reflection and transmission coefficients respectively in the case where the electric vector of the electromagnetic wave is parallel to the plane of incidence. Reflection and transmission coefficients at the metallic surface are defined as follows.

$$R_{RN} = \left(\frac{E_{OR}}{E_{OI}} \right)^2_N$$

$$R_{RP} = \left(\frac{E_{OR}}{E_{OI}} \right)^2_P$$

$$R_{TN} = \left(\frac{E_{OT}}{E_{OI}} \right)^2_N$$

$$R_{TP} = \left(\frac{E_{OT}}{E_{OI}} \right)^2_P$$

Taking the help of Eq. (6.165), the value of R_{RP} for a good conducting surface is found out to be $R_{RP} \approx 1$. Also taking the help of Eq. (6.158), the value of R_{RN} for a good conducting surface is found out to be $R_{RN} \approx 1$. Since the values of R_{RP} and R_{RN} are very close to unity in this case, the entire energy incident will be reflected back to the first medium. The values of R_{TP} and R_{TN} from Eqs (6.166) and (6.159) also complements these facts. Therefore, metals are opaque for light. The extremely small amount of energy that flows into the metal is dissipated rapidly by heat loss associated with eddy currents. The skin depths of a good conducting media (explained in detail later) are very small implying that they absorb electromagnetic waves very strongly. Therefore, we can conclude that all good conductors are good reflectors and good absorbers.

Example 6.20

The electric component vector of a plane electromagnetic wave propagating in a non-magnetic medium is given by $\vec{E} = \hat{y}\,50\cos(10^8 t + 2z)$ V/m. Find the (a) direction of propagation, (b) frequency (c) wavelength, (d) relative permittivity and (e) magnetic component vector of the electromagnetic wave.

Solution

a. By combining the given electric component vector $\vec{E} = \hat{y}\,50\cos(10^8 t + 2z)$ V/m and the equations $\vec{k} \times \vec{E} = \mu\omega\vec{H}$ and $\vec{H} \times \vec{k} = \mu\varepsilon\vec{E}$, we conclude that the direction of propagation of the electromagnetic wave is along the negative Z-axis.

b. Comparing the given wave equation with the standard wave equation, we get $\omega = 10^8$ rad/s. Hence, the frequency of the electromagnetic wave will be obtained as

$$v = \frac{\omega}{2\pi} = \frac{10^8\,\text{rad/s}}{2\pi\,\text{rad}} = 15.92 \times 10^6/\text{s} = 15.92 \times 10^6\,\text{Hz}$$

c. The magnitude of the propagation vector k is given in general as $k = \dfrac{2\pi}{\lambda}$.

   However, comparing the given wave equation with the standard wave equation, we get

   $k = 2$

   Hence, we have

   $$\frac{2\pi}{\lambda} = 2$$

   or $\lambda = \dfrac{2\pi}{2}$ m $= 3.142$ m.

d. The speed of the electromagnetic wave is

   $$v = \lambda v = 3.142 \text{ m} \times 15.92 \times 10^6/\text{sv} = 50.02 \times 10^6\,\text{m/s}$$

It is given that the medium is non-magnetic. Hence, we have $\mu_r = 1$.
We know that $\mu = \mu_0 \mu_r$ and $\varepsilon = \varepsilon_0 \varepsilon_r$. The speed of the electromagnetic wave v is given by

$$v = \frac{1}{\sqrt{\varepsilon \mu}}$$

or $v^2 = \frac{1}{\varepsilon \mu} = \frac{1}{\varepsilon_0 \varepsilon_r \mu_0 \mu_r} = \frac{c^2}{\varepsilon_r \mu_r}$

In our case, $\mu_r = 1$. Hence, from the aforementioned equation, we have

$$\varepsilon_r = \frac{c^2}{v^2} = \frac{9 \times 10^{16}}{25.02 \times 10^{14}} = 35.97.$$

e.   From the equation $\vec{k} \times \vec{E} = \mu \omega \vec{H}$, we have

$$\vec{H} = \frac{\hat{x} k E \sin 90}{\mu \omega} = \frac{\hat{x} 2\pi E}{\lambda \mu \omega} = \frac{\hat{x} 2\pi 50 \cos(10^8 t + 2z)}{3.142 \times 4\pi \times 10^{-7} 10^8}$$

or $H = \hat{x} 0.7958 \sin(10^8 t + 2z) \text{A/m}.$

Example 6.21

A material is characterized by $\varepsilon_r = 3$, $\mu_r = 3$, and $\sigma = 8 \times 10^{-5}$ Siemen/m at a frequency of 2 MHz. Calculate the values of the (a) loss tangent, (b) attenuation constant and (c) phase constant.

Solution

a.   The loss tangent, $\tan 2\theta$, is given by

$$\tan 2\theta = \frac{\sigma}{2\pi v \varepsilon} = \frac{\sigma}{2\pi v \varepsilon_r \varepsilon_0}$$

Putting the given values into this equation, we get the loss tangent = 0.240.

b.   The attenuation constant α is given by

$$\alpha = \omega \sqrt{\frac{\mu \varepsilon}{2} \left\{ \sqrt{1 + \left(\frac{\sigma}{\omega \varepsilon}\right)^2} - 1 \right\}} = 0.150 \text{Neper/m}$$

c.   The phase constant β is given by $\beta = 0.127$ rad/m

Example 6.22

A good conducting medium is characterized by conductivity $\sigma = 58 \times 10^6$ Seimen/m, relative magnetic permeability $\mu_r = 1$ at frequency 10^8 Hz. Calculate the (a) intrinsic impedance, (b) propagation constant, (c) attenuation constant, (d) phase constant (f) skin depth and (g) wave speed.

Solution

It is given that the medium is a good conductor.

a.   The intrinsic impedance of a medium is given by

$$\eta = \sqrt{\frac{i\omega\mu}{\sigma + i\omega\varepsilon}}$$

We can neglect $\omega\varepsilon$ in comparison to σ as the medium is a good conductor. Hence, this expression becomes

$$\eta = \sqrt{\frac{i\omega\mu}{\sigma}}$$

The magnitude of η is given by [Using the formula $|z| = \sqrt{a^2 + b^2}\; e^{i\tan^{-1}\frac{b}{a}}$ when $z = a + ib$]

$$\eta = \sqrt{\frac{\omega\mu}{\sigma}} e^{i\frac{\pi}{4}}\; \Omega$$

Putting the given values of ω, μ, and σ into this equation, we get the value of the intrinsic impedance of the medium as

$$\eta = 3.69 \times 10^{-3} e^{i\frac{\pi}{4}}\; \Omega$$

b.   The propagation constant γ in a medium is given by

$$\gamma = \sqrt{i\omega\mu(\sigma + i\omega\varepsilon)}$$

We can neglect $\omega\varepsilon$ in comparison to σ as the medium is a good conductor. Hence, the aforementioned expression becomes

$$\gamma = \sqrt{i\omega\mu\sigma}$$

The magnitude of γ is given by

$$\gamma = \sqrt{\omega\mu\sigma}\, e^{i\frac{\pi}{4}}$$

Putting the given values of ω, μ, and σ into this equation, we get the value of propagation constant in the medium as

$$\gamma = 2.14 \times 10^5 e^{i\frac{\pi}{4}} \text{ m}^{-1}$$

c.   The attenuation constant α in a medium is given by

$$\alpha = \omega \sqrt{\frac{\mu\varepsilon}{2}\left\{\sqrt{1+\left(\frac{\sigma}{\omega\varepsilon}\right)^2}-1\right\}}$$

We can neglect 1 in comparison to $(\sigma/\omega\varepsilon)$ as the medium is a good conductor. Hence, the aforementioned expression becomes

$$\alpha = \omega\sqrt{\frac{\mu\varepsilon}{2}\times\frac{\sigma}{\omega\varepsilon}}$$

or $\alpha = \sqrt{\dfrac{\omega\mu\sigma}{2}}$

Putting the given values of ω, μ, and σ into this equation, we get the value of the attenuation constant of the medium as

$$\alpha = 1.51 \times 10^5 \text{ Neper/m}$$

d.   The phase constant β in a medium is given by

$$\beta = \omega\sqrt{\frac{\mu\varepsilon}{2}\left\{\sqrt{1+\left(\frac{\sigma}{\omega\varepsilon}\right)^2}+1\right\}}$$

We can neglect 1 in comparison to $(\sigma/\omega\varepsilon)$ as the medium is a good conductor. Hence, the aforementioned expression becomes

$$\beta = \omega\sqrt{\frac{\mu}{2}\frac{\sigma}{\omega}} = \sqrt{\frac{\omega\mu\sigma}{2}}$$

Putting the given values of ω, μ, and σ into this equation, we get the value of the phase constant of the medium as

$$\beta = 1.51 \times 10^5 \text{ rad/m}$$

e. The skin depth δ in a medium (explained in detail later) is given by

$$\delta = \frac{1}{\alpha}$$

Putting the calculated value of α into this equation, we get the value of skin depth in the medium as

$$\delta = 6.62 \; \mu m$$

f. The wave speed v in the medium is given by

$$v = \frac{\omega}{\beta}$$

Putting the calculated value of β and the given value of ω into this equation, we get the value of the wave speed in the medium as

$$v = 4.16 \times 10^{3} \, m/s$$

Example 6.23

In a partial conducting medium, characterized by $\varepsilon_r = 18.5$, $\mu_r = 800$, $\sigma = 1$ Siemen/m, the electric field component of an electromagnetic wave is given as $E_T = 50.0 e^{-\alpha z} \cos(10^9 t - \beta z)\hat{y}$ V/m. Calculate α, β, η, v, and H.

Solution

a. We know that for a partially conducting medium

$$\alpha = \omega \sqrt{\frac{\mu\varepsilon}{2}\left\{\sqrt{1+\left(\frac{\sigma}{\omega\varepsilon}\right)^2}-1\right\}}$$

Putting the values of $\mu = \mu_r \mu_0$, $\varepsilon = \varepsilon_r \varepsilon_0$, and $\omega = 2\pi v$ into this equation, we have

or $\alpha = 1132$ Neper

b. We know that for a partially conducting medium

$$\beta = \omega \sqrt{\frac{\mu\varepsilon}{2}\left\{\sqrt{1+\left(\frac{\sigma}{\omega\varepsilon}\right)^2}+1\right\}}$$

Putting the values of $\mu = \mu_r \mu_0$, $\varepsilon = \varepsilon_r \varepsilon_0$, and $\omega = 2\pi v$ into this equation, we have

$$\beta = 2790 \text{ rad/m}$$

c. The intrinsic impedance of a medium is given by

$$\eta = \sqrt{\frac{i\omega\mu}{\sigma + i\omega\varepsilon}}$$

or $\eta = \dfrac{1}{\sqrt{\sigma^2 + \omega^2 \varepsilon^2}} \left[(\omega^2 \mu\varepsilon)^2 + (\omega\mu\sigma)^2 \right]^{\frac{1}{4}} e^{i\frac{1}{2}\tan^{-1}\frac{\sigma}{\omega\varepsilon}}$

or $\eta = \dfrac{1}{\sqrt{\sigma^2 + 4\pi^2 v^2 \varepsilon^2}} \left[(4\pi^2 v^2 \mu\varepsilon)^2 + (2\pi v\mu\sigma)^2 \right]^{\frac{1}{4}} e^{i\frac{1}{2}\tan^{-1}\frac{\sigma}{\omega\varepsilon}}$

or $\eta = \dfrac{1}{\sqrt{1 + 4\pi^2 \times 0.0268}} \Big[(4\pi^2 \times 10^{18} \times 1.65 \times 10^{-13})^2 +$

$\qquad (2\pi \times 1.005 \times 10^6)^2 \Big]^{\frac{1}{4}} e^{i\frac{1}{2}\tan^{-1} 0.972} \, \Omega$

or $\eta = 2098.4 e^{i22.09} \Omega$

or $\eta = 2100 e^{i22.1} \Omega$

d. The speed of the electromagnetic wave in the given medium will be obtained from

$$v = \frac{1}{\sqrt{\mu\varepsilon}} = \frac{1}{\sqrt{\mu_r \mu_0 \varepsilon_r \varepsilon_0}} = 2.64 \times 10^6 \text{ m/s}$$

e. The magnetic component of the electromagnetic wave is obtained

$$H_T = \frac{50.0}{\eta} e^{-\alpha z} \cos(10^9 t - \beta z)(-\hat{x})$$

Putting the value of η into this equation, we get

$$H_T = -2.38 \times 10^{-2} e^{-\alpha z} \cos(10^9 t - 0.368 - \beta z)\hat{x}\,\text{A/m}$$

Skin depth

The electromagnetic wave equation for the electric field vector in a conducting medium is given by

$$\nabla^2 \vec{E} - \mu\varepsilon \frac{\partial^2 \vec{E}}{\partial t^2} - \mu\sigma \frac{\partial \vec{E}}{\partial t} = 0 \tag{6.169}$$

The origin of the term $\mu\varepsilon \dfrac{\partial^2 \vec{E}}{\partial t^2}$ lies in the displacement current while the origin of the term $\mu\sigma \dfrac{\partial \vec{E}}{\partial t}$ lies in the conduction current. In almost all conducting media, the conduction current dominates the displacement current. Therefore, for a good conducting medium, Eq. (6.169) can be written as

$$\nabla^2 \vec{E} - \mu\sigma \frac{\partial \vec{E}}{\partial t} = 0 \tag{6.170}$$

The solution of this equation may be given as

$$E(r,t) = E_0 e^{-\beta r} e^{i(\alpha r - \omega t)} \tag{6.171}$$

The propagation vector $\vec{k}$ in a conducting medium is of the form

$$\vec{k} = \alpha + i\beta \tag{6.172}$$

where $\alpha = \sqrt{\dfrac{\mu\sigma\omega}{2}}$ and $\beta = \sqrt{\dfrac{\mu\sigma\omega}{2}}$

Let $\delta = \dfrac{1}{\alpha} = \dfrac{1}{\beta} = \sqrt{\dfrac{2}{\mu\sigma\omega}}$.

Equation (6.171) becomes

$$E(r,t) = E_0 e^{-\frac{r}{\delta}} e^{i\left(\frac{r}{\delta} - \omega t\right)} \tag{6.173}$$

Equation (6.173) shows that the amplitude of the electromagnetic wave is $E_0 e^{-\frac{r}{\delta}}$. When $r = \delta$, the amplitude of the electromagnetic wave decreases in magnitude to $1/e$ times its value at the surface. Thus, the quantity δ is a measure of the distance of penetration of electromagnetic waves into the good conducting medium. The quantity δ is called skin depth. It is given by the expression

$$\delta = \sqrt{\frac{2}{\mu \sigma \omega}} \tag{6.174}$$

Equation (6.174) shows that the skin depth is a property of the medium and also depends upon the frequency of the incident wave. If the frequency of the medium increases, the skin depth δ decreases. The more is the skin depth, the less will be the attenuation of the electromagnetic waves in the medium and vice versa. The skin depth of seawater is relatively high for radio waves. This is the reason why radio communication with submarines under several metres of seawater is difficult.

Example 6.24

An electromagnetic wave is propagating in free space with magnetic vector $\vec{H}(z,t) = e^{i(\omega t + \beta z)} \hat{x}$ A/m. Obtain an expression for the electric vector.

Solution

The direction of propagation is along the $-Z$-axis as is obvious from the given equation. The amplitude of the magnetic component is 1 ampere/metre. The amplitude of the electric component will be

$$E_0 = \mu c H_0 = 4\pi \times 10^{-7} \times 3 \times 10^8 \times 1\,\text{V/m} = 120\pi\,\text{V/m}$$

The direction of the electric vector is along the $+Y$-axis as the direction of the magnetic vector is along the $+X$-axis and the direction of propagation is along the $-Z$-axis. The expression for the electric vector will be

$$\vec{E}(z,t) = E_0 e^{i(\omega t + \beta z)} \hat{y}$$

or $\vec{E}(z,t) = 120\pi e^{i(\omega t + \beta z)} \hat{y}$

Example 6.25

Aluminum is characterized by $\sigma = 38.2 \times 10^6$ Siemen/m and $\mu_r = 1$. Determine the skin depth for aluminum at the frequency 2.6 MHz. Also calculate the propagation constant and the wave speed in the given aluminum.

Solution

a. The expression for skin depth δ is given by

$$\delta = \sqrt{\frac{2}{\omega\sigma\mu}} = \sqrt{\frac{2}{2\pi\nu\sigma\mu}} = \frac{1}{\sqrt{\pi\nu\sigma\mu}}$$

Putting the given values into this equation, we get the skin depth as

$$\delta = \frac{1}{\sqrt{3.14 \times 2.6 \times 10^6 \times 38.2 \times 10^6 \times 4 \times 3.14 \times 10^{-7}}} \text{ m}$$

$$= 50.53 \times 10^{-6} \text{ m}$$

b. The propagation constant, in this case, is given by

$$\gamma = \frac{1}{\delta} + i\frac{1}{\delta} = 2.799 \times 10^4 \times e^{i45}$$

Therefore, the phase constant β is given by $\beta = \frac{1}{\delta}$
The speed of the wave is given by

$$v = \frac{\omega}{\beta} = 2\pi\nu\delta = 825.5 \text{ m/s}$$

Example 6.26

An electromagnetic wave propagating in air is allowed to fall incident normally on the surface of a dielectric with $\varepsilon_r = 4$. Calculate the reflection and transmission coefficients.

Solution

For normal incidence of the electromagnetic wave on a dielectric surface, the reflection and transmission coefficients are the same whether the electric vector of the electromagnetic wave is parallel or perpendicular to the dielectric surface. Therefore, for normal incidence, the reflection coefficient R_R and the transmission coefficient R_T are given by

$$R_R = \left(\frac{\eta_2 - \eta_1}{\eta_2 + \eta_1}\right)^2 \quad \text{and} \quad R_T = \frac{4\dfrac{\eta_2}{\eta_1}}{\left[\dfrac{\eta_2}{\eta_1} + 1\right]^2}$$

η_1 = intrinsic impedance of first (air) medium = $\sqrt{\dfrac{\mu_0}{\varepsilon_0}} = 120\pi \ \Omega$

η_2 = intrinsic impedance of second (dielectric) medium =

$$\sqrt{\frac{\mu}{\varepsilon}} = \sqrt{\frac{\mu_r \mu_0}{\varepsilon_r \varepsilon_0}} = 120\pi\Omega \sqrt{\frac{\mu_r}{\varepsilon_r}} = 40\pi\sqrt{3}\ \Omega$$

Putting the values of η_1 and η_2 into the expressions for reflection and transmission coefficients, we get

$$R_R = \left(\frac{\eta_2 - \eta_1}{\eta_2 + \eta_1}\right)^2 = \left(\frac{40\pi\sqrt{3} - 120\pi}{40\pi\sqrt{3} + 120\pi}\right)^2 = 7.18 \times 10^{-2}$$

$$R_T = \frac{4\dfrac{\eta_2}{\eta_1}}{\left[\dfrac{\eta_2}{\eta_1} + 1\right]^2} = 0.928$$

Total internal reflection

Total internal reflection is defined as the phenomenon of returning of electromagnetic waves to the same medium while passing from an electromagnetically denser medium to a rarer medium. Suppose the electromagnetic wave in the denser medium is incident on the interface separating the denser medium from the rarer medium. Then, according to Snell's law, we have

$$\frac{\sin\theta_I}{\sin\theta_T} = \frac{n_r}{n_d} \quad \text{with} \quad \theta_T > \theta_I$$

where n_d and n_r are the absolute refractive indices of the denser medium and rarer medium respectively. If the angle of incidence goes on increasing, the angle of refraction goes on increasing. For a particular angle of incidence, the angle of refraction becomes 90° in a rarer medium. Under this condition, Snell's law becomes

$$\sin\theta_C = \frac{n_r}{n_d}$$

Here θ_C is called the critical angle. The critical angle is defined as the angle of incidence in the denser medium for which the angle of refraction is $\pi/2$ in the rarer medium. Let us see

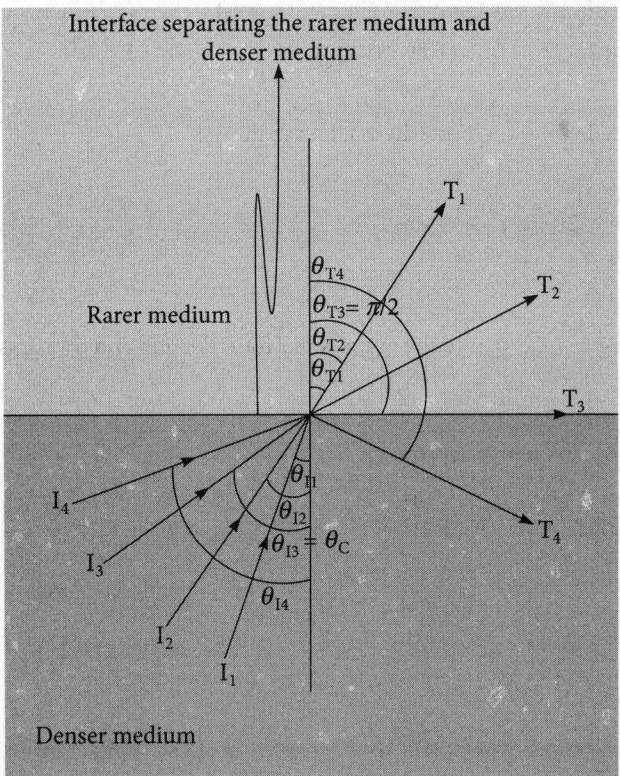

Figure 6.10 | Total internal reflection occurs when $\theta_i > \theta_C$. I_1, I_2, I_3, and I_4 are the incident waves whereas T_1, T_2, T_3, and T_4 are the corresponding refracted or transmitted waves. θ_{I1}, θ_{I2}, θ_{I3} and θ_{I4} are the angles of incidence whereas θ_{T1}, θ_{T2}, θ_{T3}, and θ_{T4} are the corresponding angles of refraction or angle of transmission. The X-axis is the interface between the denser medium and the rarer medium. The normal at the point of incidence is along the Y-axis.

what happens if the angle of incidence is more than that of the angle of refraction. When the angle of incidence in the denser medium goes on increasing, the angle of refraction/transmission increases up to $\pi/2$ in the rarer medium. With still further increase of the angle of incidence in the denser medium, the angle of refraction becomes more than $\pi/2$ and is reflected back to the denser medium obeying the laws of reflection. See Fig. 6.10. The phenomenon of total internal reflection is used when it is required to transmit electromagnetic radiation without loss in its intensity.

Example 6.27

An electromagnetic wave falls on top of a horizontal surface of a glass cuboid at an angle of incidence 45°. What must be the refractive index of the glass if total internal reflection occurs at a vertical surface

Solution

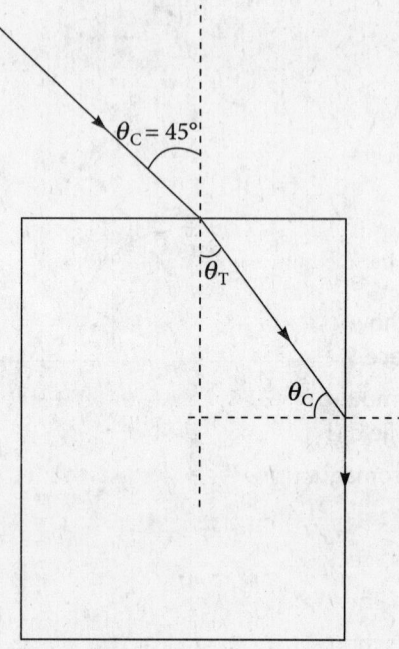

Figure 6.11 | Refraction of electromagnetic wave through a glass cuboid at an angle of incidence 45°.

Applying Snell's law to the top horizontal surface, we have

$$\frac{\sin 45}{\sin \theta_T} = n \qquad\qquad (A)$$

Total internal reflection occurs at a vertical surface. Hence, angle of incidence on the vertical surface is the critical angle. Applying Eq. (6.173) to this surface, we get

$$n = \frac{1}{\sin \theta_C} = \frac{1}{\sin(90 - \theta_T)} = \frac{1}{\cos \theta_T}$$

$$\text{or } \cos^2 \theta_T = \frac{1}{n^2} \qquad\qquad (B)$$

From Eq. (A), we get

$$\sin^2 \theta_T = \frac{1}{2n^2}$$

$$\text{or } \cos^2 \theta_T = 1 - \frac{1}{2n^2}$$

Putting this equation into Eq. (B), we get

$$\frac{1}{n^2} = 1 - \frac{1}{2n^2}$$

or $n = 1.22$

Questions

6.1 Explain in detail how $\oint_s \vec{E} \times \vec{H} \cdot \vec{ds}$ represents the power flow into/out of the volume through the surface S.

6.2 Prove that the amount of energy transferred per unit time per unit volume of the electromagnetic field is $\vec{E} \cdot \vec{J}$.

6.3 What is the electromagnetic energy density in a medium?

6.4 What is the electromagnetic energy density in free space?

6.5 What is Poynting's vector?

6.6 State and explain Poynting's theorem.

6.7 Obtain Poynting's theorem for the conservation of energy in an electromagnetic field and discuss the physical meanings of each term in the resulting equation.

6.8 Establish the law of conservation of energy for electromagnetic fields.

6.9 What does Poynting's vector represent? Explain.

6.10 Prove that the time-average value of Poynting's vector $< \vec{P} >$ is $\frac{1}{2} \text{Re} \left(\vec{E} \times \vec{H}^* \right)$, where $\vec{H}^*$ is the complex conjugate of $\vec{H}$.

6.11 Prove that the average value of magnitude of Poynting's vector is $< P > = \frac{1}{2} \sqrt{\frac{\mu}{\varepsilon}} \times H_0^2$

6.12 Dimensionally prove that $\vec{E} \times \vec{H}$ has the dimension of power per unit area.

6.13 Define scalar potential.

6.14 Define vector potential.

6.15 Express electric field in terms of vector and scalar potential.

6.16 Express magnetic field in terms of vector and scalar potential.

6.17 What is the difference between electric potential and magnetic scalar potential?

6.18 What is Laplace's equation?

6.19 What is Poisson's equation?

6.20 What is the Lorentz gauge condition?

6.21 Derive the electromagnetic wave equation in terms of electric vector $\vec{E}$.

6.22 Derive the electromagnetic wave equation in terms of magnetic vector $\vec{H}$.

6.23 Derive the electromagnetic wave equation in terms of electric vector $\vec{E}$ in vacuum.

6.24 Derive the electromagnetic wave equation in terms of magnetic vector $\vec{H}$ in vacuum.

6.25 Derive the electromagnetic wave equation in terms of magnetic induction, vector $\vec{B}$.

6.26   Derive the electromagnetic wave equation in terms of electric displacement, vector $\vec{D}$.

6.27   Derive a relation between vector potential and scalar potential.

6.28   Derive the electromagnetic wave equation in terms of vector potential, $\vec{A}$.

6.29   Derive the electromagnetic wave equation in terms of scalar potential, φ.

6.30   What do you mean by a plane electromagnetic wave?

6.31   What do you mean by propagation vector $\vec{k}$? What is its magnitude?

6.32   Prove that the propagation constant in a medium is complex.

6.33   What is the relation between the three vectors $\vec{E}$, $\vec{H}$, and $\vec{k}$?

6.34   Find the relation between electric and magnetic components of an electromagnetic wave if the wave is propagating in a conducting medium along the positive Z-axis in terms of intrinsic impedance of the medium.

$$[\text{Ans}\quad \text{If } \vec{E} = E_0 e^{-\gamma z}\hat{x}, \text{ then } \vec{H} = \sqrt{\frac{\sigma + i\omega\varepsilon}{i\omega\mu}} E_0 e^{-\gamma z}\hat{y}\,]$$

6.35   Prove that vector $\vec{E}$ and propagation vector $\vec{k}$ are perpendicular to each other.

6.36   Prove that vector $\vec{H}$ and propagation vector $\vec{k}$ are perpendicular to each other.

6.37   Prove that vector $\vec{E}$ and vector $\vec{H}$ are perpendicular to each other.

6.38   Prove that vector $\vec{E}$, vector $\vec{H}$ and vector $\vec{k}$ are mutually perpendicular to each other.

6.39   How do you conclude that an electromagnetic wave is transverse?

6.40   Derive a relation between amplitudes of vector $\vec{E}$ and vector $\vec{H}$.

6.41   What do you mean by intrinsic impedance of a medium? Derive its unit.

6.42   Express intrinsic impedance in terms of the magnitude of vector $\vec{E}$ and vector $\vec{H}$.

6.43   Prove that the speed v of an electromagnetic wave is $v = \dfrac{1}{\sqrt{\varepsilon\mu}}$.

6.44   Prove that the refractive index of a dielectric medium is equal to the square root of the dielectric constant.

6.45   Prove that for a dielectric medium, the propagation vector is real.

6.46   Prove that for any medium, the propagation constant is not real.

6.47   Prove that for a good conducting medium, the propagation vector and the propagation constant are the same.

6.48   Prove that intrinsic impedance of a medium is complex.

6.49   Prove that for a linear dielectric medium, the refractive index and intrinsic impedance are inversely proportional to each other.

6.50   Prove that $B = \dfrac{E}{v}$, where v is the speed of the electromagnetic wave. Other symbols have their usual meanings.

6.51   What result leads you to believe that light is an electromagnetic wave.

6.52   Express refractive index of a medium in terms of relative permeability and dielectric constant.

6.53 Prove that the time-average value of Poynting's vector is half the product of the amplitude of the electric vector E_0 and the amplitude of the magnetic vector H_0.

6.54 Prove that the time-average value of Poynting's vector is the product of the root-mean-square value of the electric vector and the root-mean-square value of the magnetic vector.

6.55 Prove that the time-average value of Poynting's vector is $\dfrac{1}{2}\sqrt{\dfrac{\varepsilon}{\mu}}E_O^2\ \hat{k}$.

6.56 How are the time-average value of Poynting's vector and electromagnetic energy density related to each other?

6.57 State and explain Poynting's theorem.

6.58 Explain how Poynting's theorem is the law of conservation of electromagnetic energy

6.59 Prove that energy flows in an electromagnetic wave with the same speed and same direction of the electromagnetic wave.

6.60 What is plasma angular frequency? Derive an expression for it.

6.61 Derive the electromagnetic wave in an ionized medium in terms of the electric vector $\vec{E}$.

6.62 Derive the electromagnetic wave in an ionized medium in terms of the magnetic vector $\overline{H}$.

6.63 Prove that plasma frequency is $9\sqrt{n}$, where n is the electron density in the plasma medium.

6.64 What is propagation constant? Derive an expression for it in terms of plasma angular frequency.

6.65 Discuss the electromagnetic wave propagation in an ionized medium at high frequency $\omega > \omega_p$.

6.66 Discuss the electromagnetic wave propagation in an ionized medium at very high frequency $\omega \gg \omega_p$.

6.67 Discuss the electromagnetic wave propagation in an ionized medium at low frequency $\omega < \omega_p$.

6.68 What is cut-off plasma frequency? Derive an expression for it.

6.69 What are field vectors of an electromagnetic field?

6.70 What are the boundary conditions for field vectors?

6.71 What is plane of incidence?

6.72 What is plane of reflection?

6.73 What is plane of refraction?

6.74 What is plane of transmission?

6.75 Prove the first and second law of reflection in case of an electromagnetic wave.

6.76 Prove the first and second law of refraction in case of an electromagnetic wave.

6.77 Prove that the frequency of an electromagnetic wave does not change during reflection.

6.78 Prove that the frequency of an electromagnetic wave does not change during refraction.

6.79 Derive the relation between propagation constant, speed and angular frequency of an electromagnetic wave in a non-conducting medium.

6.80 Derive the relation between propagation constant, speed and angular frequency of an electromagnetic wave in a conducting medium.

6.81 Derive an expression for reflection coefficient when an electromagnetic wave is incident on a non-conducting plane surface with its electric vector tangential to the surface.

6.82 Derive an expression for reflection coefficient when an electromagnetic wave is incident on a non-conducting plane surface with its electric vector perpendicular to the surface.

6.83 Derive an expression for the magnitude of propagation vector in a non-conducting medium.

6.84 Derive an expression for the magnitude of propagation vector in a conducting/metallic medium.

6.85 Derive an expression for reflection coefficient when an electromagnetic wave is incident on a dielectric surface at polarizing/Brewster angle. What conclusions are drawn from this equation?

6.86 Define plane of polarization and plane of vibration?

6.87 Define degree of polarization in terms of reflection coefficients.

6.88 Derive an expression for reflection coefficient when an electromagnetic wave is incident on a conducting plane surface with its electric vector tangential to the surface.

6.89 Derive an expression for reflection coefficient when an electromagnetic wave is incident on a conducting plane surface with its electric vector perpendicular to the surface.

6.90 Derive an expression for transmission coefficient when an electromagnetic wave is incident on a non-conducting plane surface with its electric vector tangential to the surface.

6.91 Derive an expression for transmission coefficient when an electromagnetic wave is incident on a non-conducting plane surface with its electric vector perpendicular to the surface.

6.92 Derive an expression for transmission coefficient when an electromagnetic wave is incident on a conducting plane surface with its electric vector tangential to the surface.

6.93 Derive an expression for transmission coefficient when an electromagnetic wave is incident on a conducting plane surface with its electric vector perpendicular to the surface.

6.94 Show that the reflected electromagnetic wave is in phase with the incident wave when reflection occurs from a dielectric surface backed by a rarer medium.

6.95 Show that the reflected electromagnetic wave is out of phase with the incident wave when reflection occurs from a dielectric surface backed by a denser medium.

6.96 Show that the refracted electromagnetic wave is in phase with the incident wave when refraction occurs from a dielectric rarer medium to a denser medium.

6.97 Show that the refracted electromagnetic wave is in phase with the incident wave when refraction occurs from a dielectric denser medium to a rarer medium.

6.98 Show that the electromagnetic wave propagates in the conducting medium along the normal whatever may be the angle of incidence.

6.99 Show that the reflected electromagnetic wave is out of phase with the incident wave when reflection occurs from a metallic surface.

6.100 Show that the metallic reflection of an electromagnetic wave is nearly 100%.

6.101 Show that the refracted electromagnetic wave is in phase with the incident wave when refraction occurs on a metallic surface.

6.102 Describe the peculiarities of metallic reflection with the theory behind it.

6.103 How does the magnitude of Poynting's vector change when a plane electromagnetic wave is refracted?

6.104 How does the magnitude of Poynting's vector change when a plane electromagnetic wave is reflected?

6.105 From the transmission coefficient of an electromagnetic wave in a conducting medium, show that attenuation of electromagnetic wave is very high in a conducting medium.

6.106 What do you mean by skin depth? On what factors does it depend?

6.107 Derive an expression for skin depth.

6.108 Explain why for a high frequency electromagnetic wave, the material becomes transparent.

6.109 Explain why radio communication with submarines under several metres of sea is difficult.

6.110 Explain why electromagnetic wave can easily penetrate into a dielectric medium but cannot do so in a conducting medium.

6.111 Maxwell's equations are the backbone of electromagnetic wave propagation. Justify the statement.

6.112 Explain: Good conductors are good reflectors of electromagnetic waves.

Problems

6.1 A material is characterized by $\varepsilon_r = 3$, $\mu_r = 1.5$, and $\sigma = 4 \times 10^{-5}$ Siemen/m at a frequency of 2 MHz. Calculate the values of (a) loss tangent, (b) attenuation constant, and (c) phase constant. [Ans (a) 0.120, (b) 5.324×10^{-3} Neper/m, (c) 8.91×10^{-2} rad/m]

6.2   In free space, the magnetic field vector is given as $H = \hat{y}\, 0.1 \times \cos(3 \times 10^8 t - \beta x)$ A/m. Calculate the value of β and the time it takes for the wave to travel a distance of $\lambda/2$. Determine the direction of wave propagation. Also calculate the wavelength.

[Ans $\beta = 1.0$ rad/s, 10.47×10^{-9} s along the positive X-axis, 6.28 m]

6.3   The electric component vector of a plane electromagnetic wave propagating in a non-magnetic medium is given by $\vec{E} = \hat{y}40\cos(10^8 t - 4z)$V/m. Find the (a) direction of propagation, (b) frequency, (c) wavelength, (d) relative permittivity and (e) magnetic component vector of the electromagnetic wave. [Ans   Along the positive Z-axis, 15.92×10^6 Hz, 1.571 metres, 143.88, $\vec{H} = \hat{x}11.12 \times 10^{-9} \sin(10^8 t + 2z)$A/m]

6.4   Calculate the value of Poynting's vector on the surface of a 200 watts spherical electric bulb having a radius of 5 cm.   [Ans   6366.20 W/m²]

6.5   The average distance between the sun and the earth is 1.5×10^{11} m. The power radiated by the sun is 3.8×10^{26} W. Show that the average solar energy incident on the earth is (2 Calorie/cm² minute) [This is called a solar constant].

6.6   The earth receives energy from the sun at the rate of (2 Calorie/cm² minute). Calculate the amplitudes of electric and magnetic field of radiation on the earth's surface.

[Ans   1027 V/m, 2.73 A/m]

6.7   A 500 watt electric bulb is glowing with full capacity illuminating the space around it uniformly. Calculate the amplitudes of the electric and magnetic field of radiation at a distance of 2 m.   [Ans   194.64 V/m, 0.5165 A/m]

6.8   A plane electromagnetic wave is travelling in a lossless dielectric medium having dielectric constant 3 and relative magnetic permeability 1. If amplitude of the electric field is 6 V/m, calculate the (a) speed of the wave, (b) intrinsic impedance of the dielectric, (c) amplitude of the magnetic field, (d) maximum value of Poynting's vector.

[Ans   (a) 1.732×10^8 m/s, (b) 217.56 Ω, (c) 0.2887A/m, (d) 0.165 W/m²]

6.9   For a dielectric medium $\sigma = 0$, $\mu_r = 10$, and $\varepsilon_r = 2.5$. Examine whether the pair of fields $\vec{E} = 100\sin 6 \times 10^7 t \sin z\hat{y}$ and $\vec{H} = -0.1328\cos 6 \times 10^7 t \cos z\hat{x}$ satisfy Maxwell's equation.   [Ans   Yes]

6.10   The electric field intensity of a uniform plane wave in air is given by $\vec{E} = 900\cos(\omega t - kz)\hat{x}$. If the wavelength of the wave is 2 m, find (a) frequency, (b) time period, (c), value of k (d) amplitude of magnetic vector H, and (e) direction of propagation.

[Ans   (a) 1.5×10^8 Hz, (b) 6.67 nanosecond, (c) π_m, (d) 2.40 A/m, (e) +Z-direction]

6.11   The electromagnetic wave is propagating in free space with electric vector $\vec{E}(z,t) = 60\cos(\omega t - kz)\hat{x}$. How much average energy crosses a circular area of radius 3 m on the X-Y plane in unit time.   [Ans   134.80 W]

6.12   The electromagnetic wave is propagating in free space with magnetic vector $\vec{H}(z,t) = 0.133\cos(4 \times 10^7 t - kz)\hat{x}$ A/m. Obtain an expression for the electric vector. Obtain an expression for E(z, t) along with the value of k.

[Ans $\vec{E}(z,t) = 50\cos(4 \times 10^7 t - kz)\hat{x}$ V/m, 0.133 rad/m]

6.13 Find the magnitude and direction of $\vec{H}(z,t) = 3\sin(\omega t - kz)\hat{x} - 4\sin(\omega t - kz)\hat{y}$ A/m at $t = 0$ and $k = \dfrac{3\lambda}{4}$.　　　　　　　　　　　　　　[Ans　5.0 A/m, $0.6\hat{x} - 0.8\hat{y}$]

6.14 The electric field component vector of a plane electromagnetic wave propagating in a non-magnetic medium is given by $E = \hat{y}\, a\, \cos(\omega t + kz)$ V/m. What are the direction of propagation and the magnetic field vector direction?　　　　[Ans　$-Z, +X$]

6.15 Aluminum is characterized by $\sigma = 38.2 \times 10^6$ Siemen/m and $\mu_r = 1$. Determine the skin depth for aluminum at the frequency 1.6 MHz. Also calculate the propagation constant and wave speed in the given aluminum.

　　　　　　　　　　[Ans　64.4 μm, $2.20 \times 10^4\, e^{i45}$/m, 647.42 m/s]

6.16 A good conducting medium is characterized by conductivity $\sigma = 68 \times 10^6$ Seimen/m, relative magnetic permeability $\mu_r = 2.5$ at frequency 10^8 Hz. Calculate the (a) intrinsic impedance, (b) propagation constant, (c) attenuation constant, (d) phase constant (f) skin depth, and (f) wave speed.

　　[Ans　(a) $\eta = 5.39 \times 10^{-3} e^{i\frac{\pi}{4}}\Omega$ (b) $\gamma = 1.59 \times 10^5 e^{i\frac{\pi}{4}} m^{-1}$ (c) $\alpha = 2.59 \times 10^5$ Neper/m
　　　　　(d) $\beta = 2.59 \times 10^5$ rad/m (e) $\delta = 3.86\,\mu m$ (f) $v = 2.43 \times 10^3$ m/s]

6.17 Calculate the frequency of an electromagnetic wave so that the electromagnetic wave will penetrate a distance of 2 mm into the surface of silver. Given for silver $\sigma = 3.0 \times 10^6$ Seimen/m.　　　　　　　　　　　　　[Ans　21.11 kHz]

6.18 The phase constant of a certain good conducting material at a certain frequency is 3.75×10^5 rad/m. Calculate the skin depth.　　　　　[Ans　2.67 μm]

6.19 The electromagnetic wave propagating in air is incident normally on the surface of a dielectric with $\varepsilon_r = 81$ and $\mu_r = 1$. Calculate the reflection and transmission coefficients.　　　　　　　　　　　　　　[Ans　0.64, 0.36]

6.20 An electromagnetic wave of frequency 10^8 Hz propagating in air is incident on a copper surface having electrical conductivity $\sigma = 58 \times 10^6$ Siemen/m. Calculate the depth of the point in the copper metal so that its amplitude becomes e^{-1} times its value on the surface.　　　　　　　　　　　　[Ans　6.61 μm]

6.21 The amount of electromagnetic energy received by the earth in the form of light from a star is 300 Watt/m². Calculate the root-mean-square value of the electric vector and the magnetic induction vector of the light wave on the earth's surface.

　　　　　　　　　　　　[Ans　336 V/m, 1.12×10^{-6} Tesla]

6.22 The fraction of energy transmitted and the fraction of energy reflected through the interface of two dielectric media in case of normal incidence are equal. Calculate the ratio of refractive indices of the two media. [Hints $R_T = R_R$].　　[Ans　$\dfrac{n_1}{n_2} = 3 + 2\sqrt{2}$]

6.23 A plane electromagnetic wave with electric vector $E_I = 30\cos(\omega t - 4z)\hat{y} V/m$ propagating in air medium is incident normally on the surface of a lossless dielectric medium characterized by $\varepsilon_r = 4$ and $\mu_r = 1$. Find the (a) λ, (b) ω, (c) $\vec{E}_T$ (transmitted electric vector), and (d) H_I (incident magnetic vector).

　　　　　　[Ans　1.57 m, 12×10^8 rad/s, $20\cos(\omega t - 4z)\hat{y}$ V/m,
　　　　　　　　　　　　$56.34 \times 10^5 \cos(\omega t - 4z)\hat{x}$ A/m]

6.24   In a partial conducting medium characterized by $\varepsilon_r = 8.5$, $\mu_r = 80$, $\sigma = 2$ Siemen/m, the electric field component of an electromagnetic wave is given as $E_T = 100.0e^{-\alpha z}\cos(10^9 t - \beta z)\hat{y}$ V/m. Calculate α, β, η, v, and $\vec{H}$.

[Ans   707 Neper, 893 rad/s, $555e^{i38.35}\Omega$, 1.15×10^7 m/s,
$$H_T = -18.02 \times 10^{-2} e^{-\alpha z}\cos(10^9 t - 1.34 - \beta z)\hat{x} \text{ A/m}]$$

6.25   A current distribution gives rise to magnetic vector potential $\vec{A} = x^2 y\hat{x} + y^2 x\hat{y} + 4xyz\hat{z}$. Calculate the magnetic induction at the point $(-1, 2 -3)$.   [Ans $12\hat{x} + 24\hat{y} + 3\hat{z}$]

6.26   A medium is characterized by relative permitivitty $\varepsilon_r = 45$ and relative permeability $\mu_r = 5$. Calculate the speed of the electromagnetic wave in the medium and the refractive index of the medium.   [Ans 2×10^7 m/s, 15]

6.27   In the photosphere of a star if there are 8.85×10^{14} electrons per unit volume, then calculate the cut-off frequency in the photosphere.   [Ans 2.7×10^8 Hz]

6.28   In a plasma medium, there are 10^{12} electrons per unit volume. Calculate the plasma frequency of the medium.   [Ans 8.97×10^6 Hz]

6.29   An electromagnetic wave with the amplitude of the electric vector as 2.5×10^{-3} volt/metre propagating in a dielectric medium having $\varepsilon_r = 6.5$ and $\mu_r = 1$ is incident normally on the interface between the dielectric medium and the air medium and simultaneously gets reflected and refracted. Calculate the amplitudes of the electric and magnetic vectors of the reflected electromagnetic wave and the refracted electromagnetic wave. Also calculate the amplitude of the incident magnetic vector.

[Ans $E_{OR} = 1.09 \times 10^{-3}$ V/m, $E_{OT} = 3.59 \times 10^{-3}$ V/m, $H_{OR} = -7.37 \times 10^{-6}$ A/m,
$$H_{OT} = 9.53 \times 10^{-6} \text{ A/m}, H_{OI} = 1.69 \times 10^{-5} \text{ A/m}]$$

6.30   An electromagnetic wave with the amplitude of the electric vector as 50 volt/meter and frequency 15 MHZ propagating in free space is incident normally on the plane conducting surface characterized by $\varepsilon_r = 1$, $\mu_r = 1$, and $\sigma = 81 \times 10^6$ Siemen/m and simultaneously gets reflected and refracted. Calculate the amplitudes of the electric vector of the reflected and refracted electromagnetic wave.

[Ans -50 V/m, $3.21 \times 10^{-4} e^{-i1}$ V/m]

6.31   Calculate the degree of polarization for ordinary light incident at an angle $57.5°$ on the glass having refractive index 1.5.   [Ans   99.8%]

6.32   Calculate the critical angle and Brewster's angle for an electromagnetic wave passing through a non-magnetic glass of relative permitivitty 9.   [Ans $19.47°, 71.56°$]

6.33   Calculate the critical angle and Brewster's angle for an electromagnetic wave passing through water of relative permitivitty 81.   [Ans $6.38°, 83.66°$]

Multiple Choice Questions

1.   Which of the following is not an electromagnetic wave?

   (i)   seismic wave

   (ii)   microwave

(iii) light wave

(iv) X-rays

(v) ultrasonic wave

(vi) gamma rays

2. Which of the following is/are not the properties of electromagnetic waves?

(i) Electromagnetic waves belong to the transverse wave category

(ii) Energy is carried by electromagnetic waves

(iii) Information is carried by electromagnetic waves

(iv) Electrons play a major role in propagation of electromagnetic waves

(v) Accelerating charges can produce electromagnetic waves

(vi) Electromagnetic waves cannot be propagated through material medium.

3. Which of the following does not mean Poynting's vector?

(i) the amount of energy flowing per second

(ii) the amount of energy flowing per second per unit area

(iii) the amount of power flowing per second per unit area

(iv) the amount of power flowing per unit area

(v) the amount of energy flowing per second per unit area

4. The conductivity of a medium for electromagnetic waves is independent of the frequency of the electromagnetic wave

(i) true (ii) false

5. The scalar potential and vector potential are completely independent of each other.

(i) yes (ii) no

6. Any function which satisfies the electromagnetic wave equation for electric and magnetic vectors must satisfy Maxwell's electromagnetic equations.

(i) yes necessarily (ii) not necessarily

7. Which of the following is/are not the properties of electromagnetic waves?

(i) Phase of a plane electromagnetic wave is constant throughout the propagation.

(ii) Wavefront of a plane electromagnetic wave is plane

(iii) Propagation vector is perpendicular to the wavefront

(iv) The plane electromagnetic wave does not have a single frequency

8. Which of the following equations are not correct for an electromagnetic wave?

(i) $H \times k = \varepsilon \omega E$

(ii) $k \times E = \mu \omega H$

(iii) $\nabla \times E = 0$

(iv) $\dfrac{E_0}{H_0} = \dfrac{\mu \omega}{k}$

(v) $\dfrac{E_0}{H_0} = \dfrac{k}{\varepsilon\omega}$

9. Plasma state of matter in undisturbed condition is neutral.

   (i) true (ii) false

10. Which of the following is not a form of the wave $E_x = \cos(\omega t - \beta z)$

   (i) $\cos(\beta z - \omega t)$

   (ii) $\sin\left(\beta z - \omega t - \dfrac{\pi}{2}\right)$

   (iii) $\cos\dfrac{2\pi t}{T} - \dfrac{2\pi z}{\lambda}$

   (iv) $\mathrm{Re}(e^{i(\omega t - \beta z)})$

   (v) $\cos\beta(z - vt)$

11. Identify which of the following functions do not satisfy the wave equation

   (i) $50e^{i(\omega t - 3z)}$

   (ii) $\sin\omega(10z + 5t)$

   (iii) $(x + 2t)^2$

   (iv) $\cos^2(y + 5t)$

   (v) $\sin x \cos t$

   (vi) $\cos(5y + 2x)$

12. Which of the following statements is not true of waves in general

   (i) It may be a function of time only

   (ii) It may be sinusoidal or co-sinusoidal

   (iii) It must be a function of time and space

   (iv) For practical reasons, it must be of finite extent

13. The electric field component of a wave in free space is given by $E = \hat{y}10\cos(10^7 t + kz)$ V/m. It can be inferred that

   (i) the wave propagates along the positive y-axis

   (ii) the wavelength = 188.5 m

   (iii) the wave amplitude = 10 V/m

   (iv) the wave number = 0.33 rad/m

   (v) the wave attenuates as it travels

14. What is the major factor for determining whether a medium is free space, lossless dielectric, lossy dielectric or a good conductor?

    (i) attenuation constants   (ii) constitutive parameters $(\sigma, \mu, \varepsilon)$

    (iii) loss tangent   (iv) reflection coefficient

15. In a certain medium, $E = \hat{x}\, 10 \cos(10^8 t - 3y)$ V/m. What type of medium is it?

    (i) free space   (ii) perfect dielectric

    (iii) lossless dielectric   (iv) conductor

16. Electromagnetic waves travel faster in conductors than in dielectrics.

    (i) true   (ii) false

17. In good conductors, E and H are in time phase.

    (i) true   (ii) false

18. Electromagnetic waves of any frequency can propagate through the ionosphere.

    (i) yes   (ii) no

19. Does scalar potential has any meaning in a region containing current?

    (i) yes   (ii) no

20. What is the shape of a wave front in case of plane electromagnetic waves?

    (i) spherical   (ii) cylindrical

    (iii) plane   (iv) cannot be said

21. Plasma in undisturbed condition is neutral.

    (i) correct   (ii) wrong

22. Out of the following in Fig. 6.12, which is the correct representation of vector $\vec{E}$ and vector $\vec{H}$ and vector $\vec{k}$?

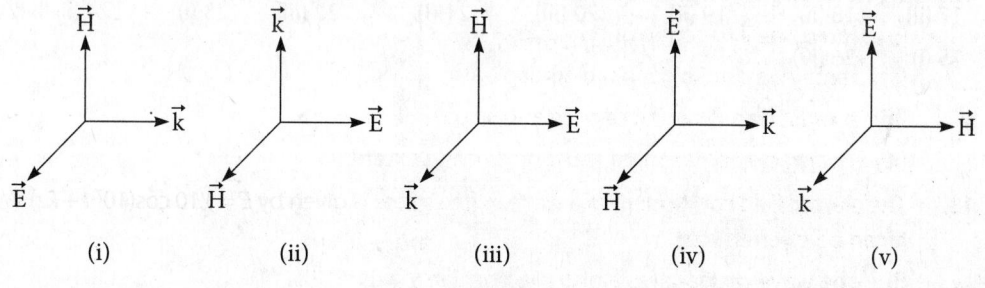

(i)   (ii)   (iii)   (iv)   (v)

Figure 6.12

23. In an ionized medium, the current density vector lags behind the electric field vector by a phase difference of $\dfrac{\pi}{2}$.

    (i) yes   (ii) no

24.   Out of the following, which are the incorrect representations of Poynting's vector $\vec{P}$?

(i) $\vec{P} = \dfrac{1}{\mu}(\vec{E} \times \vec{B})$

(ii) $\vec{P} = \dfrac{1}{\mu}(\vec{E} \times \vec{H})$

(iii) $\vec{P} = (\vec{E} \times \vec{B})$

(iv) $\vec{P} = (\vec{E} \times \vec{H})$

(v) $\vec{P} = (\vec{H} \times \vec{E})$

25.   The intrinsic impedance of vacuum is defined as the product of the speed of light in vacuum and its magnetic permeability.

(i)   yes   (ii)   no

26.   An electromagnetic wave is propagating in free space with electric vector $E(z, t) = 150 \cos(\omega t - kz) = \hat{x}$ V/m. How much average energy is passing through a rectangular hole of length 3 cm and width 1.5 cm on the Y-Z or X-Z plane in one minute.

(i)   41.67 J   (ii)   0.81 J

(iii) 1.8 J   (iv) 0.0 J

Answers

1 (i & v)	2 (iv & vi)	3 (i & iii)	4 (i)	5 (ii)	6 (ii)	7 (iv)	8 (iii)
9 (i)	10 (ii)	11 (iv & vi)	12 (i)	13 (ii & iii)	14 (iii)	15 (iii)	16 (ii)
17 (ii)	18 (ii)	19 (ii)	20 (iii)	21 (i)	22 (iii)	23 (i)	24 (ii, iii & v)
25 (i)	26 (iv)						

7 Elementary Concepts of Quantum Physics

7.1 Introduction

It is correctly told that mathematics is the queen of all sciences; in the same spirit, quantum physics or quantum mechanics may be called the king of all sciences. Our knowledge in any field of science is incomplete as long as we remain unacquainted with quantum physics. The concepts of quantum physics form the basis for our present understanding of physical phenomena on an atomic and microscopic scale. The concepts of quantum physics can be applied to most fields of science and engineering starting from biology to quantum computers to cosmology. Within engineering, important subjects of practical significance include semiconductor transistors, lasers, quantum optics, and molecular devices where quantum physics plays the most vital role. As technology advances, quantum concepts give birth to an increasing number of new electronic and opto-electronic devices. Their fabrications and functions can only be understood by using quantum physics. Within the next few years, fundamentally quantum devices such as single-electron memory cells and photonic signal processing systems may be available commercially. As nano-and atomic-scale devices become easier to manufacture, these sophisticated manufacturing units will require an increasing number of individuals with sound knowledge of quantum physics. Therefore, all universities in the world have included quantum physics as a subject in their technical course curricula. Quantum physics is no longer a theoretical subject with mathematical complexities but an engineering subject!

7.2 Need for Quantum Physics

Two time-tested proverbs are, 'Failure is the pillar of success' and 'Necessity is the mother of invention'. Classical physics based on Newtonian laws, thermodynamical laws and classical laws of electromagnetism explained successfully the macroscopic world. The macroscopic world is directly observable or can be made observable by relatively simple devices. However, classical physics failed seriously in explaining the phenomena in the realm of atoms, nucleons and elementary particles. These failures gave birth to a new branch in physics called quantum physics. In the following, we mention a few examples of the failures of classical concepts, though the list is endless.

An accelerated charge emits energy and the electron revolving around a nucleus should emit energy [its energy then should go to zero] resulting in the collapse of the atom; but atom is a stable entity! According to classical theory, the excited hydrogen atom should emit electromagnetic radiations of all the wavelengths continuously. However, the observed fact is that excited hydrogen atoms emit radiations of certain wavelengths only! When classical laws are applied to the photoelectric effect, it should take nearly 10^7 seconds [one year!] in contradiction to the fact that it is an instantaneous phenomenon. The electrical conductivity of silver is nearly 10^{24} times more than that of fused quartz, a phenomenon that cannot be reasonably explained by classical theories. The magnetic susceptibility of iron is 10^9 times more than that of other metals, a fact very hard to digest classically. If the total energy is calculated by integrating the Rayleigh–Jeans formula over the whole range of wavelength from zero to infinity, it comes out to be infinity for all temperatures except absolute zero. This is an absurd result because the energy emitted at any finite temperature should be finite. Neither the Rayleigh–Jeans formula nor Wien's displacement formula could explain the blackbody radiation phenomenon completely over the whole range of wavelengths. Only the quantum theory of radiation can explain blackbody radiation! To understand the chemical properties of matter, as summarized in the periodic table of elements, the fact that not all the states of electrons permitted by the classical model are feasible realistically must be taken into account. Even cosmological and astrophysical phenomena are not completely explainable by the laws of classical physics.

The failure of classical concepts to explain the physical phenomena completely forced the scientific community to search for the missing link. Their attempts to search for the missing link gave birth to quantum physics and the mathematical modelling of quantum physics is quantum mechanics. The finer laws of quantum physics are not far away from the reality of the macro-world or are only means to explain microscopic phenomena; they are the true laws of nature. Actually, all of physics is quantum physics. All the laws of quantum physics reduce to the laws of classical physics under certain circumstances. If quantum physics is a super set, then classical physics is a sub-set. To be specific,

$$\lim_{n \to \infty} \text{Quantum Physics} = \text{Classical Physics}$$

In this relation, n is the principal quantum number.

7.3 Particles and Waves

Everything in the world is a wave; every thing in the world is a particle. They are the manifestation of the same thing in different forms. The physical reality we perceive has its roots in the world of elementary particles. Electron has mass and charge like a particle and obeys the laws of particle mechanics as in a CRO tube or picture tube of a television set but it behaves like a perfect wave in case of an electron microscope. The wave nature of the electron was confirmed by the Davisson–Germer experiment. According to de Broglie, all material particles behave like waves. We regard electromagnetic radiations as waves in case of interference, diffraction or in polarization but in case of the photoelectric effect, they behave like particles. According to the wave theory, electromagnetic waves leave a source with their energy spread out continuously through the wave pattern and according to quantum theory, electromagnetic waves consist of individual photons, each small enough to be absorbed by an electron. The boundary between particles and waves has all but disappeared. Together with the theory of relativity, the wave–particle duality is central to the understanding of all natural phenomena from the elementary particle level to the cosmological level.

7.4 Particle Aspect of Waves

The particle aspect of radiation is exhibited in the phenomena of blackbody radiation, photoelectric effect, Compton scattering and pair production.

7.4.1 Blackbody radiation

The classical wave theory light obeying Maxwell's electromagnetic theory successfully explained the phenomena of interference, diffraction, polarizations but lasted just more than one decade. The spectrum (distribution of energy with respect to wavelength) of thermal radiation (electromagnetic radiations emitted from a body due to its temperature only) could not be explained on the basis of the classical wave theory.

A blackbody is defined as a body that absorbs all radiations of all frequencies incident on it. Since a body at finite temperature is in thermal equilibrium with its surroundings, it must absorb energy at the same rate as it emits energy. Hence, a blackbody at finite temperature emits radiation of all frequencies or wavelengths. Blackbody radiation is defined as the radiation containing all the wavelengths from zero to infinity.

A blackbody is not necessarily a black body. A hole in the wall of a hollow object having an inner polished wall is an excellent example of a blackbody. A ray of radiation entering into the object through the hole in the wall will suffer reflections on the inside polished surface and cannot come out rendering the hollow object an ideal blackbody. A blackbody has no definite shape or size. This has been pictorially explained in Fig. 7.1.

The explanation of blackbody radiations came from the German physicist Max Planck in the year 1900 with the following postulates.

i. The radiation field in a uniform temperature enclosure remains in a state of dynamic equilibrium by way of emission and absorption. For this to take place, he assumed that there were harmonic oscillators lined up along the wall of the enclosure and that they absorbed energy from the radiation field and gave it back to the field in a characteristic way. The oscillators do not radiate energy until their energy reaches

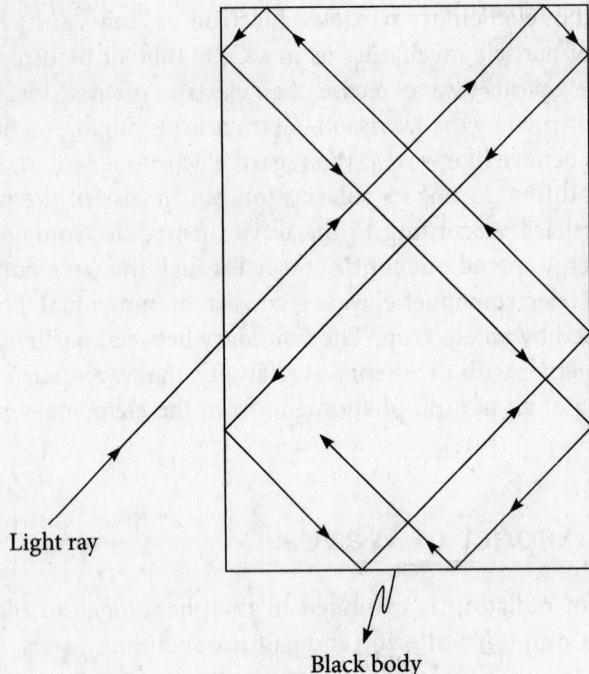

Light ray

Black body

Figure 7.1 | A perfect blackbody. A ray of radiation at any frequency incident on the hole cannot escape. A hole on the wall of a hollow object is an excellent approximation of a blackbody

an integral multiple of a certain minimum called a quantum of energy which is proportional to the frequency of the oscillators.

ii.   An oscillator absorbs energy from the radiation field and delivers it back to the field in the quanta of $0, \varepsilon, 2\varepsilon, 3\varepsilon, \ldots$, etc, where ε is a quantum of energy proportional to the frequency of the oscillator, i.e.,

$$\varepsilon \propto \nu$$

or

$$\varepsilon = h\nu \tag{7.1}$$

where the proportionality constant h is called Planck's constant with a value

$$6.626 \times 10^{-34}\,\text{Js}; \quad \hbar = \frac{h}{2\pi} = 1.0546 \times 10^{-34}\,\text{Js}; \quad hc = 12400\,\text{eV}\,\overset{\circ}{\text{A}}$$

iii.   The number of oscillators emitting a particular energy obeys the Maxwell–Boltzmann distribution law.

If N is the total number of harmonic oscillators and E is their total energy, then the energy per oscillator is given by

$$< \varepsilon > = \frac{E}{N} \tag{7.2}$$

If N_0, N_1, N_2, N_3, ..., N_r are the number of oscillators with energy 0, ε, 2ε, 3ε, ..., $r\varepsilon$, ... respectively, then

$$N = N_0 + N_1 + N_2 + N_3 + ... + N_r + ... \tag{7.3}$$

and

$$E = 0 + \varepsilon N_1 + 2\varepsilon N_2 + 3\varepsilon N_3 + ... + r\varepsilon N_r + ... \tag{7.4}$$

According to the Maxwell–Boltzmann distribution law, the number of oscillators with energy $r\varepsilon$ is given by

$$N_r = N_0 e^{-\frac{r\varepsilon}{kT}} \tag{7.5}$$

where k is called Boltzmann's constant with the value given by

$$k = 1.38 \times 10^{-23} \frac{J}{K} = 8.62 \times 10^{-5} \frac{eV}{K}$$

Substituting the values of N_0, N_1, N_2, N_3 from Eq. (7.5) into Eq. (7.3), we get

$$N = N_0 + N_0 e^{-\frac{\varepsilon}{kT}} + N_0 e^{-\frac{2\varepsilon}{kT}} + N_0 e^{-\frac{3\varepsilon}{kT}} + ... + N_0 e^{-\frac{r\varepsilon}{kT}} + ...$$

or

$$N = \frac{N_0}{1 - e^{-\frac{\varepsilon}{kT}}} \tag{7.6}$$

Substituting the values of N_0, N_1, N_2, N_3 from Eq. (7.5) into Eq. (7.4), we get

$$E = 0 + \varepsilon N_0 e^{-\frac{\varepsilon}{kT}} + 2\varepsilon N_0 e^{-\frac{2\varepsilon}{kT}} + 3\varepsilon N_0 e^{-\frac{3\varepsilon}{kT}} + ... + r\varepsilon N_0 e^{-\frac{r\varepsilon}{kT}} + ...$$

Since $\dfrac{1}{(1-x)^2} = 1 + 2x + 3x^2 + 4x^3 + ...$, this equation boils down to

$$E = \frac{N_0 \varepsilon e^{-\frac{\varepsilon}{kT}}}{\left(1 - e^{-\frac{\varepsilon}{kT}}\right)^2} \tag{7.7}$$

Putting the values of N and E from Eqs (7.6) and (7.7) into Eq. (7.2), the average energy per oscillator is obtained as

$$<\varepsilon> = \frac{\varepsilon e^{-\frac{\varepsilon}{kT}}}{\left(1 - e^{-\frac{\varepsilon}{kT}}\right)} \tag{7.8}$$

We know that the number of oscillators per unit volume in the frequency range v and $v + dv$ is given by

$$N = \frac{8\pi v^2}{c^3} dv \tag{7.9}$$

$E_v dv$ = the average radiant energy per unit volume with frequency range of $dv = N < \varepsilon >$, i.e.

$$E_v dv = N < \varepsilon >$$

Putting the value of $< \varepsilon >$ and N from Eqs (7.8) and (7.9) respectively into this equation, we get

$$E_v dv = \frac{8\pi v^2}{c^3} dv \times \frac{\varepsilon e^{-\frac{\varepsilon}{kT}}}{1 - e^{-\frac{\varepsilon}{kT}}} = \frac{8\pi v^2}{c^3} dv \times \frac{\varepsilon}{e^{\frac{\varepsilon}{kT}} - 1} = \frac{8\pi v^2}{c^3} dv \times \frac{hv}{e^{\frac{hv}{kT}} - 1}$$

or
$$E_v dv = \frac{8\pi h v^3}{c^3 \left(e^{\frac{hv}{kT}} - 1\right)} dv \tag{7.10}$$

Equation (7.10) is Planck's radiation law in terms of frequency. This expression gives the amount of radiant energy per unit volume having a frequency interval of dv.

In terms of wavelength, Planck's radiation law can be expressed in the following way.

We know that $v = \dfrac{c}{\lambda}$ or $dv = \dfrac{c}{\lambda^2} d\lambda$ in magnitude. Putting these values of v and dv into Eq. (7.10), we get

$$E_\lambda d\lambda = \frac{8\pi h \left(\dfrac{c}{\lambda}\right)^3}{c^3 \left(e^{\frac{hc}{\lambda kT}} - 1\right)} \frac{c}{\lambda^2} d\lambda = \frac{8\pi hc}{\lambda^5 \left(e^{\frac{hc}{\lambda kT}} - 1\right)} d\lambda$$

or $\qquad E_\lambda d\lambda = \frac{8\pi hc}{\lambda^5} \frac{1}{e^{\frac{hc}{\lambda kT}} - 1} d\lambda$ $\hfill (7.11)$

where E_λ = Intensity of emission. In terms of wavelength, Eq. (7.11) is Planck's radiation law. This expression gives the amount of radiant energy per unit volume having wavelength interval $d\lambda$.

Example 7.1

How many photons are there in 1.00 cm³ of radiation in thermal equilibrium at 1000 K? Calculate the average energy of one photon at this temperature.

Solution

Let there be N number of photons per unit volume. The energy of one photon is hv and so the energy of N number of photons is Nhv and the energy per unit volume will be Nhv. According to Planck's law, the energy per unit volume is

$$\frac{8\pi hv^3}{c^3 \left(e^{\frac{hv}{kT}} - 1\right)}.$$

Therefore, we can have

$$Nhv = \frac{8\pi hv^3}{c^3 \left(e^{\frac{hv}{kT}} - 1\right)}$$

or $\qquad N = \dfrac{8\pi v^2}{c^3 \left(e^{\frac{hv}{kT}} - 1\right)}$

Hence, the number of photons in volume V will be

$$n = NV = V \frac{8\pi v^2}{c^3 \left(e^{\frac{hv}{kT}} - 1 \right)}$$

and the number of photons in volume V with a frequency range dv will be

$$V \frac{8\pi v^2}{c^3 \left(e^{\frac{hv}{kT}} - 1 \right)} dv$$

The number of photons in volume V having a frequency range of 0 to ∞ will be

$$\int_0^\infty V \frac{8\pi v^2}{c^3 \left(e^{\frac{hv}{kT}} - 1 \right)} dv = \frac{8\pi V}{c^3} \int_0^\infty \frac{v^2}{\left(e^{\frac{hv}{kT}} - 1 \right)} dv$$

The integral

$$\int_0^\infty \frac{v^2}{\left(e^{\frac{hv}{kT}} - 1 \right)} dv$$

can be evaluated by following the procedure as laid down in the Section 7.4.1 (Case 4) and its value is

$$2.404 \left(\frac{kT}{h} \right)^3.$$

Therefore, the number of photons in volume V having a frequency range of 0 to ∞ will be

$$\frac{8\pi V}{c^3} \times 2.404 \left(\frac{kT}{h} \right)^3 = 19.232\pi V \left(\frac{kT}{hc} \right)^3 = 2.02 \times 10^{10}.$$

According to Planck's law, the energy per unit volume is

$$\frac{8\pi h v^3}{c^3 \left(e^{\frac{hv}{kT}} - 1 \right)}$$

and so the energy in V volume will be

$$\frac{8\pi h v^3}{c^3\left(e^{\frac{hv}{kT}}-1\right)}\times V$$

and the energy in volume V having frequency range dv will be

$$\frac{8\pi h v^3}{c^3\left(e^{\frac{hv}{kT}}-1\right)}\times V dv.$$

Thus, the energy in volume V having a frequency range of 0 to ∞ will be

$$\int_0^\infty \frac{8\pi h v^3}{c^3\left(e^{\frac{hv}{kT}}-1\right)}\times V dv = \frac{8\pi h V}{c^3}\int_0^\infty \frac{v^3 dv}{e^{\frac{hv}{kT}}-1} = \frac{8\pi(\pi kT)^4 V}{15c^3 h^3} = 7.54\times10^{-10}\,\text{J}$$

The average energy per photon will be

$$\frac{7.54\times10^{-10}\,\text{J}}{2.02\times10^{10}} = 3.73\times10^{-20}\,\text{J} = 0.233\ \text{eV}$$

Special cases of Planck's radiation law

Case 1: Wien's radiation law

λT is small for low temperatures. As a special case, Wien's radiation law follows automatically from Planck's radiation law at low temperatures.

Aa mentioned earlier, at low temperatures, λT is small and hence, $e^{\frac{hc}{\lambda kT}} \gg 1$. Hence, in Eq. (7.11), we can neglect 1 in the denominator in comparison to $e^{\frac{hc}{\lambda kT}}$ and get

$$E_\lambda d\lambda = \frac{8\pi hc}{\lambda^5}e^{-\frac{hc}{\lambda kT}}d\lambda \tag{7.12}$$

This expression is Wien's radiation law.

Case 2: Rayleigh–Jeans law

λT is large for high temperatures. As a special case, Rayleigh–Jeans radiation law follows automatically from Planck's radiation law at high temperatures.

Rayleigh–Jeans law states that the intensity of emission is directly proportional to the absolute temperature of the body and inversely proportional to the fourth power of the wavelength. Mathematically, it is given by

$$E_\lambda = \frac{AT}{\lambda^4},$$

A is the proportionality constant.

As mentioned earlier, at high temperatures, λT, i.e., λkT is large and hence, $\frac{hc}{\lambda kT} < 1$. Equation (7.11) can be expressed as

$$E_\lambda d\lambda = \frac{8\pi hc}{\lambda^5 \left(1 + \left(\frac{hc}{\lambda kT}\right) + \frac{1}{2!}\left(\frac{hc}{\lambda kT}\right)^2 + \frac{1}{3!}\left(\frac{hc}{\lambda kT}\right)^3 + \dots - 1\right)} d\lambda \text{ *}$$

Since $\frac{hc}{\lambda kT} < 1$ $\left(\frac{hc}{\lambda kT}\right)^2$ and higher powers of $\frac{hc}{\lambda kT}$ are negligibly small, we get

$$E_\lambda d\lambda = \frac{8\pi hc}{\lambda^5 \left(1 + \left(\frac{hc}{\lambda kT}\right) - 1\right)} d\lambda$$

or $\qquad E_\lambda d\lambda = \frac{8\pi kT}{\lambda^4} d\lambda$ $\hfill$ (7.13)

This above expression is Rayleigh–Jean's law.

Case 3: Wien's displacement law

Wien's displacement law states that the wavelength of radiation that is emitted with maximum intensity λ_m varies inversely with the absolute temperature T of the body. Mathematically,

$$\lambda_m = \frac{b}{T}, \quad b = \text{Wien's constant} = 2.89 \times 10^{-3} \text{ mK}$$

As a special case, Wien's displacement law also follows automatically from Planck's radiation law.

$*\quad e^x = 1 + x + \frac{1}{2!}x^2 + \frac{1}{3!}x^3 + \dots$

Applying the principle of maxima and minima function to E_λ, we have $\dfrac{dE_\lambda}{d\lambda}\bigg|_{\lambda=\lambda_m}=0$
because E_λ is to be maximum when $\lambda = \lambda_m$. Putting the value of E_λ from Planck's formula (7.11), we get

$$\frac{d}{d\lambda}8\pi hc\lambda^{-5}(e^{\frac{hc}{\lambda kT}}-1)^{-1}\bigg|_{\lambda=\lambda_m}=0$$

or $\quad (-5)\lambda_m^{-6}\left(e^{\frac{hc}{\lambda_m kT}}-1\right)^{-1}+\lambda_m^{-5}(-1)\left(e^{\frac{hc}{\lambda_m kT}}-1\right)^{-2}e^{\frac{hc}{\lambda_m kT}}\left(-\frac{hc}{\lambda_m^2 kT}\right)=0$

or $\quad \dfrac{hc}{\lambda_m kT}\dfrac{e^{\frac{hc}{\lambda_m kT}}}{e^{\frac{hc}{\lambda_m kT}}-1}=5$

This equation is in the form of $x\dfrac{e^x}{e^x-1}=5$ with $x=\dfrac{hc}{\lambda_m kT}$ whose solution is $x = 4.965$.
Hence, we have $\dfrac{hc}{\lambda_m kT}=4.965$

or $\quad \lambda_m T=\dfrac{hc}{4.965k}=b$

or $\quad \lambda_m T=2.89\times10^{-3}\,\mathrm{Km}$ \hfill (7.14)

Case 4: Stefan's fourth power law

Stefan's fourth power law states that radiant emittance, i.e., the energy radiated per second per unit area containing all the wavelengths by a perfect blackbody is directly proportional to the fourth power of its absolute temperature T. Mathematically,

Radiant emittance $=\sigma T^4$

where $\sigma=$ Stefan's constant $=5.67\times10^{-8}\dfrac{W}{m^2 K^4}$

Stefan's fourth power law is deduced from Planck's formula by integrating the expression for all the wavelengths from 0 to ∞ and hence, we obtain

$$\int_0^\infty E_\lambda d\lambda=\int_0^\infty\frac{8\pi hc}{\lambda^5\left(e^{\frac{hc}{\lambda kT}}-1\right)}d\lambda=8\pi hc\int_0^\infty\frac{e^{-\frac{hc}{\lambda kT}}}{\lambda^5\left(1-e^{-\frac{hc}{\lambda kT}}\right)}d\lambda$$

$$= 8\pi hc \int_0^\infty \lambda^{-5} e^{-\frac{hc}{\lambda kT}} \left(1 - e^{-\frac{hc}{\lambda kT}}\right)^{-1} d\lambda$$

$$= 8\pi hc \int_0^\infty \lambda^{-5} e^{-\frac{hc}{\lambda kT}} \left(1 + e^{-\frac{hc}{\lambda kT}} + e^{-2\frac{hc}{\lambda kT}} + e^{-3\frac{hc}{\lambda kT}} + \dots \infty\right) d\lambda \, **$$

$$= 8\pi hc \int_0^\infty \lambda^{-5} \left(\sum_{n=1}^\infty e^{-n\frac{hc}{\lambda kT}}\right) d\lambda$$

$$\int_0^\infty E_\lambda d\lambda = 8\pi hc \sum_{n=1}^\infty \int_0^\infty \lambda^{-5} e^{-n\frac{hc}{\lambda kT}} d\lambda \qquad\qquad (7.15)$$

Evaluation of $\int_0^\infty \lambda^{-5} e^{-n\frac{hc}{\lambda kT}} d\lambda$

$\int_0^\infty \lambda^{-5} e^{-n\frac{hc}{\lambda kT}} d\lambda$ can be evaluated by using the gamma function***.

Let $\dfrac{nhc}{\lambda kT} = x$. Here $x = 0$, when $\lambda = \infty$ and $x = \infty$, when $\lambda = 0$. Hence, we can have

$$\lambda = \frac{nhc}{xkT}$$

or   (i) $d\lambda = -\dfrac{nhc}{x^2 kT} dx$

and   (ii) $\lambda^{-5} = \left(\dfrac{nhc}{xkT}\right)^{-5}$

Putting these values of $\dfrac{nhc}{\lambda kT}$, $d\lambda$ and λ^{-5} into the integral $\int_0^\infty \lambda^{-5} e^{-n\frac{hc}{\lambda kT}} d\lambda$, we get

_____

** $\quad (1-x)^{-1} = 1 + x + x^2 + x^3 + \dots \infty, \quad (1+x)^{-1} = 1 - x + x^2 - x^3 + \dots \infty$

*** $\quad \Gamma(n) = \int_0^\infty x^{n-1} e^{-x} dx, \quad \Gamma\left(\dfrac{1}{2}\right) = \sqrt{\pi} \; ; \; \Gamma(n+1) = n\Gamma(n).$

If n is a positive integer then $\Gamma(n) = (n-1)!$

$$\int_0^\infty \lambda^{-5} e^{-n\frac{hc}{\lambda kT}} d\lambda = -\int_\infty^0 \left(\frac{nhc}{xkT}\right)^{-5} e^{-x} \frac{nhc}{x^2 kT} dx = \int_0^\infty \left(\frac{nhc}{kT}\right)^{-4} x^3 e^{-x} dx =$$

$$\left(\frac{nhc}{kT}\right)^{-4} \int_0^\infty x^{4-1} e^{-x} dx = \left(\frac{kT}{nhc}\right)^4 \Gamma(4)$$

or $\quad \int_0^\infty \lambda^{-5} e^{-n\frac{hc}{\lambda kT}} d\lambda = \left(\frac{kT}{nhc}\right)^4 \times 6$

Thus, we have $\quad \int_0^\infty \lambda^{-5} e^{-n\frac{hc}{\lambda kT}} d\lambda = 6\left(\frac{kT}{hc}\right)^4 \frac{1}{n^4}$

Putting this value of $\int_0^\infty \lambda^{-5} e^{-n\frac{hc}{\lambda kT}} d\lambda$ into Eq. (7.15), we have

$$\int_0^\infty E_\lambda d\lambda = 8\pi hc \sum_{n=1}^\infty 6\frac{k^4 T^4}{c^4 h^4}\frac{1}{n^4} = \frac{48\pi k^4 T^4}{c^3 h^3}\left(\frac{1}{1^4}+\frac{1}{2^4}+\frac{1}{3^4}+...\infty\right) = \frac{48\pi k^4 T^4}{c^3 h^3} \times \frac{\pi^4}{90}$$

or $\quad \int_0^\infty E_\lambda d\lambda = \frac{8\pi^5 k^4}{15 c^3 h^3} T^4$ $\qquad\qquad\qquad\qquad\qquad\qquad$ (7.16)

This expression gives the amount of radiant energy per unit volume having all the wavelengths.

Since the enclosure is in thermal equilibrium, according to the kinetic theory of gases, the arrival of photons per second per unit area is $\frac{c}{4}$ times the concentration of photons. Therefore, the energy arrives at the walls per second per unit area, and consequently, the energy re-radiated per second per unit area is given as

$$\text{Radiant emittance} = \frac{c}{4} \times \frac{8\pi^5 k^4}{15 c^3 h^3} T^4 = \frac{2\pi^5 k^4}{15 c^2 h^3} T^4 = \sigma T^4$$

where $\sigma = \frac{2\pi^5 k^4}{15 c^2 h^3} = 5.67 \times 10^{-8} \frac{W}{m^2 K^4}$ is called Stefan's constant. Thus, the energy radiated per second per unit area containing all the wavelengths by a perfect blackbody is given by

$$\text{Radiant emittance} = \sigma T^4. \text{ Stefan's law} \qquad\qquad\qquad\qquad (7.17)$$

The energy radiated per second per unit area containing all the wavelengths by any other body is given by

$$E = e\sigma T^4 \tag{7.18}$$

where e is called emissivity. Emissivity depends upon the nature of the radiating surface and ranges from 0 for a perfect reflector to 1 for a perfect blackbody. Stefan's law is also called the Stefan–Boltzmann law.

All these discussions show that Planck's law is perfect and all other laws follow as special cases. Thus, according to Planck's law, energy changes take place discontinuously and discretely as an integral multiple of small units of energy $h\nu$ which are called quanta or photons.

Example 7.2

Calculate the effective temperature of sun from the following data. Solar constant = $2.297 \dfrac{\text{cal}}{\text{cm}^2\text{min}}$, radius of the sun = 4.3×10^5 miles, distance of the sun from the earth = 9.3×10^7 miles. Assume that the sun is a blackbody. $\left(\sigma = 1.37 \times 10^{-12} \dfrac{\text{cal}}{\text{cm}^2\text{sec}} \right)$

Solution

$$T^4 = \left(\frac{R}{r}\right)^2 \frac{S}{\sigma} = \left(\frac{9.3 \times 10^7}{4.3 \times 10^5}\right)^2 \times \frac{2.279/60}{1.37 \times 10^{-12}}$$

or $T = 6001\text{K}$

Example 7.3

Calculate the temperature of sun from the following data. $\lambda_m T = 2.89 \times 10^{-3}$ Km, $\lambda_m = 4753$ Å.

Solution

$$T = \frac{2.89 \times 10^{-3}}{4753 \times 10^{-10}} = 6080.4\,\text{K}$$

Example 7.4

What is the maximum wavelength of thermal radiation emitted by a body at room temperature 27°C. To what temperature must we heat it so that its peak thermal radiation is in the red region of the spectrum. ($\lambda = 6500$ Å).

Solution

According to Wien's displacement law, $\lambda_m T = 2.89 \times 10^{-3}$ mK. Hence,

$$\lambda_m = \frac{2.89 \times 10^{-3} \, mK}{T}$$

$$T = 27°C = 300 \text{ K}$$

Therefore, we have $\lambda_m = \dfrac{2.89 \times 10^{-3} \, mK}{300K} = 9.63 \, \mu m$

The temperature we must heat it so that its peak thermal radiation is in the red region of the spectrum ($\lambda = 6500$ Å) is

$$T = \frac{2.89 \times 10^{-3} \, mK}{\lambda_m}$$

Hence, we have $T = \dfrac{2.89 \times 10^{-3} \, mK}{0.65 \, \mu m} = 4446 \text{ K}$

Example 7.5

At what temperature should the filament of a 100 W lamp operate so that it becomes a perfect blackbody of area 1cm².

Solution

According to Stefan's law, the energy radiated per second per unit area of a blackbody is σT^4. Hence, the amount of energy radiated per second from an area A of a perfect blackbody will be $A\sigma T^4$.

According to the data given in the question, the energy radiated per second of the lamp is 100 watt. Hence, we have

$$A\sigma T^4 = 100 \text{ W}$$

or $T^4 = \dfrac{100 \, W}{A\sigma} = \dfrac{100 \, W}{10^{-4} \, m^2 \times 5.67 \times 10^{-8} \, W/m^2 K^4} = 1.764 \times 10^{13} \, K^4$

or $T = 2049$ K

7.4.2   Photoelectric effect

Photoelectric effect is defined as the phenomenon of emission of electrons from the surface of certain substances, mainly metals when light shorter wavelength is incident on them.

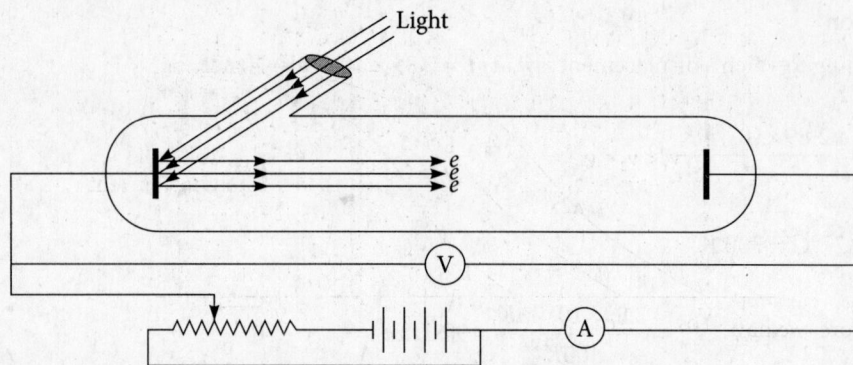

Figure 7.2 | Experimental arrangements for the study of photoelectric effect. When a light of suitable frequency is incident on the cathode, photoelectrons are emitted and move towards the anode depending upon its voltage. The voltage of the cathode and anode can be varied by the rheostat. Voltage and currents are measured at the required point. The stopping potential V_0 is the negative voltage of the anode which stops the emission of electrons from the cathode surface. The work function w_0 of the metal surface is defined as the minimum energy required to pull an electron out of the metal surface at 0 K temperature

Experimental results are represented graphically in Fig. 7.3.

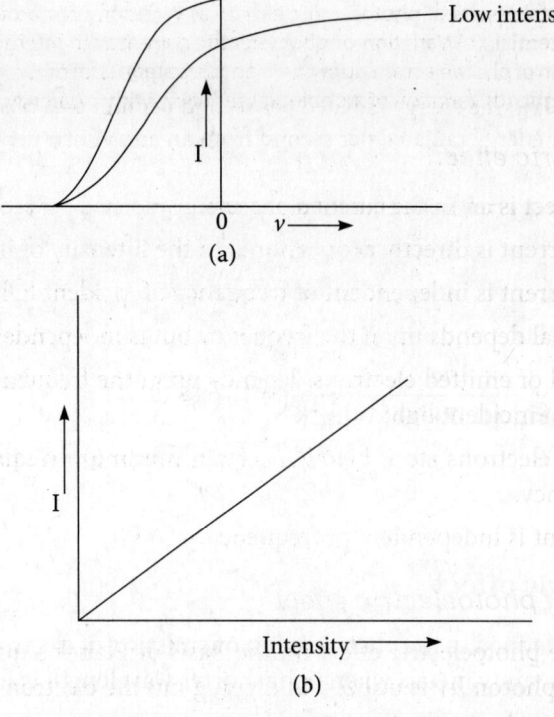

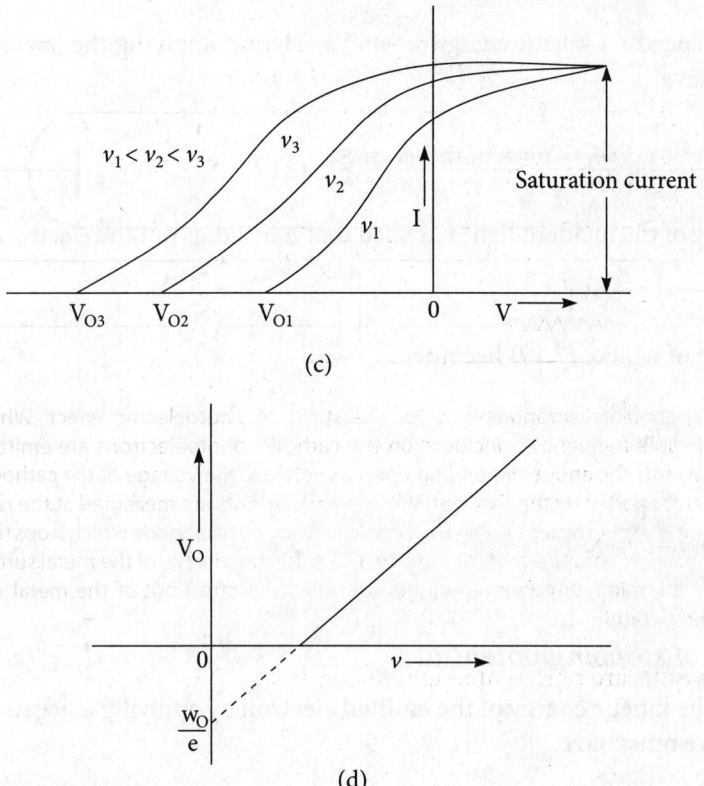

(c)

(d)

Figure 7.3 | Experimental results of photoelectric effect. (a) Variation of photoelectric current with anode potential, (b) Variation of photoelectric current with intensity of incident light, (c) Variation of photoelectric current with anode potential for different frequencies of the incident light. (d) Variation of stopping potential V_0 with frequency of the incident light

Laws of photoelectric effect

i.   Photoelectric effect is an instantaneous process.

ii.   Photoelectric current is directly proportional to the intensity of incident light.

iii.   Photoelectric current is independent of frequency of incident light.

iv.   Stopping potential depends upon the frequency but is independent of intensity.

v.   Maximum speed of emitted electrons depends upon the frequency but independent of the intensity of incident light.

vi.   The emission of electrons stops below a certain minimum frequency known as the threshold frequency.

vii.   Saturation current is independent of frequency.

Einstein's theory of photoelectric effect

Einstein explained the photoelectric effect on the basis of Planck's quantum theory. The energy of an incident photon $h\nu$ is utilized in digging out the electron (w_0) and giving the electron some speed v $\left(\text{kinetic energy} = \dfrac{1}{2}mv^2 \right)$. Hence, applying the law of conservation of energy, we have

$$hv = \frac{1}{2}m_e v^2 + w_0, \quad m_e = \text{mass of the electron} \tag{7.19}$$

If the frequency of the incident light v_0 is such that it just digs out the electron, then we have

$$hv_0 = w_0 \tag{7.20}$$

With this value of w_0, Eq. (7.19) becomes

$$hv = \frac{1}{2}mv^2 + hv_0$$

or $\quad \dfrac{1}{2}mv^2 = h(v - v_0)$ $\hfill$ (7.21)

Calculation of stopping potential V_0

To neutralize the kinetic energy of the emitted electron by applying a negative potential V_0 to the anode, we must have

$$\frac{1}{2}mv^2 = eV_0 \tag{7.22}$$

Putting Eq. (7.22) into Eq. (7.19), we get

$$hv = eV_0 + w_0$$

or $\quad V_0 = \dfrac{h}{e}v - \dfrac{w_0}{e} \quad (y = mx + c)$ $\hfill$ (7.23)

This equation has been plotted in Fig. 7.3(d)

Calculation of threshold frequency v_0

In the plot of stopping potential V_0 and frequency v [see Fig. 7.3(d)], according to Eq. (7.23), the negative Y-intercept will be $-\dfrac{w_0}{e}$. If we divide this intercept (by taking only magnitude) by h, we have

$$\frac{\dfrac{w_0}{e}}{h} = \frac{w_0}{eh}$$

Putting the value of w_0 from Eq. (7.20) into this equation, we get

$$\frac{Y - \text{intercept}}{h} = \frac{h\nu_0}{h} = \nu_0$$

or $\quad \nu_0 = \dfrac{Y - \text{intercept in } V_0 \sim \nu \text{ plot}}{h}$

Calculation of work function w_0

In the plot of stopping potential V_0 and frequency ν [see Fig. 7.3(d)], according to Eq. (7.23) the negative Y-intercept will be $-\dfrac{w_0}{e}$. If we multiply this intercept (by taking only magnitude) by e, we have

$$\frac{w_0}{e} \times e = w_0$$

Hence, we have

$$w_0 = \left(Y - \text{intercept in } V_0 \sim \nu \text{ plot}\right) \times e$$

Calculation of Planck's constant h

According to Eq. (7.23), the slope of the plot of stopping potential V_0 and frequency ν is (h/e) [see Fig. 7.3(d)]. If we multiply this slope with e, the charge of the electron, we will obtain the value of h, the Planck's constant. Mathematically,

$$h = (\text{Slope of the } V_0 \sim \nu \text{ plot}) \times e$$

Example 7.6

The photoelectric threshold wavelength for a metal is 4400 Å. What will be the maximum energy of photoelectrons emitted from this metal surface when it is irradiated by a radiation of 4000 Å?

Solution

The data given are

$$\lambda_0 = 4400 \text{ Å}$$

$$\lambda = 4000 \text{ Å}$$

According to Einstein's photoelectric equation, the maximum kinetic energy of the photoelectrons is given by

$$E_{max} = h\nu - h\nu_0 = hc\left(\frac{\lambda_0 - \lambda}{\lambda_0 \lambda}\right)$$

Putting the data given in the question into this equation, we have

$$E_{max} = 6.626 \times 10^{-34} \times 3 \times 10^8 \left(\frac{4400 \times 10^{-10} - 4000 \times 10^{-10}}{4400 \times 10^{-10} \times 4000 \times 10^{-10}}\right) J$$

$$= 4.52 \times 10^{-20} J = 0.28 eV$$

7.4.3   Compton effect

The photoelectric effect shows that the energy of an electromagnetic wave is absorbed and emitted in discrete quanta. The Compton effect gives compelling evidence of the corpuscular nature of radiation. The effect was first conceptualized theoretically and demonstrated experimentally by A. H. Compton in the year 1923.

Compton used X-rays from molybdenum with energy of approximately 20 keV. On entering the scattering material such an X-ray interacts with atomic electrons. These electrons can be considered essentially free for the X-rays because the energy of the incident X-rays is much more than the energy of the electrons of 10 eV. The electric field of the incident electromagnetic wave causes these electrons to oscillate with the frequency of the X-rays. An electron that oscillates in simple harmonic motion will radiate like an electric dipole. The frequency of the oscillations is therefore initially equal to the frequency of the incident X-rays. The incident electromagnetic wave carries momentum (E/c) since for a massless particle $E = pc$, where E is energy of the incident wave. Since the X-ray gives some of its energy to the electron, it loses momentum. This momentum is imparted to the electron, which recoils in the direction of propagation of the incident wave. As the electron recedes from the source of X-ray, it no longer sees the original frequency but a lower frequency due to Doppler effect. The electron then re-radiates with this lower frequency.

This classical argument leads to the conclusion that the frequencies of the scattered X-rays should have a continuous range of values. When the target electron is still in rest, the frequency of the scattered radiation observed in the laboratory frame should be that of incident X-rays. The frequency should then decrease continuously to a final value.

The conclusion of the Compton experiment is in direct contradiction with this conclusion drawn from classical argument. Rather than what was theorized by the classical theory, the X-rays scattered by electrons at a particular angle are found to have just one sharply defined frequency lower than that of the incident X-rays. This fact shows that scattering is not a gradual process during which electron gains momentum at a continuous rate; the interaction of X-rays and electrons is instantaneous.

The fact that the X-ray transfers its energy and momentum instantaneously in one packet suggests that the scattering process can be treated as a collision between two particles – a photon and an electron. Such instantaneous transfer does indeed occur in a two-particle collision.

Figure 7.4 shows the diagram of the Compton experiment.

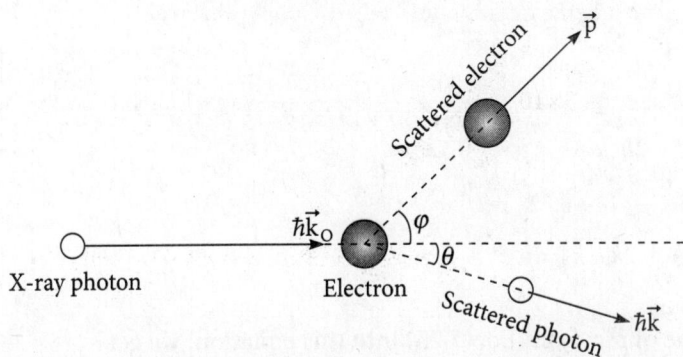

Figure 7.4 | Diagram of Compton scattering. An X-ray photon strikes the electron at rest. After collision, the incident photon and electron are scattered along the directions shown in the figure. $\hbar \vec{k}_0$ is the momentum of the incident X-ray photon, $\hbar \vec{k}$ is the momentum of the scattered X-ray photon, $\vec{P}$ is the momentum of the scattered electron

An X-ray photon strikes an electron assumed to be initially at rest in the laboratory coordinate system and is scattered away from its original direction of motion, while the electron receives some impulse and begins to move. We can think of the photon as losing an amount of energy in the collision that is the same as the kinetic energy gained by the scattered electron, although actually, separate photons are involved. If the incident photon has frequency v and the scattered photon has frequency v', then we must have $v' < v$. (Why?).

The total relativistic energy E of the moving electron is given by

$$E = \sqrt{m_0^2 c^4 + (pc)^2} \tag{7.24}$$

or $\quad p^2 c^2 = E^2 - m_0^2 c^4 \tag{7.25}$

Hence, the total relativistic energy E of the electron at rest will be $m_0 c^2$

$\hbar \vec{k}_0$ = Momentum of the incident X-ray photon.

$\vec{k}_0$ = Propagation vector of the incident photon with magnitude $\dfrac{2\pi}{\lambda_0}$, λ_0 being the wavelength of the incident photon.

$\hbar \vec{k}$ = Momentum of the scattered X-ray photon.

$\vec{k}$ = Propagation vector of the scattered photon with magnitude $\dfrac{2\pi}{\lambda_0}$, λ being the wavelength of the scattered photon.

According to the law of conservation of momentum, we have

Total momentum before collision = Total momentum after collision

or $\quad \hbar \vec{k}_0 = \hbar \vec{k} + \vec{p}$ \hfill (7.26)

or $\quad \hbar^2 k_0^2 + \hbar^2 k^2 - 2\hbar^2 \vec{k}_0 \cdot \vec{k} = p^2$

or $\quad \hbar^2 c^2 \left(k_0^2 + k^2 - 2\vec{k}_0 \cdot \vec{k} \right) = p^2 c^2$

Putting the value of $p^2 c^2$ from Eq. (7.25) into this equation, we get

$$\hbar^2 c^2 \left(k_0^2 + k^2 - 2\vec{k}_0 \cdot \vec{k} \right) = E^2 - m_0^2 c^4 \qquad (7.27)$$

The energy of the incident X-ray photon will be $\hbar \vec{k}_0 c$ since the momentum is $\hbar \vec{k}_0$.

The energy of the scattered X-ray photon will be $\hbar \vec{k} c$ since the momentum is $\hbar \vec{k}$.

According to the law of conservation of energy, we have

Total energy before collision = Total energy after collision

or $\quad \hbar k_0 c + m_0 c^2 = \hbar k c + E$ \hfill (7.28)

or $\quad \hbar c \left(k_0 - k \right) = E - m_0 c^2$ \hfill (7.29)

Squaring both sides of Eq. (7.29), we get

$$\hbar^2 c^2 \left(k_0^2 + k^2 - 2 k_0 k \right) = E^2 + m_0^2 c^4 - 2 m_0 c^2 E \qquad (7.30)$$

Subtracting Eq. (7.30) from Eq. (7.27), we get

$$\hbar^2 \left(k_0 k - \vec{k}_0 \cdot \vec{k} \right) = m_0 \left(E - m_0 c^2 \right)$$

Putting the value of $E - m_0 c^2$ from Eq. (7.29) into this equation, we obtain

$$\hbar^2 \left(k_0 k - \vec{k}_0 \cdot \vec{k} \right) = m_0 \hbar c \left(k_0 - k \right) \qquad (7.31)$$

or $\quad k_0 k - \vec{k}_0 \cdot \vec{k} = \dfrac{m_0 c}{\hbar}(k_0 - k)$ $\qquad\qquad$ (7.32)

The angle between $\vec{k}_0$, the propagation vector of the incident photon and $\vec{k}$, the propagation vector of the scattered photon as shown in Fig. 7.4 is θ. Hence,

$$\vec{k}_0 \cdot \vec{k} = k_0 k \cos\theta \qquad\qquad (7.33)$$

Putting the value of $\vec{k}_0 \cdot \vec{k}$ from Eq. (7.33) into Eq. (7.32), we get

$$k_0 k - k_0 k \cos\theta = \dfrac{m_0 c}{\hbar}(k_0 - k)$$

or $\quad 1 - \cos\theta = \dfrac{m_0 c}{\hbar}\left(\dfrac{\lambda}{2\pi} - \dfrac{\lambda_0}{2\pi}\right)$

or $\quad \lambda - \lambda_0 = \dfrac{2\pi\hbar}{m_0 c}(1 - \cos\theta)$ $\qquad\qquad$ (7.34)

or $\quad \lambda - \lambda_0 = \lambda_C(1 - \cos\theta)$ $\qquad\qquad$ (7.35)

where $\quad \lambda_c = \dfrac{2\pi\hbar}{m_0 c} = \dfrac{h}{m_0 c}$ $\qquad\qquad$ (7.36)

$\lambda_C = \dfrac{2\pi\hbar}{m_0 c} = \dfrac{h}{m_0 c}$ is called the Compton wavelength of the scattering particle. For an electron, $\lambda_C = .2426 \times 10^{-12}$ meter $= 2.426 \times 10^{-2}$ Å. Equations (7.34) or (7.35) give the change in wavelengths between the incident radiation and the scattered radiation. The change in wavelengths $\lambda - \lambda_0$ between the incident radiation and the scattered radiation is called Compton shift. Equation (7.35) shows that the Compton shift is maximum when the scattered photon goes in the opposite direction to the incident photon and is zero when the scattered photon goes in the same direction as the incident photon.

From Eq. (7.35), the wavelength of the scattered photon is

$$\lambda = \lambda_0 + \lambda_C(1 - \cos\theta) \qquad\qquad (7.37)$$

or $\quad v = \dfrac{v_0}{1 + (\lambda_C v_0 / c)(1 - \cos\theta)}$ $\qquad\qquad$ (7.38)

Equation (7.38) shows that the wavelength of the scattered photon depends upon the scattering angle θ. Experiment shows that this equation is correct. Figure 7.5 shows the effect of the scattering angle θ on the Compton shift $\lambda - \lambda_0$. The important theoretical fact is that it was obtained by considering light quantum or photon as a particle. It constitutes strong evidence in favour of the quantum theory of radiation. If the Compton effect has true features of a two-particle collision, then it must be possible to observe the scattered photon and recoiling electron simultaneously. It was experimentally verified by Bothe and Geiger in the year 1925.

The Compton effect shows clearly that photons behave like particles when they collide with electrons. It is also true that light exhibits interference phenomenon, which can only be explained by the wave nature of electromagnetic radiation. In such phenomena, light behaves like a wave. This dual wave–particle nature of electromagnetic radiation is an experimental fact.

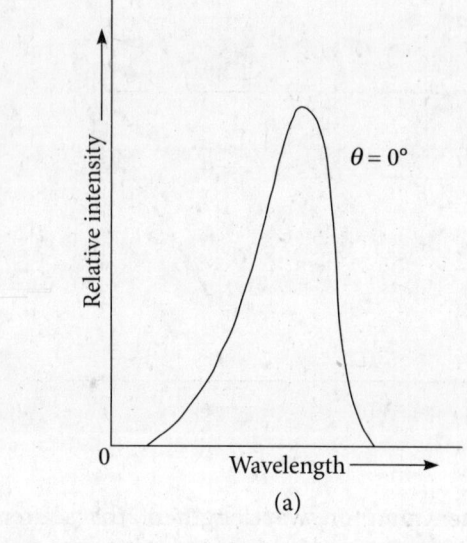

(a)

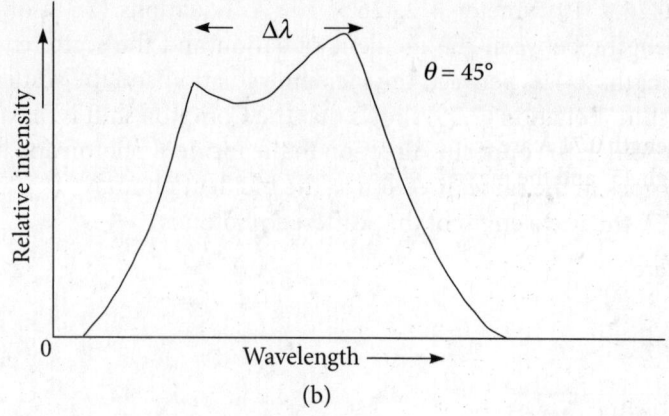

(b)

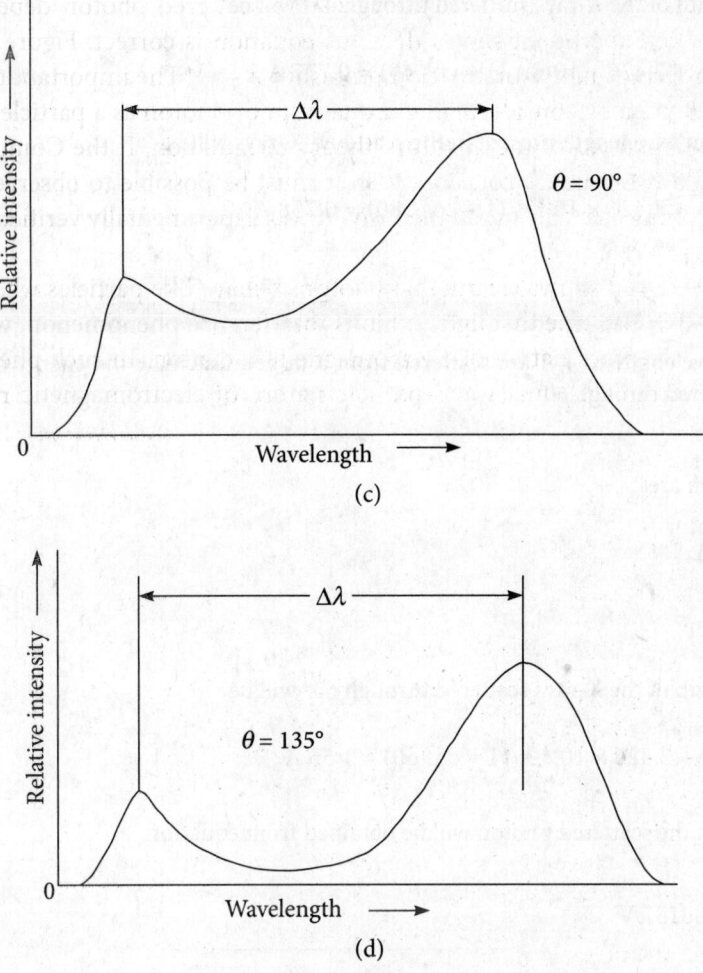

(c)

(d)

Figure 7.5 Plot of relative intensity versus wavelength. Effect of scattering angle θ on the change in wavelength between the incident photon and the scattered photon is clearly visible

Example 7.7

X-rays of wavelength 0.71 Å are scattered from a target. Calculate the wavelength of the X-rays scattered through 45° and the maximum wavelength present in the scattered X-rays.

Solution

The data given are

$$\lambda_0 = 0.71 \text{ Å}$$

$$\theta = 45°$$

The wavelength of the X-rays scattered through 45° will be

$$\lambda = 0.71 \text{ Å} + 2.426 \times 10^{-2} \text{ Å} \ (1 - \cos 45) = 0.717 \text{ Å}$$

The maximum wavelength present in the scattered X-rays will be

$$\lambda = 0.71 \text{ Å} + 2.426 \times 10^{-2} \text{ Å} \ (1 - \cos 180) = 0.758 \text{ Å}$$

Example 7.8

X-rays of wavelength 1.54 Å are scattered from a target. Calculate the energy of the X-rays photon scattered through 60°

Solution

The data given are

$$\lambda = 1.54 \text{ Å}$$

$$\theta = 60°$$

The wavelength of the X-rays scattered through 60° will be

$$\lambda = 1.54 \text{ Å} + 2.426 \times 10^{-2} \text{ Å} \ (1 - \cos 60) = 1.55 \text{ Å}$$

The energy of the scattered photon will be obtained from equation

$$E = \frac{hc}{\lambda} = 8015 \text{ eV}$$

7.4.4 Pair production

In photoelectric effect, the incident photon gives off all its energy to the electron and in the Compton effect, the incident photon gives off part of its energy to the electron. Matter can be converted to energy [$E = \Delta mc^2$]; can energy be converted to matter? That should be; because nature loves symmetry. Indeed the conversion of energy into matter exists in nature. This phenomenon is called pair production. Pair production is defined as a phenomenon in which a gamma ray photon [high energy photon] while passing near a nucleus is converted fully to an electron and a positron. A positron [positive electron] is a particle having mass equal to that of an electron and charge equal to that of a proton. It is the antiparticle of an electron. Symbolically, pair production phenomenon can be written as

$$\gamma\text{-ray photon} \xrightarrow{\text{in the vicinity of a nucleus}} \text{electron } e^- + \text{positron } e^+$$

No conservation principles are violated in pair production.

i.   *Conservation of charges*   Total charges before pair production [zero] is equal to the total charge after pair production [charge of electron + charge of positron = zero]

ii. *Conservation of energy*   Total energy before pair production is equal to the total energy including rest energy after pair production [energy itself is being converted, energy can not be destroyed]

iii. *Conservation of linear momentum*   Linear momentum of the gamma ray photon = linear momentum of the electron + linear momentum of the positron. For the process to occur near a nucleus, a part of the momentum is used. Due to the enormous mass of the nucleus, it absorbs only a negligible fraction of photon energy.

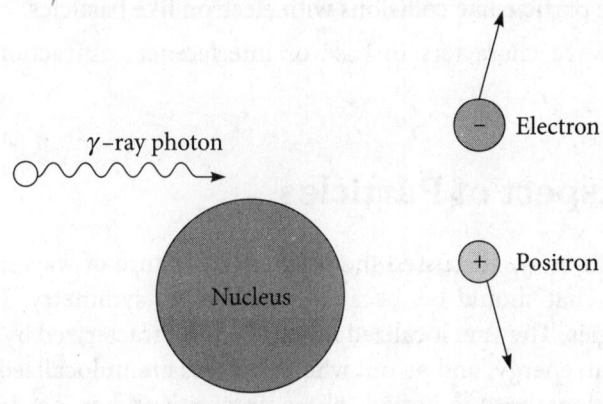

Figure 7.6 | Pair production

The rest energy $m_0 c^2$ of the electron and the positron are each equal to 0.51 Mev. Hence, for pair production to occur, the energy of the photon must be at least 1.02 Mev. The corresponding photon frequency is nearly about 2.5×10^{20} Hz. The electromagnetic wave with such a high frequency is the gamma ray.

For the successful explanation of the Compton effect, photoelectric effect and pair production phenomenon, the electromagnetic radiation is treated as consisting of particles known as photons. All the three effects occurr simultaneously in different magnitudes when radiation passes through matter. If the radiation is a gamma ray of high frequency, pair production dominates over the other two; if the radiation is an X-ray, Compton effect dominates over the other two; and if the radiation is an ultraviolet ray, photoelectric effect dominates over the other two.

Evidently, radiation has a dual character. Depending upon the situation, it reveals itself. Quantum physics accepts the dual nature of radiation where as, classical physics is unable to accept it. In classical physics, laws governing wave characteristics and particle characteristics are fundamentally different. The wave theory and quantum theory complements each other. Either theory of wave and particle can explain only certain phenomena

7.4.5 Characteristics of photon

i.   The carrier of electromagnetic field is a photon that moves with the speed of light.

ii. The rest mass and rest energy of a photon is zero.

iii. The energy of a photon is $h\nu$.

iv. The momentum of a photon is $\dfrac{h}{\lambda}$.

v.   The mass of a photon is $\dfrac{E}{c^2} = \dfrac{h\nu}{c^2}$.

vi. Photons have no electric charge.

vii. They can be created or destroyed when radiation is emitted or absorbed.

viii. They can have particle like collisions with electron like particles.

ix. They reveal wave characters in case of interference, diffraction and polarization phenomena.

7.5 Wave Aspect of Particles

In the previous section, we discussed the particle like nature of waves. Do particles show wave like nature? That should be; because nature loves symmetry. Particles are highly concentrated energies. They are localized in space and characterized by mass, momentum, angular momentum, energy, and so on, whereas waves are unlocalized and characterized by frequency, wavelength, time period, phase, intensity and so on. In the year 1924, in his PhD thesis, Louis de Broglie boldly and daringly proposed, on a theoretical basis, that moving material objects have wave as well as particle characteristics. Though initially the hypothesis was criticized severely by the scientific community, very shortly, after three years in the year 1927, the hypothesis was experimentally verified for electrons by Clinton Davisson and Lester Germer (USA) jointly and G.P. Thomson (England) independently. This opened the door to quantum mechanics. Louis de Broglie's concepts of matter waves are based completely upon the wave aspect of particles.

7.5.1 Matter waves

The two terms, matter waves and de Broglie waves, are synonymously used.

The energy of a photon is given by

$$E = h\nu$$

Relativistically, $E = \sqrt{m_0^2 c^4 + (pc)^2}$ and for a zero rest mass particle like photon $E = pc$

Hence, $pc = h\nu$

or $p = \dfrac{h\nu}{c}$

or $p = \dfrac{h}{\lambda}$ (7.39)

or $\lambda = \dfrac{h}{p}$ (7.40)

where p is the magnitude of the momentum of the particle.

Equation (7.40) connects the momentum of the photon with the wavelength of the corresponding electromagnetic radiation. Though the Eqs (7.39) or (7.40) were derived for photons, de Broglie proposed that they are completely general, applicable to material particles as well as to photons. The wave associated with a material particle obeying Eqs (7.39) or (7.40) is called a de Broglie wave. The wavelength of a material particle λ computed according to Eq. (7.38) is called a de Broglie wavelength. From Eq. (7.39), we have

$$p = \dfrac{\dfrac{h}{2\pi}}{\dfrac{\lambda}{2\pi}}$$

or $p = \hbar k$ (7.41)

where $k = \dfrac{2\pi}{\lambda}$ is called the propagation vector.

If the mass of the material particle is m and it is moving with speed v, the magnitude of the momentum p of the particle will be mv. The de Broglie wavelength associated with the particle will be given by

$$\lambda = \dfrac{h}{mv} \text{ de Broglie hypothesis.}$$ (7.42)

In Eq. (7.42) depending upon the speed of the particle m may be taken as a relativistic mass given by

$$m = \dfrac{m_0}{\sqrt{1 - \dfrac{v^2}{c^2}}} , m_0 = \text{rest mass of the particle.}$$ (7.43)

Relativistically, we know

Total energy = Kinetic energy + Rest energy

or $\quad \sqrt{(pc)^2 + (m_0 c^2)^2} = K + m_0 c^2$

or $\quad p = \dfrac{1}{c}\sqrt{K(K + 2m_0 c^2)}$

Thus, the relativistic form of the de Broglie relation is given by

$$\lambda = \frac{h}{p} = \frac{hc}{pc} = \frac{hc}{\sqrt{K(K + 2m_0 c^2)}}$$

or $\quad \lambda = \dfrac{hc}{\sqrt{K(K + 2m_0 c^2)}}$ \hfill (7.44)

where K is the kinetic energy of a particle of rest mass m_0.

Phase speed of de Broglie waves

The phase speed of a de Broglie wave is defined in the usual way as

$$v_p = \frac{\omega}{k}$$

or $\quad v_p = \dfrac{\hbar\omega}{\hbar k} = \dfrac{E}{p}$ \hfill (7.45)

Equation (7.45) is a general equation applicable to relativistic as well as non-relativistic free particles.

Relativistically free particles

For particles moving with relativistic speed Eq. (7.45) becomes

$$v_p = \frac{E}{p} = \frac{mc^2}{mv}$$

or $\quad v_p = \dfrac{c^2}{v}$ \hfill (7.46)

or $\quad v_p > c$

Thus, the phase speed of the matter wave for a relativistically freely moving particle is more than the speed of light in vacuum!

Non-relativistically free particles

For particles moving with non-relativistic speed, Eq. (7.45) becomes

$$v_p = \frac{E}{p} = \frac{\frac{1}{2}mv^2}{mv}$$

or $\quad v_p = \frac{v}{2}$ $\qquad (7.47)$

Thus, the phase speed of a matter wave for a non-relativistically freely moving particle is half of the particle speed.

Group speed of de Broglie waves

Group speed of a de Broglie wave is defined, in the usual way, as

$$v_g = \frac{d\omega}{dk}$$

or $\quad v_g = \frac{d\hbar\omega}{d\hbar k} = \frac{dE}{dp}$ $\qquad (7.48)$

Equation (7.48) is a general equation applicable to relativistic as well as non-relativistic free particles.

Relativistically free particles

For particles moving with relativistic speed, Eq. (7.48) becomes

$$v_g = \frac{dE}{dp} = \frac{d}{dp}\sqrt{m_0^2 c^4 + p^2 c^2}$$

or $\quad v_g = \frac{pc^2}{\sqrt{m_0^2 c^4 + p^2 c^2}} = \frac{mvc^2}{mc^2}$

or $\quad v_g = v$

Thus, the group speed of a matter wave for relativistically freely moving particles is equal to the speed of the particle.

Non-relativistically free particles
For particles moving with non-relativistic speed, Eq. (7.48) becomes

$$v_g = \frac{dE}{dp} = \frac{d}{dp}\frac{p^2}{2m}$$

or $$v_g = \frac{p}{m} = \frac{mv}{m} = v$$

Thus, the group speed of a matter wave for non-relativistically freely moving particles is equal to the particle speed.

Dispersive nature of matter wave

The phase speed of a matter wave is given according to Eq. (7.45) as

$$v_p = \frac{E}{p} = \frac{\sqrt{m_0^2 c^4 + p^2 c^2}}{p} = \frac{pc\sqrt{1 + \frac{m_0^2 c^2}{p^2}}}{p}$$

or $$v_p = c\sqrt{1 + \frac{m_0^2 c^2 \lambda^2}{h^2}} \tag{7.49}$$

Thus, the phase speed of a matter wave depends upon the wavelength of the matter wave even in vacuum. Therefore, even vacuum behaves as a dispersive medium for matter wave!

7.5.2 Davisson–Germer experiment

The first experimental confirmation of de Broglie's hypothesis of matter waves and quantitative confirmation of de Broglie's relation $\lambda = \dfrac{h}{mv}$ were demonstrated accidentally by Clinton Davisson and Lester Germer of USA for electrons. The wave nature of other particles such as protons, neutrons, helions and so on were demonstrated subsequently.

Davisson and Germer were studying the scattering of electrons by the surface of a nickel crystal using the apparatus shown in Fig. 7.7. The energy of electrons in the primary beam, the angle at which they strike the target and the position of the detector could be varied. During the experiment, an accident occurred and air entered into the apparatus to oxidize the crystal surface. The target was removed from its position and the oxide coating was cleaned out by heat treatments and then placed at its original position to continue the experiment.

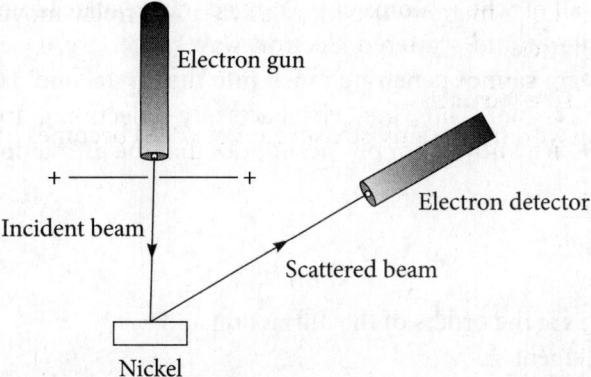

Figure 7.7 Schematic diagram of the apparatus used by Davisson and Germer to study the diffraction of electrons by Nickel crystal. Electrons leave the filament and are accelerated to wards the target under a potential difference of +V volt. The beam strikes the crystal surface and the scattered beam entering into the electron detector is measured. The detector can be moved in the range of 0° to 90°

Now the experimental observation changed abruptly. Instead of a continuous variation of scattered electron intensity with angle, distinct maxima and minima was observed. Moreover, the positions of maxima and minima depend upon the energy of the incident electron. Typical polar graphs are shown in the Fig. 7.8. The method of plotting is such that the intensity at any angle is proportional to the distance of the curve at that angle from the point of scattering. If the intensity were the same at all the scattering angles, the curves would be circles centered on the point of scattering. As shown in Fig. 7.8, when the accelerating voltage is set at 54 V, there is an intense reflection of the beam at an angle $\theta = 50°$. It is difficult to account for this intense reflection through any effect other than the superposition of waves to give an intensity maximum. Heating a block of nickel at high temperature causes the many individual crystals, of which the nickel is composed of, to combine into a

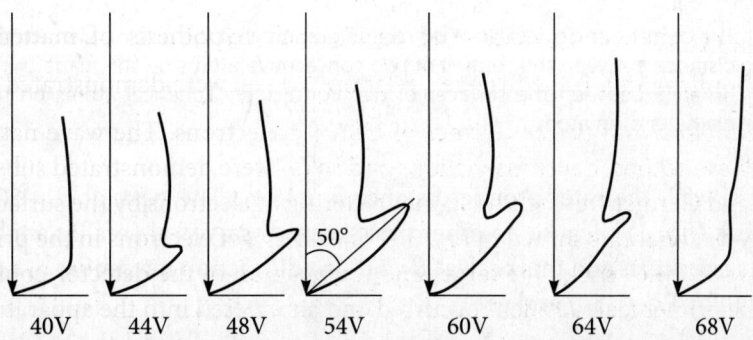

Figure 7.8 Results of Davisson–Germer experiment. When the accelerating voltage is set at 54 V, the intensity of the scattered beam is maximum at an angle $\theta = 50°$.

single large crystal, all of whose atoms are arranged in a regular lattice. Each atom of the crystal act as a scatterer and scattered electron waves interfere. Due to low energy, the incident electron beam cannot penetrate much into the crystal and diffraction takes place on the surface of the crystal. Thus, the crystal acts as a reflection grating for the electrons as shown in Fig. 7.9. We know from physical optics that the diffraction maxima occurs at an angle θ such that

$$(e+b)\sin\theta = n\lambda, \tag{7.50}$$

where $n = 1, 2, 3, \ldots$ are the orders of the diffraction and

$e + b$ = grating element

  = distance between the centres of two consecutive atoms on the plane

  = the shortest distance between the surfaces of two consecutive spherical atoms
    on the plane + diameter of an atom

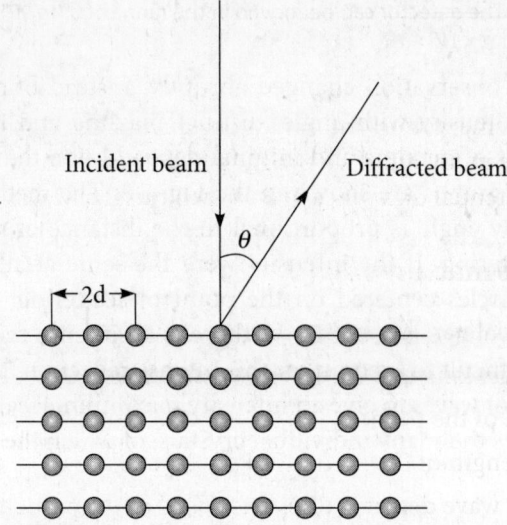

Figure 7.9 The crystal surface acts like a diffraction grating with the grating element equal to the distance between the centres of two consecutive atoms on the plane or the shortest distance between the surfaces of two consecutive spherical atoms on the plane + diameter of an atom

For a nickel crystal, the inter-atomic distance is 2.15 Å. The peak at $\theta = 50°$ must be a first order peak ($n = 1$) because no other peaks were observed at smaller angles. If indeed electrons acts like waves and this peak at $\theta = 50°$ is a diffraction maximum, we can calculate the wavelength by Eq. (7.50) and get

2.15 Å × sin 50 = 1 × λ

or $\lambda = 1.65$ Å

For confirmation of the de Broglie hypothesis, let us calculate the wavelength by applying the de Broglie relation, $\lambda = \dfrac{h}{p}$.

The kinetic energy E of electron of charge e moving under a potential difference of V volt is given by

or $\dfrac{p^2}{2m} = eV$

So $\lambda = \dfrac{h}{\sqrt{2meV}}$

Putting the values of h, m, e and V (= 54 V) into this equation, we get

$$\lambda = \frac{6.63 \times 10^{-34}\,\text{Js}}{\sqrt{2 \times 9.11 \times 10^{-31}\,\text{kg} \times 1.6 \times 10^{-19}\,C \times 54\,\text{V}}} = 1.67\,\overset{\text{O}}{\text{A}}$$

This is in excellent agreement with the value found from the diffraction maximum and provides strong evidence in favour of de Broglie's hypothesis. Thus, the beam of electrons accelerated through a potential difference of 54 V behaves like waves of wavelength 1.67 Å.

7.5.3   Properties of matter wave

i.   Only moving material particles exhibit matter waves.

ii.   Lighter is the particle, longer is the wavelength associated with it.

iii.   Smaller is the speed of the particle, longer is the wavelength associated with it.

iv.   The range of wavelength of the matter wave is from 0 to ∞.

v.   Speed of the matter wave depends upon the speed of the particles.

vi.   Matter waves are completely different from electromagnetic waves. Electromagnetic waves are produced due to acceleration of charges, whereas matter waves are produced due to the motion of material particles and are independent of the charge of the material particles

vii.   Matter waves travel with a speed more than the speed of light in vacuum as $v_p > c$.

viii. The wavelength of matter waves depends upon Planck's constant $h = 6.626 \times 10^{-34}$ Js. Hence, matter waves can be observed experimentally in case of atomic particles whose dimensions are comparable to the wavelength of matter waves.

ix.   Material particles are represented by wave packets having finite width and hence, the uncertainty principle is inherent in matter waves.

x.   The wave aspect and particle aspect of a quantum entity are both necessary for the complete description of atomic or subatomic phenomena. However, both the aspects cannot be revealed simultaneously in a single experiment. The aspects revealed depend upon the nature of the experiment being performed. This is called the principle of complementarity.

xi. Matter waves are not real waves and therefore, cannot be represented by wave displacement. Variation of the height of a water surface constitute water wave; variation of pressure constitute a sound wave; variation in electric and magnetic fields constitute electromagnetic waves. Similarly, variation of a wave function ψ constitutes a matter wave. However, wave function ψ has no direct physical significance, whereas the square of the absolute value of the wave function $|\psi|^2$ known as the probability density has direct physical significance.

xii. The superposition principle holds good for matter waves.

xiii. Matter waves produce interference, diffraction and polarization phenomena

Example 7.9

Compute the de Broglie wavelength in the following cases.

a.   A 100 kg bike travelling with at a speed of 100 km/hour(27.78 m/s).
b.   A 46 gm golf ball moving with a speed of 30 m/s.
c.   A 10 gm bullet moving with a speed of 500 m/s.
d.   A smoke particle of mass 10^{-9} gm moving with a speed 2 cm/s.
e.   An electron moving with an energy of 1 eV (1.6×10^{-19} J).
f.   An electron moving with an energy of 100 MeV (1.6×10^{-11} J).
g.   An electron moving with a speed of 10^7 m/s.

Solutions

a.   Using the classical form of the de Broglie relation, we have

$$\lambda = \frac{h}{p} = \frac{h}{mv}.$$

Putting the given data into this equation, we get

$$\lambda = \frac{6.626 \times 10^{-34}}{100 \times 27.78} \, m = 2.38 \times 10^{-27} \, \text{Å}$$

b.   Using the classical form of the de Broglie relation, and putting the given data into the equation, we get

$$\lambda = \frac{6.626 \times 10^{-34}}{46 \times 10^{-3} \times 30} \, meter = 4.8 \times 10^{-24} \, \text{Å}$$

c. Putting the given data into the classical form of the de Broglie relation, we get

$$\lambda = \frac{6.626 \times 10^{-34}}{10 \times 10^{-3} \times 500} \, \text{m} = 1.32 \times 10^{-24} \, \text{Å}$$

d. Using the classical form of the de Broglie relation, we get

$$\lambda = \frac{6.626 \times 10^{-34}}{10^{-9} \times 10^{-3} \times 0.02} \, \text{m} = 3.31 \times 10^{-10} \, \text{Å}$$

e. The rest energy of an electron is

$$m_0 c^2 = 9.11 \times 10^{-31} \times (3 \times 10^8)^2 \, \text{J} = 5.1 \times 10^5 \, \text{eV}$$

Since the given kinetic energy, 1 eV is much smaller than its rest energy of 5.1×10^5 eV, we can use non-relativistic kinematics.

Putting the given data into the classical form of the de Broglie relation, we get

$$\lambda = \frac{h}{\sqrt{2m_0 K}} = \frac{6.626 \times 10^{-34}}{\sqrt{2 \times 9.11 \times 10^{-31} \times 1.6 \times 10^{-19}}} \, \text{m} = 12.3 \, \text{Å}$$

f. The rest energy of an electron is

$$m_0 c^2 = 9.11 \times 10^{-31} \times (3 \times 10^8)^2 \, \text{J} = 5.1 \times 10^5 \, \text{eV}$$

Since the given kinetic energy, 100 MeV is much larger than to its rest energy of 5.1×10^5 eV, we can use relativistic kinematics. Taking the relativistic form of the de Broglie relation and using the given data, we have

$$\lambda = \frac{6.626 \times 10^{-34} \times 3 \times 10^8}{\sqrt{1.6 \times 10^{-11}(1.6 \times 10^{-11} + 2 \times 8.82 \times 10^{-14})}} \, \text{m} = 12.4 \times 10^{-5} \, \text{Å}$$

Alternatively,

Since the given kinetic energy, 100 MeV is much larger than in its rest energy of 5.1×10^5 eV, we can neglect $2m_0 c^2$ in comparison to K and the relativistic form of the de Broglie relation Eq. (7.44) becomes

$$\lambda = \frac{hc}{\sqrt{K^2}} = \frac{hc}{K}.$$

Putting the given data into this equation, we get $\lambda = 12.4 \times 10^{-5}$ Å

The result calculated by the alternative method does not differ much and in the present case, it is almost equal.

g. The speed of the electron 10^7 m/s is almost 30 times less than that of light. Hence, we are permitted to use the classical form of the de Broglie relation to calculate the wavelength. Thus,

$$\lambda = \frac{h}{mv} = \frac{6.626 \times 10^{-34}}{9.11 \times 10^{-31} \times 10^7} \, \text{m} = 0.73 \, \text{Å}$$

Students may calculate the de Broglie wavelength of the electron in this case by applying the relativistic formula.

Example 7.10

Calculate the kinetic energy of a proton whose de Broglie wavelength is 1.000 femtometers.

Solution

Taking the relativistic form of the de Broglie relation, we have

$$K = \frac{-2m_0 c^2 \pm \sqrt{\left(2m_0 c^2\right)^2 + \left(2hc/\lambda\right)^2}}{2} = -m_0 c^2 \pm \sqrt{\left(m_0 c^2\right)^2 + \left(hc/\lambda\right)^2}$$

Since kinetic energy cannot be negative from this equation, we have

$$K = -m_0 c^2 + \sqrt{\left(m_0 c^2\right)^2 + \left(hc/\lambda\right)^2}$$

Putting the values of $m_0 = 1.67 \times 10^{-27}$ kg (rest mass of proton) and $\lambda = 1.000$ fm $= 1.000 \times 10^{-15}$ m into this equation, we get $K = 618$ MeV.

7.6 Atom Models

Atom is the smallest unit of matter that has the characteristic properties of a chemical element. As such, the atom is the basic building block of all matter. Structure of atom had always been one of main targets of scientific communities. The decoding of atomic structure and development of quantum physics went on parallely. The failures or limitations of Thomson model (1904), Rutherford model (1911), and Bohr model (1913) in explaining the observed atomic phenomena completely, played an important role in the groundwork of quantum physics. Out of the aforementioned three models, Bohr's model was best, because interestingly he had incorporated few quantum mechanical concepts intuitively. Nowadays, the vividly clear picture of atomic or subatomic phenomena can be obtained simply by solving a second order differential equation called Schrodinger equation! Thus, quantum physics provided the best model for microscopic particles including photons.

7.6.1 Rutherford's atom model

J. J. Thomson's plum-pudding model in which electrons are immersed in a cloud of positive charge was rejected on the ground that it predicted a single spectral line for the hydrogen atom in contrary to the experimental observation of a series of spectral lines. Rutherford's atom model, proposed in 1911 from his famous α-particles scattering experiment, assumed that the positive charges were highly concentrated inside a small massive sphere called the nucleus and the electrons were revolving around the nucleus in circular paths. Rutherford's atom model was able to explain a lot of experimental observations.

Objections to Rutherford's atom model

i.   However, revolving electrons according to electromagnetic theory should lose energy by radiating electromagnetic radiation continuously at the rate

$$\frac{1}{6\pi\varepsilon_o}\frac{e^2}{c^3}a^2$$

[a is the acceleration] and finally should fall into the nucleus making the atoms collapse. It is contrary to the observed fact that atom is a stable entity.

ii.   Again according to Rutherford's model, an atom should emit continuous radiation of all frequencies which is against the experimental fact that atoms emit spectral lines of only definite frequencies.

Therefore, Rutherford's atom model based on classical physics was not fully satisfactory.

7.6.2   Bohr's atom model

Following the aforementioned serious objections to Rutherford's atom model, Bohr suggested some modifications in his model. The model is based on the following basic postulates.

i.   The central part of the atom called nucleus contains the whole of the positive charges; almost the whole of the mass of the atom is concentrated in it.

ii.   Electrons are capable of revolving around the nucleus in certain fixed orbits called stationary orbits or privileged orbits. The required centripetal force is provided by the electrostatic force of attraction between the negatively charged electron and the positively charged nucleus.

iii. Atoms do not radiate energy as long as the electrons are in these certain fixed stationary orbits.

iv.   Atoms radiate energy in the form of quanta when the electrons make transitions from one stationary orbit to another. The energy carried by one quantum or photon $h\nu$ is equal to the energy difference of the two stationary orbits, i.e., $h\nu = E_2 - E_1$.

v.   The angular momentum $\vec{L} = \vec{r} \times \vec{p}$ possessed by an electron in the stationary orbits is an integral multiple of

$$\hbar = \frac{h}{2\pi} = 1.054 \times 10^{-34} \text{ Js, i.e.,}$$

$$L = n\hbar, \quad n = 1, 2, 3, 4, \ldots$$

Here n is called the principal quantum number.

Bohr's theory of the hydrogen atom

Hydrogen atom consists of one proton (nucleus) and an electron revolving around the proton/nucleus in a fixed orbit as shown in the Fig. 7.10.

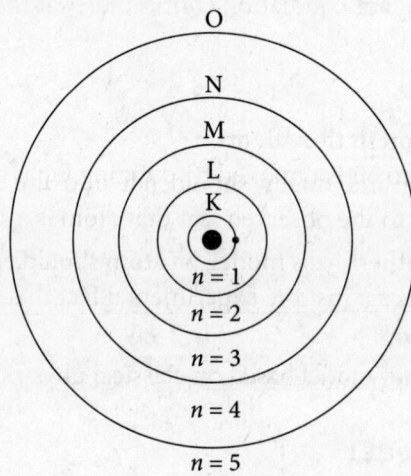

Figure 7.10 | Schematic diagram of a hydrogen atom consisting of one proton and one electron. In the ground state of the atom, the electron stays in the first orbit $n = 1$. The atom is said to be in excited state when the electron, by absorbing energy, jumps to higher orbits with $n > 1$. This diagram is in conformity with Eq. (7.55) $r_n = \dfrac{h^2}{4\pi^2 mke^2} n^2$. The variation of atomic radius with principal quantum number is parabolic

If radius of the orbit is r the charge of the proton is $+e$ and that of the electron is $-e$, the force of attraction F_e between the nucleus and the electron is given by

$$F_e = k\frac{e \times e}{r^2} = k\frac{e^2}{r^2} \tag{7.51}$$

where in SI system for free space, $k = \dfrac{1}{4\pi\varepsilon_0} = 9 \times 10^9 \, Nm^2C^{-2} =$ with $\varepsilon_0 =$ permittivity of free space $= 8.85 \times 10^{-12} \, C^2m^{-2}N^{-1}$

This force of attraction F_e produces a centripetal force $\dfrac{mv^2}{r}$, where v is the speed of the electron in the orbit and is called orbital speed. Therefore, we can have

$$\frac{mv^2}{r} = k\frac{e^2}{r^2}$$

or $\quad mv^2 = k\dfrac{e^2}{r}$ $\hfill$ (7.52)

According to Bohr's postulates, the angular momentum $L = mvr$ possessed by an electron in the stationary orbits of radius r is an integral multiple of $\dfrac{h}{2\pi}$, i.e.,

$$L = n\dfrac{h}{2\pi}$$

or $\quad mvr = n\dfrac{h}{2\pi}$ $\hfill$ (7.53)

Orbital speed of the electron in the nth orbit
The orbital speed of the electron is obtained by dividing Eq. (7.52) by Eq. (7.53) as

$$v_n = 2\pi k\dfrac{e^2}{nh}$$ $\hfill$ (7.54)

Radius of the nth orbit
The radius of the nth orbit is obtained from Eq. (7.53) as

$$r_n = \dfrac{nh}{2\pi m}\dfrac{1}{v}$$

Using the value of v from Eq. (7.54), we get

$$r_n = \dfrac{h^2}{4\pi^2 mke^2}n^2$$ $\hfill$ (7.55)

Total energy of electron
The total energy possessed by the electron moving in the nth orbit, E_n is sum of the kinetic energy, E_k and potential energy, E_p, i.e., $E_n = E_k + E_p$

a. The kinetic energy, E_k of the electron is obtained from Eq. (7.52) as

$$E_k = \dfrac{ke^2}{2r}$$

b. The potential energy E_p of the hydrogen nucleus (a single proton of charge $+e$) and the electron system is given by

$$E_p = k\dfrac{(e)\times(-e)}{r} = -\dfrac{ke^2}{r}$$

Therefore, the total energy possessed by the electron moving in the nth orbit E_n will be obtained as

$$E_n = E_k + E_p = -\frac{ke^2}{2r}$$

Using the value of r from Eq. (7.48) into this equation, we obtain

$$E_n = -\frac{2\pi^2 mk^2 e^4}{h^2 n^2} = -\frac{13.6 eV}{n^2} \qquad (7.56)$$

The total energy possessed by the electron moving in the nth orbit E_n as calculated using Eq. (7.56) is found to be negative. For the stability of any dynamical system, the total energy has to be negative. Therefore, the hydrogen atom is stable. When $n \to \infty$, according to Eq. (7.56), $E_n = 0$. Therefore, when an electron is very far away from the nucleus, its energy becomes zero and is free from the nucleus.

When electron is in the first orbit ($n = 1$), its energy is $E_1 = -13.6$ eV and the atom is said to be in the ground state. When the atom is in any other orbit, the atom is said to be in an excited state. The energy of the electron when it is in the second orbit ($n = 2$) is

$$E_2 = -\frac{13.6 eV}{2^2} = -3.4 eV.$$

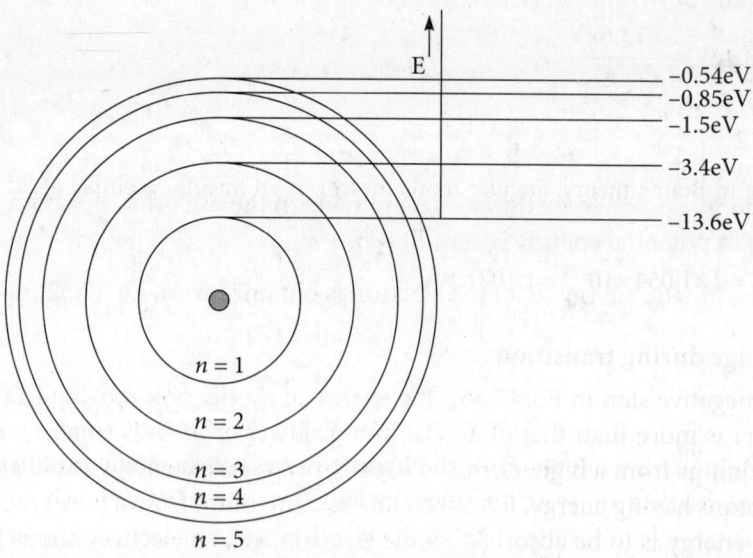

Figure 7.11 | Energy levels of hydrogen atoms in different excited state. The diagram is drawn in conformity with Eq. (7.56) $E_n = -\dfrac{2\pi^2 mk^2 e^4}{h^2 n^2} = -\dfrac{13.6 eV}{n^2}$

If the electron in the ground state of the hydrogen atom absorbs 13.6 eV energy, its energy will be –13.6 eV + 13.6 eV = 0 and is just free from the nucleus. If the electron in the ground state of the hydrogen atom absorbs 10.2 eV energy, its energy will be –13.6 eV + 10.2 eV – 3.4 eV = energy of the second level. Hence, the electron by absorbing 10.2 eV energy, will jump to the second orbit.

Example 7.11

Calculate the energy of the electron in the second orbit.

Solution

$$E_n = -\frac{13.6 \text{ eV}}{n^2} = -\frac{13.6}{4} \text{ eV} = -3.4 \text{ eV}$$

Example 7.12

Calculate the angular momentum of the electron, according to Bohr's theory, in a hydrogen atom when its energy is –3.4 eV.

Solution

The total energy of the electron when it is in the nth orbit is

$$E_n = -\frac{13.6 \text{ eV}}{n^2}$$

Hence, $-3.4\,eV = -\dfrac{13.6\,eV}{n^2}$

or $\quad n = 2$

According to Bohr's theory, angular momentum L is an integral multiple of $\hbar$. Hence, we have

$$L = n\hbar = 2 \times 1.054 \times 10^{-34} = 1.108 \times 10^{-34} \text{ Js}$$

Energy change during transition

Due to the negative sign in Eq. (7.56), the energy of an electron moving in a higher orbit (n is greater) is more than that of an electron at lower orbit (n is smaller). Hence, when an electron jumps from a higher orbit to a lower orbit, excess energy is radiated out in the form of photons having energy, $h\nu$. When an electron jumps from a lower orbit to a higher orbit, some energy is to be absorbed by the electron, i.e., an electron jumps from a lower orbit to a higher orbit by absorbing energy. If n_1 is a lower orbit of energy

$$E_1 = -\frac{2\pi^2 m k^2 e^4}{h^2 n_1^2}$$

and n_2 is a higher orbit of energy

$$E_2 = -\frac{2\pi^2 mk^2 e^4}{h^2 n_2^2},$$

the amount of energy emitted in the form of photons when an electron jumps from n_2 to n_1 will be $E_2 - E_1$. According to Bohr's postulates,

$$h\nu = E_2 - E_1 = \frac{2\pi^2 mk^2 e^4}{h^2}\left(\frac{1}{n_1^2} - \frac{1}{n_2^2}\right)$$

The frequency ν of the radiation emitted when an electron jumps from a higher energy level to a lower energy level is obtained from this equation as

$$\nu = \frac{2\pi^2 mk^2 e^4}{h^3}\left(\frac{1}{n_1^2} - \frac{1}{n_2^2}\right) \qquad (7.57)$$

or

$$\frac{1}{\lambda} = \frac{2\pi^2 mk^2 e^4}{ch^3}\left(\frac{1}{n_1^2} - \frac{1}{n_2^2}\right)$$

or

$$\frac{1}{\lambda} = R_H\left(\frac{1}{n_1^2} - \frac{1}{n_2^2}\right) \qquad (7.58)$$

where $R_H = \dfrac{2\pi^2 mk^2 e^4}{ch^3} = \dfrac{me^4}{8\varepsilon_o^2 ch^3} = 1.09737315685\times10^7\,\mathrm{m}^{-1}$ is called Rydberg's constant for hydrogen atom. Equation (7.58) gives the expression for wave number $\dfrac{1}{\lambda}$.

Example 7.13

How much energy is released when an electron in a hydrogen atom jumps from the third orbit to the second orbit? What is the wavelength of the radiation emitted?

Solution

The total energy of the electron when it is in the third orbit is

$$E_3 = -\frac{13.6\,\mathrm{eV}}{3^2} = -\frac{13.6}{9}\,\mathrm{eV} = -1.5\,\mathrm{eV}$$

The total energy of the electron when it is in the second orbit is

$$E_2 = -\frac{13.6\,\mathrm{eV}}{n^2} = -\frac{13.6}{4}\,\mathrm{eV} = -3.4\,\mathrm{eV}$$

The amount of energy released when an electron in a hydrogen atom jumps from the third orbit to the second orbit is

$$E = E_3 - E_2 = 1.9\,\text{eV}$$

If v is the frequency, we have

$$\lambda = \frac{hc}{E} = \frac{6.626 \times 10^{-34} \times 3 \times 10^8}{3.04 \times 10^{-19}} = 6539\,\text{Å}$$

Limitations of Bohr's theory

Bohr's theory successfully explained a number of experimental observations and has correctly predicted the spectral lines of hydrogen atom, single ionized helium ion and so on in terms of only the principal quantum number, n. However, Bohr's theory has the following limitations

i.   The theory does not give any information regarding the distribution and arrangements of electrons in the atoms.

ii.   Spectra of multi-electron atoms cannot be explained by Bohr's theory.

iii.   Bohr's theory does not explain the variations of intensity of spectral lines.

iv.   The theory cannot be used to calculate the rate of transition of electrons from one level to another level. It is not applicable to the selection rules that apply to them.

v.   The theory fails to explain the fine structure of spectral lines.

vi.   The theory is not applicable for the quantitative explanation of chemical bonding.

vii.   The Zeeman effect [splitting of spectral lines by magnetic fields] and the Stark effect [splitting of spectral lines by electric fields] cannot be explained by applying Bohr's theory.

In spite of all this limitations, Bohr's theory of the hydrogen atom is a mile-stone in the development of quantum physics.

Spectral lines of hydrogen atom

According to Eq. (7.56), the energy of an electron moving in a higher orbit is more than that of an electron at a lower orbit. By absorbing energy from external sources, the electron jumps to a higher energy level from a lower energy level and the atom is in excited state. Within a very short time of less than 10^{-8} s, the electron jumps back to the original lower energy level by emitting radiations and the atom returns to the ground state. The radiations of different wavelengths emitted when atoms return to their ground state from the excited state produce spectral lines. The spectral lines are the bright lines seen against the dark background or the dark lines seen against the bright background of the spectrometer. They are the characteristic of the radiation emitting atoms.

The wavelengths of different spectral lines are calculated using the relation

$$\frac{1}{\lambda} = R_H \left(\frac{1}{n_1^2} - \frac{1}{n_2^2} \right) \text{ [Eq. (7.58)].}$$

This relation explains the complete spectrum of hydrogen atoms. The wavelengths of the radiation depend upon the initial and final energy levels between which transition takes place. Accordingly, a number of series are emitted. Each series is composed of a number of lines. Detailed accounts of five important series are given here.

Lyman series ($n_1 = 1$)

The Lyman series is composed of the lines which are emitted when electronic transition takes place to the energy level with the principal quantum number $n_1 = 1$ from all the other outer energy levels. In other words, the Lyman series is composed of the lines which are emitted when an electron jumps from the outer energy levels to the first energy level with principal quantum number $n_1 = 1$. The wavelength of each line of the Lyman series can be calculated using the formula

$$\frac{1}{\lambda_{Lyman}} = R_H \left(1 - \frac{1}{n_2^2} \right) \text{ with } n_2 = 2, 3, 4, \ldots \tag{7.59}$$

The wavelength of the first line ($n_2 = 2$) of the Lyman series is obtained using

$$\frac{1}{\lambda_{Lyman-1}} = 10973731 \times 10^{-10} \left(1 - \frac{1}{2^2} \right) (\overset{o}{A})^{-1}$$

or $\lambda_{Lyman-1} = 1215 \overset{o}{A}$

The wavelength of the second line ($n_2 = 3$) of the Lyman series is obtained as $\lambda_{Lyman-2} = 1025$ Å. Similarly, the wavelengths of other lines can be calculated. The wavelength of the limiting line ($n_2 = \infty$) of the Lyman series is obtained as $\lambda_{Lyman-\infty} = 911$ Å.

The Lyman series lie in ultra-violet region of the spectrum.

Balmer series ($n_1 = 2$)

The Balmer series is composed of the lines that are emitted when electronic transition takes place to the energy level with the principal quantum number $n_1 = 2$ from all other outer energy levels. In other words, the Balmer series is composed of the lines which are emitted when an electron jumps from the outer energy levels to the second energy level with principal quantum number $n_1 = 2$. The wavelength of each line of the Balmer series can be calculated using the formula

$$\frac{1}{\lambda_{Balmer}} = R_H \left(\frac{1}{4} - \frac{1}{n_2^2} \right) \text{ with } n_2 = 3, 4, 5, \ldots \tag{7.60}$$

The wavelength of the first line ($n_2 = 3$) of the Balmer series is obtained by

$$\frac{1}{\lambda_{\text{Balmer}-1}} = 10973731 \times 10^{-10} \left(\frac{1}{4} - \frac{1}{3^2} \right) (\text{Å})^{-1}$$

or $\lambda_{\text{Balmer}-1} = 6561 \text{ Å}$

The wavelength of the second line ($n_2 = 4$) of the Balmer series is obtained as $\lambda_{\text{Balmer}-2} = 4860$ Å. Similarly, the wavelengths of other lines can be calculated. The wavelength of the limiting line ($n_2 = \infty$) of the Balmer series is obtained as $\lambda_{\text{Balmer}-\infty} = 3645$ Å. Balmer series lie in the visible and near ultra-violet region of the spectrum.

Paschen series ($n_1 = 3$)

The Paschen series is composed of the lines that are emitted when electronic transition takes place to the energy level with principal quantum number $n_1 = 3$ from all other outer energy levels. In other words, the Paschen series is composed of the lines which are emitted when the electron jumps from the outer energy levels to the third energy level with principal quantum number $n_1 = 3$. The wavelength of each line of the Paschen series can be calculated using the formula

$$\frac{1}{\lambda_{\text{Paschen}}} = R_H \left(\frac{1}{9} - \frac{1}{n_2^2} \right) \text{ with } n_2 = 4, 5, 6, 7, \ldots \qquad (7.61)$$

The wavelength of the first line ($n_2 = 4$) of the Paschen series is obtained as $\lambda_{\text{Paschen}-1} = 18746$ Å. The wavelength of the second line ($n_2 = 5$) of the Paschen series is obtained by $\lambda_{\text{Paschen}-2} = 12815$ Å. Similarly, the wavelengths of other lines can be calculated. The wavelength of the limiting line ($n_2 = \infty$) of the Paschen series is obtained as $\lambda_{\text{Paschen}-\infty} = 8201$ Å.

The Paschen series lie in the infra-red region of the spectrum.

Brackett series ($n_1 = 4$)

The Brackett series is composed of the lines that are emitted when electronic transition takes place to the energy level with principal quantum number $n_1 = 4$ from all other outer energy levels. In other words, the Brackett series is composed of the lines which are emitted when the electron jumps from the outer energy levels to the fourth energy level with principal quantum number $n_1 = 4$. The wavelength of each line of the Brackett series can be calculated using the formula

$$\frac{1}{\lambda_{\text{Brackett}}} = R_H \left(\frac{1}{16} - \frac{1}{n_2^2} \right) \text{ with } n_2 = 5, 6, 7, \ldots \qquad (7.62)$$

The wavelength of the first line ($n_2 = 5$) of the Brackett series is obtained as $\lambda_{\text{Brackett}-1} = 40501$ Å. The wavelength of the second line ($n_2 = 6$) of the Brackett series is obtained as $\lambda_{\text{Brackett-2}} = 26244$ Å. Similarly, the wavelengths of other lines can be calculated. The wavelength of the limiting line ($n_2 = \infty$) of the Brackett series is obtained as $\lambda_{\text{Brackett-}\infty} = 14580$ Å. The Brackett series lie in infra-red region of the spectrum.

Pfund series ($n_1 = 5$)

The Pfund series is composed of the lines that are emitted when electronic transition takes place to the energy level with principal quantum number $n_1 = 5$ from all other outer energy levels. In other words, the Brackett series is composed of the lines which are emitted when the electron jumps from outer energy levels to the fifth energy level with principal quantum number $n_1 = 5$. The wavelength of each line of the Pfund series can be calculated using the formula

$$\frac{1}{\lambda_{\text{Pfund}}} = R_H \left(\frac{1}{25} - \frac{1}{n_2^2} \right) \text{ with } n_2 = 6, 7, 8, 9, \ldots \tag{7.63}$$

The wavelength of the first line ($n_2 = 6$) of the Pfund series is obtained as $\lambda_{\text{Pfund-1}} = 74558$ Å. Similarly, the wavelengths of other lines can be calculated. The wavelength of the limiting line ($n_2 = \infty$) of the Pfund series is obtained as $\lambda_{\text{Pfund}} = 22782$ Å. The Pfund series lie in the infra-red region of the spectrum.

The emission of all the five important series discussed here are represented diagrammatically in Fig. 7.12.

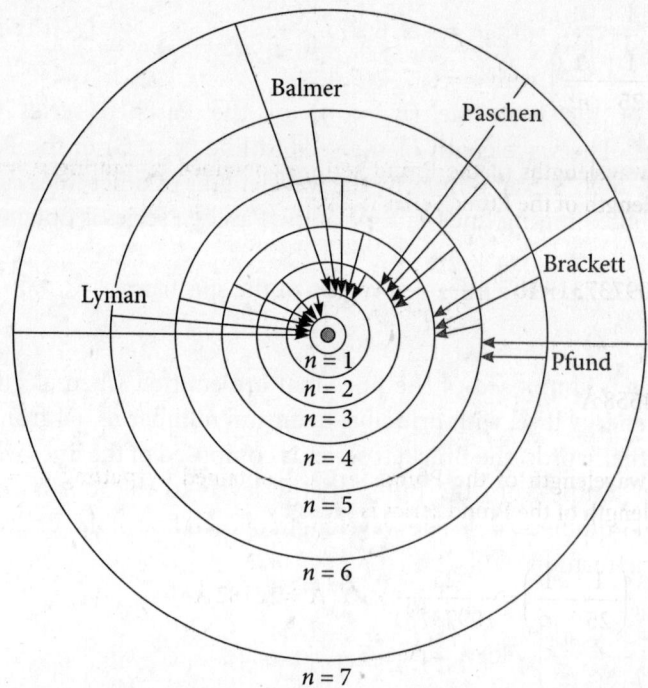

Figure 7.12 Representation of Lyman ($n_1 = 1$), Balmer ($n_1 = 2$), Paschen ($n_1 = 3$), Brackett ($n_1 = 4$) and Pfund ($n_1 = 5$) series diagrammatically

Example 7.14

The first line of the Lyman series has wavelength 1215 Å. Calculate the wavelength of the second line.

Solution

The wavelengths of the first line ($n_2 = 2$) and the second line ($n_2 = 3$) of the Lyman series are given respectively by $\lambda_1 = \dfrac{4}{3R_H}$ and $\lambda_2 = \dfrac{9}{8R_H}$

Therefore, $\dfrac{\lambda_2}{\lambda_1} = \dfrac{27}{32}$

or $\lambda_2 = \dfrac{27}{32} \times \lambda_1 = \dfrac{27}{32} \times 1215 \, \text{Å} = 1025.2 \, \text{Å}$

Example 7.15

Calculate the maximum and minimum wavelengths of the Pfund series.

Solution

The wavelength of the Pfund series is obtained from

$$\frac{1}{\lambda_{\text{Pfund}}} = R_H \left(\frac{1}{25} - \frac{1}{n_2^2} \right) \text{ with } n = 6, 7, 8, 9, \ldots$$

The maximum wavelengths of the Pfund series is obtained by putting $n_2 = 6$. Hence, the maximum wavelength of the Pfund series is given by

$$\frac{1}{\lambda_{\text{PfundMax}}} = 10973731 \times 10^{-10} \left(\frac{1}{25} - \frac{1}{6^2} \right) \left(\text{Å} \right)^{-1}$$

or $\lambda_{\text{PfundMax}} = 74558 \, \text{Å}$

The minimum wavelength of the Pfund series is obtained by putting $n_2 = \infty$. Hence, the minimum wavelength of the Pfund series is given by

$$\frac{1}{\lambda_{\text{PfundMin}}} = R_H \left(\frac{1}{25} - \frac{1}{\infty} \right) = \frac{25}{10973731} \times 10^{10} \, \text{Å} = 22782 \, \text{Å}$$

7.7 Heisenberg's Uncertainty Principle

It is not possible to know the future precisely. The uncertainty principle is a direct consequence of the wave–particle duality of nature. When we represent a particle as a wave packet, its dimension and position loses their precise meanings. The particle may be anywhere inside the wave packet. The position of the particle becomes more defined as the wave packet becomes smaller and smaller. The average value of wavelength in a smaller wave packet is less well defined as the smaller wave packet contains less number of waves. If wavelength λ is not well defined, then momentum $p = (h/\lambda)$ cannot be well defined. In summary, we observe that whenever the position of the particle is well defined, its momentum becomes ill defined and vice versa!

Basing on these concepts, Werner Heisenberg in the year 1927 stated the uncertainty principle. When Heisenberg first propounded the principle, his reasoning was based, however, on the wave–particle duality of the photon. Heisenberg's uncertainty principle is one of the most significant natural laws.

7.7.1 Statement

Heisenberg's uncertainty principle states that it is not possible to make a simultaneous determination of the position and the momentum of a particle with unlimited precision. Mathematically, Heisenberg's uncertainty principle is given by

$$\Delta x \Delta p \geq \frac{\hbar}{2} \tag{7.64}$$

(Heisenberg's uncertainty principle for linear momentum and position)
The other two uncertainty principles, mathematically, are as follows:

$$\Delta E \, \Delta t \geq \frac{\hbar}{2}. \tag{7.65}$$

(Heisenberg's uncertainty principle for energy and time)

$$\Delta J \Delta \theta \geq \frac{\hbar}{2} \tag{7.66}$$

(Heisenberg's uncertainty principle for angular momentum and angle)
In these relations

Δx = the uncertainty in measurement of the position of the particle.

Δp = the uncertainty in measurement of the momentum of the particle.

ΔE = the uncertainty in measurement of the energy of the particle.

Δt = the uncertainty in measurement of the time of the particle.

ΔJ = the uncertainty in measurement of the total angular momentum of the particle.

$\Delta\theta$ = the uncertainty in measurement of the angle made by the particle.

7.7.2 Explanation

According to Heisenberg's uncertainty principle, there is a limit to the precision with which the position and the momentum of an object can be measured at the same time. Depending upon the experimental conditions, either quantity can be measured as precisely as desired at least in principle, but the more precisely one of the quantities is measured, the less precisely the other is known. The uncertainty principle is significant only on the atomic scale because of the small value of h. If the position of a macroscopic object of mass one gram is measured with a precision of 10^{-6} m, the uncertainty principle states that its speed cannot be measured with a precision better than about 10^{-25} m/s. Such a limitation is hardly worrisome. However, if an electron is located in an atom about 10^{-10} m across, the principle gives a minimum uncertainty in the speed of about 10^6 m/s. This reasoning leading to the uncertainty principle is based on the wave–particle duality of the electron. The limits on the measurement imposed by Eq. (7.64) have nothing to do with the crudity of our measuring instruments.

The uncertainty in the momentum of the electron is proportional to the momentum of the colliding photon, which is inversely proportional to the wavelength of the photon. Hence, it is again the case that increased precision in the knowledge of the position of the electron is gained only at the expense of decreased precision in knowledge of its momentum. Heisenberg's reasoning brings out clearly the fact that the smaller the particle being observed, the more significant is the uncertainty principle.

Planck's constant h probably appears no–where that has more deep-seated significance than in Eq. (7.64). If this product had been zero instead of $\hbar/2$, classical concepts about energy, momentum, position, and so on of particles would be correct. It would then be possible to measure both momentum and position simultaneously with unlimited precision. The fact that h appears means classical concepts are not fully correct. The magnitude of h tells up to what extent the classical concepts are correct and beyond what extent the classical concepts must be replaced by quantum concepts.

7.7.3 Experimental illustration of the uncertainty principle

Due to the smallness in the value of Planck's constant h, the effects of the uncertainty principle are not observed in our day-to-day life. The smaller the particle being observed, the more significant is the uncertainty principle. The small value of Planck's constant cannot stop us from doing some thought experiments on the uncertainty principle.

Electron diffraction by a single slit

Let a parallel beam of electron be incident perpendicularly on a single slit as shown in Fig. 7.13.

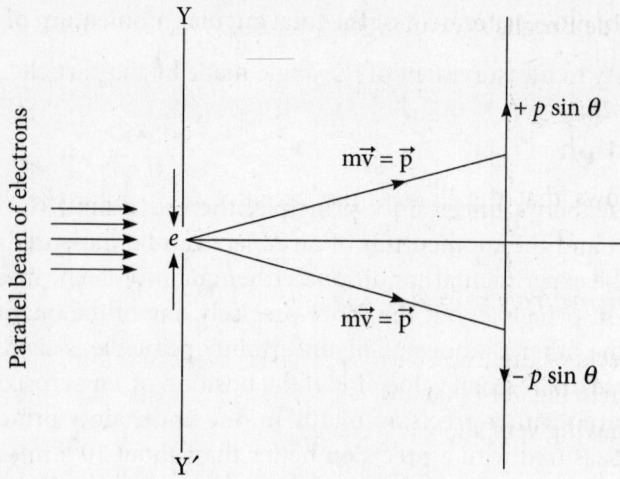

Figure 7.13 Diffraction of electron beam by a single slit of width e. The uncertainty in position along the Y-axis is e and the uncertainty in momentum along the Y-axis is 2p sin θ.

When the electron beam passes through the slit of width e the maximum uncertainty in the position Δy of the electron will be equal to the width of the slit. Hence, we have

$$\Delta y = e \tag{7.67}$$

By applying Eq. (3.64), the position of the first minima of the diffraction pattern due to a single slit is given by

$$e \sin \theta = \lambda \tag{7.68}$$

As shown in Fig. 7.13, the Y-component of the momentum of the electron is $\pm p \sin \theta$. The Y-component of the momentum of the electron cannot exceed the range from $-p \sin \theta$ to $+p \sin \theta$. Hence, when the electron beam passes through the slit, the maximum uncertainty in the momentum along the Y-axis will be $p \sin \theta + p \sin \theta = 2p \sin \theta$. Thus, we have

$$\Delta p = 2p \sin \theta \tag{7.69}$$

Multiplication of Eqs (7.67) and (7.69) gives

$$\Delta y \Delta p = e \times 2p \sin \theta \quad \text{(uncertainty product)}$$

or $\quad \Delta y \Delta p = 2p(e \sin \theta)$

Putting the value of $e \sin \theta$ from Eq. (7.68) into this equation, we get

$$\Delta y \Delta p = 2p\lambda$$

Putting the value of de Broglie's wavelength from Eq. (7.41) into this equation, we get

$$\Delta y \Delta p = 2p \frac{h}{p} = 2h \tag{7.70}$$

Equation (7.70) shows that the uncertainty product $\Delta y \Delta p$ is of the order of Planck's constant h.

Heisenberg's gamma ray microscope

To visualize the effects of the uncertainty principle, Heisenberg imagined a hypothetical microscope as shown in Fig. 7.14 that utilizes γ-rays as the probe. γ-rays are electromagnetic radiation like light having very short wavelengths.

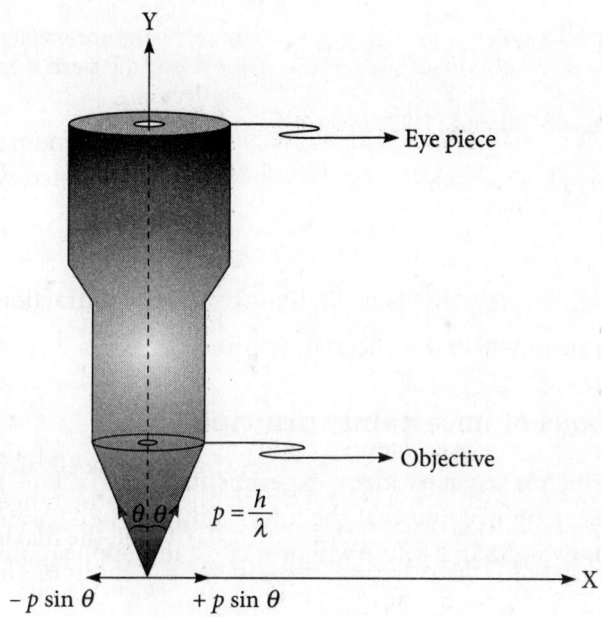

Figure 7.14 | Determination of the position of an electron using γ-ray microscope. θ is the angular aperture of the microscope. The X-component of the momentum of electron will be $p \sin \theta$

Through the γ-ray microscope, electrons can be made visible. γ-ray photons reflected by the electron will enter the objective of the γ-ray microscope through a cone of illumination. This will make the electron visible when seen through the eye-piece. It is a well-known fact that a microscope cannot resolve two points which are separated by a distance which is $\leq$ ($\lambda/2 \sin \theta$), where λ is the wavelength of the radiation and θ is the angular aperture of the microscope. Hence, the maximum possible uncertainty in the position of an electron along the X-axis is $(\lambda/2 \sin \theta)$. Thus, we have

$$\Delta x = \frac{\lambda}{2 \sin \theta} \tag{7.71}$$

Equation (7.71) shows that shorter the wavelengths (γ-rays), smaller will be the uncertainty in the position of the electron. However, a γ-ray photon will suffer Compton recoil by the electron. The recoil momentum will be h/λ. Again for the electron to be visible through the microscope, the reflected photon should be able to enter the microscope through an angle of 2θ. The X-component of the momentum of electron and hence, that of photon will be $\pm p \sin \theta$. The X-component of the momentum of the electron cannot exceed the range from $-p \sin \theta$ to $+p \sin \theta$. Hence, when the photon passes through the aperture, the maximum uncertainty in the momentum along the X-axis will be $p \sin \theta + p \sin \theta = 2p \sin \theta$. Thus, we have

$$\Delta p = 2p \sin \theta = \frac{2h}{\lambda} \sin \theta \tag{7.72}$$

Multiplication of Eqs (7.71) and (7.72) gives

$$\Delta x \Delta p = \frac{\lambda}{2 \sin \theta} \times \frac{2h}{\lambda} \sin \theta$$

or $\quad \Delta x \Delta p = h \tag{7.73}$

This is again in conformity with the uncertainty principle.

7.7.4 Applications of uncertainty principle

The uncertainty principle explains many experimental observations in the microscopic world. It is worth keeping in mind that the lower limit $\hbar/2$ of $\Delta x \Delta p$ is rarely attained. In general, $\Delta x \Delta p \geq \hbar$ or even $\Delta x \Delta p \geq h$. We will now cite a few applications of the uncertainty principle.

Ground state energy of an harmonic oscillator

The total energy of a harmonic oscillator with motion constrained along the X-axis is given by

$$E = \frac{p^2}{2m} + \frac{1}{2} m \omega^2 x^2 \tag{7.74}$$

In the ground state, uncertainties in momentum and positions, i.e., Δp and Δx are of the order of their magnitudes. That means, $\Delta p = p$ and $\Delta x = x$ and the uncertainty principle $\Delta x \Delta p = (\hbar/2)$ becomes

$$xp = \frac{\hbar}{2}$$

or $\quad p = \dfrac{\hbar}{2x}$ $\qquad\qquad$ (7.75)

Putting Eq. (7.75) into Eq. (7.74), we have

$$E = \frac{\hbar^2}{8mx^2} + \frac{1}{2}m\omega^2 x^2 \qquad\qquad (7.76)$$

Since the energy of the harmonic oscillator in the ground state is minimum, applying the principle of minima functions to this equation, we have

$$\frac{dE}{dx}\bigg|_{x=a} = 0 \text{, where } a \text{ is the value of } x \text{ in ground state}$$

Hence, for the minimum energy, we have

$$\frac{d}{dx}\left(\frac{\hbar^2}{8mx^2} + \frac{1}{2}m\omega^2 x^2 \right)\bigg|_{x=a} = 0$$

or $\quad a = \left(\dfrac{\hbar}{2m\omega} \right)^{\frac{1}{2}}$ $\qquad\qquad$ (7.77)

From Eq. (7.76), energy is minimum when $x = a$, i.e.,

$$E_{min} = \frac{\hbar^2}{8ma^2} + \frac{1}{2}m\omega^2 a^2 \qquad\qquad (7.78)$$

Putting the value of a from Eq. (7.77) into the above this equation, we get

$$E_{min} = \frac{\hbar^2}{8m} \times \frac{2m\omega}{\hbar} + \frac{1}{2}m\omega^2 \times \frac{\hbar}{2m\omega}$$

or $\quad E_{min} = \dfrac{1}{2}\hbar\omega$

According to classical physics, the ground state energy of a harmonic oscillator is zero, whereas according to quantum mechanics, it is $(1/2)\hbar\omega$. The latter is in conformity with the experimental result and other theoretical predictions.

Ground state energy of the hydrogen atom

The total energy of a hydrogen atom in its ground state is given by $E = E_k + E_p$ or

$$E = \frac{p^2}{2m} - k\frac{e^2}{x} \tag{7.79}$$

In the ground state, uncertainties in momentum and positions, i.e., Δp and Δx are of the order of their magnitudes. That means, $\Delta p = p$ and $\Delta x = x$ and the uncertainty principle $\Delta x \Delta p = \hbar$ becomes

$$xp = \hbar$$

or $\quad p = \frac{\hbar}{x}$ \hfill (7.80)

Putting Eq. (7.80) into Eq. (7.79), we have

$$E = \frac{\hbar^2}{2mx^2} - k\frac{e^2}{x} \tag{7.81}$$

Since the energy of the hydrogen atom in the ground state is minimum applying the principle of minima functions to this equation, we have

$$\left.\frac{dE}{dx}\right|_{x=a} = 0, \text{ where } a \text{ is the value of } x \text{ in ground state}$$

Hence, for the minimum energy, we have

$$\left.\frac{d}{dx}\left(\frac{\hbar^2}{2mx^2} - \frac{ke^2}{x}\right)\right|_{x=a} = 0$$

or $\quad a = \frac{\hbar^2}{mke^2}$ \hfill (7.82)

Equation (7.82) gives the radius of the atom in the ground state and is equal to the radius of the first orbit as calculated by applying Bohr's theory in Eq. (7.55)

From Eq. (7.81), energy is minimum when $x = a$, i.e.,

$$E_{min} = \frac{\hbar^2}{2ma^2} - k\frac{e^2}{a} \tag{7.83}$$

Putting the value of a from Eq. (7.82) into this equation, we get

$$E_{min} = \frac{\hbar^2}{2m} \times \frac{m^2 k^2 e^4}{\hbar^4} - ke^2 \frac{mke^2}{\hbar^2}$$

or $$E_{min} = -\frac{mk^2 e^4}{2\hbar^2} = -\frac{2\pi^2 mk^2 e^4}{h^2} \qquad (7.84)$$

Equation (7.84) gives the minimum energy of the atom in the ground state and is equal to the energy of the first orbit as calculated by applying Bohr's theory in Eq. (7.56)

Absence of electrons in the nucleus

As electrons are emitted from the nucleus during radioactive β-decay, we may assume that electrons exist inside a nucleus. If possible, let this assumption be correct.

The diameter of a typical nucleus is 1.0×10^{-14} m. Hence, the uncertainty in the position of the electron will be

$$\Delta x = 1.0 \times 10^{-14} \text{ m}$$

Applying the uncertainty principle, $\Delta x \Delta p = \hbar$, the uncertainty in momentum of the electron will be

$$\Delta p = \frac{\hbar}{1.0 \times 10^{-14}} = 1.054 \times 10^{-20} \text{ Ns}$$

If this is the uncertainty in the momentum of an electron inside a nucleus, the momentum p itself must be at least comparable in magnitude. Thus, we can have

$$p = 1.054 \times 10^{-20} \text{ Ns}$$

An electron with such a momentum has kinetic energy E_k many times greater than its rest energy $m_0 c^2$; this kinetic energy is in the relativistic range. Applying the relativistic formula, we have

$$E_k = \text{Total energy} - \text{rest energy}$$

or $$E_k = \sqrt{p^2 c^2 + m_0^2 c^4} - m_0 c^2$$

$$= 3.2 \times 10^{-12} - 8.2 \times 10^{-14} \text{ J} = 19.4 \text{ MeV}$$

Thus, the kinetic energy of an electron ($m_o = 9.11 \times 10^{-31}$ kg) inside a typical nucleus is 19.4 MeV. The electrons emitted from the nucleus in radioactive β-decay typically have kinetic energy of about 1 MeV, much smaller than the uncertainty principle requires for electrons confined inside the nucleus. This calculation thus suggests that β-decay electrons of such low energies cannot be confined in a region the size of the nucleus. The kinetic energy of an electron must exceed 19.4 MeV if it is to be inside a nucleus.

7.8 Transition from Deterministic Classical Physics to Probabilistic Quantum Physics

An essential feature of quantum physics is that it is generally impossible, even in principle, to measure a system without disturbing it; the detailed nature of this disturbance and the exact point at which it occurs are obscure and controversial. Thus, quantum mechanics has attracted some of the ablest scientists of the 20th century, and they have erected what is perhaps the finest intellectual edifice of the period.

The deterministic concept of classical physics loses its significance in quantum physics. Unlike in classical physics, point particles are treated as wave packets in quantum physics. De Broglie's concepts of matter waves or Heisenberg's uncertainty principle cannot be fit into classical physics. Explanation of the natural phenomena in terms of quantum physics necessitates the transition from deterministic classical physics to probabilistic quantum physics

Classical physics is deterministic in nature though nature itself is not deterministic. If we know the exact configuration of a system at present, we can predict the exact configuration of the system in future. For example, if the initial position and speed of a particle are known at present, its exact trajectory in future can be determined with help of the laws of classical physics with cent percent precision. Due to the smallness in the value of Planck's constant h, the exact position and momentum of a particle in the macroscopic world can be predicted in any later time by using the conditions prevalent at an earlier time. In the solar system, the exact time period, speed and so on of different planets and their interactions producing lunar and solar eclipses are predicted precisely.

According to classical physics, the exact trajectory of an electron can be determined. However, experimental observations on electrons or any other atomic particles contradict this classical concept. Even at $t = 0$, the exact position of an electron cannot be known as we have seen in the case of a single slit diffraction of an electron. The position of the electron is somewhere in the slit. In the language of probability, we can say that the probability of the presence of the electron along the length of the slit is one, whereas the probability of the presence of the electron at a particular point along the slit length is not one. Hence, comes the uncertainty in position of the electron at $t = 0$ time. How can we know the exact trajectory of the electron when we cannot even derermine the position of the electron at initial time? We cannot say exactly where an individual electron will strike the screen. We can only know the probability of arrival of the electron in a certain area of the screen.

What does it mean when we say that the probability of arrival of an electron at a certain spot on the screen is 50%? It means that out of a large number of similar electrons, only half of the electrons will arrive at the spot. The language of probability in quantum mechanics is same as that used in our day-to-day language. A surgeon who asserts that a patient has 70% probability of surviving an operation means that with a large number of similar cases, 70% of patients survive. The same language of probability is used in case of weather forecasting in our day-to-day life.

In quantum physics, particles are represented by wave packets. Where is the particle in the wave packet? A wave packet can be considered to be the superposition of a large number of waves that which interfere constructively in the vicinity of the particle, giving a resultant wave of large amplitude and interfere destructively far way from the particle giving a resultant wave of small amplitude. Thus, the probability of finding the particle is high, where the amplitude of the resultant wave is large; the probability of finding the particle is less, where the amplitude of the resultant wave is small. When we say that the wave packet is spread out in space, it does not mean that the particle itself is spread out in space. If we say that the probability of the presence of an electron at a certain point in the wave packet is 30%, it does not mean that 30% of the electron is at that point and the rest 70% of the electrons is at some other points. Therefore, we conclude that the probability of finding the particle at any point depends on the amplitude of its de Broglie wave (matter wave) at that point.

Intensity of a wave at a point

$$\propto |\text{amlitude of the wave at that poin}|^2 \quad \text{(Classical physics)}$$

Probability of finding a particle at a point

$$\propto |\text{amlitude of matter wave at that poin}|^2 \quad \text{(Quantum physics)}$$

This probabilistic description is the fundamental concept in quantum physics and is achieved by defining the wave function.

7.9 Wave Function ψ

Matter waves are not real waves and therefore, cannot be represented by wave displacement. In electromagnetic waves, the electric and magnetic field vary periodically. In sound waves, the pressure varies periodically. In water waves, the height of the water surface varies periodically. Similarly, what varies in matter waves? The quantity which varies in case of the matter wave is the wave function ψ.

In classical physics, the wave function of classical waves like mechanical waves or electromagnetic waves contains all the information about amplitude, speed and wavelength. Similarly, the wave function of a matter wave contains all the information required for the probabilistic description of a quantum mechanical system. However, unlike the wave function of classical waves, the wave function ψ of a matter wave has no direct physical significance, whereas the square of the absolute value of the wave function $|\psi|^2$ known as the probability density has a direct physical significance.

We cannot describe the motion of a particle quantum mechanically by arbitrarily defining a wave function. Only the wave functions with the following characteristics can yield physically meaningful results when used in calculations. Only such well-behaved wave functions are admissible as mathematical representations of real bodies.

7.9.1 Characteristics of the wave function of a matter wave

i.   Wave functions are the mathematical representation of particles that contain all the information required for the probabilistic description of the particles.

ii.   The wave function ψ in general is a mathematical function of space and time, i.e., $\psi = \psi(x, y, z, t)$. The wave function may be a function of the systems.

iii. The wave function ψ in general is complex so that $\psi\psi^* = |\psi|^2$, the probability density is real.

iv.   The wave function ψ must be continuous and single valued everywhere.

v. $\dfrac{\partial \psi}{\partial x}, \dfrac{\partial \psi}{\partial y}$ and $\dfrac{\partial \psi}{\partial z}$ must be continuous and single valued everywhere.

vi. $\lim\limits_{x \to \pm\infty} \psi = 0$, $\lim\limits_{y \to \pm\infty} \psi = 0$ and $\lim\limits_{z \to \pm\infty} \psi = 0$ so that $\int\limits_{-\infty}^{\infty} |\psi|^2 dv = N$, N being a finite constant.

      Under this condition, the wave function ψ is said to be normalizable.

vii. The wave function ψ must satisfy Schrödinger's equation.

7.9.2 Probability density

The probability of finding the particle [described by the wave function ψ] per unit volume is called probability density P and as discussed earlier is given by

$$P = \psi\psi^* = |\psi|^2 \tag{7.85}$$

Hence, the probability of finding the particle in the volume dv will be

$$Pdv = |\psi|^2 \, dv \tag{7.86}$$

and
    the probability of finding the particle in the entire volume will be given by

$$\int\limits_{V} Pdv = \int\limits_{V} |\psi|^2 \, dv \tag{7.87}$$

For the particle described by the wave function, ψ if $\int_V |\psi|^2 dv = 0$, then surely the particle is not present in the specified volume and if $\int_V |\psi|^2 dv = 1$, then surely the particle is present in the specified volume. If the particle is confined to move in one dimension, say along the X-axis, then $\psi \psi^* = |\psi|^2$ will be the probability of finding the particle per unit length and $|\psi|^2 dx$ will be the probability of finding the particle in dx length. The probability of finding the particle in between two points x_1 and x_2, $P_{x_1 x_2}$ will be given by

$$P_{x_1 x_2} = \int_{x_1}^{x_2} |\psi|^2 dx \qquad (7.88)$$

In this equation, ψ is the normalized wave function.

Example 7.16

A particle trapped in a one-dimensional box of length L is described by the normalized wave function

$$\psi = \sqrt{\frac{2}{L}} \sin \frac{n\pi x}{L}.$$

What is the probability that the particle is lying between $0.45\,L$ and $0.50\,L$?

Solution

According to Eq. (7.88), the probability of finding the particle in between two points x_1 and x_2, $P_{x_1 x_2}$ in one dimension is given by

$$P_{x_1 x_2} = \int_{x_1}^{x_2} |\psi|^2 dx$$

or $P_{x_1 x_2} = \int_{0.45L}^{0.5L} \frac{2}{L} \sin^2 \frac{n\pi x}{L} dx$ ****

**** $\int \sin^n x\, dx = -\frac{1}{n} \sin^{n-1} \cos x + \frac{n-1}{n} \int \sin^{n-2} dx$ and

$\int \cos^n x\, dx = \frac{1}{n} \cos^{n-1} \sin x + \frac{n-1}{n} \int \cos^{n-2} dx$ and

$\int \sin^m x \cos^n dx = -\frac{1}{m+n} \sin^{m-1} \cos^{n+1} x + \frac{m-1}{m+n} \int \sin^{m-2} \cos^n x\, dx$

$\int \sin^m x \cos^n x\, dx = \frac{1}{m+n} \sin^{m+1} \cos^{n-1} x - \frac{n-1}{m+n} \int \sin^m \cos^{n-2} x\, dx$

$$= \left(\frac{x}{L} - \frac{1}{2n\pi} \sin \frac{2n\pi x}{L} \right)\Bigg|_{0.45L}^{0.5L} = \frac{1}{40\pi n} \left(\sin \frac{9\pi n}{10} - \sin \pi n \right)$$

This equation shows that the probability of finding the particle in between two points 0.45 L to 0.50 L is a function of n. $n = 1$ corresponds to the ground state of the particle, $n = 2$ corresponds to the first excited state of the particle, $n = 3$ corresponds to the third state of the particle and so on.

The probability of finding the particle in the first excited state ($n = 1$) in between two points 0.45 L to 0.50 L is 0.25%.

Example 7.17

A particle trapped in a one dimensional box of length L is described by the wave function $\psi = ax$. What is the probability that the particle is lying between L_1 and L_2?

Solution

According to Eq. (7.88), the probability of finding the particle in between two points L_1 and L_2, $P_{L_1 L_2}$ in one dimension is given by

$$P_{L_1 L_2} = \int_{L_1}^{L_2} |\psi|^2 \, dx$$

or $P_{L_1 L_2} = \int_{L_1}^{L_2} a^2 x^2 dx = \left(\frac{a^2 x^3}{3} \right)\Bigg|_{L_1}^{L_2} = \frac{a^2}{3} \left(L_2^3 - L_1^3 \right)$

Example 7.18

The normalized wave function for a certain particle is $\psi = \sqrt{\dfrac{8}{3\pi}} \cos^2 x$ for $-\dfrac{\pi}{2} < x < \dfrac{\pi}{2}$. Calculate the probability that the particle can be found between $x = 0$ and $x = \dfrac{\pi}{4}$.

Solution

According to Eq. (7.88), the probability of finding the particle in between two points x_1 and x_2, $P_{x_1 x_2}$ is given by

$$P_{x_1 x_2} = \int_{x_1}^{x_2} |\psi|^2 \, dx$$

or $P_{0 - \frac{\pi}{4}} = \int_0^{\frac{\pi}{4}} \frac{8}{3\pi} \cos^4 x \, dx = \frac{8}{3\pi} \int_0^{\frac{\pi}{4}} \cos^4 x \, dx$

$$= \frac{2}{3\pi} + \frac{1}{4} = 0.462$$

Thus, the probability that the particle be found between $x = 0$ and $x = \frac{\pi}{4}$ is calculated to be 0.462 or 46.2%.

7.9.3 Dimensional analysis of a wave function

The probability of finding the particle described by the wave function ψ is a dimensionless quantity. If the wave function ψ describes a particle moving in space, then $|\psi|^2$ is the probability of finding the particle per unit volume, i.e., probability/volume. Hence, the dimension of probability/volume must be L^{-3}, Therefore,

$$|\psi|^2 = \frac{\text{Probability}}{\text{Volume}}$$

must have the dimension of L^{-3}, i.e.,

The dimension of $|\psi|^2$ is L^{-3}

Hence, the dimension of $|\psi|$ must be $L^{-\frac{3}{2}}$ when the wave function ψ describes a moving particle in space, i.e., the dimension of a three-dimensional wave function describing a particle moving in space is $L^{-\frac{3}{2}}$.

If the particle is confined to move in a plane, i.e., in two dimensions, then $\psi\psi^* = |\psi|^2$ will be the probability of finding the particle per unit area. Hence,

$$|\psi|^2 = \frac{\text{Probability}}{\text{Area}}$$

must have the dimension of L^{-2}, i.e.,

The dimension of $|\psi|^2$ is L^{-2}

Therefore, the dimension of $|\psi|$ must be L^{-1} when the wave function ψ describes a particle moving on a plane, i.e., the dimension of a two-dimensional wave function describing a particle moving on a plane is L^{-1}

If the particle is confined to move in one dimension, then $\psi\psi^* = |\psi|^2$ will be the probability of finding the particle per unit length. Hence,

$$|\psi|^2 = \frac{\text{Probability}}{\text{Length}}$$

must have the dimension of L^{-1}, i.e.,

The dimension of $|\psi|^2$ is L^{-1}

Therefore, the dimension of $|\psi|$ must be $L^{-\frac{1}{2}}$ when the wave function ψ describes a moving particle in a line, i.e., the dimension of a one-dimensional wave function describing a particle moving in a line is $L^{-\frac{1}{2}}$

7.10 Superposition Principle

Schrödinger's equation is a linear second order differential equation. Hence, if ψ_1, ψ_2, ψ_3, ..., ψ_n are n solutions of Schrödinger's equation, then their linear combination, i.e., $a_1\psi_1 + a_2\psi_2 + a_3\psi_3 + \ldots\ldots + a_n\psi_n$ will be a solution of Schrödinger's equation. This is an inherent property of the differential equation. The solutions of Schrödinger's equation are called wave functions. Hence, $\psi = a_1\psi_1 + a_2\psi_2 + a_3\psi_3 + \ldots\ldots + a_n\psi_n$ is a wave function satisfying Schrödinger's equation. Thus, the wave functions ψ_1, ψ_2, ψ_3, ..., ψ_n obey the superposition principle. $|a_i|^2$ is the probability of the system being in a state defined by the wave function ψ_i. For a given system, the possible allowed states are obtained by solving Schrödinger's equation with appropriate boundary conditions.

7.11 Normalization

Two wave functions differing only by a constant factor describe identical systems. The particle which is described by the wave function ψ must be inside the region of our study. This means $\int\limits_V |\psi|^2 dv = 1$. If $\int\limits_V |\psi|^2 dv \neq 1$, we can multiply the wave function ψ by a constant N to obtain another wave function ψ_1, describing an identical system so that $\int\limits_V |\psi_1|^2 dv = 1$, where $\psi_1 = N\psi$

$$\int\limits_V |N\psi|^2 dv = 1$$

or
$$N = \frac{1}{\sqrt{\int\limits_V |\psi|^2 dv}}$$
(7.89)

The constant N is called the normalization constant. The wave function ψ is said to be a normalized wave function if $\int\limits_V |\psi|^2 dv = 1$.

7.11.1 Procedures for calculation of the normalization constant

In quantum mechanics, the following procedures are followed to determine the normalization constant of any wave function ψ.

i.   Evaluate $\int_V |\psi|^2 dv$

ii.   The square root of the reciprocal of the result of (i) is the normalization constant.

To make the procedure clear, let us consider the case of evaluating the normalization constants of the following wave functions.

Let the wave function $\psi = \sin qx$ with $q = \dfrac{n\pi}{L}$ describe a particle in a one-dimensional box of length L. Now we shall normalize the wave function $\psi = \sin qx$ so that the particle is inside the box.

i.   Evaluation of $\int_0^L |\psi|^2 dx$

$$\int_0^L |\psi|^2 dx = \int_0^L \sin^2 qx\, dx = \frac{1}{2q}\int_0^L (1 - \cos 2qx)d(qx) = \frac{L}{2} - \frac{1}{4q}\sin 2qL$$

Putting the value of $q = \dfrac{n\pi}{L}$ into this equation, we have

$$\int_0^L |\psi|^2 dx = \frac{L}{2} - \frac{L}{4n\pi}\sin\left(2\frac{n\pi}{L}L\right) = \frac{L}{2}$$

ii.   The square root of the reciprocal of the result obtained is the normalization constant.

Thus, the normalization constant is $\sqrt{\dfrac{2}{L}}$ and the normalized wave function is

$$\psi = \sqrt{\frac{2}{L}}\sin\left(\frac{n\pi}{L}x\right)$$

Example 7.19

A particle trapped in a one-dimensional box of length L is described by the wave function $\psi = x$. Normalize the wave function between a and b.

Solution

i.   Evaluation of $\int_a^b |\psi|^2 dx$

$$\int_a^b |\psi|^2 dx = \int_a^b x^2 dx = \frac{1}{3}(b^3 - a^3)$$

ii.   The square root of the reciprocal of this result is the normalization constant. Thus, the normalization constant is $\dfrac{\sqrt{3}}{\sqrt{b^3 - a^3}}$ and the normalized wave function is

$$\psi = \frac{\sqrt{3}}{\sqrt{b^3 - a^3}} x$$

Example 7.20

Evaluate the normalization constant N of the wave function $\psi = Nx \exp\left(-\dfrac{x^2}{2}\right)$.

Solution

Our aim is to determine N so that particle exists in the space from $-\infty$ to $+\infty$, i.e.,

$$\int_{-\infty}^{\infty} |\psi|^2\, dx = 1$$

or $1 = \displaystyle\int_{-\infty}^{\infty} |\psi|^2\, dx = \int_{-\infty}^{\infty} N^2 x^2 e^{-x^2}\, dx = 2N^2 \int_{0}^{\infty} x^2 e^{-x^2}\, dx$ *****

This integral is evaluated by applying the gamma function in the following way.

Let $x^2 = u$ so that $x = \sqrt{u}$, and $dx = \dfrac{1}{2} u^{-\frac{1}{2}} du$

This integral becomes

$$\int_{-\infty}^{\infty} |\psi|^2\, dx = 2N^2 \int_{0}^{\infty} u e^{-u} \frac{1}{2} u^{-\frac{1}{2}} du = N^2 \int_{0}^{\infty} u^{\frac{1}{2}} e^{-u} du = N^2 \int_{0}^{\infty} u^{\frac{3}{2}-1} e^{-u} du$$

$$= N^2 \Gamma\left(\frac{3}{2}\right) = N^2 \Gamma\left(1 + \frac{1}{2}\right) = N^2 \times 1 \times \frac{1}{2} \Gamma\left(\frac{1}{2}\right) = N^2 \frac{1}{2} \sqrt{\pi}$$

Thus, we have $N^2 \dfrac{\sqrt{\pi}}{2} = 1$

or $N = \dfrac{\sqrt{2}}{\pi^{\frac{1}{4}}} = \sqrt{2} \pi^{-\frac{1}{4}}$

***** $\displaystyle\int_{-\infty}^{+\infty} x^{2n} e^{-ax^2}\, dx = \frac{\overset{n}{\underset{1}{\Pi}}(2n-1)}{2^2} \sqrt{\frac{\pi}{a}}$

Hence, the normalized wave function is given by

$$\psi = \sqrt{2}\pi^{-\frac{1}{4}} x \exp\left(-\frac{x^2}{2}\right).$$

Example 7.21

The wave function for a certain particle is $\psi = A \cos^2 x$ for $-\dfrac{\pi}{2} < x < \dfrac{\pi}{2}$. Find the value of A.

Solution

$$1 = \int_{-\frac{\pi}{2}}^{\frac{\pi}{2}} A^2 \cos^4 x \, dx = A^2 \int_{-\frac{\pi}{2}}^{\frac{\pi}{2}} \cos^4 x \, dx$$

$$= A^2 \left(\frac{\cos^2 x \sin 2x}{8} + \frac{3 \sin 2x}{16} + \frac{3x}{8} \right) \Bigg|_{-\frac{\pi}{2}}^{\frac{\pi}{2}} = A^2 \frac{3\pi}{8}$$

Thus, $A = \sqrt{\dfrac{8}{3\pi}}$. The normalized wave function for this particle will be

$$\psi = \sqrt{\frac{8}{3\pi}} \cos^2 x \quad \text{for} \quad -\frac{\pi}{2} < x < \frac{\pi}{2}.$$

Example 7.22

Normalize the wave function $\psi = e^{-r/a_0}$ for a ground state hydrogen atom (Actually a_0 is the average value of r.)

Solution

i. Evaluation of $\displaystyle\int_V |\psi|^2 \, dv$

$$\int_0^\infty |\psi|^2 \, dv = \int_0^\infty e^{-2r/a_0} 4\pi r^2 dr = 4\pi \int_0^\infty r^2 e^{-2r/a_0} dr$$

This integral is evaluated by applying the gamma function in the following way.

Let $\dfrac{2r}{a_0} = u$ so that $r = \dfrac{a_0}{2} u$, and $dr = \dfrac{a_0}{2} du$

This integral becomes

$$\int_0^\infty |\psi|^2 \, dv = 4\pi \int_0^\infty \frac{a_o^2}{4} u^2 e^{-u} \frac{a_o}{2} du = 4\pi \times \frac{a_o^3}{8} \int_0^\infty u^2 e^{-u} du$$

$$= \frac{\pi a_o^3}{2} \int_0^\infty u^{3-1} e^{-u} du = \frac{\pi a_o^3}{2} \Gamma(3) = \frac{\pi a_o^3}{2} \times 2!$$

ii. The square root of the reciprocal of this result is the normalization constant. Thus, the normalization constant is $\dfrac{1}{\sqrt{\pi a_o^3}}$ and the normalized wave function is

$$\psi = \frac{1}{\sqrt{\pi a_o^3}} e^{-r/a_o}$$

Example 7.23

The wave function given by $\psi = \exp\left(-\dfrac{x^2}{2a^2} + ikx\right)$ describes a free particle in one dimension. Normalize this wave function.

Solution

i. Evaluation of $\displaystyle\int_V |\psi|^2 \, dv$

$$\int_{-\infty}^{\infty} |\psi|^2 \, dx = \int_{-\infty}^{\infty} \psi\psi^* \, dx = \int_{-\infty}^{\infty} e^{-\frac{x^2}{2a^2}+ikx} e^{-\frac{x^2}{2a^2}-ikx} \, dx = \int_{-\infty}^{\infty} e^{-\frac{x^2}{a^2}} \, dx = 2\int_0^\infty e^{-\frac{x^2}{a^2}} \, dx$$

This integral is evaluated by applying the gamma function in the following way.

Let $x^2 = a^2 u$ so that $x = a\sqrt{u}$, and $dx = \dfrac{a}{2} u^{-\frac{1}{2}} du$

This integral becomes

$$\int_{-\infty}^{\infty} |\psi|^2 \, dx = 2\int_0^\infty e^{-\frac{x^2}{a^2}} \, dx = 2\int_0^\infty e^{-u} \frac{a}{2} u^{-\frac{1}{2}} du = a\int_0^\infty u^{\frac{1}{2}-1} e^{-u} du = a\Gamma\left(\frac{1}{2}\right) = a\sqrt{\pi}$$

ii.   The square root of the reciprocal of this above result is the normalization constant. Thus,
the normalization constant is $\dfrac{1}{\sqrt{a\sqrt{\pi}}}$ and the normalized wave function is

$$\psi = \frac{1}{\sqrt{a\sqrt{\pi}}}\exp\left(-\frac{x^2}{2a^2}+ikx\right)$$

7.12 Observables and Operators

Any dynamical quantity like position coordinates, energy, linear momentum, angular momentum and so on that can be observed or measured is known as an observable. To each observable, there is an operator. Let D be an operator. It operates on x^3 and is written as Dx^3. If the result of Dx^3 is $3x^2$, i.e., $Dx^3 = 3x^2$ then D is called a differential operator.

If the operator A satisfies the rule $A(a_1\psi_1 + a_2\psi_2) = a_1A\psi_1 + a_2A\psi_2$, the operator A is said to be a linear operator. In quantum mechanics, operators are linear. Every observable quantity of a physical system may be represented by a suitable quantum mechanical operator. Operators corresponding to some observables are given in the following table.

Observables and Operators in Three and One dimensions

Observables	Operators in three dimensions	Operators in one dimensions
Total energy (E)	$i\hbar\dfrac{\partial}{\partial t}$	$i\hbar\dfrac{\partial}{\partial t}$
Linear momentum (p)	$-i\hbar\overline{\nabla}$	$-i\hbar\dfrac{\partial}{\partial x}$
Kinetic energy $\left(\dfrac{p^2}{2m}\right)$	$-\dfrac{\hbar^2}{2m}\nabla^2$	$-\dfrac{\hbar^2}{2m}\dfrac{\partial^2}{\partial x^2}$
Position	r	x
Potential energy (V)	V	V
Hamiltonian (H) (Total energy in terms of coordinates and momenta only)	$-\dfrac{\hbar^2}{2m}\nabla^2 + V$	$-\dfrac{\hbar^2}{2m}\dfrac{\partial^2}{\partial x^2} + V$

7.13 Eigenvalues

The eigenvalues of a physical quantity of a given system are defined as the set of permitted values of the physical quantity of the given system. For example, an electron in the hydrogen atom has a number of permitted energy levels given by

$$E_n = -\frac{2\pi^2 mk^2 e^4}{h^2 n^2}.$$

These energy levels of the electron in the hydrogen atoms are called energy eigenvalues of the electron in the hydrogen atom. The energy eigenvalues of a harmonic oscillator is given by

$$E_n = \left(n + \frac{1}{2}\right)\hbar\omega_o.$$

The momentum eigenvalues of a particle trapped in a one-dimensional potential box of length L is given by

$$p_n = \pm\frac{n\pi\hbar}{L}.$$

In case of hydrogen atom, the total angular momentum eigenvalues are given by $L = \sqrt{\ell(\ell+1)}\,\hbar$ with $\ell = 0, 1, 2, 3, ..., (n-1)$. To every physical quantity, there is a set of eigenvalues. Therefore, eigenvalues are real quantities. The eigenvalues may be discrete as the energy eigenvalues of an electron in the hydrogen atom or may be continuous as in the eigenvalues of free particles.

7.14 Eigenfunctions

The solutions of Schrödinger's equation ψ_n for a physical system are called eigenfunctions. The actual wave function representing the given physical system is the linear combination of these eigenfunctions. If $\psi_1, \psi_2, \psi_3, ..., \psi_n$ are the eigenfunctions of a system with coefficients $a_1, a_2, a_3, ..., a_n$ then the wave function ψ representing the given physical system is given by

$$\psi = a_1\psi_1 + a_2\psi_2 + a_3\psi_3 + ... + a_n\psi_n \tag{7.90}$$

The squares of the coefficients give the relative probabilities of the corresponding eigenstates. This means $|a_i|^2$ is the relative probability that the system is in the state represented by the wave function ψ_i.

Example 7.24

If $\dfrac{2}{5}, \dfrac{1}{5}, \dfrac{3}{10},$ and $\dfrac{1}{10}$ are the probabilities that the system are in the four states represented by eigenfunctions $\psi_1, \psi_2, \psi_3,$ and ψ_4 respectively, what is the wave function of the system?

Solution

According to Eq. (7.90), the wave function ψ representing the given physical system is given by

$$\psi = a_1\psi_1 + a_2\psi_2 + a_3\psi_3 + \dots + a_n\psi_n \tag{A}$$

According to the question, $a_1 = \sqrt{\dfrac{2}{5}}, \; a_2 = \sqrt{\dfrac{1}{5}}, \; a_3 = \sqrt{\dfrac{3}{10}}$ and $a_4 = \sqrt{\dfrac{1}{10}}$

Putting the values of a_1, a_2, a_3 and a_4 into Eq. (A), we obtain the wave function of the system as

$$\psi = \sqrt{\dfrac{2}{5}}\psi_1 + \sqrt{\dfrac{1}{5}}\psi_2 + \sqrt{\dfrac{3}{10}}\psi_3 + \sqrt{\dfrac{1}{10}}\psi_4$$

7.15 Operators, Eigenfunctions and Eigenvalues

For some physically meaningful function ψ, it may happen that

$$A\psi = \alpha\psi \tag{7.91}$$

where A is an operator and α in general may be complex. In Eq. (7.91) ψ is the eigenfunction of the operator A, α is the eigenvalue of the operator A associated with eigenfunction ψ. Equation (7.91) is called the eigenvalue equation.

The eigenvalue equation for the momentum operator is

$$\frac{\hbar}{i}\frac{d\psi}{dx} = p\psi \tag{7.92}$$

where ψ = momentum eigenfunction and p, a real number = momentum eigenvalues.

From Eq. (7.92), we can have

$$\frac{d\psi}{\psi} = \frac{i}{\hbar}pdx$$

Integrating both sides of this equation, we have

$$\ln \psi = \frac{i}{\hbar} px + \ln c$$

or $\quad \psi = c e^{\frac{i}{\hbar} px}$

where c = normalization constant = $\frac{1}{\sqrt{2\pi\hbar}}$. The normalized wave function $\psi = \frac{1}{\sqrt{2\pi\hbar}} e^{\frac{i}{\hbar} px}$ satisfies the eigenvalue equation for the momentum operator $\frac{\hbar}{i} \frac{\partial \psi}{\partial x} = p\psi$. Hence, the normalized wave function of the particle $\psi = \frac{1}{\sqrt{2\pi\hbar}} e^{\frac{i}{\hbar} px}$ is the eigenfunction of the momentum operator.

Let us now calculate the momentum eigenfunction and momentum eigenvalues of a particle trapped in a one dimensional box of length L. As we know, this particle is described by the normalized wave function $\psi = \sqrt{\frac{2}{L}} \sin \frac{n\pi x}{L}$. We can express this normalized wave function $\psi = \sqrt{\frac{2}{L}} \sin \frac{n\pi x}{L}$ in an alternate way as

$$\psi = \sqrt{\frac{2}{L}} \sin \frac{n\pi x}{L} = \frac{1}{2i} \sqrt{\frac{2}{L}} \left(e^{\frac{in\pi}{L} x} - e^{-\frac{in\pi}{L} x} \right)$$

$$= \frac{1}{2i} \sqrt{\frac{2}{L}} e^{\frac{in\pi}{L} x} - \frac{1}{2i} \sqrt{\frac{2}{L}} e^{-\frac{in\pi}{L} x}$$

or $\quad \psi = \psi_n^+ - \psi_n^-$

where

$$\psi_n^+ = \frac{1}{2i} \sqrt{\frac{2}{L}} e^{\frac{in\pi}{L} x}$$

$$\psi_n^- = \frac{1}{2i} \sqrt{\frac{2}{L}} e^{-\frac{in\pi}{L} x}$$

These normalized wave functions ψ_n^+ and ψ_n^- satisfy the eigenvalue equation for momentum operator $\frac{\hbar}{i} \frac{\partial \psi}{\partial x} = p\psi$ with $p = \pm \frac{n\pi\hbar}{L}$. Therefore, $\psi_n^+ = \frac{1}{2i} \sqrt{\frac{2}{L}} e^{\frac{in\pi}{L} x}$ and $\psi_n^- = \frac{1}{2i} \sqrt{\frac{2}{L}} e^{-\frac{in\pi}{L} x}$ are called momentum eigenfunctions and $p = \pm \dfrac{n\pi\hbar}{L}$ are the momentum eigenvalues for the trapped particle. The energy eigenvalues of the trapped particle will be

$$E = \frac{p^2}{2m} = \frac{n^2\pi^2\hbar^2}{2mL^2} \quad \text{(Energy eigenvalues)}$$

Example 7.25

Find the eigenfunction for the operator $x + \dfrac{d}{dx}$.

Solution

The eigenvalue equation for the operator $x + \dfrac{d}{dx}$ is given by

$$\left(x + \frac{d}{dx} \right)\psi = \alpha\psi$$

where ψ is the eigenfunction and α is the eigenvalue. From this equation, we have

$$x\psi + \frac{d\psi}{dx} = \alpha\psi$$

or $\dfrac{d\psi}{dx} = (\alpha - x)\psi$

or $\dfrac{d\psi}{\psi} = (\alpha - x)dx$

Integrating both sides of this equation, we get

$$\ell n\psi = \alpha x - \frac{x^2}{2} + \ell nc, \ \ell nc \text{ is the constant of integration.}$$

or $\ell n \dfrac{\psi}{c} = \alpha x - \dfrac{x^2}{2}$

or $\psi = ce^{\alpha x - \frac{x^2}{2}}$

Example 7.26

The normalized wave function of a particle is $\psi = Ae^{-ikx}$. Calculate the energy eigenvalue of the particle.

Solution

Now $\dfrac{d\psi}{dx} = \dfrac{d}{dx} Ae^{-ikx} = (-ik)Ae^{-ikx} = (-ik)\psi$

Thus, we have $\dfrac{d\psi}{dx} = (-ik)\psi$

or $\dfrac{\hbar}{i}\dfrac{d\psi}{dx} = (-\hbar k)\psi$

Comparing this equation with the eigenvalue equation for the momentum operator as given in Eq. (7.81), we get the momentum eigenvalue of the particle as

$p = -\hbar k$

The energy eigenvalue of the particle will be given by

$$E = \frac{p^2}{2m} = \frac{\hbar^2 k^2}{2m}$$

7.16 Expectation Value

The deterministic concept of classical physics loses its significance in quantum physics. In quantum physics, every system is described in a probabilistic manner. Different eigenvalues of a physical quantity occur with different relative probabilities. The experimental value of a physical quantity may not be equal to any one of the eigenvalues; the experimental value of the physical quantity is equal to the weighted average of the eigenvalues with their relative probabilities. The expectation value of the physical quantity is defined as the weighted average of the eigenvalues with their relative probabilities.

To make the procedure clear, let us consider the case of finding the average location of a particle by measuring its x-coordinate a number of times. The observation is given in following table.

X-coordinates of the position of the particle	The number of times the X-coordinates comes out, i.e., weights of the measurements
x_1	N_1
x_2	N_2
x_3	N_3

....	....
....	....
x_n	N_n

From this table, the expectation value or the average value of the X-coordinates is

$$\langle x \rangle = \frac{N_1 x_1 + N_2 x_2 + N_3 x_1 + \ldots + N_n x_n}{N_1 + N_2 + N_3 + \ldots + N_n} = \frac{\sum\limits_{i=1}^{n} N_i x_i}{\sum\limits_{i=1}^{n} N_i} \tag{7.93}$$

The number of times, N_i that we measure each x_i is proportional to the probability $P(x_i)dx$ of finding the particle in the interval dx about x_i. Making this substitution in Eq. (7.93) and changing sums into integration, we have

$$\langle x \rangle = \frac{\int\limits_{-\infty}^{+\infty} P(x)x\,dx}{\int\limits_{-\infty}^{+\infty} P(x)\,dx} = \frac{\int\limits_{-\infty}^{+\infty} |\psi|^2 x\,dx}{\int\limits_{-\infty}^{+\infty} |\psi|^2 \,dx}$$

or $\quad \langle x \rangle = \dfrac{\int\limits_{-\infty}^{+\infty} \psi^* x\psi\,dx}{\int\limits_{-\infty}^{+\infty} \psi^* \psi\,dx} \tag{7.94}$

For a normalized wave function, $\int\limits_{-\infty}^{+\infty} |\psi|^2 \,dx = \int\limits_{-\infty}^{+\infty} \psi^*\psi\,dx = 1$. Hence, expression (9.94) becomes

$$\langle x \rangle = \int\limits_{-\infty}^{+\infty} \psi^* x\psi\,dx \tag{7.95}$$

This expression gives the expectation value or average value of the position of the particle. Similarly, the expectation value or average value of any function of x, $f(x)$ is given by

$$\langle f(x) \rangle = \int\limits_{-\infty}^{+\infty} \psi^* f(x)\psi\,dx$$

The expectation value of momentum $\langle p \rangle$ in one dimension is given by

$$\langle p \rangle = \int\limits_{-\infty}^{+\infty} \psi^* \left(-i\hbar \frac{d\psi}{dx} \right) dx \tag{7.96}$$

The expectation value of momentum $\langle p \rangle$ in three dimensions is given by

$$\langle p \rangle = \int_{-\infty}^{+\infty} \psi^* (-i\hbar \nabla \psi) dr \qquad (7.97)$$

The expectation value of potential energy $\langle V \rangle$ in one dimension is given by

$$\langle V \rangle = \int_{-\infty}^{+\infty} \psi^* V \psi \, dx \qquad (7.98)$$

The expectation value of potential energy $\langle V \rangle$ in three dimensions is given by

$$\langle V \rangle = \int_{-\infty}^{+\infty} \psi^* V \psi \, dr \qquad (7.99)$$

The expectation value of kinetic energy $\left\langle \dfrac{p^2}{2m} \right\rangle$ in one dimension is given by

$$\left\langle \frac{p^2}{2m} \right\rangle = \int_{-\infty}^{+\infty} \psi^* \left(-\frac{\hbar^2}{2m} \frac{\partial^2}{\partial x^2} \right) \psi \, dx \qquad (7.100)$$

The expectation value of kinetic energy $\left\langle \dfrac{p^2}{2m} \right\rangle$ in three dimensions is given by

$$\left\langle \frac{p^2}{2m} \right\rangle = \int_{-\infty}^{+\infty} \psi^* \left(-\frac{\hbar^2}{2m} \nabla^2 \right) \psi \, dr \qquad (7.101)$$

The expectation value of total energy $\langle E \rangle$ in one dimension is given by

$$\langle E \rangle = \int_{-\infty}^{+\infty} \psi^* \left(i\hbar \frac{\partial}{\partial t} \right) \psi \, dx \qquad (7.102)$$

The expectation value of total energy $\langle E \rangle$ in three dimensions is given by

$$\langle E \rangle = \int_{-\infty}^{+\infty} \psi^* \left(i\hbar \frac{\partial}{\partial t} \right) \psi \, dr$$

If $|a_1|^2, |a_2|^2, |a_3|^2, \ldots$ and $|a_n|^2$ are the probabilities that the system be in the states represented by eigenfunctions $\psi_1, \psi_2, \psi_3, \ldots$ and ψ_n respectively, then the wave function of the system will be given by

$$\psi = a_1\psi_1 + a_2\psi_2 + a_3\psi_3 + \ldots + a_n\psi_n. \qquad (7.103)$$

If $|a_1|^2, |a_2|^2, |a_3|^2, \ldots$ and $|a_n|^2$ are the probabilities that the system be in the states $\psi_1, \psi_2, \psi_3,$ $\ldots$ and ψ_n respectively having energy eigenvalues $E_1, E_2, E_3, \ldots$ and E_n respectively, then the energy expectation value of the system will be given by

$$\langle E \rangle = |a_1|^2 E_1 + |a_2|^2 E_2 + |a_3|^2 E_3 + \ldots + |a_n|^2 E_n. \qquad (7.104)$$

Example 7.27

$\dfrac{1}{2}, \dfrac{1}{5}$ and $\dfrac{3}{10}$ are the probabilities that the system be in the three states having energy eigenvalues 4 eV, 6 eV and 9 eV respectively. What is the energy expectation value of the system?

Solution

According to Eq. (7.104), the energy expectation value of the system is given by

$$\langle E \rangle = |a_1|^2 E_1 + |a_2|^2 E_2 + |a_3|^2 E_3 + \ldots + |a_n|^2 E_n \qquad (A)$$

According to the question, $|a_1|^2 = \dfrac{1}{2}$, $|a_2|^2 = \dfrac{1}{5}$ and $|a_3|^2 = \dfrac{3}{10}$

Putting these values of a_1, a_2 and a_3 into Eq. (A), we obtain the energy expectation value of the system as

$$\langle E \rangle = \frac{4\,\text{eV}}{2} + \frac{6\,\text{eV}}{5} + \frac{3}{10} \times 9\,\text{eV} = 5.9\,\text{eV}$$

7.16.1 Procedures for calculation of the expectation value

In quantum mechanics, the following procedures are followed to determine the expectation value of any physical quantity.

i.   Replace the physical quantity by its operator.

ii. Operator operates on ψ.

iii. ψ^* is pre-multiplied on the result of (ii).

iv. Integration of the result of (iii) over the given space gives the expectation value of the physical quantity.

To make the procedure clear, let us consider the case of finding the expectation value of square of the momentum p^2 in one dimension.

i. The operator representation of p^2 is $-\hbar^2 \dfrac{\partial^2}{\partial x^2}$.

ii. The operator $-\hbar^2 \dfrac{\partial^2}{\partial x^2}$ operates on ψ. The result is $-\hbar^2 \dfrac{\partial^2 \psi}{\partial x^2}$.

iii. $-\hbar^2 \dfrac{\partial^2 \psi}{\partial x^2}$ is pre-multiplied by ψ^*. The result is $\psi^* \times -\hbar^2 \dfrac{\partial^2 \psi}{\partial x^2} = -\hbar^2 \psi^* \dfrac{\partial^2 \psi}{\partial x^2}$

iv. The expectation value of the square of the momentum p^2 in one-dimension is

$$\left\langle p^2 \right\rangle = -\hbar^2 \int\limits_{-\infty}^{+\infty} \psi^* \frac{\partial^2 \psi}{\partial x^2} dx$$

In analogy with the definition of the standard deviation in statistics, uncertainties in position and momentum are defined respectively as

$$\Delta x = \left\langle (x - \langle x \rangle)^2 \right\rangle^{\frac{1}{2}} = \left\{ \langle x^2 \rangle - \langle x \rangle^2 \right\}^{\frac{1}{2}} \tag{7.105}$$

$$\Delta p = \left\langle (p - \langle p \rangle)^2 \right\rangle^{\frac{1}{2}} = \left\{ \langle p^2 \rangle - \langle p \rangle^2 \right\}^{\frac{1}{2}} \tag{7.106}$$

Example 7.28

A particle trapped in a one-dimensional box of length L is described by the normalized wave function $\psi = ax$. What is the expectation value of the particle's position $\langle x \rangle$?

Solution

The expectation value of the particle's position $< x >$ when the particle is described by a normalized wave function is given according to Eq. (7.95) by

$$\langle x \rangle = \int\limits_{-\infty}^{+\infty} \psi^* x \psi \, dx$$

Hence, we have $\langle x \rangle = \int\limits_{0}^{L} ax \times x \times ax \, dx = \int\limits_{0}^{L} a^2 x^3 dx = a^2 \left. \frac{x^4}{4} \right|_{0}^{L}$

or $\langle x \rangle = \dfrac{a^2}{4} L^4$

Example 7.29

The normalized wave function $\psi = \sqrt{\dfrac{2}{L}} \sin \dfrac{n\pi x}{L}$ describes a particle in a one dimensional box of length L. What is the expectation value of the particle's position $\langle x \rangle$?

Solution

The expectation value of the particle's position $\langle x \rangle$ when the particle is described by a normalized wave function is given according to Eq. (7.95) by

$$\langle x \rangle = \int_{-\infty}^{+\infty} \psi^* x \psi \, dx$$

Hence, we have

$$\langle x \rangle = \int_0^L \sqrt{\frac{2}{L}} \sin \frac{n\pi x}{L} \times x \times \sqrt{\frac{2}{L}} \sin \frac{n\pi x}{L} \, dx = \frac{2}{L} \int_0^L x \sin^2 \frac{n\pi x}{L} \, dx$$

$$= \frac{2}{L} \left(\frac{x^2}{4} - \frac{L}{4\pi n} x \sin \frac{2n\pi x}{L} - \frac{L^2}{8\pi^2 n^2} \cos \frac{2\pi n x}{L} \right) \Bigg|_0^L = \frac{2}{L} \left(\frac{L^2}{4} \right)$$

or $\langle x \rangle = \dfrac{L}{2}$

Example 7.30

The normalized wave function given by

$$\psi = \frac{\sqrt{2}}{\pi^{1/4}} x \exp\left(-\frac{x^2}{2} \right)$$

describes a particle in a one-dimensional box of length L. What is the expectation value of the particle's position $\langle x \rangle$?

Solution

The expectation value of the particle's position $\langle x \rangle$ when the particle is described by a normalized wave function is given according to Eq. (7.95) by

$$\langle x \rangle = \int_{-\infty}^{+\infty} \psi^* x \psi \, dx$$

Hence, we have

$$\langle x \rangle = \int_0^L \sqrt{\frac{2}{\sqrt{\pi}}} e^{-\frac{x^2}{2}} \times x \times \sqrt{\frac{2}{\sqrt{\pi}}} e^{-\frac{x^2}{2}} dx = \frac{2}{\sqrt{\pi}} \int_0^L x e^{-x^2} dx$$

In this integral, let $x^2 = u$ so that $x = \sqrt{u}$, $dx = \frac{1}{2} u^{-\frac{1}{2}} du$. Then integral becomes

$$\langle x \rangle = \frac{2}{\sqrt{\pi}} \int_0^L u^{\frac{1}{2}} e^{-u} \frac{1}{2} u^{-\frac{1}{2}} du = \frac{2}{\sqrt{\pi}} \int_0^L e^{-u} du = \frac{2}{\sqrt{\pi}} \left(1 - e^{-L} \right)$$

The expectation value of the particle's position $\langle x \rangle$ is obtained as

$$\langle x \rangle = \frac{2}{\sqrt{\pi}} \left(1 - e^{-L} \right)$$

Example 7.31

The normalized wave function for a certain particle is $\psi = \sqrt{\frac{2}{\pi}} \cos x$ for $-\frac{\pi}{2} < x < \frac{\pi}{2}$. What is the expectation value of the particle's position $\langle x \rangle$?

Solution

The expectation value of the particle's position $\langle x \rangle$ when the particle is described by a normalized wave function is given according to Eq. (7.95) by

$$\langle x \rangle = \int_{-\infty}^{+\infty} \psi^* x \psi \, dx$$

Hence, we have

$$\langle x \rangle = \int_{-\frac{\pi}{2}}^{\frac{\pi}{2}} \sqrt{\frac{2}{\pi}} \cos x \times x \times \sqrt{\frac{2}{\pi}} \cos x \, dx = \frac{2}{\pi} \int_{-\frac{\pi}{2}}^{\frac{\pi}{2}} x \cos^2 x \, dx \; ******$$

or $\langle x \rangle = \frac{2}{\pi} \left(\frac{x \sin 2x}{4} + \frac{\cos 2x}{8} + \frac{x^2}{4} \right) \Big|_{-\frac{\pi}{2}}^{\frac{\pi}{2}} = \frac{2}{\pi} \left(\frac{1}{4} \right) = 0.16$

$****** \quad \int uv \, dx = u \int v \, dx - \int \frac{du}{dx} \left(\int v \, dx \right) dx$

Example 7.32

The normalized wave function for ground state $2s$ electron in the hydrogen atom is given as

$$\psi = \frac{1}{4\sqrt{2\pi a_o^3}}\left(2-\frac{r}{a_o}\right)e^{-\frac{r}{2a_o}}. \text{ Calculate } \left\langle\frac{1}{r}\right\rangle.$$

Solution

The expectation value $\left\langle\frac{1}{r}\right\rangle$ when a particle is described by a normalized wave function is given according to Eq. (7.95) by

$$\left\langle\frac{1}{r}\right\rangle = \int_0^\infty \psi^*\left(\frac{1}{r}\right)\psi dV$$

Hence, we have

$$\left\langle\frac{1}{r}\right\rangle = \int_0^\infty \frac{1}{4\sqrt{2\pi a_o^3}}\left(2-\frac{r}{a_o}\right)e^{-\frac{r}{2a_o}}\left(\frac{1}{r}\right)\frac{1}{4\sqrt{2\pi a_o^3}}\left(2-\frac{r}{a_o}\right)e^{-\frac{r}{2a_o}}4\pi r^2 dr$$

$$= \frac{1}{8a_o^3}\int_0^\infty\left(2-\frac{r}{a_o}\right)^2 e^{-\frac{r}{a_o}}r dr = \frac{1}{8a_o^3}\int_0^\infty\left(4+\frac{r^2}{a_o^2}-\frac{4r}{a_o}\right)e^{-\frac{r}{a_o}}r dr$$

$$= \frac{1}{8a_o^3}\left(4\int_0^\infty re^{-\frac{r}{a_o}}dr + \frac{1}{a_o^2}\int_0^\infty r^3 e^{-\frac{r}{a_o}}dr - \frac{4}{a_o}\int_0^\infty r^2 e^{-\frac{r}{a_o}}dr\right)$$

The integrals in the RHS of this expression are evaluated by applying the gamma function. Thus, we get

$$\left\langle\frac{1}{r}\right\rangle = \frac{1}{8a_o^3}\left(4a_0^2 + \frac{1}{a_0^2}6a_0^4 - \frac{4}{a_0}2a_0^3\right) = \frac{1}{a_0}$$

Example 7.33

The normalized wave function given by

$$\psi = \frac{1}{\sqrt{a\sqrt{\pi}}}\exp\left(-\frac{x^2}{2a^2}+ikx\right)$$

describes a free particle in one dimension. In what region of space is the particle most likely found?

Solution

The expectation value of the particle's position $\langle x \rangle$ when the particle is described by a normalized wave function is given according Eq. (7.95) by

$$\langle x \rangle = \int_{-\infty}^{+\infty} \psi^* x \psi \, dx$$

Hence, we have

$$\langle x \rangle = \int_{-\infty}^{\infty} \frac{1}{\sqrt{a\sqrt{\pi}}} e^{-\frac{x^2}{2a^2} - ikx} \times x \times \frac{1}{\sqrt{a\sqrt{\pi}}} e^{-\frac{x^2}{2a^2} + ikx} \, dx = \frac{1}{a\sqrt{\pi}} \int_{-\infty}^{\infty} x e^{-\frac{x^2}{a^2}} \, dx$$

The integrand in the RHS $x e^{-\frac{x^2}{a^2}}$ is an odd function******* of x. Hence, we have $\int_{-\infty}^{\infty} x e^{-\frac{x^2}{a^2}} \, dx = 0$. Thus, we have

$$\langle x \rangle = \frac{1}{a\sqrt{\pi}} \times 0 = 0$$

The most expected value of the position of the particle $\langle x \rangle$ is zero. Thus, we can say that the particle is most likely found in the region on either side of the origin $x = 0$.

Example 7.34

The normalized wave function given by

$$\psi = \frac{1}{\sqrt{a\sqrt{\pi}}} \exp\left(-\frac{x^2}{2a^2} + ikx\right)$$

describes a free particle in one dimension. What is the expectation value of the momentum of the particle?

******* A function $f(x)$ is said to be odd if $f(-x) = -f(x)$. A function $f(x)$ is said to be even if $f(-x) = f(x)$. For odd functions,

$$\int_{-a}^{+a} f(x) \, dx = 0$$

and for even functions,

$$\int_{-a}^{+a} f(x) \, dx = 2 \int_{0}^{+a} f(x) \, dx$$

Solution

The momentum operator in one dimension is $-i\hbar \dfrac{\partial}{\partial x}$.

The expectation value of the particle's momentum $\langle p \rangle$ when the particle is described by a normalized wave function is given according to Eq. (7.96) by

$$\langle p \rangle = \int_{-\infty}^{+\infty} \psi^* \left(-i\hbar \frac{d\psi}{dx} \right) dx$$

Hence, we have

$$\langle p \rangle = \int_{-\infty}^{\infty} \frac{1}{\sqrt{a\sqrt{\pi}}} e^{-\frac{x^2}{2a^2} - ikx} \left(\frac{-i\hbar}{\sqrt{a\sqrt{\pi}}} e^{-\frac{x^2}{2a^2} + ikx} \right) \left(ik - \frac{x}{a^2} \right) dx$$

$$= \frac{-i\hbar}{a\sqrt{\pi}} \int_{-\infty}^{\infty} e^{-\frac{x^2}{a^2}} ik\, dx + \frac{i\hbar}{a^3 \sqrt{\pi}} \int_{-\infty}^{\infty} xe^{-\frac{x^2}{a^2}} dx$$

The integrand in the second integration of the RHS of this equation $xe^{-\frac{x^2}{a^2}}$ is an odd function of x. Hence, we have $\int_{-\infty}^{\infty} xe^{-\frac{x^2}{a^2}} dx = 0$. Therefore,

$$\langle p \rangle = \frac{2\hbar k}{a\sqrt{\pi}} \int_{0}^{\infty} e^{-\frac{x^2}{a^2}} dx$$

In this integral, let $x^2 = u$ so that $x = \sqrt{u}$, $dx = \dfrac{1}{2} u^{-\frac{1}{2}} du$. Then the integral becomes

$$\langle p \rangle = \frac{2\hbar k}{a\sqrt{\pi}} \int_{0}^{\infty} e^{-u} \frac{a}{2} u^{-\frac{1}{2}} du = \frac{\hbar k}{\sqrt{\pi}} \int_{0}^{\infty} u^{\frac{1}{2} - 1} e^{-u} du = \frac{\hbar k}{\sqrt{\pi}} \Gamma\left(\frac{1}{2} \right) = \hbar k$$

Example 7.35

A particle trapped in a one-dimensional box of length L is described by the normalized wave function

$$\psi = \sqrt{\frac{2}{L}} \sin \frac{n\pi x}{L}.$$

What is the expectation value of momentum of the particle?

Solution

The momentum operator in one dimension is $-i\hbar \dfrac{\partial}{\partial x}$.

The expectation value of the particle's momentum $\langle p \rangle$ when the particle is described by a normalized wave function is given according to Eq. (7.96) by

$$\langle p \rangle = \int\limits_{-\infty}^{+\infty} \psi^* \left(-i\hbar \frac{d\psi}{dx} \right) dx$$

Hence, we have

$$\langle p \rangle = \int\limits_{0}^{L} \sqrt{\frac{2}{L}} \sin \frac{n\pi x}{L} \left(-i\hbar \sqrt{\frac{2}{L}} \frac{n\pi}{L} \cos \frac{n\pi x}{L} \right) dx$$

$$= -i\hbar \frac{2}{L} \int\limits_{0}^{L} \sin \frac{n\pi x}{L} \cos \frac{n\pi x}{L} \frac{n\pi}{L} dx$$

$$= \frac{i\hbar}{2L} \left(\cos \frac{2n\pi x}{L} \right) \Bigg|_{0}^{L} = \frac{i\hbar}{2L} (\cos 2n\pi - \cos 0) = 0$$

Thus, we have $\langle p \rangle = 0$

Therefore, the expectation value of the particle's momentum $\langle p \rangle$ described by the normalized wave function $\psi = \sqrt{\dfrac{2}{L}} \sin \dfrac{n\pi x}{L}$ is zero. Incredible! (Why?)

7.17 Schrödinger's Equation

Newton's second law of motion is the fundamental equation in classical mechanics. In the same sense, Schrödinger's equation, a wave equation in the variable ψ, is the fundamental equation in quantum mechanics. Schrödinger's wave equation is a basic principle that cannot be derived from anything else.

7.17.1 Schrödinger's time-dependent equation

Schrödinger's time-dependent non-relativistic equation in three dimensions is written as

$$i\hbar \frac{\partial \psi}{\partial t} = -\frac{\hbar^2}{2m} \nabla^2 \psi + V\psi \tag{7.107}$$

or $\quad \nabla^2 \psi + i\dfrac{2m}{\hbar}\dfrac{\partial \psi}{\partial t} - \dfrac{2m}{\hbar^2}V\psi = 0$ (7.108)

In one-dimension when the particle is constrained to move along the X-axis Eqs (7.107) and (7.108) becomes respectively

$$i\hbar\frac{\partial \psi}{\partial t} = -\frac{\hbar^2}{2m}\frac{\partial^2 \psi}{\partial x^2} + V\psi$$ (7.109)

and $\quad \dfrac{\partial^2 \psi}{\partial x^2} + i\dfrac{2m}{\hbar}\dfrac{\partial \psi}{\partial t} - \dfrac{2m}{\hbar^2}V\psi = 0$ (7.110)

Potential energy in general is a function of position as well as time. If the potential energy function V is known, Eq. (7.110) can be solved for the particle wave function ψ. From the wave function ψ, all the information about the particle can be derived. For a freely moving particle, $V = 0$.

Newton's second law of motion $F = ma$, the basic principle of classical mechanics can be derived from Schrödinger's time-dependent non-relativistic equation provided the quantities it relates are the expectation values rather than the precise values.

7.17.2 Schrödinger's time-independent equation

In lot of real problems, potential energy is a function of position only and does not depend upon time. Schrödinger's time-independent equation is called the steady state equation. Schrödinger's steady state non-relativistic equation in three dimension is written as

$$\nabla^2 \psi + \frac{2m}{\hbar^2}(E - V)\psi = 0$$ (7.111)

where

E = non-relativistic total energy.

V = potential energy

In one dimension, say, if a particle is constrained to move along the X-axis, Eq. (7.111) becomes

$$\frac{\partial^2 \psi}{\partial x^2} + \frac{2m}{\hbar^2}(E - V)\psi = 0$$ (7.112)

When Schrödinger's steady state non-relativistic Eq. (7.111) or (7.112) is solved for a given system, one gets a group of solutions. These solutions ψ_n are called eigenfunctions. The the corresponding total energies E_n are called eigenvalues. Schrödinger's time-independent equation is applicable to stationary states.

In this discussion, we have not tried to derive Schrödinger's wave equation. It is impossible. At best, we can verify Schrödinger's wave equation by taking the known wave function of a system. Schrödinger's wave equation cannot be derived from other basic principles of physics; it is a basic principle itself.

7.17.3 Newton's equation and Schrödinger's equation

Newton's equation	Schrödinger's equation
$$m\frac{d^2x}{dt^2} = F$$	$$\frac{\partial^2\psi}{\partial x^2} + \frac{2m}{\hbar^2}(E-V)\psi = 0$$ $$i\hbar\frac{\partial\psi}{\partial t} = -\frac{\hbar^2}{2m}\frac{\partial^2\psi}{\partial x^2} + V\psi$$
The particle is subject to different forces	The particle is subject to different potential energies
The basic behavior of a particle is obtained by solving Newton's equation of motion.	The basic behavior of a particle is obtained by solving Schrödinger's equation.
The position of a particle is continuous across the interface; speed is also continuous as long as forces remain finite.	The wave function ψ of a particle is continuous across the interface; $\frac{\partial\psi}{\partial x}$ is also continuous as long as change in potential energy remains finite.

Questions

7.1   What phenomena led to the discovery of quantum physics?

7.2   What are the failures of classical physics?

7.3   What is the relation between quantum physics and classical physics?

7.4   Explain the statement: All natural laws belong to quantum physics.

7.5   Give few phenomena where waves behave like particles.

7.6   Give few phenomena where particles behave like waves.

7.7   What is blackbody?

7.8   What is blackbody radiation?

7.9   What is spectrum?

7.10   What are Planck's postulates regarding blackbody radiation?

7.11   What is Planck's radiation law?

7.12   Derive Planck's radiation law in terms of frequency of the radiation.

7.13   Derive Planck's radiation law in terms of wavelength of the radiation.

7.14   What is Wien's radiation law? Derive Wien's radiation law from Planck's radiation formula.

7.15   What is Wien's displacement law? Derive Wien's displacement law from Planck's radiation formula.

7.16   What is Rayleigh–Jean's law? Derive Rayleigh–Jean's radiation law from Planck's radiation formula.

7.17   What is Stefan's law? Derive Stefan's radiation law from Planck's radiation formula.

7.18   Why is quantization of energy not observed in everyday life?

7.19   What is photoelectric effect? What are the laws of photoelectric effect?

7.20   Give Einstein's theory of photoelectric effect.

7.21   What do you mean by work function of a metal? Derive an expression for it.

7.22   What do you mean by stopping potential of a metal? Derive an expression for it.

7.23   What do you mean by threshold frequency of a metal? Derive an expression for it.

7.24   Explain an experiment to determine Planck's constant.

7.25   Plot the variation of photoelectric current with anode potential.

7.26   Plot the variation of photoelectric current with intensity of incident light.

7.27   Plot the variation of photoelectric current with anode potential for different frequencies of the incident light.

7.28   What is Compton effect? Explain how the classical concept failed in explaining Compton effect.

7.29   Explain why the frequency of the scattered photon is less than that of the incident photon in case of Compton scattering.

7.30   What is Compton shift? Derive an expression for it.

7.31   What are the factors on which the Compton shift depends on?

7.32   What do you mean by the Compton wavelength of a scattered particle? What are the factors on which the Compton wavelength depends on?

7.33   Derive an expression for the wavelength of the scattered photon in the Compton effect.

7.34   Explain the phenomenon of pair production.

7.35   Explain why pair production is not possible in vacuum.

7.36   What are the characteristics of a photon?

7.37   What are the physical quantities that characterizes particle phenomenon?

7.38   What are the physical quantities that characterizes wave phenomenon?

7.39   What is the de Broglie hypothesis?

7.40   What is the de Broglie wave? Derive an expression for the wavelength of the de Broglie waves of a material particle moving with relativistic speed.

7.41   Prove that the de Broglie wavelength of a photon is equal to the wavelength of the radiation.

7.42   Two particles of masses m_1 and m_2 move with the same kinetic energy. Find the ratio of their de Broglie wavelengths.

7.43 What is the de Broglie relation for a charged particle moving under a potential difference of V volts.

7.44 Why is the wave nature of a particle not observable in our daily life?

7.45 Explain how the phase speed of matter waves for relativistically freely moving particles is more than the speed of light in vacuum.

7.46 Explain how the phase speed of matter waves for non-relativistically freely moving particles is half of the particle speed.

7.47 Prove that the group speed of a matter wave for a relativistically freely moving particle is equal to the speed of the particle.

7.48 Prove that the group speed of a matter wave for a non-relativistically freely moving particle is equal to the speed of the particle.

7.49 Explain how vacuum behaves as a dispersive medium for matter wave.

7.50 Derive an expression for the phase speed of a matter wave in vacuum for a relativistically freely moving particle.

7.51 Describe the Davisson–Germer experiment.

7.52 Plot the polar graphs of the Davisson and Germer experiment.

7.53 What conclusions were drawn from the Davisson and Germer experiment?

7.54 What are the properties of a matter wave?

7.55 On what grounds was Thompson's plum-pudding model of the atom discarded?

7.56 What is Rutherford's atom model? What are the experimental observations it can explain?

7.57 What are the experimental observations Rutherford's atom model cannot explain?

7.58 What are the postulates of Bohr's atom model?

7.59 Based on Bohr's atom model, derive an expression for the orbital speed of an electron in the hydrogen atom.

7.60 Based on Bohr's atom model, derive an expression for the diameter of an orbit of an electron in the hydrogen atom.

7.61 Based on Bohr's atom mode, prove that the variation of the atomic radius with principal quantum number is parabolic.

7.62 Based on Bohr's atom model, derive an expression for the kinetic energy of an electron in any orbit in the hydrogen atom.

7.63 Based on Bohr's atom model, derive an expression for the potential energy of an electron in any orbit in the hydrogen atom.

7.64 Based on Bohr's atom model, derive an expression for the total energy of an electron in any orbit in the hydrogen atom.

7.65 The total energy of an electron in any orbit in the hydrogen atom is found to be negative. Explain

7.66 Based on Bohr's atom model, derive an expression for the wavelength of radiation emitted when an electron jumps from one energy level to another energy level.

7.67   What are the limitations of Bohr's atom model?

7.68   What are the experimental observations Bohr's atom model can explain?

7.69   What are the experimental observations Bohr's atom model cannot explain?

7.70   What is Lyman series? Give the expression for the wavelength of this series.

7.71   What is Balmer series? Give the expression for the wavelength of this series.

7.72   What is Paschen series? Give the expression for the wavelength of this series.

7.73   What is Brackett series? Give the expression for the wavelength of this series.

7.74   What is Pfund series? Give the expression for the wavelength of this series.

7.75   State and explain Heisenberg's uncertainty principle.

7.76   Explain why we cannot see an electron even with an ideal microscope.

7.77   What is a wave packet?

7.78   Describe Heisenberg's experiment using a hypothetical gamma ray microscope.

7.79   Can the uncertainty principle be verified experimentally? If yes, explain.

7.80   Derive an expression for the minimum energy of a harmonic oscillator by using the uncertainty principle.

7.81   Derive an expression for the ground state energy of a hydrogen atom by using the uncertainty principle.

7.82   Prove by using the uncertainty principle that a nucleus cannot contain electrons.

7.83   Differentiate between classical physics and quantum physics.

7.84   What is a wave function?

7.85   What are the characteristics of a wave function?

7.86   What is probability density? Derive an expression for it.

7.87   Derive an expression for the probability of finding a particle described by the wave function ψ in a certain region.

7.88   What are the units of one, two and three-dimensional wave functions?

7.89   State and explain the superposition principle for wave functions.

7.90   Why should the wave function be normalized to a unit?

7.91   What do you mean by normalization of a wave function?

7.92   What is the meaning of a normalization constant?

7.93   How would you find the normalization constant of a wave function?

7.94   Give few observables and operators of quantum mechanics.

7.95   What do you mean by eigenvalues of a physical quantity? Explain.

7.96   What do you mean by eigenfunctions of a physical system? Explain.

7.97   Derive an expression for eigenfunctions of the momentum operator.

7.98   Calculate the momentum eigenfunctions and momentum eigenvalues of a particle trapped in a one-dimensional box of length L when the particle is described by the normalized wave function

$$\psi = \sqrt{\frac{2}{L}} \sin\frac{n\pi x}{L}.$$

Also calculate the energy eigenvalues of the particle.

7.99 What do you mean by the expectation value of a physical quantity? What does it represent?

7.100 The expectation value of the momentum of a particle trapped in a one-dimensional box of length L described by the normalized wave function

$$\psi = \sqrt{\frac{2}{L}} \sin\frac{n\pi x}{L}$$

is zero. Explain.

7.101 In determining the expectation value of any physical quantity, what procedures are to be followed?

7.102 What is the time-independent Schrödinger's equation for a particle of mass m and energy E moving in the XY-plane under a potential energy V.

7.103 What is the time-independent Schrödinger's equation for a particle of mass m and energy E moving in the YZ-plane under a potential energy V.

7.104 What is the time-independent Schrödinger's equation for a particle of mass m and energy E moving along the Y-axis under a potential energy V.

7.105 What is the time-dependent Schrödinger's equation for a particle of mass m and energy E moving in the XY-plane under a potential energy V.

7.106 What is the time-independent Schrödinger's equation for a particle of mass m and energy E moving in the YZ-plane under a potential energy V.

7.107 What is the time-independent Schrödinger's equation for a particle of mass m and energy E moving along the Z-axis under a potential energy V.

7.108 Prove that the magnitude of propagation vector is

$$\frac{\sqrt{2m(E-V)}}{\hbar}$$

where E = total energy, V = potential energy, and the other symbols have their usual meanings.

7.109 Prove that the plane wave function

$$\psi = re^{i(kx-\omega t)} \quad k = \frac{2\pi}{\lambda}$$

satisfy Schrödinger's equation.

7.110 Find the eigenfunction for the operator $-i\hbar\dfrac{d}{dx}$.

Problems

7.1 What is the energy and momentum of a photon of sodium light of wavelength 5890 Å?.

[Ans   2.11 eV, 2.11 eV/c]

7.2 How many photons are there in 1m^3 of radiation in thermal equilibrium at 1500 K? Calculate the average energy of one photon at this temperature.

[Ans 6.82×10^{16}, 0.35 eV]

7.3 What is the maximum wavelength of thermal radiation emitted by a body at room temperature 127°C. To what temperature must we heat it so that its peak thermal radiation is in the orange region of the spectrum? ($\lambda = 5900$ Å).

[Ans   7.225 μm, 4898 K]

7.4 At what temperature can the filament of a 100 watt lamp be operated so that it becomes a perfect blackbody of area 3cm^2.   [Ans   1557 K]

7.5 Radiation from the Big Bang has been Doppler shifted to longer wavelengths by the expansion of the universe and today, it has a spectrum corresponding to that of a blackbody at 2.7 K. Find the wavelength at which the energy density of this radiation is maximum. In what region of the spectrum is this radiation?

[Ans   1.1 mm, microwave region]

7.6 The rate at which sunlight reaches the earth is 1.4 (kW/m^2), the average distance between the sun and the earth is 1.5×10^8 km and the radius of the sun is 7×10^5 km. From this data, find the surface temperature of the sun.   [Ans   5800 K]

7.7 When a certain monochromatic radiation falls on a certain photosensitive metal, the maximum speed with which photoelectrons emitted is 7.75×10^5 m/s. What is the stopping potential in this case?   [Ans   1.71 V]

7.8 The work function of tungsten metal is 4.52 eV. What is the threshold wavelength of this metal?   [Ans   2740 Å]

7.9 X-rays of wavelength 1.54 Å are scattered from a target. Calculate the wavelength of the X-rays scattered through 60° and the maximum wavelength present in the scattered X-rays.   [Ans   1.55 Å. 1.59 Å]

7.10 Calculate the kinetic energy of a neutron whose de Broglie wavelength is 1.000 femtometers.   [Ans   616 MeV]

7.11 Calculate the de Broglie wavelength of a particle of mass 1 mg moving with a speed 20 m/s.   [Ans 3.3×10^{-29} m]

7.12 Through what potential difference should the electron be accelerated so that its de Broglie wavelength becomes 5500 Å.   [Ans 4.98×10^{-6} V]

7.13 How much energy is required to excite the ground state hydrogen atom to the third state.   [Ans   19.36 eV]

7.14 The second member of the Balmer series has wavelength 4861 Å. Calculate the wavelength of the first member.   [Ans   6562.4 Å]

7.15 Calculate the maximum and minimum wavelengths of the Balmer series.

[Ans   6563 Å, 3646 Å]

7.16 A particle trapped in a one-dimensional box of length L is described by the normalized wave function

$$\psi = \sqrt{\frac{2}{L}} \sin \frac{n\pi x}{L}.$$

What is the probability that the particle in the ground state is lying between (a) 0 and $L/3$ (b) $L/3$ and $2L/3$ (c) $2L/3$ and L? [Ans   19.55%, 60.90%, 19.55%]

7.17 The particle trapped in a one-dimensional box of length L is described by the normalized wave function

$$\psi = \sqrt{\frac{2}{L}} \sin \frac{n\pi x}{L}.$$

What is the probability that the particle in the nth state is lying between 0 and $\dfrac{L}{n}$?

[Ans $\dfrac{1}{n}$]

7.18 An electron trapped in a one-dimensional box of length $L = 1$ Å [a typical atomic diameter] is described by the normalized wave function

$$\psi = \sqrt{\frac{2}{L}} \sin \frac{n\pi x}{L}.$$

In the ground state, what is the probability that the electron lies between (a) 0.09 Å and 0.11 Å (b) 0.0 Å and 0.25 Å? [Ans   0.38%, 25%]

7.19 If $\dfrac{1}{2}, \dfrac{1}{3}$, and $\dfrac{1}{6}$ are the probabilities that the system be in the three states represented by eigenfunctions ψ_1, ψ_2 and ψ_3 respectively, what is the wave function of the system? [Ans $\psi = \dfrac{1}{\sqrt{2}}\psi_1 + \dfrac{1}{\sqrt{3}}\psi_2 + \dfrac{1}{\sqrt{6}}\psi_3$]

7.20 The normalized wave function of a particle is $\psi = Ae^{-inkx}$. Calculate the momentum eigenvalues and energy eigenvalues of the particle. [Ans $-nk\hbar, \dfrac{n^2k^2\hbar^2}{2m}$]

7.21 The normalized wave function for a certain particle is $\psi = \sqrt{\dfrac{8}{3\pi}}\sin^2 x$ for $-\dfrac{\pi}{2} < x < \dfrac{\pi}{2}$.

Calculate the probability that the particle be found between $x = 0$ and $x = \dfrac{\pi}{4}$.

[Ans   462]

7.22 The wave function for a certain particle is $\psi = \sin^2 x$ for $-\dfrac{\pi}{2} < x < \dfrac{\pi}{2}$. Normalize the wave function for $-\dfrac{\pi}{2} < x < \dfrac{\pi}{2}$. [Ans $\psi = \sqrt{\dfrac{8}{3\pi}}\sin^2 x$]

7.23 The wave function of a 2s electron in the hydrogen atom is $\psi = \left(2 - \dfrac{r}{a_0}\right)e^{-\frac{r}{2a_0}}$.

Normalize the wave function.

[Ans $\psi = \dfrac{1}{4\sqrt{2\pi a_0^3}}\left(2 - \dfrac{r}{a_0}\right)e^{-\frac{r}{2a_0}}$]

7.24 Normalize the Gaussian function $\psi = e^{-\frac{x^2}{2\sigma^2}}$. in one-dimension.

[Ans $\psi = \dfrac{1}{\sqrt{\sigma}\sqrt{\pi}}e^{-\frac{x^2}{2\sigma^2}}$]

7.25 Normalize the one-dimensional wave function $\psi = e^{\frac{i}{\hbar}ax}$.   [Ans $\psi = \dfrac{1}{\sqrt{2\pi\hbar}}e^{\frac{i}{\hbar}ax}$]

7.26 A particle trapped in a one-dimensional box of length 1.0 Å is described by the normalized wave function $\psi = ax$. What is the expectation value $\langle x \rangle$ of the particle's position?   [Ans $\langle x \rangle = \dfrac{a^2}{4}$]

7.27 The normalized wave function for a ground state 1s electron in the hydrogen atom is given as

$\psi = \dfrac{1}{\sqrt{\pi a_0^3}}e^{-\frac{r}{a_0}}$.

Calculate $\langle r \rangle$ and $\left\langle \dfrac{1}{r}\right\rangle$.   [Ans $\langle r \rangle = a_0$ $\left\langle \dfrac{1}{r}\right\rangle = \dfrac{1}{a_0}$]

7.28 The normalized wave function $\psi = \sqrt{\dfrac{2}{L}}\sin\dfrac{n\pi x}{L}$ describes an electron in a one-dimensional box of length L. Calculate $\langle x^2 \rangle$.   [Ans $L^2\left(\dfrac{1}{3} - \dfrac{1}{2n^2\pi^2}\right)$]

7.29 $\dfrac{2}{5}, \dfrac{1}{5}, \dfrac{3}{10}$, and $\dfrac{1}{10}$ are the probabilities that the system be in the four states having energy eigenvalues 4 eV, 6 eV, 7 eV and 9 eV respectively. What is the energy expectation value of the system?   [Ans $\langle E \rangle = 5.8$ eV]

7.30 The wave function of a system is given by $\psi = p\psi_1 + q\psi_2 + r\psi_3$. What are the probabilities of finding the system in the states ψ_1, ψ_2 and ψ_3?   [Ans $|p|^2, |q|^2, |r|^2$]

Multiple Choice Questions

1. Which of the following is correct?

   (i) $\lim\limits_{n \to \infty}$ Classical physics = Quantum physics

   (ii) $\lim\limits_{n \to \infty}$ Quantum Physics = Classical Physics

   (iii) $\lim\limits_{h \to 0}$ Quantum Physics = Classical Physics

   (iv) $\lim\limits_{h \to 0}$ Classical physics = Quantum physics

2. Classical physics cannot explain micro-phenomena and quantum physics can not explain macro-phenomena.
    (i) True   (ii) False

3. Which of the following phenomenon first confirms the wave nature of the electron?
    (i)   Compton scattering
    (ii) Thompson scattering
    (iii) Davisson–Germer experiment

4. The color of a blackbody is always black.
    (i) True   (ii) False

5. Photons obey
    (i)   Maxwell–Boltzmann statistics
    (ii) Fermi–Dirac statistics
    (iii) Bose–Einstein statistics

6. Electrons obey
    (i)   Maxwell–Boltzmann statistics
    (ii) Fermi–Dirac statistics
    (iii) Bose–Einstein statistics

7. The intensity of emission is directly proportional to the absolute temperature of the body and inversely proportional to the fourth power of the wavelength. This is
    (i)   Planck's radiation law   (ii) Wien's radiation law
    (iii) Rayleigh–Jeans law   (iv) Wein's displacement law

8. The emissivity of a blackbody is
    (i)   0   (ii) 1
    (iii) ∞ (iv) 0.5

9. What is the value of hc, h = Planck's constant, c = speed of light in vacuum,
    (i)   1240 eV Å
    (ii) 12400 eV Å
    (iii) 1240 eVnm
    (iv) 12400 eV.nm

10. Einstein's photoelectric equation is based on the law of conservation of
    (i)   momentum   (ii) angular momentum
    (iii) energy   (iv) none of the above

11. Photoelectric effect is possible with radiation having
    (i)   low wavelengths   (ii) high wavelengths
    (iii) low frequency   (iv) high frequency

12. The kinetic energy of a photoelectron is more with radiation having
    (i)   low wavelengths
    (ii) high wavelengths
    (iii) low frequency
    (iv) high frequency

13. The Compton wavelength of the scattering electron depends on
    (i)   energy of the X-ray photon
    (ii) angle of scattering of the electron
    (iii) angle of scattering of the X-ray photon
    (iv) mass of the electron

14. Name a phenomenon where energy is converted into matter.
    (i)   Compton effect
    (ii) Thompson effect
    (iii) Radioactive decay
    (iv) Pair production

15. Which of the following conservation principle is violated during pair production?
    (i)   Momentum
    (ii) Charge
    (iii) Energy
    (iv) None of the above

16. The phase speed of a matter wave for a relativistically freely moving particle in vacuum is
    (i)   equal to the speed of light in vacuum
    (ii) less than the speed of light in vacuum
    (iii) more than the speed of light vacuum
    (iv) equal to half the speed of light in vacuum

17. The phase speed of a matter wave for non-relativistically freely moving particles is
    (i)   Half of the particle speed
    (ii) Equal to the particle speed
    (iii) Less than the particle speed
    (iv) More than the particle speed

18. The group speed of a matter wave for a relativistically freely moving particle in vacuum is
    (i)   Equal to the particle speed in vacuum
    (ii) Less than the speed of light in vacuum
    (iii) More than the speed of light in vacuum
    (iv) Equal to half the speed of light in vacuum

19. The group speed of the matter wave for a non-relativistically freely moving particle is
    (i)   Half of the particle speed
    (ii) Equal to the particle speed
    (iii) Less than the particle speed
    (iv) More than the particle speed

20. Which of the following phenomena are not observed in case of matter wave?

    (i)   Interference
    (ii) Polarization
    (iii) Diffraction
    (iv) None of the above

21. What conclusion(s) was drawn from Rutherford's α-particle scattering experiment

    (i)   Electrons are revolving in circular paths
    (ii) Atom is neutral
    (iii) Positive charges are concentrated at a point in an atom
    (iv) Total mass of the atom is concentrated at a point
    (v)   Charges are concentrated at a point in an atom

22. The nucleus of the hydrogen atom consists of

    (i)   one proton and one neutron
    (ii) one proton only
    (iii) one neutron only
    (iv) one proton and two neutrons

23. The variation of the atomic radius with principal quantum number is.

    (i)   Parabolic
    (ii) Elliptic
    (iii) hyperbolic
    (iv) linear

24. Bohr's theory does not explain the variations of the intensity of spectral lines.

    (i)   True
    (ii) False

25. The spectral lines obtained when electronic transition takes place to the third energy level from other higher levels is called

    (i)   Lyman series
    (ii) Balmer series
    (iii) Paschen series
    (iv) Brackett series
    (v)   Pfund series

26. Which of the following is not an uncertainty principle?

    (i) $\Delta x \Delta p \geq \dfrac{\hbar}{2}$
    (ii) $\Delta J \Delta \theta \geq \dfrac{\hbar}{2}$
    (iii) $\Delta E \Delta t \geq \dfrac{\hbar}{2}$
    (iv) None of the above

27. Two particles of masses m_1 and m_2 move with the same momentum. The ratio of their de Broglie wavelengths will be

    (i)   0
    (ii) 1
    (iii) 3
    (iv) 4

28. What is the unit of a two-dimensional wave function?

   (i) $m^{-3/2}$

   (ii) m^{-1}

   (iii) $m^{-1/2}$

   (iv) None of the above

29. If ψ is a normalized wave function, then the value of $\int\limits_{-\infty}^{+\infty} \psi^*\psi\, dr$ will be

   (i) 0

   (ii) ∞

   (iii) 1

   (iv) None of the above

30. Which of the following relation is correct

   (i) $\Delta x = \left\langle x - \langle x \rangle^2 \right\rangle^{\frac{1}{2}} = \left\{ \langle x^2 \rangle - \langle x \rangle^2 \right\}^{\frac{1}{2}}$

   (ii) $\Delta x = \left\langle \left(x - x^2 \right)^2 \right\rangle^{\frac{1}{2}} = \left\{ \langle x^2 \rangle - \langle x \rangle^2 \right\}^{\frac{1}{2}}$

   (iii) $\Delta x = \left\langle \left(x - \langle x \rangle \right)^2 \right\rangle^{\frac{1}{2}} = \left\{ \langle x^2 \rangle - x^2 \right\}^{\frac{1}{2}}$

   (iv) $\Delta x = \left\langle \left(x - \langle x \rangle \right)^2 \right\rangle^{\frac{1}{2}} = \left\{ \langle x^2 \rangle - \langle x \rangle^2 \right\}^{\frac{1}{2}}$

Answers

1 (ii & iii)	2 (ii)	3 (iii)	4 (ii)	5 (iii)	6 (ii)	7 (iii)	8 (ii)
9 (ii & iii)	10 (iii)	11 (i & iv)	12 (i & iv)	13 (iv)	14 (iv)	15 (iv)	16 (iii)
17 (i)	18 (i)	19 (ii)	20 (iv)	21 (iii & iv)	22 (ii)	23 (i)	24 (i)
25 (iii)	26 (iv)	27 (ii)	28 (ii)	29 (iii)	30 (iv)		

8 Applications of Quantum Mechanics

8.1 Introduction

Quantum mechanics has been applied to problems in nature with great success. It has been enormously successful in explaining microscopic phenomena in all branches of science. The applications described in this chapter are examples that demonstrate the quintessence of the theory. Quantum mechanics gives the most correct results in practically every situation to which it is applied. Only for the simplicity of the mathematics, we confine ourselves to one-dimensional idealistic systems. However, most relevant features of real physical systems can be well understood by applying quantum mechanics to relatively simple idealistic systems. The complete description of the microscopic and sub-microscopic world by quantum physics requires rigorous mathematics which is beyond the scope of this book.

8.2 One-Dimensional Problems

One-dimensional systems are defined as systems which require only one space coordinate say x and time coordinate t for their analysis. Thus the wave functions ψ describing one-dimensional physical systems are functions of x and t only, i.e.,

$$\psi = \psi(x,t) \tag{8.1}$$

If the total energy of the particle has a fixed value E, the time dependence of ψ is given as

$$\psi(x,t) = \psi(x)e^{-i\omega t} \tag{8.2}$$

where the frequency ω is given by the de Broglie relationship

$$\omega = \frac{E}{\hbar} \tag{8.3}$$

Let us see how the multiplication of $\psi(x)$ by $e^{-i\omega t}$ gives a wave. The wave function of a free particle as discussed later on is given by

$$\psi(x) = Ae^{ik_0 x} + Be^{-ik_0 x}$$

Multiplying both sides of this equation by $e^{-i\omega t}$, we get

$$\psi(x)e^{-i\omega t} = Ae^{i(k_0 x - \omega t)} + Be^{-i(k_0 x + \omega t)}$$

The first term on the right represents a trigonometric function with phase $(k_0 x - \omega t)$, and thus, is a wave moving in the positive x-direction. Similarly, the second term on the right represents a trigonometric function with phase $(k_0 x + \omega t)$, and thus, is a wave moving in the negative x direction. The squared magnitudes of the coefficients of each term give intensities of the wave. Hence, the wave moving in the positive x direction has intensity $|A|^2$ and the wave moving in the negative x direction has intensity $|B|^2$.

The potential energy function in one-dimensional physical systems is only a function of x, i.e.,

$$V = V(x) \tag{8.4}$$

Equations (8.1) and (8.2) represent one-dimensional systems.

One-dimensional non-relativistic time-dependent Schrödinger's equation is given by

$$i\hbar \frac{\partial \psi}{\partial t} = -\frac{\hbar^2}{2m} \frac{\partial^2 \psi}{\partial x^2} + V\psi \tag{8.5}$$

One-dimensional non-relativistic time-independent Schrödinger's equation is given by

$$\frac{\partial^2 \psi}{\partial x^2} + \frac{2m}{\hbar^2}(E - V)\psi = 0 \tag{8.6}$$

Time-independent Schrödinger's equation is also called a steady-state equation. The solutions of Schrödinger's equations are different for different systems.

8.3 Boundary Conditions on ψ

It is worthwhile to recall herewith the essential conditions that a physically meaningful solution or wave function must satisfy. They are

i. $\int_0^\infty \psi^* \psi$ must be finite.

ii. ψ is continuous and single valued everywhere in space.

iii. $\dfrac{\partial \psi}{\partial x}$ is continuous and finite every where in space.

iv. $\lim\limits_{x \to \pm\infty} \psi(x) = 0$

The necessity of these conditions becomes clear if we look into the physical meaning of the wave function ψ. For example, the single-valuedness of the wave function is required so as to avoid ambiguity of the theoretical predictions. Any discontinuity in ψ would mean discontinuity in the probability of finding the particle in a given volume. However, the probability would vary continuously from point to point if no particles were created or destroyed. The probability of finding the particle within the volume element is always finite and hence, the first condition. Similarly, $\dfrac{\partial \psi}{\partial x}$ relates to particle momentum, which must be a continuous function of x, except at a point where potential energy is infinite. These conditions enable one to select the actual solution out of a number of possible solutions. In all our analysis, we have assumed the classical relationship $E = K + V$ and neglected the rest energy contribution to the total energy E so that we can use the classical relationship, $K = \dfrac{p^2}{2m}$.

8.4 Free Particle

A free particle is defined as a particle on which no forces are acting in a specified region of space. Since $-\dfrac{\partial V}{\partial x} = F$, for no forces acting on the particle, $V = $ constant. Thus, the particle is moving in a constant potential energy field through out the space. We are free to choose the constant V to be zero to make the mathematics simple, since potential energy is defined to be constant. In summary, for a free particle

$$V = V(x) = 0 \qquad\qquad (8.7)$$

Putting the condition of a free particle ($V = 0$) into the Eq. (8.6), one-dimensional Schrödinger's equation for the free particle is obtained as

$$\frac{\partial^2 \psi}{\partial x^2} + \frac{2mE}{\hbar^2}\psi = 0 \qquad\qquad (8.8)$$

where

$\psi(x) = $ wave function for the free particle

$E = $ allowed total energy values for the particle

$m = $ mass of the particle

or $\dfrac{\partial^2 \psi}{\partial x^2} + k_0^2 \psi = 0$ (8.9)

where $k_0 = \dfrac{\sqrt{2mE}}{\hbar} = \dfrac{2\pi}{\lambda_0}$ (8.10)

Equation (8.9) is a familiar differential equation. The solution of the second order differential Eq. (8.9) where k_0^2 is always positive is given by

$\psi(x) = A e^{ik_0 x} + B e^{-ik_0 x}$ (8.11)

where A and B are integration constants. In this solution, the function $A e^{ik_0 x}$ when multiplied by the factor $e^{-\frac{i}{\hbar} Et}$ $(= e^{-i\omega t})$ represents a plane wave going towards the right and the function $B e^{-ik_0 x}$ when multiplied by the factor $e^{-\frac{i}{\hbar} Et}$ $(= e^{-i\omega t})$ represents a plane wave going towards the left. Hence, the solution Ψ is a linear combination of plane waves. By solving Eq. (8.10), the allowed energy values are obtained as

$E = \dfrac{\hbar^2 k_0^2}{2m}$ (8.12)

Since potential energy is taken to be zero, Eq. (8.12) is the expression for kinetic energy of the particle. As Eq. (8.11) is valid for any value of k_0, kinetic energy $E = \dfrac{\hbar^2 k_0^2}{2m}$ and is permitted to have all values starting from zero to infinity. In the language of quantum mechanics, energy E is not quantized but continuous. Putting $E = \dfrac{p^2}{2m}$ into Eq. (8.12), we get an expected result

$p = \hbar k_0$ (de Broglie equation)

Example 8.1

An electron is moving freely with energy 2 eV. Calculate its de Broglie wavelength.

Solution

The datum given in the question is $E = 2$ eV.
   The de Broglie wavelength is given as

$\lambda_B = \dfrac{h}{\sqrt{2mE}} = \dfrac{6.626 \times 10^{-34}}{\sqrt{2 \times 9.11 \times 10^{-31} \times 2 \times 1.6 \times 10^{-19}}} \text{m} = 8.68\,\text{Å}$

8.5 Potential Steps

Variation of potential energy of a particle as the particle moves from one point of the space to another point constitutes potential energy steps or simply potential steps. The particle may have constant potential energy at a certain region and at a neighbourhood region, it may have a different constant potential energy. This sudden variation of constant potential energy from one constant value to another constant value constitutes sharply defined potential energy steps as shown in Fig. 8.1.

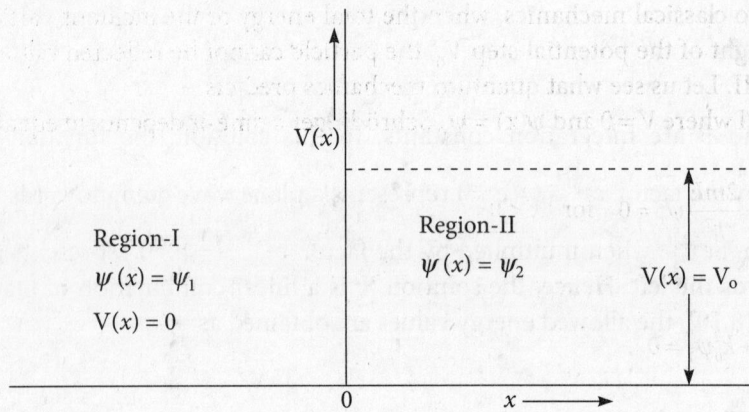

Figure 8.1 | Sharply defined one-dimensional potential energy step. In region I, the potential energy of a particle is taken to be zero for mathematical simplicity. In region II, the potential energy of the particle is taken to be a non-zero constant V_0. The height of the potential energy step is V_0. In region I, the particle is represented by the wave function ψ_1 and in region II, the particle is represented by the wave function ψ_2.

The potential energy of a charged particle changes by constant magnitude when it moves through space from one capacitor to another different capacitor. Like this there are many physical situations in nuclear and atomic science in which the potential energy of a particle changes from one constant value to another constant value as it moves from one region to another region.

The potential energy step shown in Fig. 8.1 is translated mathematically as follows.

$$V(x) = 0 \quad \text{at the region } x < 0$$

$$= V_0 \text{ at the region } x > 0 \tag{8.13}$$

Let the particles having total energy E be incident on the potential energy steps from the left. In region I, they have zero potential energy and when they enter region II, their potential energy becomes V_0. Thus, when the particles move from region I to region II, their potential energy changes by a constant value V_0. Since the total energy of the incident particles remains constant, i.e., it is independent of time, we take into consideration the steady-state form of Schrödinger's equation, which is in general one dimension, and is given by

$$\frac{\partial^2 \psi}{\partial x^2} + \frac{2m}{\hbar^2}(E - V)\psi = 0$$

Case A: $(E \geq V_0)$

According to classical mechanics, when the total energy of the incident particle E is more than the height of the potential step V_0, the particle cannot be reflected but is transmitted into region II. Let us see what quantum mechanics predicts.

In region I where $V = 0$ and $\psi(x) = \psi_1$, Schrödinger's time-independent equation becomes

$$\frac{\partial^2 \psi_1}{\partial x^2} + \frac{2mE}{\hbar^2}\psi_1 = 0 \quad \text{for} \;\; x < 0$$

or $\quad \dfrac{\partial^2 \psi_1}{\partial x^2} + k_0^2 \psi_1 = 0$ $\hfill (8.14)$

where $k_0 = \dfrac{\sqrt{2mE}}{\hbar} = \dfrac{2\pi}{\lambda_0}$ $\hfill (8.15)$

λ_0 in this equation is the de Broglie wavelength of the particle in region I. λ_0 is shown in Fig. 8.2.

Equation (8.14) is a familiar differential equation. The solution of the second order differential Eq. (8.14), where k_0^2 is always positive is given by

$$\psi_1 = Ae^{ik_0 x} + Be^{-ik_0 x}$$ $\hfill (8.16)$

where A and B are integration constants. In this solution, the function $Ae^{ik_0 x}$ when multiplied by the factor $e^{-\frac{i}{\hbar}Et}$ $(= e^{-i\omega t})$ represents a plane wave going towards the right and the function $Be^{ik_0 x}$ when multiplied by the factor $e^{-\frac{i}{\hbar}Et}$ $(= e^{-i\omega t})$ represents a plane wave going towards the left. Hence, the solution ψ_1 is a linear combination of plane waves.

Similarly, in region II where $V(x) = V_0$ and $\psi(x) = \psi_2$, Schrödinger's time-independent equation becomes

$$\frac{\partial^2 \psi_2}{\partial x^2} + \frac{2m(E - V_0)}{\hbar^2}\psi_2 = 0 \; \text{for} \; x > 0$$

or $\dfrac{\partial^2 \psi_2}{\partial x^2} + k^2 \psi_2 = 0$ (8.17)

where $k = \dfrac{\sqrt{2m(E - V_0)}}{\hbar} = \dfrac{2\pi}{\lambda_1}$ (8.18)

λ_1 in this equation is the de Broglie wavelength of the particle in region II. λ_1 is shown in Fig. 8.2.

The solution of the second order differential Eq. (8.17) is given by

$\psi_2 = Ce^{ikx} + De^{-ikx}$ (8.19)

where C and D are integration constants. In this solution, the function Ce^{ikx} when multiplied by the factor $e^{-\frac{i}{\hbar}Et}$ $(= e^{-i\omega t})$, represents a plane wave going towards the right and the function De^{-ikx} when multiplied by the factor $e^{-\frac{i}{\hbar}Et}$ $(= e^{-i\omega t})$, represents a plane wave going towards the left. Since we are considering the incidence of particles from the left on the potential step at $x = 0$, there can not be a wave in region II propagating towards the left; to overcome these difficulties, we must set $D = 0$. Therefore, Eq. (8.19) becomes

$\psi_2 = Ce^{ikx}$ (8.20)

Evaluation of integration constants A, B, C,

For ψ to be the physically meaningful, the wave function ψ is continuous every where including at $x = 0$, i.e.,

$\psi_1\big|_{x=0} = \psi_2\big|_{x=0}$

or $\left(Ae^{ik_0x} + Be^{-ik_0x}\right)\Big|_{x=0} = Ce^{ikx}\big|_{x=0}$

or $A + B = C$ (8.21)

Similarly for ψ to be the physically meaningful, wave function $\dfrac{d\psi}{dx}$ is continuous every where including at $x = 0$, i.e.,

$\dfrac{d\psi_1}{dx}\bigg|_{x=0} = \dfrac{d\psi_2}{dx}\bigg|_{x=0}$

or $\dfrac{d}{dx}\left(Ae^{ik_0x} + Be^{-ik_0x}\right)\bigg|_{x=0} = \dfrac{d}{dx}Ce^{ikx}\bigg|_{x=0}$

or $\quad k_0(A-B)=Ck$ (8.22)

Solving Eqs (8.21) and (8.22) for B and C, we get

$$B=\frac{k_0-k}{k_0+k}A$$

$$C=\frac{2k_0}{k_0+k}A$$

Putting these values of B and C into Eqs (8.16) and (8.20), we get

$$\psi_1=Ae^{ik_0x}+A\frac{k-k_0}{k+k_0}e^{-ik_0x}\quad\text{for }x<0$$ (8.23)

$$\psi_2=A\frac{2k_0}{k+k_0}e^{ikx}\quad\text{for }x>0$$ (8.24)

Equation (8.23) shows that in region I,

$$\psi_I=Ae^{ik_0x}e^{-\frac{i}{\hbar}Et}=Ae^{i\left(k_0x-\frac{E}{\hbar}t\right)}\text{ is a right going wave, i.e., incident wave.}$$ (8.25)

$$\psi_R=A\frac{k-k_0}{k+k_0}e^{-ik_0x}e^{-\frac{i}{\hbar}Et}=A\frac{k-k_0}{k+k_0}e^{-i\left(k_0x-\frac{E}{\hbar}t\right)}\text{ is a left going wave, i.e., reflected wave.}$$ (8.26)

Equation (8.24) shows that in region II,

$$\psi_T=A\frac{2k_0}{k+k_0}e^{ikx}e^{-i\frac{E}{\hbar}t}=A\frac{2k_0}{k+k_0}e^{i\left(kx-\frac{E}{\hbar}t\right)}\text{ is a right going wave, i.e., transmitted wave.}$$ (8.27)

If one particle is considered, $\psi^*\psi$ represents the probability density and if a group of particles is considered, then $\psi^*\psi$ represents the number of particles per unit volume. In this case, $\psi^*\psi\times v$ will represent the number of particles crossing unit area perpendicularly in unit time, i.e., its current density.

If v_0 is the speed of the incident particles, we have from Eq. (8.15)

$$k_0=\frac{\sqrt{2mE}}{\hbar}=\frac{\sqrt{2m\times\frac{1}{2}mv_0^2}}{\hbar}=\frac{mv_0}{\hbar}$$

or $\quad v_0 = \dfrac{\hbar k_0}{m}$ $\hfill$ (8.28)

Similarly if v_1 is the speed of the particles in region II, we have from Eq. (8.18)

$$k = \dfrac{\sqrt{2m(E - V_0)}}{\hbar} = \dfrac{\sqrt{2m \times \dfrac{1}{2}mv_1^2}}{\hbar} = \dfrac{mv_1}{\hbar}$$

or $\quad v_1 = \dfrac{\hbar k}{m}$ $\hfill$ (8.29)

From Eqs (8.28) and (8.29), we have

$$\dfrac{v_0}{v_1} = \dfrac{k_0}{k} = \dfrac{\sqrt{E}}{\sqrt{E - V_0}}$$ $\hfill$ (8.30)

Since $\dfrac{k_0}{k} > 1$, we have $v_0 > v_1$. Thus, the speed of a transmitted particle is less than that of incident particles. In a similar manner, it can be proved that the speed of reflected particles is equal to the speed of incident particles.

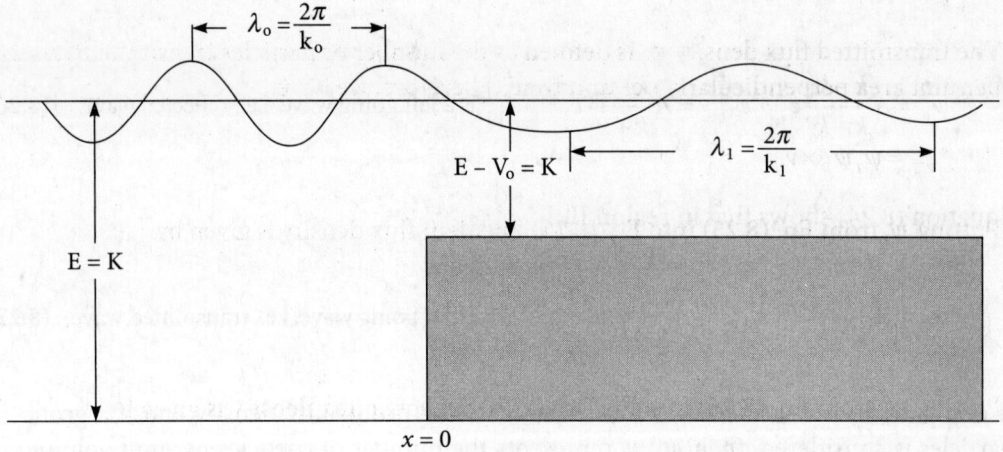

Figure 8.2 | The wave function of a particle of energy E encountering a potential step of height V_0 for the case $E \geq V_0$. The de Broglie wavelength of the particle changes from λ_0 to λ_1 when the particle enters the step, but ψ and $\dfrac{d\psi}{dx}$ are continuous at $x = 0$

8.5.1   Reflection and transmission at the boundary at $x = 0$

We have assumed, $E > V_0$. Hence, according to classical physics, all the particles should be transmitted to region II. However, according to Eq. (8.26), in the region I, there is a reflected wave given by

$$\psi_R = A \frac{k - k_0}{k + k_0} e^{-i\left(k_0 x - \frac{E}{\hbar} t\right)}$$

and the probability density of reflected particles $\psi_R^* \psi_R$ is not zero, i.e., the probability of particles being reflected is not zero. Therefore, according to quantum physics, even though $E > V_0$, the probability of particles being reflected at the boundary is not zero.

Let us calculate the fractions of particles being reflected and transmitted.

$\psi^* \psi \times v$ represents the number of particles crossing unit area perpendicularly in unit time, i.e., current density or flux density.

The incident flux density φ_I is defined as the number of particles incident per unit area perpendicularly per unit time. Hence, we have

$$\varphi_I = \psi_I^* \psi_I \times v_0 \tag{8.31}$$

The reflected flux density φ_R is defined as the number of particles reflected per unit area perpendicularly per unit time. Hence, we have

$$\varphi_R = \psi_R^* \psi_R \times v_0 \tag{8.32}$$

The transmitted flux density φ_T is defined as the number of particles transmitted/refracted per unit area perpendicularly per unit time. Therefore, we have

$$\varphi_T = \psi_T^* \psi_T \times v_1 \tag{8.33}$$

Putting ψ_I from Eq. (8.25) into Eq. (8.31), incident flux density is given by

$$\varphi_I = A^* e^{-i\left(k_0 x - \frac{E}{\hbar} t\right)} A e^{i\left(k_0 x - \frac{E}{\hbar} t\right)} v_0 = |A|^2 v_0 \tag{8.34}$$

Putting ψ_R from Eq. (8.26) into Eq. (8.32), the reflected flux density is given by

$$\varphi_R = A^* \left(\frac{k - k_0}{k + k_0}\right)^* e^{-i\left(k_0 x - \frac{E}{\hbar} t\right)} A \frac{k - k_0}{k + k_0} e^{i\left(k_0 x - \frac{E}{\hbar} t\right)} v_0 = |A|^2 \left|\frac{k - k_0}{k + k_0}\right|^2 v_0 \tag{8.35}$$

Putting ψ_T from Eq. (8.27) into Eq. (8.33), the transmitted flux density is given by

$$\varphi_T = A^* \left(\frac{2k_0}{k+k_0}\right)^* e^{-i\left(kx-\frac{E}{\hbar}t\right)} A \frac{2k_0}{k+k_0} e^{i\left(kx-\frac{E}{\hbar}t\right)} v_0 = |A|^2 \left|\frac{2k_0}{k+k_0}\right|^2 v_1 \qquad (8.36)$$

The reflection coefficient or reflection probability R, i.e., fraction of incident particles reflected is defined as the ratio of reflected flux density to incident flux density. Mathematically,

$$R = \frac{\varphi_R}{\varphi_I} \qquad (8.37)$$

Putting the values of φ_I and φ_R from Eqs (8.34) and (8.35) into these equation, we get

$$R = \frac{|A|^2 \left|\frac{k-k_0}{k+k_0}\right|^2 v_0}{|A|^2 v_0} = \left|\frac{k-k_0}{k+k_0}\right|^2 \qquad (8.38)$$

Putting the values of k_0 and k from Eqs (8.15) and (8.18) into this equation, we get

$$R = \left|\frac{\sqrt{E-V_0} - \sqrt{E}}{\sqrt{E-V_0} + \sqrt{E}}\right|^2 \qquad (8.39)$$

The transmission coefficient or transmission probability T, i.e., fraction of incident particles transmitted is defined as the ratio of the transmission flux density to the incident flux density. Mathematically,

$$T = \frac{\varphi_T}{\varphi_I}$$

Putting the values of φ_I and φ_T from Eqs (8.34) and (8.36) into this equation, we get

$$T = \frac{|A|^2 \left|\frac{2k_0}{k+k_0}\right|^2 v_1}{|A|^2 v_0} = \left|\frac{2k_0}{k+k_0}\right|^2 \frac{v_1}{v_0}$$

Putting the value of $\dfrac{v_1}{v_0} = \dfrac{k}{k_0}$ from Eq. (8.30) into this equation, we get

$$T = \left|\frac{2k_0}{k+k_0}\right|^2 \frac{k}{k_0} = \left|\frac{4kk_0}{(k+k_0)^2}\right|$$

Putting the values of k_0 and k from Eqs (8.15) and (8.18) into this equation, we get

$$T = \left|\frac{4\sqrt{E}\sqrt{E-V_0}}{\left(\sqrt{E-V_0}+\sqrt{E}\right)^2}\right| \tag{8.40}$$

Law of conservation of flux

Now we shall prove that

$$\varphi_I = \varphi_R + \varphi_T$$

Taking the values φ_R and φ_T from Eq. (8.35) and Eq. (8.36), we have

$$\varphi_R + \varphi_T = |A|^2 \left|\frac{k-k_0}{k+k_0}\right|^2 v_0 + |A|^2 \left|\frac{2k_0}{k+k_0}\right|^2 v_1$$

or

$$= \frac{A^2 v_0}{\left(k+k_0\right)^2}\left((k-k_0)^2 + 4k_0^2 \times \frac{v}{v_0}\right)$$

$$= \frac{A^2 v_0}{\left(k+k_0\right)^2}\left((k-k_0)^2 + 4k_0^2 \times \frac{k}{k_0}\right)$$

$$= \frac{A^2 v_0}{\left(k+k_0\right)^2}\left(k+k_0\right)^2 = A^2 v_0 = \varphi_I$$

Thus, $\varphi_R + \varphi_T = \varphi_I$

Refractive index μ

In the usual manner, the refractive index of the second medium ($x > 0$) with respect to the first medium ($x < 0$) is defined as

$$\mu = \frac{\text{speed in second medium}}{\text{speed in first medium}} = \frac{v_1}{v_0}$$

Putting the value of $\dfrac{v_1}{v_0}$ from Eq. (8.30) into this equation, we get

$$\mu = \frac{\sqrt{E - V_0}}{\sqrt{E}} = \sqrt{1 - \frac{V_0}{E}}$$

Example 8.2

Electrons with energy 12.0 eV are incident on a potential step 8 eV high and 0.50 nm wide. Calculate the reflection probability and transmission probability of the incident electron.

Solution

The data given are

$$E = 12 \text{ eV}$$

$$V_0 = 8.0 \text{ eV}$$

The reflection probability is given by

$$R = \left| \frac{\sqrt{E - V_0} - \sqrt{E}}{\sqrt{E - V_0} + \sqrt{E}} \right|^2$$

Putting the given data into this equation, we get

$$R = \left| \frac{\sqrt{12 - 8} - \sqrt{12}}{\sqrt{12 - 8} + \sqrt{12}} \right|^2 = 0.07$$

The transmission probability is given by

$$T = \left| \frac{4\sqrt{E}\sqrt{E - V_0}}{\left(\sqrt{E - V_0} + \sqrt{E}\right)^2} \right|$$

Putting the given data into this equation, we get

$$T = \left| \frac{4\sqrt{12}\sqrt{12-8}}{\left(\sqrt{12-8}+\sqrt{12}\right)^2} \right| = 0.93$$

Case B: $(0 < E < V_0)$

According to classical mechanics, when the total energy of the incident particle E is less than the height of the potential step V_0, the particles reflect back but cannot be transmitted into region II. Let us see what quantum mechanics tells.

In region I where $V = 0$ and $\psi(x) = \psi_1$, Schrödinger's time-independent equation becomes

$$\frac{\partial^2 \psi_1}{\partial x^2} + \frac{2mE}{\hbar^2}\psi_1 = 0 \quad \text{for } x < 0$$

or
$$\frac{\partial^2 \psi_1}{\partial x^2} + k_0^2 \psi_1 = 0 \tag{8.41}$$

where $k_0 = \dfrac{\sqrt{2mE}}{\hbar} = \dfrac{2\pi}{\lambda_0}$ \hfill (8.42)

λ_0 in this equation is the de Broglie wavelength of the particle in region I. λ_0 is shown in Fig. 8.3. The solution of the second order differential Eq. (8.41) where k_0^2 is always positive is given by

$$\psi_1 = Ae^{ik_0 x} + Be^{-ik_0 x} \tag{8.43}$$

where A and B are integration constants.

Similarly in region II, where $V(x) = V_0$ and $\psi(x) = \psi_2$, Schrödinger's time independent equation becomes

$$\frac{\partial^2 \psi_2}{\partial x^2} + \frac{2m(E-V_0)}{\hbar^2}\psi_2 = 0 \quad \text{for } x > 0$$

or
$$\frac{\partial^2 \psi_2}{\partial x^2} - k_1^2 \psi_2 = 0 \tag{8.44}$$

where $k_1 = \dfrac{\sqrt{2m(V_0-E)}}{\hbar}$ \hfill (8.45)

The solution of the second order differential Eq. (8.45) is given by

$$\psi_2 = Ce^{k_1 x} + De^{-k_1 x} \tag{8.46}$$

where C and D are integration constants. As $x \to \infty$, the first term $Ce^{k_1 x}$ does not remain finite, for which the solution ψ_2 looses its physical meanings. Hence, for the meaningful solution, we must set $C = 0$ so that ψ_2 becomes physically meaningful. Therefore, Eq. (8.46) becomes

$$\psi_2 = De^{-k_1 x} \tag{8.47}$$

Evaluation of integration constants A, B, D

For ψ to be the physically meaningful wave function ψ that is continuous everywhere including at $x = 0$, i.e.,

$$\psi_1\big|_{x=0} = \psi_2\big|_{x=0}$$

or $\quad \left(Ae^{ik_0 x} + Be^{-ik_0 x} \right)\Big|_{x=0} = De^{-k_1 x}\Big|_{x=0}$

or $\quad A + B = D \tag{8.48}$

Similarly, for ψ to be the physically meaningful wave function, $\dfrac{d\psi}{dx}$ is continuous every where including at $x = 0$, i.e.,

$$\frac{d\psi_1}{dx}\bigg|_{x=0} = \frac{d\psi_2}{dx}\bigg|_{x=0}$$

or $\quad \left(Aik_0 e^{ik_0 x} - Bik_0 e^{-ik_0 x} \right)\Big|_{x=0} = -Dk_1 e^{-k_1 x}\Big|_{x=0}$

or $\quad ik_0(A - B) = -Dk_1 \tag{8.49}$

Solving Eqs (8.48) and (8.49) for B and D, we get

$$B = \frac{ik_0 + k_1}{ik_0 - k_1} A$$

$$D = \frac{2ik_0}{ik_0 - k_1} A$$

Putting these values of B and D into the Eqs (8.43) and (8.47), we get

$$\psi_1 = Ae^{ik_0 x} + A\frac{ik_0 + k_1}{ik_0 - k_1}e^{-ik_0 x} \quad \text{for } x < 0 \tag{8.50}$$

$$\psi_2 = A\frac{2ik_0}{ik_0 - k_1}e^{-k_1 x} \quad \text{for } x > 0 \tag{8.51}$$

Equation (8.50) shows that in region I,

$$\psi_I = Ae^{ik_0 x}e^{-\frac{i}{\hbar}Et} = Ae^{i\left(k_0 x - \frac{E}{\hbar}t\right)} \text{ is a right going wave, i.e., incident wave.} \tag{8.52}$$

$$\psi_R = A\frac{ik_0 + k_1}{ik_0 - k_1}e^{-ik_0 x}e^{-\frac{i}{\hbar}Et} = A\frac{ik_0 + k_1}{ik_0 - k_1}e^{-i\left(k_0 x - \frac{E}{\hbar}t\right)} \text{ is a left going wave, i.e., reflected wave.} \tag{8.53}$$

Equation (8.51) shows that in region II,

$$\psi_T = A\frac{2ik_0}{ik_0 - k_1}e^{-k_1 x}e^{-\frac{E}{\hbar}t} = A\frac{2ik_0}{ik_0 - k_1}e^{-\left(k_1 x + \frac{E}{\hbar}t\right)} \text{ is a right going wave, i.e., transmitted wave.} \tag{8.54}$$

Transmission at the boundary (at x = 0)

We have assumed $E < V_0$. Hence, according to classical physics no particles should be transmitted to region II since its total energy is not sufficient to overcome the potential energy step. However, according to Eq. (8.54) in region II, wave function ψ_T dies away exponentially and thus, quantum physics allows the wave function and therefore, the particle to penetrate into the classically forbidden region. No experiment can ever observe the particle in region II (its kinetic energy $E - V$ would be negative there), but in certain experiments, a particle can pass through a classically forbidden region and emerge into an allowed region where it can be observed. Penetration into the forbidden region is associated with the wave nature of the particles and is inconsistent with the uncertainty principle. Detailed explanation is beyond the scope of this book.

Let us calculate the fractions of particles being reflected and transmitted. From Eq. (8.31), the incident flux density is given by

$$\varphi_I = \psi_I^* \psi_I \times v_0$$

Putting ψ_I from Eq. (8.52) into this equation, the incident flux density is given by

$$\varphi_I = |A|^2 v_0 \tag{8.55}$$

From Eq. (8.32), the incident flux density is given by

$$\varphi_R = \psi_R^* \psi_R \times v_0$$

Putting ψ_R from Eq. (8.53) into this equation, the reflected flux density is given by

$$\varphi_R = A^* \left(\frac{ik_0 + k_1}{ik_0 - k_1} \right)^* e^{i\left(k_0 x - \frac{E}{\hbar}t\right)} A \frac{ik_0 + k_1}{ik_0 - k} e^{i\left(k_0 x - \frac{E}{\hbar}t\right)} v_0$$

$$= A^* \frac{-ik_0 + k_1}{ik_0 - k_1} \times A \frac{ik_0 + k_1}{ik_0 - k_1} v_0$$

$$= A^* \frac{k_1 - ik_0}{-(k_1 + ik_0)} \times A \frac{ik_0 + k_1}{-(k_1 - ik_0)} v_0$$

or $$\varphi_R = |A|^2 v_0 \tag{8.56}$$

From Eq. (8.33), the transmitted flux density is given by

$$\varphi_T = \psi_T^* \psi_T \times v_1$$

Putting $\psi_T(x)$ from Eq. (8.54) into this equation, the transmitted flux density is given by

$$\varphi_T = A^* \left(\frac{2ik_0}{ik_0 - k_1} \right)^* e^{-kx} A \frac{2ik_0}{ik_0 - k_1} e^{-kx} v$$

$$= A^* \frac{-2ik_0}{-ik_0 - k_1} e^{-kx} A \frac{2ik_0}{ik_0 - k_1} e^{-kx} v$$

$$= A^* \frac{-2ik_0}{-(k_1 + ik_0)} \times A \frac{2ik_0}{-(k_1 - ik_0)} e^{-2kx} v$$

$$= \left| A \right|^2 \frac{4k_0^2}{\left(k_1^2 + k_0^2 \right)} e^{-2kx} v \tag{8.57}$$

Putting the values of φ_I and φ_R from Eqs (8.55) and (8.56) into the equation $R = \dfrac{\varphi_R}{\varphi_I}$, we get

$$R = \frac{\left| A \right|^2 v_0}{\left| A \right|^2 v_0} = 1 \tag{8.58}$$

Equation (8.58) shows that all incident particles are reflected back. Putting the values of φ_I and φ_T from Eqs (8.55) and (8.57) into the equation $T = \dfrac{\varphi_T}{\varphi_I}$, we get

$$T = \frac{4k_0^2}{k_1^2 + k_0^2} e^{-2k_1 x} \frac{v}{v_0} \tag{8.59}$$

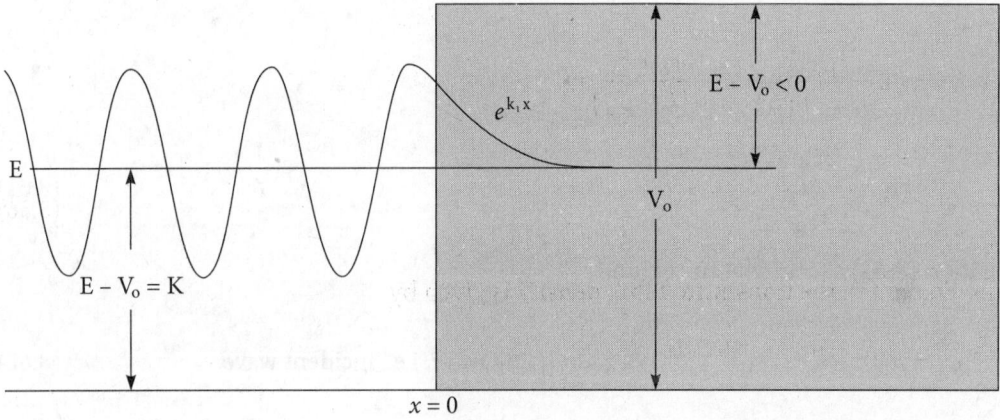

Figure 8.3 │ The wave function of a incident particle whose total energy is less than the height of the step. The wave function decreases exponentially in the classically forbidden region where the classical kinetic energy $E - V$ would be negative. At $x = 0$, the wave function ψ and $\dfrac{d\psi}{dx}$ are continuous

These discussions show that D is not zero unless k_1 and consequently, V_0 is infinity. If D is not zero, then ψ_2 is not zero for $x > 0$ and the probability density of finding the particle in region II is

$$= \left| A \right|^2 \frac{4k_0^2}{\left(k_1^2 + k_0^2 \right)} e^{-2kx}.$$

Case C: (E < 0)

In Case B, kinetic energy $E - V$ is positive in region I ($x < 0$) and negative in region II ($x > 0$). When kinetic energy $E - V$ is negative everywhere as in Case C, the wave numbers k_0 and k are both imaginary and there is no solution.

Case D: ($V \to \infty$ for x > 0) Potential step of infinity height

In region I where $V = 0$, the solution of Schrödinger's time-independent equation as calculated earlier is given by

$$\psi_1 = Ae^{ik_0 x} + Be^{-ik_0 x} \quad k_0^2 = \frac{2mE}{\hbar} \tag{8.60}$$

with $B = \dfrac{ik_0 + k_1}{ik_0 - k_1} A$. When $V \to \infty$, $k_1 \to \infty$, and $B = \dfrac{ik_0 + k_1}{ik_0 - k_1} A$ becomes $- A$ i.e., $B = - A$ when $V \to \infty$. Hence, Eq. (8.60) becomes

$$\psi_1 = Ae^{ik_0 x} - Ae^{-ik_0 x}$$

or $\quad \psi_1 = Ae^{ik_0 x} + e^{-i\frac{\pi}{2}} Ae^{-ik_0 x}$

or $\quad \psi_1 = Ae^{ik_0 x} + Ae^{-i\left(k_0 x + \frac{\pi}{2}\right)} \tag{8.61}$

Equation (8.61) shows that in region I,

$$\psi_I = Ae^{ik_0 x} e^{-\frac{i}{\hbar}Et} = Ae^{i\left(k_0 x - \frac{E}{\hbar}t\right)} \text{ is a right going wave, i.e., incident wave.} \tag{8.62}$$

$$\psi_R = Ae^{-i\left(k_0 x + \frac{\pi}{2}\right)} e^{-\frac{i}{\hbar}Et} = Ae^{-i\left(k_0 x + \frac{E}{\hbar}t + \frac{\pi}{2}\right)} \text{ is a left going wave, i.e., reflected wave.} \tag{8.63}$$

The phase difference between the incident wave and the reflected wave is $\dfrac{\pi}{2}$.

When $V \to \infty$, $k_1 \to \infty$, and $D = \dfrac{2ik_0}{ik_0 - k_1} A$ becomes zero, i.e., $D = 0$. As a result of which the wave function in region II, ψ_2 vanishes, i.e.,

$$\psi_2 = \psi_T = 0 \quad \text{for} \quad x \geq 0. \tag{8.64}$$

In this case, particles are never found in region II.

Reflection at the boundary (at x = 0)

We have assumed $V \to \infty$. Hence, according to classical physics, no particles should be transmitted to region II since the potential energy step is infinity. The same inference is drawn from quantum physics according to Eq. (8.64). Let us calculate the fractions of particles being reflected and transmitted.

From Eq. (8.62), the incident flux density is given by

$$\varphi_I = \psi_I^* \psi_I \times v_0 = |A|^2 v_0 \tag{8.65}$$

Putting ψ_R from Eq. (8.63) into this equation, the reflected flux density is given by

$$\varphi_R = \psi_R^* \psi_R \times v_0 = |A|^2 v_0 \tag{8.66}$$

From Eq. (8.64), the transmitted flux density is given by

$$\varphi_T = \psi_T^* \psi_T \times v_1 = 0 \tag{8.67}$$

Putting the values of φ_I and φ_R from Eqs (8.65) and (8.66) into equation $R = \dfrac{\varphi_R}{\varphi_I}$, the reflection coefficient or reflection probability is given by

$$R = \frac{|A|^2 v_0}{|A|^2 v_0} = 1 \tag{8.68}$$

This equation shows that all the incident particles are reflected back. Putting the values of φ_I and φ_R from Eqs (8.65) and (8.67) into equation $T = \dfrac{\varphi_T}{\varphi_I}$, we get

$$T = 0 \tag{8.69}$$

This is in conformity with Eq. (8.68).

Example 8.3

If a beam of electrons strike on any potential energy step of height 0.03 eV, find the fraction of electrons reflected and transmitted if the energy of the incident electrons is

(i)   0.04 eV,   (ii)   0.025 eV,   (iii)   0.03 eV,

Solution

i.   In this case, $E = 0.04$ and $V_0 = 0.03$ eV; $E > V_0$. Therefore, reflection probability is

$$R = \left| \frac{\sqrt{E - V_0} - \sqrt{E}}{\sqrt{E - V_0} + \sqrt{E}} \right|^2$$

and

transmission probability is $T = \left| \dfrac{4\sqrt{E}\sqrt{E-V_0}}{\left(\sqrt{E-V_0} + \sqrt{E}\right)^2} \right|$

Putting the given data into these expressions, we get

$$R = \left| \frac{\sqrt{0.04 - 0.03} - \sqrt{0.04}}{\sqrt{0.04 - 0.03} + \sqrt{0.04}} \right|^2 = 0.11$$

and $\quad T = \left| \dfrac{4\sqrt{0.04}\sqrt{0.04 - 0.03}}{\left(\sqrt{0.04 - 0.03} + \sqrt{0.04}\right)^2} \right| = 0.89$

In this case, 11% of electrons are reflected and 89% of electrons are transmitted. In this case, $R + T = 1$ is also verified.

ii.   In this case, $E = 0.025$ and $V_0 = 0.03$ eV; $E < V_0$. Since $E < V_0$, the reflection probability is $R = 1$ and the transmission probability naturally will be zero. Hence, all the incident electrons will be reflected.

iii. In this case, $E = 0.03$ and $V_0 = 0.03$ eV; $E = V_0$. Therefore, the reflection probability is

$$R = \left| \frac{\sqrt{E - V_0} - \sqrt{E}}{\sqrt{E - V_0} + \sqrt{E}} \right|^2 \quad \text{and the transmission probability is} \quad T = \left| \frac{4\sqrt{E}\sqrt{E - V_0}}{\left(\sqrt{E - V_0} + \sqrt{E}\right)^2} \right|$$

Putting the given data into these expressions, we get

$$R = \left| \frac{\sqrt{0.03 - 0.03} - \sqrt{0.03}}{\sqrt{0.03 - 0.03} + \sqrt{0.03}} \right|^2 = 1$$

and $\quad T = \left| \dfrac{4\sqrt{0.04}\sqrt{0.03 - 0.03}}{\left(\sqrt{0.03 - 0.03} + \sqrt{0.03}\right)^2} \right| = 0$

In this case also, the reflection probability is $R = 1$ and the transmission probability comes out to be zero. Hence, all the incident electrons will be reflected. In this case, $R + T = 1$ is also verified.

8.5.2 Potential energy barrier

The sharpest increase of potential energy to a certain value at a point and sharpest fall of the potential energy to zero at another point while remaining constant continuously over the interval constitute the potential barrier or potential energy barrier as shown in Fig. 8.4.

Thus, the potential barrier sharply increases to a certain value V_0, remains constant over a certain interval and then decreases again to its original value.

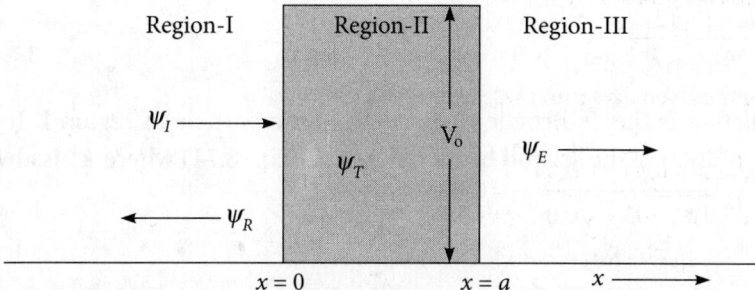

Figure 8.4 | Ideal potential energy barrier of height V_0. When a particle of energy $E < V_0$ is incident on the potential barrier, according to classical physics, the particle must be reflected back. In quantum physics, the de Broglie waves that correspond to the particle are partly reflected and partly transmitted which means that the particle has a finite probability to penetrate the barrier

Mathematically, potential barrier is defined by

$$V(x) = 0 \quad \text{for } -\infty < x < 0$$

$$= V_0 \quad \text{for } 0 \leq x \leq a \tag{8.70}$$

$$= 0 \quad \text{for } a \leq x < \infty$$

According to classical physics, if a particle is incident on the potential barrier with energy less than the height of the barrier, it will be reflected back by the barrier. However, due to the wave nature of matter, there is some probability that the particle incident on the barrier with less energy than the barrier height, penetrates out to the other side of the barrier. Experimentally it has been observed to happen; the effect is called tunnel effect.

If a particle is incident on the potential barrier with energy less than the height of the barrier, $E < V_0$, it will not necessarily be reflected by the barrier but there is always the probability that it may cross the barrier and continue its forward motion. This probability of crossing the barrier is called the tunnel effect. The tunnel effect is purely a quantum mechanical effect. The emission of electrons, β-emission and emission of α-particles are examples of tunnel effect.

Case I: $(E < V_0)$

In region I where $V = 0$ and $\psi(x) = \psi_1$, Schrödinger's time-independent equation becomes

$$\frac{\partial^2 \psi_1}{\partial x^2} + \frac{2mE}{\hbar^2}\psi_1 = 0 \quad \text{for } -\infty < x < 0$$

or
$$\frac{\partial^2 \psi_1}{\partial x^2} + k_0^2 \psi_1 = 0 \tag{8.71}$$

where
$$k_0 = \frac{\sqrt{2mE}}{\hbar} = \frac{2\pi}{\lambda_0} \tag{8.72}$$

λ_0 in this equation is the de Broglie wavelength of the particle in region I. It is shown in Fig. 8.5. The solution of the second order differential Eq. (8.71) where k_0^2 is always positive is given by

$$\psi_1 = Ae^{ik_0 x} + Be^{-ik_0 x} \text{ for } -\infty < x < 0 \tag{8.73}$$

where A and B are integration constants. In this solution, the function $Ae^{ik_0 x}$ when multiplied by the factor $e^{-\frac{i}{\hbar}Et}$ ($=e^{-i\omega t}$) represents a plane wave going towards the right and the function $Be^{-ik_0 x}$ when multiplied by the factor $e^{-\frac{i}{\hbar}Et}$ ($=e^{-i\omega t}$) represents a plane wave going towards the left. Hence, the solution ψ_1 is a linear combination of plane waves.

Similarly in region II where $V(x) = V_0$ and $\psi(x) = \psi_2$, Schrödinger's time-independent equation becomes

$$\frac{\partial^2 \psi_2}{\partial x^2} + \frac{2m(E - V_0)}{\hbar^2} \psi_2 = 0 \text{ for } 0 \le x \le a$$

$$\frac{\partial^2 \psi_2}{\partial x^2} - \frac{2m(V_0 - E)}{\hbar^2} \psi_2 = 0$$

or

$$\frac{\partial^2 \psi_2}{\partial x^2} - k_1^2 \psi_2 = 0 \tag{8.74}$$

where
$$k_1 = \frac{\sqrt{2m(V_0 - E)}}{\hbar} \text{ is real.} \tag{8.75}$$

The solution of the second order differential Eq. (8.74) is given by

$$\psi_2 = Ce^{k_1 x} + De^{-k_1 x} \text{ for } 0 \le x \le a \tag{8.76}$$

where C and D are integration constants. In this solution, the function $Ce^{k_1 x}$ when multiplied by the factor $e^{-\frac{i}{\hbar}Et}$ ($=e^{-i\omega t}$) represents a plane wave going towards the right and the function $De^{-k_1 x}$ when multiplied by the factor $e^{-\frac{i}{\hbar}Et}$ ($= e^{-i\omega t}$) represents a plane wave going towards the left.

In region III where $V = 0$ and $\psi(x) = \psi_3$, Schrödinger's time-independent equation becomes

$$\frac{\partial^2 \psi_3}{\partial x^2} + \frac{2mE}{\hbar^2} \psi_3 = 0 \text{ for } a \leq x < \infty$$

or $\qquad \dfrac{\partial^2 \psi_3}{\partial x^2} + k_0^2 \psi_3 = 0 \qquad\qquad\qquad\qquad\qquad$ (8.77)

where $\quad k_0 = \dfrac{\sqrt{2mE}}{\hbar} = \dfrac{2\pi}{\lambda_0} \qquad\qquad\qquad\qquad\qquad$ (8.78)

λ_0 in this equation is the de Broglie wavelength of the particle in region III. It is shown in Fig. 8.5.

The solution of the second order differential Eq. (8.77) where k_0^2 is always positive is given by

$$\psi_3 = Fe^{ik_0 x} + Ge^{-ik_0 x} \text{ for } a \leq x < \infty \qquad\qquad (8.79)$$

where E and F are integration constants. In this solution, the function $Ee^{ik_0 x}$ when multiplied by the factor $e^{-\frac{i}{\hbar}Et}$ $(=e^{-i\omega t})$ represents a plane wave going towards the right and the function $Fe^{-ik_0 x}$ when multiplied by the factor $e^{-\frac{i}{\hbar}Et}$ $(= e^{-i\omega t})$ represents a plane wave going towards the left. However, in region III, there cannot be a reflected wave to move towards the left. Hence, G in Eq. (8.79) is zero and Eq. (8.79) becomes

$$\psi_3 = Fe^{ik_0 x} \text{ for } a \leq x < \infty \qquad\qquad (8.80)$$

Evaluation of integration constants A, B, C, and D

For ψ to be the physically meaningful, wave function ψ is continuous everywhere including at $x = 0$, i.e.,

$$\psi_1 \big|_{x=0} = \psi_2 \big|_{x=0}$$

or $\quad \left(Ae^{ik_0 x} + Be^{-ik_0 x} \right)\Big|_{x=0} = \left(Ce^{k_1 x} + De^{-k_1 x} \right)\Big|_{x=0}$

or $\quad A + B = C + D \qquad\qquad\qquad\qquad\qquad\qquad$ (8.81)

Similarly, for ψ to be the physically meaningful, wave function $\dfrac{d\psi}{dx}$ is continuous everywhere including at $x = 0$, i.e.,

$$\left.\frac{d\psi_1}{dx}\right|_{x=0} = \left.\frac{d\psi_2}{dx}\right|_{x=0}$$

or $\quad \left(Aik_0 e^{ik_0 x} - Bik_0 e^{-ik_0 x}\right)\Big|_{x=0} = \left(Ck_1 e^{k_1 x} + Dk_1 e^{-k_1 x}\right)\Big|_{x=0}$

or $\quad ik_0(A - B) = k_1(C - D)$ $\qquad\qquad\qquad\qquad$ (8.82)

Solving Eqs (8.81) and (8.82) for A and B, we get

$$A = \left(1 - \frac{ik_1}{k_0}\right)\frac{C}{2} + \left(1 + \frac{ik_1}{k_0}\right)\frac{D}{2} \tag{8.83}$$

$$B = \left(1 + \frac{ik_1}{k_0}\right)\frac{C}{2} + \left(1 - \frac{ik_1}{k_0}\right)\frac{D}{2} \tag{8.84}$$

Similarly at $x = a$,

$$\psi_2\big|_{x=a} = \psi_3\big|_{x=a}$$

or $\quad Ce^{k_1 a} + De^{k_1 a} = Fe^{ik_0 a}$ $\qquad\qquad\qquad\qquad$ (8.85)

Similarly for ψ to be physically meaningful, wave function $\dfrac{d\psi}{dx}$ is continuous everywhere including at $x = a$, i.e.,

$$\left.\frac{d\psi_2}{dx}\right|_{x=a} = \left.\frac{d\psi_3}{dx}\right|_{x=a}$$

$$Ck_1 e^{k_1 a} - Dk_1 e^{-k_1 a} = ik_0 Fe^{ik_0 a} \tag{8.86}$$

Solving Eqs (8.85) and (8.86) for C and D, we get

$$C = \left(1 + \frac{ik_0}{k_1}\right)\frac{F}{2} e^{(ik_0 - k_1)a} \tag{8.87}$$

$$D = \left(1 - \frac{ik_0}{k_1}\right)\frac{F}{2}e^{(ik_0 + k_1)a} \tag{8.88}$$

Putting the value of C and D from Eqs (8.87) and (8.88) into Eq. (8.83), we get

$$A = \frac{F}{4}e^{ik_0 a}\left(1 - \frac{ik_1}{k_0}\right)\left(1 + i\frac{k_0}{k_1}\right)e^{-k_1 a} + \frac{F}{4}e^{ik_0 a}\left(1 + \frac{ik_1}{k_0}\right)\left(1 - \frac{ik_0}{k_1}\right)e^{k_1 a}$$

$$= \frac{F}{4}e^{ik_0 a}\left[\left(1 - i\frac{k_1}{k_0} + i\frac{k_0}{k_1} + 1\right)e^{-k_1 a} + \left(1 + i\frac{k_1}{k_0} - i\frac{k_0}{k_1} + 1\right)e^{k_1 a}\right]$$

$$= \frac{F}{4}e^{ik_0 a}\left[2\left(e^{k_1 a} + e^{-k_1 a}\right) + i\frac{k_1}{k_0}\left(e^{k_1 a} - e^{-k_1 a}\right) - i\frac{k_0}{k_1}\left(e^{k_1 a} - e^{-k_1 a}\right)\right]$$

$$= \frac{F}{4}e^{ik_0 a}\left[2 \times 2\cosh k_1 a + i2\frac{k_1}{k_0}\sinh k_1 a - i2\frac{k_0}{k_1}\sinh k_1 a\right]$$

or $\quad A = Fe^{ik_0 a}\left[\cosh k_1 a + \frac{i}{2}\left(\frac{k_1}{k_0} - \frac{k_0}{k_1}\right)\sinh k_1 a\right] \tag{8.89}$

Again putting the value of C and D from Eqs (8.87) and (8.88) into Eq. (8.84), we get

$$B = \frac{F}{4}e^{ik_0 a}\left(1 + \frac{ik_1}{k_0}\right)\left(1 + \frac{ik_0}{k_1}\right)e^{-k_1 a} + \frac{F}{4}e^{ik_0 a}\left(1 - \frac{ik_1}{k_0}\right)\left(1 - \frac{ik_0}{k_1}\right)e^{k_1 a}$$

$$= \frac{F}{4}e^{ik_0 a}\left[\left(1 + \frac{ik_1}{k_0} + \frac{ik_0}{k_1} + i^2\right)e^{-k_1 a} + \left(1 - \frac{ik_1}{k_0} - \frac{ik_0}{k_1} + i^2\right)e^{k_1 a}\right]$$

$$= \frac{F}{4}e^{ik_0 a}\left[\left(\frac{ik_1}{k_0} + \frac{ik_0}{k_1}\right)e^{-k_1 a} - \left(\frac{ik_1}{k_0} + \frac{ik_0}{k_1}\right)e^{k_1 a}\right]$$

$$= \frac{F}{4}e^{ik_0 a}\left[i\left(\frac{k_1}{k_0}+\frac{k_0}{k_1}\right)\left(e^{-k_1 a}-e^{k_1 a}\right)\right]$$

$$= \frac{F}{4}e^{ik_0 a}\left[i\left(\frac{k_1}{k_0}+\frac{k_0}{k_1}\right)2\sinh k_1 a\right]$$

or $\quad B = \dfrac{iF}{2}e^{ik_0 a}\left(\dfrac{k_1}{k_0}+\dfrac{k_0}{k_1}\right)\sinh k_1 a \qquad (8.90)$

Putting these values of A, B, C, and D into the Eqs (8.73) and (8.76), we get

$$\psi_1 = Fe^{ik_0 a}\left(\cosh k_1 a + \frac{i}{2}\left(\frac{k_1}{k_0}-\frac{k_0}{k_1}\right)\sinh k_1 a\right)e^{ik_0 x} +$$

$$\frac{iF}{2}e^{ik_0 a}\left(\frac{k_1}{k_0}+\frac{k_0}{k_1}\right)\sinh k_1 a e^{-ik_0 x} \quad \text{for } -\infty < x < 0 \qquad (8.91)$$

$$\psi_2 = \frac{F}{2}e^{(ik_0-k_1)a}\left(1+\frac{ik_0}{k_1}\right)e^{k_1 x} + \frac{F}{2}e^{(ik_0+k_1)a}\left(1-\frac{ik_0}{k_1}\right)e^{-k_1 x} \quad \text{for } 0 \le x \le a \qquad (8.92)$$

$$\psi_3 = Fe^{ik_0 x} \quad \text{for } a \le x < \infty \qquad (8.93)$$

Equation (8.91) shows that in region I,

$$\psi_I = Fe^{ik_0 a}\left(\cosh k_1 a + \frac{i}{2}\left(\frac{k_1}{k_0}-\frac{k_0}{k_1}\right)\sinh k_1 a\right)e^{ik_0 x} \text{ is a right going wave, i.e., incident wave. } (8.94)$$

$$\psi_R = \frac{iF}{2}e^{ik_0 a}\left(\frac{k_1}{k_0}+\frac{k_0}{k_1}\right)\sinh k_1 a e^{-ik_0 x} \text{ is a left going wave, i.e., reflected wave. } (8.95)$$

Equation (8.92) shows that in region II,

$$\psi_T = \frac{F}{2}e^{(ik_0-k_1)a}\left(1+\frac{ik_0}{k_1}\right)e^{k_1 x} \text{ is a right going wave, i.e., transmitted wave. } (8.96)$$

$$\psi_{R'} = \frac{F}{2} e^{(ik_0 + k_1)a} \left(1 - \frac{ik_0}{k_1}\right) e^{-k_1 x} \text{ is a left going wave, i.e., reflected wave} \tag{8.97}$$

Equation (8.93) shows that in region III,

$$\psi_E = Fe^{ik_0 x} \text{ is a right going wave, i.e., emergent wave} \tag{8.98}$$

Thus, the probability density of the particle in region III is $\psi_E \psi_E^* = Fe^{ik_0 x} Fe^{-ik_0 x} = F^2$, a non-zero quantity which implies that the probability of tunnelling of a particle through the barrier to region III is not zero even though $E < V_0$! This is one of basic characteristics of quantum physics.

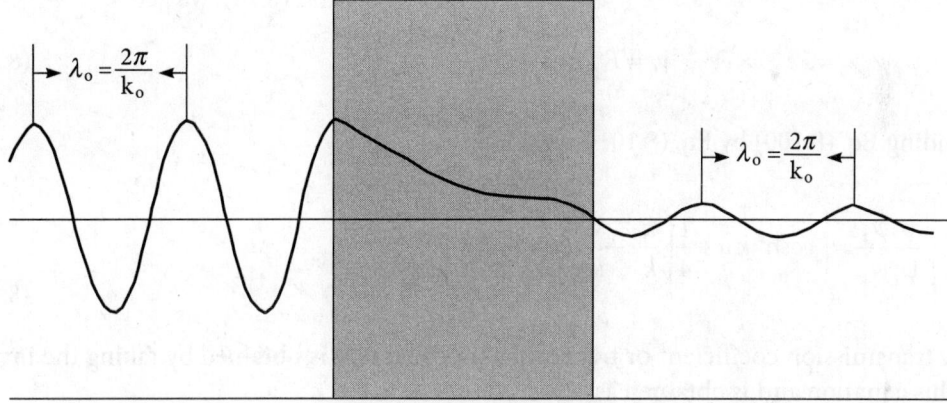

Figure 8.5 | The wave function of a particle incident on the potential energy barrier from the left side when $E < V_0$. The wavelength λ_0 is same on both sides of the barrier, whereas the amplitude beyond the barrier [region III] is much less than the original amplitude in region I. The particle can never be observed inside the barrier [region II] where it would have negative kinetic energy, but it can be observed beyond the barrier! Since a particle escapes detection by the observer in region II, the word 'tunnelling' is used. The phenomenon is called tunnelling effect or quantum mechanical tunnelling

Transmission coefficient

The transmission coefficient or transmission probability T of the barrier is defined in the usual way as

$$T = \frac{\text{Transmitted flux density through the barrier}}{\text{Incident flux density on the barrier at } x = 0} = \frac{\psi_E \psi_E^* v_0}{\psi_I \psi_I^* v_0} = \frac{\psi_E \psi_E^*}{\psi_I \psi_I^*} \tag{8.99}$$

The incident flux density on the barrier at $x = 0$ is

$$\psi_I \psi_I^* v_0 = Fe^{ik_0 a}\left(\cosh k_1 a + \frac{i}{2}\left(\frac{k_1}{k_0} - \frac{k_0}{k_1} \right) \sinh k_1 a \right) e^{ik_0 x} \times$$

$$Fe^{-ik_0 a}\left(\cosh k_1 a - \frac{i}{2}\left(\frac{k_1}{k_0} - \frac{k_0}{k_1} \right) \sinh k_1 a \right) e^{-ik_0 x} v_0$$

$$\psi_I \psi_I^* v_0 = F^2\left(\cosh^2 k_1 a + \frac{1}{4}\left(\frac{k_1}{k_0} - \frac{k_0}{k_1} \right)^2 \sinh^2 k_1 a \right) v_0 \tag{8.100}$$

The emergent flux density through the barrier is

$$\psi_E \psi_E^* v_0 = Fe^{ik_0 x} \times Fe^{-ik_0 x} v_0 = F^2 v_0 \tag{8.101}$$

Dividing Eq. (8.100) by Eq. (8.101), we obtain

$$\frac{\psi_I \psi_I^*}{\psi_E \psi_E^*} = \left(\cosh^2 k_1 a + \frac{1}{4}\left(\frac{k_1}{k_0} - \frac{k_0}{k_1} \right)^2 \sinh^2 k_1 a \right) \tag{8.102}$$

The transmission coefficient or transmission probability is obtained by taking the inverse of this equation and is obtained as

$$T = \frac{1}{\cosh^2 k_1 a + \frac{1}{4}\left(\frac{k_1}{k_0} - \frac{k_0}{k_1} \right)^2 \sinh^2 k_1 a} \tag{8.103}$$

or

$$T = \frac{1}{\cosh^2 k_1 a + \frac{1}{4}\left(\frac{k_1}{k_0} + \frac{k_0}{k_1} \right)^2 \sinh^2 k_1 a - \sinh^2 k_1 a}$$

or

$$T = \frac{1}{1 + \frac{1}{4}\left(\frac{k_1}{k_0} + \frac{k_0}{k_1} \right)^2 \sinh^2 k_1 a} \tag{8.104}$$

We can also express Eq. (8.103) in another form in the following way. From Eq. (8.103), we have

$$T = \frac{4}{4\cosh^2 k_1 a + \left(\dfrac{k_1}{k_0} - \dfrac{k_0}{k_1}\right)^2 \sinh^2 k_1 a}$$

or

$$T = \frac{4}{4\cosh^2 k_1 a + \left(\dfrac{k_1}{k_0}\right)^2 \sinh^2 k_1 a + \left(\dfrac{k_0}{k_1}\right)^2 \sinh^2 k_1 a - 2\sinh^2 k_1 a}$$

Multiplying the numerator and denominator of this equation by $\left(\dfrac{k_0}{k_1}\right)^2$, we have

$$T = \frac{4\left(\dfrac{k_0}{k_1}\right)^2}{4\left(\dfrac{k_0}{k_1}\right)^2 \cosh^2 k_1 a + \sinh^2 k_1 a + \left(\dfrac{k_0}{k_1}\right)^4 \sinh^2 k_1 a - 2\left(\dfrac{k_0}{k_1}\right)^2 \sinh^2 k_1 a}$$

$$= \frac{4\left(\dfrac{k_0}{k_1}\right)^2}{\left[1 + \left(\dfrac{k_0}{k_1}\right)^4 - 2\left(\dfrac{k_0}{k_1}\right)^2\right]\sinh^2 k_1 a + 4\left(\dfrac{k_0}{k_1}\right)^2 \cosh^2 k_1 a}$$

or

$$T = \frac{\left(\dfrac{2k_0}{k_1}\right)^2}{\left[1 - \left(\dfrac{k_0}{k_1}\right)^2\right]^2 \sinh^2 k_1 a + \left(\dfrac{2k_0}{k_1}\right)^2 \cosh^2 k_1 a} \tag{8.105}$$

Putting the value of $\left(\dfrac{k_0}{k_1}\right)^2 = \dfrac{E}{V_0 - E}$ into this equation, we have

$$T = \frac{4\left(\dfrac{E}{V_0 - E}\right)^2}{\left[1 - \left(\dfrac{E}{V_0 - E}\right)^2\right]^2 \sinh^2 k_1 a + 4\left(\dfrac{E}{V_0 - E}\right)^2 \cosh^2 k_1 a} \tag{8.106}$$

The expressions for transmission coefficient or transmission probability shows that $T = 0$ at $E = 0$ and increases steadily with E.

Reflection coefficient in region I

The reflection coefficient or reflection probability of the barrier R is defined in the usual way as

$$R = \frac{\psi_R \psi_R^* v_0}{\psi_I \psi_I^* v_0} = \frac{\psi_R \psi_R^*}{\psi_I \psi_I^*} \tag{8.107}$$

The reflected flux density on the barrier at $x = 0$ is

$$\psi_R \psi_R^* v_0 = \frac{iF}{2} e^{ik_0 a}\left(\frac{k_1}{k_0} + \frac{k_0}{k_1}\right)\sinh k_1 a e^{-ik_0 x} \times \frac{-iF}{2} e^{-ik_1 a}\left(\frac{k_1}{k_0} + \frac{k_0}{k_1}\right)\sinh k_1 a e^{-ik_0 x} v_0$$

or $\qquad \psi_R \psi_R^* v_0 = \dfrac{F^4}{4}\left(\dfrac{k_1}{k_0} + \dfrac{k_0}{k_1}\right)^2 \sinh^2 k_1 a \times v_0 \qquad\qquad$ (8.108)

Thus, the reflection coefficient or reflection probability is obtained by putting Eqs (8.100) and (8.108) into Eq. (8.107) as

$$R = \frac{\left(\dfrac{k_1}{k_0} + \dfrac{k_0}{k_1}\right)^2 \sinh^2 k_1 a}{4 + \left(\dfrac{k_1}{k_0} + \dfrac{k_0}{k_1}\right)^2 \sinh^2 k_1 a} \tag{8.109}$$

The expressions for T and R are compatible with each other. The summation of R and T gives

$$R + T = \frac{\left(\dfrac{k_1}{k_0} + \dfrac{k_0}{k_1}\right)^2 \sin h^2 k_1 a}{4 + \left(\dfrac{k_1}{k_0} + \dfrac{k_0}{k_1}\right)^2 \sin h^2 k_1 a} + \frac{4}{4 + \left(\dfrac{k_1}{k_0} + \dfrac{k_0}{k_1}\right)^2 \sinh^2 k_1 a}$$

or $$R + T = \frac{\left(\dfrac{k_1}{k_0} + \dfrac{k_0}{k_1}\right)^2 \sin h^2 k_1 a + 4}{4 + \left(\dfrac{k_1}{k_0} + \dfrac{k_0}{k_1}\right)^2 \sin h^2 k_1 a} = 1$$

Thus, we have $R + T = 1$

Approximate value of T

Let us assume that the barrier width a is wide enough for ψ_2 to be severely weakened at $0 \le x \le a$. In this case, $k_1 a \gg 1$ i.e., $e^{k_1 a} \gg e^{-k_1 a}$ and $\dfrac{k_1}{k_0} \gg \dfrac{k_0}{k_1}$. Under this conditions, Eq. (8.103) becomes

$$T = \frac{1}{\left(\dfrac{e^{k_1 a} + e^{-k_1 a}}{2}\right)^2 + \dfrac{1}{4}\left(\dfrac{k_1}{k_0} - \dfrac{k_0}{k_1}\right)^2 \left(\dfrac{e^{k_1 a} - e^{-k_1 a}}{2}\right)^2}$$

Neglecting $e^{-k_1 a}$ in comparison to $e^{k_1 a}$ and $\dfrac{k_0}{k_1}$ in comparison to $\dfrac{k_1}{k_0}$ in this equation, we get

$$T = \frac{1}{\dfrac{e^{2k_1 a}}{4} + \dfrac{1}{4}\left(\dfrac{k_1}{k_0}\right)^2 \dfrac{e^{2k_1 a}}{4}} = \frac{16}{4 + \dfrac{k_1^2}{k_0^2}} e^{-2k_1 a}$$

The quantity $\dfrac{16}{4 + \dfrac{k_1^2}{k_0^2}}$ always has the value one. Therefore, a reasonable approximation of the transmission probability is

$$T = e^{-2k_1 a} \tag{8.110}$$

Example 8.4

Electrons with energy 1.0 eV are incident on a potential barrier of 8.0 eV height and 0.50 nm width. Calculate the transmission probability of the incident electron.

Solution

The data given are

$$E = 1.0 \text{ eV} = 1.6 \times 10^{-19} \text{ J}$$

$$V_0 = 8.0 \text{ eV} = 12.8 \times 10^{-19} \text{ J}$$

$$A = 0.50 \text{ nm} = 0.50 \times 10^{-9} \text{ m}$$

$$k_1 = \frac{2\pi\sqrt{2m(V_0 - E)}}{h} = 1.4 \times 10^{10} \text{ m}^{-1}$$

The transmission coefficient is given by

$$T = e^{-2k_1 a} = \exp(-2 \times 1.4 \times 10^{10} \times 0.50 \times 10^{-9}) = 8.3 \times 10^{-7}$$

Thus, in this case, one electron out of 1.2 million electrons can tunnel through the barrier!

Example 8.5

12 million electrons with energy 3.0 eV are incident on a potential barrier of 9.0 eV height and 0.50 nm width. Calculate how many electrons will tunnel through the barrier?

Solution

The data given are

$$E = 3.0 \text{ eV} = 4.8 \times 10^{-19} \text{ J}$$

$$V_0 = 9.0 \text{ eV} = 14.4 \times 10^{-19} \text{ J}$$

$$a = 0.50 \text{ nm} = 0.50 \times 10^{-9} \text{ m}$$

$$k_1 = \frac{2\pi\sqrt{2 \times 9.11 \times 10^{-31}(14.4 - 4.8) \times 10^{-19}}}{6.626 \times 10^{-34}} = 1.2 \times 10^{10} \text{ m}^{-1}$$

The transmission probability is given by

$$T = e^{-2k_1 a} = \exp(-2 \times 1.2 \times 10^{10} \times 0.50 \times 10^{-9}) = 6.1 \times 10^{-6}$$

The number of electrons that tunnel through the barrier will be approximately

$$12 \times 10^6 \times T = 73$$

Thus, in this case, 73 electrons out of 12 million electrons can tunnel through the barrier.

Example 8.6

A beam of electrons is incident on a barrier of height 6.00 eV and 0.200 nm width. Find the energy they should have if 1.00% of them are to tunnel through the barrier.

Solution

Data given are

$$V_0 = 6.00 \text{ eV} = 9.6 \times 10^{-19} \text{ J}$$

$$a = 0.200 \text{ nm} = 0.200 \times 10^{-9} \text{ m}$$

$$T = 1.00\% = 0.01$$

The transmission probability is given by

$$T = e^{-2k_1 a}$$

or $\quad -2k_1 a = \ell n T$

or $\quad k_1 = \dfrac{-\ell n T}{2a}$

$$\sqrt{2m(V_0 - E)} = \dfrac{-h \times \ell n T}{4\pi a}$$

or $\quad E = V_0 - \dfrac{1}{2m}\left(\dfrac{-h \times \ell n T}{4\pi a}\right)^2$

Putting the given data into this equation, we get

$$E = 9.6 \times 10^{-19}\,\text{J} - \frac{1}{2 \times 9.11 \times 10^{-31}} \left(\frac{6.626 \times 10^{-34} \times \ln 0.01}{4\pi \times 0.2 \times 10^{-9}} \right)^2 \text{J} = 0.944\,\text{eV}$$

Case II: $(E \geq V_0)$

In region I where $V = 0$ and $\psi(x) = \psi_1$, Schrödinger's time-independent equation becomes

$$\frac{\partial^2 \psi_1}{\partial x^2} + \frac{2mE}{\hbar^2}\psi_1 = 0 \quad \text{for} \quad -\infty < x < 0$$

or $$\frac{\partial^2 \psi_1}{\partial x^2} + k_0^2 \psi_1 = 0 \tag{8.111}$$

where $k_0 = \dfrac{\sqrt{2mE}}{\hbar}$ \hfill (8.112)

The solution of the second order differential Eq. (8.111) where k_0^2 is always positive is given by

$$\psi_1 = A e^{ik_0 x} + B e^{-ik_0 x} \quad \text{for} \quad -\infty < x < 0 \tag{8.113}$$

where A and B are integration constants. In this solution, the function $A e^{ik_0 x}$ when multiplied by the factor $e^{-\frac{i}{\hbar}Et}$ $(= e^{-i\omega t})$ represents a plane wave going towards the right and the function $B e^{-ik_0 x}$ when multiplied by the factor $e^{-\frac{i}{\hbar}Et}$ $(= e^{-i\omega t})$ represents a plane wave going towards the left. Hence, the solution ψ_1 is a linear combination of plane waves.

Similarly, in region II where $V(x) = V_0$ and $\psi(x) = \psi_2$, Schrödinger's time-independent equation becomes

$$\frac{\partial^2 \psi_2}{\partial x^2} + \frac{2m(E - V_0)}{\hbar^2}\psi_2 = 0 \quad \text{for} \quad 0 \leq x \leq a$$

or $$\frac{\partial^2 \psi_2}{\partial x^2} + k^2 \psi_2 = 0 \tag{8.114}$$

where $k = \dfrac{\sqrt{2m(E - V_0)}}{\hbar}$ is real. $\hspace{2cm}$ (8.115)

The solution of the second order differential Eq. (8.114) is given by

$$\psi_2 = Ce^{ikx} + De^{-ikx} \quad \text{for } 0 \le x \le a \hspace{2cm} (8.116)$$

where C and D are integration constants. In this solution, the function Ce^{ikx} when multiplied by the factor $e^{-\frac{i}{\hbar}Et}$ $(= e^{-i\omega t})$ represents a plane wave going towards the right and the function De^{-ikx} when multiplied by the factor $e^{-\frac{i}{\hbar}Et}$ $(= e^{-i\omega t})$ represents a plane wave going towards the left.

In region III where $V = 0$ and $\psi(x) = \psi_3$, Schrödinger's time-independent equation becomes

$$\frac{\partial^2 \psi_3}{\partial x^2} + \frac{2mE}{\hbar^2}\psi_3 = 0 \quad \text{for } a \le x < \infty$$

or $\quad \dfrac{\partial^2 \psi_3}{\partial x^2} + k_0^2 \psi_3 = 0 \hspace{2cm} (8.117)$

where $k_0 = \dfrac{\sqrt{2mE}}{\hbar} \hspace{2cm} (8.118)$

The solution of the second order differential Eq. (8.117), where k_0^2 is always positive is given by

$$\psi_3 = Fe^{ik_0 x} + Ge^{-ik_0 x} \quad \text{for } a \le x < \infty \hspace{2cm} (8.119)$$

E and F are integration constants. In this solution, the function $Ee^{ik_0 x}$ when multiplied by the factor $e^{-\frac{i}{\hbar}Et}$ represents a plane wave going towards the right and the function $Fe^{-ik_0 x}$ when multiplied by the factor $e^{-\frac{i}{\hbar}Et}$ represents a plane wave going towards the left. However, in region III, there cannot be a reflected wave. Hence, F in Eq. (8.113) is zero and Eq. (8.113) becomes

$$\psi_3 = Fe^{ik_0 x} \quad \text{for } a \le x < \infty \hspace{2cm} (8.120)$$

Evaluation of integration constants *A, B, C,* and *D*

For ψ to be physically meaningful, wave function ψ is continuous everywhere including at $x = 0$, i.e.,

$$\psi_1\big|_{x=0} = \psi_2\big|_{x=0}$$

or $A + B = C + D$ (8.121)

Similarly, for ψ to be physically meaningful, wave function $\dfrac{d\psi}{dx}$ is continuous everywhere including at $x = 0$, i.e.,

$$\frac{d\psi_1}{dx}\bigg|_{x=0} = \frac{d\psi_2}{dx}\bigg|_{x=0}$$

or $k_0(A - B) = k(C - D)$ (8.122)

Solving Eqs (8.121) and (8.122) for *A* and *B*, we get

$$A = \left(1 + \frac{k}{k_0}\right)\frac{C}{2} + \left(1 - \frac{k}{k_0}\right)\frac{D}{2} \tag{8.123}$$

$$B = \left(1 - \frac{k}{k_0}\right)\frac{C}{2} + \left(1 + \frac{k}{k_0}\right)\frac{D}{2} \tag{8.124}$$

Similarly at $x = a$,

$$\psi_2\big|_{x=a} = \psi_3\big|_{x=L}$$

or $Ce^{ika} + De^{-ika} = Fe^{ik_0 a}$ (8.125)

Similarly, for ψ to be physically meaningful, wave function $\dfrac{d\psi}{dx}$ is continuous every where including at $x = a$, i.e.,

$$\frac{d\psi_2}{dx}\bigg|_{x=a} = \frac{d\psi_3}{dx}\bigg|_{x=a}$$

or $Cke^{ika} - Dke^{-ika} = k_0 Fe^{ik_0 a}$ (8.126)

Solving Eqs (8.125) and (8.126) for C and D, we get

$$C = \left(1 + \frac{k_0}{k}\right)\frac{F}{2}e^{i(k_0 - k)a}$$ (8.127)

$$D = \left(1 - \frac{k_0}{k}\right)\frac{F}{2}e^{i(k_0 + k)a}$$ (8.128)

Putting the value of C and D from Eq. (8.127) and (8.128) into Eq. (8.123), we get

$$A = \frac{F}{4}e^{ik_0 a}\left(1 + \frac{k}{k_0}\right)\left(1 + \frac{k_0}{k}\right)e^{-ika} + \frac{F}{4}e^{ik_0 a}\left(1 - \frac{k}{k_0}\right)\left(1 - \frac{k_0}{k}\right)e^{ika}$$

$$= \frac{F}{4}e^{ik_0 a}\left[2\left(e^{-ika} + e^{ika}\right) - \left(\frac{k}{k_0} + \frac{k_0}{k}\right)\left(e^{ika} - e^{-ika}\right)\right]$$

or $A = Fe^{ik_0 a}\left[\cos ka - \frac{i}{2}\left(\frac{k}{k_0} + \frac{k_0}{k}\right)\sin ka\right]$ (8.129)

Again putting the value of C and D from Eqs (8.127) and (8.128) into Eq. (8.124), we get

$$B = \frac{F}{4}e^{ik_0 a}\left(1 - \frac{k}{k_0}\right)\left(1 + \frac{k_0}{k}\right)e^{-ika} + \frac{F}{4}e^{ik_0 a}\left(1 + \frac{k}{k_0}\right)\left(1 - \frac{k_0}{k_1}\right)e^{ika}$$

$$= \frac{F}{4}e^{ik_0 a}\left[\left(\frac{k}{k_0} - \frac{k_0}{k}\right)\left(e^{ika} - e^{-ika}\right)\right]$$

$$= \frac{F}{4}e^{ik_0 a}\left[\left(\frac{k}{k_0} - \frac{k_0}{k}\right)2i\sin ka\right]$$

or $B = \frac{iF}{2}e^{ik_0 a}\left(\frac{k}{k_0} - \frac{k_0}{k}\right)\sin ka$ (8.130)

Putting these values of A, B, C, and D into Eqs (8.113) and (8.116), we get

$$\psi_1 = Fe^{ik_0a}\left(\cos ka - \frac{i}{2}\left(\frac{k}{k_0}+\frac{k_0}{k}\right)\sin ka\right)e^{ik_0x} + \frac{iF}{2}e^{ik_0a}\left(\frac{k}{k_0}-\frac{k_0}{k}\right)\sin ka \times e^{-ik_0x}$$

$$\text{for } -\infty < x < 0 \qquad (8.131)$$

$$\psi_2 = \frac{F}{2}e^{ik_0a}\left(1+\frac{k_0}{k}\right)e^{-ika}e^{ikx} + \frac{F}{2}e^{ik_0a}\left(1-\frac{k_0}{k}\right)e^{ika}e^{-ikx} \text{ for } 0 \le x \le a \qquad (8.132)$$

$$\psi_3 = Fe^{ik_0x} \text{ for } a \le x < \infty \qquad (8.133)$$

Equation (8.131) shows that in region I,

$$\psi_I = \frac{F}{2}e^{ik_0a}\left(2\cos ka - i\left(\frac{k}{k_0}+\frac{k_0}{k}\right)\sin ka\right)e^{ik_0x} \text{ is a right going wave, i.e., incident wave.}$$

$$(8.134)$$

$$\psi_R = \frac{iF}{2}e^{ik_0a}\left(\frac{k}{k_0}-\frac{k_0}{k}\right)\sin ka \times e^{-ik_0x} \text{ is a left going wave, i.e., reflected wave.} \qquad (8.135)$$

Equation (8.132) shows that in region II,

$$\psi_T = \frac{F}{2}e^{ik_0a}\left(1+\frac{k_0}{k}\right)e^{-ika}e^{ikx} \text{ is a right going wave, i.e., transmitted wave.} \qquad (8.136)$$

$$\psi_{R'} = \frac{F}{2}e^{ik_0a}\left(1-\frac{k_0}{k}\right)e^{ika}e^{-ikx} \text{ is a left going wave, i.e., reflected wave} \qquad (8.137)$$

Equation (8.133) shows that in region III,

$$\psi_E = Fe^{ik_0x} \text{ is a right going wave, i.e., emergent wave} \qquad (8.138)$$

Thus, the probability density of the particle in region III is $\psi_E\psi_E^* = Fe^{ik_0x}Fe^{-ik_0x} = F^2$, a non-zero quantity which implies that the probability of tunnelling of a particle through the barrier to region III is not zero when $E > V_0$; the result is in conformity with classical physics.

Transmission coefficient

The transmission coefficient or transmission probability of the barrier T is defined in the usual way as

$$T = \frac{\text{Transmitted flux density through the barrier}}{\text{Incident flux density on the barrier at } x = 0} = \frac{\psi_E \psi_E^* v_0}{\psi_I \psi_I^* v_0} = \frac{\psi_E \psi_E^*}{\psi_I \psi_I^*} \tag{8.139}$$

The incident flux density on the barrier at $x = 0$ is

$$\psi_I \psi_I^* v_0 = Fe^{ik_0 a} \left(\cos ka - \frac{i}{2} \left(\frac{k}{k_0} + \frac{k_0}{k} \right) \sinh ka \right) e^{ik_0 x} \times$$

$$Fe^{-ik_0 a} \left(\cos ka + \frac{i}{2} \left(\frac{k}{k_0} + \frac{k_0}{k} \right) \sin ka \right) e^{-ik_0 x} v_0$$

$$\psi_I \psi_I^* v_0 = F^2 \left(\cos^2 ka + \frac{1}{4} \left(\frac{k}{k_0} + \frac{k_0}{k} \right)^2 \sin^2 ka \right) v_0 \tag{8.140}$$

The transmitted flux density through the barrier is

$$\psi_E \psi_E^* v_0 = Fe^{ik_0 x} \times Fe^{-ik_0 x} v_0 = F^2 v_0 \tag{8.141}$$

Dividing Eq. (8.140) by Eq. (8.141), we obtain

$$\frac{\psi_I \psi_I^*}{\psi_E \psi_E^*} = \left(\cos^2 ka + \frac{1}{4} \left(\frac{k}{k_0} + \frac{k_0}{k} \right)^2 \sin^2 ka \right)$$

The transmission coefficient or transmission probability is obtained by taking the inverse of this equation and is obtained as

$$T = \frac{4}{4 \cos^2 ka + \left(\frac{k}{k_0} + \frac{k_0}{k} \right)^2 \sin^2 ka}$$

Putting refractive index $\mu = \dfrac{k}{k_0} = \sqrt{1 - \dfrac{V_0}{E}}$, the previous equation becomes

$$T = \frac{4}{4\cos^2 ka + \left(\mu + \dfrac{1}{\mu}\right)^2 \sin^2 ka}$$

or $\quad T = \dfrac{4\mu^2}{4\mu^2 \cos^2 ka + \left(\mu^2 + 1\right)^2 \sin^2 ka}$ (8.142)

or $\quad T = \dfrac{4\mu^2}{4\mu^2 \cos^2 \mu k_0 a + \left(\mu^2 + 1\right)^2 \sin^2 \mu k_0 a} \qquad \left(\because \quad \mu = \dfrac{k}{k_0}\right)$

Now to find out the value of T when $E = V_0$, i.e., $\mu = 0$, we have

$$\operatorname*{Lim}_{\mu \to 0} T = \operatorname*{Lim}_{\mu \to 0} \frac{4\mu^2}{4\mu^2 \cos^2 \mu k_0 a + \left(\mu^2 + 1\right)^2 \sin^2 \mu k_0 a} = \frac{4}{4 + \left(k_0 a\right)^2}$$ (8.143)

Equation (8.143) shows that

$$T = \frac{4}{4 + \left(k_0 a\right)^2}$$

at $E = V_0$.

In analogy with the total transmission of light through a thin refracting layer, perfect transmission occurs for $ka = n\pi$ because if we put $ka = n\pi$ into Eq. (8.142), $T = 1$, i.e.,

$$T = 1$$

When $ka = n\pi$ or $\dfrac{2\pi}{\lambda} a = n\pi$ or $a = n\dfrac{\lambda}{2}$, with $n = 1, 2, 3, \ldots$

Thus, total transmission occurs whenever barrier width a is an integral multiple of half the wavelengths.

Reflection coefficient in region I

The reflection coefficient of the barrier R is defined in the usual way as

$$R = \frac{\psi_R \psi_R^* v_0}{\psi_I \psi_I^* v_0} = \frac{\psi_R \psi_R^*}{\psi_I \psi_I^*} \tag{8.144}$$

The reflected flux density on the barrier at $x = 0$ is

$$\psi_R \psi_R^* v_0 = \frac{iF}{2} e^{ik_0 a} \left(\frac{k}{k_0} - \frac{k_0}{k} \right) \sin ka\, e^{-ik_0 x} \times \frac{-iF}{2} e^{-ika} \left(\frac{k}{k_0} - \frac{k_0}{k} \right) \sin ka \times e^{ik_0 x} v_0$$

or $\quad \psi_R \psi_R^* v_0 = \dfrac{F^2}{4} \left(\dfrac{k}{k_O} - \dfrac{k_0}{k} \right)^2 \sin^2 ka \times v_0 \qquad (8.145)$

Thus, the reflection coefficient is obtained by putting Eqs (8.140) and (8.145) into Eq. (8.144) as

$$R = \frac{\left(\dfrac{k}{k_0} - \dfrac{k_0}{k} \right)^2 \sin^2 ka}{4 \cos^2 ka + \left(\dfrac{k}{k_O} + \dfrac{k_0}{k} \right)^2 \sin^2 ka} \tag{8.146}$$

The expressions for T and R are compatible with each other, i.e.,

$$R + T = 1$$

All the mathematical relations in case of $E \geq V_0$ can be obtained alternately just by replacing k_1 by ik of the case $E < V_0$!

8.6 Infinity Deep Potential Well

Repulsive infinite deep potential energy well is defined mathematically as

$$V(x) = 0 \text{ for } 0 \leq x \leq a$$

$$= \infty \text{ for } x < 0 \text{ and } x > a$$

and is depicted in Fig. 8.6.

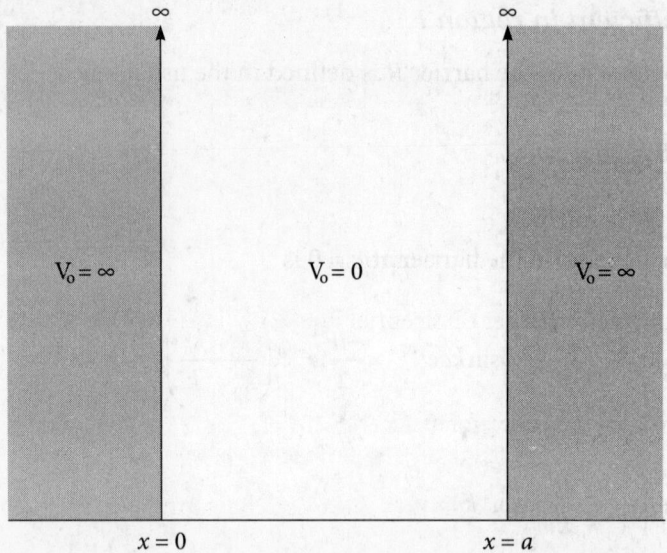

Figure 8.6 | A particle moves freely in the region $0 \leq x \leq a$ of one-dimensional repulsive infinite deep potential energy well, but is excluded completely from $x < 0$ and $x > a$

A particle moves freely in the region $0 \leq x \leq a$ of a one-dimensional repulsive infinite deep potential energy well, but is excluded completely from $x < 0$ and $x > a$. It collides with the infinite hard walls at $x = 0$ and $x = a$. A particle does not lose energy when it collides with such walls so its total energy remains constant. The potential energy $V(x)$ on both sides of the well is infinity and inside the well is constant; for convenience of mathematics, it can be taken as zero. The particle can not have infinite amount of energy. Hence, it cannot exist outside the well where potential energy is infinite. Therefore, we have

$\psi(x) = 0$ for $x < 0$ and $x > a$.

Boundary conditions on $\psi(x)$

i.　$\psi(0) = 0$ at $x = 0$ since $\psi(x) = 0$ for $x < 0$

ii.　$\psi(a) = 0$ at $x = a$ since $\psi(x) = 0$ for $x > a$

Schrödinger's time-independent equation for the particle inside the potential energy is given by

$$\frac{\partial^2 \psi}{\partial x^2} + \frac{2mE}{\hbar^2}\psi = 0 \text{ for } 0 \leq x \leq a \tag{8.147}$$

or $\dfrac{\partial^2 \psi}{\partial x^2} + k_0^2 \psi = 0$ (8.148)

where $k_0 = \dfrac{\sqrt{2mE}}{\hbar} = \dfrac{2\pi}{\lambda}$ (8.149)

λ = de Broglie wavelength of the particle.

The solution of the second order differential Eq. (8.147) where k_0^2 is always positive is given by

$$\psi(x) = A \sin k_0 x + B \cos k_0 x \text{ for } 0 \le x \le a$$ (8.150)

where A and B are constants whose values can be found out by applying boundary conditions.

Evaluation of constants A and B

According to the boundary conditions discussed earlier, $\psi(x) = 0$ for $x < 0$ and $x > a$. i.e.,

$$\psi(0) = 0 \text{ and } \psi(a) = 0$$

Hence, Eq. (8.150) becomes

$$\psi(0) = A \sin 0 + B \cos 0$$

$$\psi(0) = 0 + B$$

This equation proves that $B = 0$ because the left side of this equation is zero. Equation (8.150) becomes

$$\psi(x) = A \sin k_0 x$$ (8.151)

where A is called the normalization constant and can not be zero. If $A = 0$, $\psi = 0$ everywhere and $\psi \psi^* = 0$ every where and the wave function becomes meaningless. $\psi \psi^* = 0$ implies that there is no particle! The value of the normalization constant is found out using the procedure outlined in the previous chapter and is given directly as

$$A = \sqrt{\dfrac{2}{a}}$$

Putting this value of A into Eq. (8.151),

$$\psi(x) = \sqrt{\frac{2}{a}} \sin k_0 x \tag{8.152}$$

Again applying the boundary condition, we get

$$\psi(a) = 0$$

or $\quad \sqrt{\frac{2}{a}} \sin k_0 a = 0$

$A = \sqrt{\frac{2}{a}}$ can not be zero. Hence,

$$\sin k_0 a = 0$$

or $k_0 a = n\pi$ with $n = 1, 2, 3, 4, \ldots$ \hfill (8.153)

8.6.1   Quantization of de Broglie wavelengths

Putting the value of $k_0 = \dfrac{2\pi}{\lambda}$ into Eq. (8.153), we get

$$\frac{2\pi a}{\lambda} = n\pi \quad \text{or} \quad \lambda = \frac{2a}{n} \qquad n = 1, 2, 3, 4, \tag{8.154}$$

Thus, the solution of Schrödinger's equation for a particle trapped in a linear region of length a is a series of standing de Broglie waves. Not all the wavelengths are permitted; only certain values permitted by Eq. (8.154) may occur. Therefore, we conclude that the de Broglie wavelength is quantized.

8.6.2   Quantization of energy (energy eigenvalues)

Putting the value of $k_0 = \dfrac{\sqrt{2mE}}{\hbar}$ into Eq. (8.153), we get

$$\frac{\sqrt{2mE}}{\hbar} a = n\pi$$

Solving this equation for E, the energy eigenvalues of the particles are given by

$$E_n = \frac{\hbar^2 \pi^2 n^2}{2ma^2}, \quad n = 1, 2, 3, \ldots \tag{8.155}$$

This equation shows that the energy of the particle in a one-dimensional potential well of infinite height is quantized in contradiction to classical physics.

i $n = 1, 2, 3, 4, \ldots$ is called the principal quantum number. Each value of n corresponds to one allowed energy state of the system.

ii.   The ground state energy of the particle E_1 is obtained by putting $n = 1$ in Eq. (8.153), i.e.,

$$E_1 = \frac{\hbar^2 \pi^2}{2ma^2} \text{ (ground state energy of the particle)}$$

This equation shows that the ground state energy or zero point energy of a particle is not zero, whereas according to classical physics, the ground state energy or zero point energy of a particle is zero.

iii. The energy of any other state of the particle is given by

$$E_n = E_1 n^2 \quad n = 2, 3, 4, \ldots$$

The only allowed energy of the particle are E_1, $4E_1$, $9E_1$, $16E_1$, etc. All intermediate values such as $2E_1$, $3.2E_1$, $5E_1$, $6.7E_1$ are forbidden.

iv. $E_n = E_1 n^2$ shows that energy levels are not equispaced. The difference between n^2 for two consecutive values of n increases with increased values of n. Thus, the spacing between two successive energy levels increases for higher values of n. This has been depicted in Fig. 8.7.

Figure 8.7 | The energy levels of a particle enclosed in a one-dimensional potential well of infinite height. The energy levels are not continuous but quantized

8.6.3   Quantization of speed (speed eigenvalues)

The potential energy of a particle is assumed to be zero for mathematical simplicity. Therefore, for a non-relativistic case, the energy of the particle is purely kinetic, i.e.,

$$\frac{1}{2}mv_n^2 = E_n = \frac{\hbar^2 \pi^2 n^2}{2ma^2}$$

or $$v_n^2 = \frac{\hbar^2 \pi^2 n^2}{m^2 a^2}$$

From this equation, the speed eigenvalues of the particle under consideration is given by

$$v_n = \frac{\hbar \pi}{ma}n \qquad\qquad (8.156)$$

This equation shows that the speed of a particle in a one-dimensional potential well of infinite height is quantized in contradiction to classical physics. If a suitable bead is placed inside a suitable box [so that there is no friction and collisions are elastic ideally] and the bead is given any speed, the motion will continue infinitely. However, the particle inside the infinite deep potential well cannot be given any arbitrary initial speed for sustained states of motion; only certain initial speeds defined by Eq. (8.156) can produce sustained states of motion. The special conditioned motions are called stationary states. The states are stationary because when time dependence is included to make $\psi(x,t)$, the probability density $\psi\psi^*$ is independent of time and the expectation values calculated do not change with time.

8.6.4   Eigenfunctions

The normalized eigenfunctions $\psi_n(x)$ of a particle enclosed in a one-dimensional potential well of infinite height is obtained by putting $k_0 = \dfrac{n\pi}{a}$ [Eq. (8.153)] into Eq. (8.152) as

$$\psi_n(x) = \sqrt{\frac{2}{a}} \sin \frac{n\pi}{a} x \qquad\qquad (8.157)$$

For different values of n, normalized eigenfunctions $\psi_n(x)$ of a particle enclosed in a one-dimensional potential well of infinite height is plotted in Fig. 8.8.

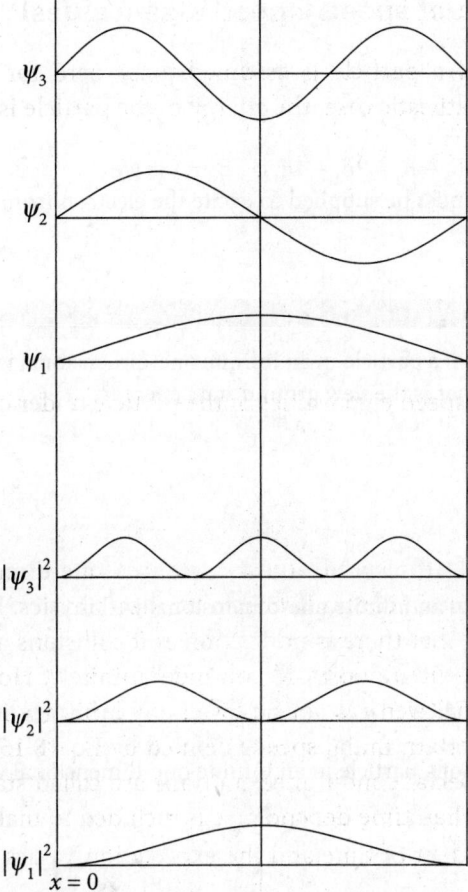

Figure 8.8 | Comparison of normalized eigenfunctions $\psi_n(x)$ and probability densities $\psi_n \psi_n^*$ of a particle in different quantum states enclosed in a one-dimensional potential well of infinite height

Example 8.7

An electron is trapped completely in a one-dimensional region of length 1 Å. How much energy must be supplied to excite the electron from the first excited state to the second excited state?

Solution

The data given are

$$a = 1 \overset{o}{A} = 10^{-10} \, \text{m}$$

$$E_1 = \frac{\hbar^2 \pi^2}{2ma^2} = 37.65 \, \text{eV}$$

$$E_2 = 4E_1$$

$$E_3 = 9E_1$$

The energy difference is $E_3 - E_2 = 9E_1 - 4E_1 = 5E_1 = 188$ eV

Thus, 188 eV energy must be supplied to excite the electron from the first excited state to the second excited state.

Example 8.8

The ground state energy of a particle in an infinite one-dimensional well is 4.4 eV. If the width of the well is doubled, what is the new ground state energy?

Solution

The data given are

$$E_1 = 4.4 \text{ eV}$$

The energy of a particle in an infinite one-dimensional well is

$$E_n = \frac{\hbar^2 \pi^2 n^2}{2ma^2}$$

The ground state energy of a particle in an infinite one-dimensional well is

$$E_1 = \frac{\hbar^2 \pi^2}{2ma^2}$$

If width of the well a is doubled, the new ground state energy will be

$$E_{1new} = \frac{\hbar^2 \pi^2}{2m4a^2} = \frac{E_1}{4} = \frac{4.4 \text{ eV}}{4} = 1.1 \text{ eV}$$

Questions

8.1   What is the scope of quantum physics?

8.2   What are the conditions for one-dimensional systems?

8.3   Explain how multiplication of $e^{-i\omega t}$ with the wave function $\psi(x)$ generates a wave.

8.4   What is the condition for a particle to be a free particle?

8.5   Write Schrödinger's time-independent equation for a free particle confined to the Y-axis.

8.6   Solve Schrödinger's time-independent equation for a free particle.

8.7   What are the boundary conditions on ψ so that it turns out a physically meaningful wave function?

8.8   What do you mean by a free particle? Derive the expression for its wave function.

8.9   Is the wave function of a free particle oscillatory or non-oscillatory? Explain.

8.10   Derive an expression for the de Broglie wavelength of a free particle in terms of its kinetic energy

8.11   Prove that the speed of a free particle in one dimension is continuous.

8.12   Prove that the momentum of a free particle in one dimension is continuous.

8.13   What is potential step? Define it mathematically.

8.14   Explain why in potential step problem, time-independent Schrödinger's equation is taken.

8.15   Prove that the speed of transmitted particles is less than that of incident particles in case of potential step problems.

8.16   What do you mean by transmission coefficient? What is its physical significance?

8.17   What do you mean by reflection coefficient? What is its physical significance?

8.18   Derive the expression for transmission probability or transmission coefficient of a particle through potential step when the energy of the incident particle is more than the height of the potential step.

8.19   Calculate the transmission probability in case of potential step if the energy of the incident particle is much more than the potential step.

8.20   Derive the expression for reflection coefficient of a particle through potential step when energy of the incident particle is more than the height of the potential step.

8.21   Prove that the sum of the transmission coefficient and the reflection coefficient of a particle through potential step when energy of the incident particle is less than the height of the potential step is unity.

8.22   Derive the expression for transmission probability or transmission coefficient of a particle through potential step when energy of the incident particle is less than the height of the potential step.

8.23   Derive the expression for transmission reflection coefficient of a particle through potential step when energy of the incident particle is less than the height of the potential step.

8.24   An electron and proton with same energy E is incident on a potential energy barrier of height V_0 such that $E < V_0$. Do they have the same probability of tunnelling through the barrier? Explain.

8.25   Prove that the sum of the transmission probability and the reflection probability of a particle through potential step when energy of the incident particle is more than the height of the potential step is unity.

8.26   Prove that in case of potential step, the speed of transmitted particle is less than that of incident particles.

8.27 What is the law of conservation of flux? Explain how the law is satisfied in case of the potential step problem.

8.28 Define refractive index of potential step when energy of the particle is more than the height potential step.

8.29 If μ is the refractive index of a potential step, then prove that

$$R = \left| \frac{1-\mu}{1+\mu} \right|^2$$

when the energy of the particle is more than the height potential step.

8.30 If μ is the refractive index of a potential step, then prove that

$$T = \left| \frac{4\mu}{(1+\mu)^2} \right|$$

when the energy of the particle is more than the height potential step.

8.31 Prove that in case of the potential energy step when the energy of the particle is less than the height potential step, the reflection probability is unity.

8.32 Solve Schrödinger's time-independent equation to obtain the wave function for incident particles in case of potential step.

8.33 Solve Schrödinger's time-independent equation to obtain the wave function for reflected particles in case of potential step.

8.34 Solve Schrödinger's time-independent equation to obtain the wave function for transmitted particles in case of potential step.

8.35 The wave function inside the potential step is oscillatory or non-oscillatory when energy of the particle is less than the height potential step. Explain

8.36 The wave function inside the potential step is oscillatory or non-oscillatory when energy of the particle is more than the height potential step. Explain

8.37 Explain how quantum physics differs from classical physics in potential step problems.

8.38 What is the potential energy barrier? Define a potential energy barrier.

8.39 Derive an expression for the transmission probability of incident particles through the potential barrier when energy of the incident particle is less than the height of the barrier.

8.40 Derive an expression for the reflection probability of incident particles through the potential barrier when energy of the incident particle is less than the height of the barrier.

8.41 Prove that the transmission probability of incident particles through the potential barrier when energy of the incident particle is less than the height of the barrier is approximately $T = e^{-2k_1 a}$.

8.42 Derive an expression for the transmission probability of incident particles through the potential barrier when energy of the incident particle is more than the height of the barrier.

8.43 Derive an expression for the reflection probability of incident particles through the potential barrier when energy of the incident particle is more than the height of the barrier.

8.44 What is quantum mechanical tunnelling? Mention two examples where this phenomenon is observed.

8.45 Explain how quantum physics differs from classical physics in potential barrier problems.

8.46 Derive an expression for the normalized wave function of a particle trapped in an infinite deep potential well.

8.47 What is the eigen functions of a particle trapped in an infinite deep potential well in its ground state and third excited state?

8.48 Prove that the energy of the particle enclosed in an infinite deep potential well is quantized.

8.49 Prove that the momentum of a particle in a one-dimensional potential well of infinity height is quantized.

8.50 Derive an expression for the energy of the particle enclosed in an infinite deep potential well.

8.51 Prove that the de Broglie wavelength of a particle enclosed in an infinite deep potential well is quantized.

8.52 A particle of mass m is enclosed inside a potential well of infinite height. What is the maximum de Broglie wavelength of the particle?

8.53 Explain how quantum physics differs from classical physics in infinite deep potential well problems.

8.54 Is the wave function inside the infinite deep potential well oscillatory or non-oscillatory? Explain

8.55 Explain how the energy of a particle in an infinite deep potential well depends on mass of the particle and width of the well.

8.56 The de Broglie wavelength of a particle inside an infinite deep potential well is quantized. Prove it.

8.57 A particle of mass m is enclosed inside a potential well of infinite height. What is the lowest speed the particle is to be given so that the particle acquires sustained states of motion?

8.58 Prove that the spacing between consecutive energy levels of a particle inside an infinite deep potential well increases with the increase of the energy.

8.59 Prove that half the de Broglie wavelength of a particle inside an infinite deep potential well cannot be less than the width of the well.

Problems

8.1 An electron is trapped completely in a one-dimensional region of length 1 Å. How much energy must be supplied to excite the electron from the ground state to the first excited state? [Ans 111 eV]

8.2 The de Broglie wavelength of a freely moving electron is 8.678 Å. Calculate its energy. [Ans 2 eV]

8.3 An electron is moving freely with energy 2 eV. Calculate its speed. [Ans 8.38 × 10^5 m/s]

8.4 The ground state energy of a particle in an infinite one-dimensional well is 5.6 eV. If the width of the well is doubled, what is the new ground state energy? [Ans 1.4 eV]

8.5 An electron is trapped completely in a one-dimensional region of length 0.132 nm. What is the energy of the electron in $n = 10$ state? [Ans 2160 eV]

8.6 What is the minimum energy of a proton confined to a region of space of nuclear dimension 10 femtometers? [Ans 2.0 MeV]

8.7 Electrons with energy 10.0 eV are incident on a potential step of 8 eV height and 0.50 nm width. Calculate the reflection probability and transmission probability of the incident electron. [Ans 0.15, 0.85]

8.8 Electrons with energy 2.0 eV are incident on a potential barrier of 10.0 eV height and 0.50 nm width. Calculate the transmission probability of the incident electrons. [Ans 2.4 × 10^{-7}]

8.9 20 × 10^6 electrons with energy 1.0 eV are incident on a potential barrier of 10.0 eV high and 0.50 nm width. Calculate how many electrons will tunnel through the barrier? [Ans only 2]

8.10 A beam of protons is incident on a barrier of height 8.00 eV and width 0.300 nm. Find the energy they should have if 10.00% of them are to tunnel through the barrier. [Ans 7.438 eV]

Multiple Choice Questions

1. Time-independent Schrödinger's equation in one-dimension is

    (i) $\dfrac{\partial^2 \psi}{\partial x^2} + \dfrac{\hbar^2}{2m}(E - V)\psi = 0$

    (ii) $\dfrac{\partial^2 \psi}{\partial x^2} + \dfrac{2m}{\hbar^2}(E - V)\psi = 0$

    (iii) $\dfrac{\partial^2 \psi}{\partial x^2} + \dfrac{2m}{\hbar^2}(V - E)\psi = 0$

    (iv) $\dfrac{\partial^2 \psi}{\partial x^2} + \dfrac{2m}{\hbar^2}(E + V)\psi = 0$

2.   Time-dependent Schrödinger's equation in one-dimension is

(i) $i\hbar \dfrac{\partial \psi}{\partial t} = \dfrac{2m}{\hbar^2} \dfrac{\partial^2 \psi}{\partial x^2} + V\psi$ (ii) $i\hbar \dfrac{\partial \psi}{\partial t} = \dfrac{\hbar^2}{2m} \dfrac{\partial^2 \psi}{\partial x^2} + V\psi$

(iii) $i\hbar \dfrac{\partial \psi}{\partial t} = -\dfrac{\hbar^2}{2m} \dfrac{\partial^2 \psi}{\partial x^2} + V\psi$ (iv) $i\hbar \dfrac{\partial \psi}{\partial t} = -\dfrac{2m}{\hbar^2} \dfrac{\partial^2 \psi}{\partial x^2} + V\psi$

3.   According to quantum physics, the energy of a free particle is

(i)   quantized   (ii)   continuous

(iii) always zero   (iv) always infinity

4.   What is the unit of transmission coefficient?

(i)   Unitless   (ii)   Weber/meter

(iii) Tesla   (iv) None of the above

5.   What is the unit of reflection coefficient?

(i)   Unitless   (ii)   Weber/meter

(iii) Tesla   (iv) None of the above

6.   What is the unit of $\vec{k}_0$?

(i)   Unitless   (ii) m^{-1}

(iii) electron volt   (iv) None of the above

7.   In case of potential step of infinite height, the phase difference between the incident wave and the reflected wave is

(i)   0   (ii) π

(iii) 2π (iv) $\pi/2$

8.   The de Broglie wavelength of a particle trapped inside an infinite deep potential well in the first excited state is

(i)   Equal to the width of the well   (ii)   Two times the width of the well

(iii) Three times the width of the well (iv) Four times the width of the well

9.   The width of an infinite deep potential well is doubled. The ground state energy of the particle will be

(i)   doubled   (ii)   halved

(iii) quartered   (iv) tripled

10.   A particle trapped in an infinite deep potential well has de Broglie wavelength λ in the ground state. What is the wavelength of the particle in the next excited state?

(i) $\dfrac{\lambda}{2}$ (ii) $\dfrac{\lambda}{3}$

(iii) $\dfrac{\lambda}{4}$ (iv) 2λ

Answers

1 (ii)   2 (iii)   3 (ii)   4 (i)   5 (i)   6 (ii)   7 (iv)   8 (ii)
9 (iii)   10 (i)

9 Special Theory of Relativity

9.1 Introduction

Newtonian mechanics describes the macro-world correctly. However, when it is applied to describe atomic particles of very high energies, or travelling at very high speeds, it gives predictions which disagree with experiments. The absoluteness of mass, length and time were discarded by the theory of relativity – it was theorized that they depend on reference frames. Though variations of mass, length and time with speed are too small to be observed in day-to-day life, their variations are clearly observable in the world of high energy atomic particles. In the special theory of relativity, gravitational effects and accelerated motions are beyond consideration and they are dealt with in the general theory of relativity. The theory of relativity is beyond the perception of common sense. It is our duty as students of science and engineering to change our minds to fit nature, not to change nature to fit the preconceptions in our minds.

9.2 Frame of Reference

To describe a physical event, we must establish a frame of reference. The motion of a body can only be described relative to some other bodies, observers, or a set of space–time coordinates. These are called frames of reference. If the coordinates are not chosen properly, the laws of physics may be unnecessarily more complicated.

9.2.1 Inertial frame of reference

The reference frame in which the law of inertia (Newton's first law) holds good is called the inertial frame of reference. All inertial frames are in a state of constant, rectilinear motion with respect to one another; they do not accelerate. They may be at rest. No experiment can be performed to know whether an isolated inertial frame of reference is in motion or at rest. In an inertial frame of reference, an object is observed to have no acceleration when no forces act on it. Furthermore, any frame moving with constant velocity with respect to an inertial frame must also be an inertial frame. Measurements in one inertial frame can be converted to measurements in another by the Galilean transformation in Newtonian physics and the Lorentz transformation in special relativity. Earth is approximately taken to be an inertial reference frame. No preferred inertial reference frame exists and all inertial reference frames are equivalent.

9.2.2 Non-inertial frame of reference

Opposite of the inertial frame of reference is the non-inertial frame of reference or accelerated frame of reference. A non-inertial reference frame is a frame of reference that is accelerating with respect to an inertial frame. An accelerometer at rest in a non-inertial frame will in general show non-zero acceleration. Unlike the inertial frames, the laws of motion in non-inertial frames take complicated forms, and the laws vary from frame to frame depending upon the acceleration. To explain the motion of bodies entirely within the viewpoint of non-inertial reference frames, fictitious forces also called inertial forces/pseudo-forces/d'Alembert forces must be introduced to account for the observed motion. One might say that $F = ma$ holds in any coordinate system provided the term "force" is re-defined to include the so-called d'Alembert forces. In a curved space–time, all frames are non-inertial. Measurements with respect to non-inertial reference frames can always be transformed to an inertial frame, directly incorporating the acceleration of the non-inertial frame as that acceleration as seen from the inertial frames.

9.3 Galilean Transformation

Let us first see how we transform from one inertial frame to another in Newtonian mechanics. Suppose an inertial frame of reference S' is moving with respect to another inertial frame of reference S with a constant velocity v along the x-direction and both S and S' coincide with each other at $t = 0$. Two observers O and O' are attached at the origin of S and S' respectively. Let a point in space–time (called an 'event') has the coordinates (x, y, z, t) in frame S as measured by O and (x', y', z', t') in S' as measured by S'. Then from our common sense/classical ideas and with reference to Fig. 9.1, the relation between the coordinates of the event in S and S' will be

$$t' = t \quad \text{or} \quad dt' = dt$$

$$x' = x - vt \quad \text{or} \quad \frac{dx'}{dt'} = \frac{dx}{dt} - v \quad \text{or} \quad u' = u - v \tag{9.1}$$

$$y' = y \quad \text{or} \quad \frac{dy'}{dt'} = \frac{dy}{dt}$$

$$z' = z \quad \text{or} \quad \frac{dz'}{dt'} = \frac{dz}{dt}$$

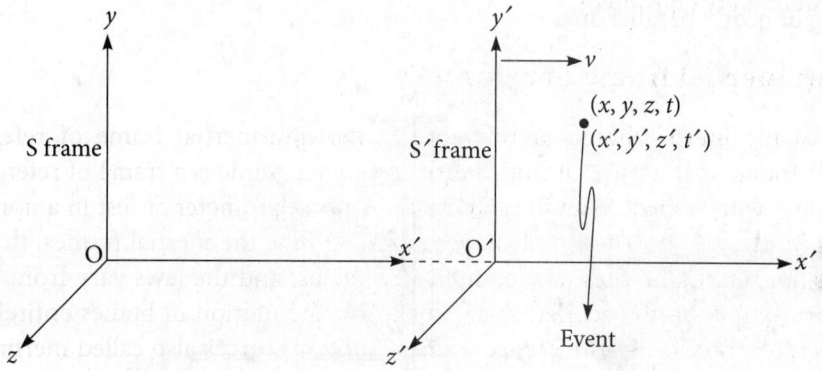

Figure 9.1 | Observers O and O' move with relative velocity v and each observer has its own set of coordinates (x, y, z, t) and (x', y', z', t')

These relations are called Galilean transformations. From this, we see that the time of occurrence of an event is the same in all inertial frames. A more precise way of stating this is that the time interval between two events is invariant under Galilean transformation.

9.4 Michelson–Morley Experiment

When we say that the speed of light in a vacuum is 2.997925×10^8 m/s $(=1/\sqrt{\varepsilon_o \mu_o})$, we do not mention any reference system. A reference system fixed in the medium of propagation of light presents difficulties because, in contrast to sound, no medium seems to exist. The 19th century physicists could not imagine that light could propagate without any medium. It seemed to be a logical step for them to postulate a medium called "ether", assigning unusual properties to it. It followed then that an observer moving through ether with velocity v would measure a velocity $c \pm v$ for a light beam according to Newtonian relativity. It was this result that the Michelson–Morley experiment was designed to test. The experimental tool used was nothing but the Michelson interferometer of which one of the two arms is aligned along the direction of Earth's motion through space. Earth moving through ether at speed v (≈ 30 km/s) is equivalent to ether flowing past Earth in the opposite direction with the same speed v. This ether wind blowing in the direction opposite the direction of Earth's motion should cause the speed of light measured in Earth's frame to be $c \pm v$ depending upon the direction of the light beam. The schematic diagram of the Michelson interferometer is shown in Fig. 9.2. The beam of light from source S is split by the partially silvered mirror M into two coherent beams by partial transmission and partial reflection. Beam 1 is reflected back to M by mirror M_1 and beam 2 by mirror M_2. Then the returning beam 1 is partially reflected and the returning beam 2 is partially transmitted by M back to a telescope at T where they interfere. Depending on the phase difference of the beams, the interference is constructive or destructive. The partially silvered mirror surface M is inclined at 45° to the beam directions so as to make beam 1 and 2 mutually perpendicular. If M_1 and M_2 are very nearly at right angles, we observe a fringe system in the telescope consisting of nearly parallel lines.

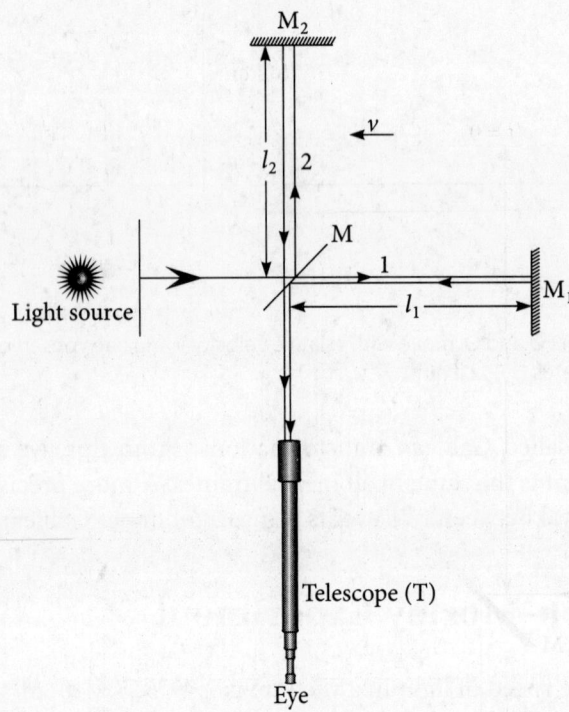

Figure 9.2 A schematic diagram of the Michelson interferometer showing splitting of light beam into two beams by mirror M. These beams are reflected by mirrors M_1 and M_2 towards M and then transmitted to the telescope T giving rise to an interference fringe pattern. Here v is the velocity of ether with respect to the interferometer

Let us compute the phase difference between the two beams 1 and 2. This phase difference is due to the different path lengths ℓ_1 and ℓ_2 that the two light rays travelled, and the different speeds of the light rays because of the "ether wind" speed v. The time for beam 1 to travel from M to M_1 and back (as shown in Fig. 9.2) is

$$t_1 = \frac{\ell_1}{c-v} + \frac{\ell_1}{c+v} = \frac{2\ell_1}{c}\left(\frac{1}{1-v^2/c^2}\right) \tag{9.2}$$

The path of beam 2, travelling from M to M_2 and back through the ether, is shown in Fig. 9.3 enabling the beam to return to the mirror M. The transit time t_2 as calculated from Fig. 9.3 is

$$\frac{ct_2}{2} = \sqrt{\ell_2^2 + (vt_2/2)^2}$$

or $\quad t_2 = \frac{2\ell_2}{c}\left(\frac{1}{\sqrt{1-v^2/c^2}}\right)$ (9.3)

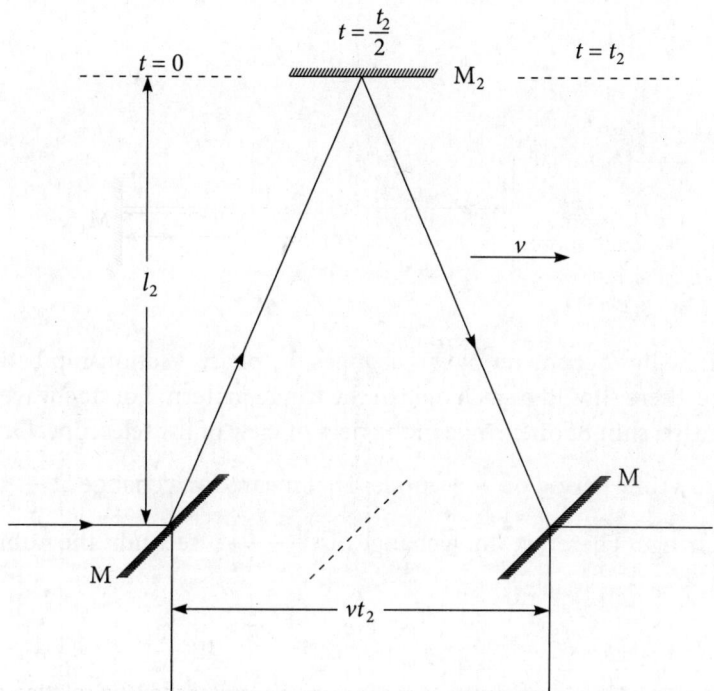

Figure 9.3 | The mirrors move through ether at a speed v, (i.e., v is the speed of the interferometer with respect to ether); the light moves through ether at speed c

The calculation of t_2 made in the ether frame and t_1 in the frame of the apparatus makes no difference because time is an absolute quantity in classical physics. The difference in transit times is

$$\Delta t = t_2 - t_1 = \frac{2}{c}\left(\frac{\ell_2}{\sqrt{1-v^2/c^2}} - \frac{\ell_1}{1-v^2/c^2}\right)$$ (9.4)

Let the instrument be rotated through 90°, thereby interchanging ℓ_1 and ℓ_2. In this case, the transit time difference will be

$$\Delta t' = \frac{2}{c}\left(\frac{\ell_2}{1-v^2/c^2} - \frac{\ell_1}{\sqrt{1-v^2/c^2}}\right) \tag{9.5}$$

Hence, the rotation through 90° changes the time differences by

$$\Delta t' - \Delta t = \frac{2}{c}\left(\frac{\ell_2+\ell_1}{1-v^2/c^2} - \frac{\ell_2+\ell_1}{\sqrt{1-v^2/c^2}}\right)$$

$$= \frac{2(\ell_1+\ell_2)}{c}\left((1-v^2/c^2)^{-1} - (1-v^2/c^2)^{-1/2}\right)$$

$$= \frac{2(\ell_1+\ell_2)}{c}\left((1+v^2/c^2) - (1+v^2/2c^2)\right)$$

or $\quad \Delta t' - \Delta t = \frac{\ell_1+\ell_2}{c}\frac{v^2}{c^2} \tag{9.6}$

The rotation of the interferometer by 90° changes the phase relationship between beams 1 and 2 and hence, there should be a change in the fringe pattern. For one wavelength λ path difference, there is a shift of one fringe in the field of view of the telescope. One wavelength λ corresponds to a time period of $\frac{\lambda}{c}$ seconds. That means for a change of $\frac{\lambda}{c}$ seconds, there is a shift of one fringe. Therefore, for a change of $\frac{\ell_1+\ell_2}{c}\frac{v^2}{c^2}$ seconds, the number of fringes shifted ΔN will be

$$\Delta N = \frac{c}{\lambda} \times \frac{\ell_1+\ell_2}{c}\frac{v^2}{c^2} = \frac{(\ell_1+\ell_2)v^2}{\lambda c^2}$$

or $\quad \Delta N = \frac{(\ell_1+\ell_2)}{\lambda}\left(\frac{30000}{3\times10^8}\right)^2 = 10^{-8}\frac{(\ell_1+\ell_2)}{\lambda} \tag{9.7}$

(We have chosen $v = 30$ km/s because it is the orbital speed of Earth).

In Michelson's 1887 experiment, $\ell_1 + \ell_2 = 22$ m and a light of wavelength $\lambda = 5.5 \times 10^{-7}$ m was used which gives

$$\Delta N = 10^{-8}\frac{22\,\text{m}}{5.5\times10^{-7}\,\text{m}} = \frac{4}{10}$$

Incredibly, a shift of four-tenths a fringe! To consider the effects of spinning and rotation of Earth (thereby interchanging ℓ_1 and ℓ_2 over long time periods), observations were made day and night throughout the year. However, the expected fringe shift was not observed. The experimental conclusion was that there was no fringe shift at all.

This null result was such a blow to the ether hypothesis that the same experiment was repeated by many researchers for more than 50 years. The null result was amply established and provided great stimuli to theoretical and experimental investigation. The Michelson–Morley experiment's null result not only contradicted the ether hypothesis, but also showed that it is impossible to measure the absolute velocity of Earth with respect to the ether frame. Einstein, in the year 1905, put forth his special theory of relativity and with it, it was possible to interpret the null results of the Michelson–Morley experiment.

9.5 Einstein's Principles of Relativity

Einstein proposed a theory that boldly removed the difficulties of measuring speed of light with respect to the earth and the contradictions in Galilean transformation. At the same time, it completely altered our notion of space and time. Einstein's special theory of relativity is based on the following two postulates:

i.   *The principle of relativity*   All the laws of physics must be the same in all inertial reference frames.

ii. *The constancy of the speed of light*   The speed of light in vacuum has the same value ($\approx 3 \times 10^8$ m/s) in all inertial frames of reference; independent of the speed of the observer or that of the source emitting the light.

9.6 Lorentz Transformation

The Lorentz transformation is the equivalent of the Galilean transformation with the added assumption that everyone measures the same speed of light no matter how fast they are travelling or the sources are travelling.

9.6.1 Mathematics of the Lorentz transformation

Consider two observers O and O' moving at a constant velocity relative to each other along the x-axis, i.e., the S' frame moves along on the x-axis with constant velocity v with respect to S. They synchronise their clocks as they pass each other so that $t = t' = 0$. They both observe the same event such as a flash of light. How will the coordinates recorded by the observers of the event that produced the flash of light be interrelated? The relationship between the coordinates can be derived on the basis of the postulates of relativity and additional assumptions of homogeneity and isotropy of space–time.

Homogeneity and isotropy assumptions

Space–time is uniform and homogeneous in all directions. If this were not the case, then when comparing lengths in different coordinate systems, the lengths would depend upon the position and orientation of the measurement. The linear equations relating coordinates in S' and S frames are:

$$x' = a_{11}x + a_{12}y + a_{13}z + a_{14}t$$

$$y' = a_{21}x + a_{22}y + a_{23}z + a_{24}t$$

$$z' = a_{31}x + a_{32}y + a_{33}z + a_{34}t$$

$$t' = a_{41}x + a_{42}y + a_{43}z + a_{44}t \tag{9.8}$$

There is no relative motion in the y or z directions. Therefore, according to the relativity postulate

$$y' = y$$

$$z' = z$$

Hence,

$$a_{22} = 1, a_{21} = a_{23} = a_{24} = 0$$

and

$$a_{33} = 1, a_{31} = a_{32} = a_{34} = 0$$

Thus, the following equations of Eq. (9.8) remain to be solved

$$x' = a_{11}x + a_{12}y + a_{13}z + a_{14}t \tag{9.9}$$

$$t' = a_{41}x + a_{42}y + a_{43}z + a_{44}t \tag{9.10}$$

If space–time is isotropic (i.e., the properties of space is the same in all directions), then the readings of clocks should be independent of the y and z coordinates. Hence,

$$a_{42} = a_{43} = 0$$

and

$$t' = a_{41}x + a_{44}t \tag{9.11}$$

The point having $x' = 0$ appears to move in the direction of the positive x-axis with speed v, so that the statement $x' = 0$ must be identical to the statement $x = vt$
   Hence, we have from Eq. (9.9),

$$0 = a_{11}vt + a_{12}y + a_{13}z + a_{14}t \tag{9.12}$$

Given that the equations are linear,

$$a_{12}y + a_{13}z = 0$$

and so from Eq. (9.12),

$$-a_{11}vt = a_{14}t \quad \text{or} \quad -a_{11}v = a_{14}$$

Therefore, the correct transformation equation for x' from Eq. (9.9) is

$$x' = a_{11}(x - vt) \tag{9.13}$$

Thus, we have the following transformation equations.

$$x' = a_{11}(x - vt) \tag{9.14}$$

$$y' = y \tag{9.15}$$

$$z' = z \tag{9.16}$$

$$t' = a_{41}x + a_{44}t \tag{9.17}$$

Assuming that the speed of light is constant (postulate ii), the coordinates of a flash of light that expands as a sphere will satisfy the following equations in each coordinate system

$$(ct)^2 = x^2 + y^2 + z^2 \tag{9.18}$$

and

$$(ct')^2 = (x')^2 + (y')^2 + (z')^2 \tag{9.19}$$

Substituting the coordinate transformation Eqs (9.14)–(9.17) into Eq. (9.19), we obtain

$$c^2(a_{41}x + a_{44}t)^2 = a_{11}^2(x - vt)^2 + y^2 + z^2 \tag{9.20}$$

$$\text{or } (c^2a_{44}^2 - v^2a_{11}^2)t^2 = (a_{11}^2 - c^2a_{41}^2)x^2 + y^2 + z^2 - 2(va_{11}^2 + c^2a_{41}a_{44})xt \tag{9.21}$$

Comparing Eq. (9.18) and (9.21), we have

$$c^2a_{44}^2 - v^2a_{11}^2 = c^2$$

$$a_{11}^2 - c^2a_{41}^2 = 1$$

$$va_{11}^2 + c^2a_{41}a_{44} = 0$$

The solution of these three simultaneous equations gives

$$a_{44} = \frac{1}{\sqrt{1 - v^2 / c^2}}$$

$$a_{11} = \frac{1}{\sqrt{1 - v^2 / c^2}}$$

$$a_{41} = \frac{v / c^2}{\sqrt{1 - v^2 / c^2}}$$

Substituting these values into Eqs (9.14)–(9.17), we have

$$x' = \frac{x - vt}{\sqrt{1 - v^2 / c^2}} \tag{9.22}$$

$$y' = y \tag{9.23}$$

$$z' = z \tag{9.24}$$

$$t' = \frac{t - (v / c^2) x}{\sqrt{1 - v^2 / c^2}} \tag{9.25}$$

Putting $\beta = v/c$ and $\gamma = \dfrac{1}{\sqrt{1 - v^2 / c^2}} = \dfrac{1}{\sqrt{1 - \beta^2}}$ into these equations, we get

$x \rightarrow x'$	$x' \rightarrow x$
$x' = \gamma(x - vt)$	$x = \gamma(x' + vt')$
$y' = y$	$y = y'$
$z' = z$	$z = z'$
$ct' = \gamma(ct - \beta x)$	$ct = \gamma(ct' + \beta x')$

These are the Lorentz transformation equations for events when the S' frame moves along the x-axis with constant velocity v with respect to S. In Fig. 9.4, we have depicted the variation of $\gamma = \dfrac{1}{\sqrt{1 - v^2 / c^2}}$ with respect to v, the velocity with which S' moves with respect to S.

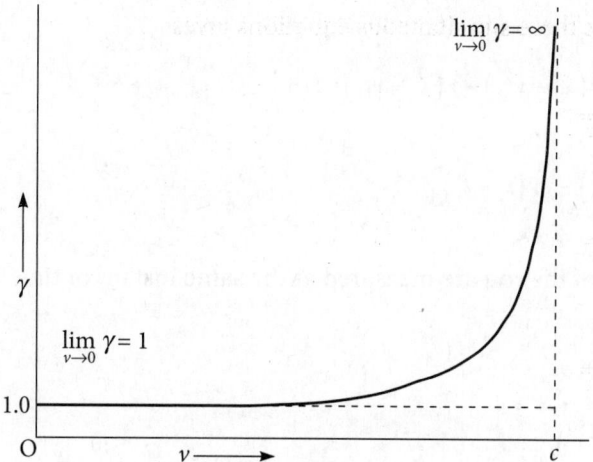

Figure 9.4 | The quantity $\gamma = \dfrac{1}{\sqrt{1 - v^2/c^2}}$ as a function of the relative speed of two frames of reference. For a particle at rest, $\gamma = 1.0$

As speed v approaches the speed of light c, γ approaches infinity.

9.6.2 Consequences of the Lorentz transformation equations

The Lorentz transformation equations give incredibly interesting results when applied to physical phenomena. Few of them are cited here.

Consequence 1

A rod's length is the greatest when it is at rest relative to an observer.

Proof

Let the rod be placed in the S' frame parallel to the x'-axis. Let its end points be at x_1' and x_2' as measured in S' at the same instant of time and hence, its rest length, i.e., proper length will be

$$\ell_0 = x_2' - x_1'$$

Here ℓ_0 is the length of the rod measured in a frame of reference in which it is at rest. Its end points are measured to be at x_1 and x_2 in S at the same instant of time and hence, its length with respect to S, i.e., improper length will be

$$\ell = x_2 - x_1$$

Therefore, we have

$$\ell_0 = x_2' - x_1' = \gamma(x_2 - vt_2) - \gamma(x_1 - vt_1)$$

or $\quad \ell_0 = \gamma[(x_2 - x_1) - \gamma v(t_2 - t_1)]$

Since the two ends of the rod are measured at the same instant of time in S, $t_1 = t_2$ and this equation gives

$$\ell_0 = \gamma(x_2 - x_1)$$

or $\quad \ell = \dfrac{\ell_0}{\gamma}$ \hfill (9.26)

Equation (9.26) shows that when a rod moves with velocity v relative to the observer, its measured length is contracted (Since $\gamma = 1/\sqrt{1 - v^2/c^2} > 1$) in the direction of motion by a factor of $\sqrt{1 - v^2/c^2}$, whereas its dimensions perpendicular to the direction of motion are unaffected. This effect is called length contraction.

Consequence 2

A clock is measured to go at its fastest rate when it is at rest relative to the observer.

Proof

Let the clock be placed in the S' frame at x'. Duration of an event in the S' frame is $t_2' = t_1'$. Since the clock is at rest in S', this time interval is called proper time. Duration of the same event in the S frame is $t_2 - t_1$. Since the clock is in motion with respect to the S frame, this time interval is called improper time. Applying Lorentz transformation equation, we have

$$ct_2 - ct_1 = \gamma(ct_2' + \beta c x_2') - \gamma(ct_1' + \beta c x_1')$$

or $\quad t_2 - t_1 = \gamma(t_2' - t_1') - \gamma\beta(x_2' - x_1')$

Since the position of the clock x' does not change in the S frame, $x_1' = x_2'$, and this equation boils down to

$$t_2 - t_1 = \gamma(t_2' - t_1')$$

or $\quad \Delta t = \gamma \Delta t_0$ \hfill (9.27)

Here $\Delta t_0 = t_2' - t_1'$ is the time interval measured in a frame of reference in which the clock is at rest. Equation (9.27) shows that when a clock moves with velocity v relative to the observer, its rate is measured to have slowed down (since $\gamma = 1/\sqrt{1-v^2/c^2} > 1$) by a factor of $\sqrt{1-v^2/c^2}$. This effect is known as time dilation. It is important to note that the time interval Δt in Eq. (9.27) involves events that occur at different space points in the frame of reference S. Moreover, any differences between Δt and the proper time Δt_0 are not caused by differences in the times required for light to travel from those space points to an observer at rest in S.

Example 9.1

A hypothetical spaceship flies at a speed of 0.990 c. The pilot in the spaceship measures its length to be 990 m. What length do observers measure on Earth?

Solution

Data given are $\ell_0 = 900$ m and $v = 0.990$ c

$$\ell = \frac{\ell_0}{\gamma} = \ell_0 \sqrt{1-v^2/c^2} = 990\sqrt{1-(0.990c)^2/c^2}\,\text{m} = 139.7\text{m}$$

Example 9.2

The proper volume of a cuboid of length, breadth, and width a_0, b_0, and c_0 along the x-, y-, z- axes respectively is $V_0 = a_0\, b_0\, c_0$. Find the volume of the cuboid if it is moving along the y-axis with a velocity v.

Solution

Since the cuboid is moving along the y-axis, only its breadth b_0 will change and become

$$b = \frac{b_0}{\gamma}$$

The volume will become

$$V = a_0 \frac{b_0}{\gamma} c_0 = \frac{V_0}{\gamma} = V_0 \sqrt{1-v^2/c^2}$$

Example 9.3

A 20-year-old woman gave birth to a girl child and then immediately left for a space journey at a speed of 0.985 c. At the age of 30 years (according to her own clock), she came back to her daughter. How old will she find her daughter?

Solution

Data given are $\Delta t_0 = 30$ yrs – 20 yrs = 10 yrs and $v = 0.985$ c

$$\Delta t = \gamma \Delta t_0 = \frac{\Delta t_0}{\sqrt{1 - v^2/c^2}} = \frac{10 \text{yrs}}{\sqrt{1 - (0.985c)^2/c^2}} = 57.95 \text{yrs}$$

Example 9.4

The period of a seconds' pendulum is 2.0 s in the reference frame of the pendulum. What is the period when measured by an observer moving at a speed of 0.950 c relative to the pendulum?

Solution

Data given are $\Delta t_0 = 2.0$ s, $v = 0.950\ c$

$$\Delta t = \gamma \Delta t_0 = \frac{\Delta t_0}{\sqrt{1 - v^2/c^2}} = \frac{2.0 \text{s}}{\sqrt{1 - (0.950c)^2/c^2}} = 6.4 \text{s}$$

Example 9.5

The life time of a particle in its own frame is 22 ns. What is its life time in the laboratory frame of reference when it is moving with a speed of 0.9 c.

Solution

Data given are $\Delta t_0 = 22$ns and $v = 0.9\ c$

$$\Delta t = \gamma \Delta t_0 = \frac{\Delta t_0}{\sqrt{1 - v^2/c^2}} = \frac{22 \text{ns}}{\sqrt{1 - (0.9c)^2/c^2}} = 50.5 \text{ns}$$

9.7 Relativity of Simultaneity

A basic premise of Newtonian mechanics is that a universal time scale exists, i.e., time is the same for all observers. Newton and his followers took simultaneity ($t' = t$) for granted. A careful analysis of simultaneity help us to develop the appropriate modifications of our notions about space and time. In the special theory of relativity, Einstein boldly abandoned this assumption.

This is a basic consequence of the special theory of relativity – that two events that are simultaneous for one observer may occur at different times for another observer in another reference frame moving relative to the first. Accordingly, it is impossible to say in an absolute sense whether two events occur at the same time if those events are separated in space. If we imagine one reference frame assigns precisely the same time to the two events at different points in space, a reference frame that is moving relative to the first may generally assign different times to the same two events even if both are inertial frames.

The relativity of simultaneity can be calculated using Lorentz transformations, which relate the coordinates used by one observer to coordinates used by another in uniform relative motion with respect to the first. Assume that the first observer uses coordinates labelled (x, y, z, t), while the second observer uses coordinates labelled (x', y', z', t'). Now suppose that the first observer sees the second observer moving in the $+x$ direction at constant speed v. Assume that both the observers' coordinate axes are parallel and that they have the same origin at initial zero time. Then according to Lorentz transformations, the two times are related by Eq. (9.25) as

$$t' = \frac{t - \left(v/c^2\right)x}{\sqrt{1 - v^2/c^2}}$$

where c is the speed of light. If two events happen at the same time in the frame of the first observer, they will have identical values of the time coordinate t. However, if they have different values of the x-coordinate, i.e., different positions in the $+x$-direction, we see that they will have different values of the t' coordinate; they will happen at different times in that frame. The quantity that accounts for the failure of absolute simultaneity is the finite speed of light in free space.

According to the principle of relativity, no inertial frame of reference is more correct than any other in the formulation of physical laws (postulate i). Each observer is correct in his/her own frame of reference. In other words, simultaneity is not an absolute concept. Whether two events are simultaneous depends on the frame of reference. Simultaneity plays an essential role in measuring time intervals. It follows that the time interval between two events may be different in different frames of reference and are related to each other by the equation

$$t_2' - t_1' = \left(t_2 - t_1\right)\sqrt{1 - v^2/c^2}.$$

9.8 Relativistic Addition of Velocity

Let the speed of an object along the x-axis be $u(= dx / dt)$ and $u'(= dx'/ dt')$ in S and S' respectively. Using Eqs (9.22) and (9.25), we get $dx' = \gamma(dx - vdt)$ and $cdt' = \gamma(cdt - \beta dx)$ respectively. Taking their ratio, we have

$$\frac{dx'}{cdt'} = \frac{dx - vdt}{cdt - \beta dx} = \frac{dx/dt - v}{c - \beta dx/dt}$$

or $\qquad u' = \dfrac{u - v}{1 - vu/c^2}$ $\hfill$ (9.28)

Similarly, by using $x = \gamma(x' + vt')$ and $ct = (ct' + \beta x')$, we get

$$\frac{dx}{cdt} = \frac{dx' + vdt'}{cdt' + \beta dx'} = \frac{dx'/dt' + v}{c + \beta dx'/dt'}$$

or $\quad u = \dfrac{u' + v}{1 + vu'/c^2}$ (9.29)

Equations (9.28) and (9.29) give the relation between u and u'. This is known as the relativistic velocity addition theorem. It applies to velocities parallel to the direction of relative motion. Equation (9.29) gives the transformation of constant velocities parallel to the direction of relative motion, i.e., the object moves along the x-axis. To signify this, we put a subscript x in u and u'. Thus, we have

$$u_x = \frac{u'_x + v}{1 + u'_x v/c^2}$$ (9.30)

and $\quad u'_x = \dfrac{u_x - v}{1 - u_x v/c^2}$ (9.31)

Let the relative velocity between S and S' be v along the common x-axis. Imagine an object moves along the vertical axis with velocity u'_y in S' and with velocity u_y in S. Thus, we have

$$u'_y = \frac{dy'}{dt'} = \frac{cdy}{\gamma(cdt - \beta dx)} = \frac{dy/dt}{\gamma(c - \beta dx/dt)}$$

or $\quad u'_y = \dfrac{u_y}{\gamma\left(1 - vu_x/c^2\right)}$ (9.32)

Again using $ct = \gamma(ct' + \beta x')$, we get

$$u_y = \frac{dy}{dt} = \frac{cdy'}{\gamma(cdt' + \beta dx')} = \frac{dy'/dt'}{\gamma(c + \beta dx'/dt')}$$

or $\quad u_y = \dfrac{u'_y}{\gamma\left(1 + vu'_x/c^2\right)}$ (9.33)

Following the exact procedure, we can obtain

$$u'_z = \frac{u_z}{\gamma(1 - u_x \beta/c)}$$ (9.34)

and $\quad u_z = \dfrac{u_z'}{\gamma\left(1+u_x'\beta/c\right)}$ $\qquad\qquad\qquad\qquad\qquad\qquad\qquad\qquad$ (9.35)

In these equations, u_x, u_y, and u_z are the velocity components as seen in the S frame and u_x', u_y', and u_z' are the corresponding velocity components as seen in the S' frame. It is also important to remember that these equations are derived by taking the S' frame to be moving with velocity v with respect to the S frame along the positive x-axis. For the convenience of the readers, all above results established till now have been summarized in the following table by slightly changing the symbols.

Inter-conversion of relativistic velocities

$v_{xos'} = \dfrac{v_{xos}-v_{s's}}{1-v_{xos}v_{s's}/c^2}$	$v_{xos} = \dfrac{v_{xos'}+v_{s's}}{1+v_{xos'}v_{s's}/c^2}$
$v_{yos'} = \dfrac{v_{yos}\sqrt{1-v_{s's}^2/c^2}}{1-v_{xos}v_{s's}/c^2}$	$v_{yos} = \dfrac{v_{yos'}\sqrt{1+v_{s's}^2/c^2}}{1+v_{xos'}v_{s's}/c^2}$
$v_{zos'} = \dfrac{v_{zos}\sqrt{1-v_{s's}^2/c^2}}{1-v_{xos}v_{s's}/c^2}$	$v_{zos} = \dfrac{v_{zos'}\sqrt{1+v_{s's}^2/c^2}}{1+v_{xos'}v_{s's}/c^2}$

Here

$v_{xos'}$ = x component of object velocity with respect to S' frame.

v_{xos} = x component of object velocity with respect to S frame.

$v_{yos'}$ = y component of object velocity with respect to S' frame.

v_{yos} = y component of object velocity with respect to S frame.

$v_{zos'}$ = z component of object velocity with respect to S' frame.

v_{zos} = z component of object velocity with respect to S frame.

$v_{s's}$ = velocity of S' frame with respect to S frame.

Example 9.6

Two space vehicles A and B are moving in opposite directions with speeds $0.756\,c$ and $0.856\,c$ as measured by an observer on Earth. Find the velocity of ship B as observed by a crew on ship A.

Solution

The problem is depicted in the schematic diagram shown in Fig. Example 6. For a crew on vehicle A, it (vehicle A) is at rest (S frame). Earth (S' frame) and vehicle B are moving away from A with speeds 0.756 c and 0.856 c respectively along the positive x-axis. The vehicle B is the object.

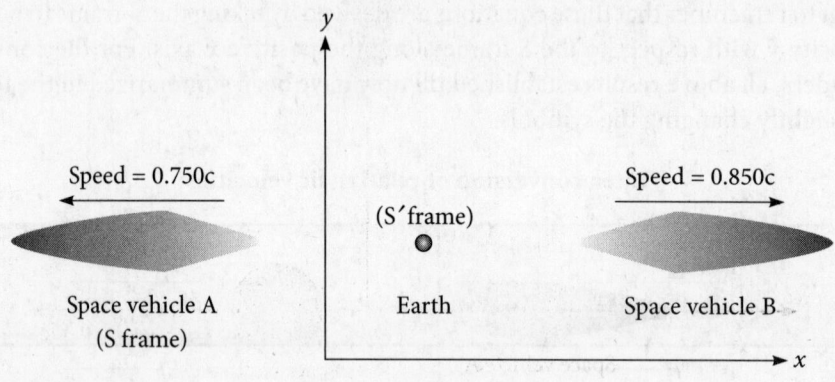

Illustration for Example 6

The data given are

$v_{xos'} = 0.856\,c$ along the positive x-axis.

$v_{s's} = 0.756\,c$ along the positive x-axis.

Let $v_{xos} = x$ component of object velocity with respect to S frame.
By using Eq. (9.28), we have

$$v_{xos} = \frac{v_{xos'} + v_{s's}}{1 + v_{xos'}v_{s's}\big/c^2} = \frac{0.856c + 0.756\,c}{1 + 0.856c \times 0.756\,c\big/c^2} = 0.979\,c \quad \text{along the positive } x\text{-axis.}$$

Example 9.7

Two space vehicles A and B are leaving their space stations along mutual perpendicular directions. The astronaut in the space station measured the velocity of A and B as 0.50 c and 0.60 c respectively. What is the velocity of B as measured by a crew in A?

Solution

The problem is depicted in the schematic diagram shown in Fig. Example 7. For a crew on vehicle A, it (vehicle A) is at rest (S frame). The vehicles A and B are moving away from the space station with speeds 0.50 c and 0.60 c along the negative x-axis and the positive y-axis respectively. The vehicle B is the object.

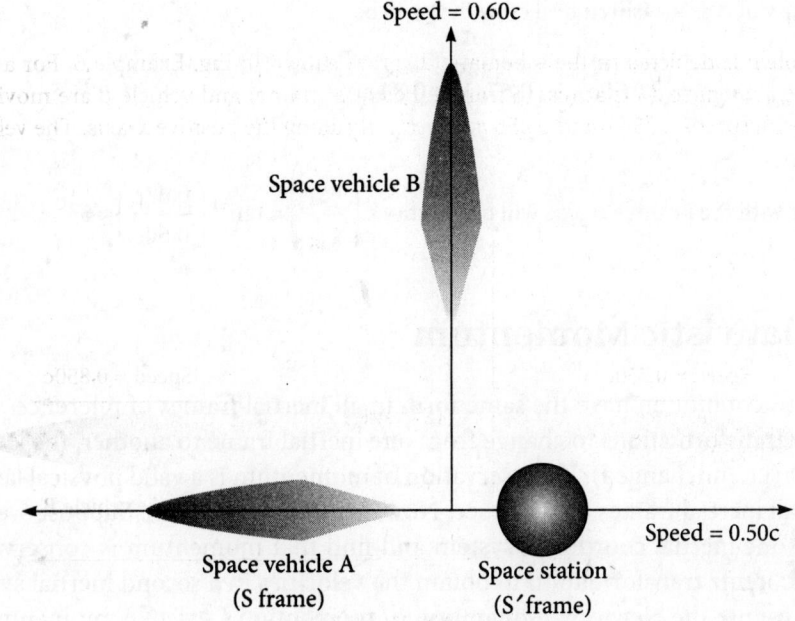

Illustration for Example 7

The data given are

$$v_{xos'} = 0$$

$$v_{yos'} = 0.60c \text{ along the positive } y\text{-axis.}$$

$$v_{s's} = 0.50c \text{ along the positive } x\text{-axis.}$$

Let
$v_{xos} = x$ component of object velocity with respect to S frame.
$v_{yos} = y$ component of object velocity with respect to S frame.

$$v_{xos} = \frac{v_{xos'} + v_{s's}}{1 + v_{xos'} \cdot v_{s's}/c^2} = \frac{0 + 0.50c}{1 + 0 \times 0.50c/c^2} = 0.50c$$

$$v_{yos} = \frac{v_{yos'}\sqrt{1 + v_{s's}^2/c^2}}{1 + v_{xos'} \cdot v_{s's}/c^2} = \frac{0.60c\sqrt{1 + (0.50c)^2/c^2}}{1 + 0 \times 0.50c/c^2} = 0.67c$$

The velocity of B as measured by a crew in A will be

$$v = \sqrt{v_{xos}^2 + v_{yos}^2} = \sqrt{(0.50c)^2 + (0.67c)^2} = 0.84c$$

The angle with the positive x-axis will be $\theta = \tan^{-1}\left(\dfrac{v_{yos}}{v_{xos}}\right) = \tan^{-1}\left(\dfrac{0.67c}{0.50c}\right) \approx 53^{\circ}$

9.9 Relativistic Momentum

Newton's laws of motion have the same form in all inertial frames of reference. When we use Lorentz transformations to change from one inertial frame to another, the laws should be invariant (i.e., unchanged). If conservation of momentum is a valid physical law, it must be valid in all inertial frames of reference. Now, here is the problem: Suppose we look at a collision in one inertial coordinate system and find that momentum is conserved. Then, we use the Lorentz transformation to obtain the velocities in a second inertial system. We find that if we use the Newtonian definition of momentum ($\vec{p} = m\vec{v}$), momentum would not be conserved in the second system. If we are convinced that the principle of relativity and the Lorentz transformation are correct, the only way to save momentum conservation is to generalize the definition of momentum.

The obvious idea for relativistic momentum is to use the same classical definition so that its dimensions remain correct, i.e.,

$$\vec{p} = m\vec{v} = m\frac{d\vec{r}}{dt}$$

By using the particles own time, i.e., proper time, we have

$$\vec{p} = m\frac{d\vec{r}}{dt'}$$

A change in proper time t' is related to a change in frame S' s time t through the time dilation formula

$$dt = \frac{dt'}{\sqrt{1 - v^2/c^2}}$$

where v is the velocity of the particle in frame S. Hence, in frame S, where the particle moves with velocity $\vec{v}$, the momentum is

$$\vec{p} = \frac{m\vec{v}}{\sqrt{1 - v^2/c^2}} \tag{9.36}$$

Thus, the components of momentum are

$$p_x = \frac{mv_x}{\sqrt{1-v^2/c^2}} \qquad (9.37)$$

$$p_y = \frac{mv_y}{\sqrt{1-v^2/c^2}} \qquad (9.38)$$

$$p_z = \frac{mv_z}{\sqrt{1-v^2/c^2}} \qquad (9.39)$$

Let us verify the correctness of this definition by applying it to law of conservation of momentum in two-body collision problems in S and S' frames. We have the following symbols.

m_A = mass of one body before collision

m_B = mass of another body before collision

m_C = mass of one of the bodies after collision

m_D = mass of one of the bodies after collision

v_A = velocity of mass m_A before collision

v_B = velocity of mass m_B before collision

v_C = velocity of mass m_C after collision

v_D = velocity of mass m_D after collision

The law of conservation of momentum in S and S' respectively states that

$$p_A + p_B = p_C + p_D$$

and

$$p_A' + p_B' = p_C' + p_D'$$

We are now proceed to show that if $p_A' + p_B' = p_C' + p_D'$, then $p_A + p_B = p_C + p_D$, i.e., the law of conservation of momentum in S follows from the law of conservation of momentum in S'.

$$p_A' + p_B' = p_C' + p_D'$$

$$\frac{m_A v_A'}{\sqrt{1-\left(v_A'/c\right)^2}} + \frac{m_B v_B'}{\sqrt{1-\left(v_B'/c\right)^2}} = \frac{m_C v_C'}{\sqrt{1-\left(v_C'/c\right)^2}} + \frac{m_D v_D'}{\sqrt{1-\left(v_D'/c\right)^2}} \qquad (9.40)$$

By using the velocity transformation relation (9.28), $v'_A = \dfrac{v_A - v}{1 - v_A v/c^2}$, we get

$$\frac{v'_A}{\sqrt{1-\left(v'_A/c\right)^2}} = \frac{v_A - v}{\sqrt{1-\left(v_A/c\right)^2}\sqrt{1-\left(v/c\right)^2}} \tag{9.41}$$

Thus, we have

$$\frac{m_A v'_A}{\sqrt{1-\left(v'_A/c\right)^2}} = \frac{m_A v_A}{\sqrt{1-\left(v_A/c\right)^2}}\frac{1}{\sqrt{1-\left(v/c\right)^2}} - \frac{m_A}{\sqrt{1-\left(v_A/c\right)^2}}\frac{v}{\sqrt{1-\left(v/c\right)^2}}$$

or $\qquad \dfrac{m_A v'_A}{\sqrt{1-\left(v'_A/c\right)^2}} = m_A v_A \gamma_A \gamma - m_A \gamma_A \gamma v$ $\qquad\qquad$ (9.42)

Similarly, we have

$$\frac{m_B v'_B}{\sqrt{1-\left(v'_B/c\right)^2}} = m_B v_B \gamma_B \gamma - m_B \gamma_B \gamma v \tag{9.43}$$

$$\frac{m_C v'_C}{\sqrt{1-\left(v'_C/c\right)^2}} = m_C v_C \gamma_C \gamma - m_C \gamma_C \gamma v \tag{9.44}$$

$$\frac{m_D v'_D}{\sqrt{1-\left(v'_D/c\right)^2}} = m_D v_D \gamma_D \gamma - m_D \gamma_D \gamma v \tag{9.45}$$

Putting Eqs (9.42)–(9.45) into Eq. (9.40), we get

$$m_A v_A \gamma_A \gamma - m_A \gamma_A \gamma v + m_B v_B \gamma_B \gamma - m_B \gamma_B \gamma v = m_C v_C \gamma_C \gamma - m_C \gamma_C \gamma v + m_D v_D \gamma_D \gamma - m_D \gamma_D \gamma v$$

or $\quad m_A v_A \gamma_A - m_A \gamma_A v + m_B v_B \gamma_B - m_B \gamma_B v = m_C v_C \gamma_C - m_C \gamma_C v + m_D v_D \gamma_D - m_D \gamma_D v$

or $\quad m_A v_A \gamma_A + m_B v_B \gamma_B = m_C v_C \gamma_C + m_D v_D \gamma_D + (m_A \gamma_A + m_B \gamma_B - m_C \gamma_C - m_D \gamma_D)v$

or $\quad m_A v_A \gamma_A + m_B v_B \gamma_B = m_C v_C \gamma_C + m_D v_D \gamma_D$

since $m_A\gamma_A + m_B\gamma_B = m_C\gamma_C + m_D\gamma_D$

Thus,

$$p_A + p_B = p_C + p_D$$

We cannot deduce $p_A + p_B = p_C + p_D$ from $p'_A + p'_B = p'_C + p'_D$ for any other definition of momentum except

$$\vec{p} = \frac{m\vec{v}}{\sqrt{1 - v^2/c^2}} = \gamma m\vec{v} \tag{9.46}$$

Newtonian mechanics correctly describes objects moving at ordinary speeds. Yet, it incorrectly predicts that momentum $m\vec{v}$ becomes infinite only if v becomes infinite which is not possible – in the Newtonian concept, there is no upper limit on the speed of a particle. Relativistic momentum becomes infinite as v approaches c. It is shown graphically in Fig. 9.5.

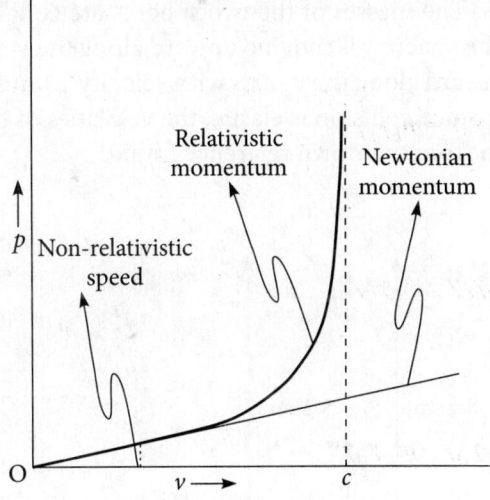

Figure 9.5 | Variation of relativistic momentum with speed. Here also variation of Newtonian momentum has been shown

Example 9.8

An electron with rest mass 9.11×10^{-31} kg and charge -1.6×10^{-19} C is moving in an electric field of magnitude 9.00×10^5 N/C. Neglecting all other forces, find the magnitudes of momentum at the instants when $v = 0.010\,c$, $0.90\,c$, and $0.95\,c$.

Solution

The data given are $m_0 = 9.11 \times 10^{-31}$ kg, $v = 0.010\,c$, $0.90\,c$, $0.95\,c$

$$\gamma = 1\Big/ \sqrt{1 - v^2/c^2} = 1.00 \qquad \text{for } v = 0.010\, c$$

$$= 2.29 \qquad \text{for } v = 0.90\, c$$

$$= 3.20 \qquad \text{for } v = 0.95\, c$$

The magnitude of the momentum $p = \gamma m_0 v = 2.7 \times 10^{-24}$ kg m/s for $v = 0.010\, c$

$$= 5.6 \times 10^{-22} \text{ kg m/s for } v = 0.90\, c$$

$$= 8.3 \times 10^{-22} \text{ kg m/s for } v = 0.95\, c$$

9.10 Variation of Mass with Speed

Let the reference frame S' move towards the reference frame S with a velocity v as shown in Fig. 9.6. Two spheres p and q are moving vertically in opposite direction with equal speed in both reference frames. The masses of the two spheres are equal when measured in the same reference frame. The sphere p is moving upward along the y-axis with velocity u_y and sphere q is moving downward along the y'-axis with velocity u_y' in such a manner that they have an elastic collision. Since collision is elastic, the velocities of the spheres are reversed and interchanged as seen from their own reference frame.

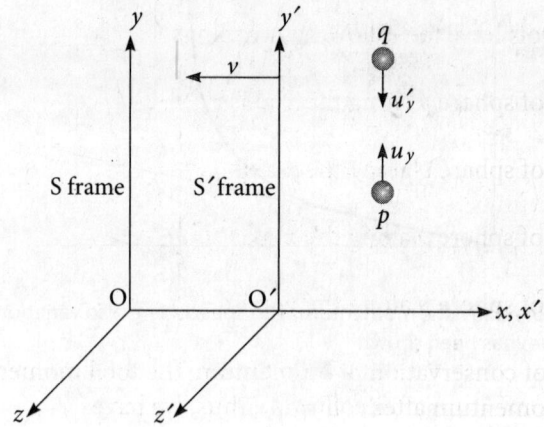

Figure 9.6 | Derivation of $m = \dfrac{m_0}{\sqrt{1 - v^2/c^2}}$

Thus, we have the following table.

Spheres and mass	Before collision	After collision
p, m	u_y	$-u_y$ and $\lvert -u_y \rvert = u_y$
q, m	$-u_y'$ and $\lvert -u_y' \rvert = u_y$	u_y

In the reference frame S, we have the following table after collision. (according to the Lorentz transformation)

Spheres and mass	Before collision	After collision
p	$u_{xp} = 0,\ u_{yp} = u_y$	$u_{xp} = 0,\ u_{yp} = -u_y$
q	$u_{xq} = \dfrac{u_x' + v}{1 + u_x' v / c^2} = v,$ $u_{yq} = \dfrac{u_y'\sqrt{\left(1 - v^2/c^2\right)}}{1 + u_x' v / c^2} = -u_y'\sqrt{\left(1 - v^2/c^2\right)}$	$u_{xq} = \dfrac{u_x' + v}{1 + u_x' v / c^2} = v$ $u_{yq} = \dfrac{u_y'\sqrt{\left(1 - v^2/c^2\right)}}{1 + u_x' v / c^2} = u_y'\sqrt{\left(1 - v^2/c^2\right)}$
Momentum	$m_0 u_y + \left(-m u_y'\sqrt{\left(1 - v^2/c^2\right)}\right)$	$-m_0 u_y + \left(m u_y'\sqrt{\left(1 - v^2/c^2\right)}\right)$

In this table, the symbols have the following meanings.

u_{xp} = velocity of sphere p along the x-axis

u_{yp} = velocity of sphere p along the y-axis

u_{xq} = velocity of sphere q along the x-axis

u_{yq} = velocity of sphere p along the y-axis

According to the law of conservation of momentum, the total momentum before collision is equal to the total momentum after collision. Thus, we have

$$m_0 u_y + \left(-m u_y'\sqrt{\left(1 - v^2/c^2\right)}\right) = -m_0 u_y + \left(m u_y'\sqrt{\left(1 - v^2/c^2\right)}\right)$$

or $$m_0 u_y + \left(-m u_y\sqrt{\left(1 - v^2/c^2\right)}\right) = -m_0 u_y + \left(m u_y\sqrt{\left(1 - v^2/c^2\right)}\right)$$

Simplifying this equation, we get

$$m = \frac{m_0}{\sqrt{1 - v^2/c^2}} \tag{9.47}$$

This expression gives the relativistic mass m in terms of the rest mass/invariant mass m_0. It is fascinating as we come to know from this discussion that mass variation with speed is a natural consequence of the Lorentz transformation. The invariant mass of a particle is independent of its velocity v, whereas relativistic mass increases with velocity and tends to infinity as the velocity approaches the speed of light c. At zero velocity, the relativistic mass is equal to the invariant mass.

The word 'relativistic mass' was not used originally by Einstein. Later on, two pioneer physicists of the 20th century, namely S. Hawking and R. Feynman, popularized its use in their popular science books written for the common people. Although the word 'relativistic mass' is not wrong, it often leads to confusion and is less useful in advanced applications such as quantum field theory and general relativity; gradually its use is limited to special relativity at an elementary level. It is better to say that mass increases with energy content of the body than to argue with relativistic mass, rest mass or invariant mass!

Example 9.9

At what relativistic speed, is the mass of a moving body equal to 3 times its rest mass.

Solution

Data given is $m = 3m_0$

$$\frac{1}{\sqrt{(1 - v^2/c^2}} = \frac{m}{m_0} = 3$$

or $\quad v = c\sqrt{1 - \frac{1}{9}} = 0.943\,c$

9.11 Mass–Energy Equivalence

Energy is defined as the work done in moving a body from one place to another. Infinitesimal energy due to displacement dx is given by

$$dE = Fdx$$

Total energy will be

$$E = \int Fdx \tag{9.48}$$

Kinetic energy (K) is the energy used to move a body from a velocity of 0 to a velocity u. Hence:

$$K = \int_0^u F dx$$

$$K = \int_0^u \frac{dp}{dt} dx = \int_0^u \frac{dmu}{dt} dx = \int_0^u (mdu + udm)u$$

or $$K = \int_0^u (mudu + u^2 dm) \qquad (9.49)$$

In relativity, mass is given by

$$m = \frac{m_0}{\sqrt{1 - u^2/c^2}}$$

or $mc^2 - m_0 c^2 = mu^2$

Differentiating this equation, we get

$$mudu + u^2 dm = c^2 dm \qquad (9.50)$$

Substituting Eq. (9.50) into Eq. (9.49), we get

$$K = \int_{m_0}^m c^2 dm = mc^2 - m_0 c^2 \qquad (9.51)$$

or $$K = (\gamma - 1)m_0 c^2 \qquad (9.52)$$

The quantity mc^2 is known as the total energy of the particle. The quantity $m_0 c^2$ is known as the rest energy of the particle, i.e., energy of the particle at rest. If the total energy of the particle is given by the symbol E, we have

$$E = K + m_0 c^2 = mc^2 = \gamma m_0 c^2 \qquad (9.53)$$

This is the origin of the famous formula '$E = mc^2$' that is iconic of the nuclear age.

At low speeds, this expression boils down to the classical expression of kinetic energy as obtained from Newton's laws.

$$K = m_0 c^2 \left(1 -= u^2/c^2\right)^{-1/2} - m_0 c^2$$

Expanding the factor $(1 - u^2 / c^2)^{-1/2}$ by binomial expression and taking the first two terms, we get

$$K \approx m_0 c^2 \left(1 + \frac{u^2}{2c^2}\right) - m_0 c^2 = \frac{1}{2} m_0 u^2$$

which is equal to the classical expression for kinetic energy of a particle moving with a speed u. If the difference between relativistic and non-relativistic kinetic energy is more than 1%, the relativistic kinetic energy formula is to be used. The variation of kinetic energy with speed is illustrated in Fig. 9.7.

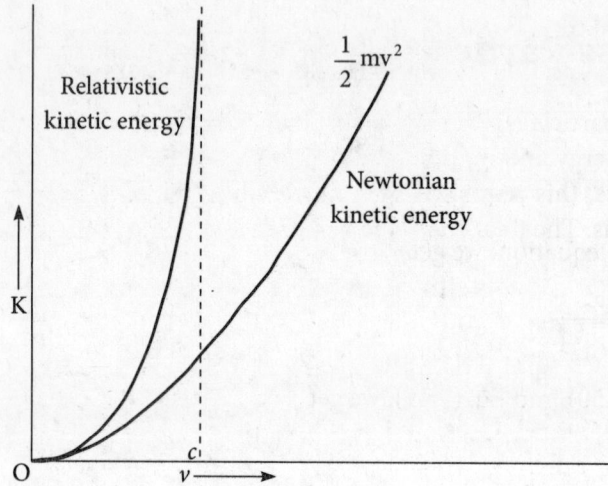

Figure 9.7 | The variation of the kinetic energy of a particle of rest mass m_0 as a function of speed v. The kinetic energy grows more rapidly as v increases than classical physics would say. Here also we have shown that the Newtonian prediction gives correct results only at speeds much less than c

Example 9.10

Find the speed of an electron of rest mass 9.11×10^{-31} kg and charge -1.6×10^{-19} C that has been accelerated through a potential difference of 5.00 MV.

Solution

The data given are $m_0 = 9.11 \times 10^{-31}$ kg, $V = 5.00 \times 10^6$ Volt

The rest energy $E_0 = m_0 c^2 = 9.11 \times 10^{-31} \left(2.998 \times 10^8\right)^2 \, \text{J} = 8.19 \times 10^{-14} \, \text{J}$

The kinetic energy of the electron

$$K = e \times V = 1.6 \times 10^{-19} \times 5 \times 10^6 \, J = 8.0 \times 10^{-13} \, J$$

$$\gamma = \frac{K + m_0 c^2}{m_0 c^2} = 1 + \frac{8.0 \times 10^{-13}}{8.19 \times 10^{-14}} = 10.77$$

The speed of the electron $v = c\sqrt{1 - \left(1/\gamma\right)^2}$

$$= c \times \sqrt{1 - \left(1/10.77\right)^2} \, m/s = 0.996 \, c$$

9.12 Massless Particles ($m_0 = 0$)

If the rest mass of a particle is zero, we call it a massless particle. In Newtonian mechanics, particles can have energy and momentum if they have rest mass, i.e., $m_0 \neq 0$. However, in relativistic mechanics, this requirement is not binding. This follows from the following mathematical analysis. The total energy of a particle according to Eq. (9.53) is given by

$$E = mc^2 = \frac{m_0 c^2}{\sqrt{1 - u^2/c^2}}$$

or $$E^2 = \frac{m_0^2 c^4}{1 - u^2/c^2} \qquad (9.54)$$

Momentum of a particle according Eq. (9.46) is given by

$$p = \frac{m_0 u}{\sqrt{1 - u^2/c^2}}$$

or $$p^2 c^2 = \frac{m_0^2 u^2 c^2}{1 - u^2/c^2} \qquad (9.55)$$

Combining Eqs (9.54) and (9.55), we get

$$E^2 - p^2 c^2 = \frac{m_0^2 c^4}{1 - u^2/c^2} - \frac{m_0^2 u^2 c^2}{1 - u^2/c^2} = \frac{m_0^2 c^4 \left(1 - u^2/c^2\right)}{1 - u^2/c^2} = m_0^2 c^4$$

or $E = \sqrt{m_0^2 c^4 + p^2 c^2}$ (9.56)

This above equation shows that if a particle with $m_0 = 0$ exists, then its speed will be c and its total energy will be given by

$E = pc$ (9.57)

Indeed massless particles exist; as of 2014, gauge bosons, photons and gluons are examples!

9.13 Generalization of Newton's Second Law

The net force on a particle equals the time rate of change of its momentum, i.e.,

$$\vec{F} = \frac{d\vec{P}}{dt} = m \frac{d\vec{v}}{dt} + \vec{v} \frac{dm}{dt}$$ (9.58)

However, we know from Eq. (9.53) that

$K + m_0 c^2 = mc^2 = E$

The differentiation of this above equation with respect to time gives

$$\frac{dm}{dt} = \frac{1}{c^2} \frac{dE}{dt} = \frac{1}{c^2} \frac{d(K + m_0 c^2)}{dt} = \frac{1}{c^2} \frac{dK}{dt}$$

or $\frac{dm}{dt} = \frac{1}{c^2} \frac{(\vec{F}.\vec{dl})}{dt} = \frac{1}{c^2} \vec{F}.\frac{(\vec{dl})}{dt} = \frac{1}{c^2} \vec{F}.\vec{v}$ (9.59)

By substituting Eqs (9.59) into (9.58), we have

$$\vec{F} = m \frac{d\vec{v}}{dt} + \frac{\vec{v}(\vec{F}.\vec{v})}{c^2}$$ (9.60)

Thus, the general expression for acceleration $\vec{a}$ of the body defined by $\vec{a} = \frac{d\vec{v}}{dt}$ is given as

$$\vec{a} = \frac{d\vec{v}}{dt} = \frac{\vec{F}}{m} - \frac{\vec{v}}{mc^2}(\vec{F}.\vec{v})$$ (9.61)

This equation tells that in general, acceleration $\vec{a}$ is neither in the direction of $\vec{F}$ nor in the direction of $\vec{v}$!

In one special case, when $\vec{a}$, $\vec{F}$, and $\vec{v}$ are in the same direction, we can have

$$\vec{F} = \frac{d\vec{P}}{dt} = \frac{m_0\vec{a}}{\sqrt{1-v^2/c^2}} + \vec{v}\frac{d}{dt}\left(\frac{m_0}{\sqrt{1-v^2/c^2}}\right)$$

or $$\vec{F} = \frac{m_0}{\left(1-v^2/c^2\right)^{3/2}}\vec{a} \qquad (9.62)$$

The quantity $\dfrac{m_0}{\left(1-v^2/c^2\right)^{3/2}}$ is sometimes called longitudinal mass.

Similarly, in another special case, when $\vec{a}$, and $\vec{F}$ are in the same direction, but $\vec{v}$ is perpendicular to $\vec{F}$ (i.e., $\vec{F}.\vec{v}=0$), we have

$$F = m\frac{dv}{dt} + 0 = \frac{m_0}{\sqrt{1-v^2/c^2}}a \qquad (9.63)$$

The quantity $\dfrac{m_0}{\sqrt{1-v^2/c^2}}$ is sometimes called transversal mass.

Questions

9.1   What are the necessities of frame reference? Distinguish between inertial frame of reference and non-inertial frame of reference.

9.2   Describe with necessary theory the Michelson–Morley experiment. What are its inferences?

9.3   Derive the Lorentz transformation equations.

9.4   Derive the Galilean transformation equations from the Lorentz transformation equations.

9.5   Mention a few consequences of the Lorentz transformation equations.

9.6   List three ways our day-to-day lives would change if the speed of light were only 60 m/s.

9.7   What do you mean by relativity of simultaneity? Explain

9.8   How is it possible that photons of light, which have zero mass, have momentum?

9.9   Two identical clocks are in the same house, one upstairs in a bedroom and the other downstairs in the kitchen. Which clock runs slower? Explain

9.10 Show the variation of momentum of a particle with respect to velocity mathematically.

9.11 The speed of an object along the x-axis is $u(= dx / dt)$ and $u'(= dx' / dt')$ in S and S' respectively. The two reference frames S and S' have a relative velocity v along the x-axis. Derive the relation connecting u and u'.

9.12 Explain how the classical concept of momentum was generalized in the theory of relativity.

9.13 Derive the expression for the variation of mass with speed.

9.14 Explain how the special theory of relativity gives birth to the idea of massless particles.

9.15 Derive the mass–energy relation.

9.16 Derive the classical expression for kinetic energy of a moving body from its relativistic expression.

9.17 The theory of relativity sets an upper limit on the speed that a particle can have. Are there also limits on the energy and momentum of a particle? Explain.

9.18 Show that the addition of any velocity to c leaves the velocity as c only.

9.19 Using the special theory of relativity concepts, show that the rest mass of a photon is zero.

9.20 The speed of light in water is 2.3×10^8 m/s. Suppose an electron is moving through water at 2.4×10^8 m/s. Does that violate the principle of relativity? Explain.

Problems

9.1 The length of a meter stick is measured to be 0.3048 m by an observer at rest. The meter stick is moving at a high speed parallel to its long axis. Find the speed of the meter stick. [Ans 0.952 c]

9.2 A moving rod is observed to have a length of 2.00 m and to be oriented at an angle of 30.0° with respect to the direction of motion. If the speed of the rod is 0.995c, calculate the proper length of the rod. [Ans 17.34 m]

9.3 Two rods of proper length ℓ_0 are moving towards each other lengthwise with velocity v. Calculate the length of one rod as measured by an observer on the other rod.

$$[\text{Ans} \quad \ell = \ell_0 \frac{1 - v^2/c^2}{1 + v^2/c^2}]$$

9.4 When at rest, a box is a cube with sides 30 cm. What is its volume when it moves with a speed of 0.9c parallel to one of its sides. [Ans 11769 cm³]

9.5 A proton moves at a speed of 0.950c. Calculate its total energy. [Ans 3.00 Gev]

9.6 The total energy of a proton is thrice its rest energy. Find the momentum of the proton in MeV/c units. [Ans 2.66×10^3 MeV/c]

9.7 The rest energy of an electron is 0.511 MeV and that of a proton is 938 MeV. Assume that both particles have kinetic energies of 4.00 MeV. Find the speed of (a) the electron and (b) the proton. [Ans 0.994 c, 0.092 c]

9.8   An elementary particle (imaginary) is observed to move at 0.96c and have a lifetime of 2.0×10^{-8} s. Calculate the lifetime of the particle in its own reference frame?

[Ans 0.56×10^{-8} s]

9.9   An observer on the earth observes a spacecraft and finds that, between 5.00 pm and 6.00 pm according to his clock, 3610 s elapse on the spacecraft's clock. What is the spacecraft's speed relative to the earth?   [Ans   0.074 c]

9.10   A spaceship is measured to be 120.0 m long at rest relative to an observer. If this spaceship now flies with a speed of 0.99c, what length does the observer on the earth measure? What length does the observer on the earth measure if speed of the spaceship is 3×10^{7} m/s?   [Ans   17 m, 119.4 m]

9.11   How fast must a meter stick be travelling relative to your reference frame for you to observe a 2% contraction?   [Ans   0.20 c]

9.12   An object is moving with relativistic speed and its has mass equals to two times its rest mass. Calculate its velocity.   [Ans   0.867 c]

9.13   A spaceship moving away from the earth at 0.850 c fires a space probe in the same direction as its motion at 0.700 c relative to the spaceship. What is the probe's velocity relative to the earth?   [Ans   0 872 c]

9.14   An electron in a typical television picture tube moves with a kinetic energy 0.017 MeV Find its speed.   [Ans   0.250 c]

9.15   Two protons moving in opposite direction with equal speed suffer a head on elastic collision and as a result, a neutral pion of mass 2.40×10^{-28} kg is produced. Calculate the initial speed of the protons if after collision, the three particles are at rest.

[Ans   0.360 c]

Multiple Choice Questions

1.   Which of the following are beyond the scope of the special theory of relativity?

(i)   Effect of speed on mass   (ii)   Effect of gravity on velocity

(iii)   Effect of speed on length, and   (iv)   Effect of speed on time

2.   In special theory of relativity, which type of reference frame is chosen?

(i)   Inertial frame of reference   (ii)   Non-inertial frame of reference

(iii)   Both of the above   (iv)   None of the above

3.   Galilean transformation can be obtained from the Lorentz transformation by setting

(i) $v = c$ (ii) $v < c$

(iii) $v > c$ (iv) $v << c$; $v =$ speed of object

4.   In special theory of relativity, mass, length and time are invariant.

(i)   True   (ii)   False

5.   The Michelson–Morley experiment was conducted to

(i)   Disapprove Galilean transformations

(ii)   Test the invariance nature of time

(iii) Know the shape of interference fringes during rotation of Earth

(iv) Test the presence of ether medium

6. Which of the following is not an Einstein's principle of relativity?

(i) All the laws of physics must be the same in all frames of reference

(ii) The speed of light in vacuum has the same value, in all frames of reference

(iii) The mass–energy relation is $E = mc^2$

(iv) All of the above

7. Which of the following is not a conclusion of the Michelson–Morley experiment?

(i) Constancy of the speed of light

(ii) Absence of ether medium

(iii) Variation of time with speed

(iv) Non-existence of an absolute frame of reference

8. Which of the following relations is incorrect?

(i) $x' = \gamma(x - vt)$

(ii) $ct' = \gamma(ct + \beta x)$

(iii) $m = \dfrac{m_0}{\sqrt{1 - v^2/c^2}}$

(iv) Total energy = kinetic energy + $m_0 c^2$. Symbols have their usual meanings.

9. What is the minimum possible values of $\gamma = \dfrac{1}{\sqrt{1 - v^2/c^2}}$?

(i) 0 (ii) 1

(iii) 0.5 (iv) ∞

10. A rod's length is greatest when it is

(i) At rest relative to the observer (ii) In motion relative to the observer

(iii) Independent of the observer (iv) None of above

11. A clock is measured to go at its slowest rate when it is

(i) At rest relative to the observer (ii) In motion relative to the observer

(iii) Independent of the observer (iv) None of the above

12. Weight of a body is measured to be minimum when it is

(i) At rest relative to the observer (ii) In motion relative to the observer

(iii) Independent of the observer (iv) None of above

13. Length contraction effect in special theory of relativity is

(i) The contraction of the body in all directions

(ii) Contraction of the rod in the direction perpendicular to the direction of motion

(iii) Contraction of the body in the direction of motion

(iv) None of the above

14. A clock is measured to go at its fastest rate when it is at

(i)   Rest relative to the observer   (ii) Motion towards the observer

(iii) Motion away from the observer   (iv) Independent of the relative motion

15. Which of the following is not a consequence of the special theory of relativity?

(i)   Existence of massless particle

(ii) Change of mass of a body with relative speed between body and observer

(iii) Change of length of a rod with relative speed between rod and observer

(iv) Change of color of light with speed of observer

16. Temperature of a body increases. Then, according to the special theory of relativity, its mass

(i)   Increases   (ii) Decreases

(iii) Remains the same   (iv) None of the above

17. You are moving in the direction of the light beam with velocity $0.9\,c$. You will measure the speed of light to be

(i) $0.1\,c$ (ii) $1.9\,c$

(iii) c (iv) $0.44\,c$

18. Two events are simultaneous for two observers if

(i)   Two observers are in the same reference frame

(ii) Two observers are in different reference frames

(iii) Two observers have constant relative velocity

(iv) Simultaneity of events is independent of observers

19. The speed of a non-zero rest mass body can be

(i)   More than the light speed in vacuum   (ii) Equal to the light speed in vacuum

(iii) Less than the light speed in vacuum   (iv) All of the above

20. Energy can be possessed by a body if it has

(i)   Non-zero rest mass   (ii) Zero rest mass

(iii) Momentum   (iv) All the above

Answers

1 (ii)	2 (i)	3 (iv)	4 (ii)	5 (iv)	6 (iv)	7 (iii)	8 (ii)
9 (ii)	10 (i)	11 (ii)	12 (i)	13 (iii)	14 (i)	15 (iv)	16 (i)
17 (iii)	18 (i)	19 (iii)	20 (iv)				

10 Architectural Acoustics

10.1 Introduction

Sound is a disturbance in an elastic medium resulting in an audible sensation. From the acoustics point of view, sound and noise constitute the same phenomenon of atmospheric pressure fluctuations about the mean atmospheric pressure; the differentiation is greatly subjective. What is sound to one person can very well be noise to somebody else. It should be mentioned that the field of sound in a real sense is so complicated that it is not open to exact mathematical treatment. Actually, the term "acoustics" is derived from the Greek verb "$\alpha\kappa o\acute{\gamma}\nu\varepsilon\iota\nu$" (akúɪn) which means to hear. The branch of science which deals with the planning of a building or hall with a view to provide the best audible sound to the audience is called acoustics of a building or architectural acoustics. The recognition of noise as a serious health hazard is a development of modern times.

10.2 Basic Requirements of an Acoustically Good Hall

We all know that a concert hall, theatre, or lecture room may have good or poor "acoustics". As far as speech in these rooms is concerned, it is relatively simple to make some sort of judgement on their quality by rating the ease with which the spoken word is understood. An everyday experience is that living rooms, offices, restaurants and all kinds of rooms for work can be acoustically satisfactory or unsatisfactory.

In 1895, the original Fogg Art Museum Lecture Hall (now known as Hunt Hall) of Harvard University was dedicated as a memorial to William Hayes Fogg. It was soon discovered that the acoustics of the main lecture hall was so bad that a good orator could hardly make his words intelligible to the audience and the space had to be abandoned as unusable. The then Harvard's president, Charles W. Eliot, turned to Wallace Clement

Sabine, a 27-year-old assistant professor of physics, for help in resolving the difficulty. Sabine undertook the challenge despite the discouragement of his senior faculty colleagues who considered it beyond solution. The acoustical difficulties of the lecture hall are best described in Sabine's own words: "The rate of absorption was so small that a word spoken in an ordinary tone of voice was audible for five and a half seconds afterwards. During this time, even a very deliberate speaker would have uttered twelve or fifteen successive syllables". It was he who first scientifically tackled the problem of satisfactory speech and music in a hall. He is often credited with transforming the understanding of acoustics from a mysterious art to a respected discipline, and is considered by many to be the "father of modern architectural acoustics". According to Sabine, the following are a few basic requirements of an acoustically good hall.

i.   The sound heard must be sufficiently loud in every part of the hall and no echoes should be present.

ii.   The total quality of the speech and music must be unchanged, i.e., the relative intensities of the several components of a complex sound must be maintained.

iii. For the sake of clarity, successive syllables spoken must be clear and distinct, i.e., there must be no confusion due to overlapping of syllables.

iv.   Reverberation should be quite proper – neither too large nor too small. The reverberation time should be 1 to 2 seconds for music and 0.5 to 1 second for speech.

v.   There should be no concentration of sound in any part of the hall.

vi.   The boundaries should be sufficiently sound proof to exclude extraneous noise from outside.

vii. There should be no Echelon effect.

viii. There should be no resonance within the building.

10.3 Reverberation and Reverberation Time

Whenever a sound pulse is produced in an auditorium, it is partially reflected from the walls, ceiling and other materials in the room. The waves received by the listener are: (i) direct waves from the source and (ii) multiple reflected waves of decreasing intensity from various surfaces. The quality of the note received by the listener will be the combined effect of these two sets of waves. There is also a time gap between the direct wave received by the listener and the waves received by successive reflection. Due to this, the sound waves persist for sometime even after the source has stopped emitting sound. This persistence of sound in a hall after the source has stopped is termed as reverberation. The duration for which the audible sound persists after source is cut off is called reverberation time. It is the time taken for the sound to fall below the minimum audibility measured from the instant when the source stopped sounding. Sabine defined the standard reverberation time quantitatively as the time taken by sound to fall to one-millionth of its power which is equivalent to a sound level of 60 dB just before the source is cut off. The reverberation time will depend on the volume and shape of the auditorium, the nature and area of the reflecting materials in the hall. In a good auditorium, it is necessary to keep the reverberation time to an optimum value. Typical values of reverberation times vary from about 0.3 s (living rooms) up to 10 s (empty reverberation chambers). Most large halls have reverberation times between 0.2 s to 0.7 s.

10.3.1 Sabine's formula for reverberation time

Sabine developed the reverberation time formula to express the rise and fall of sound in an auditorium. The main assumptions are:

i.   The density of the sound wave (sound energy per unit volume) is uniform throughout the hall.

ii. The decrease in energy is only due to the absorption by the materials present in the room and also due to transmission through the walls, windows and ventilators. Both these factors are included in the term "absorption" of energy.

Suppose a point source is emitting sound continuously in all directions. Consider the incidence of sound energy on a small element ds of a plane wall AB as shown in Fig. 10.1.

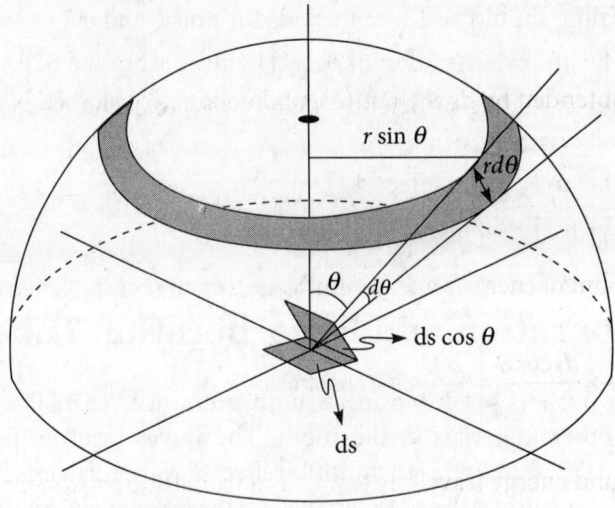

Figure 10.1 | Derivation of equation (10.1)

The sound energy from all points above the plane of ds is incident on an elemental area ds. Therefore, a hemisphere of radius r is imagined with ds at its centre. If dr is the thickness of the hemisphere, then volume of an elemental portion of the strip shown in Fig. 10.1 is

$$2\pi r \sin\theta \times r d\theta \times dr, \ 0 \le \theta \le \pi/2 \tag{10.1}$$

where

$2\pi r \sin \theta$ = elemental length of the strip

$rd\theta$ = width of the strip

dr = thickness of the strip

If ρ is the density of the sound energy in the room at any instant of time, then the amount of energy contained in this strip at that instant is

$$2\pi\rho r^2 \sin \theta d\theta dr \qquad (10.2)$$

This sound energy is travelling through the element equally in all directions. Since the solid angle subtended around a point is 4π, the energy travelling per unit solid angle along any direction is

$$\frac{2\pi\rho r^2 \sin \theta d\theta dr}{4\pi} = \frac{\rho r^2 \sin \theta d\theta dr}{2} \qquad (10.3)$$

The solid angle subtended by ds at the elemental volume considered is

$$\frac{ds \cos \theta}{r^2} \qquad (10.4)$$

Therefore, the amount of energy travelling towards ds from the circular strip will be given by

$$\frac{\rho r^2 \sin \theta d\theta dr}{2} \times \frac{ds \cos \theta}{r^2} = \frac{\rho ds}{2} \sin \theta \cos \theta d\theta dr \qquad (10.5)$$

The amount of sound energy travelling towards ds from the hemisphere is thus obtained by

$$\frac{\rho ds}{2} dr \int_0^{\pi/2} \sin \theta \cos \theta d\theta = \frac{\rho ds}{4} dr \qquad (10.6)$$

If c is the distance travelled by the sound in one second (i.e., c = speed of sound), then the amount of sound energy incident on ds per unit time will be obtained by replacing dr by c as

$$\frac{\rho ds}{4} c \qquad (10.7)$$

This expression gives the amount of sound energy incident on ds per unit time from all directions.

If α be the absorption coefficient (α = energy absorbed/energy incident) of the portion ds which is a part of AB, then the energy absorbed by ds per unit time will be

$$\frac{\rho ds}{4} c \times \alpha = \frac{\rho c}{4} \alpha ds \qquad (10.8)$$

If different portions of the wall AB have different absorption coefficients, then the sound energy absorbed by AB per unit time will be

$$\frac{\rho c}{4} \sum \alpha_i ds_i = \frac{\rho c A}{4} \qquad (10.9)$$

where $A = \sum \alpha_i ds_i$ is the total absorption on all the surfaces on which sound falls. Here α_t and ds_t are the absorption coefficient and the surface area of the ith surface respectively. Expression (10.9) gives the rate of absorption of the sound energy. If P is the power of the sound source, i.e., the rate at which the sound source emits sound energy into the room, then the net growth of sound energy in the room per second will be

$$P - \frac{\rho c A}{4} \qquad (10.10)$$

In Eq. (10.10), ρ is the sound energy density at any instant of time and so at that instant, the sound energy content of the room will be ρV. Therefore, the net rate of growth of sound energy will be

$$\frac{d\rho V}{dt} = V \frac{d\rho}{dt} \qquad (10.11)$$

Taking into account the rate of emission and rate of absorption at any instant of time we have

$$V \frac{d\rho}{dt} = P - \frac{\rho c A}{4} \qquad (10.12)$$

The sound content of the room cannot increase indefinitely. When the steady state is attained, $V \frac{d\rho}{dt} = 0$. If the steady-state energy density is represented by ρ_0, then its value as obtained from Eq. (10.12) is given as

$$\rho_0 = \frac{4P}{cA} \qquad (10.13)$$

From Eq. (10.12), we have

$$V \frac{d\rho}{dt} = P - \frac{\rho c A}{4}$$

or $\qquad \dfrac{d\rho}{dt} = \dfrac{P}{V} - \dfrac{\rho c A}{4V}$ $\hfill (10.14)$

Putting $\beta = \dfrac{cA}{4V}$ into this equation, we have

$$\frac{d\rho}{dt} + \rho\beta = \frac{4P}{cA}\beta$$

or $\qquad e^{\beta t} \dfrac{d\rho}{dt} + e^{\beta t} \rho\beta = e^{\beta t} \dfrac{4P}{cA}\beta$ (Multiplying both sides by $e^{\beta t}$)

or $\qquad \dfrac{de^{\beta t}\rho}{dt} = \dfrac{4P}{cA}\beta e^{\beta t}$

or $\qquad e^{\beta t}\rho = \displaystyle\int \dfrac{4P}{cA}\beta e^{\beta t}\, dt = \dfrac{4P}{cA}e^{\beta t} + K$ $\hfill (10.15)$

Using the boundary conditions, we can find the value of the integration constant, K.

Growth of the sound energy density

If t is measured from the instant the sound source starts emitting sound, then at $t = 0$, $\rho = 0$. Applying this condition to Eq. (10.15), we have

$$0 = \frac{4P}{cA} + K$$

or $\qquad K = -\dfrac{4P}{cA}$

or $\qquad \rho = \dfrac{4P}{cA}\left(1.0 - e^{-\beta t}\right) = \rho_0 \left(1.0 - e^{-\beta t}\right)$ $\hfill (10.16)$

Equation (10.16) shows the growth of sound energy density with time. After infinite time, $\rho = \rho_0$. The growth is of exponential nature as shown in Fig. 10.2.

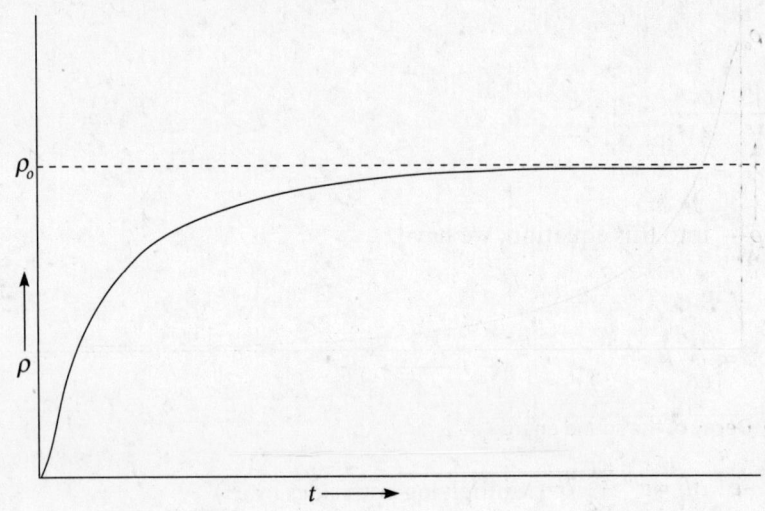

Figure 10.2 | Growth of the sound energy density

Decay of sound energy density

Let the source be cut off after steady state is reached. At this instant,

$t = 0, P = 0$ and $\rho = \rho_0$

Putting these boundary conditions into Eq. (10.15), we have

$\rho_0 = K$

$\rho = \rho_0 e^{-\beta t}$ since $P = 0$ (10.17)

Equation (10.17) shows that the decay of the sound energy density with time after the source is cut off is exponential in nature. The plotting of this equation gives the exponential curves shown in Fig. 10.3.

Standard reverberation time

The standard time of reverberation T is defined as the time taken for the sound energy density to fall to one-millionth of its maximum audible value. Hence, to calculate T, we set

$$\frac{\rho}{\rho_0} = 10^{-6} = e^{-\beta T}$$

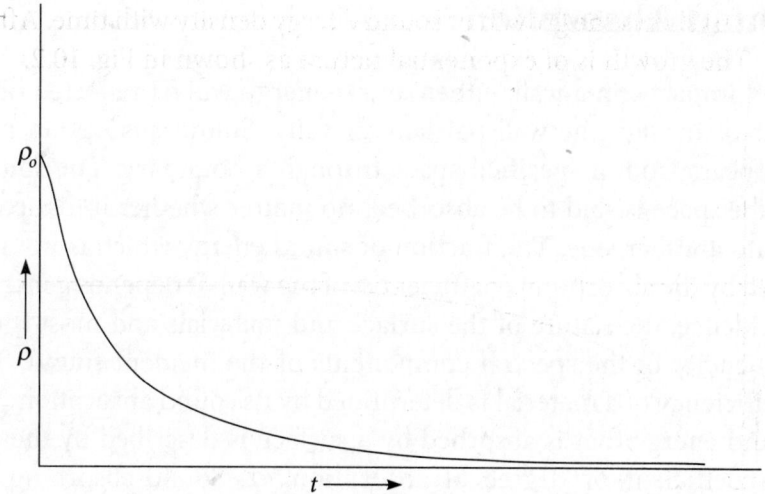

Figure 10.3 Decay of the sound energy density

or $\quad T = \dfrac{6\ell n 10}{\beta} = \dfrac{24 V \ell n 10}{cA}$ $\hspace{4cm}$ (10.18)

Putting the value of the speed of sound in air medium as 340 m/s into Eq. (10.18), we have

$$T = \dfrac{0.165 V}{\sum \alpha_i ds_i}$$ $\hspace{4cm}$ (10.19)

This equation is in good agreement with the experimental values obtained by Sabine. Sabine's reverberation time is still the best known and most important quantity in architectural acoustics.

Example 10.1

An auditorium has a volume of 30600 m³. It is required to have a reverberation time of 0.8 s. What should be the total absorption in the auditorium?

Solution

Data given are $V = 30600$ m³, $T = 0.8$ s

$$\text{The total absorption} = \sum \alpha_i ds_i = \frac{0.165 V}{T} = \frac{0.165 \times 3060}{0.8} = 631 \text{ O.U.W.}$$

10.4 Sound Absorption

When sound impacts on a wall, either sound energy will be reflected or absorbed or transmitted through the wall partially or fully. Sound absorption means that sound disappears from a specified space through a boundary. The sound energy that leaves the space is said to be absorbed, no matter whether it dissipates or just transmits into another side. The fraction of sound energy which is not reflected is characterised by the absorption coefficient α of the wall. It depends generally on the angle of incidence, the nature of the surface and materials and most importantly, on the frequencies of the spectral components of the incident sound. The sound absorbing efficiency of a material is determined by its sound absorption coefficient, α. The sound energy that is absorbed by a surface is described by the parameter absorption coefficient or degree of absorption, α. Sound-absorbing materials, carpets, acoustical tiles, and other specially fabricated absorbing products can absorb appreciable amounts of sound energy. Rooms whose surfaces absorb all the sound energy ($\alpha = 1$) are called anechoic rooms or dead rooms. Rooms opposite in nature ($\alpha = 0$) are called reverberant rooms, live rooms or hard rooms.

The coefficient of absorption α of a material is defined as the ratio of the sound energy absorbed by the surface to that of the total incident sound energy on the surface. Mathematically, it is given by

$$\alpha = \frac{\text{Sound energy absorbed}}{\text{Sound energy incident}} = \frac{\text{Sound energy incident-Sound energy reflected}}{\text{Sound energy incident}}$$

or $\quad \alpha = 1 - \dfrac{\text{Sound energy reflected}}{\text{Sound energy incident}} = 1 - R^2$ \hfill (10.20)

The ratio of the sound energy reflected and the sound energy incident is called coefficient of reflection or degree of reflection R. For a wall with zero reflectivity ($R = 0$), the absorption coefficient has its maximum value 1. Then, the wall is said to be totally absorbent. If $R = 1$ and there is in-phase reflection, the wall is said to be "rigid" or "hard". In the case of $R = -1$ and there is out-of-phase reflection, we speak of a "soft" wall. In both cases, there is no sound absorption ($\alpha = 0$). It may vary from 0 (no absorption) to 1 (complete absorption).

The absorption of sound by a material depends upon its exposed surface area. As sound leaves through open windows without returning, an open window behaves as a perfect absorber of sound and hence, the unit area of an open window is taken as a standard unit of absorption. The absorption of all other substances can be measured in this unit. Thus, the absorption coefficient of a material is defined as the ratio of the sound energy absorbed by a certain area of the surface to that of an open window of the same area. In other words, the absorption coefficient of a surface is defined as the reciprocal of its area which absorbs the same sound energy as absorbed at a unit area (1.0 m²) of an open window. The absorption coefficient is measured in open window unit written as O.W.U. or Sabin in honour of W. C. Sabine (1867–1919). For example, if 5.0 m² of a surface absorbs the same amount of sound energy as absorbed by 1.0 m² of the open window, then the absorption coefficient of the surface is $\dfrac{1.0}{5.0} = 0.2$ Sabin. 1.0 m² window has an absorption coefficient of 1 sabin.

10.4.1 Room averaged sound absorption coefficient

No rooms have the same walls on all sides – there are different types of walls having different absorption coefficients. The room averaged sound absorption coefficient can be used for different types of materials and areas of walls averaged together. It is defined as

$$\bar{\alpha} = \frac{\displaystyle\sum_{i=1}^{n} \alpha_i s_1}{\displaystyle\sum_{i=1}^{n} s_1} = \frac{\displaystyle\sum_{i=1}^{n} \alpha_i s_1}{S} \tag{10.21}$$

Here,

α_t = absorption coefficient of the ith surface

s_t = area of the ith surface

$S = \displaystyle\sum_{i=1}^{n} s_1$ = total surface area of the room

n = total number of absorptive surfaces in the room

Example 10.2

A conference hall has a volume of 30000 m³. It has a reverberation time of 1.2 s and the total sound absorbing surface is 6000 m². Calculate the average absorbing power of the surfaces.

Solution

Data given are $V = 30000$ m³, $T = 1.2$s, $S = 6000$ m²

The average absorbing power of the surfaces

$$\frac{\sum \alpha_i ds_i}{S} = \frac{0.165V}{TS} = \frac{0.165 \times 30000}{1.2 \times 6000} = 0.688$$

Example 10.3

The wall areas of a room are 3000 m², the floor area is 1500 m² and the ceiling area is 1500 m². The average sound absorption coefficient for the walls is 0·03, for the ceiling is 0·8, and for the floor is 0·06. Calculate the average sound absorption coefficient.

Solution

Data given are $s_1 = 3000$ m², $\alpha_1 = 0.03$; $s_2 = 1500$ m², $\alpha_2 = 0.08$; $s_3 = 1500$ m², $\alpha_3 = 0.06$.

The average sound absorption coefficient $\bar{\alpha} = \dfrac{\sum\limits_{i=1}^{n} \alpha_i s_i}{\sum\limits_{i=1}^{n} s_i} = \dfrac{\alpha_1 s_1 + \alpha_2 s_2 + \alpha_3 s_3}{s_1 + s_2 + s_3}$

$$= \frac{3000 \times 0.03 + 1500 \times 0.08 + 1500 \times 0.06}{3000 + 1500 + 1500} = 0.05$$

10.4.2 Measurement of absorption coefficient

The following methods are used to measure the of absorption coefficient.

i.   **Method 1**

In this method, we can calculate the coefficient of absorption of a sample material. This method is based on the determination of standard times of reverberation in the room without and with a standard large sample of the material inside the chamber. If T_1 is the reverberation time without the materials and T_2 is the reverberation time with the material having absorption coefficient a and surface area S, then by applying Sabine's formula, we have

$$T_1 = \frac{0.165V}{\sum \alpha_i ds_i} \text{ and } T_2 = \frac{0.165V}{\sum \alpha_i ds_i + \alpha S}$$

or $\dfrac{1}{T_2} - \dfrac{1}{T_1} = \dfrac{\sum \alpha_i ds_i + \alpha S}{0.165V} - \dfrac{\sum \alpha_i ds_i}{0.165V} = \dfrac{\alpha S}{0.165V}$

or $\quad \alpha = \dfrac{0.165V}{S} \left(\dfrac{1}{T_2} - \dfrac{1}{T_1} \right)$ \hfill (10.22)

Hence by measuring the reverberation times T_1, T_2, the surface area of the sample material S and volume of the hall, we can calculate the absorption coefficient of the sample material.

ii. **Method 2**

In this method, we can calculate the average coefficient of absorption of a hall. This method consists of finding the times of decay of the steady energy density of two sources of power outputs P_1 and P_2, respectively, to the same audibility level. Let t_1 and t_2 be the time duration in which the sound energy density levels from two sources of power outputs P_1 and P_2 attain the same value ρ'. From the equation of the decay of energy density, we have the following equation for one source

$$\rho' = \rho_{01}e^{-\beta t_1} \text{ with } \rho_{01} = \frac{4P_1}{cA} \text{ and } \beta = \frac{cA}{4V}$$

and for the other source

$$\rho' = \rho_{02}e^{-\beta t_2} \text{ with } \rho_{02} = \frac{4P_2}{cA} \text{ and } \beta = \frac{cA}{4V}$$

From these relations, we have

$$\frac{P_1}{P_2} = \frac{\rho_{01}}{\rho_{02}} = \frac{e^{\beta t_1}}{e^{\beta t_2}} = e^{\beta(t_1 - t_2)}$$

or $\quad \dfrac{cA}{4V}(t_1 - t_2) = \ell n \dfrac{P_1}{P_2}$

or $\quad A = \bar{\alpha}S = \dfrac{4V}{c(t_1 - t_2)} \ell n \dfrac{P_1}{P_2}$

or $\quad \bar{\alpha} = \dfrac{4V}{cS(t_1 - t_2)} \ell n \dfrac{P_1}{P_2}$ $\hfill (10.23)$

Hence, by measuring the times t_1, t_2, surface area of the sample material S, volume of the hall and knowing the output powers of the two sources, we can calculate the average absorption coefficient of the room.

10.5 Factors Affecting the Acoustics of Buildings

Reverberation is the most important single factor that affects the acoustics of a room or a hall. Besides reverberation, factors that affect the acoustics of a room or a hall are loudness, focussing, echelon effect, extraneous noise, resonance and so on.

i. **Loudness**

The speech of a person in a hall can be heard by an audience consisting of about 1000 persons. However, to ensure uniform distribution of the sound intensity in the hall, electrically amplified loud speakers are used. These speakers are kept at different places in the auditorium and are located generally a little higher than the speaker's head. Amplifiers, however, make low frequency tones more prominent and hence, the amplification has to be kept low. The presence of low artificial ceilings improves the audibility in general.

ii. **Focussing**

As shown in Fig. 10.4, the presence of cylindrical or spherical surfaces on the walls of the ceiling gives rise to undesirable focussing.

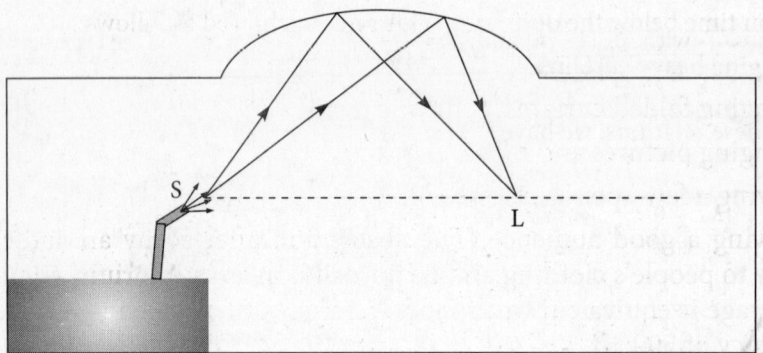

Figure 10.4 | Focussing of reflected sound waves

In Fig. 10.4, the listener at L receives sound from the speaker along the direct path SL. The observer also receives the sound waves after reflection from the ceiling. The direct and reflected sound wave may interfere constructively (large sound) or destructively (low sound). Again the direct and the reflected sound waves may form a stationary wave pattern. These cause uneven distribution of sound intensity in the hall.

iii. **Echelon effect**

If there is a regular structure similar to a flight of stairs or a set of railings in the hall, the sound produced in front of such a structure may produce a musical note due to the regular successive echoes of sound reaching the observer. Such an effect is called the echelon effect. To avoid echelon effect, staircases are covered with soft carpets so that instead of reflection, sound is absorbed by them.

iv. **Extraneous noise**

Extraneous noise may be due to (a) sound received from outside the room and (b) the sound produced by fans, AC, and so on inside the auditorium. External sound cannot be completely eliminated but can be minimised. Proper attention must also be paid to see that fans run at the optimum permissible speed and there is optimum rate of air circulation in the room. Air conditioning pipes should be covered with cork and insulated acoustically from the main building.

v. **Resonance**

The acoustics of a building may also be affected by resonance. If there is resonance for any audio frequency note, the intensity of the note will be entirely different from the intensity desired. In halls of large size, the resonance frequency is much below the audible limit.

10.5.1 Requisites for good acoustics

The reverberation of sound in an auditorium is due to multiple reflections taking place at various surfaces present within the auditorium. The acoustics of an auditorium can be improved by using surfaces with high absorption coefficient. This will reduce the reverberation time below the optimum value and is achieved as follows:

i. By hanging heavy curtains.

ii. By hanging folded curtains.

iii. By hanging pictures and maps.

iv. By having a few open windows.

v. By having a good audience. The absorption affected by an audience is due mainly to people's clothing and its porosity. In an auditorium, each person on an average is equivalent to about 0.75 m^2 area of an open window at a sound frequency of 500 Hz.

vi. Curved walls and corners bounded by two walls should be avoided as these cause (a) concentration of sound and (b) dead spaces. If domes or other concave surfaces are desired, they must be treated with absorptive materials to reduce unwanted reflections.

vii. Upholstered seats should be provided so that the absorption is approximately the same with or without the audience.

viii. The walls and the ceiling should be covered with materials having high absorption coefficient, i.e., with perforated cardboards, felts, fibers, carpets and the like.

ix. The walls should be engraved and made rough with decorative materials to increase absorption.

x. If possible, noisy equipment and activities should be placed in remote areas.

10.6 Decibel Scale

A linear scale based on the square of the sound pressure would require 10 unit divisions to cover the range of human experience; however, the human brain is not organised to encompass such a range. The remarkable dynamic range of the ear suggests that some kind of compressed scale should be used. A scale suitable for expressing the square of the sound pressure in units best matched to subjective response is logarithmic rather than linear. Thus, the bel was introduced which is the logarithm of the ratio of two quantities, one of which is a reference quantity. To avoid a scale which is too compressed over the sensitivity range of the ear, a factor of 10 is introduced, giving rise to the decibel. For sound pressure level measurements, a reference value of $2 \times 10^{-5}\,\mathrm{Nm^{-2}}$ is chosen internationally for air. This is the threshold of hearing for a typical healthy young person. The sound pressure level (SPL) in decibels for any sound for which the pressure is known, is given internationally by the following expression:

$$L_p = 10 \times \log\left(\frac{p_{\mathrm{rms}}}{p_0}\right)^2 = 20\log\frac{p_{\mathrm{rms}}}{p_0} \tag{10.24}$$

where

L_p = sound pressure level (SPL) in decibels (dB)

p_{rms} = root mean square of the acoustic pressure fluctuations

= measured acoustic pressure

p_0 =   reference sound pressure in air, usually taken to be $2 \times 10^{-5}\,\mathrm{Nm^{-2}}$ internationally; for other media, it is taken as $10^{-6}\,\mathrm{Nm^{-2}}$. The measured acoustic pressure will be obtained from Eq. (10.24) as

$$p = p_0 \times 10^{\frac{L_p}{20}} = 2 \times 10^{-5} \times 10^{\frac{L_p}{20}}\,\mathrm{Nm^{-2}} \tag{10.25}$$

The sound intensity and the root mean square of the acoustic pressure fluctuations p_{rms} in a medium are related by the relationship

$$I = \frac{p_{\mathrm{rms}}^2}{\rho v} \tag{10.26}$$

where

ρ = density of the medium

v = speed of sound in the medium

With the help of Eqs (10.24) and (10.26), the sound intensity level (SIL) in decibels for any sound for which the intensity is known, is given by the following expression

$$L_I = 10\log\frac{I}{I_0} \tag{10.27}$$

where

I_0 = reference sound intensity, usually taken to be $10^{-12}\,Wm^{-2}$ internationally.

The measured acoustic intensity will be obtained from Eq. (10.27) as

$$I = I_0 \times 10^{\frac{L_I}{10}} = 10^{-12} \times 10^{\frac{L_I}{10}}\,Wm^{-2} \tag{10.28}$$

Similarly, from Eq. (10.27), we can have an expression for sound power level as

$$L_p = 10 \log \frac{P}{P_0} \tag{10.29}$$

where

P_0 = reference sound power, usually taken to be $10^{-12}\,W$ internationally.

It is to be noted that sound levels are not additive in nature.

Example 10.4

Calculate, in air, the acoustic pressure of a plane acoustic wave of pressure level of 150 decibels with reference to the standard acoustic pressure level of $2 \times 10^{-5}\,Nm^{-2}$.

Solution

Data given are $p = 150\,db$, $p_0 = 2 \times 10^{-5}\,Nm^{-2}$
The acoustic pressure

$$p = 2 \times 10^{-5} \times 10^{\frac{L_p}{20}}\,Nm^{-2} = 2 \times 10^{-5} \times 10^{\frac{150}{20}}\,Nm^{-2} = 632.46\,Nm^{-2}$$

Example 10.5

Calculate the increase in sound pressure level when the sound pressure is doubled.

Solution

Data given is $p_2 = 2 \times p_1$

$$L_{p1} = 20 \log \frac{p_1}{p_0} \quad \text{and} \quad L_{p2} = 20 \log \frac{p_2}{p_0}$$

The increase in sound pressure level is given as

$$L_{p2} - L_{p1} = 20 \log \frac{p_2}{p_0} - 20 \log \frac{p_1}{p_0} = 20 \times \log \frac{p_2}{p_1} = 20 \times \log 2 = 6.02 db$$

Example 10.6

Calculate the increase in acoustic intensity level when the sound intensity is doubled.

Solution

Data given are $I_2 = 2 \times I_1$

$$L_{I1} = 10 \log \frac{I_1}{I_0} \quad \text{and} \quad L_{I2} = 10 \log \frac{I_2}{I_0}$$

$$L_{I2} - L_{I1} = 10 \log \frac{I_2}{I_0} - 10 \log \frac{I_1}{I_0} = 10 \log \frac{I_2}{I_1} = 10 \log \frac{2I_1}{I_1} = 3.01 db$$

Example 10.7

Calculate the acoustic intensity of a plane acoustic wave in air of intensity level 120 decibels.

Solution

Data given are $L = 80 \, db$

$$I = I_0 \times 10^{\frac{L_I}{10}} = 10^{-12} \times 10^{\frac{L_I}{10}} \, \text{Wm}^{-2} = 1.0 \, \text{Wm}^{-1}$$

Example 10.8

Calculate the acoustic intensity level in each case, at a distance of 10 metres from a point source which radiates energy at the rate of 3.14 W.

Solution

Data given are $r = 10$ m, $P = 3.14$ W
Since the source is a point source, it radiates sound energy uniformly in all directions. It can be imagined to be at the centre of a sphere of radius 10 m.
Thus, on a $4\pi \times 10^2$ m^2 area, energy incidence is 3.14 W. Hence, we have

$$I = \frac{\text{Power}}{\text{area}} = \frac{3.14}{4\pi r^2} = \frac{1}{4 \times 10^2} \, \text{Wm}^{-2} = 2.5 \times 10^{-3} \, \text{Wm}^{-2}$$

$$L_I = 10 \log \frac{I}{I_0} = 10 \log \frac{2.5 \times 10^{-3}}{10^{-12}} \, bd = 93.98 db$$

Example 10.9

A ceiling fan operates at a sound intensity level of 75 db. If it is operated in a room with an existing sound intensity level of 70 db, what will be the resultant intensity level?

Solution

Data given are $L_{I1} = 60$ bd and $L_{I2} = 50$ bd

$$I_1 = 10^{-12} \times 10^{\frac{L_{I1}}{10}} \, \text{Wm}^{-2} = 10^{-12} \times 10^{\frac{60}{10}} \, \text{Wm}^{-2} = 10^{-6} \, \text{Wm}^{-2}$$

$$I_2 = 10^{-12} \times 10^{\frac{L_{I2}}{10}} \, \text{Wm}^{-2} = 10^{-12} \times 10^{\frac{50}{10}} \, \text{Wm}^{-2} = 10^{-7} \, \text{Wm}^{-2}$$

Total intensity $I = I_1 + I_2 = 10^{-6} \, \text{Wm}^{-2} + 10^{-7} \, \text{Wm}^{-2} = 1.1 \times 10^{-6} \, \text{Wm}^{-2}$

Resultant intensity level $L_I = 10 \log \dfrac{I}{I_0} = 10 \log \dfrac{1.1 \times 10^{-6}}{10^{-12}} bd = 60.41 db$

10.7 Acoustic Quieting

Acoustic quieting is the process of making machinery quieter by damping vibrations. Machinery vibrates, causing sound waves in air, hydroacoustic waves in water, and mechanical stresses in solid matter. Quieting is achieved by absorbing the vibrational energy or minimizing the source of the vibration. It may also be re-directed away from the observer. In the late 20th century, acoustic quieting techniques were developed to make submarines difficult to detect by sonar. Later on, this technology was adapted to many industries and products.

10.7.1 Aspects of acoustic quieting

A number of different aspects might be considered to achieve acoustic quietening. Each aspect of acoustics can be taken alone or in concert so that the end result is that the reception of noise by the observer is minimized. Acoustic quieting engineers consider the following factors.

i. Noise generation: by limiting the noise at its source.

ii. Sympathetic vibrations: by acoustic decoupling.

iii. Resonations: by acoustic damping or changing the size of the resonator.

iv. Sound transmissions: by reducing transmission using many methods depending upon whether the transmission is through air, liquid, or solid.

v. Sound reflections: by limiting the reflections using many methods, for example, by using acoustic absorption (deadening) materials, trapping the sound, opening a "window" to let sound out, and so on.

By analyzing the entire sequence of events, from the source to the observer, an acoustic engineer can provide many ways to quieten the machine or structure. The challenge is to do this in a practical and inexpensive way. The engineer might focus on changing materials, using a damping material, isolating the machine, running the machine in partial vacuum, or running the machine slower.

10.7.2 Methods of quieting

Different methods are adopted to achieve acoustic quietening in a machine. Each method works independently or in collaboration with other methods so that the reception of noise by the observer is minimum.

Mechanical acoustic quieting

i.   *Sound isolation*   Noise isolation refers to isolating noise source to prevent noise from transferring out of one area, using barriers like deadening materials to trap sound. In advanced countries, in home and office construction, many builders place sound-control barriers such as fiberglass batting in walls to deaden the transmission of noise through them.

ii. *Noise absorption*   Unwanted sounds, i.e., noise can be absorbed rather than reflected inside the room. This is useful when a listener needs to hear sounds only from a point source and not echo reflections. In a recording studio, sound proofing is accomplished with bass traps and anechoic chambers. Another example is the ubiquitous use of dropped ceilings and acoustical tiles in modern office buildings with high ceilings. Submarine hulls have special coatings that absorb noise.

iii. *Acoustic damping*   Vibration isolation prevents vibration from transferring beyond the device into another material. Damping mounts have been developed to offer vibrational resistance in many degrees of freedom. Recent advances in the field include shock isolators damping in at least six degrees of freedom. Acoustic damping also has uses in seismic shock protection of buildings. Motors and rotating shafts are commonly fitted with these mounts at the points where they have contact with the building or the chassis of a large machine.

iv. *Acoustic decoupling*   Certain parts of a machine can be built to keep the frame, chassis, or external shafts from receiving unwanted vibrations from a moving part.

v.   *Preventing stalls*   Whenever a machine undergoes an aerodynamic stall, it will abruptly vibrate. To prevent stalls first of all, we should avoid conditions that lead to an aerodynamic stall. Also, we should be able to recognize the warning signs of a stall so that precautionary measures can be taken in time. If already happened, appropriate recovery techniques should be applied as soon as possible.

vi. *Preventing cavitation*   When a machine is in contact with a fluid, it may be susceptible to cavitation. The sounds of gas bubbles imploding is the source of the noise. Ships and submarines that cavitate are more vulnerable to detection by sonar. The excessive pump head height is one the main causes of causes of cavitation. So the pump head height should be in the medium range. The Disconnection of suction strainers from the suction line of the hydraulic pump also decreases cavitation. The rags in the liquid may be responsible for cavitation. The proper diameter of inlet end plays role in control of cavitation. The cavitation can also be decreased by decreasing the liquid temperature.

vii. *Preventing water hammer*   In hydraulics and plumbing, water hammer is a known cause for the failure of piping systems. It also generates considerable noise. A valve that abruptly opens or shuts at the open end of the pipeline is the most common cause for water hammer. So the operator should close or open valves slowly to prevent water hammer. By decreasing the pressure head of the water supply we can foil the effects of water hammer.

viii. *Shock absorption*   Just as automobile shock absorbers are used to prevent mechanical shocks from reaching the passengers in a car, they are also important for quieting shocks.

ix. *Reduction of resonance*   At certain frequencies, certain material components of a machine/structure are susceptible to resonance. A machine that resonates would make tremendous noise. Resonance also occurs in enclosures, such as when echoes reverberate in musical instruments like the ocarina, organ pipe, and so on. Addition of removal of masses to the material components helps in reducing resonance. By stiffening the components and by changing the dampness of the medium, we can control the occurrence of resonance. (Chapter-1 may be seen)

x. *Material selection*   By choosing non-metallic components, the transmission of sound and vibrations can be minimized. Instead of using rigid brass fittings, a machine using flexible plastic pipe fittings may be much quieter. In some cases, air can be evacuated from a machine and sealed hermetically, the vacuum inside becoming a barrier to sound transmission.

Electronic quieting

i. *Electronic vibration control*   Electronics, sensors, and computers are now employed to reduce vibration. Using high speed logic, vibrations can be damped quickly and effectively by counteracting the motion before it exceeds a certain threshold.

ii. *Electronic noise control*   Electronics, sensors, and computers are also employed to cancel noise by using phase cancellation which matches the sound amplitude with a wave of the opposite polarity. This method employs the use of an active sound generating device, such as a loudspeaker to counteract ambient noise in an area.

iii. *Noise reduction*   In sound and video equipment, noise reduction is the process of removing noise from a signal. This is strictly for electronic noise or noise which has been detected and put into electronic form.

iv. *Noise cancellation*   If both the noise and the signal are received by an electronic or digital medium, noise can be filtered from the signal electronically and re-transmitted without the noise. Helicopter pilots rely on this technology to speak on the radio.

10.7.3 Quieting for specific observers

i.   *Underwater acoustics*   All of types of acoustic quieting discussed earlier apply to submarines. Additionally, a submarine may employ a tactic that prevents sounds from reaching a listener at a particular ocean depth. Operating below the depth of the sound channel axis, where the speed of sound in water is the lowest, submarines can prevent detection by surface ships.

ii. *Sound refraction*   Just as a submarine can use refraction to hide its acoustic signature from surface vessels, the same principle of sound refraction can be used to prevent certain observers from hearing the noise. For example, an outdoor observer close to the ground will have sound waves refracted towards him when the ground is cooler than the ambient air and away from him when the ground is hotter than the air.

iii. *Sound re-direction*   One of the obvious ways to reduce the received sound level of an observer is to place the observer out of the path of the highest amplitude sounds. For example, if we mark off a circle around a jet engine and make sound power level observations along that circle, we would expect that the sound is loudest directly in line with the jet's exhaust. Observations perpendicular to the exhaust would be significantly quieter.

iv. *Hearing protection*   An observer may be forced to wear ear plugs in areas of high ambient noise levels. This may be the only quieting method available in areas of noise pollution, such as an open-air firing range or an airport.

10.7.4 Mufflers

Mufflers/silencers/acoustic filters are used in a number of devices requiring the suppression or attenuation of sound. Acoustic mufflers make everyday life much more pleasant. Many common appliances, such as refrigerators and air conditioners, use acoustic mufflers to produce a minimal working noise. Acoustic mufflers are used where there is a large amount of radiated sound such as high pressure exhaust pipes, internal combustion engines, gas turbines, and rotary pumps for reducing the amount of noise emitted. Mufflers are designed to reduce the loudness of the sound pressure created by the engine. Basically, such a muffler consists of a pipeline with sudden changes in the cross-sectional area to reflect the sound energy. An unavoidable side effect of using mufflers is an increase of back pressure which decreases engine efficiency. Performance-oriented mufflers and exhaust systems thus strive to minimize back pressure by employing numerous technologies and methods to attenuate the sound. In some special type of mufflers, inner linings are coated with sound absorbing materials to absorb sound energy. For the majority of such systems, however, the general rule of "more power, more noise" applies.

Though there are varieties of acoustic mufflers used in a variety of devices, they are broadly classified into two main types. These are (i) absorptive mufflers and (ii) reactive mufflers. Many of the more complex mufflers today incorporate both methods to optimize sound attenuation and provide realistic specifications.

Absorptive mufflers

Absorptive mufflers incorporate sound absorbing materials to attenuate the sound energy. These mufflers also transform acoustic energy into heat. Absorptive mufflers are typically straight pipes lined with multiple layers of absorptive materials to reduce radiated sound power. The most important property of absorptive mufflers is the attenuation constant. Higher attenuation constants lead to higher sound energy dissipation. The advantages of absorptive mufflers are (i) high amount of absorption at higher frequencies, (ii) good for applications involving broadband and narrowband noise, (iii) reduced amount of back pressure as compared to reactive mufflers. Disadvantages of absorptive mufflers are (i) poor performance at low frequencies and (ii) material can degrade under certain circumstances such as high heat condition, and so on. A typical absorptive muffler is shown in Fig. 10.5 illustrating the different absorptive layers.

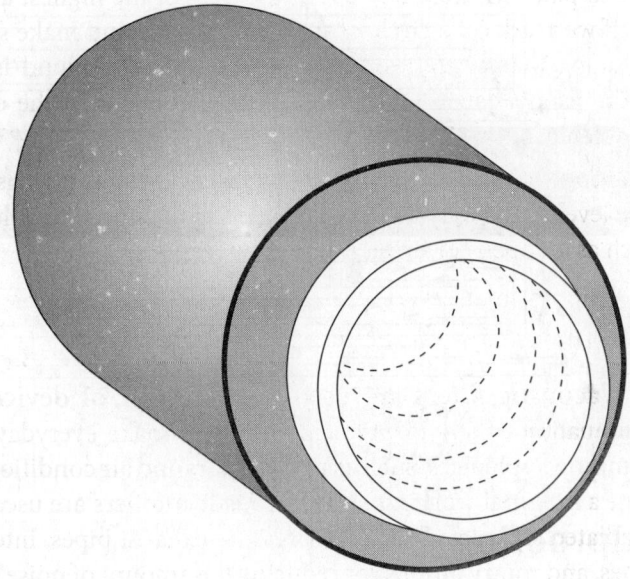

Figure 10.5 | Absorptive muffler

There are a number of applications for absorptive mufflers. The most well-known application is in racing cars, where engine performance is highly desired. Absorptive mufflers do not create a large amount of back pressure when attenuating the sound – this leads to higher muffler performance. It should be noted however, that the radiated sound is much higher. Other applications include plenum chambers, i.e., large chambers lined with absorptive materials, lined ducts, and ventilation systems.

Reactive mufflers

Reactive mufflers use a number of complex passages or lumped elements to reduce the amount of acoustic energy transmitted. These mufflers maximize sound attenuation while meeting a set of specifications, such as pressure drop, volume flow, and so on. This is accomplished by a change in impedance at the intersections, which gives rise to reflected waves and hence, effectively reduces the amount of transmitted acoustic energy.

Since the amount of energy transmitted is minimized, the energy reflected back to the source is quite high. This can actually degrade the performance of engines and other sources. Unlike absorptive mufflers, which dissipate the acoustic energy in the form of heat, reactive mufflers keep the energy constant within the system. Advantages of reactive mufflers are (i) high performance at low frequencies, (ii) they typically give high insertion loss, for stationary tones, (iii) they are useful in harsh conditions. The main disadvantage of reactive mufflers is poor performance at high frequencies. A typical reactive muffler is shown in Fig. 10.6.

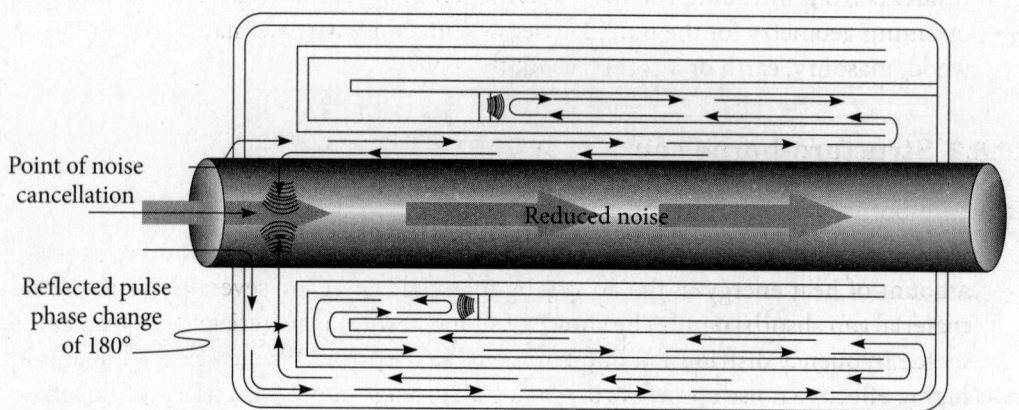

Point of noise cancellation

Reduced noise

Reflected pulse phase change of 180°

Figure 10.6 | Reactive muffler

10.8 Soundproofing

Soundproofing are the measures that reduce the transmission of sound from the source of the sound to the receiver. There are several basic approaches for soundproofing. Few of them are (i) increasing the distance between source and receiver, (ii) using noise barriers to reflect or absorb the energy of the sound waves, (iii) using damping structures such as sound baffles, and (iv) using active anti-noise sound generators. Two distinct soundproofing problems need to be considered when designing acoustic treatments – (i) improving sound quality within a room and (ii) reducing sound leakage to/from adjacent rooms or outdoors.

When speaking of soundproofing, a distinction is made between structure-borne soundproofing and airborne soundproofing. Structure-borne soundproofing is protection against sound within a hall and airborne soundproofing is protection against sound from outside. Airborne sound enters the room through walls, ceilings, windows and doors. Structure-borne soundproofing is essential in cases where sound transmission between rooms of a building proves to be disturbing. Structure-borne sound is transmitted through pipes, posts and beams that run through the building.

10.8.1 Airborne soundproofing

Airborne soundproofing measures are described in the following.

i.   The energy density of sound waves decreases following the inverse square law as they spread out. Hence, increasing the distance between the receiver and the source results in lesser intensity of sound reaching the receiver.

ii. In an outdoor environment such as highways, embankments or panelling are often used to reflect sound upwards into the sky.

iii. In advanced countries, noise barriers along major highways are used to protect adjacent residents from intruding roadway noise. Technology exists to predict accurately the optimum geometry for the noise barrier design. Noise barriers may be constructed of wood, masonry, earth or a combination thereof.

10.8.2 Structure-borne soundproofing

Structure-borne soundproofing measures are described in the following.

i.   A part of the absorbed sound energy spontaneously gets converted into a very small amount of heat energy in the absorbing material. There are several ways in which a material can absorb sound. The choice of sound absorbing material will be determined by the frequency distribution of the noise to be absorbed. Porous open cell foams are highly effective noise absorbers across a broad range of medium to high frequencies. The absorption is low at low frequencies. The exact absorption profile of a porous open cell foam is determined by a number of factors like (i) cell size, (ii) tortuosity, (iii) porosity, (iv) material thickness, and (v) material density and so on.

ii. Damping means to reduce resonance in the room, by absorption or re-direction using methods of reflection or diffusion. It can reduce the acoustic resonance in the air, or mechanical resonance in the structure of the room itself or things in the room. Resonant panels, Helmholtz resonators and other resonant absorbers work by damping a sound wave using the method of reflection. Unlike porous absorbers, resonant absorbers are most effective at low to medium frequencies.

iii. If a specular reflection from a hard flat surface is giving problematic echos, then an acoustic diffuser may be applied to the surface. It will scatter sound in all directions.

iv. Honeycombed curtains can be used in windows to damp sound. Single, double and triple honeycomb designs achieve relatively greater degrees of sound damping. The primary soundproofing limit of curtains is low since the edge of the curtains are not sealed. Double pane windows achieve somewhat greater sound damping than single pane windows. Noise reduction can be achieved to a good extent by installing a second interior window.

v.   Noise cancellation generator for active noise control is a modern device. A microphone is used to pick up the sound and then analyzed by a computer; then, sound waves with opposite polarity are thrown through a speaker. Both sound waves cause destructive interference and reduce much of the noise.

vi. A room within a room is one method of isolating sound and stopping it from transmitting to the outside world where it would be undesirable. Sound transfer from a room to the outside or other rooms occurs through mechanical means. The sound vibration passes directly through the brick, woodwork, and other solid structural elements. The wall, ceiling, floor or closed window acts as a sounding board when sound vibrations strike them. Thus, vibration is amplified and heard in the second space. The use of acoustic foam and other absorbent means is less effective against this transmitted vibration. The user is advised to break the connection between the room that contains the noise source and the outside world by a new wall. This is called acoustic de-coupling. Ideal de-coupling involves eliminating vibration transfer in both solid materials and in the air.

All or few of the techniques discussed here are applied in one way or the other in fighter planes to submarines to motor propelled yachts to road vehicles to reduce noise levels in them.

Questions

10.1   What are the basic requirements of an acoustically good hall?

10.2   What is reverberation time? Derive Sabine's formula for reverberation time.

10.3   What are absorption coefficient and reflection coefficient? Derive the relation connecting the two.

10.4   What are the methods used for measuring of absorption coefficient?

10.5   Besides reverberation, what are the other factors that affect the acoustics of buildings?

10.6   What are the requisites for good acoustics?

10.7   What is acoustic quieting? What are the factors taken into consideration to achieve it?

10.8   What are the mechanical methods followed in acoustic quieting?

10.9   What are the electronic methods followed in acoustic quieting?

10.10 What is a muffler? Distinguish between absorptive muffler and reactive mufflers.

10.11 What is soundproofing? Explain different methods to achieve it.

10.12 Distinguish between air-borne soundproofing and structure-borne soundproofing.

Problems

10.1   An auditorium has a volume of 8000 m³. It is required to have a reverberation time of 1.6 s. What should be the total absorption in the auditorium?   [Ans   825 O.U.W.]

10.2   A theatre hall has a volume of 25000 m³. It has a reverberation time of 1·5 s and the total sound absorbing surface is 5000 m². Calculate the average absorbing power of the surfaces.   [Ans   0.55]

10.3   A hall of volume 5500 m³ is found to have a reverberation time of 2.3 s. The sound absorbing surface of the hall has an area of 750 m². Calculate the average absorption coefficient.   [Ans   0.504]

10.4   Calculate the acoustic intensity level in each case, at a distance of 2.51 metres from a point source which radiates energy at the rate of 6.3 W.   [Ans   109 db]

10.5   Calculate the acoustic intensity of a plane acoustic wave in air of intensity level 130 decibels.   [Ans   10 Wm⁻²]

10.6   A grinder operates at a sound intensity level of 90 db. If it is operated in a room with an existing sound intensity level of 75 db, what will be the resultant intensity level?   [Ans   90.14 db]

10.7   Two sources of sound A and B emit sound waves of different frequencies. The two sound pressure levels, as recorded at a place, are 70 db and 60 db respectively. Calculate the resultant sound pressure level at the same point due to the combined effect.   [Ans   72.39 db]

10.8   The noise from a helicopter engine 100 m from an observer is 40 db in intensity level. What will be the intensity when the helicopter flies overhead at a height of 2 km?   [Ans   13.98 db]

Multiple Choice Questions

1.   The term "acoustics" is derived from the Greek verb which means
   (i)   to sound   (ii)   to speak
   (iii) to hear   (iv)   to cry

2.   Who is the father of modern architectural acoustics?
   (i)   Maxwell   (ii)   Hawkings
   (iii) Einstein   (iv)   Sabine

3.   What should be the range of reverberation time for music?
   (i)   0 – 1 s   (ii)   1 – 2 s
   (iii) 0.5 – 1 s   (iv)   1.5 – 2 s

4.   What should be the range of reverberation time for speech?
   (i)   0 – 1 s   (ii)   1 – 2 s
   (iii) 0.5 – 1 s   (iv)   1.5 – 2 s

5. The reverberation time of a hall does not depend on
   (i) color of the hall
   (ii) shape of the hall
   (iii) the nature of the reflecting materials in the hall
   (iv) the volume of the hall

6. The growth and decay of the sound energy density is
   (i) linear   (ii) parabolic
   (iii) hyperbolic   (iv) exponential

7. Sound absorption means that
   (i) the sound disappears from a specified space
   (ii) sound energy leaves the space
   (iii) it dissipates in it
   (iv) all the above

8. The sound absorption coefficient of a material is zero if
   (i) reflectivity $= 0$ (ii) reflectivity $= 1$
   (iii) reflectivity $= -1$ (iv) reflectivity $= \infty$

9. Give an example of a perfect absorber of sound.
   (i) open door   (ii) loosely bound cotton
   (iii) foam   (iv) powdered material

10. Which of the following is not a factor affecting the acoustics of buildings?
    (i) brick quality   (ii) loudness
    (iii) extraneous noise   (iv) resonance

11. Which of the following is not a requisite for good acoustics of buildings?
    (i) hanging heavy curtains   (ii) hanging folded curtains
    (iii) hanging pictures and maps   (iv) having closed windows

12. What is the reference value of sound pressure level measurements in air?
    (i) $2 \times 10^{-4}\,\text{Nm}^{-2}$ (ii) $2 \times 10^{-5}\,\text{Nm}^{-2}$
    (iii) $1.011 \times 10^{-5}\,\text{Nm}^{-2}$ (iv) $6.023 \times 10^{23}\,\text{Nm}^{-2}$

13. To achieve acoustic quietening in a machine, we cannot
    (i) use a damping material in the machine
    (ii) use diesel as fuel
    (iii) isolate the machine
    (iv) run the machine in partial vacuum

14. In which of the following machines, acoustic mufflers are not used.
    (i) Air conditioner   (ii) fridge
    (iii) grinder   (iv) gas turbines

15. Which of the following is not a measure that reduces the transmission of sound from the source to the receiver?
    (i)   increasing the distance between source and receiver
    (ii) using noise barriers to reflect or absorb the sound energy
    (iii) using active anti-noise sound generators
    (iv) using electromagnetic wave reflectors

Answers

1 (iii)	2 (iv)	3 (ii)	4 (iii)	5 (i)	6 (iv)	7 (i)	8 (ii)
9 (i)	10 (i)	11 (iv)	12 (ii)	13 (ii)	14 (iii)	15 (iv)	

11 Ultrasonics

11.1 Introduction

Ultrasonics is the branch of science and technology concerned with the study and use of ultrasonic waves. Sound waves of frequency more than 20 kHz are called ultrasound. Though these sound waves are not sensed by a normal human ear, they are sensed by a few lower creatures like cats, fox, puppies, nocturnal insects and animals, dolphins, a few variety of whales, and fishes. Sound waves of a frequency less than 20 Hz are called infrasonic. Though infrasonic sound waves are not audible to us, it sometimes gives us a sensation of "ghost vision". Elephants use infrasound for their communication. In addition to the properties of audible sound waves, ultrasonic waves exhibit other new phenomena. Ultrasonic waves have a large number of applications in all fields of science and technology. The ultrasound is used in many different fields, typically to penetrate a medium and measure the reflection signature or supply focused energy. The reflection signature can reveal details about the inner structure of the medium.

11.2 Production of Ultrasonic Waves

Unlike audible sound waves (20 Hz–20 kHz), production of ultrasonic waves requires special devices and methods. In the following, we shall discuss few methods of production of ultrasonic waves.

11.2.1 Galton's whistle

Galton's whistle (also known as silent whistle or dog whistle) is a type of whistle that emits sound in the ultrasonic range. It was invented in 1876 by Francis Galton and is used in the training of dogs and cats. It is believed that the wild ancestors of cats and dogs evolved this hearing range in order to hear high frequency sounds made by their preferred prey, small rodents.

Principle   Galton's whistle works on the principle of the organ pipe. In a close ended organ pipe, resonance occurs when the length of the pipe is one-fourth times the length of the wavelength of sound in the air medium, i.e., $\lambda = 4\ell$.

Construction   It consists of a closed end air column A whose length can be adjusted with the help of a movable piston. The piston P can be moved to the desired position with the help of a screw S_2. The open end of the pipe A is fitted with a lip L. The position of the pipe C can be adjusted with the help of the screw S_1. The gap between the ends of A and C can be adjusted with the help of the screw S_1. The construction details are shown in the Fig. 11.1.

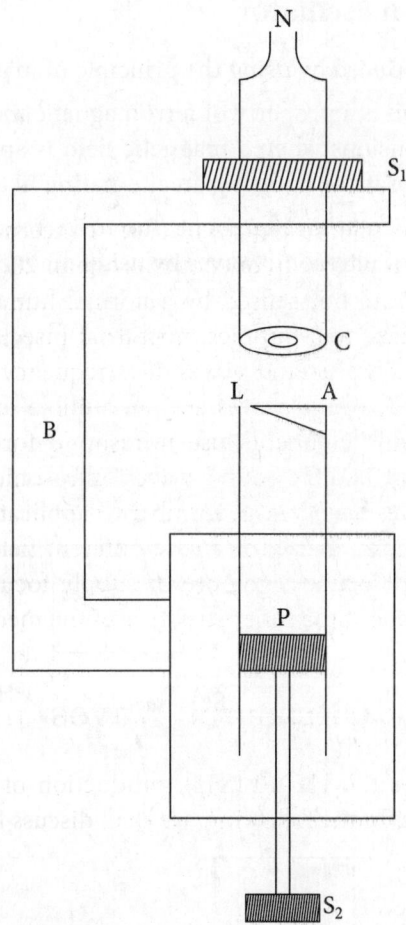

Figure 11.1 | Construction details of Galton's whistle

Working An air blast is blown through the nozzle N at the top. The blast of air coming out of C strikes against the lip L and the column of air in the pipe is set into vibration. By adjusting the length of the air column in A by screw S_2, it is brought to the resonant position. The resonant frequency will depend on the length and diameter of the pipe A. If ℓ is the length of the air column in A, x the end correction, then the wavelength at resonance will be $\lambda = 4(\ell + x)$ and the corresponding frequency will be

$$v = \frac{v}{\lambda} = \frac{v}{4(\ell + x)} \tag{11.1}$$

With the help of this whistle, frequencies in the range of ultrasonics can be produced. The micrometer screw S_2 can also be calibrated to directly give the frequency of the sound.

11.2.2 Magnetostriction oscillator

Ultrasonic waves can be produced by using the principle of magnetostriction.

Principle Magnetostriction is a property of ferromagnetic materials that causes them to change their shape or dimensions when a magnetic field is applied. Thus, ferromagnetic materials can be made to vibrate by applying an alternating magnetic field.

Construction An experimental arrangement due to Pierce for explaining the physics involved in the production of ultrasonic waves by using magnetostriction effects is shown in Fig. 11.2.

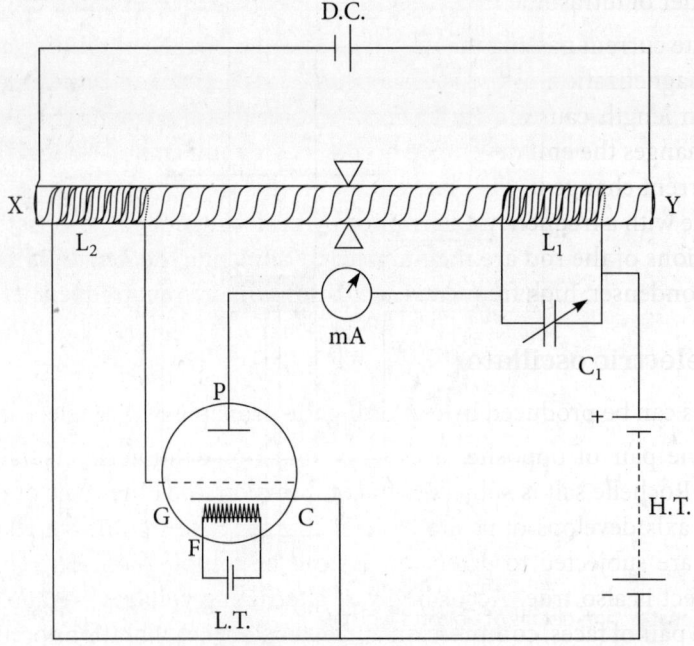

Figure 11.2 | Schematic diagram for the production of ultrasonic waves by magnetostriction effect

XY is a bar of ferromagnetic material say of iron or nickel. The bar is clamped in the middle. A coil which is wrapped around the coil and rod is permanently magnetised in the beginning by passing direct current through this coil. L_1 and L_2 are two coils surrounding the bar *XY*. L_1 and C_1 are connected in parallel and the combination is connected in the plate circuit to which a mili-ammeter is connected. L_2 is connected between the grid and the cathode. The values of L_1 and C_1 determine the frequency

$$\left(= \frac{1}{2\pi\sqrt{L_1 C_1}} \right)$$

of the oscillatory circuit. The vibrations are maintained due to the coupling provided by the coil L_2.

Working If a ferromagnetic material in the form of a bar is subjected to an alternating magnetic field along the length, it expands and contracts in length, alternately. This change of length is independent of the sign of the field and only depends upon the magnitude of the field and nature of the material. The frequency of contraction or expansion is twice the frequency of the alternating magnetic field. The alternating magnetic field is produced with the help of an oscillatory circuit. Ordinarily, the amplitude of the vibrations of the rod is small. If, however, the frequency of the alternating current is the same as the natural frequency of the rod, then resonance occurs and the amplitude of vibration is considerably increased. Due to the longitudinal contraction and expansion of the bar, sound waves are now emitted from the ends of the rod. If the applied frequency of the alternating magnetic field is of the order of ultrasonic frequency, the rod sends out ultrasonic waves.

When the plate current passing through the coil L_1 is changed, it causes a corresponding change in the magnetization of the rod. Thus, there is a change in the length of the rod. This variation in length causes a variation in the magnetic flux through the grid coil L_2 which in turn changes the emf developed across it. This emf acts on the grid and produces an amplified current change in the plate circuit. In this way, the plate current builds up to a large amplitude with a frequency determined by the frequency of the vibration of the rod. Thus, the vibrations of the rod are maintained. By adjusting the length of the rod and the capacity of the condenser, high frequency oscillations of different frequencies are obtained.

11.2.3 Piezoelectric oscillator

Ultrasonic waves can be produced by exploiting the principle of piezoelectric effects.

Principle If one pair of opposite faces of certain non-symmetric crystals like quartz, tourmaline and Rochelle salt is subjected to compression, the other pair of opposite faces along the polar axis develops opposite electric charges. The sign of the charges changes when the faces are subjected to extension instead of compression. The converse of the Piezoelectric effect is also true. Accordingly, if alternating voltages are applied along the polar axis to one pair of faces, compression and extension, i.e., vibrations occur at the other pair of faces of the crystal. The direction of the cut of the crystal with reference to the polar axis is quite important.

Construction An experimental arrangement for explaining the physics involved in the production of ultrasonic waves by using piezoelectric effects is shown in Fig. 11.3.

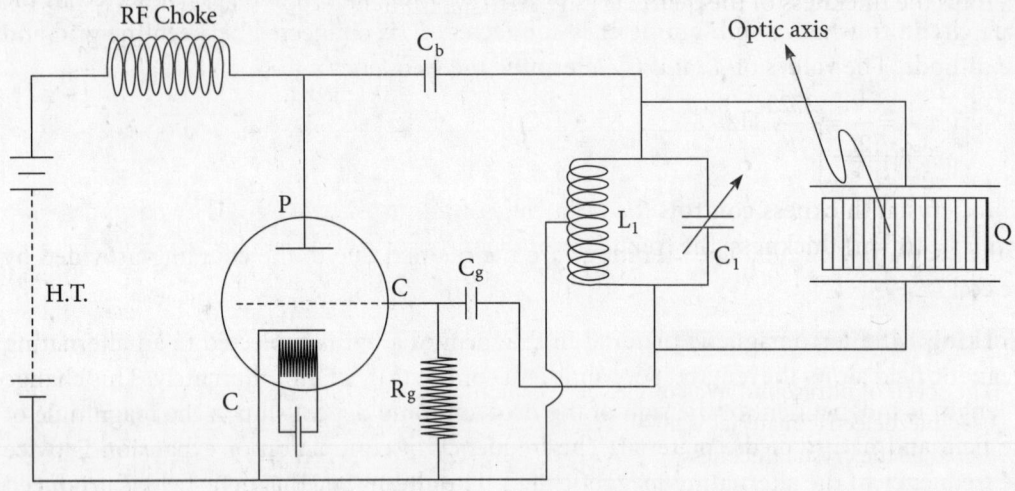

Figure 11.3 | Schematic diagram for the production of ultrasonic wave by piezoelectric effect

A quartz crystal is cut in such a manner that the polar axis is perpendicular to the surface. A thin slice of this. The quartz crystal is connected parallel to the tank circuit of the oscillatory circuit. Thus, a high frequency alternating voltage is applied to the quartz crystal along the length of the polar axis. The high frequency alternating voltage which is applied to the crystal is obtained by a Hartley oscillatory circuit.

Working The high frequency alternating voltage obtained from the Hartley oscillator is applied to the crystal. Ordinarily, the amplitude of the vibrations of crystal is small. Now the variable capacitor's capacitance C_1 is adjusted so that the frequency of the oscillating circuit

$$\left(= \frac{1}{2\pi\sqrt{L_1 C_1}} \right)$$

is tuned to the natural frequency of the crystal and resonance is produced. The quartz crystal is set into mechanical vibrations and ultrasonic waves are produced.

The velocity of the ultrasonic wave in quartz in a direction perpendicular to the polar axis is given by

$$v = \sqrt{\frac{Y}{\rho}} = 5450\,\text{m/s}$$

with Y = Young's modulus of quartz along the polar axis = 7.9×10^{10} N/m² and ρ = density of quartz = 2650 kg/m².

If t is the thickness of the quartz crystal, then at resonance $t = \dfrac{\lambda}{2}$ (closed organ pipe) and we have

$$\nu = \frac{v}{\lambda} = \frac{v}{2t} = \frac{2725}{t}\, \text{Hz} \tag{11.2}$$

Thus, crystal thickness controls the frequency of ultrasonic waves. For a quartz crystal of 1mm (= 10^{-3} m) thickness, the frequency of the ultrasonic wave will be 2725 kHz.

Example 11.1

The speed of ultrasound in a commercial preparation of lead zirconate titanate, a commonly used piezoelectric material is 4000 m/s. If a vibration frequency of 5 MHz is required, what would be the crystal thickness?

Solution

$$t = \frac{v}{2f} = \frac{4000}{2 \times 5 \times 10^6}\, \text{m} = 0.4\,\text{mm}$$

11.3 Detection of Ultrasonic Waves

Since ultrasonic waves are beyond human audible range, we human beings cannot directly detect them. However, ultrasonic waves propagated through a medium can be detected in a number of ways; some of the methods employed are as follows.

i.   **Piezo-electric detector**

     A quartz crystal can also be used to detect ultrasonic waves. When one pair of faces of the quartz crystal is subjected to ultrasonic waves, feeble voltages are developed on the other faces. By amplifying this voltage using suitable electronic devices, we can determine the presence of ultrasonic waves.

ii. **Kundt's tube method**

     Ultrasonic waves can be detected with the help of Kundt's tube. This method is suitable for detecting ultrasonic waves of large wavelength. When ultrasonic waves pass through the tube, the lycopodium powder sprinkled in the tube collects in the form of heaps at the nodal points and is blown off at the antinodal points. The average distance between two adjacent heaps is equal to half the wavelength.

iii. **Sensitive flame method**

     A narrow sensitive flame is moved slowly through the medium in which the ultrasonic wave is supposed to be present. At the position of the antinode, the flame will be steady and at the node position, the flame will flicker because there will be change in pressure. The average distance between two adjacent nodes is equal to half the wavelength.

iv. **Thermal detector method**

In this method, a fine platinum wire connected to a sensitive bridge arrangement is moved in the medium in which ultrasonic wave is supposed to be present. The temperature changes at the nodes and remains constant at the antinode. Hence, the resistance of the platinum wire changes at the nodes and remain constant at the antinode. This change in the resistance of the platinum wire can be detected by using a sensitive bridge arrangement. The bridge will be in the balanced position when the platinum wire is at the antinodes.

11.4 Properties of Ultrasonic Waves

i. Ultrasonic waves are highly directional, i.e., unlike ordinary sound, ultrasound travels in a straight line like light waves. Bending of high frequency ultrasonic waves around obstacles is negligibly small.

ii. Due to their high frequency, they are highly energetic and hence highly penetrating.

iii. Their speed of propagation increases with increase in frequency.

iv. Due to their small wavelength, their scattering is negligibly small. Hence, they can be transmitted over long distances without any appreciable loss of energy.

v. The temperature of tissues or water increases by absorbing ultrasound energy.

vi. Intense ultrasonic radiation has a disruptive effect in liquids by causing cavitation.

vii. When ultrasonic waves are propagated in a liquid bath, a stationary wave pattern is formed due to the reflection of the wave from the opposite end. The density of the liquid thus varies from layer to layer along the direction of propagation of the wave. In this way, an acoustic grating is formed which can give a diffraction pattern of light as we will discuss later.

viii. All these properties of ultrasonic waves are exploited technologically for use in industry, the medical field, scientific researchand the like.

11.5 Wavelength Determination of Ultrasonic Waves

The phenomenon of diffraction of light while passing through a liquid subjected to ultrasonic waves was first observed by Debye and Sears in 1932. The formation of stationary ultrasonic waves in a liquid, gives rise to fixed positions of nodal and antinodal planes. The liquid density is maximum at nodal planes and minimum at antinodal planes due to the difference of pressure at these planes. If monochromatic light is passed through the liquid in this condition at right angles to the waves, the liquid behaves as a diffraction grating.

Such a grating is known as an acoustical grating. This grating behaves in the same way as a ruled optical grating. The distance d between two consecutive nodal or antinodal planes is half the wavelength $(\lambda/2)$ of ultrasonic waves (i.e., $d = \lambda/2$). Hence, the method can be used for finding the wavelength and velocity of ultrasonic waves in a liquid in a similar way to that of light. The experimental arrangement is shown in Fig. 11.4.

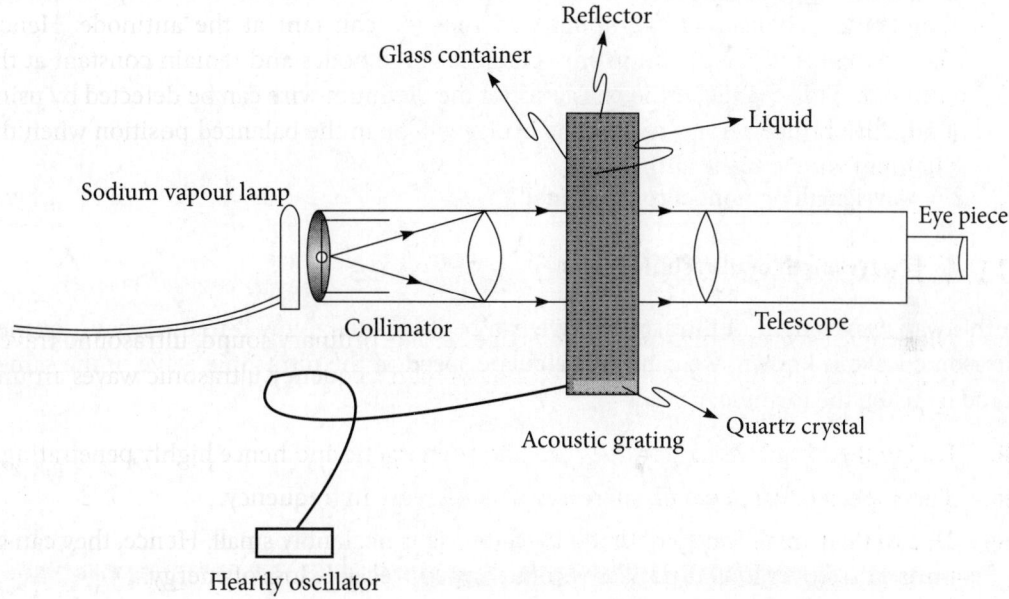

Figure 11.4 | Wavelength determination of ultrasonic waves by acoustic grating

The light from a monochromatic source of light S, passes through a collimator which makes it a parallel beam. This parallel beam of light then passes through the ultrasonic cell. The cell consists of a rectangular glass tank containing the liquid and a piezoelectric crystal connected to the Hartley oscillator at the bottom. The top surface of the glass container is an ultrasonic wave reflector. Now, this acoustic grating is placed on the prism table of the levelled spectrometer.

The piezoelectric crystal produces and directs ultrasonic waves towards the reflector and the reflector in turn reflects the waves. Due to superposition of direct and reflected waves, stationary waves are produced in the liquid. These stationary waves give rise to fixed positions of nodal and antinodal planes parallel to the direction of propagation of light forming an acoustical grating.

Now light emerging from the acoustic grating is seen through a well focussed telescope. When the crystal is arrested, a single image of the collimator slit is observed. However, when ultrasonic waves are produced in the liquid by resonant excitation of the crystal, a number of diffracted images appear on either side of the central maximum with diminishing intensity. The angular separation θ_n between the direct image of the slit and the diffracted image of any order n is measured. Applying the theory of diffraction grating, the wavelength of ultrasonic waves can be calculated in the following way. From the theory of light diffraction, we have

$$d \sin \theta_n = n\lambda_L$$

or $$\lambda_{uw} = \frac{2n\lambda_L}{\sin \theta_n} \quad \text{since } d = \frac{\lambda_{uw}}{2}$$ (11.3)

Here

$n = 1, 2, 3, \ldots$

λ_L = wavelength of monochromatic light

λ_{uw} = wavelength of ultrasonic waves

In this way, wavelength of ultrasonic wave can be calculated. Now if frequency v_{uw} of the ultrasonic wave is known, we can also calculate speed of the ultrasonic wave in the same liquid by using the formula

$$v = \lambda_{uw} v_{uw}.$$ (11.4)

Example 11.2

The speed of ultrasonic wave in a certain medium is 5450 m/s. If the frequency of the ultrasonic wave is 2725 kHz, find the wavelength of the ultrasonic wave.

Solution

The data given are $v = 5450$ m/s, $v_{uw} = 2725 \times 10^3$ Hz

$$\lambda_{uw} = \frac{5450 \text{m/s}}{2725 \times 10^3 /\text{s}} = 2 \times 10^{-3} \text{m}$$

Example 11.3

Acoustic grating is formed in the water medium by passing monochromatic ultrasonic waves through it. The third order diffraction maximum is observed at an angle 2° for sodium light of wavelength 5890 Å. Calculate the wavelength of the ultrasonic wave.

Solution

The data given are $n = 1$, $\theta = 2°$, $\lambda_L = 5890 \times 10^{-10}$ m

$$\lambda_{uw} = \frac{2\lambda_L}{\sin \theta} = \frac{2 \times 3 \times 5890 \times 10^{-10} \text{m}}{\sin 2°} = 1.01 \times 10^{-4} \text{m}$$

11.6 Ultrasound Cavitation

Cavitation is the formation and then immediate implosion of bubbles in a liquid. It usually occurs when a liquid is subjected to rapid changes of pressure that cause the formation of bubbles where the pressure is relatively low.

Sound, including ultrasound, propagates through any physical medium in the form of compression and rarefaction. It stretches and compresses the molecular spacing of the medium. As the ultrasound crosses the liquid medium, the average distance between the molecules will vary very rapidly in accordance with its frequency. At the ultrasonic frequency, cohesive forces of the liquid media are overcome and voids/cavitations are created. At the position of rarefaction, the distance between the molecules of the liquid exceeds the minimum molecular distance required to hold the liquid intact. Hence, the liquid breaks down and voids are created. These voids are the so-called cavitation bubbles. As the liquid compresses and stretches, cavitation bubbles behave in the following two ways. At fairly low ultrasonic intensities (1.0 – 3.0 watt/cm²), stable cavitation bubbles form and oscillate about some equilibrium size for many acoustic cycles. At ultrasonic intensities exceeding 10.0 watt/cm², transient cavitation bubbles are formed which expand through a few acoustic cycles to a radius of at least twice their initial size, before collapsing violently on compression. Transient bubble collapsing is considered to be the main source of the chemical and mechanical effects of ultrasonic sound energy. Each collapsing bubble can be considered as a microreactor in which temperatures of several thousand degrees (≈ 5000 K) and pressures higher than one thousand atmospheres (≈ 2000 atm) are created instantaneously. The implosion of the cavitation bubble also results in liquid jets of up to 280 m/s velocity.

11.6.1 Parameters affecting ultrasonic cavitation

The following parameters affect the ultrasonic cavitation phenomenon.

i.　**Frequency**

At high ultrasonic frequencies, in the order of MHz, the production of cavitation bubbles becomes more difficult than at low ultrasonic frequencies, in the order of kHz. Ten times more power is required to induce cavitation in water at 400 kHz than at 30 kHz. The physical explanation for this lies in the fact that at very high frequencies, the cycle of compression and rarefaction caused by the ultrasonic waves becomes so short that the molecules of the liquid cannot be separated to form a void and, thus, cavitation is no longer obtained.

ii.　**Intensity**

The intensity of ultrasonic wave is proportional to the square of the amplitude of ultrasonic waves and, as such, an increment in the amplitude of vibration will lead to an increase in the intensity of vibration and to an increase in the sono-chemical effects. To achieve the cavitation threshold, a minimum intensity is required. This means that higher amplitudes are not always necessary to obtain the desired results. In addition, high amplitudes can lead to rapid deterioration of the ultrasonic transducer, resulting in liquid agitation instead of cavitation and poor transmission of the ultrasound through the liquid media. However, the amplitude should be increased when working with samples of high viscosity, such as blood.

iii. **Solvent**

The solvent used to perform sample treatment with ultrasonic waves must be carefully chosen. As a general rule, most applications are performed in water. However, other less polar liquids, such as some organic solvents, can also be used, depending on the intended purpose. Both solvent viscosity and surface tension are expected to inhibit cavitation. The higher the natural cohesive forces acting within a liquid (e.g., high viscosity and high surface tension), the more difficult it is to attain cavitation.

iv. **Temperature**

Solvent temperature plays two roles in ultrasonic cavitation. On the one hand, the use of high temperatures helps to disrupt strong solute–matrix interactions such as the Van der Waals forces, hydrogen bonding and dipole attractions. On the other hand, cavitation is better attained at lower temperatures because faster diffusion rates occur at higher temperatures as a result of which vapour fills the cavitation bubbles, which then tend to collapse less violently and thus the ultrasonic effects are less intense than expected. Hence, a compromise between temperature and cavitation must be achieved.

v. **External pressure**

If the external pressure is increased, then a greater ultrasonic energy is required to induce cavitation, that is, to break the solvent molecular forces. There is also there is an increment in the intensity of the cavitational bubble collapse and, consequently, an enhancement in sono-chemical effects is obtained. For a specific frequency, there is a particular external pressure that will provide an optimum sono-chemical reaction.

vi. **Bubbled gas**

It must be stressed that most ultrasonic applications are performed under atmospheric pressure. Dissolved gas bubbles in a fluid can act as nuclei for cavitation, favouring the ultrasonic process. Mono-atomic gases such as He, Ar, and Ne are bubbled continuously into the solvent to increase cavitation effects.

11.6.2 Consequences of ultrasonic cavitation

High-intensity ultrasound produces violent agitation in low-viscosity liquids, which can be used in dispersion process of solid in liquid, solid in gas and immiscible liquids. At liquid/solid or gas/solid interfaces, the asymmetric implosion of cavitation bubbles can cause extreme turbulences that reduce the diffusion boundary layer, increase the convection mass transfer, and considerably accelerate diffusion in systems where ordinary mixing is not possible.

In general, cavitation in liquids may produce the following effects:

i. It causes fast and complete de-gassing.

ii. It initiates various chemical reactions by generating free chemical ions.

iii. It accelerates chemical reactions by facilitating the mixing of reactants.

iv. It enhances polymerization and depolymerization reactions by temporarily dispersing aggregates or by permanently breaking chemical bonds in polymeric chains.

v. It increases emulsification rates.

vi. It improves diffusion rates.

vii. It produces highly concentrated emulsions or uniform dispersions of micron-size or nano-size materials.

viii. It assists the extraction of substances such as enzymes from animal, plant, yeast, or bacterial cells.

ix. It removes viruses from infected tissue.

x. It erodes and breaks down susceptible particles, including micro-organisms.

11.7 Applications of Ultrasonic Waves

Ultrasound has evolved from an emerging technology and developed into a fully commercial processing technology in the last ten years. High reliability and scaleablility as well as low maintenance costs and high energy efficiency make ultrasound a promising player in the field of science and technology. Ultrasound offers additional exciting opportunities like cavitation, the basic ultrasonic effect, which allows for new results in biological, chemical and physical processes. While low-intensity ultrasound is mainly used for analysis, non-destructive testing and imaging, high-intensity ultrasound is used for the processing of liquids such as mixing, emulsifying, dispersing and de-agglomeration, cell disintegration of enzyme deactivation. In the following, we enlist a few applications of ultrasonic waves in different fields though the list is endless.

a. **Depth of sea** Ultrasonic waves of high frequency are used to determine the depth of the sea. A piezoelectric quartz oscillator is used for this purpose. The crystal is placed between two metal plates and the plates are connected to a spark oscillator, producing damped oscillations. The frequency of the damped oscillator is tuned to be the same as the natural frequency of the quartz crystal. The quartz crystal itself acts as a transmitter and a receiver of the ultrasonic waves. The ultrasonic waves transmitted by the crystal are directed towards the bed of the sea. These waves are reflected back from the bed and the echo is detected by the crystal itself. In this case, the metal plates are automatically connected to an amplifier and to a cathode ray oscillograph. The time interval between the emitted signal and the echo is determined with the help of the oscillograph. Knowing the velocity of sound through sea water and the time interval, the depth of the sea can be calculated. Suppose, t is the time interval between the transmission of the ultrasonic wave and receipt of the echo and v, the velocity of sound waves through sea water, then depth of the sea,

$$h = \frac{vt}{2}$$ (11.5)

This method is also suitable to detect the presence and depth of submarines, rocks and so on from the surface of seawater. The instrument directly calibrated to know the depth of sea is called a fathometer or echometer.

b. **Signalling**   Ultrasonic waves are used for directional signalling. The frequency of ultrasonic waves is higher than that of audible sound waves. Therefore, the wavelength is comparatively small. Due to the small wavelength, ultrasonic waves can be sent in the form of a short beam. If a quartz crystal, taken in the form of a disc of radius r, is used as a source of ultrasonic waves, an angle of the cone containing these waves is given by

$$\sin\theta = \frac{0.61\lambda}{r} \tag{11.6}$$

For small wavelengths λ, θ is small. Even for a small amplitude of the vibrating crystal, a large amount of energy is radiated, whereas it is not possible in the case of audio frequency waves. Recently, ultrasonic microscope has been invented. It is used to detect concealed objects. The frequency is very high so that the wavelength is of the order of the visible light.

c. **Heating effects**   When a beam of ultrasonic waves is passed through a substance, it gets heated. If ultrasonic waves pass through water at 0°C, the water can be made to boil.

d. **Mechanical effects**

i. *Ultrasonic drills*   Ultrasonic drills are used to bore holes in steel and other metals or their alloys. Here the drill oscillates with ultrasonic frequency and can bore any hard metal.

ii. *Crack in metals*   Ultrasonic waves can be used to detect cracks or discontinuity in metal structures. In this case, an emitter and detector of ultrasonic waves are used. Ultrasonic waves from the emitter are directed towards the metal. The reflected beam is detected by the detector. If there is a crack or discontinuity, there will be rise in energy received by the detector, if the emitter and the detector are on the same side. If the emitter and the detector are on opposite sides of the metal, there will be fall in energy at the regions of cracks or discontinuity.

iii. *Formation of alloys*   Alloys of uniform composition are obtained by subjecting the constituents to an ultrasonic beam. The two constituents are well mixed by the ultrasonic waves, even though the constituents differ in density.

iv. *Ultrasonic wet-milling and grinding*   Ultrasonic waves can be used for wet-milling and micro-grinding of particles. In particular, for the manufacture of superfine-size slurries, ultrasound has many advantages, when compared with common size reduction equipments.

e. **Chemical effects**

Ultrasonic waves act like catalytic agents and accelerate chemical reactions. They bring about a number of chemical changes. Some of the chemical applications are as follows:

i.   When potassium iodide is subjected to ultrasonic waves, it liberates iodine.

ii.   Water is decomposed into hydrogen and hydroxyl ions, by the action of ultrasonic waves.

iii. Ultrasonic waves reduce mercuric chloride into mercurous chloride.

iv.   Emulsions are dispersions of two or more immiscible liquids. Water and oil are immiscible. An emulsion of water and oil is obtained when the mixture is subjected to ultrasonic waves. Similarly, an emulsion of water and mercury can be prepared.

v.   The dispersion and de-agglomeration of solids into liquids is an important application of ultrasonic devices. Ultrasonic cavitation generates high shear forces that break particle agglomerates into single dispersed particles. Ultrasonic dispersion and de-agglomeration is a common process in paint, ink, shampoo, beverages, or polishing media factories.

vi.   Ultrasonic waves accelerate crystallization.

vii. Ultrasonic waves explode nitrogen iodide.

viii. *Soldering*   Aluminium cannot be soldered by the ordinary soldering method. To solder aluminium, ultrasonic waves are used in addition to the electrical soldering iron. The ultrasonic waves remove the oxide film and facilitate soldering.

ix.   *Ultrasonic wire, cable and strip cleaning*   Ultrasonic cleaning is an environmentally-friendly alternative for cleaning continuous materials, such as wire and cable, tape or tubes. The effect of the cavitation generated by the ultrasonic power removes lubrication residues like oil or grease, soaps or dust. These waves can also be used for cleaning liquid tanks, utensils, washing clothes, removing dust and soot from chimneys.

x.   *Ultrasonic Trans-esterification of oil to biodiesel*   Ultrasonication increases the chemical reaction speed and yield of the trans-esterification of vegetable oils and animal fats into biodiesel. This allows changing the production from batch processing into continuous flow processing and reduces investment and operational costs. Ultrasonication can achieve a biodiesel yield in excess of 99%. Ultrasound reduces the processing time and the separation time significantly.

xi.   *Ultrasonic De-gassing of liquids*   De-gassing of liquids is an interesting application of ultrasonic devices. In this case, the ultrasound removes small suspended gas bubbles from the liquid and reduces the level of dissolved gas below the natural equilibrium level.

xii. *Sonication of bottles and cans for leak detection* Ultrasound is used in bottling and filling machines to check cans and bottles for leaks. The instantaneous release of carbon dioxide is the decisive effect of ultrasonic leakage tests of containers filled with carbonated beverages.

xiii. *Cell disintegration* Ultrasonic treatment can disintegrate fibrous, cellulosic material into fine particles and break the walls of the cell structure. This releases more of the intra-cellular material, such as starch or sugar into the liquid.

xiv. *Ultrasonic cell extraction* The extraction of enzymes and proteins stored in cells and sub-cellular particles is an effective application of high-intensity ultrasound. Ultrasound has a potential benefit in the extraction and isolation of novel potentially bioactive components.

f. **Medical applications**

Ultrasonic waves have a large number of applications in the field of medicine. Some of the important applications are as follows.

i. *Neuralgic pain* Ultrasonic waves are useful for relieving neuralgic and rheumatic pains. The affected portion of the body is exposed to ultrasonic waves. The waves produce a soothing massaging action and relieve pain.

ii. *Arthritis* Ultrasonic waves are used to relieve pain due to arthritis. Here a small metal head, vibrating with a frequency of more than 10^6 Hz is moved over the skin of the patient. These vibrations after passing through the tissues produce a deep massaging action. The patient is relieved of the pain.

iii. *Contracted fingers* Ultrasonic waves are used to restore contracted fingers. They are also used to loosen up the scar tissues in various parts of the human body.

iv. *Broken teeth* Now-a-days, ultrasonic waves are used by dentists for the proper treatment of broken teeth.

v. *Bloodless surgery* Ultrasonic waves are used in bloodless surgery. Here the ultrasonic waves are focussed on a sharp instrument and the tissues are destroyed without any loss of blood. The doctors have used such instruments for conducting bloodless brain operations.

vi. *Sterilization* Ultrasonic waves can destroy unicellular organisms. Bacteria perish under the action of ultrasonic waves. Ultrasonic waves are used in the sterilization of water and milk.

vii. *Enemy of lower life* When some lower animals like rats, frogs, fish, and so on, are exposed to ultrasonic waves, they become lame.

viii. *Detection of abnormal growth* Abnormal growth in the brain, certain tumours which cannot be detected by X-rays can be detected by ultrasonic waves.

ix. *Body Shape* Ultrasound cavitation technology can be used to destroy fat cells in the body and as a result, can reduce the body weight keeping the body in perfect shape.

x. *Kidney stones* Lithotripsy is a technique to remove kidney stones by ultrasound.

In addition to the aforementioned medical applications of ultrasound, the most important and popular application of ultrasound is sonography which requires special mention.

11.8 Sonograms

A sonogram, also known as an ultrasound in layman language, is a computerized picture taken by bouncing sound waves off organs and other interior body parts. A wand called a transducer is glided along the outside of the body over a centralized area or organ. As it glides, it introduces sound waves into the body. These sound waves are reflected back into the transducer by the intended area. The transducer feeds the wave into a computer. The computer software decodes all the information concealed in the reflected wave. The picture then appears on a special computer screen. The sonogram is most often used to monitor pregnancy.

As the sonogram uses sound waves and not radiation, it is mostly safe. In addition, a sonogram can offer details X-rays cannot. Doctors can also discover a tubal pregnancy early and take proper measures to ensure the mother's safety. Again a sonogram can also detect a multiple pregnancy, giving the doctor as well as the parents enough time for preparation Birth defects can now be discovered early. Sonograms can also identify the causes of pelvic bleeding and discomfort, find the source of menstrual problems, identify cysts and locate cancerous cells. A sonogram is not just for women, either. It can also be used to help treat prostate and other cancers in men. Parents can even learn the gender of their unborn child months before delivery. Unfortunately, in India, sex determination has become a menace; the male–female ratio has decreased to alarmingly dangerous levels in different states.

There is not too much preparation involved for taking a sonogram. It is all dependent on the area to be examined. For instance, those who are to have an abdominal sonogram may be asked not to eat or drink anything for 24 hours so that their doctor can better examine the stomach. A pregnant woman is usually asked to drink lots of water before her sonogram, as it helps the doctor to see the foetus a little clearer. Loose, comfortable clothing should be worn in order to make the procedure run a little smoother.

11.9 Sonar

Importance of ultrasonics will remain incomplete without the mention of "sonar". Sonar (originally an acronym for Sound Navigation And Ranging) is a technique that uses sound propagation to navigate, communicate with or detect objects on or under the surface of the water. The sonar is of two types – passive sonar that receives sound made by distant vessels and active sonar that simultaneously emits sound pulses and receives echoes. The sinking of the *Titanic* in 1912 most probably laid the foundation stone for the use of sonar and war time demands make it technologically advanced. The acoustic frequencies used in sonar systems vary from very low (infrasonic) to extremely high (ultrasonic). The study of underwater sound is known as underwater acoustics or hydro-acoustics. Acoustic technology is especially well suited for underwater applications since sound travels farther and faster underwater than in air.

Active sonar creates a pulse of sound, often called a "ping", and then the reflections (echo) of the pulse are analyzed. This pulse of sound is generally created electronically using a sonar projector consisting of a signal generator, power amplifier and electro-acoustic transducer. The complication is the active sonar receives echoes from other objects in the sea also such as whales, rocks and so on. Active sonar works the same way as a radar. Passive sonar detects the target's radiated noise characteristics. The radiated spectrum comprises a continuous spectrum of noise with peaks at certain frequencies which can be used to extract the profile of the object. Passive sonar has several advantages. Most importantly, it is silent. If the target radiated noise level is high enough, it can have a greater range than active sonar, and allows the target to be identified. Another use of passive sonar is to determine the target's trajectory. This process is called target motion analysis (TMA), by which a target's range, course, and speed can be determined.

11.9.1 Applications of sonar

Military applications

Modern naval warfare makes extensive use of both passive and active sonar from water-borne vessels, aircraft and fixed installations. The relative usefulness of active versus passive sonar depends on the radiated noise characteristics of the target, generally a submarine. Few sonar applications in military are given here.

i.   **Anti-submarine warfare**   Ship sonars were usually with hull mounted arrays, either amidships or at the bow. It was found after their initial use that a means of reducing flow noise was required to escape enemy surveillance. For this reason, now-a-days domes are usually made of reinforced plastic or pressurised rubber.

ii.   **Torpedoes**   Modern torpedoes are generally fitted with an active/passive sonar to directly attack the target.

iii.   **Mines**   Mines may be fitted with sonar to detect, localize and recognize the required target.

iv.   **Submarine navigation**   Submarines rely on sonar to a greater extent than surface ships as they cannot use radar at depth. Sonar arrays may be hull mounted.

v.   **Aircraft**   Helicopters can be used for anti-submarine warfare by deploying fields of active/passive sonobuoys. Fixed wing aircraft can also deploy sonobuoys and have greater endurance and capacity to deploy them. (A sonobuoy is a relatively small and expendable sonar system that is dropped/ejected from aircraft or ships conducting anti-submarine warfare or underwater acoustic research). Helicopters have also been used for mine counter-measure missions using towed sonars.

vi.   **Underwater communications**   Dedicated sonars can be fitted to ships and submarines for underwater communication.

vii. **Ocean surveillance**   For many years, the United States America operated a large set of passive sonar arrays at various points in the world's oceans for ocean surveillance.

viii. **Underwater security**   Sonar system can be used to detect frogmen and other scuba divers. This can be applicable around ships or at port entrances. Active sonar can also be used as a deterrent and disablement mechanism.

ix. **Intercept sonar**   This sonar system is designed to detect and locate transmissions from hostile active sonars.

Civilian applications

Few civilian applications of sonar are listed here:

i. **Fisheries**   Today, commercial fishing vessels rely almost completely on acoustic sonar and sounders to detect fish. Acoustic technology has been one of the most important driving forces behind the development of modern commercial fisheries. Sound waves travel differently through schools of fish than through water because a fish's air-filled swim bladder has a different density than seawater. This density difference allows the detection of schools of fish by using reflected sound. In advanced countries, fishermen also use active sonar and echo sounder technology to determine water depth, bottom contour, and bottom composition.

ii. **Depth determination**   An echo sounder sends an acoustic pulse directly downwards to the sea bed and records the returned echo. As the speed of sound in water is around 1,500 m/s, the time interval between the pulse being transmitted and the echo being received allows bottom depth and targets to be measured.

iii. **Ship velocity measurement**   Sonars have been developed for measuring a ship's velocity either relative to the water or to the bottom.

iv. **ROV and UUV**   Small sonars have been fitted to Remotely Operated Vehicles (ROV) and Unmanned Underwater Vehicles (UUV) to allow their operation in murky conditions. The sonars are used to supply information about their route conditions.

v. **Vehicle location**   Sonars which act as beacons are fitted to aircraft to allow their location in the event of a crash in the sea.

Scientific applications

Few of the scientific applications of sonar are mentioned here.

i. **Biomass estimation**   Detection of fish, and other marine and aquatic life, and estimation of their individual sizes or total biomass use active sonar techniques. As the sound pulse travels through water, it encounters objects that are of different density or acoustic characteristics than the surrounding medium, such as fish, that reflect sound back towards the sound source. These echoes provide information on fish size, location, abundance and behavior.

ii. **Wave measurement**   An upward looking echo sounder (echo producing device) mounted on the bottom or on a platform may be used to make measurements of wave height and period. From this, statistics of the surface conditions at a location can be derived.

iii. **Water velocity measurement** Special short range sonars have been developed to measure water velocity.

iv. **Bottom type assessment** Sonars have been developed that can be used to determine water depth, bottom contour, and bottom composition of sea at a point.

v. **Bottom topography measurement** Side-scan sonars can be used to derive maps of the topography of an area by moving the sonar across it just above the bottom. Powerful low frequency echo-sounders have been developed for providing profiles of the upper layers of the ocean bottom.

vi. **Synthetic aperture sonar** Various synthetic aperture sonars have been built in the laboratory and some have entered use in mine hunting and search systems.

Therefore, we conclude that ultrasonics have more and more practical applications in all fields. Active research work is still in progress to study and exploit the effect of ultrasonic waves in mechanical, biological, chemical, physical and industrial fields.

11.10 Hazards of Ultrasound

It would be against scientific temper if we do not mention the demerits of ultrasound. In spite of all the advantages of ultrasound, there are some disadvantages of it, though they are very minimal in comparison to other techniques. They are:

i. Tissues or water of the human body absorb the ultrasound energy which increases their temperature locally.

ii. Cavitations are formed when dissolved gases come out of solution due to local heat caused by ultrasound.

iii. High intensity systems are actually used for therapy.

Questions

11.1 Distinguish between ultrasound and infrasound.

11.2 Describe with necessary theory how ultrasound is produced by Galton's whistle.

11.3 Describe how ultrasound can be produced using the magnetostriction principle.

11.4 Describe how ultrasound can be produced using the piezoelectric principle.

11.5 What are the different methods of detection of ultrasounds?

11.6 What are the properties of ultrasonic waves?

11.7 Describe a method of determination of wavelength of ultrasonic waves.

11.8 What is ultrasound cavitation? How it is formed in liquid? Mention parameters that affect ultrasonic cavitation.

11.9   What are the consequences of ultrasonic cavitation?

11.10 Explain how the statistics of big objects in deep sea can be determined using ultrasonics.

11.11 Explain how ultrasonics can be used in signalling.

11.12 Explain how ultrasound is helpful to a mechanical engineer.

11.13 Explain how ultrasound is helpful to a chemical engineer.

11.14 Explain how ultrasound is helpful to a metallurgist.

11.15 Explain how ultrasound is helpful to a doctor.

11.16 Explain how ultrasound is helpful to defence personnel.

11.17 What are the dangers of ultrasound?

Problems

11.1   A piezoelectric X-cut quartz plate has a thickness of 1.2 mm. If the velocity of propagation of longitudinal sound waves along the x direction is 5460 m/s, calculate the fundamental frequency of the crystal. [Ans   2275 kHz]

11.2   A quartz crystal of thickness 0.14 cm is vibrating at resonance. Calculate the fundamental frequency of vibration if Young's modulus and density for quartz is 8.79×10^{10} N/m^2 and 2650 kg/m^3 respectively. [Ans   2057 kHz]

11.3   The speed of ultrasonic wave in a certain medium is 5050 m/s. If the wavelength of the ultrasonic wave is 2.5×10^{-3} m, find the frequency of the ultrasonic wave. [Ans   2.02 MHz]

11.4   Acoustic grating is formed in a water medium by passing monochromatic ultrasonic wave at a speed 1500 m/s. The second order diffraction maximum is observed at an angle 1.5° for sodium light of wavelength 5890Å. Calculate the frequency of the ultrasonic wave. [Ans   16.67 MHz]

11.5   A marine engineer measured the depth of sea at a point by sending ultrasonic wave towards the sea bed. He received the reflected wave after three seconds. What will be the depth if the speed of ultrasonic wave in sea water is 1750 m/s? [Ans   2625 m]

Multiple Choice Questions

1.   The frequency range of hearing for a normal human being is

   (i)   20 Hz–20 kHz   (ii)   20 kHz–20 MHz

   (iii) more than 20 kHz   (iv) less than 20 Hz

2.   The frequency range of ultrasound is

   (i)   20 Hz–20 kHz   (ii)   20 kHz–20 MHz

   (iii) more than 20 kHz   (iv) less than 20 Hz

3.   The frequency range of infrasound is
      (i)   20 Hz–20 kHz   (ii)   20 kHz–20 MHz
      (iii) more than 20 kHz   (iv)   less than 20 Hz

4.   Galton's whistle works on the principle of
      (i)   Hartley oscillator   (ii)   red-shift
      (iii) organ pipe   (iv)   Kundt's tube

5.   By which of the following methods, can the presence of ultrasonic waves not be detected?
      (i)   Kundt's tube method   (ii)   Sensitive flame method
      (iii) Thermal detector method   (iv)   Position sensitive method

6.   Magnetostriction is a property of
      (i)   pyroelectric material   (ii)   piezoelectric materials
      (iii) ferromagnetic materials   (iv)   ferroelectric material

7.   High frequency magnetic field is applied to the crystal to produce ultrasound by magnetostriction method using
      (i)   rotating bar magnet   (ii)   rotating electromagnet
      (iii) atomic currents of the crystals (iv) inductively coupled circuits

8.   Acoustic grating can be formed in
      (i)   solid medium   (ii)   liquid medium
      (iii) vacuum   (iv)   plasma medium

9.   Acoustic grating can diffract
      (i)   sound wave   (ii)   mechanical wave
      (iii) light wave   (iv)   none of the above

10.   What is the full form of sonar?
      (i)   sound out navigation audible ranging
      (ii)   sound navigation audible ranging
      (iii) sound of navigation and ranging
      (iv)   sound navigation and ranging

Answers

| 1 (i) | 2 (iii) | 3 (iv) | 4 (iii) | 5 (iv) | 6 (iii) | 7 (iv) | 8 (ii) |
| 9 (iii) | 10 (iv) | | | | | | |

12 Non-Destructive Testing

12.1 Introduction

The field of non-destructive testing (NDT) is a very broad area. It plays a critical role in assuring that structural components and systems perform their function in a reliable and cost effective fashion. NDT is a part of the quality control process. Non-destructive testing (NDT)/non-destructive inspection (NDI)/non-destructive evaluation (NDE) is a wide group of analysis techniques used in science and industry to detect surface or internal flaws of a material, component or system without causing damage. Since NDT does not permanently alter the article being inspected, it is a highly valuable method that can save both money and time in product evaluation, troubleshooting and research. NDT is not just a method for rejecting sub-standard materials; it is an assurance that the supposedly good is really good. NDT is a commonly used tool in forensic engineering, mechanical engineering, electrical engineering, civil engineering, systems engineering, aeronautical engineering, metallurgical engineering, electronic engineering, medicine, and art. In this chapter, we shall discuss the basic principle involved in NDT and very briefly, the methods of NDT.

12.2 Objectives of NDT

NDT provides an excellent balance between quality control and cost-effectiveness. There are NDE applications at almost any stage in the production or life cycle of a component.

i.   To assist in product development.

ii.   To screen or sort out incoming materials.

iii. To assist in product development.

iv.   To monitor, improve or control manufacturing processes.

v. To verify proper processing such as heat treatment.

vi. To verify proper assembly.

vii. To inspect in-service damage.

NDE may be used to determine material properties such as fracture toughness, ductility, conductivity, and other physical characteristics. NDT are used in

i. Flaw detection and evaluation.

ii. Leak detection.

iii. Location determination.

iv. Dimensional measurements.

v. Structure and microstructure characterization.

vi. Estimation of mechanical and physical properties.

vii. Stress/Strain and dynamic response measurements.

viii. Material sorting and chemical composition determination.

12.3 Methods of NDT

The NDT technique uses a variety of scientific principles. The list of NDT methods that can be used to inspect components and make measurements is large and continues to grow. Researchers continue to find new ways of applying principles of physics and other scientific disciplines to develop better NDT methods. However, there are six NDT methods that are used most often. These methods are

i. Visual inspection.

ii. Dye penetrant testing.

iii. Magnetic Particle Testing.

iv. Electromagnetic or eddy current testing.

v. Radiography.

vi. Ultrasonic testing.

However, these are by no means the total of the principles available to the NDT engineer. Electrical potential drop, sonics, infra-red, acoustic emission, and spectrography, to name but a few, have been used to provide information that the aforementioned techniques have been unable to yield.

12.3.1 Visual and optical testing (VOT)

This is the most basic and common inspection method. Visual inspection involves using an inspector's eyes to look for defects. The inspector may also use special tools such as magnifying glasses, mirrors, or borescopes to gain access and more closely inspect the subject area. Portable video inspection units with zoom allow inspection of large tanks and vessels, railroad tank cars, sewer lines. Robotic crawlers permit observation in hazardous or tight areas, such as air ducts, nuclear reactors, pipelines. Visual examiners follow procedures that range from simple to very complex.

12.3.2 Dye penetrant testing (DPT)

This method is frequently used for the detection of surface breaking flaws in non-ferromagnetic materials. The object to be examined is first of all chemically cleaned so that all traces of foreign material, grease, dirt, and so on from the surfaces, and also from within the cracks are removed. Next the penetrant (which is a very fine thin oil usually dyed bright red or ultra-violet fluorescent) is applied and allowed to remain in contact with the surface for approximately fifteen minutes. Capillary action draws the penetrant into the crack during this period. The surplus penetrant on the surface is then removed completely and a thin coating of developer (a simple example of developer is chalk powder) is applied. After a further period (development time), the developer acts as a blotter, drawing the trapped penetrant out of imperfections open to the surface. With visible dyes, vivid color contrasts between the penetrant and the developer making a "bleedout". With fluorescent dyes, ultraviolet light is used to make the bleedout fluoresce brightly, thus allowing imperfections to be readily seen. The process is purely a mechanical and chemical one.

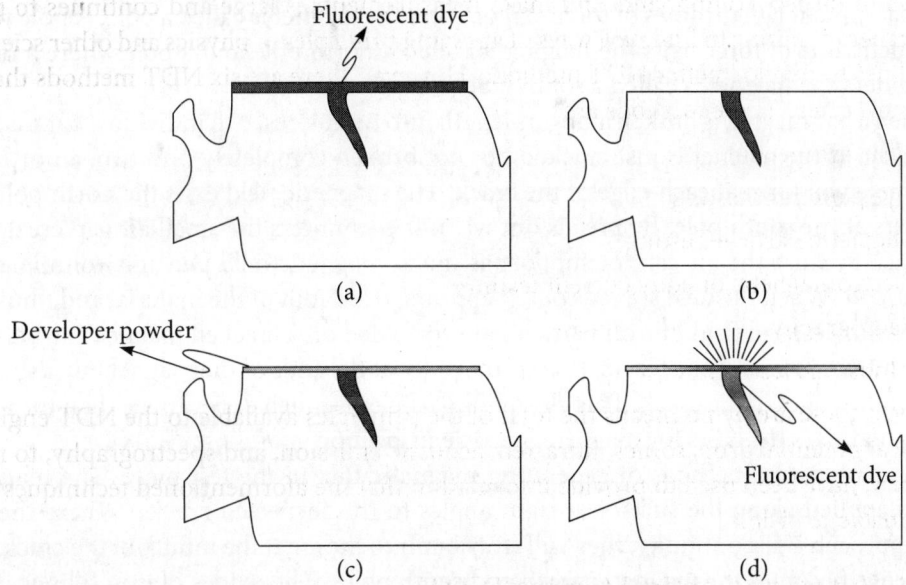

Figure 12.1 | (a) Penetrant applied to the surface enters the defect, (b) Excess penetrant removed from the surface, (c) Developer powder applied to the surface to draw the penetrant out, (d) The crack become prominent under ultraviolet light as the penetrant spreads out of the crack

Advantages

i.   Simple in operation.

ii.   It is the best method for surface breaking cracks in non-ferrous metals.

iii. It is quantitative.

Disadvantages

i.   Restricted to surface breaking defects only

ii.   Decreased sensitivity

iii. Uses a considerable amount of consumables.

12.3.3 Magnetic particle testing

This method is suitable for the detection of surface and near surface discontinuities in ferromagnetic material. It is carried out by inducing a magnetic field in a ferromagnetic material and then dusting the surface with iron particles (either dry or suspended in liquid). Surface and near-surface imperfections distort the magnetic field and concentrate iron particles near imperfections, previewing a visual indication of the flaw.

Basic principles

Any place that magnetic lines of force exit or enter the magnet is called a pole. A pole where a magnetic line of force exits the magnet is called a north pole and a pole where a line of force enters the magnet is called a south pole.

    When a bar magnet is broken along its length into two pieces, two individual bar magnets will result. If the magnet is just cracked but not broken completely into two, a north and south pole will form at each edge of the crack. The magnetic field exits the north pole and re-enters at the south pole. It spreads out when it encounters the small air gap created by the crack because the air cannot support as much magnetic field per unit volume as the magnet can. When the field spreads out, it appears to leak out of the material and, thus, it is called a flux leakage field. If iron particles are sprinkled on a cracked magnet, the particles will be attracted to and cluster not only at the poles at the ends of the magnet but also at the poles at the edges of the crack. This cluster of particles is much easier to see than the actual crack and this is the basis for magnetic particle inspection.

    There are many methods of generating magnetic flux in the test piece. The magnetic field is applied along the surface at right angles to the suspected cracks. Where the flux lines approach a discontinuity, they will stray out into the air at the mouth of the crack. The crack edge becomes the magnetic north and south poles. The oxides of iron fillings in the size range 20 to 30 microns suspended in a liquid are attracted towards the crack edges. In some instances, they can be applied in a dry powder form. The particles can be coated with a fluorescent dye which fluoresces brilliantly under ultraviolet illumination. The technique not only detects those defects which are not normally visible to the unaided eye, but also renders easily visible those defects which would otherwise require close scrutiny of the

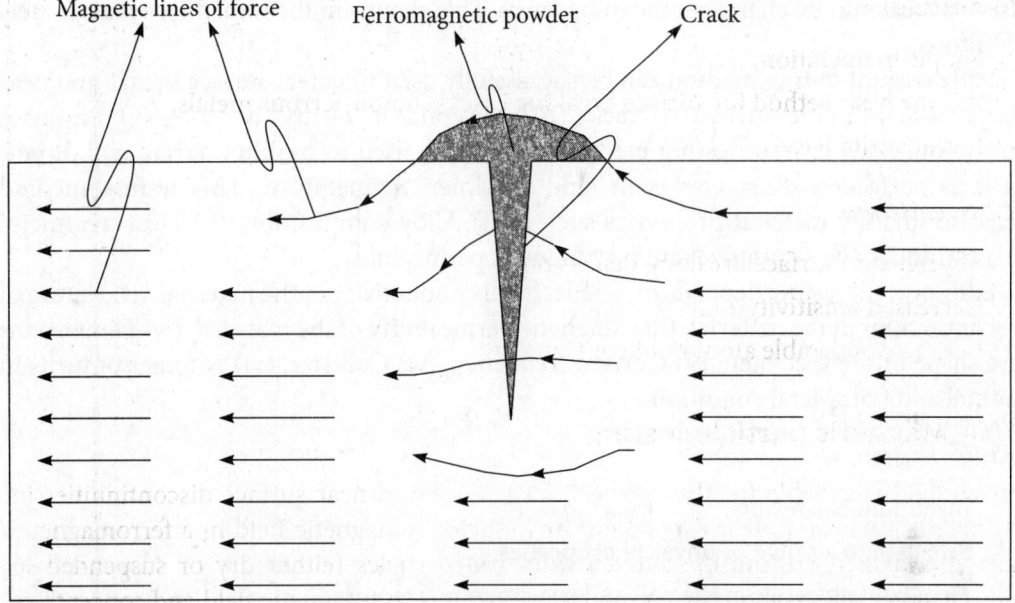

Magnetic lines of force   Ferromagnetic powder   Crack

Figure 12.2 | An illustration of the principle of magnetic particle inspection

surface. Normally, to ensure that a test piece has no cracks, it is necessary to magnetize it in at least two directions.

Advantages

i.   Simplicity of operation and application

ii.   It is quantitative

iii. It can be automated

Disadvantages

i.   It is restricted to ferromagnetic materials

ii.   It is also restricted to surface or near surface flaws

12.3.4 Electromagnetic or eddy current testing

A varying electric current flowing in a coil gives rise to a varying magnetic field. A nearby conductor resists this magnetic field by producing a current in it. This current flows in circles just below the surface of the material and hence is called eddy current. Eddy current in the conductor produces a magnetic field that opposes the magnetic field produced by the coil, resulting in a change of impedance of the coil. It is this impedance change that is to be detected with a high degree of accuracy by the measuring equipment. Cracks and other material conditions change the magnetic field of the eddy currents and hence give rise to a local minute change in the impedance. This change in the impedance is accurately monitored.

Eddy current testing method can be successfully used to detect surface breaks and near surface discontinuities such as (i) Cracks, (ii) Inclusions, (iii) Dents, (iv) Holes, (v) Scratches and so on. Eddy current testing method can also be used to monitor surface conditions such as surface coating, corrosion, and specimen temperature. This testing method can also identify material properties such as (i) Alloy composition, (ii) Heat treatment, (iii) Hardness, (iv) Grain size and (v) Magnetic permeability.

Eddy current testing depends on (i) Electrical conductivity of the material, (ii) Nature of discontinuities in the material, (iii) Magnetic permeability of the material, (iv) Dimensions and shape of the specimen, (v) Current frequency, (vi) Coil size, (vi) Number of turns in the coil and (vii) Metal condition.

Advantages:

i.   Instantaneous results.
ii.   Sensitive to a range of physical properties.
iii. Firm contact between the coil and specimen not required.
iv.   Equipment is small and self-contained.
v.   Can detect very small discontinuities.
vi.   Defects in tubes and other circular parts can be detected using special probes.

Disadvantages:

i.   It can be used on electrical conductors only.
ii.   Depth of penetration is restricted.
iii. Interpretation needs skill.
iv.   Defects parallel to coil surface can be missed.
v.   Ends of the parts cannot be tested.

12.3.5 Radiographic testing

In this method of non-destructive testing, the penetration property of X-rays and gamma rays is exploited to detect the discontinuities in the materials. X-rays and gamma rays are the shorter wavelength (less than 100 nm) part of the electromagnetic spectrum. The object to be inspected is placed between the radiation source and a piece of film. The energetic X-rays or gamma rays pass through the object. X-rays and gamma rays are differentially absorbed by different material. Also, greater the thickness, greater is the absorption. Furthermore, denser the material, greater is the absorption. Thicker and denser areas will stop more of the radiation and show on the film lighter than thinner or less dense areas. Most weld defects will show on the film darker than the surrounding area. When materials with internal voids are tested by this method, the voids appear as darkened areas on the film, where more radiation has reached the film, on a lighted background. The principles are the same for both X-ray and gamma-ray radiography.

Recent developments in radiography permit "real time" analysis. Such techniques as computerized tomography yield much important information, though these methods maybe suitable for only investigative purposes and not generally employed in production quality control.

Advantages

i.   Information is presented pictorially.

ii.   Suitable for most materials.

iii.   Gives a permanent record which may be viewed at a time and place distant from the test.

iv.   Detects internal flaws.

v.   Detects volumetric flaws readily.

vi.   Can be used on most materials.

vii.   Can check for correct assembly.

viii. Gives direct images.

ix.   Real time image is possible.

Disadvantages

i.   Radiation health hazards.

ii.   Not suitable for surface defects.

iii.   Has limited ability to detect fine cracks.

iv.   Access is required to both sides of the object.

v.   Limited thickness of the material can be penetrated.

vi.   Skilled radiographic interpretation is required.

vii.   Film processing and viewing facilities are necessary.

viii. Require high capital cost.

ix.   Relatively slow process.

x.   Require high running cost due to use of consumables.

12.3.6 Ultrasonic testing

Sound waves of frequency more than 20 kHz are called ultrasound and these sound waves are not sensed by a normal human ear. Nevertheless, the ultrasound is used in many different fields, typically to penetrate a medium and measure the reflection signature or supply focused energy. The reflection signature can reveal details about the inner structure of the medium. The most well-known application of ultrasound is screening of the uterus during pregnancy to check the well-being of the baby. There are a vast number of other applications as well.

Ultrasonic testing (UT) uses high frequency sound energy to conduct internal inspection and make measurements. Ultrasonic inspection can be used for flaw detection/evaluation, dimensional measurements, material characterization, and more. Frequencies of 0.5 MHz to 25 MHz are used in this method so that the resulting wavelengths are in mm (wavelength = speed/frequency). Inspection may be manual or automated and is an essential part of modern manufacturing processes. Most metals, plastics and aerospace composites can be inspected. Lower frequency ultrasound (50 kHz–500 kHz) can also be used to inspect less dense materials such as wood, concrete and cement.

Basic principles of ultrasonic testing

Sound energy propagates through the materials in the form of waves. When there is a discontinuity (such as a crack) in the wave path, part of the energy will be reflected back (from the flaw surfaces). The reflected wave signal is transformed into an electrical signal by the transducer and is displayed on a screen. The principle is illustrated schematically in Fig. 12.3.

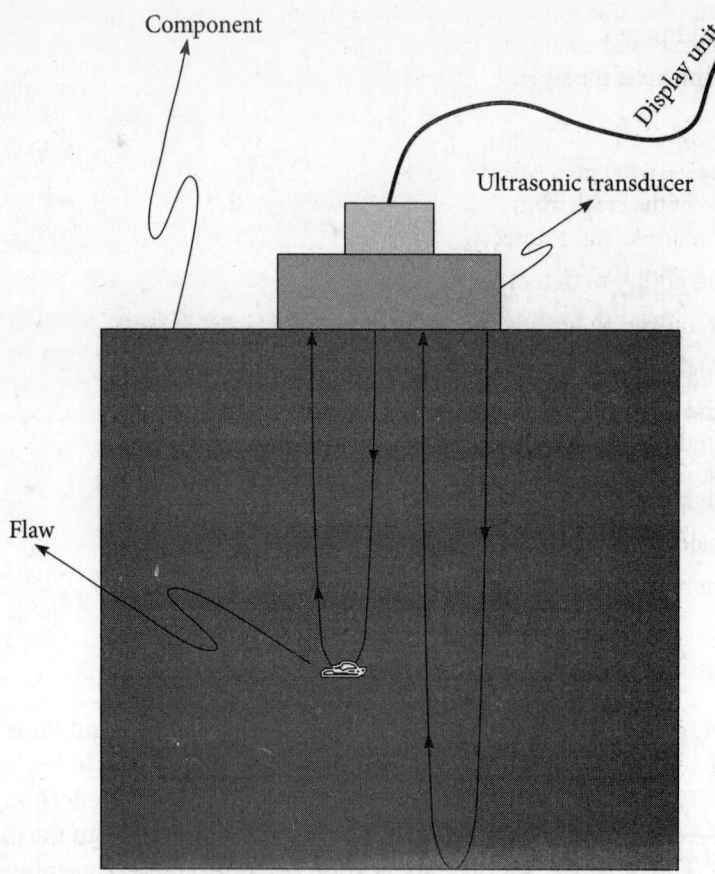

Figure 12.3 Basic principles of ultrasonic testing

The principle is in some respects similar to echo sounding. A short pulse of ultrasound is generated by a piezoelectric crystal, which vibrates for a very short period at an ultrasonic frequency range. In flaw detection, this frequency is usually in the range of 1 MHz to 6 MHz. Vibrations or sound waves at this frequency have the ability to travel a considerable distance in homogeneous elastic material, such as many metals with little attenuation. The velocity at which these waves propagate is related to Young's modulus and Poisson's ratio for the material in the following way and is characteristic of that material. The pulse/echo velocity, v, of a material is calculated by the formula

$$v = \frac{2\ell}{\Delta t}$$

(12.1)

Here ℓ = material thickness and Δt = time of round trip.

Example 12.1

An aluminium plate was tested for internal cracks by the ultrasonic method. The reflection signature was received after 10^{-4} seconds. If speed of the ultrasound in the material is 6400 m/s, then what is the location of the crack from the surface?

Solution

Data given are $v = 6400$ m/s, $\Delta t = 10$ s
   The position of the crack from the surface is calculated as

$$\ell = \frac{v\Delta t}{2} = \frac{6400 \times 10^{-4}}{2} \, m = 32 \, cm$$

In solids, sound waves can propagate in different modes that are based on the way the particles oscillate. Sound can propagate as longitudinal waves, shear waves, surface waves, and as plate waves. Longitudinal and shear waves are the two modes of propagation most widely used in ultrasonic testing. Thus, the velocity measurement depends on the density, elastic modulus, parallelism of the front and back surface of the specimen and specimen thickness.

   Thin rod velocity

$$v_t = \sqrt{\frac{Y}{\rho}}$$

(12.2)

Longitudinal wave velocity

$$v_L = \sqrt{\frac{Y(1-\sigma)}{\rho(1+\sigma).(1-2\sigma)}}$$

(12.3)

Transverse wave velocity

$$v_T = \sqrt{\frac{Y}{2\rho(1+\sigma)}} = \sqrt{\frac{G}{\rho}} \qquad (12.4)$$

where

Y = Young's modulus of elasticity

G = shear modulus of elasticity

ρ = density

σ = Poisson's ratio

Example 12.2

Young's modulus of brass is $9.1 \times 10^{10}\,\mathrm{Nm^{-2}}$. Calculate the speed of ultrasonic wave in a brass wire if its density is 8500 kgm^{-3}.

Solution

Data given are $Y = 9.1 \times 10^{10}\,\mathrm{Nm^{-2}}$, $\rho = 8500$ kgm^{-3}

The speed of ultrasonic wave is $v = \sqrt{\dfrac{Y}{\rho}} = \sqrt{\dfrac{9.1 \times 10^{10}\,\mathrm{Nm^{-2}}}{8500\,\mathrm{kgm^{-3}}}} = 3272.0\,\mathrm{m/s}$

Example 12.3

Young's modulus, shear modulus and density of brass are $9.1 \times 10^{10}\,\mathrm{Nm^{-2}}$, $3.5 \times 10^{10}\,\mathrm{Nm^{-2}}$ and 8500 kgm^{-3} respectively. Calculate the speed of ultrasonic longitudinal wave and transverse wave in it.

Solution

The data given are $Y = 9.1 \times 10^{10}$ Nm^{-2}, $\eta = 3.5 \times 10^{10}\,\mathrm{Nm^{-2}}$, and $\rho = 8500$ kgm^{-3}
Poisson's ratio σ is obtained as

$$\sigma = \frac{Y}{2\eta} - 1 = 0.3$$

The longitudinal wave speed

$$v_L = \sqrt{\frac{Y(1-\sigma)}{\rho(1+\sigma)(1-2\sigma)}} = \sqrt{\frac{9.1 \times 10^{10} \times (1-0.3)}{8500 \times (1+0.3) \times (1-0.6)}}\,\mathrm{m/s}$$

$$= 3796.3\ \mathrm{m/s}$$

The transverse wave speed

$$v_T = \sqrt{\frac{Y}{2\rho(1+\sigma)}} = \sqrt{\frac{9.1\times10^{10}}{2\times8500(1+0.3)}} = 2029.2\,\text{m/s}$$

12.3.7 Pulse–echo system

The most common system used in ultrasonic thickness measurement and ultrasonic flaw detection is the pulse–echo system. In the pulse–echo ultrasonic testing technique, an ultrasound transducer generates an ultrasonic pulse and receives its "echo". The ultrasonic transducer functions as both a transmitter and a receiver in one unit. Most ultrasonic transducer units use an electronic pulse to generate a corresponding sound pulse, using the piezoelectric effect. Here the piezoelectric transducer is repeatedly excited for a short duration to generate ultrasonic pulses. There is a delay of nano seconds between each pulse.

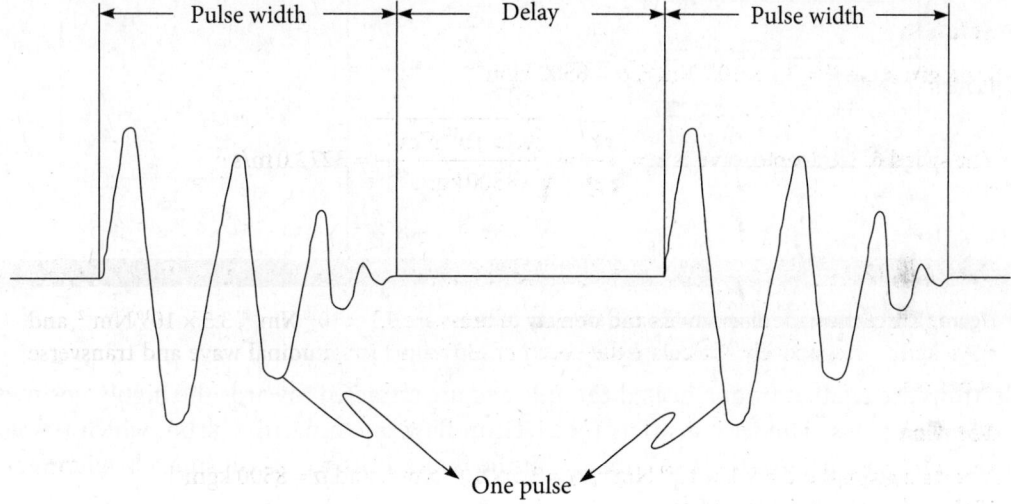

Figure 12.4 | The pulse–echo system

Sound wave pulses travel through the material under test until they meet an interface or boundary (Fig. 12.4), where they are reflected back. If the sound hits the interface at right angles, then the reflected sound travels back to the transducer as an echo. Echoes coming back to the transducer are re-converted into electrical signals and the time between transmitting the pulse and receiving the echo is electronically measured.

The time between any two echoes Δt is the length of time required for the pulse to travel through the specimen and back to the transducer. The speed of sound in the solid can be derived from the observed round trip transit time, Δt, and the measured thickness of the specimen, ℓ because

$$\ell = \frac{1}{2} v\, \Delta t \tag{12.5}$$

By calibrating the ultrasonic equipment for the speed of sound in the test material, the equipment is able to display the time taken for the pulse–echo to travel through the material as a distance.

Visual display units

The display of information can take several forms depending on the type of flaw detector, but all units use a cathode ray tube as shown in Fig. 12.5.

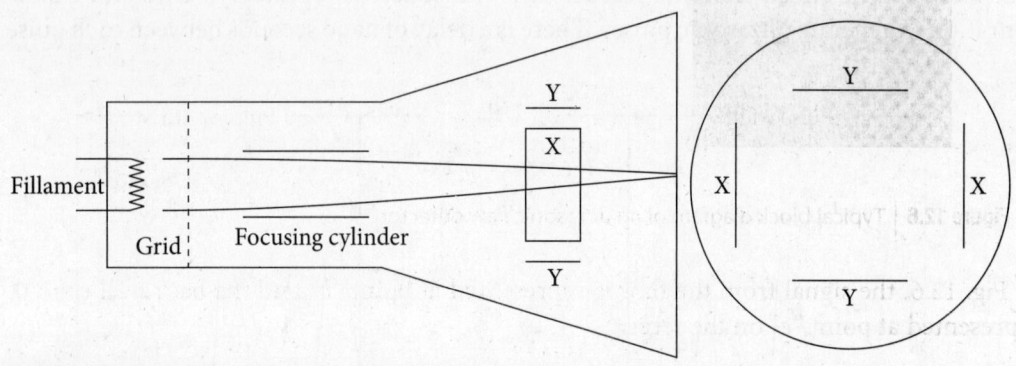

Figure 12.5 | Schematic diagram of a cathode ray tube

Electrons are emitted from a heated cathode and are attracted towards the highly positive anode. The focus cylinder constricts the electron flow into a narrow beam, which passes through the anode cylinder to eventually hit the fluorescent screen causing a bright green display.

The brightness of the display is controlled by the grid that, which controls flow rate of electrons. The horizontal and vertical movements of the electrons are controlled by the X and Y plates respectively by applying potentials across the plates. Changing the potential between the X plates, for example, causes the electron beam to traverse the screen. Figure 12.6 gives a general block diagram of an ultrasonic flaw detector.

The pulse generator sends a pulse to the probe and also triggers the time base generator. The time base generator causes the electron beam to cross the screen at the same rate as the ultrasonic pulse emitted from the probe crosses the steel block and back. The initial pulse appears at point "*a*" on Fig. 12.6. The electrical signals from the receiver transducer are amplified, and fed to the Y plates where they cause deflections in the electron beam.

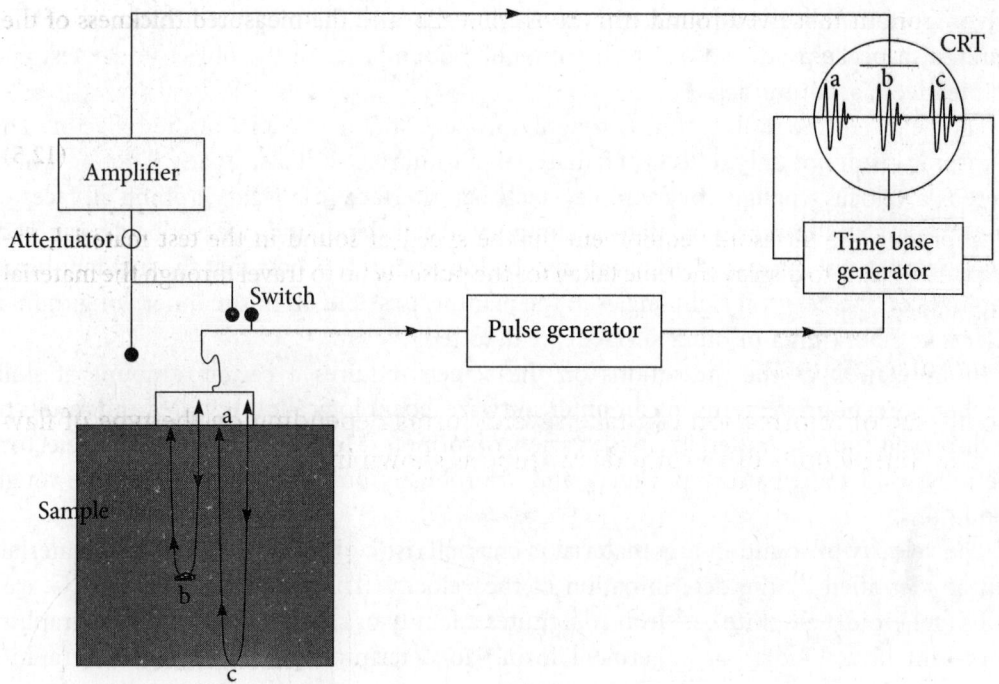

Figure 12.6 | Typical block diagram of an ultrasonic flaw detector

In Fig. 12.6, the signal from the flaw is represented at point "b" and the back wall echo is represented at point "c" on the screen.

Calibration of flaw detectors

The horizontal and vertical scales on a flaw detector display are only quantitative when they have been calibrated properly. The horizontal scale, more commonly known as the time base, can be calibrated to give depth values for different materials and sound velocities. The vertical or amplitude scale can be calibrated to give information on defect size. The method normally employed to obtain quantitative information about a test piece is to compare the screen signals with those from specially machined blocks.

For contact testing, the oscillating crystal is incorporated in a hand-held probe, which is applied to the surface of the material to be tested. Ultrasonic energy is considerably attenuated in air. Moreover, considerable amount of ultrasonic energy is reflected back at the interface between the material and air. To facilitate the transfer of ultrasonic energy efficiently across the small air gap between the probe and the test piece, the probe must be coupled to the material surface by means of a liquid, like glycerine, oil or grease. These are called "couplants". Piezoelectric materials can be used as not only a generator of sound waves but also a detector of returned pulses. The crystal is in a quiescent state to detect returned pulses. The pulse takes a finite time to travel through the material to the interface and to be reflected back to the probe. Therefore, it is possible not only to discover a defect between the surface and the back wall, but also to measure its distance below the surface.

It is important that the equipment is properly calibrated. The operator must be able to distinguish peaks produced due to the intended boundaries of the object under test and unintended discontinuities.

The height of the peak (echo) is roughly proportional to the area of the reflector. The ultrasonic beam not only reflects at a material–air interface but also at any junction where there is a velocity change, for example, steel/slag interface in a weld. Probing all faces of a test piece, three-dimensional defects along with their depth and size can be determined. Two-dimensional (planar) defects can also be found out. It is best that the incident beam impinges on the defect at right angles to the plane as possible. In this manner, longitudinal defects in tubes (inner or outer surface) are detected.

Interpretation of the indications on the screen requires a certain amount of skill. Furthermore, improvements in computer software technology allow test data and results to be displayed and out-putted in a wide variety of formats. Modern ultrasonic flaw detectors are fully solid state, battery powered, and are robustly built to withstand on-site rough conditions.

The velocity of sound in any material is characteristic of that material; some materials can be identified by the determination of the velocity. This can be applied in S.G. cast irons (spheroidal graphite cast iron constitutes a family of cast irons in which the graphite is present in a nodular or spheroidal form.) to determine the percentage of graphite nodularity.

The process can also be automated and is now in use in many foundries. Typical equipment is the qualiron. It automatically clamps the casting between the transducer and the anvil and provides a reading of the sound velocity. In a wide range of steels velocity is constant. Hence, the time taken for the pulse to travel through the material is proportional to its thickness. Therefore, with a properly calibrated instrument, it is possible to measure thickness from one side with an accuracy in thousandths of an inch. This technique is now in very common use. A development of the standard flaw detector is the digital wall thickness gauge. A typical ultrasonic flaw detection is illustrated in Fig. 12.7(a) and (b).

Applications

i. Detection of internal flaws in metal parts and alloys

ii. Bonds produced by welding, brazing, soldering and adhesive bonding can also be ultrasonically inspected

iii. For quality control and material inspection in all industries

iv. For measuring thickness of metal sections

v. Special ultrasonic techniques and equipments have been used for determination of rate of growth of fatigue cracks, measurement of elastic moduli, nodularity in cast iron

Advantages

i. Its superior penetrating power as compared to other non-destructive (NDT) methods allows the detection of flaws several feet deep in the part.

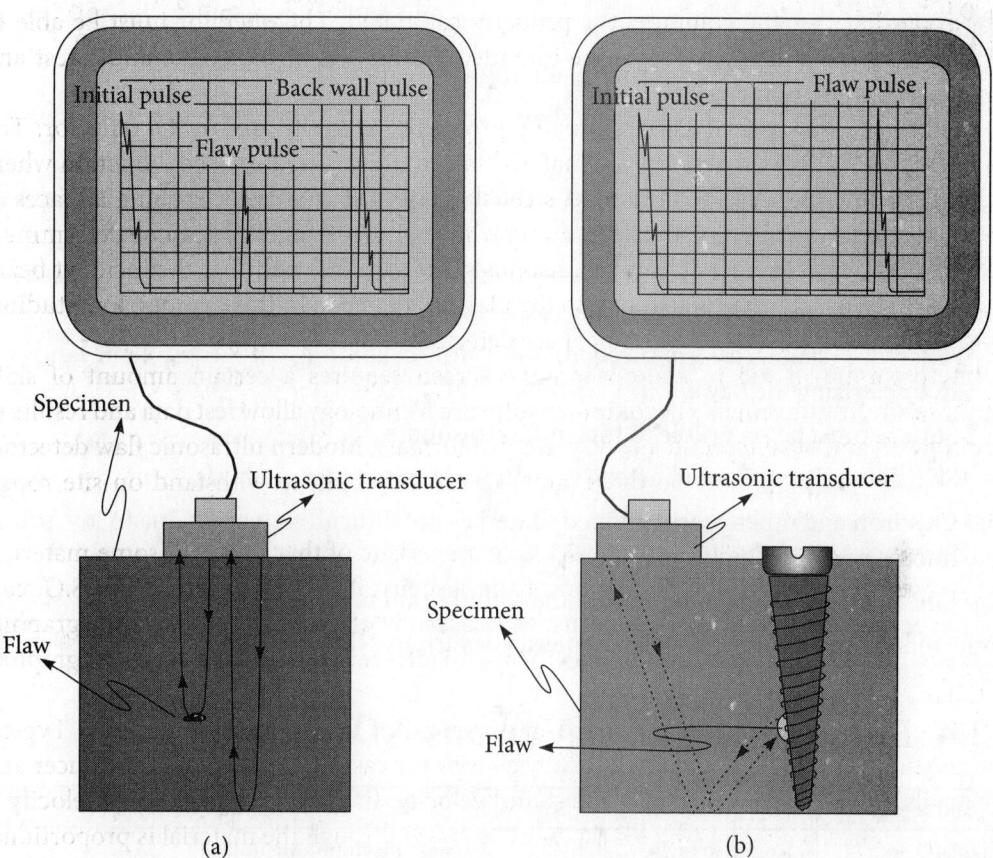

Figure 12.7 | (a) Schematic diagram of an ultrasonic detection slag in steel section using a normal probe. (b) Schematic diagram of the use of an angle probe to detect defects not directly under the probe as in weld inspection

ii. It is sensitive to both surface and sub-surface discontinuities. It is more accurate than other NDT methods in determining the position of internal flaws, estimating their size and characterizing their orientation, shape and nature.

iii. Only single-sided access is needed when the pulse–echo technique is used.

iv. Operation is electronic which provides instantaneous indication of flaws. This makes the method suitable for immediate interpretation, automation, rapid scanning and online production monitoring and process control. With most systems, a permanent record of inspection results can be made for future reference.

v. Minimal part preparation is required.

vi. It is not hazardous to operations or to nearby personnel, and has no effect on equipment and materials in the vicinity.

vii. Detailed images can be produced with automated systems.

viii. It has other uses such as thickness measurements, in addition to flaw detection.

Disadvantages

i.   Manual operation requires careful attention by experienced technicians

ii.   Parts that are rough, irregular in shape, very small or thin or not homogeneous are difficult to inspect.

iii. Discontinuities that are present in a shallow layer just below the surface may not be detectable.

iv.   Couplants are needed to provide effective transfer of ultrasonic wave energy between transducers and parts being inspected.

v.   Reference standards are needed, both for calibrating the equipment and for characterizing the flaws.

vi.   Surface must be accessible to transmit ultrasound.

vii. Skill and training is more extensive than with some other methods.

viii. Cast iron and other coarse grained materials are difficult to inspect due to low sound transmission and high signal noise.

ix.   Linear defects oriented parallel to the sound beam may go undetected.

In the following, we show the relative merits of various NDT methods in tabular form.

12.4 Relative Merits of Various NDT Methods

Test method	U.T.	R.T.	E.T.	M.T.	P.T.
Capital cost	Medium to high	High	Low to medium	Medium	Low
Consumable cost	Very low	High	Low	Medium	Medium
Time of results	Immediate	Delayed	Immediate	Short delay	Short delay
Effect of geometry	Important	Important	Important	Not too important	Not too important
Access problems	Important	Important	Important	Important	Important
Type of defect	Internal	Most	External	External near	Surface breaking
Relative sensitivity	High	Medium	High	Low	Low
Operator skill	High	High	Medium	Low	Low
Operator training	Important	Important	Important	Important	Not important
Training needs	High	High	Medium	Low	Low
Portability of equipment	High	Low	High to medium	High to medium	High
Capabilities	Thickness gauging, composition testing	Thickness gauging	Thickness gauging, grade sorting	Defects only	Defects only

R.T.: X- or Gamma radiography; M.T.: Magnetic particle inspection; P.T.: Dye penetrant; U.T.: Ultrasonic; E.T.: Eddy current.

In the following, we show the various NDT methods and their applications in tabular form.

12.5 Non-Destructive Testing Methods and Applications

Material	Flaw type						
	Surface Cracks & Flaws	Sub-Surface Cracks & Flaws	Internal Flaws & Discont-inuities	Lack of Bond or Lack of Fusion	Non-Metallic Inclusions – Slag, Porosity	Material Quality	Lamina-tions, Thickness Measur-ement
Ferrous Forgings & Stampings	M.T.	M.T. & U.T.	R.T. & U.T.		R.T. & U.T.		U.T.
Ferrous raw Materials & Rolled Products	M.T.	M.T. & U.T.	U.T.		M.T. & U.T.		U.T.
Ferrous Tube & Pipe	M.T. & E.T.	M.T. & U.T.	U.T.	U.T.	M.T. & U.T.		U.T.
Ferrous Welds	M.T. & U.T.	U.T.	R.T. & U.T.	R.T. & U.T.	R.T. & U.T.		U.T.
Steel Castings	M.T.	M.T. & U.T.	R.T. & U.T.		R.T. & U.T.		U.T.
Iron Castings	M.T.	U.T. & E.T.	U.T.		R.T. & U.T.	U.T.	U.T.
Non-Ferrous Components & Materials	P.T. & E.T.		R.T. & U.T.	U.T.	P.T. & U.T.		U.T.
Ferrous Components Finished	M.T.	U.T. & E.T.	R.T. & U.T.	U.T.	M.T. & U.T.		U.T.
Non-Ferrous Components Finished	P.T. & E.T.	U.T. & E.T.	R.T.& U.T.		U.T. & E.T.		U.T.
Aircraft Ferrous Components	R.T. & M.T.	M.T. & U.T.	R.T. & U.T.	U.T.	M.T. & U.T.		U.T.
Aircraft Non-Ferrous Components	R.T., P.T. & E.T.	R.T. & U.T.	R.T. & U.T.	U.T.	P.T. & U.T.		U.T.

Questions

12.1 What are the objectives and methods of NDT?

12.2 Describe the dye penetrant testing used in detection of cracks in metals.

12.3 Describe magnetic particle testing along with its basic principle. Mention some of its advantages and disadvantages.

12.4 Describe electromagnetic or eddy current testing. Mention some of its advantages and disadvantages.

12.5 What is radiography testing? Mention some of its advantages and disadvantages.

12.6 Explain how ultrasound is used in flaw detection. Mention some of its advantages and disadvantages.

12.7 Describe the ultrasound pulse–echo system used in flaw detection in materials.

12.8 Mention various NDT methods and their relative merits.

12.9 A manufacture unit has supplied non-ferrous components of aircraft. What are different methods to ensure their quality?

12.10 A manufacture unit has supplied ferrous components of aircraft. What are different methods to ensure their quality?

12.11 What are the methods to check the quality of concrete slabs?

12.12 Mention different methods to ensure the quality of steel and iron castings.

Problems

12.1 A brass ingot was tested for internal cracks by ultrasonic method. The reflection signature was received after 10^{-4} seconds. If speed of ultrasound in the material is 4490 m/s, then what is the location of the crack from the surface?

[Ans   22.45 cm]

12.2 Young's modulus of aluminium is 7.0×10^{10} Nm^{-2}. Calculate the speed of ultrasonic wave in a brass wire if its density is 2700 kgm^{-3}.   [Ans   5091.8 m/s]

12.3 The Young's modulus, shear modulus and density of aluminium are 7.0×10^{10} Nm^{-2}, 2.5×10^{10} Nm^{-2}, and 2700 kgm^{-3} respectively. Calculate the speed of ultrasonic longitudinal wave and transverse wave in it.

[Ans   7453.6 m/s, 3242.9 m/s]

Multiple Choice Questions

1.   Which of the following is not an NDT method?

(i)   Visual inspection   (ii) Radiography

(iii) Filtration   (iv) Penetrant testing

2.   Which of the following are not used in visual and optical testings methods?

   (i)   Lens   (ii)   mirrors

   (iii) electromagnetic field   (iv) magnetic field

3.   Which of the following NDT methods is used frequently for the detection of surface breaking flaws in non-ferromagnetic materials?

   (i)   Dye penetrant testing   (ii)   Magnetic particle testing

   (iii) Eddy current testing   (iv) Radiography

4.   Which of the following NDT methods is used frequently for the detection of surface breaking flaws in ferromagnetic materials?

   (i)   Dye penetrant testing

   (ii)   Magnetic particle testing

   (iii) Eddy current testing,

   (iv) Radiography

5.   Eddy current testing method to detect flaws in materials cannot be applied to

   (i)   Ferromagnetic materials   (ii)   diamagnetic materials

   (iii) paramagnetic materials   (iv) insulating materials

6.   In which of the following method to detect flaws in materials, X-rays or gamma rays is used

   (i)   Magnetic particle testing   (ii)   Eddy current testing

   (iii) Radiography   (iv) Ultrasonic testing

7.   Which of the following NDT methods can be used to detect internal flaws in a material?

   (i)   Visual inspection   (ii)   Penetrant testing

   (iii) Magnetic particle testing   (iv) Radiography

8.   In which of the following NDT methods, highly energetic sound energy is used to detect internal flaws in a material?

   (i)   Ultrasonics   (ii)   Eddy current testing

   (iii) Magnetic particle testing   (iv) Radiography

9.   Sound energy cannot propagate as

   (i)   Transverse waves   (ii)   longitudinal waves

   (iii) shear waves   (iv) surface waves

10. By which of the following NDT methods, nodularity in cast iron can be determined?

   (i)   Magnetic particle testing

   (ii)   Electromagnetic or eddy current testing

   (iii) Radiography

   (iv) Ultrasonic testing

Answers

1 (iii)	2 (iv)	3 (i)	4 (ii)	5 (iv)	6 (iii)	7 (iv)	8 (i)
9 (i)	10 (iv)						

13 Nuclear Accelerators

13.1 Introduction

Any device that produces a beam of fast-moving, electrically charged atomic or sub-atomic particles having high energy is called a particle accelerator or nuclear accelerator or particle collider. The effectiveness of an accelerator is usually characterized by the kinetic energy, rather than the speed, of the particles. Compared with the quantities of energy encountered in everyday experience, even the teraelectron volt (10^{12} eV) is a very small amount, about that of a mosquito in flight. The masses of the particles accelerated are so small, that kinetic energies in this range correspond to very high speeds. The particles that are accelerated most often are electrons or protons (ionized hydrogen), and their antiparticles, or heavier ionized atoms like α-particles. Physicists use accelerators in fundamental research on the structure of nuclei, the nature of nuclear forces and the properties of nuclei not found in nature, such as the transuranic (heavier than uranium) elements and other unstable elements. This is the motivation for most of the development of the various types of particle accelerators. Accelerators are also used for radioisotope production, industrial radiography, cancer therapy, sterilization of biological materials, polymerization of plastics and a certain form of radiocarbon dating. The largest accelerators are used in research on the fundamental interactions of the elementary sub-atomic particles. Last year (2015), we saw the role of the large hadron collider in decoding nature's mystery of creation. Twenty-first century accelerators are still direct descendants of the first accelerators conceived in the twentieth century.

13.2 Need of Nuclear Accelerators

In particle accelerators, electric and magnetic fields are used to guide and accelerate a beam of charged particles to high speed in order to gain high energy. Many new discoveries were possible from the α=particle bombardment experiment of Ernest Rutherford performed in 1911. At that time, it was only possible to obtain the low speed α-particles naturally from radium. The distance of the closest approach δ of α-particles to the target nucleus is given by

$$\delta = \frac{Ze^2}{\pi\varepsilon_0 mv^2} = \frac{1}{2}\frac{ze^2}{\pi\varepsilon_0 E_k} \tag{13.1}$$

where $E_k = \frac{1}{2}mv^2$

and $2e$ are kinetic energy and charge of the α -particle, Ze is the charge of the target nucleus. Nuclei can be disintegrated only when $\delta \rightarrow 0$. Therefore, only light elements were possible to be disintegrated by the low energy α-particles obtained from natural radium. To disintegrate heavier elements in order to research into the properties of atomic nuclei, sub-atomic particles and elementary particles, high energy incident particles under complete human control are required. The projectiles obtained from radioactive sources cannot be fully utilized in nuclear research as they have low energy, low flux beyond human control. Also the desired type of projectiles cannot be obtained from natural sources. Because of all these constraints in using projectiles obtained from radioactive sources or cosmic rays, there was a frantic search for projectile sources under full human control. The search ended, at the construction of different types of particle accelerators or nuclear accelerators.

13.3 Basic Mechanism of a Nuclear Accelerator

Charged particles attain high energy when they are accelerated through strong magnetic fields or high potential differences. The basic mechanism of acceleration of charged particles involves two basic principles of the electric and magnetic field.

1. A charged particle of charge q when placed in an electric field gets accelerated by an electric force $\overline{Eq}$. If V is the potential difference through which the charged particle is accelerated, the kinetic energy ($1/2$ mv^2) gained by the charged particle will be given by the expression

$$\frac{1}{2}mv^2 = qV \tag{13.2}$$

Equation (13.2) shows that to get high energy, the potential difference through which the charged particle gets accelerated should be of very high value.

2. A charged particle of charge q when projected into a magnetic field of induction $\vec{B}$ with velocity $\vec{v}$ gets accelerated by a magnetic force $q(\vec{v} \times \vec{B})$. If the angle between $\vec{v}$ and $\vec{B}$ is 90°, the trajectory of the charged particle will be circular of with radius R. The kinetic energy $(1/2\ mv^2)$ gained by the charged particle will be given by the expression

$$\frac{1}{2}mv^2 = \frac{1}{2m}q^2B^2R^2 \tag{13.3}$$

Equation (13.3) shows that to get high energy, the magnetic induction as well as the circular path along which the charged particle gets accelerated should be of very high value. By increasing the magnitude of magnetic induction and radius of the circular path, we can increase the kinetic energy of the charged particle.

13.4 Main Components

Every accelerator has three essential parts: (i) source of the ions to be accelerated, (ii) accelerating tube to accelerate the ions, and (iii) source of the electric fields and magnetic fields needed to cause the acceleration.

13.4.1 Ion sources

The characteristics of ion sources are

i. It should be a large well-focused ion source.

ii. It should be compact, rugged, dependable and have a long life.

iii. It should be designed in such a manner that it can be controlled by external manipulations.

iv. The energy of all the ions emitted should have approximately the same energy.

v. The efficiency of the ion source should be high so as to simplify the problem of maintaining a vacuum and to reduce gas consumption.

To obtain different types of ions/projectiles, there are various types of ion sources such as cold cathode canal ray tube, spark discharge sources, hot cathode arc, capillary arc, magnetic ion source and so on.

13.4.2 Accelerating tube

High D.C. tension is not applied directly to the ion source as it will damage the ion source itself. It is applied in several small steps as shown in the Fig. 13.1. Accelerating electrodes are in the form of metallic cylinders Cs and are placed inside an evacuated glass tube. The ion beam travels along the axis of these cylinders and is suitably deflected by a magnetic field. It heats the target to produce transmutation.

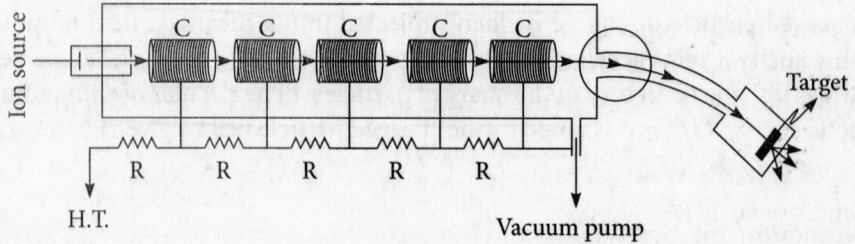

Figure 13.1 Schematic diagram of an accelerating tube. The Cs are the deflecting metallic cylinders

The region in which the ions are accelerated must be highly evacuated to keep the particles from being scattered out of the beam, or even stopped, by collisions with molecules of air.

13.5 Performance Index

The following factors determine the performance index of a nuclear accelerator.

i. The type of accelerated ions.

ii. Operation of the accelerator (pulsed or continuous)

iii. Intensity of the output beam.

iv. Maximum energy attainable.

v. Stability of energy.

vi. Homogeneity/sharpness of energy beam.

vii. Collimation of the energy beam.

13.6 Types of Accelerators

Nuclear accelerators can be broadly classified into two categories: D.C accelerators and R.F accelerators. The Cockcroft–Walton accelerators and the Van de Graaff accelerators are belong to D.C (direct current) accelerators. Cyclotrons, betatrons and linear accelerators are R.F (radio frequency) accelerators. Depending upon the path followed by the ions, accelerators are of two types: linear accelerators (ions follow a linear path) and cyclic accelerators (ions follow a circular or spiral path).

13.7 D.C. Accelerators

D.C. accelerators/electrostatics accelerators are those accelerators in which a high D.C voltage V is built up between two terminals and the charge particles are accelerated in their passage across the space. If a charged particle of charge Ze is released at one of the terminals, it will reach the other terminal with a kinetic energy ZeV. There are few D.C. accelerators produce potential difference in the order of 10^6 volts. Two of them – the Cockcroft–Walton accelerator and the Van de Graaff accelerator have been in use for many years as particle accelerators.

13.7.1 Cockcroft–Walton accelerator (D.C. accelerator)

The operation of the first successful accelerator which artificially accelerated ions was demonstrated at Cambridge University, England, by John Douglas Cockcroft and E.T.S. Walton in 1932. Using a voltage multiplier, they accelerated protons to energies as high as 710 keV and showed that these react with the lithium nucleus, the products being two energetic α-particles. The device was named in their honour as the Cockcroft–Walton accelerator.

The Cockcroft–Walton accelerator is based on the voltage multiplier principle. A number of capacitors of equal capacitance and same number of rectifiers are arranged as shown in the Fig. 13.2.

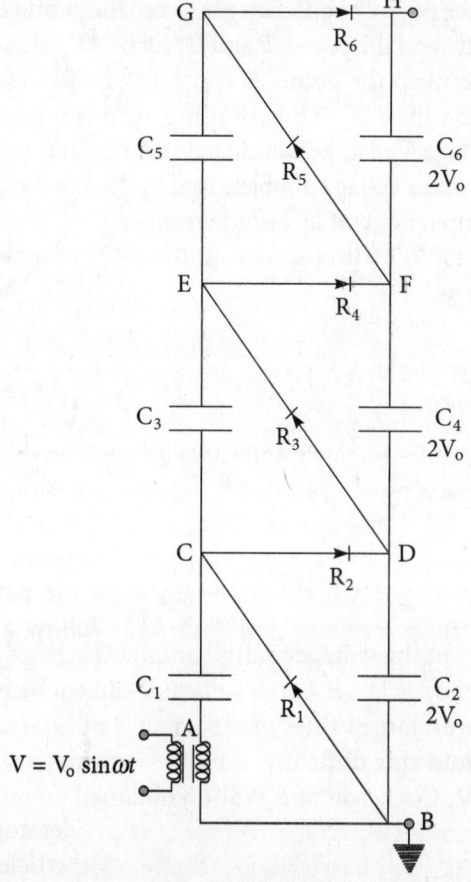

Figure 13.2 | The arrangement is the voltage multiplier circuit. $C_1, C_2, C_3, \dots$ and so on are the capacitors each having capacitance C. $R_1, R_2, R_3, \dots$ and so on are the rectifiers

Working of the voltage multiplier circuit

We assume that there is no current loss at any point in the circuit. A radio frequency transformer with output peak voltage V_0 ranging from 25 kV to 100 kV is connected to the voltage multiplier circuit as shown in the Fig. 13.2. The point B is earthed so that its potential will always be zero. The potential of the point A oscillates with peak voltage $+V_0$ and $-V_0$. Suppose in the first half cycle A is negative, then B will be positive (with respect to A). At this instant, rectifier R_1 conducts and capacitor C_1 is charged to a potential of V_0. In the second half cycle, R_1 does not conduct and the point C having potential V_0 becomes isolated. In the third half cycle R_1 conducts and the point C attains potential $2V_0$. Also, during this cycle, the point C is positive with respect to B and hence, R_2 conducts. Hence, the charge accumulated in the capacitor C_1 is shared with capacitor C_2 increasing its potential to V_0. Again during the fourth half cycle, the potential of capacitor C_2 increases. After a few cycles, an equilibrium is reached in which there is no current in either R_1 or R_2 and the potentials of the capacitors C_1 and C_2 attain a steady state value of $2V_0$.

Now the potential of D is $2V_0$ with respect to B. Same logic is also applicable to the circuit CDEF. After a few cycles, the potential difference between the points F and D will be $2V_0$. Hence, the potential difference between the points F and B will be $4V_0$. Again after a few more cycles, the potential difference between the points H and F will be $2V_0$ and the potential difference between the points H and B will be $6V_0$. Thus, by using such arrangements of say, two capacitors and two rectifiers, the voltage V_0 can be multiplied to any value V. Each circuit BCD, DCEF, FEGH, and so on are known as voltage doublers as they double the voltage. The total circuit is known as the voltage multiplier circuit or cascade rectifier.

If a voltage multiplier circuit with load current I has N number of voltage doublers the output voltage V is given as

$$V = 2NV_0 - \frac{I}{12\nu C}\left(8N^3 + 9N^2 + N\right)!$$ (13.4)

or $V = 2NV_0$ if the values of frequency ν and capacitance C are high.

The ripple voltage is given by

$$\pm \Delta V = \frac{IN(N+1)}{4\nu C}$$ (13.5)

Hence, the output voltage of the voltage multiplier circuit is large if the capacitance of each capacitor and the frequency is large. Large values of capacitance and frequency decrease the ripple voltage. However, large values of capacitance of each capacitor increase the size of the accelerator. To avoid this difficulty, a radio frequency transformer has been used. Starting with $V_0 = 100$ kV, Cockcroft and Walton obtained an output voltage of 400 kV in their first attempt.

Working of Cockcroft–Walton accelerator

The high voltage between the terminals of the voltage multiplier circuit is applied to the charged particles which are to be accelerated through an evacuated accelerating tube. One end of the accelerating tube contains the ion source and the target is attached to the other end. In the Cockcroft–Walton accelerator, energy of 4 MeV has been realized in practice. Now-a-days, the Cockcroft–Walton accelerator is used as a pre-accelerator, giving enough energy to ions so that they can be injected into a larger accelerator.

Advantages

The advantages of the Cockcroft–Walton accelerator are the following.

i.   Simple in design and construction.

ii.   It provides large ion current at constant voltage.

iii. It can be used to accelerate both positive and negative charges or ions to moderate energy.

iv.   Since there is no moving part in this accelerator, there is no wear and tear.

Disadvantages

i.   The maximum attainable voltage is limited by insulation problems.

ii.   The sensitivity and control over voltage is low.

iii. The charged particles can be accelerated only to moderate energies.

13.7.2 Van de Graaff accelerator (D.C. accelerator)

Robert Jemison Van de Graaff an American physicist, in the year 1931, designed a direct current accelerator which can develop a potential difference of several million volt and when used with positive ion tubes can impart energies of several MeV to the ions. This device, called the Van de Graaff accelerator, has found widespread use not only in atomic research but also in medicine and industry.

Principle

If a charged conductor is placed inside a hollow conductor and connected internally with the hollow conductor, all the charge of the conductor is transferred to the hollow conductor no matter how high the potential of the hollow conductor may be. The potential of the hollow conductor can be raised to any desired value by adding charges to the internal conductor successively.

In Fig. 13.3 a small sphere of radius r is charged by a positive charge q and kept inside a large hollow sphere of radius R containing positive charge Q. The outer large sphere is inside the potential field of the small sphere. Hence the total potential of the outer sphere V_R will be given as

$$V_R = \frac{kq}{R} + \frac{kQ}{R}$$

where kq/R is the potential on the surface of the large sphere due to the small sphere and kQ/R is the potential on the surface of the large sphere due to its own charge Q. Again the

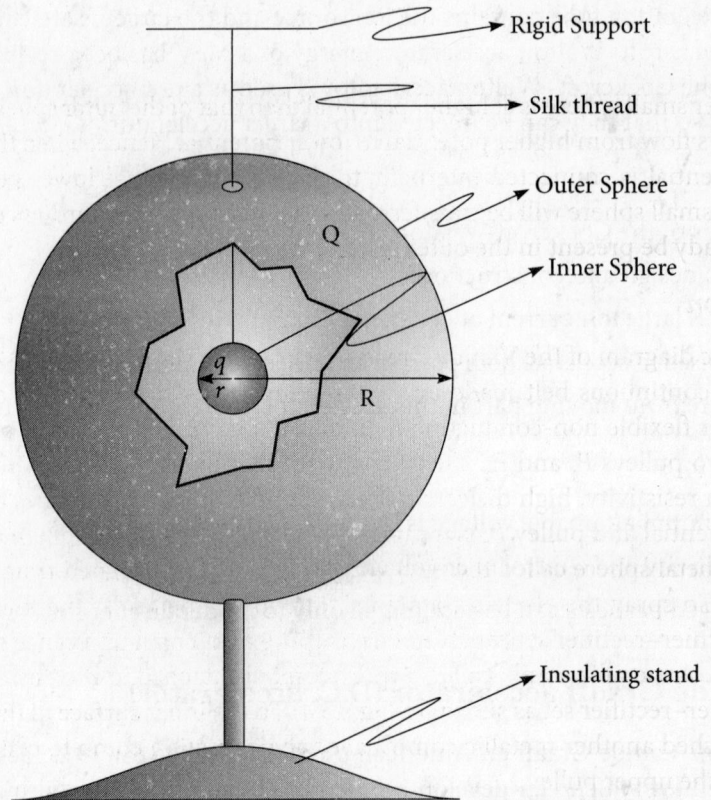

Figure 13.3 A small charged sphere of radius r is kept inside a large hollow sphere of radius R. When the small sphere is connected internally to the outer sphere, all its charge is transferred to the outer sphere

small sphere is inside the potential field of the large sphere. Hence, the total potential of the small sphere V_r will be given as

$$V_r = \frac{kq}{r} + \frac{kQ}{R}$$

where kq/r is the potential on the surface of the small sphere due to its own charge and kQ/R is the potential on the surface of the small sphere due the large sphere containing charge Q. Therefore the potential difference $V_r - V_R$ between the two spheres will be given as

$$V_r - V_R = \frac{kq}{r} + \frac{kQ}{R} - \frac{kq}{R} - \frac{kQ}{R} = kq\left(\frac{1}{r} - \frac{1}{R}\right) \tag{13.6}$$

Since q is positive and $\frac{1}{r} > \frac{1}{R}$, we conclude

$$V_r - V_R > 0$$

or $V_r > V_R$

Thus, the inner small sphere is at higher potential than that of the outer sphere. The positive charges always flow from higher potential to lower potential.Hence, when the small sphere at higher potential is connected internally to the outer sphere at lower potential, all the charge of the small sphere will be transferred to the outer sphere regardless of the charge Q that may already be present in the outer sphere.

Construction

The schematic diagram of the Van de Graaff accelerator is shown in Fig. 13.4. It essentially consists of a continuous belt made up of some insulating material such as rubber, silk, linen, or other flexible non-conducting materials that is run by means of a motor at high speed over two pulleys P_1 and P_2. This charge-carrying belt should have high mechanical strength, high resistivity, high dielectric strength and high fire resistance. The pulley P_1 is at ground potential and pulley P_2 is mounted inside a hollow metal sphere of large radius. This hollow metal sphere called the high voltage terminal is insulated from the rest of the apparatuses. To spray the electric charge on the moving belt near the lower pulley P_1, a small transformer–rectifier set capable of developing a potential difference of 10 to 100 kV is kept near it. A metallic comb, called a spray comb is attached to the positive terminal of the transformer–rectifier set as shown in Fig. 13.4. To the inner surface of the hollow metal sphere is attached another metallic comb, called the collection comb to collect the electric charges near the upper pulley.

The acceleration tube is placed parallel to the charge-carrying belt so that one end of the tube is inside the high voltage terminal. The accelerating tube is an evacuated tube of insulating material, commonly porcelain or glass cylinders several inches long and of large diameters with vacuum tight seals to metal plate electrodes between sections. These electrodes are connected to corresponding equipotential rings in the column to maintain a uniform distribution of potential along the tube. The entire set up, except the target end of the acceleration tube, the transformer–rectifier set and high vacuum pump, is enclosed by a high pressure tank as shown in the Fig. 13.4. The high pressure tank contains dried up CCl_2F_2–air or SF_6–air mixtures under high pressure. Their role is to increase the voltage of the high voltage terminal to a very high value without corona discharge.

Working

The belt near the spray comb moves upward with the help of the electric motor attached to the pulleys. Since the spray comb is attached to the positive terminal of the high D.C. voltage, a high electric field is produced at the corona points, i.e., tips of the spray comb as a result of which the nearby air is ionized. Since the spray comb is positively charged by the high D.C. voltage, the positive ions produced there by ionization are repelled towards the upward moving belt. These positive ions attach themselves to the surface of the moving belt and are

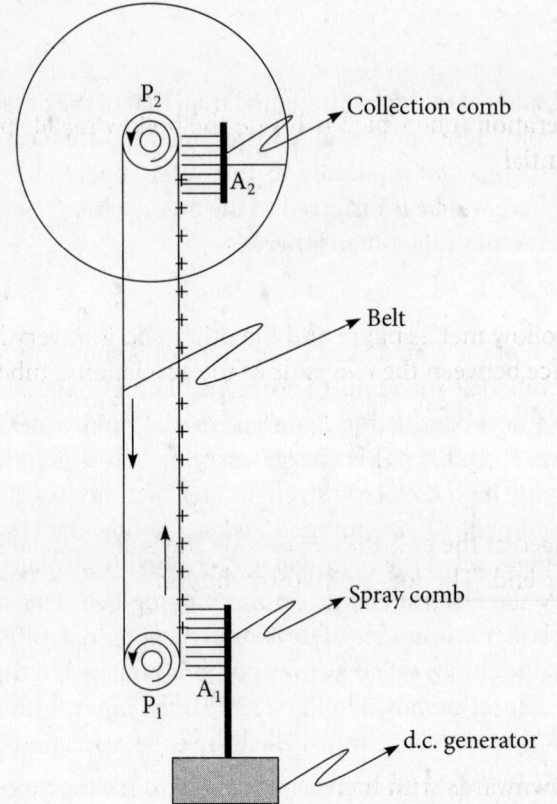

P$_2$

Collection comb

A$_2$

Belt

Spray comb

P$_1$ A$_1$

d.c. generator

Figure 13.4 | Schematic diagram of a Van de Graaff accelerator showing the different parts. P_1 and P_2 are two pulleys, A_1 and A_1 are the spray comb and collection comb respectively, B is the insulating belt

carried upward mechanically by the belt. They are removed from the belt by the collection comb and are passed to the inner surface of the high voltage terminal as a result of which its potential increases. This process goes on continuously and so the potential of the high voltage terminal goes on increasing continuously to a certain maximum value called the equilibrium value. Beyond the equilibrium value, the excess charge on the high voltage terminal will be lost to the surrounding by corona discharge and leakage of supports. Thus, the potential of the high voltage terminal can be increased to a certain maximum value by this method and this high potential is applied to the acceleration tube by the induction method.

If Q is the charge on this surface at any instant of time, its potential with respect to the ground is simply given by

$$V = \frac{Q}{C} \tag{13.7}$$

where C is the capacitance of the system. The rate of increase of potential $\dfrac{dV}{dt}$ is given by

$$\frac{dV}{dt} = \frac{1}{C}\frac{dQ}{dt} = \frac{i}{C} \tag{13.8}$$

The end of the acceleration tube which is inside the hollow metal sphere is automatically subjected to the potential

$$\left(= \frac{9 \times 10^9 Q}{R} V \right)$$

due to the charged hollow metal sphere and the other end is at very low potential. Hence, the potential difference between the two ends of the acceleration tube is

$$\frac{9 \times 10^9 Q}{R} V.$$

The ion source is placed at the end which is inside the hollow metal sphere and the target is placed at the other end. The ion emitted by the ion source is subjected to a potential difference of

$$\frac{9 \times 10^9 Q}{R} V$$

and is accelerated downwards with increasing energy to hit the target. It is used in X-ray generation and as a pre-accelerator, giving enough energy to ions so that they can be injected into a larger accelerator. Commercial Van de Graaff accelerators are available now-a-days, to provide a voltage up to 10million volts and proton beam currents of the order of 6 to 8 μA.

Advantages

i.   The potential difference attained is much more than the Cockcroft–Walton accelerator.

ii.   The potential difference and hence, ion energy is completely controllable.

iii. The homogeneity of the beam is very high. At energies below 3 MeV it is possible to obtain a resolution better than 400 eV.

iv.   Both positive and negative charges or ions can be accelerated.

v.   The sensitivity and control over the voltage is very high.

Disadvantages

i.   The maximum particle energies that can be realized by this device are relatively small in comparison with most other types of accelerators.

ii.   The charge-carrying belt gets damaged frequently and maintenance is cumbersome.

iii. The wear and tear of the charge-carrying belt continuously produces dust which is to be cleaned quickly.

13.7.3 Tandem accelerator (D.C. accelerator)

A modified version of the Van de Graaff accelerator is the tandem accelerator. In a tandem accelerator without increasing the operating voltage, the energy of an ion can be increased two folds indigenously. The tandem accelerator is also called the tandem Van de Graaff accelerator.

Principle

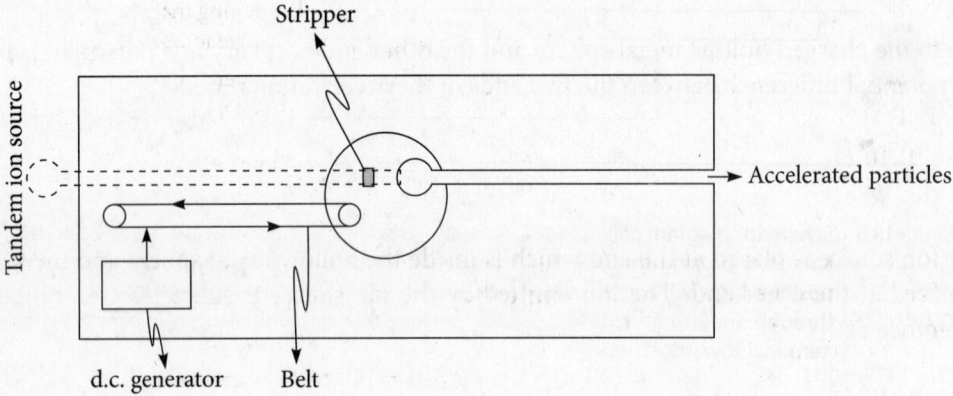

Figure 13.5 | Schematic diagram of a Van de Graaff tandem accelerator

In Fig. 13.5, the metallic hollow sphere of the Van de Graaff accelerator is supposed to be positively charged. The singly ionized negative ion from the tandem ion source will move under Coulomb attraction towards the metallic hollow sphere of the Van de Graaff accelerator. If some how, at the later part of the hollow metallic sphere, two electrons are stripped off from the singly ionized negative ion, it becomes a positive ion and will move in the forward direction due to inertia of motion and will simultaneously be repelled in the direction of motion by the Coulomb force. Thus, at two-stages, the ion gets accelerated almost by equal amounts, doubling its energy finally. The first instrument of this type called the two-stage tandem accelerator was built in U.S.A. and was put into operation at the Chalk River Laboratory, Canada in 1959.

Construction

The constructional details of a two-stage tandem accelerator are shown schematically in Fig. 13.6.

The main components of a two-stage tandem accelerator are the ion source, electron adding canal, analyzing magnet, stripping canal, Van de Graaff accelerator, switching magnet, and the target. The arrangements of the components are depicted in Fig. 13.6. In a Van de Graaff accelerator, the ion source is placed inside the high voltage terminal at one end of the acceleration tube. However, in a tandem accelerator, the two ends of the acceleration tube are at the ground potential with one end containing the target and the other end containing the positive ion source.

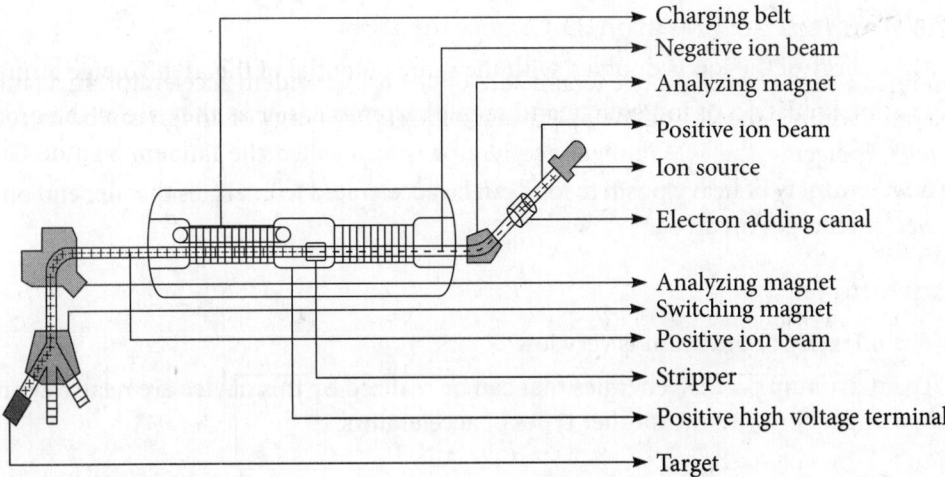

- Charging belt
- Negative ion beam
- Analyzing magnet
- Positive ion beam
- Ion source
- Electron adding canal
- Analyzing magnet
- Switching magnet
- Positive ion beam
- Stripper
- Positive high voltage terminal
- Target

Figure 13.6 | Schematic diagram of a 12 MeV tandem accelerator. The positive ions become negative ions after passing through an electron-adding canal and are accelerated towards the positive high voltage terminal, where they again become positive ions by passing through the stripping canal. These positive ions are repelled by the high voltage terminal towards the target

Working

Positive ions are produced in the ion source and are accelerated to about 50 keV. In passing through the electron-adding canal containing gas at low pressure, approximately 1% of the positive ions are converted into negative ions by picking up two electrons. The negative ion beam is separated from the unwanted neutral and positive ions by means of an analyzing magnet. The negative ions are attracted towards the high voltage terminal and gain energy of V eV, where V is the voltage of the high voltage terminal. Inside the high voltage terminal, the negative ions are passed through a stripper canal consisting of a thin carbon foil or a gas at low pressure. The energy of the negative ions is already much higher than that required to remove the most tightly bound electrons in a negative ion. Therefore, all the ions emerging from the stripper will have lost all the electrons to the carbon foil and have become positive ions. These positive ions will gain again V eV energy from the potential of the high voltage terminal to the ground potential.

The working of the tandem accelerator shows that second stage is extremely powerful when heavy ions are to be accelerated. Take for example, oxygen atom ($Z = 8$) and let the potential of the high voltage terminal be $V = 5 \times 10^6$V. In the first stage, the oxygen atoms are singly charged ($q = 1.6 \times 10^{-19}$ C) and will gain energy of ($W = qV$ J) 5 MeV. However, in the second stage, all the eight electrons are removed from the oxygen atom rendering it an oxygen nucleus having a positive charge of $8 \times 1.6 \times 10^{-19}$ C which will gain energy of 40 MeV. Thus, the total energy gain is 45 MeV!

Tandem accelerators are now the standard tool for investigating of proton scattering in the region of 6 to 20 MeV. A wide variety of positive ions can be accelerated to energies that depend upon the electric charges involved.

Advantages

i. The energy of the ion is doubled with the same potential of the high voltage terminal

ii. The manipulation of ion source and target becomes easier as they are at the ground potential.

iii. A wide variety of heavy positive ions can be accelerated to energies that depend on the electric charges involved.

Disadvantages

i. The intensity of the beam is very low

ii. The maximum particle energies that can be realized by this device are relatively small in comparison with most other types of accelerators.

13.8 R.F. Accelerators

R.F. accelerators are those accelerators in which a high R.F. voltage $V_0 \sin \omega t$ is built up between terminals and charge particles are accelerated in steps in their passage across the spaces. In a few R.F. accelerators, a magnetic field is also used to accelerate charged particles. Instead of using high potential as in electrostatic accelerators, moderate potential is used repeatedly to add energy to the ions in R.F. accelerators. Linear accelerators, cyclotrons, betatrons are a few R.F. accelerators that have been in use as particle accelerators for many years. Cyclotrons and betatrons are low energy cyclic accelerators.

13.8.1 Linear accelerators

In a linear accelerator, the energy of an ion is increased steadily or in steps as it travels in a straight line. The small accelerations the ions get in each step add up together to give the particles more energy than could be achieved by the voltage used in one section alone. In 1924, Gustaf Ising, a Swedish physicist, proposed accelerating particles using alternating electric fields, with "drift tubes" positioned at appropriate intervals to shield the particles during the half-cycle when the field is in the wrong direction for acceleration. Four years later, the Norwegian engineer Rolf Wideröe built the first machine of this kind, successfully accelerating potassium ions to an energy of 50 keV. Linear accelerators are also called linacs and are of two types one is a drift tube linear accelerator, which we shall discuss, and the other is a wave guide linear accelerator, which is beyond the scope of the book.

Construction of a drift tube linear accelerator

The drift tube linear accelerator consists of a number of cylindrical electrodes of increasing length arranged in a straight line as shown in Fig. 13.7. These cylindrical electrodes are called drift tubes. The gap or separation between any two consecutive drift tube is the same. Alternate drift tubes, i.e., drift tubes numbered 1, 3, 5, 7, ... are joined to one terminal and drift tubes numbered 2, 4, 6, 8, ... are joined to the other terminal of the radio frequency (r.f) power supply. Hence, at any instant, alternate electrodes carry opposite potentials. All the electrodes which are at a positive potential in a particular half cycle become negative in the next half cycle.

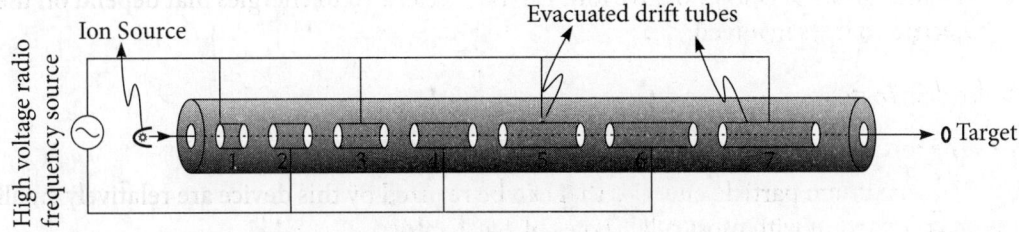

Figure 13.7 | Schematic diagram of a drift tube linear accelerator

Principle of the *drift tube linear accelerator*

The drift tubes are connected alternately as described earlier to an alternating voltage source as a result of which the electric field inside the drift tube is zero and thus, the electric force acting on the moving ion is zero. Hence, the ions are not accelerated when they are moving inside the drift tube. When the ions move in the gap between two consecutive drift tubes, they are accelerated since the successive tubes are of opposite polarity.

Working of a *drift tube linear accelerator*

Suppose positive ions from the ion source move from left to right along the common axis of the drift tubes. While passing through the first electrode, the ions receive no acceleration, since the moving ions are within the uniform potential of the first electrode. If at this instant, the first drift tube is positive, the second drift tube must be at negative potential, the positive ions will be accelerated in the gap between these two drift tubes. The positive ions then enter the second drift tube and travel through it at a constant but higher speed than in the first drift tube. The length of the second drift tube is such that just as the ions reach the gap between it and the third drift tube, the potential of these drift tubes are reversed. Now the second drift tube becomes positive and the third drift tube negative, thus the positive ions are again accelerated in this gap. Therefore, at each gap of the accelerator, the positive ions get accelerated.

Theory behind the *drift tube linear accelerator*

Let

L_n = length of the nth drift tube plus a gap

= separation between the $(n - 1)$th and nth gap.

v_n = speed of the ion in the nth drift tube.

The speed v_n of the ion inside a drift tube does not change because the potential is uniform inside the drift tube, i.e., the electric field intensity inside the drift tube is zero.

The length/distance L_n is travelled by the positive ions during the time duration of one half cycle $\frac{T}{2}$ of the applied frequency v. Therefore, we have

$$L_n = v_n \times \frac{T}{2} = \frac{v_n}{2v} \tag{13.9}$$

or $\quad v_n = 2L_n v \tag{13.10}$

Now if V_0 is the peak value of the potential difference between any two consecutive drift tubes, then the kinetic energy gained by the ion of charge q in passing any gap will be qV_0. Thus, the total kinetic energy possessed by the ion after passing the 1st gap will be qV_0, after passing the 2nd gap will be $2qV_0$, after passing the 3rd gap will be $3qV_0$ and so on. Therefore, the total kinetic energy possessed by the ion after passing the nth gap will be nqV_0. If v_n is the speed of the ion in the nth drift tube, its kinetic energy in the nth gap will be $\frac{1}{2}mv_n^2$. Thus, we have

$$\frac{1}{2}mv_n^2 = nqV_0$$

or $\quad v_n = \sqrt{\frac{2nqV_0}{m}} \tag{13.11}$

Equation (13.11) shows that for a particular experimental set up,

$$v_n \propto \sqrt{n}, \quad n = 1, 2, 3, 4, \ldots \tag{13.12}$$

i.e., $v_1 : v_2 : v_3 : \ldots = 1 : \sqrt{2} : \sqrt{3} : \ldots$

Putting Eq. (13.11) into Eq. (13.9), we get

$$L_n = \frac{1}{2v}\sqrt{\frac{2nqV_0}{m}} \tag{13.13}$$

Eq. (13.13) indicates that for a particular experimental set up,

$$L_n = \sqrt{n}, \quad n = 1, 2, 3, 4, \ldots \tag{13.14}$$

i.e., $\quad L_1 : L_2 : L_3 : ..... = 1 : \sqrt{2} : \sqrt{3} : .....$ (13.15)

Combining Eqs (13.11) and (13.13), we have

$$\frac{v_n}{L_n} = 2\nu$$ (13.16)

Energy of the ion

If there are n number of drift tubes in a drift tube linear accelerator, the total energy E possessed by the output positive ion will be given by

$$E = nqV_0$$

The ion hits the target in pulses. The number of pulses per unit time depends on the frequency of the r.f. source.

Advantages

i.   As $E = nqV_0$, the energy of ions can be increased simply by increasing the number of drift tubes.
ii. It produces high intensity ion beam.
iii. It can be used to accelerate both positive and negative ions.
iv. The peak voltage of the r.f. source is not very high as a result of which the problems associated with high insulation is minimized.
v.   In linear accelerators, the cost per MeV is less than that of cyclic accelerators if ions are electrons.

Disadvantages

i.   As the total length of the accelerator is a few kilometers, it is cumbersome to maintain vacuum and other arrangements.
ii. The ion beam is pulsed rather than continuous.
iii. In linear accelerators, the cost per MeV is more than that of cyclic accelerators, if ions are protons.

Example 13.1

One of oldest linear accelerator at Berkeley has 46 drift tubes. If the length of the shortest tube is 1.0 m, what is the length of the longest tube?

Solution

In a linear accelerator, the first tube is the shortest tube and the last tube is the longest tube. The length of the nth number tube is given as

$$L_n = \frac{1}{2v}\sqrt{\frac{2nqV_0}{m}}.$$

The length of the first tube L_1 will be

$$L_1 = \frac{1}{2v}\sqrt{\frac{2qV_0}{m}}$$

and we obtain

$$L_n = L_1\sqrt{n}$$

The data given are $L_1 = 0.5$ m, $n = 46$. Thus, we have

$$L_{46} = 2\sqrt{46}\,\text{m} = 13.56\,\text{m}$$

13.8.2 Cyclotron

The commonly used cyclic particle accelerator, the cyclotron was developed by E.O. Lawrence and M.S. Livingston and put into operation in the year 1932. In the cyclotron, generally deuterons and α-particles are accelerated to high energies.

Principle

If a charge q is projected perpendicularly into a uniform magnetic field of induction B with speed v, the charge q will trace a circular path of radius r. The kinetic energy of the charge q is given by

$$\frac{1}{2}mv^2 == \frac{B^2R^2q^2}{2m} \quad \text{since } Brq = mv$$

Example 13.2

Calculate the energy of a proton that is moving in a circle of radius 60 cm under the influence of a magnetic field 10000 gauss.

Solution

The data given are $R = 60$ cm $= 0.6$ m, $B = 10000$ gauss $= 1$ Tesla

The energy of a proton that is moving in a circular path under the influence of a magnetic field is given by

$$\frac{1}{2}mv^2 = \frac{B^2R^2q^2}{2m} = \frac{1^2 \times 0.6^2 (1.6 \times 10^{-19})^2}{2 \times 1.67 \times 10^{-27}} J = 17.2\,Mev$$

Construction

The schematic diagram of a cyclotron is shown in Fig. 13.8. In its simplest form, it consists of two hollow flat semicircular metal boxes D_1 and D_2 called dees on account of their shape like the letter D.

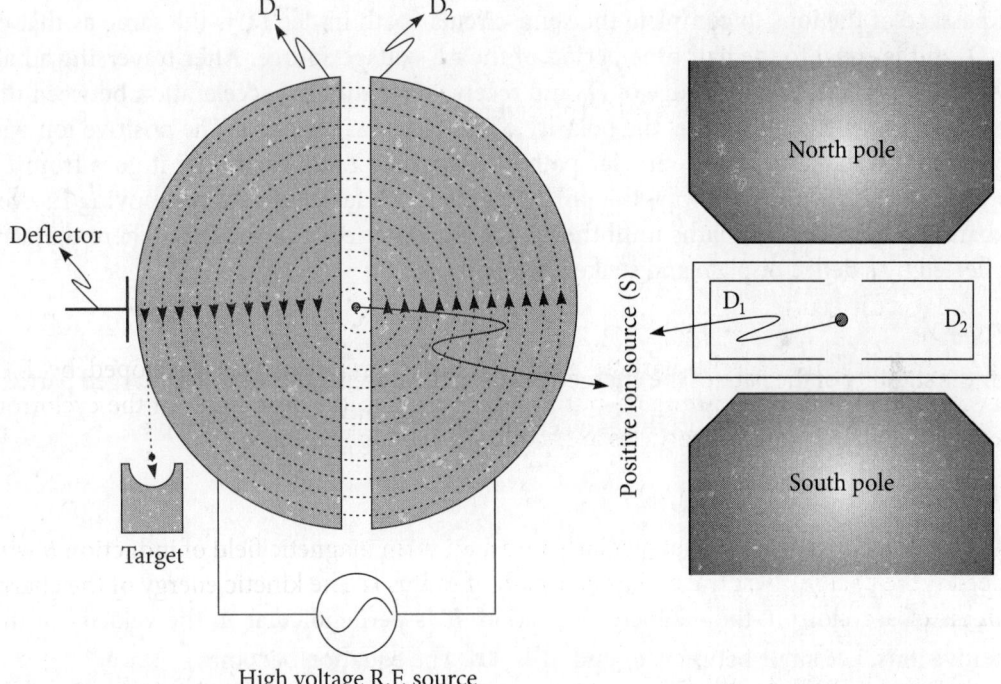

Figure 13.8 | Schematic diagram of a cyclotron

These two dees have their diametric edge parallel and slightly separated from each other. A radio frequency alternating potential of peak value of the order of 10^5 volts and frequency of the order of megacycles is applied the dees which act as electrodes. Thus, an alternating electric field is established in the gap between the two dees, i.e., an electric field is directed towards D_1 and D_2 alternately. D_1 and D_2 alternately become positive and negative at the same rate as the frequency of the r.f voltage source. An ion source is placed at the centre of the gap between the two dees which supplies the positive ions to be accelerated. The whole se-tup is placed between the poles of a strong electromagnet which provides a magnetic field of induction of the order of 1Tesla perpendicular to the plane of the dees.

Working

Suppose at any particular instant, D_1 is positive and D_2 negative. A positive ion starting from the source S will be attracted by the dee D_2. Since there is no electric field inside the dees, the positive ions move with constant speed along a circle of constant radius $r = (mv/qB)$ under the influence of a magnetic field B which is perpendicular to the plane of dees. By the time, the positive ions emerge from D_2, the polarity of the applied potential is reversed, i.e., D_2 becomes positive and D_1 becomes negative. The positive ion will again face the negative dee D_1 and thus again will be accelerated by the electric field in the gap and ion speed will be increased. Since the speed of positive ions has been increased, they will move through D_1 along a circular arc of greater radius as shown in Fig. 13.8. The time of passage of the ions to complete the semi-circular path in dee D_1 is the same as that of in D_2 and is equal to the half time period of the r.f. voltage source. After traversing a half cycle in D_1, it will reach the edge of D_1 and receive an additional acceleration between the gaps because in the meantime, the polarity of the dees has changed. The positive ion will continue travelling in a semi-circular path of increasing radii, each time it goes from D_1 to D_2 and D_2 to D_1. In this way, the positive ions move faster and faster moving in ever expanding semi-circular paths until they reach the outer edge of the dees where they are deflected by a deflector plate and strike the target.

Theory

For the stability of the path of the charged particle, the Lorentz force on the charged particle is $q(\vec{v} \times \vec{B}) = qvB\sin\theta$ which produces a centripetal force $\dfrac{mv^2}{r}$. Thus, we have

$$qvB\sin\theta = \frac{mv^2}{r}$$

In case of a cyclotron, the magnetic induction $\vec{B}$ is perpendicular to the velocity of the positive ions, i.e., angle between $\vec{v}$, and $\vec{B}$ is 90°. The equation becomes

$$qvB = \frac{mv^2}{r}$$

or $$v = \frac{Brq}{m}$$ (13.17)

Here m = mass of the positive ion and is assumed to be constant non-relativistically.

The time of passage of the positive ions to complete the semi-circular path in dee D_1 is same as that of in D_2 and is equal to the half time period of the r.f. voltage source. If T is the time period of the r.f. voltage source, then within $T/2$ seconds, the positive ions travel a semi-circle of radius r. Thus, we have

$$\pi r = v \times \frac{T}{2} = \frac{v}{2v}$$

or $$v = \frac{v}{2\pi r} \qquad (13.18)$$

Here v is the frequency of the r.f. voltage source.

Putting Eqs (13.17) into (13.18), we get

$$v = \frac{qB}{2\pi m} \qquad (13.19)$$

Since for a particular ion, q/m is constant, the value of v is adjusted corresponding to B and vice versa so that when ions complete the semi-circular path in D_1 and enters the gap between the two dees, D_1 becomes positive and D_2 becomes negative. Due to this reason, the speed and the radii of the semi-circular paths go on increasing. The Eq. (13.19) is called the cyclotron resonance condition and $v = \dfrac{qB}{2\pi m}$ is called the cyclotron frequency.

Example 13.3

The applied magnetic induction in a cyclotron is 15000 gauss. Calculate the frequency of the rf voltage source to accelerate deuterons.

Solution

The datum given is B = 15000 gauss = 1.5 Tesla

The frequency of the rf voltage source to accelerate particles is given as

$$v = \frac{qB}{2\pi m} = \frac{1.6 \times 10^{-19} \times 1.5}{2\pi \times 3.33 \times 10^{-27}} \, \text{Hz} = 11.5 \, \text{MHz}$$

Energy of the ion

The positive ion will have maximum energy at the ime when it reaches the periphery of the dees having radius R. According to Eq. (13.17), the maximum speed v_m at the periphery will be given by

$$v_m = \frac{BRq}{m} \qquad (13.20)$$

Therefore, the maximum kinetic energy $E_m = \frac{1}{2}mv_m^2$ of the ion will be obtained as

$$E_m = \frac{1}{2}mv_m^2 = \frac{B^2 R^2 q^2}{2m} = 2\pi^2 mv^2 R^2 \tag{13.21}$$

Equation (13.21) shows that the ion energy is limited due to the following factors:

i.   Due to limited power and frequency of the r.f. voltage source.

ii.   Due to maximum strength of the magnetic field which can be produced.

iii. The expression for maximum energy contains mass of the ion. When speed of ion approaches the speed of light, the energy gained by the ion while passing the gap between the two dees is used to increase the mass of the ion rather than its speed. Due to this reason, heavy ions such as protons, deuterons, α-particles are accelerated in the cyclotron instead of electrons.

iv.   The maximum energy acquired by the ion does not depend on the voltage of the r.f. voltage source.

v.   As the radius of the semi-circular path increases, the successive paths become closer and closer.

Advantages

i.   Cyclotron is capable of producing high energetic ions that are more intense than that of electrostatic accelerators.

ii.   It occupies less area.

iii. It can accelerate both positive and negative ions.

iv.   It can accelerate heavy ions.

Disadvantages

i.   The output ion beam is not continuous but pulsed.

ii.   The energy of the ions are much less constant. This limits its the scope of its wide applications.

iii. Due to relativistic effects, lighter ions like electrons cannot be accelerated to very high energy.

iv   Cyclotrons cannot be operated in the relativistic range.

v.   It is not cost effective.

Example 13.4

The beam in a cyclotron has a maximum diameter of 1.6 m. The magnetic induction is 0.75 Tesla. Calculate the kinetic energy of a proton coming out of the cyclotron. Given: mass of proton = 1.67×10^{-27} kg

Solution

The data given are (i)

$$R = \frac{1.6}{2}\,m = 0.8\,m, \quad B = 0.75.$$

The kinetic energy of a proton coming out of the cyclotron is calculated by

$$E_k = \frac{B^2 R^2 q^2}{2m} = \frac{0.75^2 \times 0.8^2 \times \left(1.6 \times 10^{-19}\right)^2}{2 \times 1.67 \times 10^{-19}}\,J = 17.3\,Mev$$

Example 13.5

Deuterons in a cyclotron describe a circle of radius 32.0 cm just before emerging from the dees. The frequency of the applied alternating voltage is 12 MHz. Neglecting relativistic effects, find the energy and speed of a deuteron on emergence.

Solution

The data given are (i) $R = 32.0$ cm $= 0.32$ m, $v = 12$ MHz $= 12 \times 10^6$ Hz.
  The kinetic energy of a deuteron coming out of the cyclotron is calculated by

$$E_k = 2\pi^2 m v^2 R^2 = 9.69 \times 10^{-13}\,J = 6.06\,Mev$$

The speed of a deuteron coming out of the cyclotron is directly calculated by

$$v = 2\pi v R = 2.413 \times 10^7\,m/s$$

13.9 Electron Accelerators

The cyclotron we have discussed earlier is not suitable for accelerating electrons to high kinetic energy. Since the electron is a very light particle (mass $= 9.11 \times 10^{-31}$ kg), even at 1 MeV energy, its mass 'm' becomes very high according to the relativistic equation

$$m = m_0 \left(1 - \frac{v^2}{c^2}\right)^{-\frac{1}{2}}.$$

Therefore, for 1 MeV or less energy electrons, frequency v given by Eq. (13.19), $v = (qB/2\pi m)$, decreases with increasing speed of electrons because m increases with increase of speed. Thus, the electrons get out of step with the r.f voltage source and eventually, the energy of the circulating electron stops increasing.

When the frequency of the r.f. voltage source remains constant throughout the operation time, the cyclotron discussed earlier has to be modified so that the equality of the LHS and the RHS of the Eq. (13.19), $v = (qB/2\pi m)$, can be maintained and electrons can be accelerated to high energy. This modification of the cyclotron with a view to accelerate electrons to high energy gave birth to synchrocyclotron [with increase of the speed of the electrons, v is decreased] and synchrotrons [with increase of the speed of electrons, B is increased].

i. *Synchrocyclotron:* As the speed of electrons increases, the relativistic mass m also increases. To maintain the equality of the LHS and the RHS of the Eq. (13.19) $v = (qB/2\pi m)$, and thus to ensure resonance, one may decrease the frequency of the r.f voltage source as the electrons accelerate in such a way that the product vm remains constant. In this case, the magnetic induction B is kept constant. The accelerators that use this technique are called synchrocyclotrons. Synchrocyclotrons are also called frequency-modulated cyclotrons.

ii. *Synchrotrons:* As the speed of electrons increases, the relativistic mass m also increases. To maintain the equality of the LHS and the RHS of Eq. (13.19), $v = (qB/2\pi m)$, and thus to ensure resonance, one may increase the magnetic induction B as the electrons accelerate in such a way that B/m remains constant. In this case, the magnetic induction B increases with increase of speed of the electrons. The accelerators that use this technique are called synchrotrons.

In addition to these techniques, we can increase the speed of electrons along a circular path by subjecting them to a space-varying and time-varying magnetic field. The device employing this technique was first constructed in 1940 and is called a betatron.

13.9.1 Betatron

The betatron is an accelerator used to accelerate β–particles, i.e., electrons to high speeds by allowing them to be acted upon by induced electric fields that are set up by a changing magnetic flux. The energetic electrons can be used for fundamental research or to produce penetrating X-rays which are useful in cancer therapy, in industry, biological and medical research. The first betatron was constructed by D.W. Kerst at the University of Illinois in 1940, to accelerate electrons to an energy of 2.3 MeV. Since then, several other betatrons have been constructed, the largest being the one at the University of Illinois, completed in 1950. It yielded electrons of 300 MeV.

When an alternating magnetic field is applied parallel to the axis of a circular tube containing electrons, an electromotive force is produced in the electrons' orbit by changing the magnetic flux that gives an additional energy to the electrons. A radial force is also produced by the action of the magnetic field whose direction is perpendicular to the electron's velocity which keeps the electron moving in a circular path. Instead of spiralling, as in a cyclotron, the conditions are arranged such that the increasing magnetic field keeps the electrons in a circular orbit of constant radius. This is the basic law behind the functioning of a betatron. A betatron provides an excellent illustration of the physical reality of the equation

$$\oint_C \vec{E}.\vec{d\ell} = -\frac{d\varphi_B}{dt}.$$

Construction

The construction of a betatron is shown in Fig. 13.9. It consists of a highly evacuated circular tube called a doughnut chamber. This chamber is placed between the two poles of an electromagnet so that magnetic induction is along the axis of the circular doughnut, i.e., magnetic flux is perpendicular to the plane of the circular doughnut. The electrons produced by the electron gun are injected into the doughnut at the beginning of each cycle of alternate current in the electromagnet.

The poles of the electromagnet are constructed in such a way as to provide certain stronger field in the central place and certain weaker field at the orbit of the electron. Thus, the magnetic field is non-uniform in space and varies with time. This variable magnetic field performs several functions in the betatron. Those functions are enlisted as:

i. The variable magnetic flux guides the electrons in a circular path.

ii. It gives rise to an induced electric field around the doughnut which accelerates the orbiting electrons to high speed as a result of which their energies increases.

iii. It keeps the radius of the orbit in which electrons are moving constant.

iv It introduces the electrons into the orbit initially and removes them from the orbit after they have reached the desired energy.

v. It provides a restoring force that resists any tendency for the electrons to leave their orbit either vertically or radially.

The direction of variable magnetic field at a particular instant of time is shown in Fig. 13.9(a). The electric field produced by this variable magnetic field at this time is into the paper at A so that the force on the electrons is out of the paper at A. The magnetic field shown in this figure is not to scale.

Working

The electrons from the electron gun are injected into the doughnut shape vacuum chamber when the magnetic field is just rising from its zero value in the first quarter cycle ab as shown in Fig. 13.10.

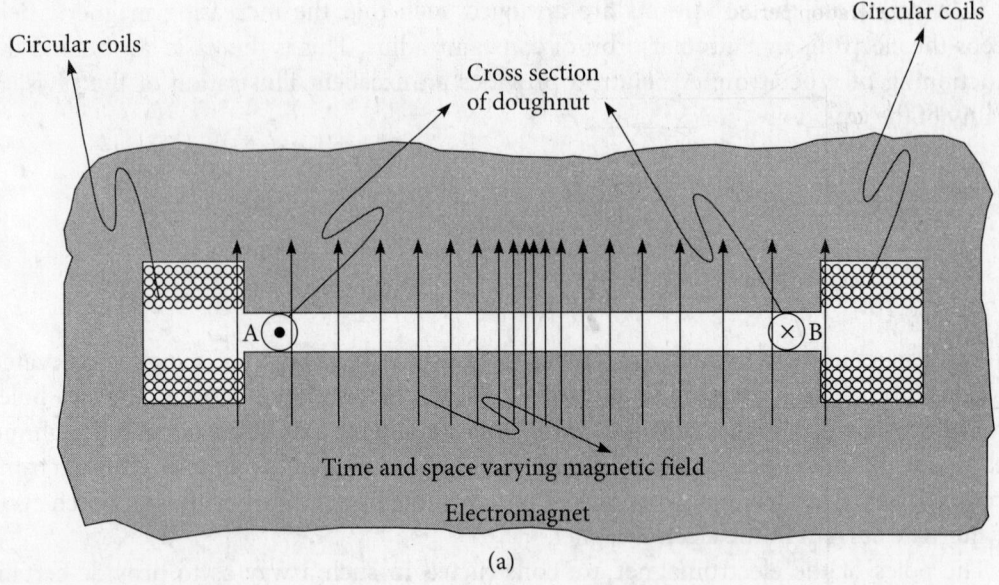

Circular coils

Cross section
of doughnut

Circular coils

A ⊙

⊗ B

Time and space varying magnetic field

Electromagnet

(a)

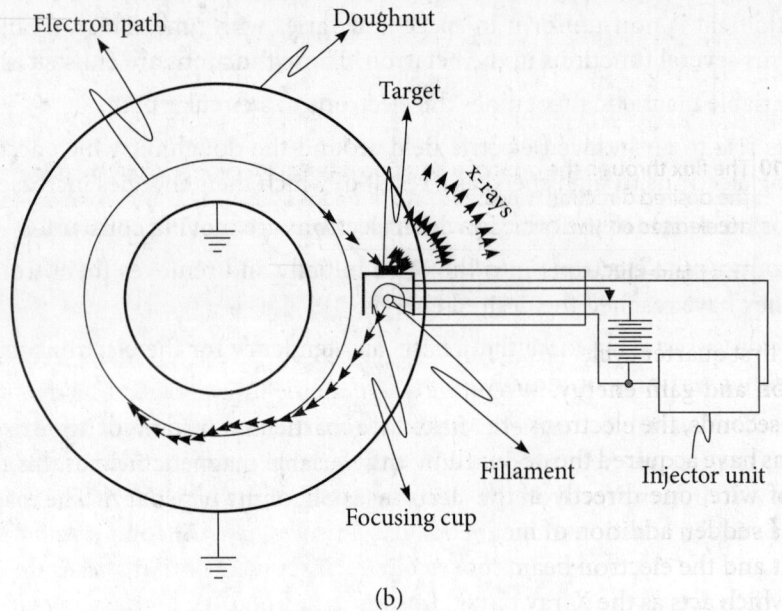

Electron path

Doughnut

Target

X-rays

Fillament

Injector unit

Focusing cup

(b)

Figure 13.9 (a) Vertical section of a betatron. Cross-sections of the coil and the doughnut are shown. The plane of the coil and the doughnut are the same and perpendicular to the plane of the page. Electrons come out of the doughnut at *A* and enter into it at *B*. The induced electric field is into the paper at *A*. (b) Horizontal section of a betatron

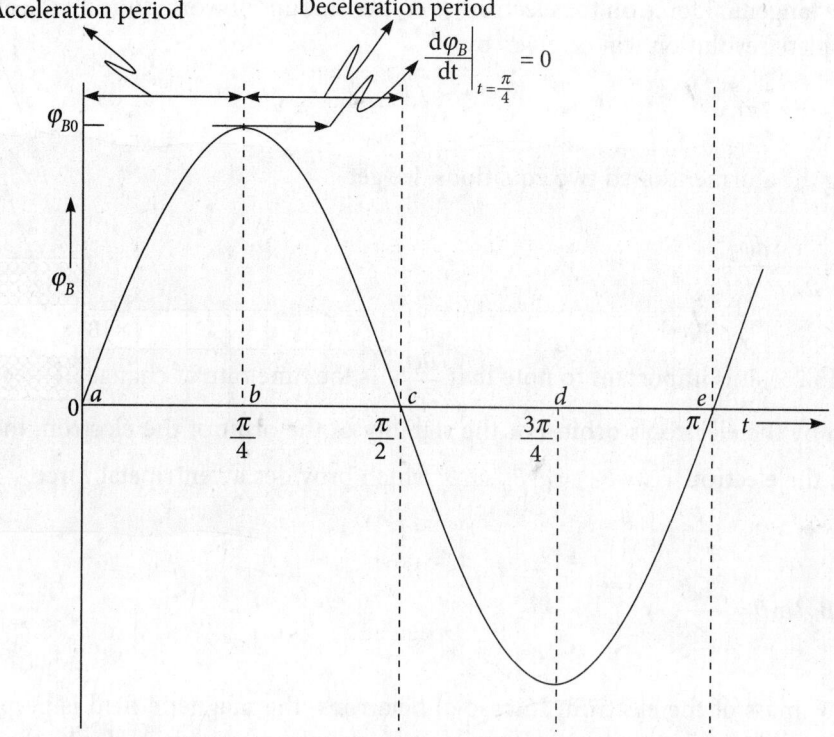

Figure 13.10 The flux through the orbit of a betatron during one cycle. Circulation of the electron in the desired direction is possible only during the half cycle ac out of which electrons are accelerated only in first quarter cycle ab and are decelerated in the second quarter cycle bc.

This figure shows that $F = \dfrac{e}{2\pi r}\dfrac{d\varphi_B}{dt} = 0$ at $t = \dfrac{\pi}{4}$.

During the first quarter cycle, i.e., in time $T/4$ seconds, the electrons make several thousands of revolution and gain energy. When the magnetic field has reached its maximum value within $T/4$ seconds, the electrons are removed from their orbit in the following way. When the electrons have acquired the desired amount of energy, a capacitor is discharged through two coils of wire, one directly above and the other directly below the stable orbit, thus producing a sudden addition of magnetic flux. This destroys the condition for the stability of this orbit and the electron beam moves out to larger radii until it strikes the back of the injector P which acts as the X-ray target. In some betatrons, the high energy electron beam emerges from the apparatus through a window for other specific purposes.

Theory

By using Faraday's law of induction, the work done on an electron in one complete revolution W is obtained as

$$W = e\frac{d\varphi_B}{dt} \qquad (13.22)$$

If F is the tangential force on the electron, then the amount of work done on the electron in one complete revolution will be given by

$$W = F \times 2\pi r$$

Equating the aforementioned two equations, we get

$$F = \frac{e}{2\pi r} \frac{d\varphi_B}{dt} \qquad (13.23)$$

In Eq. (13.23), it is important to note that $\dfrac{d\varphi_B}{dt}$ is the time rate of change of magnetic flux enclosed by the electron's orbit. For the stability of the orbit of the electron, the Lorentz force on the electron $e(\vec{v} \times \vec{B}_G) = evB_G \sin\theta$ which provides a centripetal force $\dfrac{mv^2}{r}$. Thus, we have

$$evB_G \sin\theta = \frac{mv^2}{r}$$

Here m = mass of the electron. In case of betatrons, the magnetic field is perpendicular to the plane of circulation of the electron in the doughnut chamber. Hence, the magnetic induction $\vec{B}_G$ is perpendicular to the velocity of the electron, i.e., the angle between v and $\vec{B}_G$ is 90°. This equation becomes

$$evB_G = \frac{mv^2}{r}$$

or $\qquad mv = B_G re \qquad (13.24)$

In Eq. (13.24), it should be understood that $\vec{B}_G$ is the magnetic induction on the electron's orbit. It is the guiding field on the orbit of the electron. Equation (13.24) gives the momentum (obviously tangential!) of the electron. The tangential force acting on the electron is obtained by differentiating Eq. (13.24) with respect to time. Hence,

$$F = \frac{d(mv)}{dt} = re \frac{dB_G}{dt} \qquad (13.25)$$

Combining Eqs (13.23) and (13.25), we get

$$\frac{e}{2\pi r} \frac{d\varphi_B}{dt} = re \frac{dB_G}{dt}$$

or $\quad \dfrac{d}{dt}\dfrac{\varphi_B}{\pi r^2}=2\dfrac{dB_G}{dt}$

or $\quad \dfrac{d}{dt}<B>=2\dfrac{dB_G}{dt}$

or $\quad B_G=\dfrac{1}{2}<B>$ $\hspace{5cm}$ (13.26)

$<B>=\dfrac{\varphi_B}{\pi r^2}$ is the average value of the magnetic induction over the orbit. Equation (13.26) is called the betatron condition. The betatron condition states that an electron can be accelerated into a circular orbit of constant radius by the action of a varying magnetic field only if the variation of the magnetic field over the pane of the electron's orbit is such that the magnetic field at the orbit $\vec{B_G}$ is equal to half the average value of the magnetic induction $<B>$ over the orbit.

Energy of the electron

The energy of the electrons can be calculated from the average induced emf and the total number of revolutions made by the electrons during the first quarter cycle of the applied magnetic field. Here it is assumed that the betatron condition is satisfied. Let the time-varying magnetic flux be given as $\varphi=\varphi_{BO}\sin\omega t$. The time period of the variation of this magnetic flux is $\dfrac{2\pi}{\omega}$. As we have discussed earlier, the electrons are accelerated only for

$$\frac{1}{4}\times\frac{2\pi}{\omega}=\frac{\pi}{2\omega}\text{ seconds.}$$

Using Eq. (13.22), the energy gained by an electron per revolution is obtained as

$$E'=e\frac{d}{dt}\varphi_{BO}\sin\omega t=e\omega\varphi_{BO}\cos\omega t \hspace{3cm} (13.27)$$

The average value of the electron energy per revolution during acceleration period $\dfrac{\pi}{2\omega}$ will be

$$<E'>=\frac{\displaystyle\int_0^{\frac{\pi}{2\omega}} e\omega\varphi_{BO}\cos\omega t\,dt}{\displaystyle\int_0^{\frac{\pi}{2\omega}} dt}=\frac{2}{\pi}e\omega\varphi_{BO} \hspace{2cm} (13.28)$$

For most of the time, electron travels with the speed of light c. Therefore, during acceleration time, $\dfrac{\pi}{2\omega}$, the total distance travelled by the electron will be $\dfrac{\pi c}{2\omega}$. For one complete revolution, the electron travels a distance of $2\pi r$. Therefore, the number of revolutions the electron completes in travelling a distance $\dfrac{\pi c}{2\omega}$ will be

$$\frac{\dfrac{\pi c}{2\omega}}{2\pi r} = \frac{c}{4\omega r}.$$

According to Eq. (13.28), the average energy gained by the electron per revolution is $\dfrac{2}{\pi} e\omega\varphi_{BO}$. Therefore, the average energy gained by the electron in $\dfrac{c}{4\omega r}$ number of revolutions $<E>$ will be obtained as

$$<E> = \frac{2}{\pi} e\omega\varphi_{BO} \times \frac{c}{4\omega r}$$

or $$<E> = \frac{e\varphi_{BO}c}{2\pi r} \tag{13.29}$$

Equation (13.29) shows that the energy obtainable is limited by the radius and peak magnetic flux of the applied magnetic field.

The energy of the electron can also be obtained with the help of the relativistic equation for energy as

$$K.E. = pc = B_{GO}rce \tag{13.30}$$

Here B_{GO} is the peak value of the magnetic induction at the orbit.

Example 13.6

In a certain betatron, the radius of the stable electron orbit is 30inches and operates at the maximum magnetic induction of 4000 gauss and 50 Hz. Calculate the (a) average energy gained per revolution and (b) the final energy of the electrons.

Solution

The data given are $R = 30$ inches $= 0.762$ m, $B = 4000$ gauss $= 0.4$ Tesla, $v = 50$ Hz.

(b) The energy of an emerging electron in a betatron is given by

$$E = BRce = 0.4 \times 0.762 \times 3 \times 10^8 \times 1.6 \times 10^{-19} \text{ J} = 91.4 \text{ MeV}$$

(a) 91.4 MeV is obtained by the electron in

$$\frac{c}{4\omega R} = \frac{c}{8\pi v R}$$

number of revolutions. Hence, the average energy gained per revolution will be

$$\frac{E}{\dfrac{c}{8\pi v R}} = \frac{8\pi v R E}{c} = \frac{8\pi \times 50 \times 0.762 \times 91.4}{3 \times 10^8} = 291.74\,\text{eV}$$

Example 13.7

In a 100 MeV betatron the radius of a stable electron orbit is 35 cm and the electron gains energy at the rate of 480 eV per revolution. How far will the electron travel in attaining full energy?

Solution

The data given are $E = 100$ MeV, $R = 35$ cm $= 0.35$ m, and the energy gain per revolution $= 480$ eV The electron gains an energy of 480 eV per revolution; to gain 100 MeV of energy, the electron will complete

$$\frac{100 \times 10^6\,\text{eV}}{480\,\text{eV}} = 2.083 \times 10^5$$

revolutions. The distance traveled by the electron in attaining 100 MeV will be

$$\frac{100 \times 10^6\,\text{eV}}{480\,\text{eV}} \times 0.35\,\text{m} = 4.58 \times 10^5\,\text{m}.$$

Since Eqs (13.29) and (13.30) are independent of the mass of the electron, one might think that there would be no upper limit of energy obtainable by means of the betatron; however, beyond a certain upper limit radiation, loss offsets the gain. Since the electrons are accelerating in a circular path, their energy loss by radiation is more in comparison to a linear accelerator. The total energy radiated U by a circulating electron of energy E in one revolution is calculated by using the formula

$$U = \frac{e^2}{3r\varepsilon_0} \frac{E^4}{m_0^4 c^8} = 88.5 \frac{E^4}{r}\,\text{keV} \tag{13.31}$$

In Eq. (13.31), U has been measured in keV and E in geV (1 geV = 1000 MeV). Equation (13.31) shows that the energy radiated per revolution increases with the fourth power of the electron energy and thus becomes very serious at very high energies.

Example 13.8

In the 70 MeV betatron the radius of a stable electron orbit is 28 cm. Calculate the (a) value of magnetic induction at the orbit for this energy, (b) the frequency of the applied electric field, (c) the energy loss by radiation during a single revolution of an electron.

Solution

The data given are $E = 70$ MeV, $R = 28$ cm $= 0.28$ m.

a. The energy of an emerging electron in the betatron is given by

$$E = BRce$$

or $\quad B = \dfrac{E}{Rce} = \dfrac{70 \times 10^6 \times 1.6 \times 10^{-19}}{0.28 \times 3 \times 10^8 \times 1.6 \times 10^{-19}} \text{T} = 0.83\,\text{T}$

b. The frequency v of the applied electric field is given by

$$v = \dfrac{eBc^2}{2\pi E} = \dfrac{1.6 \times 10^{-19} \times 0.83 \times (3 \times 10^8)^2}{2\pi \times 70 \times 10^6 \times 1.6 \times 10^{-19}} \text{MHz} = 170.6\,\text{MHz}$$

c. The total energy U radiated by a circulating electron of energy E in one revolution is given by

$$U = \dfrac{e^2}{3R\varepsilon_0} \dfrac{E^4}{m_0^4 c^8} = 88.5 \dfrac{E^4}{R} = \dfrac{88.5 \times (0.07)^4}{0.28} \text{keV} \doteq 7.5\,\text{eV}$$

Advantages

i.   Betatrons can accelerate electrons up to an energy of 340 MeV.

ii. Most modern betatrons are operated from a 60–Hz ac source.

iii. It is cost effective.

Disadvantages

i.   It is difficult to maintain the stability of the electron's orbit.

ii. Since the electrons are accelerating in a circular path, their energy loss by radiation is more in comparison to a linear accelerator.

13.10 Applications of Accelerators

As the definition shows, the fundamental and only direct application of a nuclear accelerator is to accelerate charged particles to high energies. These high energy charged particles are used extensively scientifically in nuclear research and commercially in the medical and material processing field and in agriculture. A brief summary of the uses of nuclear accelerators both direct and indirect, in the production of isotopes, radiation processing of materials and medicals are given in this section.

13.10.1 Radiation processing of materials

The knowledge of the ffects of nuclear radiations on various kinds of materials is important both for practical problems connected with the operation of the various devices under irradiation and for the numerous aspects of solid state physics. The properties of many materials can be changed by exposing them to nuclear radiations. Thus, many new valuable materials having optimum properties can be obtained from the conventional materials with the help of nuclear radiation. The controllable modification of the properties of materials is accomplished in general by implantation of heavy ions in the materials. It may be proved detrimental to use materials and devices in a radiation environment without knowing the effects of radiation on the materials or devices.

a. Ionizing radiations passing through a gas increase its electrical conductivity in textile industries. Electrified fibers are made neutral by irradiating them by α–active $_{94}Pu^{239}$ or β–active $_{61}Pm^{147}$ as a result of which the productivity increases by 30%.

b. Nuclear radiations are used to combine monomers to form certain polymers.

c. The cross-linking of certain polymers to build specific three-dimensional structures is performed with the help of nuclear radiations.

d. Nuclear radiations are used to degrade certain polymers. Destruction is accompanied by an increase in solubility and plasticity and decrease in strength and rupture elongation. Thus, materials liable to destruction should not be used as structural materials intended to work in conditions of irradiation.

e. Nuclear radiations, now-a-days are in use in developing methods for the production of new valuable materials like linked polyethylene.

f. Using radiation in the vulcanization process of rubber, the lifetime of tyres increases by scores of percent.

g. The strength of woods has been increased to several times when exposed to nuclear radiation of certain intensity for a certain time.

h. High temperature epoxy resins have been produced using nuclear radiation.

i. Scientists are able to synthesize, by nuclear radiation methods in laboratory conditions, such materials as cyanic acid and hydrazine whose production by conventional methods is an intricate process requiring much money and energy.

j. Different types of crystal imperfections are produced when crystals are exposed to nuclear radiations resulting in modified electrical, optical, mechanical and thermal properties of the crystals. These changes are being exploited technologically.

k. Radiation changes in materials are quite stable.

l The density of quartz decreases by 15% when exposed to nuclear radiation of a particular intensity for a certain time.

m.   Nuclear radiation changes the lattice types in some crystals.

n.   The color centres in cooking salt crystals are produced by using nuclear radiations.

o.   Depending upon the types of dielectrics, the electrical conductivity may increase or decrease when irradiated by nuclear radiations. In certain applications, it is important to know the behavior of dielectrics under nuclear radiations.

p.   The electrical conductivity of semiconductors varies in a complicated manner when irradiated by nuclear radiations. In certain applications, it is important to know the behavior of semiconductors under nuclear radiations.

q.   The implantation of boron, phosphorous, and tantalum considerably improves the behavior of silicon and germanium detectors.

r.   *p–n* junctions can be produced from a singe semiconductor by using nuclear radiations.

s.   Thermal conductivity of non-metallic materials decreases when irradiated by nuclear radiations.

t.   The implantation of heavy ions opens up great possibilities for the manufacture and study of the properties of new alloys, which cannot be fabricated by other methods because of the chemical incompatibility of the components.

u.   The excellent thin film filters of certain pore diameters ($\approx 0.4\,\mu m$) can be manufactured by using nuclear radiations. Purification of gases and water, the sorting of micro-impurities according to their size, the study of the size and shape of blood cells, the sterilization of biological media, the filtration and separation of viruses and molecules, the purification of wine and beer are done by this nuclear filters.

v.   The use of nuclear radiation to kill microorganisms has found applications in preservation of food stuffs. Irradiated potatoes and onions can be kept without deterioration or sprouting for nearly two years! This is a better substitute to refrigeration.

w.   Sterilization of surgical instruments and bandages are done by nuclear radiation. Disposable needles, for example, are ordinarily sterilized by exposure to radioactive sources.

x.   Genetically modified plants have been developed by exposing the seeds to a particular nuclear radiation.

y.   Instead of chemical pesticides to increase the resistance of sprouts against pests, seeds are being de-insecticized by nuclear radiation.

z.   In medicine, the ionizing power of nuclear radiation is used for destruction of malignant tumours in the body.

13.10.2 Uses of isotopes

By choosing the intensity and energy of the particle beam from the nuclear accelerators, the desired nuclear reaction can be studied under controlled conditions and the nuclei can be modified to produce artificial isotopes (same Z-value), isobars (same A-value), isotones (same N-value) and isodiapheres (same $(A\text{-}2Z)$value). For example, $_{17}Cl^{37}$ is an isotope of $_{17}Cl^{35}$, an isobar of $_{16}S^{37}$, an isotone of $_{19}K^{39}$, an isodiaphere of $_{18}Ar^{39}$. At present, more than 150 various radioactive isotopes are being produced in specially designed cyclotrons. Separated stable isotopes are being produced on an industrial scale. Both the radioactive and stable isotopes are produced not only in elementary form but also in the form of various chemical compounds. Depending on the properties and half life of these isotopes, they are used as tracers in various fields such as medicine, chemistry, engineering, biology, agriculture, metallurgy, hydrology, oceanography, ecology, and so on. It has found practical applications in medical diagnosis, criminology, leak detection, fuel transport and many other technological problems.

a. The wear on a piston ring is rapidly and sensitively measured by incorporating iron isotopes.

b. The degree of oil burning in diesel engines is checked by using isotopes.

c. Different types of oil filters are tested by radioisotope tracers.

d. Isotopes are used to detect mechanical flaws in materials.

e. The volume of mercury in air is monitored by $_{80}H^{197}$ isotopes in fluorescent lamp production units.

f. The efficiencies of washing machines and detergents are checked by soaking rags in oil containing $_{15}P^{32}$ and measuring the radioactivity after washing.

g. The methods of applying pesticides in agriculture are devised by using radioisotopes.

h. The methods of destroying pests are devised by feeding them with radioactive materials.

i. Deuterium, an isotope of hydrogen, is used for the investigation of fat metabolism.

j. Many radioisotopes are used in medical diagnosis. Abnormality of thyroid gland is tested by $_{53}I^{131}$.

k. Anemia (deficiency of RBC) and its converse, polycythemia, are studied by injecting iron isotopes into the blood stream.

l. The size and location of brain tumours can be known with high precision by introducing certain radioisotopes into the body.

m. The blockage of blood flow is detected introducing $_{11}Na^{24}$ into the body.

n. The failure of the body to absorb vitamin B_{12}, causing pernicious anemia, is diagnosed by introducing $_{27}Co^{60}$ into the body.

o. The radiations from radioisotopes have been used in controlling cancerous cells.

p. Radioisotope $_{6}C^{14}$ is used to study the kinetics of photosynthesis.

q. Metallurgists use radioisotopes to detect flaws in metals.

r. Chemical engineers use radioisotopes in the study of plastics and polymers and other synthetic materials.

s. Engineers use radioisotopes in the investigation of wear and lubrications.

t. Information regarding the thickness of coatings, thin films, papers are obtained by using radioisotopes.

u. Information about the level of liquid in a tank, the surface of separation between immiscible liquids, velocity of flow, degree of intermixing of oil stocks flowing in a pipe line can be obtained with the help of radioisotopes.

v. The age of meteors, rocks, and minerals containing uranium can be determined from the known half life of $_{92}U^{238}$ and the quantity of helium accumulated from the emitted α–particle.

w. Radio carbon dating has been used for the age measurements of specimens of archeological and geological importance.

x. Radio isotopes are used to study atomic and nuclear properties.

Questions

13.1 What is a nuclear accelerator?

13.2 What is distance of closest approach?

13.3 Low energy α–particles can disintegrate only lighter elements. Why?

13.4 What are the limitations in using radiations from natural radioactive elements?

13.5 Why are need of nuclear accelerators needed?

13.6 Why is vacuum maintained inside an acceleration tube?

13.7 What are the basic principles of a nuclear accelerator?

13.8 Under what conditions, can a charge not be accelerated by using magnetic field?

13.9 What are the characteristics of an ion source in nuclear accelerators?

13.10 What are the factors on which the performance index of a nuclear accelerator depends?

13.11 Explain the working of a voltage multiplier circuit.

13.12 Write an expression for the output voltage of a voltage multiplier circuit.

13.13 What are the factors on which the output voltage of a voltage multiplier circuit depend?

13.14 Explain the construction of the Cockcroft–Walton accelerator.

13.15 Explain the working of the Cockcroft–Walton accelerator.

13.16 What are the advantages ofthe Cockcroft–Walton accelerator?

13.17 What are the disadvantages of the Cockcroft–Walton accelerator?

13.18 What is the principle of the Van de Graaff accelerator?

13.19 Explain the principle of the Van de Graaff accelerator.

13.20 What are the properties of the belt used in the Van de Graaff accelerator?

13.21 What is the function of the spray comb used in the Van de Graaff accelerator?

13.22 What is the function of collection comb used in the Van de Graaff accelerator?

13.23 Describe the construction of the Van de Graaff accelerator.

13.24 Under what conditions, does corona discharge occur in the Van de Graaff accelerator?

13.25 Describe the working of the Van de Graaff accelerator.

13.26 What are the advantages of the Van de Graaff accelerator?

13.27 What are the disadvantages of the Van de Graaff accelerator?

13.28 What is the literal meaning of the word 'tandem'?

13.29 Explain how the tandem accelerator is a modified version of the Van de Graaff accelerator

13.30 Explain the principle of the tandem accelerator.

13.31 Describe the construction of the tandem accelerator.

13.32 Explain the working of the tandem accelerator giving one example.

13.33 What are the advantages of the tandem accelerator?

13.34 What are the disadvantages of the tandem accelerator?

13.35 What is the basic principle in the operation of a radio frequency accelerator?

13.36 Give the schematic diagram of a drift tube linear accelerator.

13.37 Why in the drift tube linear accelerator, are the alternate drift tubes connected?

13.38 Why is the electric force on a moving ion inside a drift tube zero?

13.39 Give the principle of a drift tube linear accelerator.

13.40 Explain how the drift tube linear accelerator works?

13.41 Derive an expression for the length of the nth number drift tube.

13.42 Derive an expression for the speed of an ion in the nth number drift tube.

13.43 What are the advantages of a drift tube linear accelerator?

13.44 What are the disadvantages of a drift tube linear accelerator?

13.45 Give the principle of a cyclotron.

13.46 Give the construction of a cyclotron.

13.47 What is the shape of the path of ions in a cyclotron? Justify your answer.

13.48 What are the factors on which the frequency of the r.f. voltage source used in cyclotron depends? Derive an expression for it.

13.49 Derive an expression for the speed of the ion in a cyclotron at any instant of time.

13.50 Does the frequency of the r.f. voltage source used in a cyclotron remain constant throughout the operation time? Justify your answer.

13.51 Derive an expression for the maximum energy of the ion in a cyclotron.

13.52 What are the factors on which the maximum energy of the ion in a cyclotron depends?

13.53 Derive an expression for the cyclotron frequency.

13.54 What are the advantages of a cyclotron?

13.55 What are the disadvantages of a cyclotron?

13.56 Why can an electron not be accelerated by a cyclotron to a very high energy?

13.57 Show that the radius of curvature R of the path of a particle inside the dees of a cyclotron is proportional to $\sqrt{n}$, where n is the number of times the particle has been accelerated across the space between the dees.

13.58 What is a synchrocyclotron?

13.59 What is a synchrotron?

13.60 What are the properties of β-particles?

13.61 What is the principle of a betatron?

13.62 Distinguish between the path of an electron in a betatron and the path of a proton in a cyclotron.

13.63 Name an accelerator in which the validity of the equation $\oint_C \vec{E}.\vec{dl} = -\dfrac{d\varphi_B}{dt}$ is demonstrated.

13.64 What are the roles of applied magnetic flux in a betatron?

13.65 Describe the construction of a betatron.

13.66 Describe the working of a betatron by a neat diagram.

13.67 What is the betatron condition?

13.68 Derive an expression for the betatron condition.

13.69 Derive an expression for the average energy of an electron in a betatron.

13.70 What are the factors on which the average energy of an electron in betatron depends?

13.71 What are the advantages of a cyclotron?

13.72 What are the disadvantages of a cyclotron?

13.73 What are the uses of nuclear accelerators in medical science?

13.74 Give a few uses of nuclear accelerators in agriculture.

13.75 Mention a few uses of nuclear accelerators in material processing.

13.76 Give a few uses of nuclear accelerators in the textile industry.

13.77 What is the role of nuclear accelerators in food stuff preservation?

13.78 How is nuclear radiation used in producing few chemicals?

13.79 How is nuclear radiation used in automobile plants?

13.80 Write an essay on 'role of nuclear accelerator in the development of physics'.

13.81 How is nuclear radiation is helpful to the farmers?

13.82 How is nuclear radiation helpful to the paper industry and print ing press?

13.83 How is nuclear radiation helpful in filling up of ketchup in an opaque container?

Problems

13.1   A linear accelerator has 960 drift tubes. If the length of the 1st tube is 1.5 m, what is the length of the last tube?   [Ans   46.48 m]

13.2 There is a proton linear accelerator with energy 800 MeV and a current of 1milliAmp. How many protons are produced per second? [Ans 6.25×10^{25} protons/s]

13.3 Calculate the energy of a deuteron that is moving in a circle of radius 60 cm under the influence of a magnetic field 10000 gauss. Given: mass of deuteron = 3.33×10^{-27} kg. [Ans 8.65 MeV]

13.4 The applied magnetic induction in a cyclotron is 15000 gauss. Calculate the frequency of the r.f. voltage source to accelerate protons. [Ans 22.9 MHz]

13.5 The beam in a cyclotron has a maximum diameter of 1.6 m. Given that the magnetic induction is 0.75 Tesla. Calculate the kinetic energy of a deuteron coming out of the cyclotron. Given: mass of deuteron = 3.33×10^{-27} kg. [Ans 8.65 MeV]

13.6 Protons in a cyclotron describe a circle of radius 40 cm just before emerging from the dees. The frequency of the applied alternating voltage is 15 MHz. Neglecting relativistic effects, find the energy and speed of a deuteron on emergence. [Ans 7.42 MeV, 3.77×10^7 m/s]

13.7 A deuteron is moving in a circular path in a magnetic field of 17000 gauss in resonance with the applied dee frequency of 12 MHz. Calculate its energy, momentum, linear speed, radius of the deuteron path. [Ans 153 MeV, 766 MeV/c, 0.42c, 165 cm]

13.8 In a certain betatron, the radius of the stable electron orbit is 100 cm and it operates at the maximum magnetic induction of 5000 gauss and 60 Hz. Calculate the (a) final energy of the electrons and (b) average energy gained per revolution. [Ans 150 MeV, 754 eV]

13.9 In a 110 MeV betatron, the radius of the stable electron orbit is 50 cm. The electron gains energy at the rate of 500 eV per revolution. How far will the electron travel in attaining full energy? [Ans 6.91×10^5 m]

13.10 The radius of the stable electron orbit of a small betatron–synchrotron is 0.12m. The maximum magnetic field at this position is 0.9 Tesla. Calculate the maximum electron produced. The arrangement is the voltage multiplier [Ans 32.4 MeV]

Multiple Choice Questions

1. The effectiveness of a nuclear accelerator is characterized by
   (i) momentum
   (ii) speed
   (iii) kinetic energy
   (iv) mass

2. The gain in kinetic energy of a charge q in an electric field may be given as

   (i) $\dfrac{1}{2}qV$

   (ii) $\dfrac{1}{2}qV^2$

   (iii) qV^2

   (iv) qV

3. The gain in kinetic energy of a charge q moving in circular path of radius R in a magnetic field may be given as

    (i) $\dfrac{1}{2} q^2 B^2 R^2$ (ii) $\dfrac{1}{2m} q^2 B^2 R^2$

    (iii) $\dfrac{1}{2} mq^2 B^2 R^2$ (iv) $\dfrac{1}{2m} qBR$

4. Which of the following is not an R.F. accelerator?

    (i)   cyclotron   (ii)   betatron

    (iii)   Cockcroft–Walton accelerators   (iv)   linear accelerators

5. Which of the following is a D.C. accelerator?

    (i)   cyclotron   (ii)   betatron

    (iii)   linear accelerators   (iv)   Van de Graaff accelerator

6. The belt used in the Van de Graaff accelerator is made of

    (i)   conducting material   (ii)   insulating material

    (iii)   semiconducting material   (iv)   any one of the above

7. The stripper canal in the tandem accelerator is made up of

    (i)   thin carbon foil   (ii)   thin aluminum foil

    (iii)   thin platinum foil   (iv)   a gas at low pressure

8. The relation between the length of the nth number drift tube and the ion speed in it is

    (i) $\dfrac{v_n}{L_n} = 2v$ (ii) $\dfrac{L_n}{v_n} = 2v$

    (iii) $\dfrac{L_n}{v_n} = v$ (iv) $\dfrac{v_n}{L_n} = v$

9. The energy of an ion in the nth number drift tube of a drift tube linear accelerator is

    (i) $E = qV_0^n$ (ii) $E = nqV_0$

    (iii) $E = \dfrac{qV_0}{n}$ (iv) $E = n^2 qV_0$

10. Which of the following charged particles cannot be accelerated by a cyclotron?

    (i) α-particles   (ii)   deuterons

    (iii)   electrons   (iv)   tritons

11. Which of the following accelerators is used to accelerate electrons?

    (i)   cyclotron   (ii)   drift tube linear accelerator

    (iii)   Cockcroft–Walton accelerator   (iv)   betatron

12.   Which of the following is not an accelerator?

    (i)   Van de Graaff generator   (ii) cyclotron

    (iii) drift tube linear accelerator   (iv) none of the above

Answers

1 (iii)   2 (iv)   3 (ii)   4 (iii)   5 (iv)   6 (ii)   7 (i & iv)   8 (i)

9 (ii)   10 (iii)   11 (iv)   12 (iv)

14 Holography

14.1 Introduction

Electromagnetic radiation scattered by an object contains all the information about the object. In photography, distribution of only the intensities of light scattered from the object are recorded on photographic plates. It is a 2-D representation of 2-D or 3-D object; it gives only a single view. The term "holography" is derived from Greek by combining *Holos* (whole) and graphy (writing) meaning complete recording of information on images in terms of scattered intensities and phases. In holography, distribution of intensities of light scattered from the object along with is phases are recorded on photographic plates. It is a 3-D representation of a 3-D object on a 2-D sheet; it gives a 3-D view. Here three-dimensional images using coherent laser light are produced on photographic plates. Therefore, though holography was invented by Dennis Gabor in 1948 (it earned him th Nobel prize in Physics in 1971), it gained momentum only after the invention of the laser in 1960, the powerful source of coherent light. In this chapter, the concept of holography is presented very briefly.

14.2 Basic Principles of Holography

The basic principle of holography was first put forward by Dennis Gabor in 1948 while working on a project to improve the resolving power of the electron microscope. In holography, three-dimensional perspective images (the clarity of vision changes with change in eye position) are produced on a photographic film which gives a sensation of the exact original 3-D object. Here the intensities and the local propagation direction or phase is recorded on the photographic film. However, all recording films record only the variation of the intensity of the object wave (i.e., light wave scattered by the object); therefore, to record phases, we must convert the phase information into variations of intensity. This does not happen in photographic images. However, it can be achieved through the process of optical interference of the light waves. In the interference and diffraction patterns of light waves, both intensities and phases are recorded.

The basic principle of holography can be understood by referring to Fig. 14.1

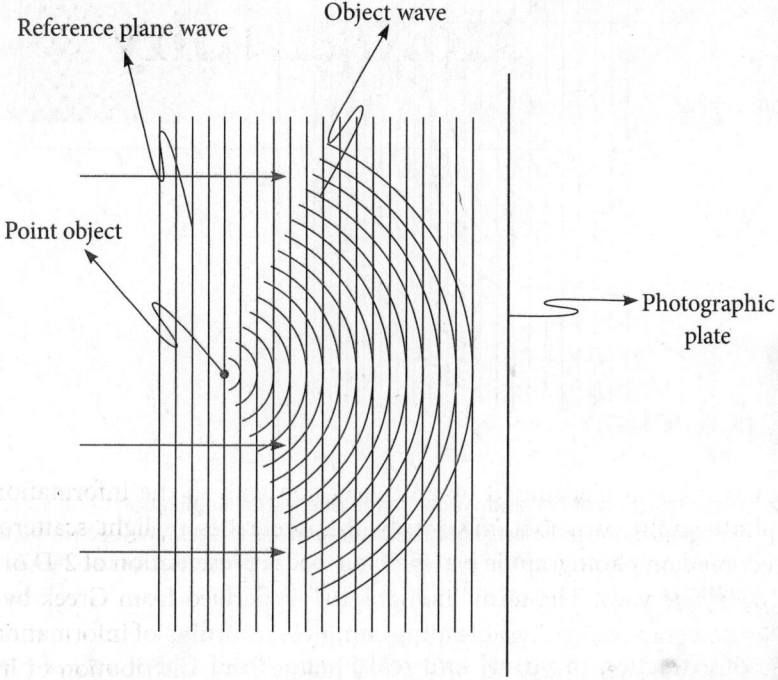

Figure 14.1 | The interference between directly incident laser light (reference wave) and corresponding scattered laser light (object wave) by a point object

The plane monochromatic wave of laser light, known as reference wave, is incident on a point object and is scattered by it. Now the scattered waves known as object waves and the reference waves of laser light advance and interfere with each other producing concentric circular interference fringes on a high resolution photographic plate placed in front of them. On the plane of the photographic plate, bright and dark concentric rings will be formed due to constructive and destructive interference between the scattered object waves and the direct reference waves. The light and dark partially absorbing circular fringes become visible in a reverse way when it is developed. This interference pattern (here in this case, it appears like Newton's rings) on the developed photographic plate is called Gabor's zone plate or Gabor's hologram. It was the first hologram constructed by Dennis Gabor and it laid the foundation for holography. Gabor's zone plate is very similar to Fresnel's zone plate except (i) the light and dark fringes shade continuously into each other and (ii) has a single focus.

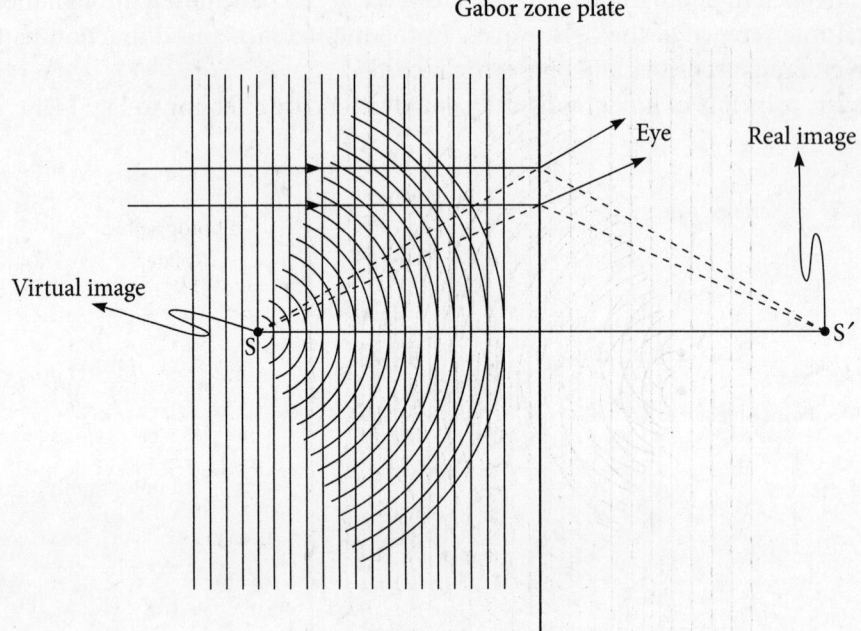

Figure 14.2 Reconstruction of real and virtual image due to the illumination of Gabor's zone plate hologram by the same laser light The virtual and real images have been formed at S and S' respectively as seen by the eye. The virtual and real images look exactly look like point scatterer

The process of extraction of virtual and real images from Gabor's zone plate is called reconstruction. For reconstruction, the plate is now illuminated by the same original plane monochromatic wave of laser (i.e., reference wave) light as shown in Fig. 14.2, but without the point scatterer.

The light passing through Gabor's zone plate hologram will now produce a first-order interference maximum at S'. To the eye, this light will therefore appear to diverge from S. All the points on the hologram will diffract the reference wave fronts and thus, virtual and real images are formed in the absence of the real scatterer! Thus, the reconstruction of virtual and real images from Gabor's zone plate made for a point object is explained.

Let us now discuss the construction of Gabor's zone plate for two point objects and the corresponding image reconstruction. It has been delineated in Fig. 14.3. Suppose instead of one scatterer, there are two scatterers on a vertical line separated by a small distance as shown in Fig. 14.3. Each scatterer will create a Gabor's zone plate. If the photographic response of the photographic plate is linear, the modulation of intensity of each concentric fringe pattern on the zone plate will be proportional to the scattered light intensity. Reconstruction in this case is done in the same method as that of a single point. It will produce two virtual and real images of both scattering centres, each with its proportionate intensity. This has been illustrated in Fig. 14.4.

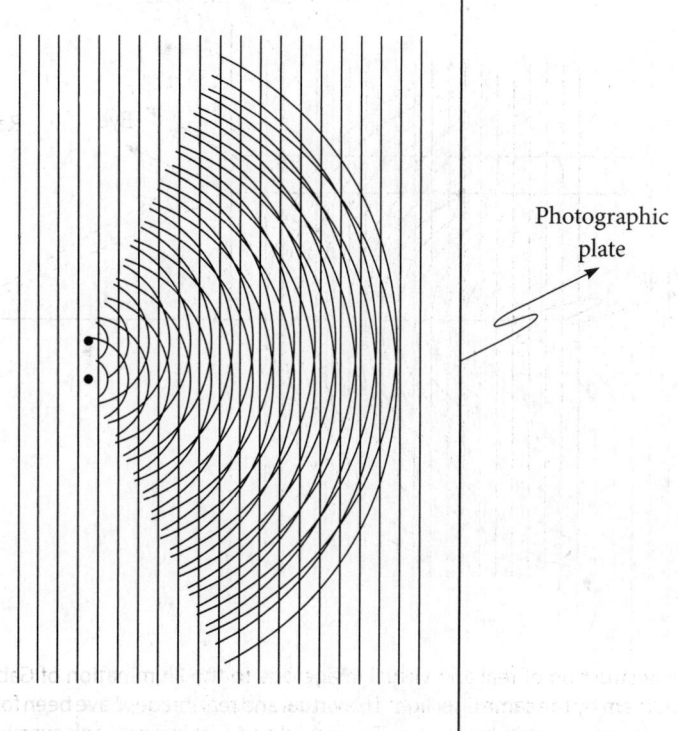

Photographic
plate

Figure 14.3 The interference between a directly incident laser light (reference wave) and the corresponding scattered laser light (object waves) by two point objects

The argument can now be extended to a 3-D object. A 3-D object consists of infinitely many points and each point acts like a scatterer or scattering center. A 3-D object is nothing but a continuous distribution of scattering centres. When it is illuminated by laser light, the scattered lights interfere producing a continuous interference pattern on the high resolution photographic plate placed in front of it. This interference pattern on the photographic plate is now called Gabor's zone plate. The reconstruction of this zone plate is shown in Fig. 14.5.

Now when Gabor's zone plate is illuminated by a reference laser beam, all the continuous points on Gabor's zone plate will diffract the reference wave fronts, producing virtual and real images on the left and right side of the hologram in the absence of a real 3-D object. Upon reconstruction, the distributed virtual image and real image should appear exactly like the original 3-D object as viewed from the right of the hologram.

We have presented the basic principles of holography in a simple way as envisioned by Dennis Gabor. However, the application of these principles suffered from poor quality of the reconstructed image. The invention of off-axis holography by Leith and Upatnieks in 1962 removed this difficulty.

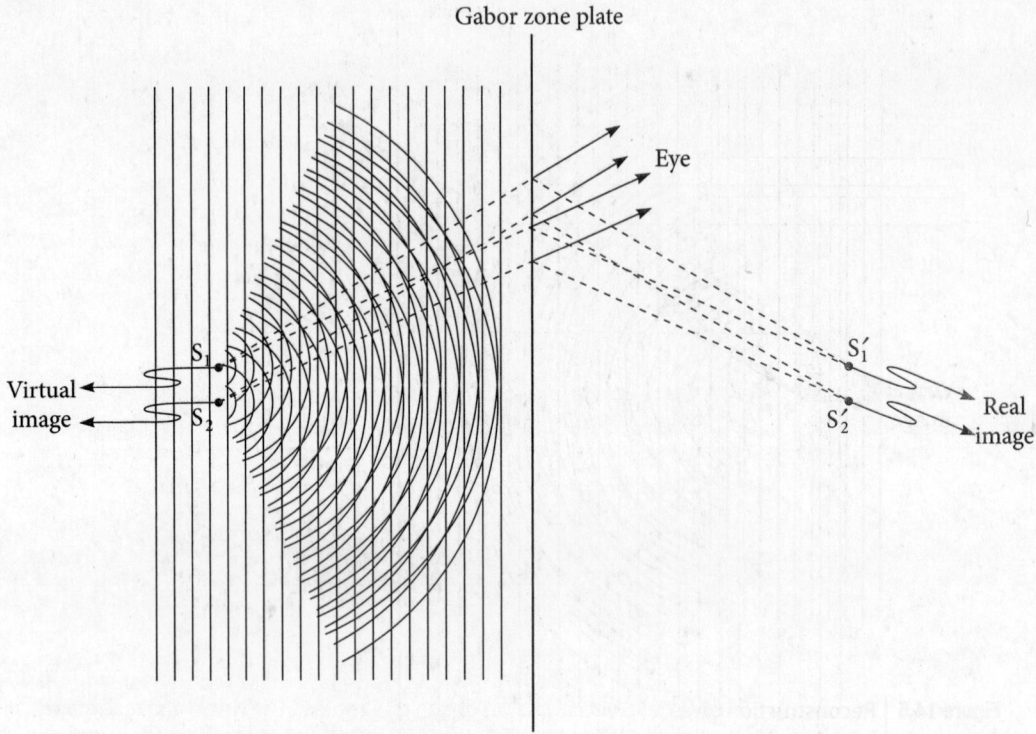

Figure 14.4 Reconstruction of real and virtual images due to two point scatterers. The virtual and real images have been formed at the left side and right side of the Gabor's zone plate respectively as seen by the eye

Depending upon the method of recording the holograms, holography is of two types, namely (i) in-line holography and (ii) off-axis holography.

i.   In-line holography

The previous discussions pertain to in-line holography. In Figs 14.2, 14.4, and 14.5, Gabor's zone plate is illuminated by two light waves along an axis perpendicular to it – one is the reference wave and the other is the object wave. The quality of the reconstructed image is poor due to superposition of the conjugate image on the reconstructed image.

ii.   Off-axis holography

Emmett Leith and Juris Upatnieks of the University of Michigan invented the off-axis holography in 1962 with a laser source which increased coherence length. The conjugate image problem of in-line holography was corrected in off-axis holography. A typical diagram showing off axis holography recording is shown in Fig. 14.6.

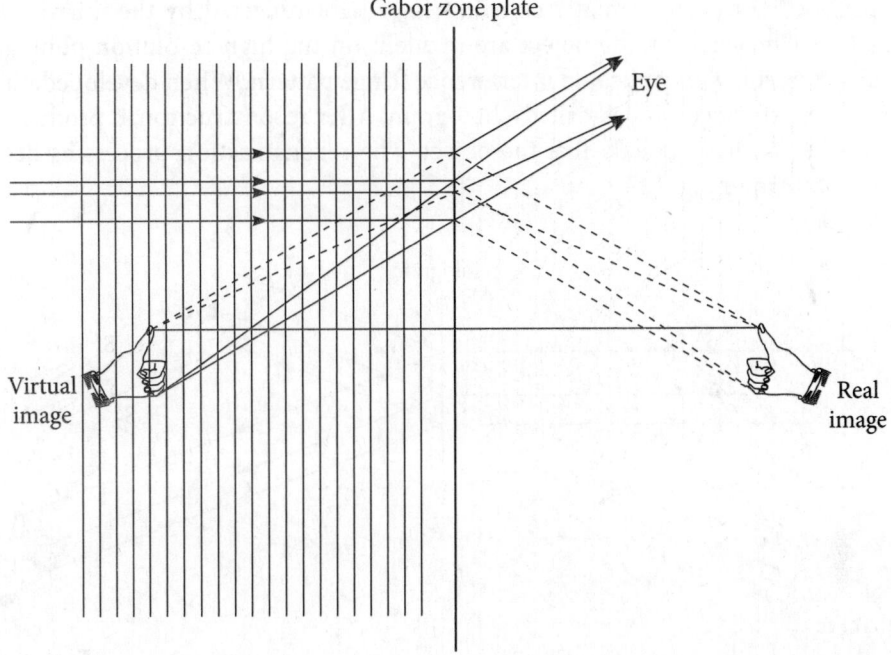

Figure 14.5 | Reconstruction of real and virtual images due to a 3-D object. The virtual and real images have been formed at the left side and right side of the Gabor's zone plate respectively as seen by the eye

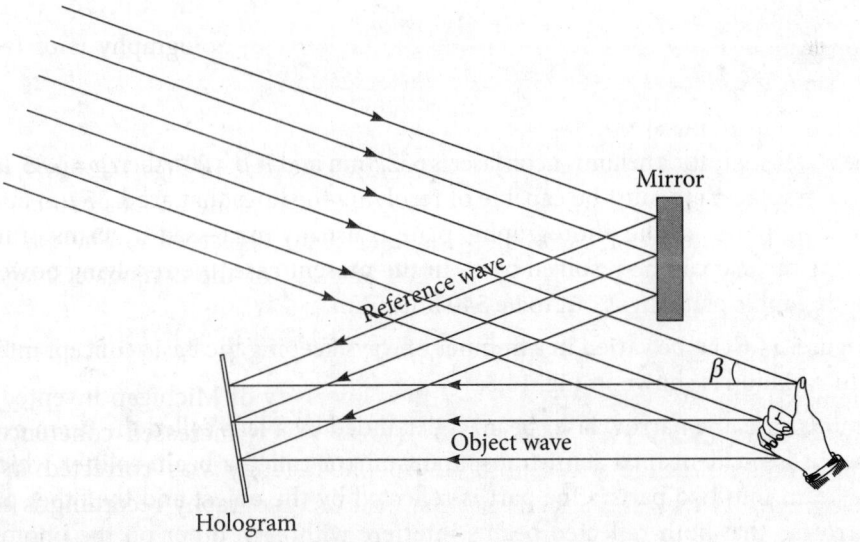

Figure 14.6 | Typical illustration of the basics of off-axis holography

A part of the monochromatic coherent laser light reflected by the mirror and the other part reflected by the object are incident on the high resolution photographic plate producing a complicated interference fringe pattern. When developed, the plate reveals interference fringes called a hologram. After reconstruction, it produces a 3-D image which looks exactly like the object. The reconstruction process as described here is shown in Fig. 14.7.

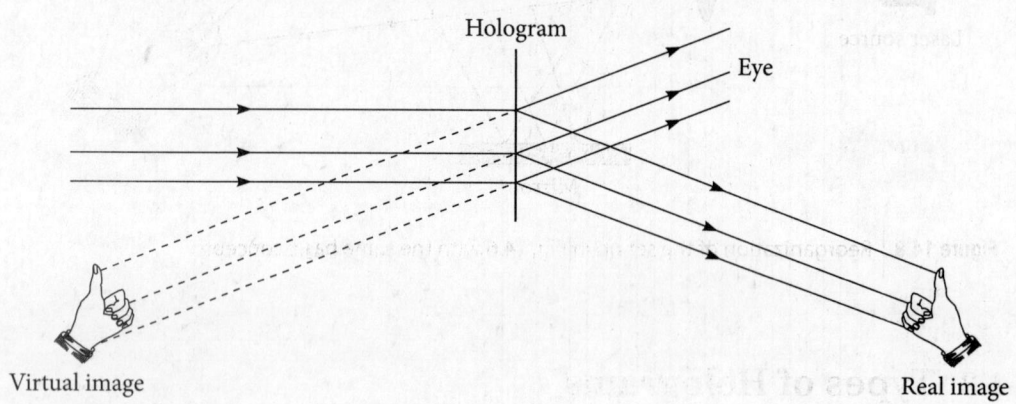

Figure 14.7 | The reconstruction process

The angle β between the scattered light and the reference beam will determine the quality of the reconstructed image. The distance between consecutive maxima or minima is known as the pitch p and is given as

$$p = \frac{\lambda}{\sin \beta}$$

The wavelength for a helium–neon laser is 632.8nm and if $\beta = 20°$, then $p = 1.85$ μm. The photographic plate must be capable of resolving the lines that are 1.85 μm apart. The resolving power of the photographic plate is usually expressed in terms of lines per millimeter that can be resolved by it. In the present case, the resolving power of the photographic plate turns out to be 540 lines/mm.

Figure. 14.6 can be varied in a number of ways keeping the basic concept intact. One such variation is shown in Fig. 14.8.

In this case, the narrow laser beam is expanded by a lens called the beam expander. Then it is incident on a semi-transparent mirror called a beam splitter which splits the beam into two parts. One part is reflected by the object and the other part by a mirror so that both reflected beams interfere with each other on the photographic plate producing a complicated interference pattern. Thus, an off-axis hologram of the object is produced.

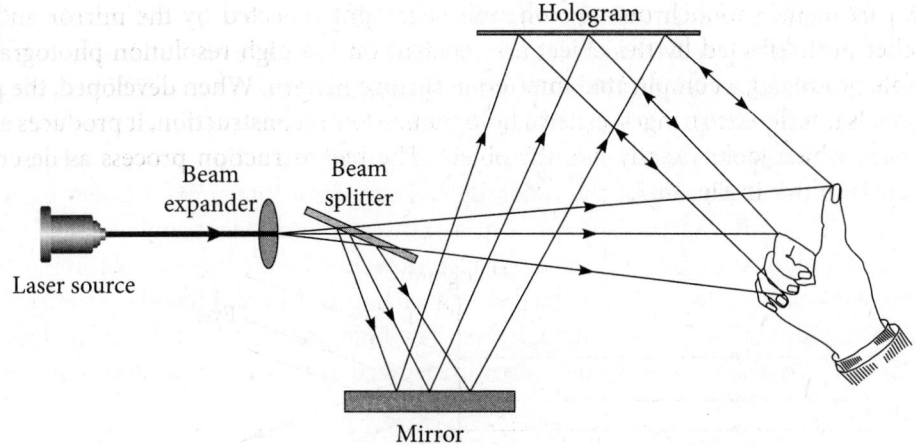

Figure 14.8 | Reorganization of the set up in Fig. 14.6 with the same basic concepts

14.3 Types of Holograms

There are basically two types of holograms: (i) Reflection holograms and (ii) transmission holograms.

14.3.1 Reflection holograms

Reflection holograms are the most common types of holograms. White ordinary light is all that is needed for viewing them. The hologram is illuminated by white light at a specific angle and distance and located on the viewer's side of the hologram. Thus, the image is formed by the light reflected by the hologram. Reflection holograms are made with the reference beam on the opposite side of the object beam. The fringes that form the image of a reflection hologram are very closely spaced. The fringe spacing is very less compared to that of transmission holograms. The very closely spaced fringe patterns require the recording medium and processing method to be of very high standard. Excellent stability of the optical system is another essential requirement. Reflection holograms produce images that are usually dimmer than that of off-axis transmission holograms. Recently, a particular type of reflection holograms has been made which produces holographic images that are optically indistinguishable from the original objects even in color. In these types of reflection holograms, the holographic image of the mirror would reflect white light and the holographic image of the diamond would sparkle.

14.3.2 Transmission holograms

The explanations for the basic principles of holography that have been discussed till now are based on the transmission type of holograms. Normally, fringes are spaced 1.0 μm to

10.0 μm apart and are wider than that of reflection holograms. The processing is therefore less stringent than in a reflection hologram. Transmission holograms are made with both reference and object waves on the same side. The typical transmission off-axis hologram can be viewed only with laser light, usually of the same type used to make the recording. The transmission of reference beams through the holograms makes images visible to the eye placed on the other side of the holograms. The virtual image can be very clear and perspective. The vision of the hologram of a full-size room with people in it, gives the sensation of looking into the room through a window. If this hologram is broken into small pieces, one can still see the entire scene through each piece. A real image of a transmission hologram can also be projected onto a screen. The holograms used in different electronic cards are actually transmission holograms mirrored with a layer of aluminum on the back.

14.3.3 Comparison of transmission and reflection holograms

In the following table, we compare transmission and reflection holograms.

	Transmission holograms	Reflection holograms
1	Transmission off-axis holograms are not viewable in white light.	Reflection holograms are viewable in white sun light.
2	The optical system usually involves at least a beam splitter and a beam spreading lens to obtain object waves and reference waves.	The optical system is simple. The laser beam is widened with a beam spreading lens and is incident on the object after passing through a transparent photosensitive plate.
3	Fringes that form the image of a transmission hologram are typically 1.0 μm to 10.0 μm apart.	The fringes that form the image of a reflection hologram are spaced about half a wavelength apart, or of the order of $3 \times 10^{-1} \mu$m.
4	The recording medium and processing methods are less stringent because fringes are loosely spaced.	The recording medium and processing methods are very stringent. Closely spaced fringes demand high quality recording medium and processing.
5	The image can be very deep if the original coherent laser light is used for viewing. White light cannot be used.	Reflection hologram images are usually much dimmer than off-axis transmission holograms. These are visible in white light.
6	Transmission holograms may be possible even when old developer is used.	If the developer is old, reflection holograms will be murky.
7	In this case, it is not highly essential that a system is stable against the slightest of motions.	In this case, stability of the system against slightest motions is highly essential.
8	The two waves (reference waves and object waves) incident on the photographic plate comes from the same side of the plate at a relatively small angle.	Here, the two waves may come from the opposite sides of the photographic plate at angles close to 180°.

14.4 White Light Holograms

Till now we have described holograms that produce sharp images only when viewed with highly monochromatic light. For the sake of completion of the chapter, we need to include a brief description of white light holograms. The holograms discussed earlier are assumed to have negligible thickness and are referred to as plane holograms. Reflection holograms can be viewed in white light.

During the recording of reflection holograms, reference waves and object waves come from opposite sides of the photographic plate at angles close to 180°. Thus, the resulting interference pattern is nearly perpendicular to the plane of the film. Therefore, one has to use a thicker (many times the fringe spacing) photographic emulsion and record the surfaces of maxima within the volume of the photographic emulsion. In this case, the interference pattern recorded is truly three-dimensional, with twenty or more hyperboloid interference surfaces within the emulsion. These surfaces are semi-transparent and semi-reflecting and are separated consecutively by $\lambda/2$ distance. Upon development, dark fringe and bright surfaces become reversed and a variation of refractive index is created within it. This has been illustrated in Fig. 14.9.

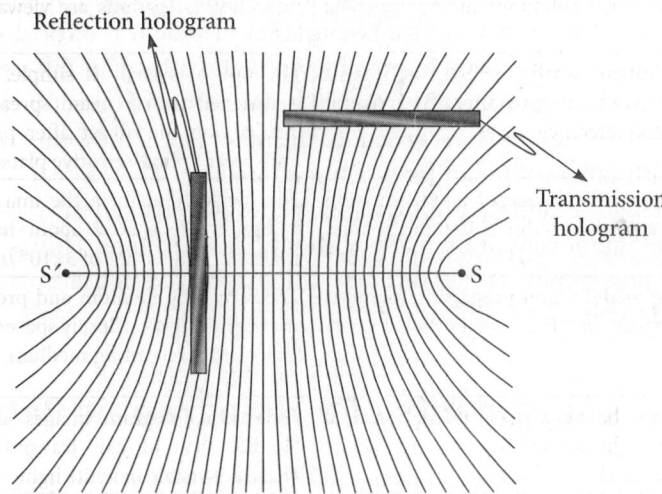

Figure 14.9 | Formation of a set of semi-transparent and semi-reflecting hyperboloidal interference surfaces within thick holograms. The interference fringe pattern is produced by two monochromatic coherent point sources Q and Q'. Here the difference between transmission and reflection holography has also been shown

However, in real life, after developing and drying the plate, the emulsion shrinks from its original dimensions and the interference surfaces get closer. Therefore, the reflected light generally has a smaller wavelength.

When it is illuminated by white light beam, some of the light is diffracted backward from the multiple layers. Due to the multiple reflections from these layers, a complicated interference pattern is formed which carries the whole 3-D visual information about the object. For each particular angle of incidence satisfying the conditions of interference, out of the whole continuous spectrum of the white light, only a narrow range of wavelengths is reflected into the observer's eye. The proper wavelength to be reflected for each angle of incidence is automatically selected by the hologram. Thus, we can view the 3-D reflection holograms with white light, direct sunlight or light coming from a small light bulb. Therefore, viewing of reflection holograms requires white light, not the same laser light as the one used to record the hologram.

14.5 Necessity of Laser Source

We know that in holography, sustainable high quality interference fringe patterns are produced due to the interference of reference waves and object waves and are recorded on high resolution recording media. The recording and production of sustained stable interference patterns require that the path difference between various interfering light waves should always be less than the longitudinal coherence length. The longitudinal coherence length is very small for ordinary light. In this case, for formation of Gabor's zone plate, the object should be placed very close to the photographic plate so that the path difference is very small. It will give rise to a lot of practical difficulties which make the formation of Gabor's zone plate impossible. In reality, the path difference introduced between different light waves reflected from different points of object is not small and so the interference pattern cannot be recorded. In the other hand, the longitudinal coherence length for laser light can be of the order of a few kilometers. Therefore, Gabor's zone plate can be formed on high resolution photographic plates placed at a larger distance from the object. Thus, a hologram cannot be made without a laser source. This is the reason that laser is needed in holography. It explains why holography did not receive much attention until lasers were developed.

14.6 Basic Requirements of a Holographic Laboratory

Unlike a photographic laboratory, a holographic laboratory requires special attention. In the following, the basic requirements of a holographic laboratory have been described.

i. The reconstruction of interference patterns in Gabor's zone plate is possible if they are stable. Therefore, the maximum path difference between the object wave and the reference wave should not exceed the coherence length.

ii. The spatial coherence is to be maintained so that the object waves scattered from different points of the object and reference waves interfere to produce high quality interference patterns on the photographic plates.

iii. The reconstruction depends on both the wavelength and the position of the reconstructing source. High resolution of reconstruction is very essential. Therefore, the source must not be broad and must emit a narrow band of wavelengths.

iv. Like ordinary photography, holography also suffers from aberrations. The aberration in reconstruction is least if the reconstruction source is of the same wavelength and is situated at the same relative position as that during recording time.

v. The stability of all the recording components is highly essential in making fine holograms. Therefore, the photographic plate, the object, and any mirrors used in producing the reference beam must be motionless with respect to one another during exposure. This is possible by mounting the components rigidly to supports attached to massive granite and marble plates placed on vibration-absorbing base like inflated rubber tubes or sand bags. Even slightest vibrations, even the fraction of wavelength of light, affect the recording of fringe patterns on the photographic plate by changing the optical path lengths. Here in this case, the interference fringes would vibrate during the exposure time causing a blurring of the recording hologram.

vi. It is natural that the output results would be best if the resolution of the photographic plate is high, because, it can record the very closely spaced fringe patterns.

vii. The highly coherent light laser for holographic applications should be powerful enough so that the exposure time is very short.

14.7 Viewing a Hologram

The photographic plate containing the interference fringes is kept under the same identical surroundings as that during the recording time. When we look at the plate, as shown in the Figs 14.5 or 14.7, the diffracted waves appear to diverge from the virtual image. The eye adjusts itself so that a real image is formed on the retina. The image is three-dimensional as well as perspective. When the observer moves his eyes to different positions, the image displays different parts of the object because the rays of light entering each eye come from different points of the fringe pattern on the hologram. Thus, the hologram enables us to see the object in a different perspective. The observer can even see the object behind an obstacle by moving his head! When a hologram is broken into many small pieces, the perspective will be limited and there may be a loss in resolution, but each piece will be a hologram of the complete object scene!

The real image is formed on the observer's side at equal distance from the plate as that of the virtual image. This real image can be captured on a photographic plate placed there. We can also see it by placing our eyes at a distance distinct from it.

14.8 Difference between Photography and Holography

A hologram differs from a photograph in several ways. A few differences are cited in the following table.

	Holography	Photography
1	Each point in the holographic recording includes light scattered from every point in the object.	Each point in a photograph has light scattered only from a single point in the object.
2	The hologram allows the recorded scene to be viewed from a wide range of angles.	The photograph gives only a single view.
3	The reproduced range of a hologram adds many of the same depth perception cues that were present in the original scene, which are again recognized by the human brain and translated into the same perception of a three-dimensional image as when the original scene might have been viewed.	The photograph is a flat 2-D representation of 2-D or 3-D objects.
4	The developed hologram surface itself consists of a very fine, seemingly random pattern, which appears to bear no relationship to the scene which it has recorded.	A photograph clearly maps out the light field of the original scene.
5	When a hologram is cut into pieces, each part is capable of reconstructing the entire object (Accordingly, the perspective will be limited and there may be a loss in resolution).	When a photograph is cut into pieces, each piece shows only part of the scene.
6	Holograms can only be viewed with very specific forms of illumination.	A photograph can be viewed in a wide range of lighting conditions.
7	The hologram cannot be constructed without a laser source.	For photograph recording, laser source is not required at all.
8	By principle, in holography, no lens is required for focusing purpose; sometimes, holography is called lensless photography.	A lens for focusing purpose is required in photography to record the image.

14.9 Applications of Holography

Holography is a very useful tool which finds applications in many areas, such as commerce, telecommunication, computers, scientific research, aviation, medicine, and industry; the list is not exhaustive.

14.9.1 Common applications of holography

In the following, a few common applications of holography have been discussed briefly.

i.   Commerce:

Supermarket and department store scanners use a holographic lens system that directs laser light onto the bar codes of the merchandise. Holograms are used in advertisements and consumer packaging of products to attract potential buyers. They have been used on covers of magazine publications

ii.   Telecommunication

Holographic telepresence is an evolving technology for full-motion, real time 3-D video conferencing. The technology will reduce the necessity of travel for meetings and facilitate distance education. Work in the direction of holographic telepresence to bring digital participants and remote locations into 3-D virtual space in real time is going on. Though holographic telepresence is still in its infancy and cost prohibitive, it may become a reality within the next few decades.

iii. Computers

Holograms can also be used to store, retrieve, and process information optically. The holographic storing device operates in three dimensions rather than two dimensions, thus increasing the storing capacity. The holographic memory cards of the size of a sugar cube can have the memory of a few hundred terabytes. Moreover, even though part of a hologram becomes defective or is destroyed, while the remaining part will still retain all the data intact. The technology could be available in the coming decade. Holographic optical computers will be capable of delivering trillions of bits of information faster than the latest computers. Color liquid crystal displays (LCD) can be brighter and whiter as a result of incorporation of a holographic reflector that will reflect ambient light to produce whiter background.

iv. Scientific research

Holographic techniques are used in the characterization of materials. A hologram of an abject can be made before and after it is subjected to stress. The change in the shape of the object due to that stress is recorded in the holograms. Particle physicists make holographic records of bubble chambers from which accurate measurements can be made. Electron holography is a method used to create images from materials at atomic dimensions. Now-a-days, it is applied in the fields of nanostructured materials, DRAM capacitor design, high temperature superconductors, characterization of complex molecules, including catalysts, polymers and different types of electronic materials.

Holographic technology in combination with actual data of Mars will give scientists a strolling experience on the planet. Interaction becomes more natural facilitating the exploration and understanding of theRed Planet.

v. Aviation

Holographic optical elements are used for navigation by airplane pilots. A holographic image of the cockpit instruments appears to float in front of the windshield. This allows the pilot to keep his eyes on the runway or the sky while reading the instruments. The feature is available on some models of automobiles.

vi. Health science

Medical doctors can use three-dimensional holographic CAT scans to make measurements without invasive surgery. This technique is also used in medical education.

vii. Industry

All types of industries use holograms to enhance the image of their brands in the market as genuine and authenticated. The holograms are almost impossible to counterfeit. They are used for attractive product packaging, security applications and so on. Holographic techniques can be incorporated in non-destructive testing methods to test the product quality. In this method, the hologram of the test specimen is compared with that of the flawless master specimen. In industrial situations, this can be used in component testing and quality control. Holographic night vision goggles and fancy items have been made using of this technology.

viii. Security

The use of holograms on credit cards and debit cards provide added security to minimize counterfeiting. They are used on tamper-resistant packaging. Few countries have started using holograms on their currency to stop counterfeiting by fraudulent people.

ix. Preservation of antiques

Holography has been used to make archival recordings of valuable and/or fragile museum artifacts. Many museums have made holograms of valuable articles in their collections, both for insurance purposes and to check for deterioration. The holograms of national treasures, antiques can be made and send to remote areas, enabling people there to see and appreciate their national heritage in their local areas without long travelling. The original items are thus properly preserved, without fear of theft.

x. Photography

Holographic techniques applied to the field of photography are found to be far superior to ordinary photographic perspective scenery. Holographic technology can be incorporated in digital cameras to detect the edge of the subject and differentiate between it and the background. As a result, the camera is able to focus accurately in dark conditions. Holography has been in use by artists to create pulsed holographic portraits as well as other works of art.

xi. Television

Holographic motion picture technology was implemented practically in the late 1970s was vastly successful. Work on holographic video is going on. Holographic television may become a reality within the next few years. Many researchers believe that truly holographic commercial televisions will become available in the coming decade. Though true holographic projection that can be viewed from any angle in air is impossible today, it may be possible in the future.

14.9.2 Application of holographic interferometry

Holographic interferometry is an important technique in experimental physics. The technique can be applied to the study of quasi-static and quasi-dynamic behavior of objects. It has been developed rapidly in recent years and will probably find more applications in engineering in the future. Therefore, the application of holographic interferometry needs special mention.

Optical interferometry is a powerful tool for measuring very small displacements of the order of wavelength of light. In the field of interferometry, the concept of holography can be exploited both for scientific research and industrial process. It is possible to record several wave fronts on the same plate and obtain several holograms. We can take the interference patterns of the object waves and reference waves at different real times, before and after the object is subjected to external stimuli. The technique is called double-exposure holographic interferometry. In double-exposure holographic interferometry, two successive holograms are recorded on the same film with little apparent deterioration in quality. Here the photographic plate is first partially exposed to the object wave and the reference wave before the application of external stimuli such as stress, magnetic field, electric field, heat and so on. Then the interference pattern of the same reference wave and the object wave from the same object under identical conditions taken on the same photographic plate is again exposed along with the same reference wave. The photographic plate after development forms the hologram consisting of two interference patterns. On reconstruction, it gives two images – one corresponds to the unstressed object and the other to the stressed object. These changes can be decoded and recorded for future use.

14.9.3 Application of holographic microscopy

The application of holographic microscopy needs special mention. It is also called digital holographic microscopy or digital holography. Digital holographic microscopy opens door for quantitative phase contrast imaging, suitable for high resolution investigations. Though digital holography is based on the normal holographic principle, the recording is performed by digital image sensors like CCD (charge coupled device) or CMOS camera instead of photographic plate. The reconstruction process is carried out numerically with a computer. Readymade softwares are available for the purpose.

The ability of holography to record information about depth finds application in studying transient microscopic events. These events can be captured on the hologram and one can leisurely observe through the depth of the reconstructed image at a later time. In traditional microscopy, it is hard to get a sense of the three-dimensional shape of objects like living cells. In holographic microscopy, it possible to record the 3-D shape of tiny objects such as cells in high resolution. This method is clearly powerful and also very cheap. Holographic microscopy belongs mostly to off-axis holography. Digital holography provides a new method for surface analysis and dynamic life cell imaging. Therefore, digital holography finds a wide range of applications in medical science such as cell cultures, cell cycle analysis, morphology analysis of cells, nerve cell studies, red blood cell analysis, cell division and migration, tomography studies deep in living tissue and so on.

This is not the be all and end all of holography; it is the beginning. Different types of holograms are available now-a-days. All the 3-D images available commercially are not truly holographic. The visual effects produced by lenticular printing, the Pepper's Ghost illusion, tomography and volumetric displays give the sensation of holographic images. The principle of holography is not only limited to electromagnetic waves but can also be applied to any waves like sound waves, electron waves, and so on. Acoustic holography using sound waves has become increasingly popular in the field of medical science. Electron holography has been developed for use in the field of scientific and industrial research. In recent years, atomic holography has been developed which produces much higher resolution holographic images. This is because the de Broglie wavelength of the atoms is much smaller than that of light. Holographic telepresence may revolutionize the future world.

Questions

14.1 What is holography? Describe its basic principles

14.2 What is Gabor's zone plate?

14.3 Describe the construction and reconstruction of images in holography.

14.4 Compare ordinary photography with holography.

14.5 Give a comparison between Gabor's zone plate and Fresnel's zone plate.

14.6 Distinguish between in-line holography and off-axis holography.

14.7 Describe the construction and reconstruction of images in off-axis holography.

14.8 What are reflection holograms?

14.9 What are transmission holograms?

14.10 Distinguish between reflection holograms and transmission holograms.

14.11 What are the basic requirements of a holographic laboratory?

14.12 Holograms cannot be produced without laser. Comment on the statement.

14.13 Explain how holography technology can be used in memory cards

14.14 Mention a few examples of the application of holography techznology in scientific research.

14.15 Write a few lines about holographic telepresence.

14.16 Explain how the use of holograms can help to prevent counterfeiting of currency.

14.17 Explain the basis principle of holographic interferometry.

14.18 Distinguish between digital holography and ordinary holography.

14.19 What are the advantages of digital holography over ordinary holography

14.20 Write an essay about the future use of holography technique.

Multiple Choice Questions

1. Which parameters of light are recorded in holography
   (i) Intensities
   (ii) Phases
   (iii) Speed
   (iv) Both (i) and (ii)

2. Holography was discovered by
   (i) Dennis Gabor
   (ii) Huygens
   (iii) Newton
   (iv) Einstein

3. Through holography technology, we can produce
    (i)   One-dimensional image
    (ii) Two-dimensional image
    (iii) Three-dimensional image
    (iv) Four-dimensional image

4. Holography produces
    (i)   real images
    (ii) virtual images
    (iii) Both (i) and (ii)
    (iv) none of the above

5. A recorded holographic plate contains information in the patterns of
    (i)   Reflection–refraction
    (ii) Interference
    (iii) Diffraction
    (iv) Polarization

6. In holography, interference pattern is produced due to superposition of
    (i)   Object wave
    (ii) Reference wave
    (iii) Both (i) and (ii)
    (iv) none of the above

7. If a hologram breaks into pieces, then each piece contains
    (i)   Total image
    (ii) No image
    (iii) Parts of the image
    (iv) Nothing

8. Holography is capturing pictorial details of 3-D objects by using the principle of
    (i)   Diffraction
    (ii) Interference
    (iii) Refraction
    (iv) Total internal reflection

9. In holography, there-dimensional image of an object may be reproduced
    (i)   By using convex lens
    (ii) By using concave lens
    (iii) By using double convex lens
    (iv) Without using lens

10. In holography, we record
    (i)   Only the phase of different parts of the object
    (ii) Only the intensity from different parts of the object
    (iii) Both the phase and intensity from different parts of the object
    (iv) None of the above

Answers

1 (iv)   2 (i)   3 (iii)   4 (iii)   5 (ii)   6 (iii)   7 (i)   8 (ii)

9 (iv)   10 (iii)

Bibliography

Symon, Keith R. 1971. *Mechanics* 3rd Edition. Massachusetts: Addison-Wesley Publishing Company Inc.

Resnick, R. and D Halliday. 1962. *Physics (Part I and II)* 2nd Edition. New Delhi: Wiley Eastern Limited.

Jenkins, F. A. and H. E. White. 1981. *Fundamentals of Optics* 4th Edition. Tokyo: McGraw-Hill Ltd.

Ghatak, A. 1992. *Optics* 2nd Edition. New Delhi: Tata McGraw-Hilsl Publishing Company Limited.

Mathur, B. K. 1982. *Principle of Optics* 3rd Edition. Kanpur: Gopal Printing Press.

Hayt, W. H. 1981. *Engineering Electromagnetics* 4th Edition. , Tokyo: McGraw-Hill Book Company.

Sadiku, M. N. O. 2006. *Elements of Electromagnetics* 3rd Edition. , New Delhi: Oxford University Press.

Theory and Problems of Electromagnetics. 1979 Schaum's Outline series, New York: McGraw-Hill Inc.

Reitz, J. R., F. J. Milford, R. W. Christy. 1979. *Foundation of Electromagnetic Theory* 3rd Edition. New Delhi: Narosa Publishing House.

Beiser, A. 1995. *Concepts of Modern Physics* 5th Edition. New Delhi: Tata McGraw-Hill Publishing Company Limited.

Powell, J. L. and B. Crasemann. 1961. *Quantum Mechanics.* New Delhi: Oxford & IBH Publishing Company.

Weber, R. L., K. V. Manning, M. W. White, G. A. Weygand. 1977. *College Physics* 5th Edition. New Delhi: Tata McGraw-Hill Publishing Company Limited.

Index